SOLID-STATE ELECTRONICS

The late F. K. Richtmyer was Consulting Editor of the series from its inception in 1929 to his death in 1939. Lee A. DuBridge was Consulting Editor from 1939 to 1946; and G. P. Harnwell from 1947 to 1954.

SHYH WANG

Professor of
Electrical Engineering
University of California,
Berkeley

SOLID-STATE ELECTRONICS

McGRAW-HILL BOOK COMPANY

New York	*San Francisco*	*St. Louis*
Toronto	*London*	*Sydney*

TO DILA,
MY WIFE

PREFACE

In the field of solid-state physics, progress toward understanding the basic properties of solid materials has stimulated a parallel progress in the development and use of solid-state electronic devices. The invention of transistors and lasers has opened a vast new territory for technological exploitation and has encouraged electrical engineers and physicists to work closely together. The need for a text which treats the properties of solid-state materials from a device standpoint has become evident; this book, therefore, is written for electrical engineers and applied physicists who will have to translate our basic knowledge in the sciences and new discoveries of physical phenomena into technological advances.

Based on the lecture notes of courses taught by the author at the University of California, Berkeley, this book deals with solid-state electronic devices that utilize the conductive, dielectric, magnetic, and optical properties of materials. The main objective is to introduce the student to the modern theory of solid-state devices, starting with a discussion of material properties. In a rapidly expanding field such as solid-state electronics, it is important that the student obtain an over-all view of the whole field. Therefore this book, which attempts to bridge the gap between physics and electrical engineering, is written as an introductory text at the advanced senior and graduate level.

Because of the wide range of subjects covered, a systematic and unified presentation is essential. Introductory discussion of subject matter is simple and understandable to students with varied backgrounds; qualitative discussion, based on simple physical models, is followed by a detailed quantitative analysis. Depth of treatment is achieved by concentrating discussions on subjects for which physical models are well established.

The first two chapters of the book are devoted to a general review of the fundamentals of atomic physics and aspects concerning the crystalline state. Chapter 1 discusses basic concepts leading to quantization (and, hence, to the periodic system of atomic elements) and statistics governing the occupancy of atomic energy states. Chapter 2 discusses interatomic forces, reviews various crystal structures and their symmetry properties, and introduces lattice imperfections, lattice vibrations, and diffusion and alloying processes.

Chapter 3 is devoted to the theory of electronic conduction in semiconductors. Fundamental concepts such as mobility, diffusion, and effective masses are introduced and the electrical properties of elemental, III-V, and II-VI semiconductors are compared. Formal treatment of the transport equation is presented, electronic mobilities due to different scattering processes are evaluated, and thermoelectric and galvanomagnetic effects are discussed in Chapter 4, as well as experiments concerning electron-phonon interaction and ultrasonic amplification in semiconductors and semimetals. Chapter 5 primarily concerns the transport and recombination of excess carriers; different processes of bulk and surface recombination are discussed. The solution of the time-dependent diffusion equation is obtained and then applied to Haynes-Shockley and photoconductive experiments for measurement of mobility, diffusion constant, bulk lifetime, and surface recombination velocity of free carriers. Theories of semiconductor junction devices are developed in Chapter 6, emphasizing a-c response and the over-all design considerations. Discussion of the diode parametric amplifier introduces the reader to the concepts of energy and momentum conservation. Considerable attention is given to developing the theory for tunnel diodes, a theory important in learning such fundamental properties as band structure and the role of lattice waves in the tunneling process.

Dielectric and ferroelectric properties are discussed in Chapter 7. The correlation between the dielectric behavior of a substance and its molecular structure is stressed. Both the Lorentz and Onsager models are discussed in the calculation of internal fields. Piezoelectric and ferroelectric crystals are also included in Chapter 7. Finally, recent development in the theory of dielectric constant and ferroelectricity is presented.

Chapter 8 serves as an introduction to the magnetic properties of solids. The phenomenological theory of magnetism is supplemented by a discussion based on the modern theory of atomic structure. Chapter 9 treats the theories of paramagnetic resonance phenomena and masers. The concepts of spin-spin and spin-lattice relaxation times are presented. A discussion of the quantum-mechanical treatment, together with the classical treatment, introduces the student to the nature of the resonance process. Various schemes for achieving a population inversion are presented. Finally, Einstein's interpretation of Planck's theory of black-body radiation forms the basis for the discussion of noise in masers. Chapter 10 is devoted to the presentation of various phenomena and effects associated with ferromagnetism. The magnetization process is analyzed in terms of magnetization rotation and domain wall movement. The treatment of the fundamental physical processes is followed by a discussion of ferromagnetic and ferrimagnetic devices. Finally, the concept of spin waves is introduced, and the excitation of spin and magneto-elastic waves in a magnetic resonance experiment is discussed.

The optical properties of solids are treated in Chapter 11. The absorption process in covalent semiconductors is discussed in detail, and the use of the Kramers-Kronig relation in the optical region is stressed. Chapter 12 discusses lasers and related optical effects. Energy levels involved in a laser transition are analyzed, and the roles of coulomb interaction, spin-orbit interaction, and crystalline-field splitting in determining atomic energy levels are discussed. At the end of Chapter 12, various schemes for modulation are introduced, and nonlinear optical effects are presented.

According to their relative degrees of sophistication, the material in this book is divided into three categories, as indicated in the table of contents by the following: (1) unmarked sections are for undergraduate instruction (advanced senior, the first course), (2) sections marked ** are for graduate instruction (the second course), and (3) sections marked * are of an intermediate level. Those sections marked * can be used by the bright undergraduate student as further reading material and by graduate students who lack sufficient background. These sections also provide flexibility for effective teaching; the instructor may choose suitable subjects and methods of presentation by carefully mixing the proper amount of quantitative analysis with descriptive discussion. The author feels that it is his obligation to make some concrete recommendations for the selection of topics. In doing so, he fully realizes that such a selection depends on one's own personal view; therefore the following list of suggested topics represents only the author's opinion.

AN ADVANCED UNDERGRADUATE COURSE EMPHASIZING THE PROPERTIES OF SOLID-STATE MATERIALS. The sections in *parentheses* are of an intermediate level; a qualitative discussion of these topics is appropriate.

Chapter 1: Sections 1–3, (4–8), 9, (10–12)
Chapter 2: Sections 1–2, 8–9
Chapter 3: Sections 1–4, 6, (7, 8, 10)
Chapter 5: Sections (3-6), 8, (9)
Chapter 6: Sections 1-3, (4, 6, 7, 9)
Chapter 7: Sections 1-4, (5, 8, 9)
Chapter 8: Sections 1–3, (4–7), 8–11
Chapter 9: Sections 1–2, (10)
Chapter 10: Sections 1–5
Chapter 11: Sections (1–5)
Chapter 12: Sections (1–2)

AN ADVANCED SENIOR OR GRADUATE COURSE IN SEMICONDUCTOR ELECTRONICS. The amount of coverage and the depth of presentation for sections in *parentheses* are up to the individual instructor.

Chapter 1: Sections 1–3, (4–8), 9, (10–12)
Chapter 2: Sections 1–2, 8–9
Chapter 3: All sections
Chapter 4: Sections (1–9)
Chapter 5: Sections (1–2), 3–10
Chapter 6: Sections 1–5, (6), 7–8, (9)
Chapter 11: Sections 1–5, (6–9)
Chapter 12: Sections (1–2)

A GRADUATE COURSE IN SOLID-STATE ELECTRONICS EMPHASIZING THE THEORY OF ELECTRONIC DEVICES. The sections in *parentheses* are for review purposes and home-reading material for graduate students with insufficient background. The instructor can choose among topics in *brackets* to maintain a proper balance in the presentation of the three main subfields of solid-state electronics: semiconductor electronics, microwave magnetic devices, and quantum electronics.

Chapter 1: Sections (1–12)
Chapter 2: Sections (1–10)
Chapter 3: Sections (1–4), 5, (6), 7–10
Chapter 4: Sections [1–9]
Chapter 5: Sections (1–10)
Chapter 6: Sections (1–5), 6–9
Chapter 7: Sections (1–10), 11
Chapter 8: Sections (1–3), 4–7, (8–11)
Chapter 9: Sections (1–3), [4–11]
Chapter 10: Sections (1–5), [6–12]
Chapter 11: Sections 1–5, [6–9]
Chapter 12: Sections [1–7]

A GRADUATE COURSE IN QUANTUM ELECTRONICS. The sections in *parentheses* are for review purposes and home-reading material for graduate students with insufficient background.

Chapter 1: Sections (1–12)
Chapter 2: Sections (1–10)
Chapter 3: Sections (1–4), 5, (6), 7–10
Chapter 5: Sections (1–2), 3–5, (6–10)
Chapter 8: Sections (1–3), 4–7
Chapter 9: Sections (1–3), 4–11
Chapter 11: Sections (1–3), 4–8
Chapter 12: Sections 1–7

The author wishes to thank Professors C. Kittel and B. Lax for reviewing the manuscript and for their valuable comments and suggestions which led to the addition of several sections. Thanks are also due to Professor M. Sparks for his many helpful comments. It is a pleasure to thank Professor R. S. Muller for his thorough and helpful scrutiny of the entire manuscript, and his colleagues, especially Professor A. C. English, for their generous criticisms of the manuscript during the early stages of its development. In addition, the author benefited from a series of lectures on nonlinear optics given by Professor N. Bloembergen at the University of California, Berkeley, in the spring of 1965. The support of a John Simon Guggenheim Memorial Foundation Fellowship and the hospitality of the Institut für angewandte Physik, Universität Bern, provided the environment in which to read the final proofs. Finally, the efforts of many graduate students are gratefully acknowledged; their inquisitive minds have helped eliminate some of the obscurities and errors in the class notes upon which this book is based.

Bern Switzerland   *Shyh Wang*

CONTENTS

* For an explanation of the asterisks, see p. ix in the preface.

12   LASERS AND RELATED OPTICAL EFFECTS   687

PHYSICAL CONSTANTS

h = Planck's constant = 6.625×10^{-27} erg-sec

$\hbar$ = $h/2\pi$ = 1.054×10^{-27} erg-sec

k = Boltzmann's constant = 1.380×10^{-16} erg/°K

N = Avogadro's number = 6.025×10^{23} molecules/mole

R = ideal gas constant = kN = 8.316×10^{7} erg/deg/mole

e = electronic charge = 1.601×10^{-19} coul

m_0 = electronic rest mass = 9.108×10^{-28} gram

M = proton mass = 1.672×10^{-24} gram

ϵ_0 = dielectric constant of free space

$$= 8.854 \times 10^{-12} \left(\approx \frac{1}{36\pi} \times 10^{-9} \right) \text{farad/m}$$

μ_0 = permeability of free space = 1.257×10^{-6} $(= 4\pi \times 10^{-7})$ henry/m

c = velocity of light = 2.998×10^{8} m/sec

u_B = Bohr magneton = 0.9273×10^{-23} joule-m^2/weber

a_1 = first Bohr radius = 5.292×10^{-11} m

$\dfrac{kT}{e}$ = 0.02586 ev (at 300°K)

Mechanical equivalent of heat = 4.186 joules/calorie

SYMBOLS

Script and greek (scalar quantities)

$\mathcal{E}$	energy
$\mathcal{E}_f$	Fermi level
$\mathcal{E}_g$	energy gap
$\mathcal{F}$	quasi Fermi level
$\mathcal{H}$	Hamiltonian
α (alpha)	polarizability
γ (gamma)	force constant; gyromagnetic ratio
ϵ (epsilon)	dielectric constant
Θ (theta)	temperature
λ (lambda)	wavelength
μ (mu)	permeability; mobility
ν (nu)	frequency
ρ (rho)	mass density; resistivity
ϱ (rho)	charge density
σ (sigma)	conductivity
τ (tau)	relaxation or mean free time
$\boldsymbol{\tau}$ (tau)	lifetime
χ (chi)	susceptibility
ω (omega)	angular frequency

Roman (symbols set in boldface type represent vector quantities)

a	Bohr radius; lattice constant
$\mathbf{B}, B$	magnetic induction or flux density
c	velocity of light
c_{ij}	elastic constant
$\mathbf{D}, D$	electric displacement or flux density
D	diffusion constant
e	electronic charge
$\mathbf{E}, E$	electric field
$\mathbf{F}, F$	force
g	Landé splitting factor
h	Planck's constant
$\mathbf{H}, H$	magnetic field

i	$= \sqrt{-1}$
I	current
j	inner quantum number
$\mathbf{J}, J$	current density; total angular momentum; total angular momentum quantum number
k	Boltzmann constant
$\mathbf{k}, k$	wave number
K	thermal conductivity; compressibility; absorption coefficient
l	azimuthal quantum number for a single electron; mean free path
$\mathbf{L}, L$	total orbital angular momentum; resultant azimuthal quantum number
L	diffusion length
m	magnetic quantum number for a single electron; mass of an electron
$\mathbf{M}, M$	magnetization
M	mass of an atom or ion; resultant magnetic quantum number
n	index of refraction; electron concentration
$\mathbf{p}, p$	electric dipole moment; linear momentum
p	hole concentration
$\mathbf{P}, P$	polarization
Q	total charge
s	spin quantum number for a single electron
$\mathbf{S}, S$	strain; total spin angular momentum; resultant spin quantum number
S	entropy
t	time
$\mathbf{T}, T$	stress; torque
T	temperature
$\mathbf{u}, u$	magnetic dipole moment
U	internal energy
$\mathbf{v}, v$	velocity
V	potential energy; electric potential; voltage
z	atomic number
$\mathbf{Z}$	coordination number

1 ATOMIC STRUCTURE AND QUANTUM THEORY

1.1 INTRODUCTION

This text is primarily concerned with the conductive, the dielectric, and the magnetic properties of solid-state materials. It may help us to realize the immense scope of the subject if we merely look at the problem from one particular aspect, the expansion of our frontiers in the domain of the electromagnetic wave spectrum shown in Fig. 1.1. In the old days, applications in electrical engineering were limited to low-frequency domains and electric devices were few, notably electrical machines and electronic tubes. At that time not only was our fundamental knowledge about material properties rather limited, our motivation for the pursuit of exact theories was meager because empirical approaches often yielded satisfactory results. The development of klystrons and magnetrons during World War II pushed the frontier of useful (coherent) spectrum into the microwave region. The breakthrough in the field of quantum electronics, which occurred recently with the advent of lasers, brings us

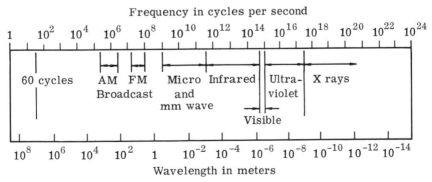

Figure 1.1

The electromagnetic wave spectrum. The frequency ranges commonly assigned for use in engineering applications are: (a) for power generation, 50 or 60 cps; (b) for radio broadcast, around 10^6 cps (amplitude modulation) and 10^8 cps (frequency modulation); and (c) for transcontinental telephone relay, in the microwave region. The recent invention of lasers has pushed our frontier into the infrared and visible region, a vast territory now open for future use in communication.

further into the infrared and visible regions of the spectrum. The rapid advance toward higher and higher frequencies requires a rather fundamental and necessary change in our thinking.

At low frequencies we are accustomed to the use of lumped-circuit elements, resistance, inductance, and capacitance. When the wavelength of an electromagnetic signal becomes comparable to or smaller than the physical dimensions of the circuit elements involved, we can no longer uniquely define a reference voltage or current without knowing in full detail electromagnetic-wave propagation in these circuit elements. Therefore, at high frequencies, we must use Maxwell's field equations. It may be worthwhile to see how this shift can affect our views toward material properties. The lumped-circuit approach is simple indeed compared with the field approach. Once the values of circuit elements are known, the process of finding the solution of circuit equations based on Kirchhoff's laws is purely mathematical. On the other hand, with Maxwell's field equations, the first problem facing us is to determine the various parameters which appear in the auxiliary relations known as *conductivity* σ, *dielectric constant* ϵ, and *permeability* μ of a medium. From an engineering point of view, many questions which are hidden in the lumped-circuit approach must now be considered.

First, the aforementioned parameters are the unique properties of a given substance. In lumped circuits it is of no consequence whether two capacitors are made of the same material if they have the same capacitance. In Maxwell's equations not only are the three parameters an integral part of material properties but their magnitudes are also independent of physical dimensions and hence can be used for direct comparison. Furthermore, unlike voltage and current, the fields appearing in Maxwell's equations are vector quantities. The auxiliary relations

$$\mathbf{J} = \sigma\mathbf{E} \qquad \mathbf{D} = \epsilon\mathbf{E} \qquad \mathbf{B} = \mu\mathbf{H} \tag{1.1}$$

are also vector equations. In Eq. (1.1) the symbols $\mathbf{J}$, $\mathbf{D}$, $\mathbf{E}$, $\mathbf{B}$, and $\mathbf{H}$ represent the current density, electric flux density, electric field, magnetic flux density, and magnetic field, respectively. It is quite possible that the parameter ϵ relating the x component of D to E_x may be different from that relating D_y to E_y. In a completely general sense, we should not even exclude the possibility that an electric field in one direction may give rise to a nonvanishing electric flux density in another direction. In other words, there may be nine different values of ϵ in Eq. (1.1). It is evident that a precise knowledge of σ, ϵ, and μ is needed in dealing with Maxwell's equations.

The foregoing discussion also implies that at high frequencies our choice of materials becomes more critical. As the usable electromagnetic spectrum expands, not only will we look for new materials but we must also know material properties over a wide frequency range. A practical example is the modulation of a laser beam. The frequency of a laser beam can be in the infrared region, whereas the modulating frequency is in the microwave region. It turns out that most significant changes in the behavior of σ, ϵ, and μ occur at frequencies in the microwave region and higher. To provide guidance in the choice of materials we must understand the fundamental mechanisms governing the conductive, the dielectric, and the magnetic properties of materials.

The situation with active elements (electronic devices) is similar to that with passive elements. As new electronic devices are invented, the underlying theories of these devices get progressively more complicated. In vacuum tubes the work function of a cathode is the only quantity which involves material properties. Next come semiconductor devices, such as transistors and diodes. In order to grasp the mechanism of electric conduction in semiconductors, we are introduced to new concepts, such as energy bands and effective masses. We then realize that crystal structure plays an important role and that electrons in a crystal do not behave like a classical particle. Needless to say, knowledge of atomic states and quantum theory is essential to the understanding of the behavior of electrons in semiconductors. For quantum electronic devices such as masers and lasers the situation is even more striking. In these devices an electron emits a photon of energy $h\nu$ in going from a higher to a lower energy state, and quantum mechanical transitions are directly involved in the photon emission process.

The preceding discussion serves to stress that a systematic approach must be undertaken in analyzing material properties. At first it may seem an almost impossible task for electrical engineers to cover such a vast field over such a wide frequency spectrum. Fortunately, information is available to us, largely through the efforts of solid-state physicists. The modern theory of solids, of which the electrical and magnetic properties are only a part, is based on the physics of atomic and crystalline structures. Therefore the first two chapters of this book will be devoted to a general review of the elements of modern physics and the fundamentals of crystals. In reviewing this material we realize that different subject matter may require varying amounts of background knowledge. A point in question is the quantum theory.

In treating interactions which involve masses and orbits of atomic dimensions, a working knowledge of quantum mechanics is essential. Such is the case with masers and lasers. On the other hand, in dealing with the conductive properties of electrons in a solid, only the consequences of quantum theory are important. The concepts of energy bands and effective masses in semiconductors are good examples.

Once we incorporate necessary modifications in the equation of motion, however, the behavior of electrons can be described in the framework of classical mechanics. Thus we see that the treatment to be given in this chapter can vary a great deal, depending on the emphasis (for example, transistor electronics versus quantum electronics) and the orientation (theory rather than application) of the book. We have tried to maintain a proper balance. Only the essential quantum concepts are covered in this chapter; the quantum mechanical tools will be developed in later chapters when needed.

1.2 HISTORICAL BACKGROUND

The classical viewpoints. The first thing we learn about electrons in introductory physics courses is the motion of an electron in electric and magnetic fields. If **E** is the electric field and **B** the magnetic flux density, then the acceleration experienced by an electron having velocity **v** is equal to

$$\mathbf{a} = \frac{-e(\mathbf{E} + \mathbf{v} \times \mathbf{B})}{m} \tag{1.2}$$

where e is the charge and m is the mass of an electron. Many important electronic devices are based on Eq. (1.2) for their operation. A few examples include the deflection of an electron beam in a cathode ray tube, the orbiting of electrons in a magnetron, and the velocity modulation of electrons in a klystron. There is no doubt that Eq. (1.2) correctly predicts the behavior of electrons when they are essentially free. The classical particle mechanics, on which Eq. (1.2) is based, is totally incapable, however, of accounting for any phenomenon involving atomic interactions. The atomic-emission spectrum of hydrogen is a good example. According to the classical picture, an electron revolves in an elliptical orbit around the nucleus. We know from the theory of electrodynamics that an accelerated electron behaves like a dipole antenna, radiating electromagnetic energy. Therefore the classical theory would have predicted not only a continuously changing emission spectrum but also an unstable atom. This is contrary to experimental facts.

Before pursuing further discussions concerning electrons, we would like to discuss the wave theory of light in contrast to classical particle mechanics. For a plane wave of angular frequency ω and propagating in the z direction, the electric field **E** can be written as

$$\mathbf{E} = \mathbf{E}_0 \exp\left[i(\omega t - kz)\right] \tag{1.3}$$

where $k = 2\pi/\lambda$ is the wave number associated with the wave, λ being the wavelength, and $\mathbf{E}_0$ is a vector in a plane perpendicular to the direction of propagation. A similar expression for **H** can be obtained from Maxwell's equations relating **H** to **E**. It is obvious from Eq. (1.3) that two light waves of the same angular frequency may reinforce or interfere with each other, depending on their relative phase. Figure 1.2a shows an interference experiment in which two parallel rays arrive with the same phase at the two slits, S_1 and S_2. At a point A on the screen, the two diffracted rays will be out of phase. Reinforcement of the rays occurs when

$$k(AS_1 - AS_2) = 2n\pi \tag{1.4}$$

The difference in the optical path can be approximated by $d \sin \phi$ if $d/L \ll 1$; thus Eq. (1.4) becomes

$$d \sin \phi = n\lambda \tag{1.5}$$

where n is an integer, d is the distance separating the two slits, and ϕ is the angle between the normal to the slit plane and the line AS.

The diffraction of an X-ray beam by a crystal (Fig. 1.2b) is another manifestation of wave phenomenon. Consider a beam of monochromatic X rays of wavelength λ incident on a crystal and making an angle θ with the plane of atoms. Part of the beam will be reflected by the atomic plane. The reflected waves from two adjacent parallel planes interfere constructively only when their optical paths differ by an integral multiple of λ. According to Fig. 1.2b, we have

$$2d \sin \theta = n\lambda \tag{1.6}$$

where d is the spacing between two parallel atomic planes and n is an integer. Equation (1.6) is known as the *Bragg law* of X-ray diffraction. If more and more atomic planes are considered, a slight deviation from Eq. (1.6) will result in a cancellation of the reflected waves from different planes. Therefore a strong diffracted beam is detected only when Eq. (1.6) is observed.

S_1,S_2 Slits
S midway point between slits

Figure 1.2

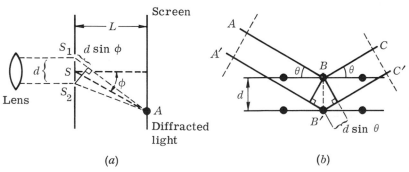

(a)   (b)

Experiments that demonstrate the wave nature of light: (a) the interference experiment and (b) the X-ray diffraction experiment.

Wave-corpuscle experiments. From the above discussion we see that corpuscles and waves are two distinct concepts in the classical theory. In the following material we shall discuss experiments which are beyond comprehension within the realm of classical mechanics and classical electromagnetic theory. One of the most important experiments in modern physics is the study of photoelectric effect with an ordinary vacuum photocell. The experimental setup for such an investigation is illustrated in Fig. 1.3a, and the energy-barrier diagram at the surface of the cathode for zero-bias voltage is shown in Fig. 1.3b. In the classical analysis electrons are accelerated in the electromagnetic fields of a light wave according to Eqs. (1.2) and (1.3). The stronger the light intensity, the greater the energy acquired by an electron. Therefore the classical theory would have predicted that photoemission began when the light intensity exceeded a certain critical strength to give electrons sufficient energy to overcome the surface barrier $\mathcal{E}_w$. This did not happen, however, to be the case. Experimentally we find that the energy $\mathcal{E}$ of photoelectrons after leaving the cathode depends on the frequency ν of the light as follows:

$$\mathcal{E} = h\nu - \mathcal{E}_w \qquad (1.7)$$

where h is Planck's constant. We also find that the magnitude of the photocurrent is directly proportional to the light intensity. To explain these results, new concepts are required.

According to the hypothesis of *photons* (light quanta) advanced by Einstein, light consists of quanta (corpuscles) of energy $h\nu$. Every light quantum colliding with electrons inside the cathode gives away its whole energy to the electron. On the basis of photon hypothesis Eq. (1.7) is self-evident; furthermore, the number of electrons liberated from the cathode will be proportional to the number of impinging photons and, therefore, to the light intensity.

Figure 1.3

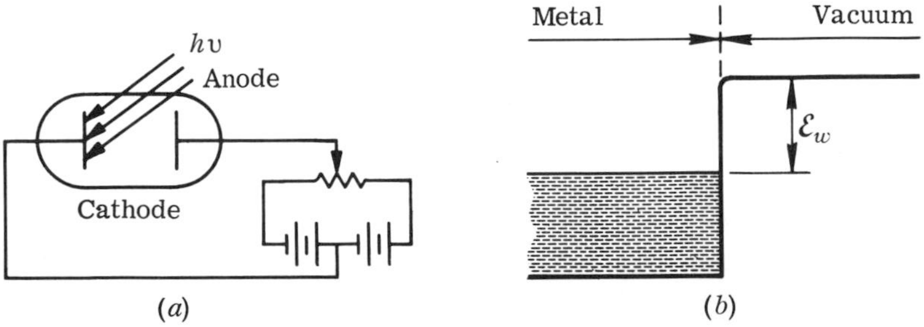

(a) The photoelectric experiment: A beam of light of frequency ν is incident on the photoemissive cathode of a vacuum-tube diode. (b) The potential-barrier diagram at the surface of a cathode. In order for an electron to escape from the cathode, it must receive an energy of the amount $\mathcal{E}_w$ from the impinging light.

Our discussion so far has been concerned with the energy of a photon. To complete the corpuscular nature of light, considerations should also be given to the momentum associated with a photon. We shall now discuss an experiment which clearly demonstrates that a photon has momentum. Compton, in investigating the scattering of X rays by paraffin, found that the frequency of the scattered light was lower than that of the incident radiation. Before attempting to explain the Compton effect, we must have some idea of the energy involved. For comparison with atomic binding energy, it is often convenient to express the energy of a photon in electron volts as follows:

$$\varepsilon \text{ (in ev)} = h\nu/e = 4.13 \times 10^{-15}\nu = 1.24 \times 10^4/\lambda \text{ (in Å)} \quad (1.8)$$

where λ is in angstroms (1 Å $= 10^{-10}$ m). For X rays from the molybdenum 0.71 Å line used by Compton, a photon energy about 18 kev is involved. Since the binding energy of electrons in paraffin is only of the order of 10 ev, the electrons can be considered as free.

In view of the foregoing discussion, the scattering process in Compton's experiment can be treated as a collision between two particles, one photon and one free electron at rest with rest mass m_0. Figure 1.4 shows a schematic illustration of the event. Conservation of energy requires that

$$h\nu_0 + m_0c^2 = h\nu + (m_0{}^2c^4 + p^2c^2)^{1/2} \quad (1.9)$$

where ν_0 and ν are the frequencies of the incident and scattered rays, respectively, c is the velocity of light, and p is the momentum of the electron after collision. Since the energy 18 kev is sufficient to push electrons into the relativistic region, relativistic energy expressions are used for electrons in Eq. (1.9). The momentum of a photon can be derived from the relativistic energy expression by demanding that the rest mass of a photon must be zero. Thus, for the momen-

Figure 1.4

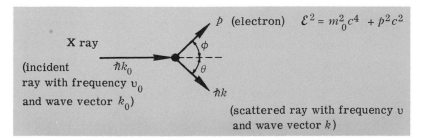

The Compton effect. In colliding with an electron, a light quantum transfers part of its energy and momentum to the electron. Consequently, both the wave vector **k** and the frequency ν of the light quantum change in the collision process.

tum P of a photon, we have

$$P = \frac{\mathcal{E}}{c} = \frac{h\nu}{c} = \hbar k \tag{1.10}$$

where $k = 2\pi/\lambda$. Applying the law of cosines to Fig. 1.4, we find

$$\hbar^2(k_0{}^2 + k^2 - 2k_0 k \cos \theta) = p^2 \tag{1.11}$$

Eliminating p^2 from Eqs. (1.9) and (1.11) gives

$$\lambda - \lambda_0 = \frac{h(1 - \cos \theta)}{m_0 c} \tag{1.12}$$

Equation (1.12) has been confirmed by experiments.

Although the photoelectric effect and the Compton effect clearly demonstrate that light exhibits particle behavior in its interaction with matter, there is also experimental evidence indicating that particles exhibit wavelike behavior. In the experiment of Davisson and Germer, the angular distribution of a beam of monoenergetic electrons scattered from a nickel target was investigated. The experiment showed that the angular distribution was analogous to optical diffraction patterns. Previous to this experiment, de Broglie not only postulated the hypothesis that the same dualism of wave and particle as exhibited by light might also occur in matter but also suggested the relation

$$\lambda = \frac{h}{p} \tag{1.13}$$

between the wavelength λ of the matter wave and the momentum p of a particle. For electrons with a well-defined kinetic energy, the de Broglie wavelength can be calculated from Eq. (1.13), and this value of λ can be checked against the one calculated from Eq. (1.6) with known values of d, θ, and n. Since the work done by Davisson and Germer, a large number of experiments on the wave nature of elementary particles, such as electrons and neutrons, has been conducted and the experimental results have confirmed de Broglie's hypothesis.

It may be worthwhile to summarize the essential points[1] in our discussion. In the classical picture, matter is considered as consisting of particles whose motion can be predicted by classical mechanics, Eq. (1.2). Associated with the motion of each particle, there are definite energy and momentum. In contrast to this corpuscle viewpoint, the electromagnetic wave theory, based on Maxwell's field equations, is generally believed to give a complete description of

---

[1] For a more comprehensive discussion the reader is referred to the book "Atomic Physics" by M. Born, Ch. 4, pp. 78–102, Hafner Publishing Co., New York, 1957.

wave phenomenon. Referring to Eq. (1.3), a wave is defined in terms of an amplitude E_0 and a phase angle $(\omega t - kz)$. The photoelectric and Compton experiments are significant because they clearly show that wave also possesses particle properties, an energy $\mathcal{E} = \hbar\omega$ and a momentum $p = \hbar k$. The experimental confirmation of de Broglie's hypothesis, that a particle with energy $\mathcal{E}$ and momentum p can be described as a matter wave having an angular frequency $\omega = \mathcal{E}/\hbar$ and a wave number $k = p/\hbar$, completes the other half of wave-particle dualism. It is this dualism which led to the development of quantum theory.

1.3 LINE SPECTRA AND THE BOHR ATOM

In the previous section we presented a series of arguments to prove that the classical laws of motion are incapable of explaining phenomena which involve interactions with atoms. The arguments are based on experiments which clearly demonstrate a dual wave-particle property for light and matter. The recognition of this wave-particle duality has been essential to the development of quantum theory. However, other works are equally important in shaping the theory of modern physics. For example, when an electrical discharge is produced in a low-pressure gas, emission of radiation is observed with a series of sharp, discrete wavelengths. The existence of sharp spectral lines is completely incomprehensible from the classical viewpoint. The first step toward an explanation of the phenomenon was taken by Bohr. He postulated that an atom could exist only in definite stationary states with discrete energies. An atom can change from a lower energy state $(\mathcal{E}_0)$ to a higher one $(\mathcal{E}_1)$ by absorbing a photon of energy $h\nu = \mathcal{E}_1 - \mathcal{E}_0$. Similarly, emission of a photon of the same energy results if the atom returns from the higher $(\mathcal{E}_1)$ to the lower energy $(\mathcal{E}_0)$ state.

At this point we should discuss the experiment of Franck and Hertz, which gives direct support to Bohr's theory. The experimental arrangement is shown in Fig. 1.5a. As electrons are accelerated toward the grid, they collide with the mercury atom. If the collision is elastic, the electron loses very little energy because of the large mass of the mercury atom. Thus the anode current is expected to increase with the grid voltage, as shown in the low-voltage region of Fig. 1.5b. However, as the accelerating voltage reaches 4.9 volts, inelastic collisions take place. Electrons, after losing all their energy to mercury atoms, can no longer reach the anode because the anode is negatively biased with respect to the grid. The sharp drop in current shown in Fig. 1.5b is a direct consequence of inelastic collisions. When the voltage reaches 9.8 volts, electrons can make two inelastic collisions, and a second dip in the current occurs. The important point of the experiment is that inelastic collisions take place only at a definite value of the electron kinetic energy, and this amount of energy is taken up in whole by mercury atoms.

Another important result of the Franck-Hertz experiment is that a strong ultraviolet radiation appears at 2536 Å when the voltage exceeds 4.9 volts. Using Eq. (1.8), we find that 4.9 ev corresponds to a wavelength of 2530 Å. Because of experimental uncertainty in locating the value of V from Fig. 1.5b, this calculated value agrees well with the observed value at which the ultraviolet radiation appears. Now, an explanation of the experiment based on Bohr's hypothesis is quite obvious. During an inelastic collision process, a mercury atom goes from a lower to a higher state by absorbing an energy 4.9 ev from an electron. When the excited mercury atom returns to the lower state, it emits a quantum of the same energy. The discrete nature of this energy, 4.9 ev in the case under discussion, is another important aspect of the quantum theory. In the following, we shall give a detailed presentation of Bohr's theory for two reasons. First, the work of Niels Bohr played a preeminent role in the development of the quantum theory. Second, in the discussion we may see clearly where quantization comes into play.

Consider an atom in which a single electron of mass m_0 and charge $-e$ revolves in a circular orbit of radius a around the nucleus with charge $+ze$. This is Bohr's model of an atom. A balance of the centrifugal force and the coulomb force gives the following:

$$\frac{m_0 v^2}{a} = \frac{ze^2}{4\pi\epsilon_0 a^2} \tag{1.14}$$

where v is the velocity of the electron and ϵ_0 is the dielectric constant of free space. To add to this classical formulation, Wilson and Sommerfeld postulated the quantum condition that the angular momentum of the electron takes on only certain definite values

Figure 1.5

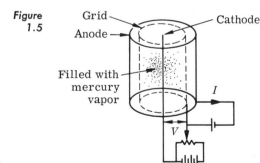

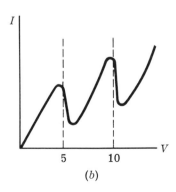

The experiment of Franck and Hertz: (a) the experimental setup and (b) the experimental curve.

such that[2]

$$m_0 a v = \frac{nh}{2\pi} \tag{1.15}$$

where n is an integer. Eliminating v from Eqs. (1.14) and (1.15) gives

$$a = \frac{a_1 n^2}{z} \quad \text{and} \quad a_1 = \frac{4\pi\epsilon_0 \hbar^2}{m_0 e^2} = 0.529 \text{ Å} \tag{1.16}$$

where $\hbar = h/2\pi$ is generally referred to as "h-bar" and a_1 represents the radius of the first Bohr orbit, commonly called the *Bohr radius*. The total energy $\mathcal{E}$ is a sum of the kinetic and the potential energy, or

$$\mathcal{E} = \frac{1}{2} m_0 v^2 - \frac{ze^2}{4\pi\epsilon_0 a} = -\frac{ze^2}{8\pi\epsilon_0 a} = -\frac{m_0 z^2 e^4}{2(4\pi\epsilon_0 \hbar)^2} \frac{1}{n^2} \tag{1.17}$$

because of Eqs. (1.14) and (1.16). Our presentation here may not follow the historical development but it is given in this order for smoothness. The circular orbit rather than the general elliptical orbit is also chosen for ease of discussion.

In 1885 Balmer found that the emission spectra of hydrogen obeyed the following empirical relation:

$$\frac{1}{\lambda} = R_\mathrm{H} \left(\frac{1}{2^2} - \frac{1}{n^2} \right) \tag{1.18}$$

where λ is the wavelength and R_H is known as the Rydberg constant equal to 109,677 cm^{-1} for hydrogen. Later, other series were observed. Using the Bohr frequency rule $h\nu = \mathcal{E}_k - \mathcal{E}_n$ mentioned earlier in this section, and the value of $\mathcal{E}$ from Eq. (1.17), we find

$$\frac{1}{\lambda} = \frac{\nu}{c} = \frac{m_0 z^2 e^4}{4\pi\hbar^3 c} \frac{1}{(4\pi\epsilon_0)^2} \left(\frac{1}{k^2} - \frac{1}{n^2} \right) \tag{1.19}$$

The five hydrogen series, named after their discoverers, are (1) $k = 1$, Lyman series, (2) $k = 2$, Balmer series, (3) $k = 3$, Paschen series, (4) $k = 4$, Brackett series, and (5) $k = 5$, Pfund series. The atomic transitions involved in these series are shown in Fig. 1.6. Since the diagram is for illustration only, it is not drawn to the energy scale. The numerical factor in front of the parenthesis in Eq. (1.19) has a value of 109,737 cm^{-1} in hydrogen with $z = 1$. The discrepancy between this calculated value and the observed value of R_H is due to the fact that an infinite nuclear mass is assumed in our calculation. During an ionization process, a hydrogen atom is excited from the ground state ($k = 1$) to an ionized state ($n = \infty$).

[2] The quantum condition $\oint p \, dx = nh$ was originally postulated by Planck in his analysis of the black-body radiation, and hence the constant h is known as the Planck constant. The application of the quantum condition was later extended to the hydrogen atom by Wilson and Sommerfeld.

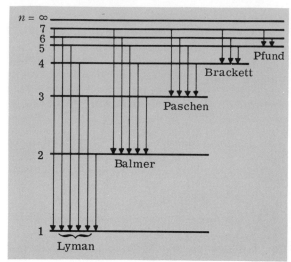

Energy levels of the hydrogen atom and the atomic transitions involved in the five hydrogen series.

Using these values in Eq. (1.19) and substituting the value of $1/\lambda$ into Eq. (1.8), we find the ionization potential for hydrogen atoms to be 13.6 volts, in agreement with experiments.

1.4 THE SCHRÖDINGER WAVE EQUATION

In the two preceding sections we discussed experimental evidences which demand a new concept to supplement the classical viewpoint. This new physics is the quantum physics or wave mechanics. The relationship between the quantum and the classical physics is best described by Bohr's *correspondence principle*. The classical Newtonian mechanics and classical electromagnetic theory, based on a wealth of experimental evidence, have proved their correctness in all processes of motion, macroscopic and microscopic, down to the motion of atoms as a whole. Therefore it is required that, in the limiting cases of large masses and of orbits of large dimensions, the quantum theory yield results identical to those of classical physics. Known as the correspondence principle, this is a guiding principle in the formulation of the new quantum physics.

The heart of the quantum theory is the Schrödinger wave equation

$$-\frac{\hbar^2}{2m_0}\left(\frac{\partial^2\psi}{\partial x^2} + \frac{\partial^2\psi}{\partial y^2} + \frac{\partial^2\psi}{\partial z^2}\right) + V(x,y,z)\psi = \mathcal{E}\psi = \frac{-\hbar}{i}\frac{\partial\psi}{\partial t} \quad (1.20)$$

Since the primary objective of this book is to discuss the use of quantum concepts in analyzing material properties, our approach to the Schrödinger equation is pragmatic. In the following, we shall use examples to show that Eq. (1.20) is compatible with ideas developed in the last two sections rather than present lengthy arguments leading to the logical development of Eq. (1.20). With this background in mind, we simply identify the terms in the Schrödinger equation by comparing it with the classical energy equation

$$K \text{ (kinetic energy)} + V \text{ (potential energy)} = \mathcal{E} \text{ (total energy)} \quad (1.21)$$

The physical meanings of the various differential operators in Eq. (1.20) will be explained later. At this point we shall state without proof that the first and second terms on the left-hand side of Eq. (1.20) arise from the kinetic and potential energies, respectively, whereas the term on the right-hand side of Eq. (1.20) has something to do with the total energy.

To show that the Schrödinger wave equation is consistent with wave-corpuscle dualism, we shall apply Eq. (1.20) to a free electron of momentum p and energy $\mathcal{E}$. According to de Broglie's wave theory of matter, a wave of the form

$$\psi = A \exp\left[-2\pi i \left(\nu t - \frac{x}{\lambda} \right) \right] \quad (1.22)$$

is associated with the electron motion with $\nu = \mathcal{E}/h$ and $1/\lambda = p/h$. For ease of discussion, a one-dimensional case (p in the x direction) is considered here. Substituting Eq. (1.22) into Eq. (1.20), we find

$$-\frac{\hbar^2}{2m_0} \frac{\partial^2 \psi}{\partial x^2} = \frac{p^2}{2m_0} \psi = K\psi \quad \text{and} \quad -\frac{\hbar}{i} \frac{\partial \psi}{\partial t} = h\nu\psi = \mathcal{E}\psi \quad (1.23)$$

Note that for free electrons, $V = 0$ and $K = \mathcal{E}$. In view of Eq. (1.23), the Schrödinger equation, Eq. (1.20), does seem to bear resemblance to the classical energy equation, Eq. (1.21), except for the physical meaning of ψ, which is left undefined. Since Eq. (1.22) is a solution of Eq. (1.20), de Broglie's wave theory of matter has naturally been incorporated into the formulation of the Schrödinger wave equation. Besides its energy, the other physical attributes of a particle are its position and velocity. To complete our discussion of the Schrödinger equation concerning these properties, we shall examine the physical significance of Eq. (1.22) more carefully.

Equation (1.22) represents a monochromatic plane wave propagating in the positive x direction. If a wave is to be associated with a particle, it seems reasonable to demand that the wavelike phenomena be confined to the vicinity of the particle. In circuit analysis we are quite familiar with the fact that a short pulse in time t can be expressed in terms of a distribution function in the frequency (ν) domain through the Fourier transform integral as illustrated in Fig. 1.7. In the same way, the wave motion of a localized (in space x) particle is related to the distribution function

$\varphi(p)$ in the momentum space by a transform equation

$$\psi(x,t) = \frac{1}{\sqrt{h}} \int_{-\infty}^{\infty} \varphi(p) \exp\left[\frac{i(px - \mathcal{E}t)}{\hbar}\right] dp \qquad (1.24)$$

We also know that an infinitely short pulse has an infinitely wide frequency spectrum. Similarly, a point charge requires an infinite spread in p. This is known as Heisenberg's *uncertainty principle*. From Eq. (1.24) we see that to keep ψ localized there must be a finite spread in p. In other words, a localized electron must be represented by a train of plane waves [Eq. (1.24)] instead of a monochromatic wave [Eq. (1.22)].

The wave represented by Eq. (1.24) is generally called a *wave packet*. A case of common interest is the wave packet having a Gaussian distribution. For this distribution it can be shown mathematically by Eq. (1.24) that the uncertainty in momentum Δp and that in position Δx are such that $\Delta p\, \Delta x \geqq \hbar/2$, provided that the standard deviations in the ψ and φ functions are taken as the proper representation for uncertainties Δx and Δp. In the classical theory, the state of a particle is completely determined by knowing the initial position and velocity of the particle and the force acting on

Figure 1.7

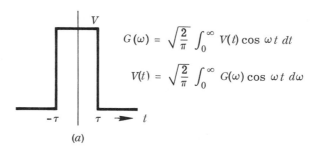

$$G(\omega) = \sqrt{\frac{2}{\pi}} \int_{0}^{\infty} V(t) \cos \omega t \, dt$$

$$V(t) = \sqrt{\frac{2}{\pi}} \int_{0}^{\infty} G(\omega) \cos \omega t \, d\omega$$

(a)

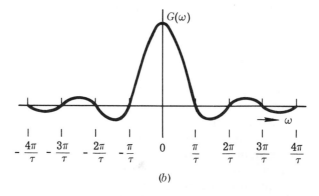

(b)

The impulse function $V(t)$ and the distribution function $G(\omega)$ in the frequency domain are related to each other through the Fourier integral. Note that a pulse of finite width $\Delta \tau$ has a finite bandwidth $\Delta \omega$.

the particle. In order to combine the wave and corpuscle views in a theoretical formalism, this deterministic view must be abandoned. In the quantum theory, the quantity $\hbar/2$ represents an absolute limit to the simultaneous measurement of coordinate and momentum. As mentioned earlier, the classical theory is completely incapable of dealing with interactions which involve masses and orbits of atomic dimensions. A quick reference to Eq. (1.15) shows that for the lowest Bohr orbit, the product of momentum and orbital radius is equal to $\hbar$. In view of Heisenberg's uncertainty principle, the failure of the classical theory is not only evident, but a new interpretation of the physical meaning of $m_0 v$ and a in Eq. (1.15) seems necessary. A quantum mechanical interpretation of these two quantities must wait, however, until we define the physical meaning of the wave function ψ and φ in the next section.

Let us now return to Eq. (1.22) and examine the meaning of velocity from both the wave and corpuscle viewpoints. For clarity and ease of discussion, we shall consider the superposition of two plane waves[3] of equal amplitude but with slightly different frequency and wave number.

$$
\begin{aligned}
\psi_1 + \psi_2 &= A(\exp\{i[(k + \Delta k)x - (\omega + \Delta\omega)t]\} \\
&\qquad + \exp\{i[(k - \Delta k)x - (\omega - \Delta\omega)t]\}) \\
&= 2A\cos(\Delta kx - \Delta\omega t)\exp[i(kx - \omega t)]
\end{aligned}
\tag{1.25}
$$

where $k = 2\pi/\lambda$ and $\omega = 2\pi\nu$. Two velocities can be defined from Eq. (1.25). The *phase velocity* v_p is the speed with which the point with constant phase angle moves as the wave propagates. Thus

$$
kx - \omega t = \text{const} \qquad \text{or} \qquad v_p = \frac{dx}{dt} = \frac{\omega}{k}
\tag{1.26}
$$

The envelope of the wave, however, moves with a different velocity which is called the *group velocity* v_g. In the limit Δk goes to zero,

$$
v_g = \lim_{\Delta k \to 0} \frac{\Delta\omega}{\Delta k} = \frac{d\omega}{dk} = \frac{d\nu}{d(1/\lambda)}
\tag{1.27}
$$

We shall now apply the foregoing results to a free electron. According to the theory of relativity, the energy and momentum of a particle with velocity v and rest mass m_0 are given by

$$
\mathcal{E} = \frac{m_0 c^2}{\sqrt{1 - \beta^2}} \qquad \text{and} \qquad p = \frac{m_0 c \beta}{\sqrt{1 - \beta^2}}
\tag{1.28}
$$

where $\beta = v/c$ and c is the velocity of light. From Eqs. (1.27) and

---

[3] For a more general discussion, the reader is referred to "Atomic Physics" by M. Born, Appendix XI, pp. 336–337, Hafner Publishing Co., New York, 1957.

(1.28) the phase velocity of the electron is given by

$$v_p = \frac{\omega}{k} = \frac{\mathcal{E}}{p} = \frac{c^2}{v} \tag{1.29}$$

and the group velocity by

$$v_g = \frac{\partial \mathcal{E}}{\partial p} = \frac{\partial \mathcal{E}/\partial \beta}{\partial p/\partial \beta} = v \tag{1.30}$$

Apparently the phase velocity, which is greater than c for $v < c$, has no physical significance except for the fact it is the velocity of propagation of points of constant phase. On the other hand, the group velocity of a wave packet is actually associated with the velocity of a particle. From this simple illustration we have shown that a wave packet, of which Eq. (1.25) is a special case, has the propagation property of a wave and the group behavior of a particle. Since Eq. (1.24) is a solution to the Schrödinger equation, we have shown that the equation is consistent with wave-corpuscle dualism, which is the cornerstone for the theoretic formulation of modern physics.

1.5 PHYSICAL INTERPRETATION OF THE WAVE FUNCTION

In the previous section, although we have shown that the wave function $\psi(x,t)$ has dual wave-particle properties, no attempt has been made to attach any physical meaning to ψ. As concepts, wave and particle usually appear far apart in our thinking, we must seek a deeper and better understanding of the relationship between these two concepts. This understanding lies in the interpretation of the physical meaning of ψ, which is to supplement the mathematical treatment given in the previous section. To make our points clear and definite, we refer again to the electron-diffraction experiment. If the same experiment is carried out with single electrons, two definite possibilities exist regarding the outcome. If each individual electron acted like a wave, the same diffraction pattern would appear whole but with reduced intensity. Then the particle concept would be completely destroyed, but this is not what happens. Individual electrons arrive at different spots on the screen, and these spots are scattered over the whole screen. The important points which need be stressed are the following. Individually each electron does behave like a particle, but it is not possible for us to predict where any one individual electron would land on the screen.

In a way the situation with the electron-diffraction experiment is similar to that of throwing a pair of dice. The appearance of the number two in a given throw is a single event and is unpredictable. When numerous throwings are made, however, the probability of getting two is precisely one out of thirty-six. An analogous situation applies to the electron-diffraction experiment. When electrons arrive in large numbers, they will appear on the screen with a certain statistical distribution. It is this statistical distribution which is observed in the experiment of Davisson and Germer, and it is the same statistical distribution with which we shall associate the wave function $\psi(x,t)$. It is obvious that if wave mechanics are to be of any use at all, we must be able to deduce the outcome of an experiment from knowledge of ψ. Let us now turn to the wave aspect of ψ. The intensity of a wave is proportional to the square of the wave amplitude or $|\psi|^2$, and hence $|\psi|^2$, not ψ, gives the statistical distribution of electrons on the screen. The above discussion leads us to interpret $|\psi|^2$ as the probability of a single electron hitting the screen at a certain point.

This probabilistic interpretation of $|\psi|^2$ was originally advanced by Born. If a particle is described by a certain wave function ψ, the probability dW of finding the particle in a volume element $dr = dx\,dy\,dz$ at the point r is proportional to

$$dW(r) = |\psi|^2\,dr = \psi\psi^*\,dr \qquad (1.31)$$

Since, in general, ψ is complex, the asterisk indicates the complex conjugate. If a total number of N particles is involved, integrating Eq. (1.31) gives $\int|\psi|^2\,dr = N$. However, generally it is convenient to use a normalized wave function $\psi_1 = \psi/\sqrt{N}$ such that $\int\psi_1\psi_1^*\,dr = 1$. Such one-electron wave function will be used extensively in later chapters when we discuss the behavior of electrons in semiconductors. Obviously many wave functions are solutions of the Schrödinger wave equation. Those wave functions which satisfy the physical conditions of a given problem are called *proper* or *eigen functions*. We shall discuss the properties of eigen functions when we treat a specific problem in later sections.

The statistical interpretation of $|\psi|^2$ permits us to make other predictions. According to the theory of probability, the expectation value $\langle f \rangle$ of a function $f(r)$ is given by

$$\langle f \rangle = \frac{\int f(r)\,dW}{\int dW} = \frac{\int f(r)|\psi|^2\,dr}{\int|\psi|^2\,dr} \qquad (1.32)$$

If we make an even bet with a dollar that either the number two or the number eleven will appear, the expectation value of our bet is $\frac{1}{36} + \frac{1}{18} = \frac{1}{12}$ of a dollar, the summation being used for a discrete distribution. In other words, the quantity $\langle f \rangle$ is a weighted average of the possible values of the function $f(r)$, the weighting factor being the probability that a particle turns up at position r. An obvious application of Eq. (1.32) is to find the expectation value

$\langle x \rangle$ of the position of a particle

$$\langle x \rangle = \int_{-\infty}^{\infty} x |\psi|^2 \, dr = \int_{-\infty}^{\infty} \psi^* x \psi \, dx \, dy \, dz \qquad (1.33)$$

For simplicity, the wave function ψ in Eq. (1.33) is taken to be normalized and the integral represents a triple integration.

Besides position, other quantities which may be of interest are the total energy and the momentum of the particle. We may multiply both sides of Eq. (1.23) by ψ^* and then integrate the products over dr to obtain the expectation values of the total and the kinetic energy, $\langle \mathcal{E} \rangle$ and $\langle p^2/2m_0 \rangle$. Note that in the Schrödinger equation, Eq. (1.20), differential operators such as $(\hbar/i)(\partial/\partial x)$ are used in place of the classical momentum p_x. Since the momentum operator will be used frequently in later chapters, we shall justify the use of such an operator in the following discussion. If the correspondence principle is a guide, it seems reasonable to suggest that the expectation value of momentum $\langle p \rangle$ should be related to $\langle x \rangle$ by

$$\langle p \rangle = m_0 \frac{d\langle x \rangle}{dt} = m_0 \int_{-\infty}^{\infty} x \left(\psi \frac{\partial \psi^*}{\partial t} + \psi^* \frac{\partial \psi}{\partial t} \right) dx \qquad (1.34)$$

For simplicity, a one-dimensional case is assumed. We should also point out that in the quantum mechanical description, information about the motion of a particle is carried in the time variation of the wave function ψ and not in the space variable x.

Using Eq. (1.20) and its conjugate equation in Eq. (1.34), we find

$$\langle p \rangle = \frac{i\hbar}{2} \int_{-\infty}^{\infty} x \left(\psi^* \frac{\partial^2 \psi}{\partial x^2} - \psi \frac{\partial^2 \psi^*}{\partial x^2} \right) dx \qquad (1.35)$$

Integrating Eq. (1.35) by parts, with ψ and $\partial \psi/\partial x$ assumed to be zero at $x = \pm \infty$, gives

$$\langle p \rangle = \frac{\hbar}{2i} \int_{-\infty}^{\infty} \left(\psi^* \frac{\partial \psi}{\partial x} - \psi \frac{\partial \psi^*}{\partial x} \right) dx \qquad (1.36)$$

Since $\langle p \rangle$ is an experimentally measurable quantity, it must be real. Furthermore, $\psi^* \, \partial \psi/\partial x$ and $\psi \, \partial \psi^*/\partial x$ are complex conjugate to each other; hence the real part of $\psi^* \, \partial \psi/\partial x$ is given by

$$\text{Re} \int_{-\infty}^{\infty} \psi^* \frac{\partial \psi}{\partial x} \, dx = \frac{1}{2} \int_{-\infty}^{\infty} \left(\psi^* \frac{\partial \psi}{\partial x} + \psi \frac{\partial \psi^*}{\partial x} \right) dx = \frac{1}{2} |\psi|^2 \Big|_{-\infty}^{\infty} = 0 \tag{1.37}$$

In view of Eq. (1.37), we can write Eq. (1.36) as

$$\langle p \rangle = \int_{-\infty}^{\infty} \psi^* \left(\frac{\hbar}{i} \frac{\partial}{\partial x} \right) \psi \, dx \qquad (1.38)$$

and $\langle p \rangle$ is proved to be real.

Several comments about Eq. (1.38) are in order. In comparing Eq. (1.38) with Eq. (1.32), we see that a differential operator $(\hbar/i)(\partial/\partial x)$ appears in place of p_x in Eq. (1.38). This result is consistent with the fact that p_x, p_y, and p_z in Eq. (1.20) are replaced by the same differential operators. Since a differential operator is involved in Eq. (1.38), the order of the terms in the integrand must be maintained. Finally, according to Eq. (1.24), the wave functions $\psi(x)$ and $\varphi(p)$ are Fourier transforms of one another. In terms of $\varphi(p)$, the expectation values of x and p become

$$\langle x \rangle = \int \varphi^*(p) i\hbar \frac{\partial \varphi(p)}{\partial p} \, dp \tag{1.39}$$

$$\langle p \rangle = \int \varphi^*(p) p \varphi(p) \, dp \tag{1.40}$$

where $dp = dp_x \, dp_y \, dp_z$. The quantity $|\varphi|^2 \, dp$ represents the probability of finding a particle in a volume element $dp_x \, dp_y \, dp_z$ at the point $p(p_x, p_y, p_z)$ in momentum space. To find $\varphi(p)$ directly from the Schrödinger equation, the space variables x, y, and z in Eq. (1.20) should be replaced by their respective differential operators $i\hbar\partial/\partial p$. However, for later applications, we shall use the wave function $\psi(x)$ exclusively. Therefore only Eqs. (1.20), (1.33), and (1.38) will be referred to in later chapters.

1.6 THE HARMONIC OSCILLATOR

A particle bound to its equilibrium position by a force which increases linearly with its displacement from the equilibrium position is a *harmonic oscillator*. The classical expression for the energy of a particle in one-dimensional harmonic motion is given by

$$\mathcal{E} = \text{KE} + \text{PE} = \frac{p^2}{2M} + \frac{\gamma x^2}{2} \tag{1.41}$$

where γ is the force constant or Hooke's-law constant. The equation of motion of the particle may be found by using Newton's second law:

$$M \frac{d^2x}{dt^2} + \gamma x = 0 \tag{1.42}$$

From Eq. (1.42) we find the angular frequency of oscillation $\omega = \sqrt{\gamma/M}$ and from Eq. (1.41) the amplitude of oscillation $a = \sqrt{2\mathcal{E}/\gamma}$. The quantum mechanical or Schrödinger equation for a harmonic oscillator takes the form

$$\frac{\hbar^2}{2M} \frac{d^2\psi}{dx^2} + \left(\mathcal{E} - \frac{\gamma x^2}{2} \right) \psi = 0 \tag{1.43}$$

when p^2 in Eq. (1.41) is replaced by quantum mechanical operator $-\hbar^2 d^2/dx^2$. Not only has Eq. (1.43) played an important part in the development of quantum mechanics but the result from Eq. (1.43) can also be applied to many similar problems in later applications.

It is convenient to express Eq. (1.43) in terms of dimensionless quantities. In our previous discussion, the energy of a photon, which can be considered classically as resulting from an oscillating dipole, is quantized in units of $\hbar\omega$. It seems natural, therefore, to use

$$y = \frac{x}{a} = \sqrt{\frac{M\omega}{\hbar}}\,x \qquad \text{and} \qquad \mathcal{E} = \frac{\lambda\hbar\omega}{2} \qquad (1.44)$$

where $a = \sqrt{\hbar\omega/\gamma}$. The factor $\frac{1}{2}$ used in $\mathcal{E}$ will become obvious later. In terms of y and λ, Eq. (1.43) becomes

$$\frac{d^2\psi}{dy^2} + (\lambda - y^2)\psi = 0 \qquad (1.45)$$

Equation (1.45) can be solved by proposing a solution of the kind

$$\psi = \exp\left(-\frac{y^2}{2}\right) \sum_{n=0}^{\infty} a_n y^{n+\alpha} \qquad (1.46)$$

Since the procedure of finding the explicit form for $\psi(x)$ can be found in most texts on quantum mechanics,[4] only an outline of the mathematics will be given here.

Substituting Eq. (1.46) into Eq. (1.45) and collecting terms with equal powers of y, we find that for a nonzero solution $\alpha = 0$ or 1. Further, a recurrence relation exists:

$$a_{n+2} = \frac{2n - (\lambda - 1)}{(n + 1)(n + 2)}\, a_n \qquad (1.47)$$

For large y, the series in Eq. (1.46) behaves like $\exp y^2$ if it does not terminate. Obviously, under such circumstances $\psi(x)$ will become infinitely large. Yet we know that the motion of a classical oscillator is bounded. That means the series must terminate and become a polynomial. From Eq. (1.47) the condition for $|\psi|^2$ being bounded gives

$$\lambda = 2n + 1 \qquad (1.48)$$

where n is an integer, and the energy of the oscillator is

$$\mathcal{E} = \left(n + \frac{1}{2}\right)\hbar\omega \qquad (1.49)$$

---

[4]   See for example, L. I. Schiff, "Quantum Mechanics," McGraw-Hill Book Company, New York, 1949; N. F. Mott and I. N. Sneddon, "Wave Mechanics and Its Applications," Oxford University Press, Fair Lawn, N.J., 1948; J. I. Powell and B. Crasemann, "Quantum Mechanics," Addison Wesley Publishing Company, Inc., Reading, Mass., 1963.

The reader is asked to show that according to the old quantum theory based on the celebrated Planck rule of quantization $\oint p\,dx = nh$, $\mathcal{E} = n\hbar\omega$. In the new quantum theory the minimum amount of energy is not zero but $\hbar\omega/2$ and is called the zero-point energy; it is a direct consequence of the uncertainty principle. This zero-point energy plays an important role in explaining the origin of van der Waals forces between molecules.

The polynomials $\psi_n(y)$ obtained from evaluating a_n are known as *Hermite polynomials*. The first few polynomials are listed below.

$$
\begin{aligned}
\psi_0 &= A_0 \exp\left(\frac{-y^2}{2}\right) & \psi_2 &= A_2(4y^2 - 2)\exp\left(\frac{-y^2}{2}\right) \\
\psi_1 &= A_1 2y \exp\left(\frac{-y^2}{2}\right) & \psi_3 &= A_3(8y^3 - 12y)\exp\left(\frac{-y^2}{2}\right)
\end{aligned}
\tag{1.50}
$$

The function ψ_1 and the classical turning point $a_1 = \sqrt{3}\,a$ are shown in Fig. 1.8 for comparison. Note that the wave function

Figure 1.8

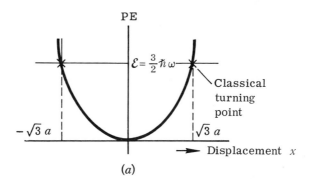

(a)

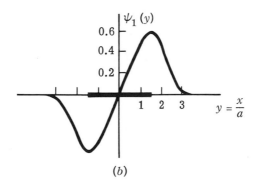

(b)

(a) *The potential energy against displacement curve. For* $\mathcal{E} = \tfrac{3}{2}\hbar\omega$, *the classical turning point is at* $x = \sqrt{3}a$ *as predicted from Eq. (1.41) by setting* KE = 0. (b) *The quantum-mechanical wave function for a harmonic oscillator with* $\mathcal{E} = \tfrac{3}{2}\hbar\omega$. *The heavy horizontal line indicates the region within which the motion of the particle is confined according to the classical theory.*

extends well into the region forbidden classically. The various poly-nomials are normalized orthogonal functions, and the constant A_n in ψ_n is a normalization constant such that

$$\int_{-\infty}^{\infty} \psi_m \psi_n \, dy = \delta_{mn} \tag{1.51}$$

where $\delta_{mn} = 1$ if $m = n$ and $\delta_{mn} = 0$ if $m \neq n$ and $A_n^{-2} = 2^n n! \sqrt{\pi}$. The function δ_{mn} is known as the *Kronecker delta*. A recurrence rela-tion exists between different ψ_n's:

$$\sqrt{n+1}\, \psi_{n+1} - \sqrt{2}\, y\psi_n + \sqrt{n}\, \psi_{n-1} = 0 \tag{1.52}$$

If Eq. (1.52) is multiplied by either ψ_{n+1} or ψ_{n-1} and integrated, the following equations result:

$$\int_{-\infty}^{\infty} \psi_{n+1} y\psi_n \, dy = \sqrt{\frac{n+1}{2}} \quad \text{and} \quad \int_{-\infty}^{\infty} \psi_{n-1} y\psi_n \, dy = \sqrt{\frac{n}{2}} \tag{1.53}$$

Since we shall use the result of Eq. (1.53) quite often later in discussing the interactions of radiation field and vibrational motion with matter, we shall derive Eq. (1.53) from a different approach so that we can see the physical meaning much more clearly. Let us rewrite Eq. (1.45) as

$$\mathcal{3C}\psi = \left(\frac{d^2}{dy^2} - y^2\right)\psi = \lambda\psi \tag{1.54}$$

where $\mathcal{3C}$ stands for $(d^2/dy^2 - y^2)$. The symbol $\mathcal{3C}$ is called the *Ham-iltonian operator* and is derived from the classical Hamiltonian by replacing p by the corresponding operator. The $\mathcal{3C}$ in Eq. (1.54) differs, however, from the ordinary $\mathcal{3C}$ by a constant as a result of the change of variables expressed by Eq. (1.44).

Next, we define two operators

$$A = \left(y + \frac{d}{dy}\right) \quad \text{and} \quad B = \left(y - \frac{d}{dy}\right) \tag{1.55}$$

of which the physical meaning will become quite clear a little later. Note that A and B involve differential operators and thus the order of A and B must be preserved in a product of the two, for example AB. Through Eq. (1.55) it is easy to show that

$$AB\psi = (\mathcal{3C} + 1)\psi \quad \text{and} \quad BA\psi = (\mathcal{3C} - 1)\psi \tag{1.56}$$

Other useful expressions for the following discussion are

$$A\mathcal{3C} - \mathcal{3C}A = A(AB - 1) - (AB - 1)A = A(AB - BA) = 2A$$
$$B\mathcal{3C} - \mathcal{3C}B = B(BA + 1) - (BA + 1)B = B(BA - AB) = -2B \tag{1.57}$$

For brevity, the wave function ψ is omitted in Eq. (1.57). Suppose that there exists a certain state ψ_n with energy $\mathcal{E}_n$ which satisfies the equation

$$\mathcal{H}\psi_n = \lambda\psi_n \tag{1.58}$$

Then we can show that the functions $A\psi_n$ and $B\psi_n$, respectively, obey

$$\mathcal{H}(A\psi_n) = (A\mathcal{H} - 2A)\psi_n = (\lambda - 2)(A\psi_n)$$
$$\mathcal{H}(B\psi_n) = (B\mathcal{H} + 2B)\psi_n = (\lambda + 2)(B\psi_n) \tag{1.59}$$

The physical meaning of operators A and B becomes clear by comparing Eq. (1.59) with Eq. (1.58). The functions $A\psi_n$ and $B\psi_n$ now have eigenvalues $\lambda - 2$ and $\lambda + 2$, respectively, as compared to the eigenvalue λ for ψ_n. Since λ is related to the energy of a harmonic oscillator, operating A on ψ_n lowers the energy of the oscillator by $\Delta\lambda = -2$ or $\Delta\mathcal{E} = -\hbar\omega$. Therefore the operators A and B are often referred to as *annihilation* and *creation operators*, respectively, because they destroy or create a quantum of energy $\hbar\omega$. If we repeat the application of A on $A\psi_n$ and its resultant terms, the energy of the system is lowered indefinitely. There must be a limit to this. The energy of the system is given by Eq. (1.41). According to the uncertainty principle the lowest energy possible is given by

$$\mathcal{E} = \frac{(\Delta p)^2}{2M} + \frac{\gamma(\Delta x)^2}{2} = \left(\frac{\Delta p}{\sqrt{2M}} - \sqrt{\frac{\gamma}{2}}\,\Delta x\right)^2 + \sqrt{\frac{\gamma}{M}}\,\Delta p\,\Delta x \tag{1.60}$$

where Δp and Δx represent, respectively, the uncertainties in the values of the momentum and position of a particle. Since $\Delta p\,\Delta x \geq \hbar/2$, the lowest value for $\mathcal{E}$ is $\hbar\omega/2$. According to Eq. (1.44) this corresponds to a lowest possible value for λ of 1.

If we assign $\mathcal{E}_0 = \hbar\omega/2$ for ψ_0, then it is apparent that for ψ_n $\mathcal{E}_n = (n + \frac{1}{2})\hbar\omega$ in agreement with Eq. (1.49). Furthermore, the change in energy must obey the quantum condition

$$\Delta\mathcal{E} = \pm\hbar\omega \qquad \text{or} \qquad \Delta n = \pm 1 \tag{1.61}$$

in view of Eq. (1.59). In other words,

$$\int_{-\infty}^{\infty} \psi_m A\psi_n \, dy = 0$$
$$\int_{-\infty}^{\infty} \psi_n B\psi_m \, dy = 0 \qquad \text{unless } n = m + 1 \tag{1.62}$$

The rule $\Delta n = \pm 1$ is a quantum-mechanical *selection rule*. We shall have other but similar selection rules later when we discuss transitions between different atomic levels.

Operators A and B also have other interesting properties. According to Eq. (1.59) we can write

$$A\psi_n = a_n\psi_{n-1} \qquad \text{and} \qquad B\psi_n = b_{n+1}\psi_{n+1} \qquad (1.63)$$

The coefficients a_n and b_{n+1} are found by considering the following:

$$BA\psi_n = Ba_n\psi_{n-1} = a_nb_n\psi_n \qquad (1.64)$$

$$\int_{-\infty}^{\infty} \psi_n B(A\psi_n)\,dy = \int_{-\infty}^{\infty} (A\psi_n)(A\psi_n)\,dy \qquad (1.65)$$

Equation (1.65) is obtained by integration by parts, remembering that $\psi_n = 0$ at infinity. Since ψ_n is normalized, the left- and right-hand sides of Eq. (1.65) are given by a_nb_n and a_n^2, respectively. Hence we conclude that $a_n = b_n$. Furthermore, from Eqs. (1.56) and (1.64), realizing that $(\mathcal{H} - 1)\psi_n = 2n\psi_n$, we have $a_n^2 = a_nb_n = 2n$. Thus Eq. (1.63) becomes

$$A\psi_n = \sqrt{2n}\,\psi_{n-1} \qquad \text{and} \qquad B\psi_n = \sqrt{2(n+1)}\,\psi_{n+1} \qquad (1.66)$$

Note that Eq. (1.66) will lead to results identical to Eq. (1.53).

In summary, we see from the preceding analysis that the energy of a harmonic oscillator is quantized according to Eq. (1.49) in steps of $\hbar\omega$. An interesting example to which Eq. (1.49) can directly apply is the mechanical vibration (elastic wave) associated with atoms in a crystal. As will be discussed in Chapter 2, the lattice vibration can be analyzed in terms of its Fourier components in a way analogous to that applied to an electric signal. Each Fourier component has a definite frequency and wave number. According to our discussion of wave-particle dualism in Sec. 1.2, any wave phenomenon must have its particle counterpart. For electromagnetic waves, the quantum particle is called *photon*, whereas for elastic waves it is called *phonon*.

Let us discuss a simple example in which phonons are involved. When an electric current passes through a conductor, an ohmic loss develops in the form of heat. In quantum mechanical terms, a thermally excited solid can be thought of as a gas of phonons and the mechanism of energy transfer as a process involving the interaction of electrons and phonons. The ohmic loss and the subsequent heating up of the solid are a direct consequence of the fact that during the process of interaction more phonons are created than destroyed. In a formal treatment of such a process we shall use the annihilation and creation operators A and B defined in Eq. (1.55) and their operational relations given in Eq. (1.66). The detailed mathematics presented in this section serves mainly to provide the necessary background for our discussion of electron-phonon interaction in Chapter 4. In a first reading, however, the reader need not follow the detailed mathematics.

1.7 THE HYDROGEN ATOM[5]

The Hamiltonian of a hydrogen atom is given by

$$\mathcal{H} = \frac{1}{2m_0}(p_x{}^2 + p_y{}^2 + p_z{}^2) - \frac{e^2}{4\pi\epsilon_0 r} \tag{1.67}$$

if the mass of the proton is assumed infinitely large. The Schrödinger equation derived from $\mathcal{H}$ is

$$\left(-\frac{\hbar^2}{2m_0}\nabla^2 - \frac{e^2}{4\pi\epsilon_0 r}\right)\psi = \mathcal{E}\psi \tag{1.68}$$

where ∇^2 is the Laplacian or Laplace's operator. In polar coordinates, r, θ, and ϕ, Eq. (1.68) becomes

$$\frac{\hbar^2}{2m_0}\left(\frac{\partial^2}{\partial r^2} + \frac{2}{r}\frac{\partial}{\partial r} + \frac{1}{r^2}\frac{\partial^2}{\partial\theta^2} + \frac{\cot\theta}{r^2}\frac{\partial}{\partial\theta} + \frac{1}{r^2\sin^2\theta}\frac{\partial^2}{\partial\phi^2}\right)\psi$$

$$+ \left(\mathcal{E} + \frac{e^2}{4\pi\epsilon_0 r}\right)\psi = 0 \quad (1.69)$$

This equation can be solved by separation of variables. If we let

$$\psi = R(r)\Theta(\theta)\Phi(\phi) \tag{1.70}$$

then Eq. (1.69) can be split into three separate equations:

$$\left(\frac{d^2}{d\phi^2} + m^2\right)\Phi = 0 \tag{1.71}$$

$$\left(\frac{d^2}{d\theta^2} + \cot\theta\frac{d}{d\theta} + \lambda - \frac{m^2}{\sin^2\theta}\right)\Theta = 0 \tag{1.72}$$

$$\left[\frac{d^2}{dr^2} + \frac{2}{r}\frac{d}{dr} + \frac{2m_0}{\hbar^2}\left(\mathcal{E} + \frac{e^2}{4\pi\epsilon_0 r}\right) - \frac{\lambda}{r^2}\right]R = 0 \tag{1.73}$$

The solution of Eq. (1.71) is given by

$$\Phi = \exp\left(\pm im\phi\right) \qquad \text{or} \qquad \sin m\phi \text{ and } \cos m\phi \tag{1.74}$$

For the function Φ to be single valued, m must be an integer. Note that in the above equations we designate m_0 as the mass of an electron and m as the quantum number to save confusion. Equation (1.72) can be transformed into the associated Legendre equation

$$\frac{d}{du}\left[(1 - u^2)\frac{dP}{du}\right] + \left(\lambda - \frac{m^2}{1 - u^2}\right)P = 0 \tag{1.75}$$

by the substitution $u = \cos\theta$ and $\Theta(\theta) = P(u)$. The preceding equation can be solved again by the method of series in a way similar to

[5] For a first reading, the reader need not follow the detailed mathematics presented early in this section. He may proceed directly to the paragraph following Eq. (1.94) for the concluding remarks.

that outlined for Eq. (1.45). Similarly, for a bounded solution at $u = \pm 1$, the parameter λ in Eq. (1.75) must be

$$\lambda = l(l + 1)$$

and $$\qquad m = -l, -(l - 1), \ldots, 0, \ldots, (l - 1), l \qquad (1.76)$$

where l is a positive integer. If these conditions are satisfied, the solutions of Eq. (1.75) are *associated Legendre polynomials*, $P_l{}^m(\cos \theta)$. The functions $P_l{}^m$ can be computed from the Legendre polynomials P_l by the following relation:

$$P_l{}^m(\cos \theta) = \sin^m \theta \frac{d^m}{du^m} P_l \qquad (1.77)$$

where P_l is given by

$$P_l = \frac{1}{2^l l!} \frac{d^l}{du^l} (u^2 - 1)^l \qquad (1.78)$$

The combination of $\Theta(\theta)$ and $\Phi(\phi)$ is generally called the *spherical harmonics* $Y_l{}^m(\theta, \phi)$, where

$$Y_l{}^m(\theta, \phi) = (-1)^m \sqrt{\frac{2l + 1}{4\pi} \frac{(l - m)!}{(l + m)!}} \exp{(im\phi)} P_l{}^m(\cos \theta) \qquad (1.79)$$

and m is a positive integer. The case for m being negative will be given later in Eq. (1.99). The various spherical harmonics form a complete orthonormal set such that

$$\int_0^\pi \int_0^{2\pi} Y_l{}^m(\theta, \phi) \, Y_{l'}^{m'}(\theta, \phi) \sin \theta \, d\theta \, d\phi = \delta_{ll'} \delta_{mm'} \qquad (1.80)$$

where the Kronecker delta ($\delta_{ll'}$ or $\delta_{mm'}$) has the same meaning as in Eq. (1.51). For future references we list below some of the spherical harmonics:

For $l = 0$: $\qquad Y_0{}^0 = \dfrac{1}{\sqrt{4\pi}}$

For $l = 1$: $\qquad Y_1{}^0 = \sqrt{\dfrac{3}{4\pi}} \cos \theta$

$$Y_1{}^1 = -\sqrt{\frac{3}{8\pi}} \exp{(i\phi)} \sin \theta$$

For $l = 2$: $\qquad Y_2{}^0 = \sqrt{\dfrac{5}{16\pi}} (3 \cos^2 \theta - 1)$

$$(1.81)$$

$$Y_2{}^1 = -\sqrt{\frac{15}{8\pi}} \exp{(i\phi)} \sin \theta \cos \theta$$

$$Y_2{}^2 = \sqrt{\frac{15}{32\pi}} \exp{(2i\phi)} \sin^2 \theta$$

In many physical problems it is useful for analytical purposes to expand a given function $f(\theta,\phi)$ in terms of Y_l^m as follows:

$$f(\theta,\phi) = \sum_{l=0}^{\infty} \sum_{m=-l}^{l} a_{lm} Y_l^m(\theta,\phi) \qquad (1.82)$$

A simple example is the electrostatic interaction e^2/r_{12} between two charged particles, r_{12} being the distance of separation.

In treating Eq. (1.73) it is again convenient to express ε and r in terms of dimensionless quantities. In view of our discussion in Sec. 1.3 we choose

$$\varepsilon = \frac{m_0 e^4}{2(4\pi\epsilon_0\hbar)^2}\,\varepsilon \quad \text{and} \quad r = \frac{4\pi\epsilon_0\hbar^2}{m_0 e^2}\,\rho = a_1\rho \qquad (1.83)$$

and then Eq. (1.73) becomes

$$\frac{d^2R}{d\rho^2} + \frac{2}{\rho}\frac{dR}{d\rho} + \left[\varepsilon + \frac{2}{\rho} - \frac{l(l+1)}{\rho^2}\right]R = 0 \qquad (1.84)$$

For $\varepsilon < 0$, we know that the electron is bound to the nucleus. For large ρ, the last two terms in the brackets of Eq. (1.84) can be neglected in comparison with ε; hence the solution should behave asymptotically as

$$R \sim \rho^{-1} \exp\left(-\sqrt{-\varepsilon}\,\rho\right) \qquad (1.85)$$

Thus it seems reasonable to assume a solution of the form

$$R = [\rho^{-1} \exp\left(-\sqrt{-\varepsilon}\,\rho\right)]\rho^\alpha \sum_\nu a_\nu\rho^\nu \qquad (1.86)$$

where ν is a positive integer. Substituting Eq. (1.86) into Eq. (1.84) and collecting terms involving ρ^α, we find

$$[\alpha(\alpha-1) - l(l+1)]a_0 = 0 \qquad (1.87)$$

which has two possible solutions, $\alpha = -l$ and $\alpha = l+1$.

Since the integral $\int R^2 r^2\,dr\int|Y_l^m|^2 \sin\theta\,d\theta\,d\phi$ is interpreted as being the probability of finding an electron, it must be finite. Therefore, to keep $\int R^2 r^2\,dr$ finite, $R^2 r^2$ must also behave properly at $r = 0$. Consequently, only $\alpha = l+1$ is allowed. Since a_0 is nonzero, the other coefficients a_ν are all related to a_0 through the following general relation:

$$a_{\nu+1} = 2\,\frac{(\nu+l+1)\sqrt{-\varepsilon}-1}{2(\nu+1)(l+1)+\nu(\nu+1)}\,a_\nu \qquad (1.88)$$

For large ν, the ratio $a_{\nu+1}/a_\nu$ approaches $2\sqrt{-\varepsilon}/\nu$; thus the terms of the series behave very much like those in the expansion of $\exp(2\sqrt{-\varepsilon}\,\rho)$. Consequently, for a bounded R at large ρ, the series must terminate, requiring

$$\varepsilon = -\frac{1}{(\nu+l+1)^2} = -\frac{1}{n^2} \qquad (1.89)$$

Since ν and l are integers, so is n, with $n = \nu + l + 1$. Furthermore, Eq. (1.86) becomes

$$R(\rho) = N\rho^l \exp\left(-\frac{\rho}{n}\right) \mathcal{L}_{n-l-1}^{2l+1}\left(\frac{2\rho}{n}\right)$$

(1.90)

where N is a normalization constant.

The polynomial $\mathcal{L}_{n-l-1}^{2l+1}(2\rho/n)$ is known as the *associated Laguerre polynomial* of degree $n - l - 1$ and of order $2l + 1$, and is related to the Laguerre polynomial $\mathcal{L}_{n-l-1}(2\rho/n)$ by

$$\mathcal{L}_{n-l-1}^{2l+1}(t) = \frac{d^{2l+1}}{dt^{2l+1}} \mathcal{L}_{n+l}(t)$$

(1.91)

where $t = 2\rho/n$. The Laguerre polynomial of variable t may be defined by means of the generating function

$$\left(\frac{1}{1-u}\right) \exp\left(\frac{-ut}{1-u}\right) = \sum_{p=0}^{\infty} \frac{\mathcal{L}_p(t)}{p!} u^p$$

(1.92)

Since a formal treatment of the radial function as well as of the Laguerre polynomials can be found in most texts on quantum mechanics, we shall only give a summary of the results. If the orthonormal wave functions for the bound states (with $\varepsilon < 0$) of hydrogen are written as

$$\psi_{n,l,m}(\rho,\theta,\phi) = R_{n,l}(\rho) Y_l^m(\theta,\phi)$$

(1.93)

then the functions $R_{n,l}(\rho)$ are

For $n = 1$: $\quad R_{10} = 2e^{-\rho}$

For $n = 2$: $\quad R_{20} = \dfrac{1}{\sqrt{8}} e^{-\rho/2}(2 - \rho)$

$$R_{21} = \frac{1}{\sqrt{24}} \rho e^{-\rho/2}$$

(1.94)

For $n = 3$: $\quad R_{30} = \dfrac{2}{81\sqrt{3}} e^{-\rho/3}(27 - 18\rho + 2\rho^2)$

$$R_{31} = \frac{4}{81\sqrt{6}} \rho e^{-\rho/3}(6 - \rho)$$

$$R_{32} = \frac{4}{81\sqrt{30}} \rho^2 e^{-\rho/3}$$

where $\rho = r/a_1$ and $a_1 = 4\pi\epsilon_0\hbar^2/(m_0 e^2)$.

The following is a summary of the results of the quantum mechanical treatment of a hydrogen atom obtained thus far. In the separation of variables, the Schrödinger equation, Eq. (1.69), is split into three separate differential equations which contain three parameters,

m, λ, and $\mathcal{E}$. Conditions for the proper behavior of $\Phi_m(\phi)$, $\Theta_l^m(\theta)$, and $R_n{}^l(r)$ are (1) m must be an integer, (2) $\lambda = l(l+1)$ and l is a positive integer and $|m| \leq l$, and (3) in a bound state

$$\mathcal{E} = -\frac{m_0 e^4}{2(4\pi\epsilon_0\hbar)^2}\frac{1}{n^2} \tag{1.95}$$

where n is an integer and $n \geq l + 1$. The three integers n, l, and m are called, respectively, the *principal*, the *azimuthal*, and the *magnetic quantum numbers*. The allowed values of the principal quantum number are

$$n = 1, 2, 3, 4 \ldots \tag{1.96}$$

Take $n = 3$ as an example. According to Eq. (1.86), if $\nu = 0$ there is only one term in the series, the constant term. Furthermore, the value of $l = n - \nu - 1 = 2$ from Eq. (1.89), reducing R of Eq. (1.90) to R_{32} of Eq. (1.94). The other two functions R_{30} and R_{31} are obtained for $\nu = 2$ and 1, respectively. Thus the allowed values for the azimuthal quantum number are

$$l = 0, 1, 2, \ldots, n - 1 \tag{1.97}$$

Both n and l are positive integers; however, for a given n, the maximum value of l is $n - 1$. Now let us turn to Eqs. (1.77) and (1.78). The term with highest power in P_l is u^l. Thus, for nonzero P_l^m, the maximum value for m is l. However, both $+m$ and $-m$ are allowed in Eq. (1.74). To avoid confusion we say that the allowed values for m are

$$m = -l, -(l-1), \ldots, 0, \ldots, (l-1), l \tag{1.98}$$

For negative values of m the sign in front of the spherical harmonics is so chosen that

$$Y_l^{-|m|}(\theta,\phi) = (-1)^{|m|} Y_l^{|m|*}(\theta,\phi) \tag{1.99}$$

Thus, for example, $Y_1^{-1} = \sqrt{3/8\pi}\, e^{-i\phi} \sin\theta$. The Y's given in Eq. (1.81) are for positive values of m.

We shall try to give some physical meaning to the various quantum numbers. The principal quantum number n determines the energy of a hydrogen atom as in the Bohr theory. The azimuthal quantum number l describes the orbital motion of the electron. According to the classical definition of angular momentum $\mathbf{L} = \mathbf{r} \times \mathbf{p}$, the corresponding quantum mechanical operators are

$$L_x = yp_z - zp_y = \frac{\hbar}{i}\left(y\frac{\partial}{\partial z} - z\frac{\partial}{\partial y}\right) \tag{1.100}$$

for the component of angular momentum about the x axis and similar expressions for L_y and L_z. It will be shown in Chapter 8 that the expectation value for

$$\langle L^2 \rangle = \langle L_x{}^2 + L_y{}^2 + L_z{}^2 \rangle$$

is given by

$$\langle L^2 \rangle = \iiint \psi_{n,l,m}^*(L_x^2 + L_y^2 + L_z^2)\psi_{n,l,m}r^2\,dr\,\sin\theta\,d\theta\,d\phi$$
$$= \hbar^2(l)(l+1) \tag{1.101}$$

and that for $\langle L_z \rangle$ by

$$\langle L_z \rangle = \iiint \psi_{n,l,m}^* L_z \psi_{n,l,m} r^2\,dr\,\sin\theta\,d\theta\,d\phi = \hbar m \tag{1.102}$$

Since the quantum numbers l and m have important bearings on the magnetic property of a substance, we shall discuss these quantities more thoroughly in Chapter 8.

The existence of quantum numbers l and m can be verified experimentally from spectroscopic studies. If the relativistic variation of mass is taken into consideration, the energy of a hydrogen atom has an additional term $\mathcal{E}(n,l)$ dependent on l or

$$\mathcal{E} = -\frac{m_0 e^4}{2(4\pi\epsilon_0\hbar)^2}\frac{1}{n^2}\left[1 + \frac{\alpha^2}{n^2}\left(\frac{n}{l+\frac{1}{2}} - \frac{3}{4}\right)\right] \tag{1.103}$$

where $\alpha = e^2/(4\pi\epsilon_0\hbar c)$. Thus transition between different atomic levels involves not only n but l also. Furthermore, concerning l, there is a quantum mechanical selection rule which limits the allowable transitions to those involving $\Delta l = \pm 1$. Therefore the only transitions which can occur are those from $l = 0$ to $l = 1$ and from $l = 1$ to $l = 0$ or $l = 2$, and so on. The following letter code is generally used by spectroscopists for the values of l:

$$l = 0, 1, 2, 3, 4, 5, 6, 7, \ldots$$
$$s, p, d, f, g, h, i, k, \ldots$$

The various possible allowed transitions are shown in Fig. 1.9 by dotted lines:

$$(1)\ 1s \leftrightarrow np \qquad (2)\ 2p \leftrightarrow ns \qquad (3)\ 2p \leftrightarrow nd$$

Figure 1.9

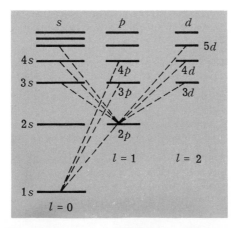

Fine structures of the hydrogen atom. Dashed lines represent the allowable transitions. Since the energy level depends on the azimuthal quantum number l, transitions (for example, 2p ↔ 3s and 2p ↔ 3d) involving different l's will have different energy separations.

The state which an electron occupies is called the *configuration* of the electron and is specified by two quantum numbers n and l. For example, a $2p$ electron has $n = 2$ and $l = 1$. Because of the term $\varepsilon(n,l)$, the transition between $2p$ and $3s$ levels and that between $2p$ and $3d$ levels involve different energies. This is called the *fine structure* of the spectral lines.

Further splitting of the spectral line is possible when a magnetic field is applied. The effect was first observed by Zeeman and is known as the *Zeeman effect*. It will be shown in Chapter 8 that the orbital motion of an electron constitutes a magnetic dipole. The energy of a magnetic dipole changes in a magnetic field, depending on its orientation with respect to the field direction. In Eq. (1.102) the z direction is chosen as the direction of the magnetic field. This explains why the quantum number m is called the magnetic quantum number. In a magnetic field different m values correspond to different energies. Since m can be any one of the values given by Eq. (1.98), there exist $2l + 1$ energy levels. However, quantum mechanical selection rules are such that only transitions involving $\Delta m = \pm 1$ and $\Delta m = 0$ are allowed. This restriction limits the number of spectral lines. These selection rules also play a very important role in the operation of masers to be discussed in great detail in Chapter 9.

Finally, in view of our discussion in Sec. 1.5 it may be worthwhile to give some pictorial illustrations of the wave function. First, let us examine the directional properties and refer to Eq. (1.81). Obviously, for s electrons ($l = 0$), the distribution is spherically symmetrical with respect to the nucleus. For p electrons ($l = 1$), we have three independent wave functions. For purposes of illustration, the three functions are chosen to be real by a proper linear combination of the Y's in Eq. (1.81). Realizing that $\cos \theta = z/r$, $\sin \phi \sin \theta = y$ and $\cos \phi \sin \theta = x$, we have

$$f_1 = A \frac{z}{r} \qquad f_2 = A \frac{y}{r} \qquad f_3 = A \frac{x}{r} \qquad (1.104)$$

where $A = \sqrt{3/(4\pi)}$, $f_1 = Y_1{}^0$, $f_2 = i(Y_1{}^1 + Y_1{}^{-1})/\sqrt{2}$, and $f_3 = (Y_1{}^{-1} - Y_1{}^1)/\sqrt{2}$. Hence the three real wave functions have electron distributions directed along three different axes as illustrated in Fig. 1.10. In a solid, the constituent atoms may be charged, creating an electrostatic field generally referred to as the *crystalline field*. If the form of the crystalline field is known, Fig. 1.10 immediately tells us which of the three distributions is favored energetically. The effect of the crystalline field on the wave functions and that of the electrostatic interaction in general will be discussed in greater detail in later chapters. Figure 1.11 shows the radial distribution function $4\pi r^2 |R_{n,l}(r)|^2$ plotted as a function of $\rho = r/(na_1)$ for

**Figure
1.10**

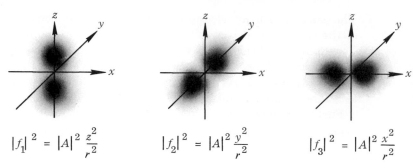

$$|f_1|^2 = |A|^2 \frac{z^2}{r^2} \qquad |f_2|^2 = |A|^2 \frac{y^2}{r^2} \qquad |f_3|^2 = |A|^2 \frac{x^2}{r^2}$$

Schematic illustration of the distribution of electrons for the three p-orbital wave functions. (After L. Pauling and E. B. Wilson, Jr., "Introduction to Quantum Mechanics," p. 150, McGraw-Hill Book Company, 1935.)

various atomic states. The expectation value $\langle r \rangle$ of the distance of the electron from the nucleus, using the definition given in Eq. (1.33), is found to be

$$\langle r_{n,l,m} \rangle = n^2 a_1 \left\{ 1 + \frac{1}{2} \left[1 - \frac{l(l+1)}{n^2} \right] \right\} \tag{1.105}$$

On the scale of ρ, these values are nearly proportional to n as represented by vertical lines in Fig. 1.11. This result agrees with the result based on the old quantum theory of Bohr.

**Figure
1.11**

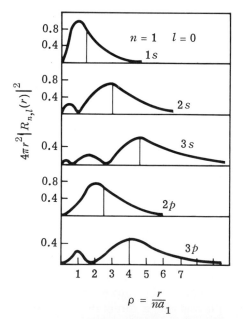

Radial distribution functions $4\pi r^2 |R_{n,l}(r)|^2$ for the hydrogen atom. (After L. Pauling and E. B. Wilson, Jr., "Introduction to Quantum Mechanics," p. 143, McGraw-Hill Book Company, 1935.) The expectation values of $\rho = r/(na_1)$ for the various atomic states are marked by vertical lines.

1.8 ELECTRON SPIN AND PAULI'S EXCLUSION PRINCIPLE

The quantum theory that we have developed so far is still incomplete because it can not account for some very fine details of the spectral lines and results from other experiments. In the Stern-Gerlach experiment, a beam of neutral atoms is sent into an inhomogeneous magnetic field. An atom is deflected by a force

$$F_z = u_z \frac{dB_z}{dz} \tag{1.106}$$

if it possesses a magnetic dipole of moment u_z in the direction of field inhomogeneity. The ground state of a neutral hydrogen atom is in 1s state with $l = 0$. Hence m must also be zero and so would be u_z in view of our discussion in the previous section. Therefore no deflection of a hydrogen beam would be expected. The observed beam not only shows a splitting of the beam into two beams but also indicates that the value of u_z is quantized.

Further evidences of the need for a new concept come from a careful study of the spectral lines. Besides hydrogen, a neutral alkali atom has particularly simple electronic structure. It consists of an inert ion core and one valence electron, and hence the hydrogenic model should be applicable to alkali atoms. According to Eq. (1.103) the energy of an atomic level is fully determined by two quantum numbers n and l. Spectroscopic studies show, however, that the levels associated with $l \neq 0$ (p, d, f states etc.) are split into two levels (doublet levels). The existence of such doublets again cannot be explained by the quantum theory based on three quantum numbers n, l, and m.

The answer to these puzzling facts was given by Uhlenbeck and Goudsmit, who postulated that an electron could also possess rotational degrees of freedom about an axis through its charged body. Such an additional property of the electron is called the *spin*. The electron spin comes out naturally from Dirac's relativistic treatment of the electron. Since a discussion of this kind is clearly beyond the scope of this book, we shall only present the result from such a treatment. Similar to the result of Eq. (1.101) for orbital motion, the magnitude of the angular momentum due to spinning motion is given by $\hbar \sqrt{s(s + 1)}$, where s is the *spin quantum number*. Instead of being an integer as for l, the value of s is found to be $\frac{1}{2}$. In a magnetic field, the spin vector is quantized in the field direction such that

$$s_z = +\frac{1}{2} \quad \text{or} \quad s_z = -\frac{1}{2} \tag{1.107}$$

This produces two values (positive and negative) of u_z in Eq. (1.106), causing a split of the beam. Furthermore, the spin moment s and the orbital moment l can combine vectorially, giving a resultant moment j:

$$\mathbf{j} = \mathbf{l} + \mathbf{s} \tag{1.108}$$

The quantity j, which represents the total angular momentum, is called the *inner quantum number*. For s electrons with $l = 0$, only one possibility exists with $j = s$. For other electrons with $l \neq 0$, two values of j may result, with $j_1 = l + \frac{1}{2}$ and $j_2 = l - \frac{1}{2}$. These two states have different energies because of the magnetic coupling of spin and orbit, the energy difference corresponding to the energy separation between the doublet.

In order to understand the shell structure of atoms, it is important to know the assignment of the four quantum numbers n, l, m, and s. The Pauli *exclusion principle* states that no two electrons can have the same values of the four quantum numbers. For example, the helium atom, next to hydrogen in the periodic table, has two $1s$ electrons. Since the two electrons have the same $n = 1$, $l = 0$, and $m = 0$, the spin quantum numbers must be different. This principle is indispensable not only in constructing the periodic system but also in determining the type of quantum statistics which a particle obeys. These topics will be the subject of discussion for later sections. At present, we shall discuss briefly the nature of the electron wave function involving two electrons and the role of the Pauli exclusion principle.

As a zero-order approximation, the electrostatic interaction energy between the two electrons is assumed small so the total wave function can be taken as a product of the one-electron wave functions. Because of the indistinguishability of electrons, however, the probability of finding electron 1 in state a and electron 2 in state b must be equal to that of finding electron 1 in state b and electron 2 in state a. In other words, the interchange of 1 and 2 should result in the same probability or $|\psi(1,2)|^2 = |\psi(2,1)|^2$. This means

$$\psi(1,2) = \psi(2,1) \qquad \text{or} \qquad \psi(1,2) = -\psi(2,1) \tag{1.109}$$

The wave functions which satisfy the two alternative conditions are called the *symmetrical* and *antisymmetrical wave functions*, respectively. For the case just discussed, the symmetrical wave function is given by

$$\psi_S = \frac{1}{\sqrt{2}} \left[\psi_a(1)\psi_b(2) + \psi_a(2)\psi_b(1) \right] \tag{1.110}$$

and the antisymmetrical wave function by

$$\psi_A = \frac{1}{\sqrt{2}} \left[\psi_a(1)\psi_b(2) - \psi_a(2)\psi_b(1) \right] \tag{1.111}$$

the factor $1/\sqrt{2}$ being the normalization constant. For the total spin wave function, the proper combinations are

$$
\left.\begin{array}{c}
\alpha(1)\alpha(2) \\[4pt]
\dfrac{1}{\sqrt{2}}\,[\alpha(1)\beta(2) + \alpha(2)\beta(1)] \\[4pt]
\beta(1)\beta(2)
\end{array}\right\} \qquad \text{symmetrical}
$$

$$
\frac{1}{\sqrt{2}}\,[\alpha(1)\beta(2) - \alpha(2)\beta(1)] \qquad \text{antisymmetrical}
$$

(1.112)

where α and β denote the one-electron spin wave functions corresponding to the value $s_z = +\tfrac{1}{2}$ and $-\tfrac{1}{2}$, respectively.

The question as to which combinations of wave functions actually occur in nature can only be answered by experimental facts. It is found that the total wave function is antisymmetrical, that is, a combination of symmetrical orbit and antisymmetrical spin wave functions or that of antisymmetrical orbit and symmetrical spin wave functions. In other words, out of all possible combinations of Eqs. (1.110) and (1.111) with Eq. (1.112), only the following combinations are allowed:

$$
\frac{1}{\sqrt{2}}\,[\psi_a(1)\psi_b(2) + \psi_a(2)\psi_b(1)]\ \frac{1}{\sqrt{2}}\,[\alpha(1)\beta(2) - \alpha(2)\beta(1)]
$$

$$
\frac{1}{\sqrt{2}}\,[\psi_a(1)\psi_b(2) - \psi_a(2)\psi_b(1)]
\left\{\begin{array}{c}
\alpha(1)\alpha(2) \\[4pt]
\dfrac{1}{\sqrt{2}}\,[\alpha(1)\beta(2) + \alpha(2)\beta(1)] \\[4pt]
\beta(1)\beta(2)
\end{array}\right\}
$$

(1.113)

Note that the total wave function is identically zero if $\alpha = \beta$ and $a = b$. Therefore, in quantum mechanical language, the demand for an antisymmetrical total wave function is just another way of stating the Pauli exclusion principle. In discussing the property of a solid, however, we often face the situation in which the interaction between electrons on neighboring atoms is so strong that the electrons can be thought of as being shared by the whole crystal. Such is the case with conductors and semiconductors. For such a system, the assignment of the quantum numbers is not so simple as in the case involving only two electrons. The exclusion principle then can be stated simply as follows. No two (or more) electrons can occupy the same quantum state. We shall define what we mean by a quantum state in such cases when we apply Fermi-Dirac statistics to conduction electrons in Sec. 1.12.

1.9 THE PERIODIC SYSTEM

The periodic table of chemical elements was developed long before the quantum theory was established. Chemical elements occupying

TABLE 1.1 *Periodic table of the elements**

Period	Group I a	Group I b	Group II a	Group II b	Group III a	Group III b	Group IV a	Group IV b	Group V a	Group V b	Group VI a	Group VI b	Group VII a	Group VII b	Group VIII a	Group VIII a	Group VIII a	Group VIII b
I														1 H 1.0080				2 He 4.003
II	3 Li 6.940		4 Be 9.02			5 B 10.82		6 C 12.010		7 N 14.008		8 O 16.0000		9 F 19.00					10 Ne 20.183
III	11 Na 22.997		12 Mg 24.32			13 Al 26.97		14 Si 28.06		15 P 30.98		16 S 32.066		17 Cl 35.457					18 A 39.944
IV	19 K 39.096	29 Cu 63.54	20 Ca 40.08	30 Zn 65.38	21 Sc 45.10	31 Ga 69.72	22 Ti 47.90	32 Ge 72.60	23 V 50.95	33 As 74.91	24 Cr 52.01	34 Se 78.96	25 Mn 54.93	35 Br 79.916	26 Fe 55.85	27 Co 58.94	28 Ni 58.69	36 Kr 83.7
V	37 Rb 85.48	47 Ag 107.880	38 Sr 87.63	48 Cd 112.41	39 Y 88.92	49 In 114.76	40 Zr 91.22	50 Sn 118.70	41 Nb 92.91	51 Sb 121.76	42 Mo 95.95	52 Te 127.61	43 Tc 99	53 I 126.92	44 Ru 101.7	45 Rh 102.91	46 Pd 106.7	54 Xe 131.3
VI	55 Cs 132.91	79 Au 197.2	56 Ba 137.36	80 Hg 200.61	57–71 Rare earths	81 Tl 204.39	72 Hf 178.6	82 Pb 207.21	73 Ta 180.88	83 Bi 209.00	74 W 183.92	84 Po 210	75 Re 186.31	85 At 211	76 Os 190.2	77 Ir 193.1	78 Pt 195.23	86 Rn 222
VII	87 Fr 223		88 Ra 226.05		89 Ac 227		90 Th 232.12		91 Pa 231		92 U 238.07	93 Np 237	94 Pu 239	95 Am 241	96 Cm 242 97 Bk 246 98 Ct 249 99 E 254 100 Fm 256			101 Mv 256

Rare Earths

Period															
VI 57–71	57 La 138.92	58 Ce 140.13	59 Pr 140.92	60 Nd 144.27	61 Pm 147	62 Sm 150.43	63 Eu 152.0	64 Gd 156.9	65 Tb 159.2	66 Dy 162.46	67 Ho 164.90	68 Er 167.2	69 Tm 169.4	70 Yb 173.04	71 Lu 174.99

* The numbers in front of the symbols of the elements denote the atomic numbers; the numbers underneath are the atomic weights.

similar places to a large extent have similar chemical properties, notably the alkali metals, the halogens, and the noble gases, as can be seen from Table 1.1. The arrangement of elements in the table was originally in the order of increasing atomic weights; however, it was later discovered that elements should be arranged according to their *atomic number z*, which is an integer. To understand the origin of such a periodic arrangement, we must study the periodic system in conjunction with the electronic structure of atoms as shown in Table 1.2. The atoms of a given element consist of a nucleus of positive charge ze, surrounded by z electrons. A nucleus, however, is composed of both protons and neutrons; the former contributes to both the charge and mass, whereas the latter contributes only to the mass of the nucleus. The chemical behavior of a given element depends on the number of revolving electrons; therefore the determining factor in the periodic arrangement is the atomic number, not the atomic weight.

What determines the chemical behavior is the distribution of these electrons among the various available quantum states defined by the four quantum numbers. If we neglect the electrostatic interaction energy between the electrons, each electron moves independently in a nuclear potential of $ze^2/4\pi\epsilon_0 r$, and the energy levels for different quantum states can be obtained by a slight modification of Eq. (1.103).

$$\mathcal{E} = -\frac{m_0 e^4 z^2}{2(4\pi\epsilon_0\hbar)^2}\frac{1}{n^2}\left[1 + \frac{\alpha^2 z^2}{n^2}\left(\frac{n}{l + \frac{1}{2}} - \frac{3}{4}\right)\right] \qquad (1.114)$$

Equation (1.114) indicates that the $2p$ level should be slightly above the $2s$ level and that the $3d$ level should be slightly above the $3p$ level and so on. Therefore, except for a change of energy scale, the energy levels shown in Fig. 1.9 can still be used in our present discussion. Let us first discuss the maximum number of electrons which each energy level can accommodate in view of the exclusion principle. For a given l there are $(2l + 1)$ values of m. Counting the two possible choices for the spin orientation, there are altogether $2(2l + 1)$ places in each level, that is, for given (n,l). Take the $3d$ level as an example; the maximum number of electrons in $3d$ is 10. This number is attained in copper as shown in Table 1.2.

Now we shall redraw Fig. 1.9 as Fig. 1.12 and distribute electrons of various elements among the available energy levels. For helium it is obvious that the lowest energy configuration has two electrons in the $1s$ level. Since two is the maximum number possible for the $1s$ level (K shell in X-ray nomenclature), the K shell is full or closed. Elements having their shells completed are chemically inert, such as He, Ne, A, and Kr. After He comes Li. The third electron of Li must go to a higher energy state (the $2s$ level) as indicated in Fig. 1.12, the arrows indicating the possible spin orientations. The chem-

TABLE 1.2 *Atomic structure of the elements*

Principal quantum number n Shell designation (X ray)			1 K shell	2 L shell		3 M shell			4 N shell	
Azimuthal quantum number l Letter designation			0 $1s$	0 $2s$	1 $2p$	0 $3s$	1 $3p$	2 $3d$	0 $4s$	1 $4p$
z	Element	ε_i(ev)								
1	Hydrogen	(H) 13.59	1							
2	Helium	(He) 24.56	2							
3	Lithium	(Li) 5.40	2	1	—					
4	Beryllium	(Be) 9.32	2	2	—					
5	Boron	(B) 8.28	2	2	1					
6	Carbon	(C) 11.27	2	2	2					
7	Nitrogen	(N) 14.55	2	2	3					
8	Oxygen	(O) 13.62	2	2	4					
9	Fluorine	(F) 17.43	2	2	5					
10	Neon	(Ne) 21.56	2	2	6					
11	Sodium	(Na) 5.14		Neon configuration		1	—			
12	Magnesium	(Mg) 7.64				2	—			
13	Aluminum	(Al) 5.97				2	1			
14	Silicon	(Si) 8.15				2	2			
15	Phosphorus	(P) 10.9				2	3			
16	Sulphur	(S) 10.36				2	4			
17	Chlorine	(Cl) 12.90				2	5			
18	Argon	(A) 15.76				2	6			
19	Potassium	(K) 4.34						—	1	
20	Calcium	(Ca) 6.11						—	2	
21	Scandium	(Sc) 6.7						1	2	
22	Titanium	(Ti) 6.84						2	2	
23	Vanadium	(V) 6.71						3	2	
24	Chromium	(Cr) 6.74						5	1	
25	Manganese	(Mn) 7.43						5	2	
26	Iron	(Fe) 7.83		Argon configuration				6	2	
27	Cobalt	(Co) 7.84						7	2	
28	Nickel	(Ni) 7.63						8	2	
29	Copper	(Cu) 7.72						10	1	—
30	Zinc	(Zn) 9.39						10	2	—
31	Gallium	(Ga) 5.97						10	2	1
32	Germanium	(Ge) 8.13						10	2	2
33	Arsenic	(As) 10.5						10	2	3
34	Selenium	(Se) 9.73						10	2	4
35	Bromine	(Br) 11.76						10	2	5
36	Krypton	(Kr) 14.00						10	2	6

TABLE 1.2 (continued)

Principal quantum number n Shell designation (X ray)					4 N shell		5 O shell			6 P shell
Azimuthal quantum number l Letter designation					2 4d	3 4f	0 5s	1 5p	2 5d	0 6s
z	Element		ε_i(ev)	Configuration of inner shells						
37	Rubidium	(Rb)	4.17		—	—	1			
38	Strontium	(Sr)	5.69		—	—	2			
39	Yttrium	(Y)	6.5		1	—	2			
40	Zirconium	(Zr)	6.95	Krypton	2	—	2			
41	Niobium	(Nb)	6.77	configuration	4	—	1			
42	Molybdenum	(Mo)	7.06		5	—	1			
43	Technetium	(Tc)	7.1		6	—	1			
44	Ruthenium	(Ru)	7.5		7	—	1			
45	Rhodium	(Rh)	7.7		8	—	1			
46	Palladium	(Pd)	8.1		10	—	—			
47	Silver	(Ag)	7.58			—	1	—		
48	Cadmium	(Cd)	8.99			—	2	—		
49	Indium	(In)	5.79	Palladium		—	2	1		
50	Tin	(Sn)	7.30	configuration		—	2	2		
51	Antimony	(Sb)	8.64			—	2	3		
52	Tellurium	(Te)	8.96			—	2	4		
53	Iodine	(I)	10.44			—	2	5		
54	Xenon	(Xe)	12.13			—	2	6		
55	Cesium	(Cs)	3.89			—			—	1
56	Barium	(Ba)	5.21			—			—	2
57	Lanthanum	(La)	5.61			—			1	2
58	Cerium	(Ce)	(6.54)			2			—	2
59	Praseodymium	(Pr)	(5.76)			3			—	2
60	Neodymium	(Nd)	(6.31)			4			—	2
61	Promethium	(Pm)	(6.3)	The shells 1s to		5	The shells		—	2
62	Samarium	(Sm)	(5.6)	4d contain 46		6	5s to 5p		—	2
63	Europium	(Eu)	5.64	electrons		7	contain 8		—	2
64	Gadolinium	(Gd)	6.7			7	electrons		1	2
65	Terbium	(Tb)	(6.74)			8			1	2
66	Dysprosium	(Dy)	(6.82)			9			1	2
67	Holmium	(Ho)				10			1	2
68	Erbium	(Er)				11			1	2
69	Thulium	(Tm)				12			1	2
70	Ytterbium	(Yb)	6.2			13			1	2
71	Lutetium	(Lu)	5.0			14			1	2

TABLE 1.2 (continued)

Principal quantum number n Shell designation (X ray)				5 O shell		6 P shell			7 Q shell
Azimuthal quantum number l Letter designation				2 5d	3 5f	0 6s	1 6p	2 6d	0 7s
z	Element	ε_i(ev)	Configuration of inner shells						
72 Hafnium	(Hf)	5.5		2	—	2			
73 Tantalum	(Ta)	6.0	The shells	3	—	2			
74 Tungsten	(W)	7.94	1s to 5p	4	—	2			
75 Rhenium	(Re)	7.87	contain	5	—	2			
76 Osmium	(Os)	8.7	68	6	—	2			
77 Iridium	(Ir)	9.2	electrons	9	—	0			
78 Platinum	(Pt)	8.96		9	—	1			
79 Gold	(Au)	9.23			—	1	—		
80 Mercury	(Hg)	10.44			—	2	—		
81 Thallium	(Tl)	6.12			—	2	1		
82 Lead	(Pb)	7.42			—	2	2		
83 Bismuth	(Bi)	(8.8)			—	2	3		
84 Polonium	(Po)	(8.2)			—	2	4		
85 Astatine	(At)	(9.6)	The shells		—	2	5		
86 Radon	(Rn)	10.75	1s to 5d		—	2	6		
87 Francium	(Fr)	(4.0)	contain		—	2	6	—	1
88 Radium	(Ra)	5.27	78		—	2	6	—	2
89 Actinium	(Ac)		electrons		—	2·	6	1	2
90 Thorium	(Th)				1(—)	2	6	1(2)	2
91 Protactinium	(Pa)				2(1)	2	6	1(2)	2
92 Uranium	(U)	4.0			3	2	6	1	2
93 Neptunium	(Np)				5(4)	2	6	—(1)	2
94 Plutonium	(Pu)				6(5)	2	6	—(1)	2
95 Americium	(Am)				7	2	6	—	2
96 Curium	(Cm)				7	2	6	1	2

Figure 1.12

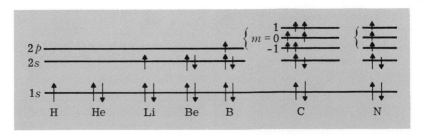

Distribution of electrons among available quantum states in a selected number of chemical elements.

ical *valency* of an element is determined by the number of free electrons or electrons needed to complete a shell. Table 1.2 shows that Li, Na, and K should have similar chemical behavior, and so should F, Cl, and Br.

Our one-electron energy scheme, or Eq. (1.114), works well qualitatively up to argon in explaining the distribution of electrons among different energy levels. But potassium shows a deviation from the normal pattern. A state of lower energy is achieved by filling the $4s$ level first instead of the $3d$ level which belongs to a lower shell. Actually, the deficiency of the one-electron approach is already obvious from the ionization energy of helium. From Eq. (1.114) it is expected that ε_i for He would be about four times larger than that for H because of z^2. The actual value for He is only about twice larger than that for H. This discrepancy is due to the fact that the charge cloud of one electron partly screens the coulomb field of the nucleus seen by the other electron. Obviously the degree of screening or the value of ze felt by an electron depends on the depth of penetration of the electron-wave function. For low values of n, the energy separation between levels with different n is so large that the effect due to different degrees of screening is secondary. For large values of n, as in the case of potassium, the $4s$ wave function penetrates much more deeply than the $3d$ wave function into the region of low potential energy near to the nucleus. The difference in the effective ze for the two wave functions actually overcompensates the difference in n in the energy expression. The same thing happens again in the N shell ($n = 4$). Electrons begin to settle in the O shell ($5s$ and $5p$ levels) before they fill up the N shell ($4d$ and $4f$ levels). The incomplete $3d$ shell in the iron group and the incomplete $4f$ shell of the rare earth group will be discussed later in connection with their magnetic properties.

The information given in Table 1.2 concerns only the electron configuration of the ground state (the lowest energy state). Since the core electrons do not play a role in determining material properties, they are usually omitted for brevity in writing the electron configuration. For example, in Fe, the electron configuration can be written simply as $3d^6 4s^2$, designating only the six $3d$ electrons and the two $4s$ electrons.

Referring again to Table 1.2, we see that electrons are assigned quantum numbers n and l, but no detail is given as to the states of the magnetic quantum number m and the spin quantum number s. Take the electron configuration $2s^2 2p^2$ of carbon as an example. Figure 1.12 shows that the scheme for distributing electrons in carbon is somewhat more complicated than that in the elements preceding carbon. The complication arises because of the following factors. First, the $2p$ level actually consists of three levels with $m = 1, 0,$ and -1. The two $2p$ electrons can be distributed in three different ways among these levels. In addition, the electrons can be arranged with spins parallel or antiparallel.

If the electrostatic interaction energy between the electrons is taken into account, the state with parallel spins has the lowest energy. In view of the Pauli exclusion principle, only three possible arrangements are possible in carbon, as shown in Fig. 1.12. Note that the resultant value of $S(= s_1 + s_2)$ is 1, whereas that for $M(= m_1 + m_2)$ is 1, 0, and -1. Since m can be considered as the projection of the l vector on a prescribed axis, the resultant vectors M and L are related to each other in the same way. Therefore the ground state of carbon has a resultant $S = 1$ and a resultant $L = 1$. The same spin-parallel arrangement requires that the ground state of nitrogen have $S = \frac{3}{2}$ and $L = 0$ as shown in Fig. 1.12. A letter code similar to that for individual electrons is often used with capital letters. For carbon, the ground state is ^3P with the understanding that for P, $L = 1$ and the number in the superscript is equal to $(2S + 1)$. In discussing transitions in masers and lasers, we need a precise knowledge of the atomic state, especially the ground state. Such information is made available from spectroscopic studies and often quoted in literature. The brief discussion presented here serves merely to introduce to the reader the designation of electronic states adopted by spectroscopists so that the reader may be able to translate the information.

1.10 THE MAXWELL-BOLTZMANN DISTRIBUTION LAW

Every one of us has certainly had some contacts with statistics or statistical methods. Simple, common examples of cases in which statistics play a role are flipping a coin and throwing a pair of dice. If the coin and dice are not crooked, the chances of getting a head or tail or a certain combination in dice throwing are well known. Admittedly, if we only make ten flips or throws, the number of occurrences of certain events may not come out exactly according to the predicted odds. However, as we increase the number of samples, that means we keep on flipping coins and throwing dice, the outcome of the game deviates less and less from the predicted odds. If we have a million persons simultaneously flipping coins or throwing dice, it is pointless to ask the outcome of each flipping or throwing because the outcome of the game on the whole is well predicted by statistical methods. The same applies to the behavior of a physical system. In dealing with a large collection of many small particles, such as molecules in a gas or electrons in a solid, it is useless as well as impossible to follow the motion of each and every individual particle because these particles are constantly and rapidly interchanging their states through collision processes. As long as these interchanges are internal, they do not affect the properties of the system as a whole. As a matter of fact, because of these rapid interchanges the whole system is in thermodynamic equilibrium; only under the thermal equilibrium condition do such parameters as temperature, energy, and composition represent meaningfully the physical properties of the system.

The most important thing in determining the outcome of a game is the rule by which the game is played. For example, if a rule is set that any repetitious number after each throwing must be discounted, the outcome of the game with and the outcome without this rule must be quite different. Similarly, molecules in a gas and electrons in a solid obey different statistical laws because they are governed by different rules. The three laws which govern the distribution of particles among available energy states are (1) the Maxwell-Boltzmann, (2) the Bose-Einstein, and (3) the Fermi-Dirac distribution laws. Each distribution law will be treated separately, and at the end of the discussion a general comparison of the three distribution laws and the assumptions involved will be made.

Consider the distribution of n balls into q boxes of different sizes with the ratio of the size of the kth box to the size of the whole q boxes being equal to g_k. The probability of getting one ball into the number 1 box is of course g_1 and that of getting n_1 balls into the number 1 box is $g_1 g_1$ (n_1 times). The question is to find the probability of a given distribution, say $n_1, n_2 \cdots n_q$ balls in the various (number 1 to number q) boxes. Because all balls are assumed similar, a permutation of the n balls among themselves does not present a new distribution. The number of permutations is given by $n!/(n_1! \cdots n_q!)$. Therefore the total probability of the given distribution is

$$W = \frac{n!}{n_1! n_2! \cdots n_q!} g_1{}^{n_1} g_2{}^{n_2} \cdots g_q{}^{n_q} \qquad (1.115)$$

Using the Stirling formula for large n [$\log n! = n(\log n) - n$], we find[6]

$$\log W = n \log n - \sum n_1 \log \left(\frac{n_1}{g_1}\right) \qquad (1.116)$$

Now we shall apply the above equation to an ideal gas of noninteracting molecules. If n is the total number of molecules, then

$$n_1 + n_2 + \cdots + n_q = n \qquad (1.117)$$

Let the different boxes correspond to different energy states; thus the total energy U of the system is given by

$$n_1 \varepsilon_1 + n_2 \varepsilon_2 + \cdots + n_q \varepsilon_q = U \qquad (1.118)$$

---

[6] Since the common logarithm to the base 10 is not used in this text, the symbols "log" and "ln" both stand for the natural logarithm to the base $e = 2.718$.

By using Lagrange's multipliers, we obtain the condition for maximum W as

$$\frac{\partial \log W}{\partial n_i} = -\log\left(\frac{n_i}{g_i}\right) - 1 = \beta + \alpha\mathcal{E}_i \qquad (1.119)$$

which leads to the *Boltzmann distribution law*

$$n_i = g_i A \exp\left(-\alpha\mathcal{E}_i\right) = g_i A \exp\left(-\frac{\mathcal{E}_i}{kT}\right) \qquad (1.120)$$

We shall later show that the value of α is equal to $1/(kT)$ as it is implied in Eq. (1.120). We should again emphasize that the number of molecules involved in any volume of conventional size, say 1 cm³, is enormous; therefore the probability W is negligibly small for any other distribution which deviates from Eq. (1.120). In other words, the most probable distribution given by Eq. (1.120) represents the physical state of the system.

To find the constants g_i, A, and α, we shall use monatomic gases as an example. The energy $\mathcal{E}_i$ is given by

$$\mathcal{E}_i = \frac{mv_i^2}{2} = \frac{m(v_{xi}^2 + v_{yi}^2 + v_{zi}^2)}{2} \qquad (1.121)$$

since only translational motions of atoms are involved. In Eq. (1.121) m is the mass of an atom. To determine the relative sizes of boxes or the values of g_i, we need to consider the volume in velocity space. For velocities between v and $v + dv$, the volume in velocity space is $4\pi v^2\, dv$. Since n_1, n_2, and so on are continuous functions of v, we can remove the subscript and further replace the summation of Eq. (1.117) by an integral

$$n = \int_0^\infty \rho(v)\, dv \qquad (1.122)$$

where $\rho(v)\, dv$ represents the number of atoms having velocity between v and $v + dv$ and

$$\rho(v) = 4\pi A v^2 \exp\left(-\alpha\mathcal{E}\right) \qquad (1.123)$$

In terms of $\rho(v)$, the total energy of the system is given by

$$U = \int_0^\infty \tfrac{1}{2}mv^2\rho(v)\, dv \qquad (1.124)$$

It is well known that the specific heat for a monatomic gas is $3nk/2$ or $U = 3nkT/2$, where k is the Boltzmann constant. From Eqs. (1.122) and (1.124), the constants A and α can be expressed in terms of n and T as $A = n(m/2\pi kT)^{3/2}$ and $\alpha = 1/kT$. Thus Eq. (1.123) becomes

$$\rho(v) = 4\pi n \left(\frac{m}{2\pi kT}\right)^{3/2} v^2 \exp\left(\frac{-mv^2}{2kT}\right) \qquad (1.125)$$

This relation is known as the Maxwell velocity distribution.

We should add that although Eq. (1.125) applies to a system having a continuous distribution of energy, Eq. (1.120) should be used for a system of discrete energy levels. For example, according to Table 1.2, the ground level of a hydrogen atom has $1s$ electron configuration. At high temperatures, an appreciable percentage of hydrogen atoms may have its electron in one of its excited levels. From Eq. (1.120), the population of the electrons in the various levels is in the ratio $n_i/n_j = (g_i/g_j) \exp[-(\mathcal{E}_i - \mathcal{E}_j)/kT]$. In this case, the sizes of the boxes and hence the values of the g's are proportional to the number of quantum states available to each energy level. To illustrate, take the $2s$ and $2p$ levels. At the $2s$ level the electron has two possible choices, whereas at the $2p$ level it has six possible choices. Hence the ratio $g(2p)/g(2s)$ is equal to 3 in the case of a hydrogen atom.

1.11 THE BOSE-EINSTEIN DISTRIBUTION LAW

The essential difference between quantum and classical statistics is that no distinction between individual atoms is made in quantum statistics. The indistinguishability of atoms reduces the number of possible ways of arriving at a given distribution. Suppose we wish to distribute three balls (α, β, and γ) into two boxes (I and II). According to the Boltzmann statistics, the number of ways of having two balls in box I and one ball in box II is $3!/(2!1!) = 3$, as illustrated in Fig. 1.13a. In Bose-Einstein statistics no distinction between α, β, and γ is made, and a distribution of the kind shown in Fig. 1.13b is called a microstate. A macrostate may be represented by various microstates and the number of these microstates gives the probability of the macrostate. The four microstates are shown in Fig. 1.14a and the way to get the four arrangements is illustrated in Fig. 1.14b. Suppose we mix the numbers I and II with the balls and arrange them in any arbitrary order with the understanding that the balls on the right of a letter belong to the box bearing that

Figure 1.13

Box	Ball				Box	Ball
I	$\alpha\ \beta$	$\gamma\ \alpha$	$\beta\ \gamma$		I	X X
II	γ	β	α		II	X

Classical counting	Quantum-mechanical counting
(a)	(b)

Schematic illustration showing the difference between (a) classical and (b) quantum-mechanical counting.

letter. Under this rule, a letter must come first, and hence there are g choices, where g is the number of boxes. This leaves a total of $(n + g - 1)$ remaining objects for arrangements, where n is the number of balls. The total number of possible arrangements is therefore $g(n + g - 1)!/n!g!$. The factor $g!$ takes care of the fact that a permutation of g boxes, together with the balls after them, does not represent a new arrangement.

Suppose now that we wish to distribute n_s particles among g_s states, the subscript s referring to a subgroup having energy $\mathcal{E}_s$. The number of possible arrangements within the subgroup is equal to $(n_s + g_s - 1)!/n_s!(g_s - 1)!$. If there are r such subgroups, the total number of distinguishable arrangements is

$$W = \prod_{s=1}^{r} \frac{(n_s + g_s - 1)!}{n_s!(g_s - 1)!} \tag{1.126}$$

Again using the Stirling formula, we obtain

$$\log W = \sum_s [(n_s + g_s) \log (n_s + g_s) - n_s \log n_s - g_s \log g_s] \tag{1.127}$$

which can be maximized under the same auxiliary conditions imposed on the total energy of the system and the total number of particles of the system. In deriving Eq. (1.127) we have neglected one in comparison with large numbers g's and n's. Following the same procedure as for the Boltzmann statistics, we find

$$\frac{\partial \log W}{\partial n_s} = \log \left(\frac{g_s + n_s}{n_s} \right) = \beta + \alpha \mathcal{E}_s \tag{1.128}$$

or the *Bose-Einstein distribution function*

$$n_s = \frac{g_s}{\exp (\beta + \alpha \mathcal{E}_s) - 1} = \frac{g_s}{\exp (\mathcal{E}_s/kT) - 1} \tag{1.129}$$

The values of β and α will be shown later to be $\beta = 0$ and $\alpha = 1/kT$.

Figure 1.14

Box	No. of balls						
I	3	0	2	1
II	0	3	1	2

(a)

I X X X II	same as	II I X X X
I II X X X	same as	II X X X I
I X X II X	same as	II X I X X
I X II X X	same as	II X X I X

(b)

Diagrams illustrating the arrangements of three balls into two boxes. The balls are assumed indistinguishable, but no limit is placed on the number of balls allowed in each box; (a) shows the four possible arrangements and (b) illustrates how these arrangements can be counted.

Now we shall discuss Planck's law of black-body radiation in connection with the Bose-Einstein statistics. The radiation energy density of a black body in the frequency range between ν and $\nu + d\nu$, according to Planck's radiation law, is given by

$$\mathcal{E}(\nu)\, d\nu = \frac{8\pi h\nu^3\, d\nu}{c^3[\exp{(h\nu/kT)} - 1]} \tag{1.130}$$

where T is the temperature of the black body. Consider a uniform medium within an enclosure having reflecting walls. Under thermodynamic equilibrium, every element of the volume in the medium radiates a certain amount of energy by converting heat energy into incoherent electromagnetic radiation and receives an equal amount of radiation energy from the environment, converting it back into heat. By a black body, we mean a body which absorbs the whole of radiation falling upon it. An enclosure with a small opening in all practical reality is a black body because any radiation passing through the opening into the enclosure will be completely absorbed by the medium. Similarly, the radiation which emerges from the opening has a spectral distribution given by Eq. (1.130) which is characteristic of black-body radiation and independent of the nature of the medium.

For simplicity, the enclosure is assumed cubic with a linear dimension a. It can be shown by solving Maxwell's wave equation that the electric field inside a cubic box varies spatially as $\sin k_x x \sin k_y y \sin k_z z$, where $k_x^2 + k_y^2 + k_z^2 = k^2 = (2\pi/\lambda)^2$, λ being the wavelength of the electromagnetic radiation. Because the electric field must vanish at the boundary, the allowable values of k_x, k_y, and k_z are such that $k_x a$, $k_y a$, and $k_z a$ are integral multiples of π. Letting

$$k_{x,y,z}a = n_{1,2,3}\pi \qquad \text{and} \qquad n^2 = n_1^2 + n_2^2 + n_3^2 \tag{1.131}$$

we note that

$$n^2 = n_1^2 + n_2^2 + n_3^2 = \frac{k^2 a^2}{\pi^2} = \frac{4a^2}{\lambda^2} \tag{1.132}$$

where n_1, n_2, and n_3 are integers. To calculate the number of proper modes of oscillation with wavelength $\lambda > \lambda_0$, let us consider a two-dimensional case with $n_3 = 0$ and draw a two-dimensional lattice as shown in Fig. 1.15. All possible values of n_1 and n_2 for $\lambda > \lambda_0$ are indicated by lattice points inside the positive quadrant of a circle of radius $2a/\lambda_0$. For large values of $2a/\lambda_0$, the area of the quadrant approaches the number of lattice points. Similarly, in a three-dimensional case, the number of lattice points with positive n_1, n_2, and n_3 inside a sphere of radius $2a/\lambda_0$ is equal to one-eighth of the volume of the sphere. Since each lattice point represents a mode of electromagnetic radiation, the total number of proper modes inside a volume $V\ (=a^3)$ with wavelength $\lambda > \lambda_0$ is given by

$$N = \frac{1}{8}\frac{4\pi}{3}n^3 = \frac{4\pi}{3}\frac{V}{\lambda^3} = \frac{4\pi}{3}\left(\frac{k}{2\pi}\right)^3 V = \frac{4\pi}{3}\frac{\nu^3}{c^3}V \qquad (1.133)$$

where c and ν are the velocity and the frequency of the electromagnetic radiation.

The number of proper modes of oscillation having frequency between ν and $\nu + d\nu$, according to Eq. (1.133), is

$$dN = 2\frac{4\pi\nu^2\,d\nu}{c^3}V \qquad (1.134)$$

the extra factor 2 being for two possible directions of polarization. This number dN should be equated with g_s in Eq. (1.129). Noting that the energy associated with each mode with frequency ν is $h\nu$, we have from Eqs. (1.134) and (1.129) the radiation energy density between ν and $\nu + d\nu$

$$\mathcal{E}(\nu)\,d\nu = \frac{n_s h\nu}{V} = \frac{8\pi h\nu^3\,d\nu}{c^3[\exp{(\beta + \alpha h\nu)} - 1]} \qquad (1.135)$$

A comparison of Eq. (1.135) with Eq. (1.130) shows that $\beta = 0$ and $\alpha = 1/kT$.

Besides photons, lattice phonons also obey the Bose-Einstein statistics. As mentioned earlier, both photons and phonons are quantum particles and hence they must obey quantum statistics.

Figure
1.15

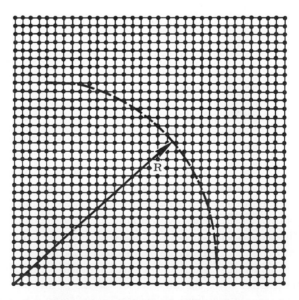

Counting the normal modes of oscillation in a two-dimensional box. Referring to Eq. (1.132), each lattice point represents a set of n_1 and n_2. The total number of lattice points inside the circular arc is equal to the number of normal modes.

Furthermore, the number of phonons and photons in any given normal mode is essentially unlimited by theory (the density of phonons and photons being proportional to the intensities of mechanical vibration and electromagnetic oscillation, respectively). In contrast, the conduction electrons in metals and semiconductors must observe the Pauli exclusion principle. This essential difference is reflected in the quantum statistics which apply to these particles: Bose-Einstein to photons and phonons, and Fermi-Dirac to conduction electrons. The result that $\beta = 0$ in Eq. (1.135) is also connected with the nature of photons and phonons. Since phonons and photons can be created or destroyed, no restriction is imposed on the constancy of their total number n. In other words, an equation similar to Eq. (1.117) should not enter into the Lagrange's multipliers in the case of photons and phonons, resulting in $\beta = 0$ in Eqs. (1.128), (1.129), and (1.135). Finally, we would like to add that in applying Eq. (1.129) to a system of phonons, the value of α in Eq. (1.129) can be determined by measuring the phonon contribution to the specific heat in a way similar to the evaluation of α for the Boltzmann statistics. The value of α is again found to be $1/kT$.

1.12 THE FERMI-DIRAC DISTRIBUTION LAW

The Fermi-Dirac statistics, like the Bose-Einstein statistics, assumes indistinguishability of particles, but it differs from the latter in that it has an added restriction imposed by the Pauli exclusion principle. Suppose that there are five states in a subgroup s, and we wish to distribute three balls among these five states. Since only one ball is allowed in a given state, the number of allowable distributions is further limited. Typical allowable and nonallowable distributions are shown in Fig. 1.16. For the situation shown in

Figure 1.16

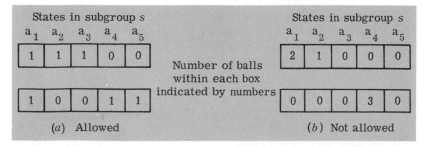

Diagrams illustrating the distribution of three balls (indistinguishable) into five boxes in observance of the rule that only one or zero ball is allowed in a given box. This rule corresponds to the Pauli exclusion principle. Under the rule, the distributions shown in (a) are allowed but the distributions shown in (b) are not allowed.

Fig. 1.16a, the first ball has 5 choices, the second 4 choices, and the third 3 choices. Since the three balls are indistinguishable, there are altogether $5 \times 4 \times 3/3! = 5!/3!2!$ distributions. For a subgroup s having g_s states and n_s balls, the number of allowable distributions is equal to $g_s!/n_s!(g_s - n_s)!$. It is understood that the states in a subgroup have the same energy $\mathcal{E}_s$. For a system consisting of r subgroups, the total number of allowable distributions is given by

$$W = \prod_{s=1}^{r} \frac{g_s!}{n_s!(g_s - n_s)!} \tag{1.136}$$

Equation (1.136) now can be maximized under the same auxiliary conditions imposed on the total number of particles and the total energy of the system. Following the same procedure as for the Bose-Einstein statistics, we find

$$\frac{\partial \log W}{\partial n_s} = \log\left(\frac{g_s - n_s}{n_s}\right) = \beta + \alpha \mathcal{E}_s \tag{1.137}$$

or the *Fermi-Dirac distribution law*

$$n_s = \frac{g_s}{\exp(\beta + \alpha\mathcal{E}_s) + 1} = g_s f(\mathcal{E}_s) \tag{1.138}$$

Let us refer to Fig. 1.16a. We see that because of the Pauli exclusion principle, a given box is either occupied by or devoid of an electron. Within the subgroup s, some boxes will be occupied by an electron and some will not be occupied by an electron. The total number of occupied states (balls) is n_s, while the total number of available states (boxes) is g_s. Therefore the ratio n_s/g_s, and hence the function $f(\mathcal{E}_s)$, generally referred to as the *Fermi function*, can be interpreted as the average probability that a quantum state having energy $\mathcal{E}_s$ is being occupied by an electron. Since the maximum value of n_s is equal to g_s, corresponding to the state that every box is occupied by a ball, the function $f(\mathcal{E}_s)$ varies from a maximum value of one to a minimum value of zero.

Now we shall apply Eq. (1.138) to electrons inside a metal. As mentioned earlier (Sec. 1.8), the electrons inside a conductor or semiconductor interact so strongly that they are shared by the whole crystal. Under such circumstances, the behavior of electrons can best be described as a wave propagating throughout the crystal. According to de Broglie hypothesis, matter possesses both wavelike and particlelike properties, and the connection between these properties is through the well-known de Broglie relation, or Eq. (1.13)

$$\lambda = \frac{h}{p} \quad \text{or} \quad p = \hbar k \tag{1.139}$$

where p is the momentum of the electron. Therefore the number of quantum states having momentum between p and $p + dp$ can be found from Eqs. (1.133) and (1.139) and is given by

$$dg_s = 2 \left(\frac{4\pi p^2 \, dp}{h^3} \right) V \qquad (1.140)$$

The factor 2 in front is for two possible orientations of electron spin. Since $\mathcal{E}_s$ is a continuous function of p, the subscript s can be dropped. The density of electrons having momentum between p and $p + dp$ is equal to

$$dn = f(\mathcal{E}) \frac{dg}{V} = \frac{8\pi p^2 \, dp}{h^3} \frac{1}{\exp\,(\beta + \alpha\mathcal{E}) + 1} \qquad (1.141)$$

If electrons are assumed free inside a metal, $\mathcal{E} = p^2/2m_0$, where m_0 is the mass of a free electron. Equation (1.141) can be physically interpreted as follows. The density of electrons (occupied states) dn is equal to the density of available states dg/V multiplied by the average probability $f(\mathcal{E})$ that a given state is occupied.

A direct application of Eq. (1.141) can be found in metals by studying the temperature dependence of thermionic emission from metals. First, the value of α in Eq. (1.141) is again $1/kT$ as we shall see later. Second, the value of β in Eq. (1.141) has a special significance. The Fermi function $f(\mathcal{E})$ can be written as

$$f(\mathcal{E}) = \frac{1}{1 + \exp\,(\beta + \alpha\mathcal{E})} = \frac{1}{1 + \exp\,(\mathcal{E} - \mathcal{E}_f)/kT} \qquad (1.142)$$

where $\mathcal{E}_f = -\beta kT$, and $f(\mathcal{E})$ has the general shape shown in Fig. 1.17. For states having energy $\mathcal{E} < \mathcal{E}_f$ such that $(\mathcal{E}_f - \mathcal{E})/kT \gg 1$, the function $f(\mathcal{E})$ reaches a maximum value of one, meaning that such states are occupied by electrons. The meaning of the energy $\mathcal{E}_f$, usually referred to as the *Fermi level*, is now clear. At absolute zero temperature, the states with energy $\mathcal{E} < \mathcal{E}_f$ are completely occupied by electrons and those with $\mathcal{E} > \mathcal{E}_f$ are completely void of electrons. At a finite temperature, the states with $\mathcal{E} = \mathcal{E}_f$ are half-filled and half-empty according to Fig. 1.17. Note that $f(\mathcal{E})$ changes very rapidly from one to zero within a very narrow energy range of the order of kT, which is only $\frac{1}{40}$ ev at room temperature. Therefore

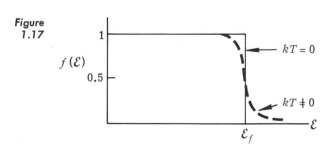

$f(\mathcal{E})$

1

0.5

$\leftarrow kT = 0$

$\leftarrow kT \neq 0$

$\mathcal{E}$

$\mathcal{E}_f$

The Fermi function $f(\mathcal{E})$ of Eq. (1.142), plotted as a function of energy $\mathcal{E}$.

even at a finite temperature it is a good approximation to say that almost all electrons inside a metal are confined to states with $\mathcal{E} < \mathcal{E}_f$.

Figure 1.18a shows the energy of electrons near the surface of a metal. The difference between the references of potential energy inside and outside the metal is known as the potential barrier $\mathcal{E}_b$, and the energy $\mathcal{E}_w = \mathcal{E}_b - \mathcal{E}_f$ is known as the *work function* of the metal. Let us define a new quantity $\rho(\mathcal{E})$ such that $\rho(\mathcal{E}) \, d\mathcal{E}$ represents the density of electrons having energy between $\mathcal{E}$ and $\mathcal{E} + d\mathcal{E}$. Thus, letting $p = (2m_0\mathcal{E})^{1/2}$ in Eq. (1.141) gives

$$\rho(\mathcal{E}) = \frac{dn}{d\mathcal{E}} = \frac{4\pi(2m_0)^{3/2}\sqrt{\mathcal{E}}\, f(\mathcal{E})}{h^3} = g(\mathcal{E})f(\mathcal{E}) \qquad (1.143)$$

where $g(\mathcal{E})$ is the density-of-states function such that $g(\mathcal{E}) \, d\mathcal{E}$ represents the available states per unit volume having energy between $\mathcal{E}$ and $\mathcal{E} + d\mathcal{E}$. Figure 1.18b shows $\rho(\mathcal{E})$ as a function of $\mathcal{E}$. Only electrons in the tail end of $\rho(\mathcal{E})$ with $\mathcal{E} > \mathcal{E}_b$ have sufficient energy to escape from the solid, and hence they are responsible for thermionic emission. The current density resulting from thermionic emission is given by

$$J = \int ev_x \, dn = \int_{\mathcal{E}_b}^{\infty} \frac{ep_x\rho(\mathcal{E}) \, d\mathcal{E}}{m_0} \qquad \text{amp/m}^2 \qquad (1.144)$$

where v_x is the component of velocity normal to the surface of the emitting solid. For electrons with $(\mathcal{E} - \mathcal{E}_f)/kT \gg 1$, $f(\mathcal{E})$ can be replaced by $\exp\left[-(\mathcal{E} - \mathcal{E}_f)/kT\right]$. The integration can easily be carried out by converting $\sqrt{\mathcal{E}} \, d\mathcal{E}$ back to $dp_x \, dp_y \, dp_z$, and the final result is given as follows:

$$J = \frac{4\pi e m_0}{h^3}(kT)^2 \exp\left(\frac{-\mathcal{E}_w}{kT}\right) \qquad \text{amp/m}^2 \qquad (1.145)$$

which is known as the Dushman equation. A derivation of Eq.

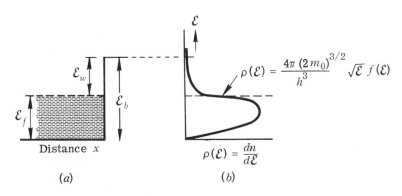

Figure 1.18

(a) Distance x

(b) $\rho(\mathcal{E}) = \dfrac{dn}{d\mathcal{E}}$

$\rho(\mathcal{E}) = \dfrac{4\pi(2m_0)^{3/2}}{h^3}\sqrt{\mathcal{E}}\, f(\mathcal{E})$

Energy diagram used to calculate the thermionic current: (a) energy diagram at the surface of a metal and (b) density of occupied states as a function of energy.

(1.145) and a similar expression based on the classical Boltzmann distribution can be found in the problems at the end of this chapter.

In concluding this section we shall compare the three distributions[7] and discuss conditions under which a given distribution is applicable. Note that both Eqs. (1.143) and (1.135) reduce to the classical Boltzmann distribution, Eq. (1.125), if the exponential term in Eqs. (1.141) and (1.129) are much greater than unity. Take Eq. (1.143) as an example. If $\exp(\beta + \alpha\mathcal{E}) \gg 1$, then $f(\mathcal{E}) \simeq \exp(-\beta - \alpha\mathcal{E})$ and Eq. (1.143) after integration becomes

$$
\begin{aligned}
n &= \frac{4\pi(2m_0)^{3/2}}{h^3} \exp\left(-\frac{\mathcal{E}_f}{kT}\right) \int_0^\infty \sqrt{\mathcal{E}} \exp\left(-\frac{\mathcal{E}}{kT}\right) d\mathcal{E} \\
&= \frac{2}{h^3}(2\pi m_0 kT)^{3/2} \exp\left(\frac{\mathcal{E}_f}{kT}\right)
\end{aligned}
\tag{1.146}
$$

noting that $\int_0^\infty \sqrt{x}\, e^{-\lambda x}\, dx = \frac{1}{2}\sqrt{\pi/\lambda}$. If the value of $\exp(\mathcal{E}_f/kT)$ is substituted back into Eq. (1.143), we find

$$
\rho(\mathcal{E}) = 2\pi n \frac{\sqrt{\mathcal{E}}}{(\pi kT)^{3/2}} \exp\left(-\frac{\mathcal{E}}{kT}\right)
\tag{1.147}
$$

which is exactly equivalent to Eq. (1.125) if $\rho(\mathcal{E})\, d\mathcal{E}$ is changed to $\rho(v)\, dv$.

According to Eq. (1.146) the Fermi level $\mathcal{E}_f$ has a new meaning. For $\mathcal{E}_f > 0$, which is the case with metals as shown in Fig. 1.18a, Eq. (1.141) must be used. For $\mathcal{E}_f < 0$, which exists in most semiconductors, Eq. (1.141) can be approximated by Eq. (1.146); or, in other words, the Fermi-Dirac statistics can be replaced by the Boltzmann statistics. In the classical theory of electric conduction in metals and semiconductors, the electrons are considered essentially free as molecules in a gas, and hence a metal or semiconductor can be thought of as an electron gas. Unlike gas of molecules, however, the statistics of the electron gas changes as the density of electrons increases. When the density of electrons is low, the Fermi-Dirac statistics can be approximated by the Boltzmann statistics, and hence the electron gas behaves like a classical gas (nondegenerate gas). When the density of electrons is high, the Fermi-Dirac statistics must be used, and the electron gas is then a degenerate gas. A parameter D, given below, is usually referred to as the *degeneracy parameter*:

$$
D = \frac{nh^3}{2(2\pi m_0 kT)^{3/2}}
\tag{1.148}
$$

---

[7] The reader may find it instructive to read an alternate derivation of the Bose-Einstein and Fermi-Dirac distributions due originally to F. Bloch. Reference: C. Kittel, "Elementary Solid State Physics," p. 125, John Wiley & Sons, Inc., New York, 1957.

By degeneracy we mean that if $D > 1$ the quantum Fermi-Dirac or Bose-Einstein statistics must be used, and if $D \ll 1$ both cases pass over to the classical statistics.

Now we shall give some physical interpretation to this degeneracy parameter. According to Heisenberg's uncertainty principle,

$$\Delta p \, \Delta q \geq \frac{h}{4\pi} \tag{1.149}$$

where Δp and Δq, respectively, are the root-mean-square deviations of the momentum and coordinate of a particle. Since electrons are in constant thermal motion, we can take $(\Delta p_x)^2/2m_0 = kT/2$. Thus

$$\Delta x \, \Delta y \, \Delta z \geq \left(\frac{h}{4\pi}\right)^3 \frac{1}{(m_0 kT)^{3/2}} \tag{1.150}$$

For electrons not to interfere with each other, $\Delta x \, \Delta y \, \Delta z$ must be smaller than the average volume allowed for each electron, which can be taken as $1/n$. Thus for nondegeneracy we have the condition

$$n \leq \frac{(4\pi)^3 (m_0 kT)^{3/2}}{h^3} \tag{1.151}$$

The conditions stated in Eqs. (1.151) and (1.148) are the same except for a numerical factor which arises because of a slightly different definition of degeneracy in terms of the degree of overlapping of the electron wave functions in these two equations. Therefore, for a system of weakly interacting particles, that is, $D \ll 1$, the quantum statistics reduces to the classical statistics or, in other words, the Boltzmann statistics is still applicable. The difference between the quantum Bose-Einstein and Fermi-Dirac statistics has its origin in the symmetry property of the total wave function. For an antisymmetrical wave function which satisfies the requirement of the Pauli exclusion principle, the Fermi-Dirac statistics should be used. The particles which obey the Fermi-Dirac statistics are known as *Fermions;* the electron is an example. By an antisymmetrical and a symmetrical wave function, we mean a function which, respectively, changes its sign or remains the same if the states of two particles are interchanged. For symmetrical wave functions, the Bose-Einstein statistics should be used. The particles which obey the Bose-Einstein statistics are called *Bosons;* photons and phonons are examples.

PROBLEMS

1.1 Consider a plane wave whose electric field can be represented by

$$\mathbf{E} = \mathbf{E}_0 \sin{(\omega t - kz)}$$

Plot E as a function of z as kz changes from -2π to 2π for three different values of t such that $\omega t = 0$, $\pi/2$, and π. Find the velocity and wavelength associated with the wave. In free space, $k = \omega \sqrt{\mu_0 \epsilon_0}$ where μ_0 and ϵ_0 are the permeability and dielectric constant respectively. Calculate the velocity of propagation.

1.2 In the diffraction experiment (Fig. 1.2a), the wave arriving at point A may be considered as consisting of two components as given below

$$\mathbf{E} = \mathbf{E}_0 \sin (\omega t - \theta_1) + \mathbf{E}_0 \sin (\omega t - \theta_2)$$

where $\theta_1 = k(AS_1)$ and $\theta_2 = k(AS_2)$ are two phase angles associated with the two components because of the difference in their optical path lengths. Find the intensity of the diffracted beam at point A for $\theta_1 - \theta_2 = \pi/2$, π, and 2π.

1.3 The NaCl crystal has a cubic structure shown in Fig. 2.11a and a density of 2165 kg/m³. Calculate the distance d_0 between Na and Cl atoms. If the molybdenum 0.71 Å line is used in the X-ray diffraction experiment, find the Bragg angle θ.

1.4 In a photo-emission experiment, a cesium cathode is illuminated with an ultraviolet light of wavelength $\lambda = 2500$ Å. The retarding potential needed to stop the photo-emission is found to be 3.06 volts. Find the work function of cesium.

1.5 A vibrating charged particle acts like an antenna, sending out electromagnetic radiation. The rate of energy radiated is given by

$$\frac{d\mathcal{E}}{dt} = - \frac{2e^2(\ddot{r})^2}{3c^3} \quad \text{(cgs)} \qquad - \frac{2}{3} \frac{e^2(\ddot{r})^2}{c^2} \frac{1}{4\pi} \sqrt{\frac{\mu}{\epsilon}} \quad \text{(mks)}$$

Consider a one-dimensional harmonic oscillator with

$$r = x = a \sin \omega t$$

Show that the average rate of energy loss by radiation is

$$\frac{d\bar{\mathcal{E}}}{dt} = - \frac{2e^2 \omega^2}{3m_0 c^3} \mathcal{E} \quad \text{(cgs)}$$

where $\mathcal{E}$ is the total energy of the harmonic oscillator. Calculate the initial decay constant for $\omega = 10^{15}$ rad/sec and $\omega = 10^{10}$ rad/sec.

1.6 Using Eq. (1.28) or

$$\mathcal{E} = \frac{m_0 c^2}{\sqrt{1 - \beta^2}} \qquad \text{and} \qquad p = \frac{m_0 c \beta}{\sqrt{1 - \beta^2}}$$

show that the group velocity of an electron is $v = c\beta$.

1.7 Find the ionization potential of hydrogen from the Rydberg constant. In a hydrogen discharge tube, a current drop is observed at 1.89 volts. Identify the transition.

1.8 In an electron diffraction experiment, electrons acquire energy from an accelerating voltage V. Show that the electron momentum (considering the relativistic correction) is given by

$$p = \sqrt{1 + eV/(2m_0c^2)} \sqrt{2m_0eV}$$

Calculate the wavelength in Å for $V = 10^3$ and 10^5 volts.

1.9 Particle diffraction experiments are not confined to electrons. Find the de Broglie wavelength of thermal neutrons which have an energy corresponding to a temperature of 300°K and a mass equal to that of a proton. Calculate the photon energy of an X-ray beam which has the same wavelength.

1.10 If the molybdenum 0.71 Å line is used in the Compton effect experiment, calculate the energy and the angle ϕ of the recoil electron for $\theta = \pi/2$.

1.11 Derive Eq. (1.12) from Eqs. (1.10) and (1.11) by filling in the necessary steps.

1.12 Show that for free particles, that is $V = 0$, the Schrödinger equation has a solution of the form

$$\psi(x,y,z) = N \sin (k_x x + \phi_x) \sin (k_y y + \phi_y) \sin (k_z z + \phi_z)$$

provided that

$$k^2 = k_x{}^2 + k_y{}^2 + k_z{}^2 = (2m\mathcal{E}/\hbar^2)$$

The form of solution given above is known as standing waves which can easily be decomposed into two traveling waves of the type given by Eq. (1.22) but propagating in opposite directions. Draw a vector diagram defining the direction of propagation of the wave in terms of k_x, k_y, and k_z.

1.13 Consider a one-dimensional problem in which the potential $V(x)$ takes the form $V = 0$ for $-a < x < a$, and $V = \infty$ elsewhere. Since the energy of a particle is finite, the wave function $\psi(x)$ is essentially zero for $|x| \geq a$. Show that the energy of the particle has quantized values $\mathcal{E} = (h^2 n^2)/(32ma^2)$ where n is an integer. Sketch the wave functions $\psi_n(x)$ for $n = 1, 2,$ and 3.

1.14 Consider two resonant circuits: one being resonant at a radio frequency of $\nu_0 = 10^6$ cps and having a Q of 1000 and the other being resonant at an optical frequency $\nu_0 = 5 \times 10^{14}$ cps and having an optical Q of 50. Calculate the number of quanta delivered per second by the resonator and the number of quanta stored in the resonator for an output of 1 watt.

1.15 For an even function $V(t)$, the frequency spectrum can be found by the use of the Fourier integral theorem

$$G(\omega) = \sqrt{\frac{2}{\pi}} \int_0^\infty V(t) \cos \omega t \, dt$$

Show that for the pulse shown in Fig. 1.7, that is $V(t) = V_0$ for $-\tau < t < \tau$ and zero elsewhere,

$$G(\omega) = V_0 \left(\frac{2}{\pi}\right)^{\frac{1}{2}} \frac{\sin \omega\tau}{\omega}$$

Plot $G(\omega)$ as a function of $\omega\tau$. We define the bandwidth $\Delta\omega$ such that at $\omega = \Delta\omega/2$, the spectral function $G(\omega)$ drops to $2/\pi$ of its maximum value. Show that the product $\Delta\nu\,\Delta t = 1$, Δt being the duration of the pulse and $\nu = \omega/2\pi$.

1.16 Consider a one-dimensional problem in which an electron with energy $\mathcal{E}$ moves in a potential V with $V = V_0$ for $x > 0$ and $V = 0$ for $x < 0$. The motion of the electron can be represented by a wave of the following form for $\mathcal{E} > V_0$

$$\psi_1 = [A \exp (ik_0x) + B \exp (-ik_0x)] \exp \left(-i\frac{\mathcal{E}}{\hbar} t\right) \qquad \text{for } x < 0$$

$$\psi_2 = [C \exp (ikx)] \exp \left(-i\frac{\mathcal{E}}{\hbar} t\right) \qquad \text{for } x > 0$$

Express k and k_0 in terms of $\mathcal{E}$ and V_0. Comment on the directions of propagation of the various wave components. Show that by equating $\psi_1 = \psi_2$ and the derivatives of ψ_1 and ψ_2 at $x = 0$, the following results are obtained:

$$R = \frac{B}{A} = \frac{k_0 - k}{k_0 + k} \qquad \text{and} \qquad T = \frac{C}{A} = \frac{2k_0}{k_0 + k}$$

which may be called the reflection and transmission coefficient of the wave respectively.

1.17 Consider another one-dimensional problem in which a potential barrier V_0 exists for $0 < x < a$ in the path of an electron having energy $\mathcal{E} < V_0$. Set up the solutions of the Schrödinger equation in the three respective regions: (1) $x < 0$, (2) $0 < x < a$ and (3) $x > a$. Assuming that $V_0 \gg \mathcal{E}$ so the reflected wave is negligibly small in region 2, find the ratio $\psi(x = a)/\psi(x = 0)$ and the probability with which an electron may appear in region 3 when it hits the potential barrier at $x = 0$. Take the electron wave function in region 1 as a sinusoidal wave, then complete the sketch of the electron wave function as a function of x by making ψ and its derivative continuous at the boundaries between the three regions, that is at $x = 0$ and $x = a$.

1.18 The wave function for the lowest energy state $(n = 0)$ of a one-dimensional harmonic oscillator is

$$\psi_0 = \frac{1}{\pi^{\frac{1}{4}}} \exp \left(-\frac{y^2}{2}\right)$$

where y is a dimensionless quantity and is related to the space coordinate x by $y = \sqrt{M\omega/\hbar}\, x$. Using the following relations

$$\int_{-\infty}^{\infty} \psi_0{}^2 \, dy = 1 \qquad \text{and} \qquad \int_{-\infty}^{\infty} \psi_0{}^2 y^2 \, dy = \frac{1}{2}$$

show that

(1)
$$\langle x^2 \rangle = \frac{\int_{-\infty}^{\infty} \psi_0^2 x^2 \, dx}{\int_{-\infty}^{\infty} \psi_0^2 \, dx} = \frac{\hbar}{2M\omega}$$

(2)
$$\hbar^2 \frac{\partial^2 \psi_0}{\partial x^2} = (-\psi_0 + y^2 \psi_0) M \hbar \omega$$

(3)
$$(\Delta p)(\Delta x) = \sqrt{\langle x^2 \rangle \langle p^2 \rangle} = \frac{\hbar}{2}$$

(4)
$$\mathcal{E} = \hbar\omega/2$$

1.19 Using Eqs. (1.81) and (1.99), show that the angular dependence of the wave functions can be represented by real functions such as

$$f_1 = Y_1^0 = A\frac{z}{r} \qquad f_2 = \frac{i(Y_1^1 + Y_1^{-1})}{\sqrt{2}} = A\frac{y}{r}$$

$$f_3 = \frac{Y_1^{-1} - Y_1^1}{\sqrt{2}} = A\frac{x}{r}$$

for p electrons and

$$f_1 = \frac{i(Y_2^1 + Y_2^{-1})}{\sqrt{2}} = B\frac{yz}{r^2} \qquad f_2 = \frac{Y_2^{-1} - Y_2^1}{\sqrt{2}} = B\frac{zx}{r^2}$$

$$f_3 = \frac{i(Y_2^{-2} - Y_2^2)}{\sqrt{2}} = B\frac{xy}{r^2} \qquad f_4 = -\frac{Y_2^2 + Y_2^{-2}}{\sqrt{2}} = \frac{C(y^2 - x^2)}{r^2}$$

$$f_5 = \sqrt{3}\, Y_2^0 = \frac{C(2z^2 - x^2 - y^2)}{r^2}$$

for d electrons with $A = \sqrt{3/(4\pi)}$, $B = \sqrt{15/(4\pi)}$ and $C = \sqrt{15/(16\pi)}$.

1.20 Consider the ground ($n = 1$) and the first excited ($n = 2$) states of hydrogen. In the Boltzmann distribution law

$$N_i = g_i A \exp(-\mathcal{E}_i/kT)$$

the weighting factor g_i is defined in the text as being the size of the box for state i. Determine the value of g_i for the ground and first excited states of hydrogen by considering the number of quantum states available to electrons for these states. Calculate the percentage of hydrogen atoms in the excited state if a hydrogen gas is heated to a temperature of $3000°K$.

1.21 Show that

$$\iint_{-\infty}^{\infty} \exp\left(\frac{-mv_x^2}{2kT}\right) \exp\left(\frac{-mv_y^2}{2kT}\right) dv_x \, dv_y = \frac{2\pi kT}{m}$$

by considering a two-dimensional space with $v^2 = v_x^2 + v_y^2$ and converting the elementary area $dv_x dv_y$ into polar coordinates v and θ. Also verify the following expression

$$\int_{-\infty}^{\infty} \tfrac{1}{2} m v_x{}^2 \exp\left(\frac{-m v_x{}^2}{2kT}\right) dv_x = \tfrac{1}{2} kT \sqrt{\frac{2\pi kT}{m}}$$

by integration by parts. Using the above result, prove that the average kinetic energy associated with each degree of freedom is given by $kT/2$. This is known as the *equipartition theorem*.

1.22 We shall derive the ideal gas law from the results obtained from the Boltzmann statistics. First, let us assume that gas molecules have a uniform velocity v. Verify the following statements step by step.

1. If n is the total number of molecules/cm³, the density of molecules having the direction of their velocity between θ and $\theta + d\theta$ is given by

$$dn_1 = (n/2) \sin \theta \, d\theta$$

2. The number of molecules striking the wall in time dt is

$$dn_2 = v \cos \theta (dt)(dn_1) A$$

where A is the total area of the wall.

3. Assume that the collision of molecules with the wall is elastic; show that the pressure exerted on the wall is equal to

$$p = \frac{F}{A} = \frac{1}{A}\frac{dP}{dt} = \frac{nmv^2}{3}$$

by considering the change of momentum P at the wall.

Now, we shall treat the case in which gas molecules have a velocity distribution governed by the Boltzmann statistics. Using the equipartition theorem stated in the previous problem, show that

$$PV = RT$$

which is known as the *ideal gas law*. What is R?

1.23 Find the average value of velocity squared for the molecules which escape in unit time through a small hole in the wall of a container. For simplicity, the container is assumed cubic and the wall with the hole is parallel to the yz plane. It is further assumed that the molecules leaving the hole obey the Boltzmann-Maxwellian velocity distribution. Show that the average kinetic energy of the molecules which leave the container is $2kT$, not $\tfrac{3}{2}kT$. [*Hint:* Use the results of parts (1) and (2) of Problem 1.22.]

1.24 The energy of a harmonic oscillator is given by

$$\mathcal{E} = (n + \tfrac{1}{2})\hbar\omega$$

where n is a positive integer. According to the Boltzmann distribution law, the population of a given energy level is proportional to $\exp(-\mathcal{E}/kT)$. Show that the average energy of the oscillator is equal to

$$\bar{\mathcal{E}} = \sum_n \mathcal{E} P(\mathcal{E}) = \frac{\hbar\omega}{\exp(\hbar\omega/kT) - 1}$$

where $P(\mathcal{E})$ is the probability that the oscillator happens to have the quantum number n. *Hint:*

$$1 + e^{-x} + e^{-2x} + \cdots = \frac{1}{1 - e^{-x}}$$

and

$$e^{-x} + 2e^{-2x} + \cdots = -\frac{d}{dx}\frac{1}{1 - e^{-x}}$$

Interpret the above result in view of the Bose-Einstein distribution law. Show that at high temperatures such that $kT > \hbar\omega$, $\mathcal{E} = kT$. How many degrees of freedom does a linear harmonic oscillator possess? Under what condition does the classical equipartition theorem yield a correct value for $\bar{\mathcal{E}}$? What happens to the zero-point energy term?

1.25 Consider a transmission line of length L terminated at both ends by a resistor of resistance R which is equal to the characteristic impedance of the line. It is obvious that the lowest mode of oscillation which can be supported by the line has a wavelength $\lambda = 2L$. For higher modes, $2L = n\lambda$ where n is an integer. Show that within a frequency range $\Delta\nu$, there are $2L\,\Delta\nu/c$ modes where c is velocity of propagation of the electromagnetic wave along the line.

An electric resonant circuit is the electric analogy of a harmonic oscillator. By analogy, what is the average energy of the circuit or in more general terms what is the average energy associated with each mode of oscillation? (Refer to Prob. 1.24.)

It is known that the power flow P along the line is given by $P = \mathcal{E}c$ where $\mathcal{E}$ is the energy stored per unit length of the line. Show that by combining results in parts (1) and (2)

$$P = \frac{\hbar\omega\,\Delta\nu}{\exp(\hbar\omega/kT) - 1}$$

The Johnson or thermal noise voltage squared V_N^2 generated across a resistor R within a frequency range $\Delta\nu$ is given by $4RkT\,\Delta\nu$. Discuss how the above expression can lead to the thermal noise voltage.

1.26 Using the de Broglie equation $\lambda = h/(mv)$ and arguing that each electron at least occupies a volume λ^3, show that the least volume in momentum space needed to accommodate N electrons in a volume V is

$$2 \times v_x v_y v_z = 2 \times \frac{4\pi v^3}{3} \geq \left(\frac{h}{m}\right)^3 \frac{N}{V}$$

the factor 2 being for two possible spin orientations. In a metal, the electrons can be considered as piling up on top of each other. Which sign, greater or equal, in the above equation should we choose? The Fermi-level $\mathcal{E}_f$ can be taken as the maximum energy of electrons in a metal. Show that

$$\mathcal{E}_f = \frac{m}{2} v_{\max}^2 = \frac{h^2}{2m}\left(\frac{3N}{8\pi V}\right)^{2/3}$$

1.27 Find the average kinetic energy of electrons in a metal in terms of the maximum kinetic energy. Also find the internal pressure ($p = nmv^2/3$

from Prob. 1.22) caused by electrons in a metal. For simplicity, the temperature is taken to be very low so that all states above $\mathcal{E}_f$ are empty and all states below $\mathcal{E}_f$ are occupied.

REFERENCES

Born, M.: "Atomic Physics," Hafner Publishing Co., New York, 1962.

Eisberg, R. M.: "Fundamentals of Modern Physics," John Wiley & Sons, Inc., New York, 1961.

Harnwell, G. P., and W. E. Stephens: "Atomic Physics," McGraw-Hill Book Company, New York, 1955.

Herzberg, G.: "Atomic Spectra and Atomic Structure," 2d ed., Dover Publications, Inc., New York, 1959.

Hume-Rothery, W.: "Atomic Theory," The Institute of Metals, Monograph Series No. 3, London, 1962.

Leighton, R. B.: "Principles of Modern Physics," McGraw-Hill Book Company, New York, 1959.

Mott, N. F., and I. N. Sneddon: "Wave Mechanics and Its Applications," The Clarendon Press, Oxford, 1948.

Pauling, L., and E. B. Wilson: "Introduction to Quantum Mechanics," McGraw-Hill Book Company, New York, 1935.

Powell, J. L., and B. Crasemann: "Quantum Mechanics," Addison-Wesley Publishing Company, Reading, Mass., 1961.

Schiff, L. I.: "Quantum Mechanics," McGraw-Hill Book Company, New York, 1955.

Sproull, R. L.: "Modern Physics," 2d ed., John Wiley & Sons, Inc., New York, 1963.

2 INTERATOMIC FORCES AND CRYSTAL STRUCTURES

2.1 CLASSIFICATION OF BOND TYPES

In the previous chapter we discussed the periodic system of chemical elements based on shell structures of atoms. The chemical bond between atoms in a solid can be understood on a similar basis. The strength of a chemical bond is measured by the energy required to dissociate a solid into isolated atoms or molecules as appropriate, and this energy is generally known as the *cohesive energy*. For example, the value is 7.8 ev/molecule for NaCl and 0.078 ev/atom for argon. The large difference in the cohesive energy indicates that chemical bonds of very different nature are involved in the two solids. Generally speaking, there are four principal types of chemical binding: ionic (NaCl, CsCl, CaO, etc.), covalent (Ge, Si, GaAs, etc.), metallic (Na, Ni, Ag, etc.), and molecular (A, Kr, Xe, etc.). In making this general classification we must realize that the lines of demarcation are not rigidly drawn in boundary areas. For example, ZnS can be considered as having a mixture of ionic and covalent bonding, with the character of the former being stronger than that of the latter. The situation in SiC is just the opposite.

The ionic bond. In the following discussion we shall give some general characterizations of the four types of chemical bonds. The simplest among these is the ionic bond. As discussed in Chapter 1, alkali metals such as sodium have a single valence electron outside a closed shell, whereas halogens such as chlorine are one electron short of having a complete outer shell. An electron transfer from the alkali metal X to the halogen Y will result in closed shells for both X^+ and Y^- ions. This happens in a salt XY. For example, in NaCl, Na^+ has an electronic configuration $1s^2$, $2s^2$, $2p^6$, which is the same as that of neon, whereas Cl^- has $1s^2$, $2s^2$, $2p^6$, $3s^2$, $3p^6$, which is the same as that of argon. Since these electronic configurations are very stable, we can say that an ionic crystal consists of positive and negative ions. As an order-of-magnitude estimate of the binding energy, we take electronic charge e as the charge carried by each ion and the interatomic distance r in NaCl as 2.8 Å. Thus electrostatic attraction between the ions yields an energy $e^2/(4\pi\epsilon_0 r)$, about 5 ev compared to the measured value of 7.8 ev for cohesive energy.

Besides alkali halides, ionic crystals can also be made of divalent and trivalent elements. Examples include CaO, $MgCl_2$, Al_2O_3, Na_2S, and ZnO, to name a few.

The covalent bond. Besides electron transfer, an incomplete shell can be filled by sharing electrons with neighboring atoms. The hydrogen molecule is a good example. The Schrödinger equation for a hydrogen molecule is

$$\frac{\hbar^2}{2m_0}(\nabla_1^2\psi + \nabla_2^2\psi)$$

$$+ \left[\varepsilon + \frac{1}{4\pi\epsilon_0}\left(\frac{e^2}{r_{1b}} + \frac{e^2}{r_{1a}} + \frac{e^2}{r_{2b}} + \frac{e^2}{r_{2a}} - \frac{e^2}{r_{12}} - \frac{e^2}{r_{ab}}\right)\right]\psi = 0 \quad (2.1)$$

where the subscripts a and b refer to the two nuclei, the subscripts 1 and 2 refer to the two electrons, and r is the distance between them, as illustrated in Fig. 2.1a. If it were assumed that the two hydrogen atoms were far apart and that electrons 1 and 2 belonged to nuclei a and b, respectively, then $\psi = \psi_a(1)\psi_b(2)$ would be the proper total wave function and ψ would satisfy the following equation:

$$\frac{\hbar^2}{2m_0}(\nabla_1^2\psi + \nabla_2^2\psi) + \left[\varepsilon + \frac{1}{4\pi\epsilon_0}\left(\frac{e^2}{r_{1a}} + \frac{e^2}{r_{2b}}\right)\right]\psi = 0 \quad (2.2)$$

where ε would be the sum of the energies of two isolated hydrogen atoms. Comparing Eq. (2.2) with Eq. (2.1) shows that the terms

$$V = \frac{1}{4\pi\epsilon_0}\left(\frac{e^2}{r_{ab}} + \frac{e^2}{r_{12}} - \frac{e^2}{r_{1b}} - \frac{e^2}{r_{2a}}\right) \quad (2.3)$$

represent classically the additional electrostatic interaction energy of two hydrogen atoms separated at a finite distance, over and above the energies of two isolated hydrogen atoms; and the quantum

Figure 2.1

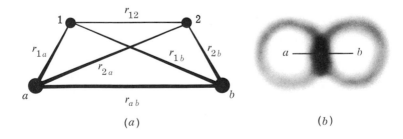

(a) $\qquad\qquad\qquad\qquad (b)$

Diagrams used to analyze the bonding in a hydrogen molecule. (a) Points a and b represent the positions of the two hydrogen nuclei while points 1 and 2 represent those of the electrons; (b) a schematic illustration of the electron clouds associated with the two electrons.

mechanical counterpart of this energy is the expectation value of V, or, in other words,

$$\mathcal{E}_a = \langle V \rangle = \int \psi_a^*(1)\psi_b^*(2)V\psi_a(1)\psi_b(2)\ dx_1\ dy_1\ dz_1\ dx_2\ dy_2\ dz_2 \quad (2.4)$$

x_1, y_1, and z_1 and x_2, y_2, and z_2 being the cartesian coordinates for the two electrons.

The same result would be obtained if it were assumed that electron 1 belonged to nucleus b and electron 2 to nucleus a by interchanging the subscripts a and b in Eqs. (2.3) and (2.4). However, it is pointed out in Sec. 1.8 that because of the indistinguishability of electrons the total wave function must be a linear combination of the products of the one-electron wave functions, the symmetrical wave function being given by ψ_S of Eq. (1.110) and the antisymmetrical wave function by ψ_A of Eq. (1.111). If the same process is repeated, we find that the additional energy $\mathcal{E}$ now becomes

$$\mathcal{E} = \int \psi^* V \psi\ dx_1\ dy_1\ dz_1\ dx_2\ dy_2\ dz_2 = \mathcal{E}_a \pm \mathcal{E}_b \quad (2.5)$$

where $\mathcal{E}_b$ is equal to

$$\mathcal{E}_b = \int \psi_a^*(2)\psi_b^*(1)V\psi_a(1)\psi_b(2)\ dx_1\ dy_1\ dz_1\ dx_2\ dy_2\ dz_2 \quad (2.6)$$

and the plus and minus signs are for ψ_S and ψ_A, respectively. From an energetic viewpoint, we see that the H atom-H atom structure if it existed would be unstable and that either the state ψ_S or the state ψ_A must be favored. Knowing ψ_a and ψ_b, the quantities $\mathcal{E}_a$ and $\mathcal{E}_b$ can be calculated. The result of such a calculation for the two 1s electrons is shown in Fig. 2.2, with the state ψ_S having the lowest energy.

If we recall, the square of the wave function $|\psi_a(1)|^2$ is interpreted as the charge density surrounding nucleus a so that $|\psi_a(1)|^2\ dv$ represents the probability of finding an electron in volume element dv around nucleus a. In Eq. (2.6) a different charge density, called the *exchange charge density*, must be defined such that $\varrho_1 = \psi_a(1)\psi_b(1)$ and $\varrho_2 = \psi_a(2)\psi_b(2)$. This charge density can be viewed as being shared by both nuclei a and b. The energy $\mathcal{E}_b$ is often called the *exchange energy* and it plays a decisive role in a covalent bond. First, we note that the product $\varrho_1\varrho_2$ has values significantly different from zero only in regions where the wave functions overlap, as indicated by the dark shaded area of Fig. 2.1b. This happens, however, only when the two nuclei are brought within a distance around one to two times the Bohr radius as is evident from Fig. 1.11. Further, for this distance, the contribution of the last two terms of Eq. (2.3) to $\mathcal{E}_b$ outweighs that of the first two terms, giving rise to a negative $\mathcal{E}_b$. For smaller r_{ab}, however, not only the term e^2/r_{ab} in Eq. (2.3) dominates, but $\mathcal{E}_a \gg \mathcal{E}_b$, resulting in a sharp rise of energy $\mathcal{E}$.

From the preceding discussion we can see that a stable hydrogen molecule forms in the state ψ_S, and with a separation between the two atoms about 1.5 times the Bohr radius. Since the state ψ_S has a minimum energy lower than the total energy of two isolated hydrogen atoms shown as $\mathcal{E}_t$ in Fig. 2.2, the two atoms attract each other in the state ψ_S. The opposite is true for the two atoms in the state ψ_A. We must not forget that associated with ψ_S and ψ_A are the spin-wave functions of Eq. (1.112), antiparallel spins with ψ_S and parallel spins with ψ_A. In summary, a covalent bond is a result of minimizing the energy by sharing two electrons with antiparallel spins between two neighboring nuclei. Exchanging two electrons with parallel spins, on the other hand, will result in a strong repulsion of the two hydrogen atoms. Other examples of covalent bond include the following:

$$\ddot{:}\overset{\cdot\cdot}{\underset{\cdot\cdot}{Cl}}\overset{\cdot\cdot}{\underset{\cdot\cdot}{:Cl}}\ddot{:} \qquad H\overset{\overset{\textstyle H}{\cdot\cdot}}{\underset{\underset{\textstyle H}{\cdot\cdot}}{:C:}}H \qquad :\overset{\cdot\cdot}{\underset{\cdot}{O}}\vdots\overset{\cdot}{\underset{\cdot}{O}}: \quad \text{etc.}$$

the two dots between the two atoms indicating the electron-pair bond.

The foregoing material leads to an important property of a covalent bond, the saturation property. Consider the interaction of a hydrogen atom with a helium atom as shown in Fig. 2.3. An exchange of electrons between H and He with parallel spins (Fig. 2.3a) results in repulsion, whereas an exchange with antiparallel

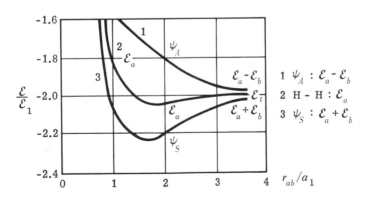

Figure 2.2

The energy of a hydrogen molecule plotted as a function of interatomic separation. The three curves are for three different states of the molecule. (1) The two electrons are being shared by the two nuclei with antisymmetrical orbital and symmetrical spin wave functions (spin parallel arrangement). (2) The two electrons belong to each nucleus separately and hence the molecule is just two atoms kept at a close distance. (3) The two electrons are being shared by the two nuclei with symmetrical orbital and antisymmetrical spin wave functions (spin antiparallel arrangement). The ordinate and abscissa are normalized with respect to the ionization energy and the Bohr radius of a hydrogen atom. (After L. Pauling and E. B. Wilson, Jr., "Introduction to Quantum Mechanics," p. 344, McGraw-Hill Book Company, New York, 1935.)

spins (Fig. 2.3b) violates the Pauli exclusion principle. Therefore covalent bonding exists only between atoms with unpaired spins. An example is CH_4. For carbon to have unpaired spins, the four valence electrons must have the configuration $2s2p^3$ in contrast to the ground-state configuration $2s^22p^2$, the superscript indicating the number of electrons in the $2s$ and $2p$ states. The four orbital wave functions are mixtures of the one $2s$ and the three $2p$ wave functions called the *sp^3-hybrid orbital* and they have electron distributions directed toward the four corners of a tetrahedron as shown in Fig. 2.4a. Typical examples of crystals with nearly pure covalent bond are diamond, silicon, and germanium, and these crystals all have the diamond structure. In this structure each atom has four nearest neighbors; a representative unit is shown in Fig. 2.4b. Each of the four orbitals of the sp^3 hybrid of each atom in the diamond structure forms an electron-pair bond with a corresponding one from one of its nearest neighbors. These covalent crystals are important semiconductors, the electrical properties of which will be discussed in detail in later chapters.

The molecular bond.   The difference in the bond property of Ge and CH_4 is worth exploring. Figure 2.5a shows a schematic representation of the bond pairs of Ge in a diamond structure. Note that germanium atoms can always be added to the existing structure because there exist unpaired bonds at the boundary of the structure. These unpaired bonds are generally referred to as the dangling bonds, and they are responsible for the fact that germanium and silicon surfaces are quite active chemically. The important point of our present discussion is the following one. Covalent bonding is all that is needed for Ge atoms to form a Ge crystal. This is not the case with CH_4. The bonds in CH_4 are all used up; thus no additional bond pairs can be formed with neighboring CH_4 molecules. The bond between such molecules comes from interactions of a different nature, interactions known as van der Waals forces. It turns out

Figure 2.3

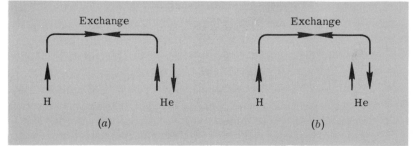

Consideration of exchange coupling between a hydrogen and helium atom. The diagram is used to illustrate the saturation property of a covalent bond.

Figure
2.4

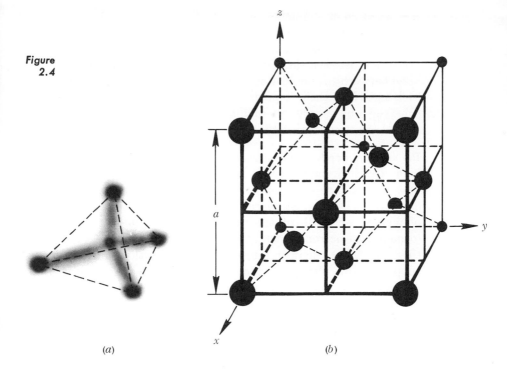

(a) *The four sp³ hybrid wave functions of carbon in* CH_4 *having electron distributions directed toward the four corners of a tetrahedron.* (b) *The tetrahedron as the basic unit in the diamond structure. Important covalent semiconductors such as germanium and silicon crystallize in this structure. The electron wave functions in these materials can also be considered as being made of the sp³ hybrid orbitals.*

Figure
2.5

Ge Ge Ge
Ge Ge Ge
Ge Ge Ge
(a)

H C H   H C H

(b)

Schematic representation of covalent bonds in (a) *Ge and* (b) CH_4. *Note that while new germanium atoms can be added to the existing structure, the bonds in* CH_4 *are all used up so that no additional bond pairs can be formed with neighboring* CH_4 *molecules.*

that an atom or a molecule is polarizable not only under an external electric field but under the field of other atoms or molecules as well. As a result, a very small electric dipoie develops, and the weak attraction between these mutually induced dipoles produces a weak bond between the noble atoms such as A and the saturated molecules such as CH_4 in their respective liquid phase. The fact that all these substances have very low boiling and melting points is a manifestation of the very weak bonds.

The metallic bond.   The outstanding characteristic of a metal is its high electrical conductivity. It can be inferred from this property that the valence electrons such as the $2s$ electrons in Li and the $3s$ electrons in Na must be free to move about. In the classical theory of metals, we think of the valence electrons in these crystals as forming an electron sea and the positive ions as being embedded in this sea. Later, quantum mechanical calculations have generally supported this picture. Figure 2.6 shows the wave function of the valence electron in a sodium crystal (solid line) as compared to the $3s$ wave function of a sodium atom (dashed line). The wave function is not disturbed near the ion core, but in the outer region it is considerably flattened and squeezed in by neighboring valence electrons. Since the distribution of electrons in a volume element $4\pi r^2 \, dr$ is given by $|\psi|^2 4\pi r^2 \, dr$, over 90 per cent of the electron distribution is in this flat region. The potential energy of the electron is obviously lowered because the average electron distribution is now closer to the nucleus. The kinetic energy, being proportional to $|\hbar \, \partial\psi/\partial r|^2$, is also reduced because of a smaller $\partial\psi/\partial r$ in the flattened region. This reduction in the total energy constitutes a bond between atoms in a metallic crystal.

The reader may wonder whether the picture just presented is in conflict with the theory of covalent binding. In the case of a hydrogen molecule, the addition of a third hydrogen atom would either result in repulsion or would violate the Pauli exclusion principle.

Figure 2.6

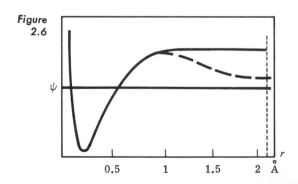

The electron wave function in a sodium crystal (solid curve) as compared to that in an isolated sodium atom (dashed curve). The vertical line indicates the boundary between neighboring atoms in a sodium crystal. (After F. Seitz, "The Modern Theory of Solids," p. 351, McGraw-Hill Book Company, New York, 1940.)

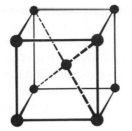

Figure
2.7

*The body-centered
cubic structure of
sodium.*

In the case of metallic sodium, the situation is changed for two essential reasons. First, each sodium atom has eight nearest neighbors in a body-centered cubic structure shown in Fig. 2.7; therefore, on the average, each atom forms only one-eighth of an electron-pair bond with one particular neighbor. Second, as a result of electrostatic interactions, the energy levels of the atom broaden out to form a band as the interatomic distance decreases. This process is illustrated in Fig. 2.8. The number of quantum states in each band is exactly equal to the number of atomic states multiplied by the number of atoms involved. If the model of covalent binding is adopted, we see that quantum states are now available to the third sodium atom. From the viewpoint of covalent bond, the metallic bond is an unsaturated one, and hence there is no contradiction between the two viewpoints.

Figure
2.8

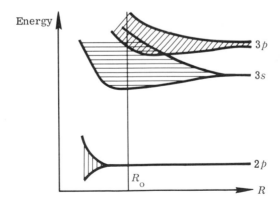

The energy-band diagram of metallic sodium as a function of interatomic distance. Because of strong interactions among electrons, the energy levels broaden into a band. The number of quantum levels in each band, for example, the 3s band, is exactly equal to the number of atomic states (2 for the 3s electrons) multiplied by the number of sodium atoms. These energy levels are discrete, but their separation in energy is so minute that for all practical purposes, the levels can be considered as forming a continuum.

In conclusion, Table 2.1 compares the strengths of the various bonds. As stated earlier, the transition between crystal types is not sharp at all. For example, tin can exist in two crystal forms, one being metallic and the other covalent. The reader should not think that all semiconductors are covalently bound. Crystals such as CuO, ZnO, and CdS are more ionic than covalent, but they also have energy bands as do Ge and Si. When sufficient electrons are in the upper band, these crystals will behave like semiconductors. Other examples of transition cases include: some metal alloys such as Mg_3Sb_2 and Zn_3As_2, which can be considered partly metallic and partly ionic, and some molecular crystals such as S and P, which are partly molecular and partly covalent.

TABLE 2.1   Cohesive energies (in ev) of different crystal types*
(Energy required to dissociate a solid into monatomic gas in the case of elemental solids and to molecular gas in the case of compound solids.)

Ionic		*Covalent*		*Metallic*		*Molecular*	
Material	*Energy*	*Material*	*Energy*	*Material*	*Energy*	*Material*	*Energy*
LiCl	8.6	Diamond	7.4	Na	1.13	A	0.08
NaCl	7.9	Si	3.7	Au	3.58	Kr	0.11
KI	6.6	Ge	3.7	Mg	1.56	CH_4	0.10
CsI	6.3	SiC	12.3	Ga	2.86	H_2	0.01
CuCl	9.6	SiO_2	17.6	Fe	4.20	He (liquid)	0.001

* References: C. A. Wert and R. M. Thomson, "Physics of Solids," McGraw-Hill Book Company, New York, 1964; F. Seitz, "The Modern Theory of Solids," McGraw-Hill Book Company, New York, 1940; A. H. Cottrell, "The Mechanical Properties of Matter," John Wiley & Sons, Inc., New York, 1964.

2.2   BRAVAIS LATTICES AND SIMPLE CRYSTAL STRUCTURES

A substance is said to be in a *crystalline state* if its constituent atoms are arranged in a regular and repetitious pattern. In order to specify the pattern of the arrangement, a system of axes called the crystal axes must be chosen. In Fig. 2.9 a parallelepipedon is shown, with the length of each side designated by a, b, and c and the angle between the axes by α, β, and γ. Depending on the specifications about the lengths and the angles, crystals can be grouped into seven systems as shown in Table 2.2. The simplest representative unit of each system has lattice points at the corners of the parallelepipedon; these units are shown in Fig. 2.10 and are referred to as *primitive units*. Bravais showed, however, that various symmetry operations possible for a three-dimensional point lattice require 14 different

Figure
2.9

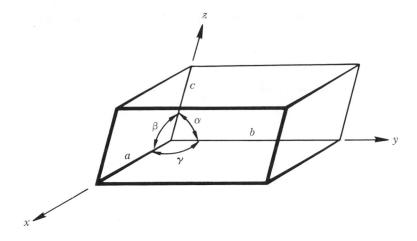

The parallelepipedon which defines three lengths a, b, and c and three angles α, β, and γ. Crystal structures are classified into seven systems depending on the specifications about the lengths and the angles.

space or *Bravais lattices.* Considerations of these symmetry operations will be discussed later. The other seven units shown in Fig. 2.10 have additional lattice points at the base centers, the face centers, or the body center of the basic unit.

Many important materials have simple structures. In the case of metals, the alkali metals (Li, Na, K, Rb, Cs) form body-centered cubic lattices, while the noble metals (Cu, Ag, Au) form face-centered cubic lattices. The diamond structure of covalent semiconductors shown in Fig. 2.4b is actually a combination of two interpenetrating face-centered cubic lattices, one sublattice having its origin at (000) and the other at (a/4, a/4, a/4). The zinc blende structure differs from the diamond structure only in that the two sublattices are occupied by two different elements. Thus C (diamond), Si, and Ge have the diamond structure, while III–V compound semiconductors such as InSb, GaAs, and GaSb have the zinc blende structure. For ionic crystals, the two common types are the

TABLE 2.2 *Crystal systems*

Triclinic	$a \neq b \neq c$	$\alpha \neq \beta \neq \gamma$
Monoclinic	$a \neq b \neq c$	$\alpha = \beta = 90° \neq \gamma$
Orthorhombic	$a \neq b \neq c$	$\alpha = \beta = \gamma = 90°$
Tetragonal	$a = b \neq c$	$\alpha = \beta = \gamma = 90°$
Cubic	$a = b = c$	$\alpha = \beta = \gamma = 90°$
Hexagonal	$a = b \neq c$	$\alpha = \beta = 90°, \gamma = 120°$
Trigonal	$a = b = c$	$\alpha = \beta = \gamma \neq 90°$

Figure 2.10

Triclinic

Monoclinic

Primitive
P

Base-centered
C

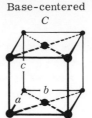

Orthorhombic

Primitive
P

Base-centered
C

Body-centered
I

Face-centered
F

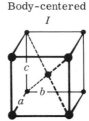

Cubic

Primitive
P

Body-centered
I

Face-centered
F

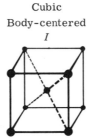
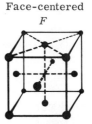

Tetragonal

Primitive
P

Body-centered
I

Trigonal
R

Hexagonal
P

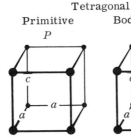
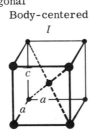
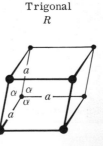

120°

a_3 — a_2

120°

a_1

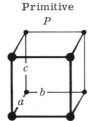

Representative units of the 14 Bravais lattices. The primitive units in each system have atoms at the corners of the parallelepipedon only and are designated by the letter P. The letter C indicates that the structure has extra atoms at the centers of its bases. The letters I (German Innenzentrierte) and F refer, respectively, to body-centered and face-centered structures.

sodium chloride and the cesium chloride structures shown in Fig. 2.11*a* and *b*, respectively. The sodium chloride structure consists of two face-centered cubic sublattices, one occupied by Na and the other by Cl atoms. The two sublattices are displaced from each other by a distance $a/2$ along any one of the three cubic edges. The cesium chloride structure looks very much like a body-centered cubic lattice except that one element occupies the eight corners and the other the center of the cube. Actually, the structure consists of two interpenetrating simple cubic sublattices, displaced $(a/2, a/2, a/2)$ apart and occupied by two different elements. Among ionic crystals, the alkali halides have the sodium chloride structure except for CsCl, CsBr, and CsI, which have the cesium chloride structure.

In calculating the interatomic forces we must know the arrangement of neighboring atoms, especially that of the nearest neighbors. The number of nearest neighbors is generally known as the *coordination number* denoted by Z. In the diamond and zinc blende structures, each atom is surrounded by four nearest neighbors; so the coordination number is 4. In the sodium chloride and cesium chloride structures, the coordination numbers are 6 and 8, respectively. Two crystal structures, the face-centered cubic structure and the hexagonal close-packed structure, have the maximum coordination

*Figure
2.11*

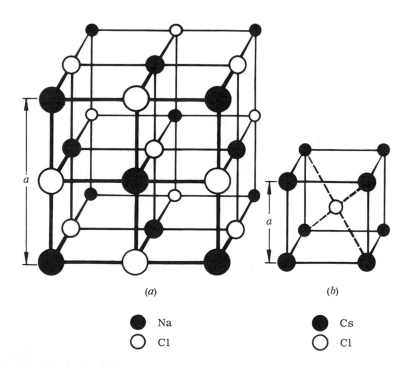

(a)   (b)

● Na   ● Cs
○ Cl   ○ Cl

Common structures for ionic crystals: (a) NaCl structure and (b) CsCl structure.

number of 12; they are most densely packed. In Fig. 2.12 the arrangement of atoms in the face-centered cubic structure is shown in detail. If we compare the arrangement in three different horizontal planes viewed from the top (Fig. 2.12a) with that in three different vertical planes viewed from the front (Fig. 2.12c), we find that not only is the pattern the same but atoms 5, 6, 7, and 8 in the front view and atoms 9, 10, 11, and 12 in the side view are similarly situated with respect to atom 0 as atoms 1, 2, 3, and 4 in the top view. We further see that in the face-centered cubic structure there is very little space left between the atoms. Another way of arranging the atoms to minimize the interstitial space leads to a hexagonal close-packed structure of Fig. 2.13 in which a middle layer is added to the regular hexagonal structure. The atoms in the middle layer rest in the depression against three atoms in the basal plane. If circles are drawn in a two-dimensional space to represent the atoms, it is clear that there is only room for three atoms, shown as atoms 7, 8, and 9 in Fig. 2.13b. However, there are two sides to the basal plane. Atoms 10, 11, and 12 represent atoms from the bottom, which have a similar arrangement to atoms 7, 8, and 9. Metals such as Be, Mg, Zn, and Cd have the hexagonal close-packed structure. Since the purpose of this section is to acquaint

Figure 2.12

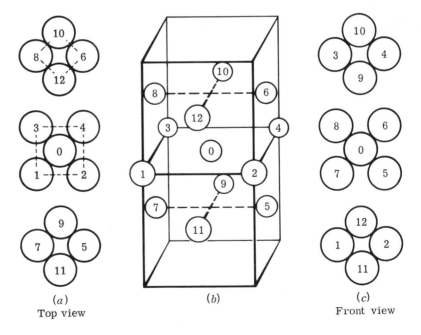

(a)
Top view

(b)

(c)
Front view

The face-centered (close-packed) cubic structure with (a) the arrangement of atoms in three different horizontal planes as viewed from the top and (c) that in three different vertical planes as viewed from the front.

Figure 2.13

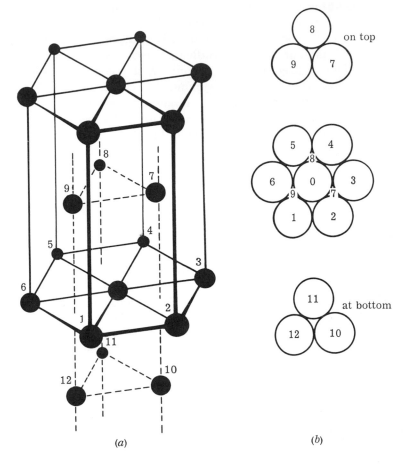

on top

at bottom

(a)   (b)

The hexagonal close-packed structure. The 12 nearest neighbors of atom O have their spatial relations to atom O as shown in (b).

the reader with some of the simplest and most common crystal structures, more complicated structures will be discussed in later chapters when appropriate. The lattice constants as well as structures of some common substances are given in Table 2.3.

2.3 IONIC CRYSTALS

In the two preceding sections we presented a general background for four principal crystal types: the ionic, the covalent, the molecular, and the metallic crystals. Of the four types, ionic and metallic crystals will be discussed further. The foundations of the classical theory of ionic and metallic crystals were laid before the quantum theory was advanced. Therefore these discussions are quite appropriate at the level of our present understanding about material properties. Furthermore, these examples will serve to illustrate that the classical theory can be remarkably successful if effects of quantum mechanical origins are handled properly.

The idealized model of an ionic crystal assumes that the constituent ions have spherically symmetrical charge distributions as in the rare gas atoms and that the charges of these ions are multiples of the electronic charge. For definiteness, we shall use NaCl as an example, the structure of which is shown in Fig. 2.11a. According to the theory of electrostatics, the interaction energy between two nonoverlapping spherical charge distributions is the same as that between two point charges; thus the total electrostatic energy of the crystal is given by

$$\mathcal{E}_1 = \sum (\pm) \frac{e^2}{4\pi\epsilon_0 r_{ij}} \tag{2.7}$$

where r_{ij} is the distance between the centers of the two charge distributions involved, ϵ_0 is the dielectric constant of free space, and the $+$ and $-$ signs are for interactions between charges of like and opposite polarities, respectively. The summation in Eq. (2.7) extends over all charge pairs. Take the center Cl ion in Fig. 2.11a as the reference ion. The contribution to Eq. (2.7) from the 6 nearest Na ions is $6e^2/4\pi\epsilon_0 R$, R being the distance between nearest neighbors. Next are the 12 nearest Cl ions which are $\sqrt{2}\,R$ apart from the

TABLE 2.3 *Common lattice structures of selected substances*

Substance	Structure	Lattice constant (Å)		Nearest neighbor distance (Å)
Carbon	Diamond	3.56		1.54
Cesium	bcc	6.05 (92°K)		5.24
Copper	fcc	3.61		2.55
Germanium	Diamond	5.65		2.44
Iron	bcc	2.86		2.48
Magnesium	hcp	3.20	5.20 (*c* axis)	3.20
Silicon	Diamond	5.43		2.35
Sodium	bcc	4.28		3.71
Xenon	fcc	6.24 (92°K)		4.41
Zinc	hcp	2.66	4.94 (*c* axis)	2.66
NaCl	NaCl	5.63		2.82
MgO	NaCl	4.20		2.10
CsCl	CsCl	4.11		3.56
NH$_4$Cl	CsCl	3.87		3.35
ZnS	Zinc blende	5.41		2.34
CdS	Zinc blende	5.82		2.51
InSb	Zinc blende	6.46		2.79

center Cl ion. Then there are 8 next to the nearest Na ions at a distance $\sqrt{3}\,R$ apart and 6 next to the nearest Cl ions at a distance $\sqrt{4}\,R$ apart. If there are $2N$ ions or N pairs of NaCl ions in a crystal, Eq. (2.7) becomes

$$\varepsilon_1 = \frac{Ne^2}{4\pi\epsilon_0 R}\left(\frac{6}{1} - \frac{12}{\sqrt{2}} + \frac{8}{\sqrt{3}} - \frac{6}{\sqrt{4}} + \frac{24}{\sqrt{5}} - \cdots\right) = \frac{ANe^2}{4\pi\epsilon_0 R}$$

(2.8)

The constant A is known as the *Madelung constant*. For NaCl, $A = 1.75$. Note that the factor N, not $2N$, appears in Eq. (2.8) because the electrostatic energy between two charged particles $e_i e_j / 4\pi\epsilon_0 r_{ij}$ should be counted only once. Since the series in Eq. (2.8) converges rather slowly, special techniques have been developed in the method of counting to improve the convergence. The reader is referred to the book[1] by Kittel for references as well as for a discussion of the techniques and to the review paper[2] by Sherman for values of Madelung constants in different crystal structures.

As shown in Fig. 2.2, the potential energy of a hydrogen molecule increases very rapidly as the distance between the two atoms gets below the Bohr radius. Repulsive potential of a similar origin exists also in an ionic crystal. When the distance between two neighboring ions gets smaller and smaller, the two ions may come into contact. In quantum mechanical language, the two charge clouds of Na and Cl ions are said to overlap each other. When this happens, the repulsive potential becomes very large and increases much more rapidly than the coulomb energy. The situation is illustrated in Fig. 2.14. The actual form of this repulsive potential can be calcu-

[1]   C. Kittel, "Introduction to Solid State Physics," John Wiley & Sons, Inc., New York, 1956.
[2]   J. Sherman, *Chem. Rev.*, vol. 11, p. 93, 1932.

Figure 2.14

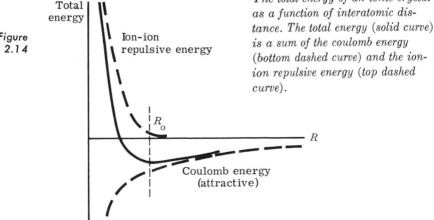

The total energy of an ionic crystal as a function of interatomic distance. The total energy (solid curve) is a sum of the coulomb energy (bottom dashed curve) and the ion-ion repulsive energy (top dashed curve).

lated quantum mechanically as a function of R. However, we are interested only in the variation of this potential near R_0, the equilibrium value of R. For this purpose, the repulsive potential can be approximated by

$$\mathcal{E}_2 = \frac{BN}{R^n} \tag{2.9}$$

Since the two energies $\mathcal{E}_1$ and $\mathcal{E}_2$ have opposite signs, the total energy of a NaCl crystal is given by $\mathcal{E}_2 - \mathcal{E}_1$.

At the equilibrium separation R_0, $\mathcal{E}_2 - \mathcal{E}_1$ should be a minimum; thus

$$\frac{-nBN}{R_0^{n+1}} + \frac{ANe^2}{4\pi\epsilon_0 R_0^2} = 0 \quad \text{or} \quad R_0^{n-1} = \frac{4\pi\epsilon_0 nB}{Ae^2} \tag{2.10}$$

Substituting R_0 from Eq. (2.10) into Eqs. (2.8) and (2.9), we find the total energy

$$\mathcal{E} = \mathcal{E}_2 - \mathcal{E}_1 = -\frac{ANe^2}{4\pi\epsilon_0 R_0}\left(1 - \frac{1}{n}\right) \tag{2.11}$$

The binding energy of the crystal per molecule, $\mathcal{E}/N$, is known as the cohesive energy of an ionic crystal, and it is found experimentally to be 183 kcal/mole or 7.95 ev/molecule for NaCl. The energy $\mathcal{E}_1$ due to coulomb attraction is 8.94 ev, taking $R_0 = 2.82$ Å in NaCl. Substituting these values into Eq. (2.11), we find $n = 9$. The value of n can also be obtained from the compressibility K. The reader is asked to show that

$$\frac{1}{K} = -V\frac{dp}{dV} = V\frac{d^2\mathcal{E}}{dV^2} = \frac{(n-1)e^2 A}{72\pi\epsilon_0 R_0^4} \tag{2.12}$$

where V and p represent, respectively, the volume and pressure. Equation (2.12) is derived by taking $V = 2NR^3$. For $K = 3.3 \times 10^{-12}$ cm^2/dyne in NaCl, Eq. (2.12) gives a value of $n = 9.1$. The discrepancy between the two values of n is due to a very small contribution from van der Waals forces to the total energy. From the above example we see that the cohesive energy of an ionic crystal is mostly due to coulomb attraction between the ions.

At this point we shall add the electrical conductive property of an ionic crystal to our discussion. Take NaCl as an example. In a solid, the $3p$ energy levels of Cl$^-$ ions broaden into a band as illustrated in Fig. 2.15. Normally this $3p$ band is completely occupied by the six $3p$ electrons of Cl$^-$ ions. In an ideal crystal, if an electron is lifted from the $3p$ band, that is, if it is taken away from one Cl$^-$ ion, this electron must go to a Na$^+$ ion, thus becoming a $3s$ electron of Na. The $3s$ levels of Na also form a band in a solid. The lifting of an electron from the $3p$ band of Cl into the $3s$ band of Na requires a very large amount of energy because the energy gap in NaCl is around 8 ev. Therefore electronic conduction is almost nonexistent.

For all practical purposes, crystals with purely ionic bonding can be considered as insulators. However, at high temperatures ionic, but not electronic, conduction is possible in ionic crystals. This ionic conduction will be discussed in connection with lattice imperfections later in this chapter.

Figure 2.15

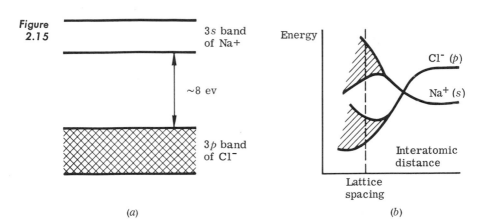

(a)   *(b)*

The energy-band diagram of NaCl. *In a solid, the atomic levels broaden to form bands. The 3s band of* Na *is raised and the 3p band of* Cl *is depressed because of the Madelung potential of Eq. (2.8). (W. Shockley, Phys. Rev. vol. 50, p. 754, 1936.)*

2.4 METALLIC CRYSTALS

According to the classical description mentioned earlier in Sec. 2.1, a metallic crystal is generally considered as being composed of arrays of positive ions embedded in a uniform sea of conduction electrons. For concreteness, we take sodium as an example. As shown in Fig. 2.6, the distribution of electrons in sodium is essentially uniform over 90 per cent of the volume occupied by the electrons. In calculating the electrostatic interaction energy in such a crystal, it seems reasonable to assume that a point charge of $+e$ is surrounded by a charge $-e$ uniformly distributed over a sphere of volume V. Referring to Fig. 2.16, we see that the charge in a volume element $4\pi r^2\, dr$ is given by $e4\pi r^2\, dr/V$. Furthermore, the coulomb potentials seen by the same charge due to the positive point charge and the negative charge cloud are, respectively, equal to $-e/(4\pi\epsilon_0 r)$ and $er^2/(3\epsilon_0 V)$. The total coulomb energy is, therefore, given by

$$\mathcal{E}_1 = \int_0^R \left(-\frac{e}{4\pi\epsilon_0 r} + \frac{er^2}{3\epsilon_0 V} \right) \frac{e4\pi r^2\, dr}{V} = \frac{-9e^2}{40\pi\epsilon_0 R} = -\frac{A}{R} \quad (2.13)$$

with $A = 9e^2/(40\pi\epsilon_0)$.

Figure 2.16

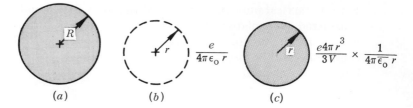

(a) *(b)* *(c)*

Diagrams used in calculating the electrostatic energy in a metallic crystal according to the classical theory. (a) The metal atom is considered as consisting of a positive ion core with a charge $+e$ surrounded by an electron cloud of charge $-e$ uniformly distributed over a volume V. The potential at a point r from the nucleus is given by (b) $-e/(4\pi\epsilon_0 r)$ due to the ion core and (c) $er^2/(3\epsilon_0 V)$ due to the electron cloud. The negative and positive signs are used for attractive and repulsive potentials, respectively.

To calculate the kinetic energy of electrons we must consider the energy level diagram of sodium. The ground state of the valence electron is the $3s$ level and its first excited state the $3p$ level. In a crystal, both levels broaden out to form bands. Since the atomic $3s$ and $3p$ levels are not far apart in energy, the two bands, the $3s$ and $3p$ bands, overlap as shown in Fig. 2.17. However, the overlap is so slight that only states in the $3s$ band are normally occupied. For simplicity, we shall take the temperature to be absolute zero so that only states below the Fermi level $\mathcal{E}_f$ are occupied. Thus, according

Figure 2.17

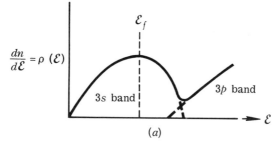

(a)

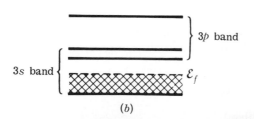

(b)

Energy-band diagrams for sodium. The density of states $\rho(\mathcal{E})$ is plotted as a function of energy in diagram (a) and the states with $\mathcal{E} < \mathcal{E}_f$ (the Fermi energy being indicated by the dashed line) are occupied by electrons.

to Eq. (1.143), we find the density of electrons n to be

$$n = \int_0^{\varepsilon_f} 4\pi(2m_0)^{3/2} \sqrt{\varepsilon} \frac{d\varepsilon}{h^3} = \frac{8\pi}{3} \frac{(2m_0)^{3/2}}{h^3} \varepsilon_f^{3/2} \qquad (2.14)$$

The most important characteristic of a metallic crystal is the ability of the conduction (valence) electrons to move about in the crystal. As shown in Fig. 2.6, the electron wave function of a sodium atom joins smoothly that of a neighboring atom at their geometric boundary, and hence the movement of electrons from one atom to another is greatly facilitated. As a matter of fact, if the conduction electrons are free as in vacuo, the Schrödinger equation simply reduces to

$$-\frac{\hbar^2}{2m_0} \nabla^2 \psi = \varepsilon \psi \qquad (2.15)$$

and the solution of Eq. (2.15) is given by

$$\psi = \left(\frac{1}{V}\right)^{1/2} \exp(i\mathbf{k} \cdot \mathbf{r}) \qquad \text{with } \frac{\hbar^2 k^2}{2m_0} = \varepsilon \qquad (2.16)$$

where $(1/V)^{1/2}$ is the normalization factor and V is the volume of the crystal. The energy ε is entirely kinetic of course, and the wave number k is related to the wavelength λ associated with the electron motion by the famous de Broglie relation of Eq. (1.139) or Eq. (1.13). The free electron theory of metals, therefore, predicts an average kinetic energy ε_3 at $0°K$:

$$\varepsilon_3 = \frac{1}{n} \int_0^{\varepsilon_f} \varepsilon 4\pi(2m_0)^{3/2} \sqrt{\varepsilon} \frac{d\varepsilon}{h^3} = \frac{8\pi}{5} \frac{(2m_0)^{3/2}}{h^3 n} \varepsilon_f^{5/2} \qquad (2.17)$$

In a solid, electrons are subject to a periodic crystalline potential arising from the nucleus. As will be shown in the next chapter, the effect of the crystalline potential can be incorporated into the free electron theory by proposing an effective mass m^* such that the kinetic energy of an electron is given by $\varepsilon = \hbar^2 k^2/2m^*$. Using m^* instead of m_0 in Eqs. (2.17) and (2.14) and realizing that $n4\pi R^3/3 = 1$, we can rewrite Eq. (2.17) in terms of m^* and R as

$$\varepsilon_3 = \frac{3}{10} \frac{h^2}{m^*} \left(\frac{9}{32\pi^2}\right)^{2/3} \frac{1}{R^2} = \frac{B}{R^2} \qquad \text{with } B = \frac{3}{10} \frac{h^2}{m^*} \left(\frac{9}{32\pi^2}\right)^{2/3} \qquad (2.18)$$

The total energy of sodium is shown schematically in Fig. 2.18 as a function of R, which is one-half the distance between the nuclei of two nearest neighbors. Since the ion-core repulsive energy of the form given in Eq. (2.9) is still not significant near the equilibrium R_0, the value of R_0 and the total energy can be found by minimizing $\varepsilon_1 + \varepsilon_3$, yielding $R_0 = 2B/A$ and $\varepsilon = -A^2/4B$. Using $m^*/m = 0.93$ for sodium, we find $R_0 = 1.3$ Å and $\varepsilon = -5$ ev. In a body-centered cubic structure the eight atoms at the corner of a given cube are shared by eight cubes; hence the cube is occupied by two electrons, one from the center atom and the other being contributed by the eight corner atoms of one-eighth each. If a is the lattice constant, then the radius R_0 should be such that $4\pi R_0{}^3/3 = a^3/2$. For sodium, $a = 4.28$ Å, giving a value of $R_0 = 0.49a = 2.1$ Å in contrast to the calculated value of 1.3 Å. Apparently either Eq. (2.13) or Eq. (2.18) must be in error.

As pointed out earlier in Sec. 1.9, the potential seen by the valence electron depends on the degree of screening of the nuclear potential by the inner closed-shell electrons. In the actual computation of the wave function, we assume a radial potential $V_c(r)$ in the Schrödinger equation

$$-\frac{\hbar^2}{2m_0}\nabla^2\psi + V_c(r)\psi = \mathcal{E}\psi \qquad (2.19)$$

such that the eigenvalues of $\mathcal{E}$ give the correct value of the ionization energy which can be obtained experimentally. If we used the atomic 3s wave function for sodium in Eq. (2.19), we should obtain a value of 5.18 ev for the ionization potential of a sodium atom. In a crystal, the wave function is squeezed in by the valence electrons of the neighboring atoms, and hence each electron is confined to its own private quarter, which is taken to be a sphere of radius R. Since every sphere is exactly the same as every other sphere, the boundary condition at adjacent spheres is $\partial\psi/\partial R = 0$ in contrast to the condition $\psi = 0$ at $R = \infty$ for the atomic wave function. The wave function shown in Fig. 2.6, which obviously satisfies the boundary condi-

Figure 2.18

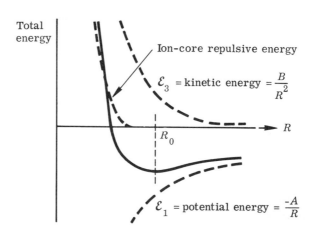

The total energy of a sodium atom in a crystal plotted as a function of R, which is one-half the distance between nearest-neighbor atoms. According to the semi-classical treatment presented in the text, the total energy (solid curve) is a sum of the potential energy, the ion-core repulsive energy, and the kinetic energy.

tion $\partial\psi/\partial R = 0$, is actually obtained by following the procedure just outlined.

Using the crystal wave function of Fig. 2.6, the energy $\mathcal{E}_c$ of sodium in crystal form can be computed from Eq. (2.19), and the result of such a calculation is plotted in Fig. 2.19a as a function of R. For comparison, the energy $\mathcal{E}_1$ calculated from Eq. (2.13) is also shown. The quantity $\mathcal{E}_1$ differs from $\mathcal{E}_c$ largely because the assumed ion-core potential $e/(4\pi\epsilon_0 r)$ in Eq. (2.13) is not a good enough approximation. For $R = 2.1$ Å, the value of $\mathcal{E}_c$ is found to be -8.4 ev from Fig. 2.19a and that of $\mathcal{E}_3$ to be 2.2 ev from Eq. (2.18). To calculate the cohesive energy we note that the energy $\mathcal{E}_c$ in Fig. 2.19a refers to the energy of the ionized state (Na$^+$ + electron). According

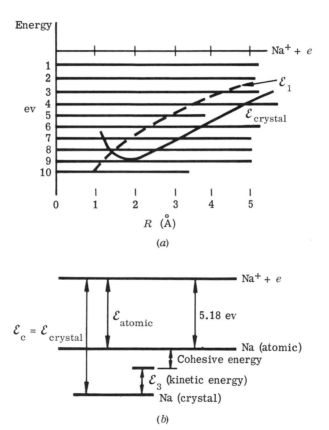

Figure 2.19

Diagrams used to calculate the cohesive energy of a sodium crystal. In diagram (a), the energies $\mathcal{E}_c$ (based on quantum-mechanical calculations) and $\mathcal{E}_1$ (based on classical models) are plotted as a function of R. (After F. Seitz, "The Modern Theory of Solids," p. 351, McGraw-Hill Book Company, New York, 1940.) Diagram (b) shows the various energy levels involved in the calculation of the cohesive energy.

to Fig. 2.19b, the cohesive energy of a sodium crystal is equal to $|\mathcal{E}_c + \mathcal{E}_3| - 5.2 = 1.0$ ev in agreement with the measured value of 1.1 ev.

From the above discussion we see that the source of error in the semiclassical model is in the calculation of the potential energy $\mathcal{E}_1$ of Eq. (2.13), not in that of the kinetic energy $\mathcal{E}_3$ of Eq. (2.18). The free electron theory, which is based on de Broglie's theory of matter wave, predicts correctly not only the value of $\mathcal{E}_3$ but also the dynamic behavior of electrons in metals and semiconductors. As will be discussed in Chapters 3 and 4, conduction electrons in these materials can be treated essentially as a classical particle with an effective mass m^*. In obtaining the value of m^*, however, we must use the definition of the group velocity v_g given in Eq. (1.27). In the wave-corpuscle dualism, Eq. (1.27) is the link connecting the particle concept (the use of an effective mass) and the wave concept (the use of wave number k in counting the available quantum states). The success which we have had with Eq. (2.18) in predicting the value of $\mathcal{E}_3$ proves the correctness of the free electron theory, that is, Eq. (2.16), for conduction electrons in metals and semi-conductors. This leads to a great simplification of the theory concerning the conductive property of these materials.

To conclude this section, we should like to add the following comments. The relative importance of the various energy terms changes from metal to metal; therefore sodium should not be taken as a typical example for other metals. If the size of the inner shell is large, ion-core repulsive force may be significant. The wide variation of cohesive energy from 0.8 ev in Cs (alkali metal) to 4.2 ev in Fe (transition metal) is an indication that forces of different physical origins may be involved. However, all metallic crystals have one outstanding property in common, an extremely low electrical resistivity. Typical metallic resistivities at room temperature range from 1.50×10^{-6} Ω-cm in copper and silver to 53×10^{-6} Ω-cm in gallium. These values may be compared with the resistivity of pure germanium at room temperature, which is 48 Ω-cm. The factor of order of 10^6 is mainly due to the difference in the concentration of free carriers in these materials. Note that Ag has only one valence electron, whereas Ge has four. Therefore the availability of electrons for electric conduction depends very critically on the nature of the chemical bond which distinguishes Ge from Ag. The electrical properties of semiconductors will be discussed at length in the next chapter.

2.5 CRYSTALLOGRAPHIC NOTATION

In Sec. 2.2 we presented the 14 Bravais lattices, the representative units of which are shown in Fig. 2.10. The reason that these units are representative of the whole lattice is that we can generate the whole lattice by translations of the representative unit along the three crystal axes. If a translation vector **t** is defined in terms of the crystal basis vectors **a**, **b**, and **c** as

$$t = u\mathbf{a} + v\mathbf{b} + w\mathbf{c} \qquad (2.20)$$

periodicity of the lattice requires that u, v, and w be integers. Besides the geometric arrangement of atoms, a function $f(\mathbf{r})$ describing any property of the crystal must also be periodic or $f(\mathbf{r}) = f(\mathbf{r} + \mathbf{t})$. For example, the wave function ψ of Fig. 2.6 is periodic in a sodium crystal. At this point the reader may ask whether the crystal axes of Fig. 2.10 represent a unique choice and, if not, what determines the choice of a particular set of axes. To answer this question we shall use the base-centered monoclinic and face-centered cubic structures as examples. Since each corner atom is shared by eight neighboring cells and each atom in the face by two neighboring cells, the representative units of Fig. 2.10 for the two structures have two and four atoms, respectively. Therefore it should be possible to reduce the size of these representative units.

In the monoclinic structure, if the two base diagonals are chosen as two crystal axes, a primitive cell is formed which contains only corner atoms or, in other words, only one atom per cell. The relation between the primitive and the unit cell is illustrated in Fig. 2.20a. In the face-centered cubic structure, a primitive cell is formed by using the three face diagonals as the three axes. The reader can easily show that the volumes of the primitive cells shown in Fig. 2.20a and b are, respectively, one-half and one-quarter of the volume of the corresponding units shown in Fig. 2.10. If a unit cell contains more than one atom per cell, it is called a nonprimitive or multiple cell. The three axes **a'**, **b'**, and **c'** used in defining a translation operation in Eq. (2.20) must be primitive vectors, or, in other

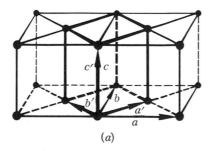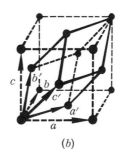

Figure
2.20

(a)   (b)

Diagrams showing the relationship between a primitive cell (heavy lines) and a unit cell for (a) the monoclinic structure and (b) the face-centered cubic structure.

words, basis vectors of the primitive cell. However, in many applications it is convenient to use the conventional crystal axes of Fig. 2.10 as the reference axes. The most important reason for doing this is that the representative units of Fig. 2.10 in each system have common symmetry properties. For example, the fact that a face-centered cubic crystal belongs to the cubic system is more easily recognized in the conventional system of axes $\mathbf{a}$, $\mathbf{b}$, and $\mathbf{c}$ than in the primitive system of axes $\mathbf{a}'$, $\mathbf{b}'$, and $\mathbf{c}'$ (Fig. 2.20b). To avoid confusion we shall use the conventional axes exclusively in the following discussion unless it is specifically stated that the primitive axes are to be used.

A basic necessity for adopting a unique system of reference axes is the need to describe directions and specific planes in space. In terms of vectors $\mathbf{a}$, $\mathbf{b}$, and $\mathbf{c}$, any vector in space can be specified by three numbers u, v, and w, conventionally arranged in brackets as $[uvw]$. Since any multiple of the vector by a number does not change the direction of the vector, the indices of a direction are given by the lowest set of integers. For example, the three face diagonals of a cube ($\mathbf{a}'$, $\mathbf{b}'$, and $\mathbf{c}'$ of Fig. 2.20b) have indices $[110]$, $[011]$, and $[101]$. A negative index is indicated by placing a minus sign above the index. Thus the negative x axis has indices $[\bar{1}00]$. Further, a full set of equivalent directions is denoted by $\langle uvw \rangle$. The eight body diagonals of a cube, $[111]$, $[11\bar{1}]$, $[1\bar{1}1]$, $[\bar{1}11]$, $[\bar{1}\bar{1}1]$, $[\bar{1}1\bar{1}]$, and $[\bar{1}1\bar{1}]$, are equivalent from symmetry considerations and hence can be designated as a group by $\langle 111 \rangle$. This is no longer true, however, with the eight body diagonals of a monoclinic structure.

The orientation of a crystal plane can similarly be determined by three numbers called the *Miller indices* of a plane. These numbers are defined in terms of the intercepts of the plane on the coordinate axes expressed in units of lattice constants. For example, to obtain the Miller indices of a plane that has intercepts $3a$, $6b$, and $2c$, we take the reciprocals of these numbers ($\frac{1}{3}$, $\frac{1}{6}$, and $\frac{1}{2}$, respectively) and then reduce the resultant fractions to the lowest set of three integers having the same ratio, that is, 2, 1, and 3. The Miller indices of a plane are written in parentheses (213). In general, the indices (hkl) may denote a single plane or a set of parallel planes, while indices written in braces $\{hkl\}$ denote a set of planes which are equivalent from symmetry considerations. As examples, several planes (110), (111), $(\bar{1}\bar{1}1)$, and (221) in a cubic crystal are shown in Fig. 2.21 with (111) and $(\bar{1}\bar{1}1)$ belonging to the set $\{111\}$. In a hexagonal system it is conventional to use four indices $(hkjl)$, the first three indices being determined by the intercepts on the three axes $\mathbf{a}_1$, $\mathbf{a}_2$, and $\mathbf{a}_3$ as shown in Fig. 2.22. Since $\mathbf{a}_3 = -(\mathbf{a}_1 + \mathbf{a}_2)$, a relationship exists between h, k, and j such that $j = -(h + k)$.

The concept of the reciprocal lattice is sometimes extremely useful in describing the periodic property of a crystal and will be referred

Figure
2.21

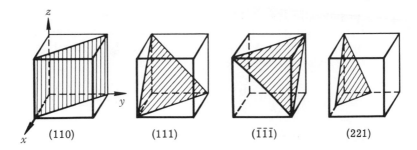

(110) (111) ($\bar{1}\bar{1}\bar{1}$) (221)

Several planes chosen for illustration in a cubic crystal and their Miller indices.

to in later applications. If **a**, **b**, and **c** are the basis vectors of a primitive cell, three new vectors **a***, **b***, and **c*** defined as

$$\mathbf{a}^* = \frac{\mathbf{b} \times \mathbf{c}}{\mathbf{a} \cdot [\mathbf{b} \times \mathbf{c}]} \qquad \mathbf{b}^* = \frac{\mathbf{c} \times \mathbf{a}}{\mathbf{a} \cdot [\mathbf{b} \times \mathbf{c}]} \qquad \mathbf{c}^* = \frac{\mathbf{a} \times \mathbf{b}}{\mathbf{a} \cdot [\mathbf{b} \times \mathbf{c}]} \qquad (2.21)$$

have the property that $\mathbf{a}^* \cdot \mathbf{a} = \mathbf{b}^* \cdot \mathbf{b} = \mathbf{c}^* \cdot \mathbf{c} = 1$ and $\mathbf{a}^*$, $\mathbf{b}^*$, and $\mathbf{c}^*$ are, respectively, perpendicular to the planes containing vectors **b** and **c**, vectors **c** and **a**, and vectors **a** and **b**. The set of reciprocal basis vectors $\mathbf{a}^*$, $\mathbf{b}^*$, and $\mathbf{c}^*$ defines a new primitive cell, and the new lattice based on the reciprocal vectors is called the *reciprocal lattice*. Let us now make use of the new concept. A vector $\mathbf{r}^*$ of the form

$$\mathbf{r}^* = h\mathbf{a}^* + k\mathbf{b}^* + l\mathbf{c}^* \qquad (2.22)$$

is perpendicular to the (hkl) plane. Since any vector in a plane can be expressed as a linear combination of two nonparallel vectors in the same plane, we can choose any two vectors in the (hkl) plane.

Figure
2.22

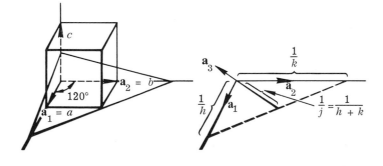

Illustrations to show how Miller indices are found in a hexagonal structure. The three intercepts along the $\mathbf{a}_1$, $\mathbf{a}_2$, and $\mathbf{a}_3$ axes are in the ratio $1/h:1/k:1/j$.

As shown in Fig. 2.23a, the two lines which represent the intersections of the (hkl) plane with ab and bc planes are given vectorially by $\mathbf{a}/h - \mathbf{b}/k$ and $\mathbf{b}/k - \mathbf{c}/l$, respectively. Using Eqs. (2.21) and (2.22), the reader can easily show that

$$\mathbf{r^*} \cdot \left(\frac{\mathbf{a}}{h} - \frac{\mathbf{b}}{k} \right) = 0 \qquad \text{and} \qquad \mathbf{r^*} \cdot \left(\frac{\mathbf{b}}{k} - \frac{\mathbf{c}}{l} \right) = 0 \qquad (2.23)$$

proving that $\mathbf{r^*}$ is a normal to the (hkl) plane.

The distance d between adjacent planes can also be found from $\mathbf{r^*}$. As illustrated in Fig. 2.23b and c, the distance d is the projection of any one of the three intercepts in the normal direction, or, in other words,

$$d = \frac{\mathbf{b}}{k} \cdot \frac{\mathbf{r^*}}{|\mathbf{r^*}|} = \frac{1}{|\mathbf{r^*}|} \qquad (2.24)$$

In a simple cubic lattice, the three primitive vectors are mutually

Figure 2.23

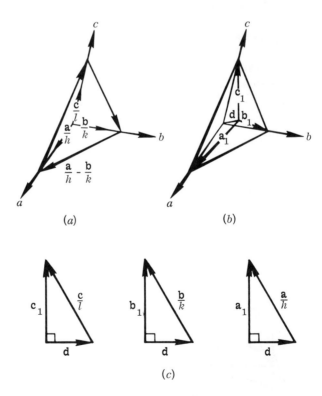

(a) (b)

(c)

Diagrams showing (a) the intercepts of the (hkl) plane with the crystal axes $\mathbf{a}$, $\mathbf{b}$, $\mathbf{c}$; (b) the normal vector $\mathbf{d}$ to the (hkl) plane; and (c) the vectorial relationship between the vector $\mathbf{d}$ and other vectors.

perpendicular and so are the reciprocal vectors. Therefore Eq. (2.24) reduces to

$$d = \frac{a}{(h^2 + k^2 + l^2)^{1/2}} \tag{2.25}$$

Figure 2.24 shows the projection of a cubic lattice in the xy plane. Any line thus represents a $(hk0)$ plane in the three-dimensional

Figure 2.24

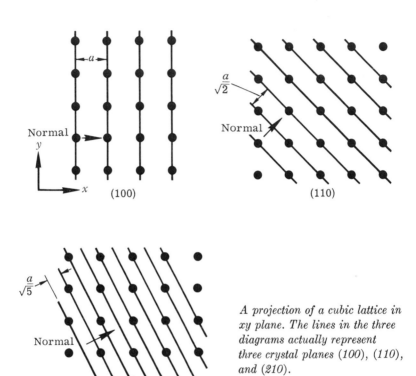

(100)

(110)

(210)

A projection of a cubic lattice in xy plane. The lines in the three diagrams actually represent three crystal planes (100), (110), and (210).

drawing. The reader can easily verify that the distance d between (100), (110), and (210) planes are a, $a/\sqrt{2}$, and $a/\sqrt{5}$, as shown in Fig. 2.24, and that the normals to these planes are given by $h\mathbf{a}^* + k\mathbf{b}^*$, which is equal to $(h\mathbf{a} + k\mathbf{b})/a^2$ for a simple cubic lattice. We also see that planes with lower index numbers not only have a wider interplanar spacing but also a higher density of atoms in the plane.

2.6 X-RAY DIFFRACTION

In Sec. 1.2 we based our derivation of the Bragg law of X-ray diffraction, that is, Eq. (1.6), on a very simple model in which adjacent atomic planes have atoms lined up one on top of the other as shown in Fig. 1.2b. Even in a simple cubic system the model is realized only for planes with lowest index numbers, that is, {100}, as can be seen from Fig. 2.24. Hence a generalization of the law is necessary. Consider two lattice points B and C separated in space by a displacement vector $\mathbf{r}$. Figure 2.25 shows the wave vectors $\mathbf{p}_0$ and $\mathbf{p}$ of the X-ray beams incident on and diffracted by lattice points B and C. Since the wavelengths of the incident and diffracted beams are the same, so are the magnitudes of $\mathbf{p}$ and $\mathbf{p}_0$. Therefore, in a vectorial presentation (Fig. 2.25b), the vector $\mathbf{P} = \mathbf{p} - \mathbf{p}_0$ is perpendicular to a line EF which bisects the angle 2θ between the directions of $\mathbf{p}_0$ and $\mathbf{p}$. In a three-dimensional plot, the line EF becomes a plane, and we assume that the plane has indices (hkl). If $\mathbf{n}$ is a unit vector in a direction normal to the (hkl) plane, then

$$\mathbf{P} = (2|\mathbf{p}| \sin \theta)\mathbf{n} \tag{2.26}$$

Referring to Fig. 2.25a, the path difference between the two diffracted waves is

$$AC - BD = \frac{\mathbf{r} \cdot \mathbf{p}_0 - \mathbf{r} \cdot \mathbf{p}}{|\mathbf{p}|} = -(2 \sin \theta)\mathbf{r} \cdot \mathbf{n} \tag{2.27}$$

Let α, β, and γ be the direction cosines of the vector $\mathbf{n}$ with respect to the three primitive vectors $\mathbf{a}$, $\mathbf{b}$, and $\mathbf{c}$. Furthermore, we realize that by the definition of primitive vectors, the smallest possible values for $\mathbf{r}$ are $\mathbf{a}$, $\mathbf{b}$, and $\mathbf{c}$. The phase differences are, therefore, given by

$$\phi_a = (2 \sin \theta)\mathbf{a} \cdot \mathbf{n} \frac{2\pi}{\lambda} = (2 \sin \theta)(a\alpha) \frac{2\pi}{\lambda} \tag{2.28}$$

and similar expressions for ϕ_b and ϕ_c. Referring to Fig. 2.23c and realizing that $\mathbf{d}$ is in the same direction as $\mathbf{n}$, we have

$$\alpha = \frac{dh}{a} \qquad \beta = \frac{dk}{b} \qquad \gamma = \frac{dl}{c} \tag{2.29}$$

Figure 2.25

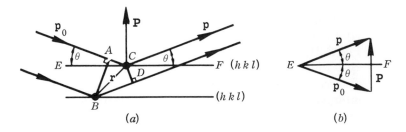

(a)   (b)

Diffraction of an X-ray beam from the (hkl) plane. The figure differs from Fig. 1.2b in that the two atoms B and C are no longer lined up one on top of the other.

The amplitude of the diffracted wave is a maximum when ϕ_a, ϕ_b, and ϕ_c are integral multiples of 2π. In view of Eq. (2.29), we have

$$\phi_a = (2d \sin \theta) \left(\frac{2\pi h}{\lambda}\right) = h_1 2\pi$$

$$\phi_b = (2d \sin \theta) \left(\frac{2\pi k}{\lambda}\right) = k_1 2\pi \qquad (2.30)$$

$$\phi_c = (2d \sin\theta) \left(\frac{2\pi l}{\lambda}\right) = l_1 2\pi$$

where h_1, k_1, and l_1 are integers and $d(hkl)$ is the spacing between adjacent (hkl) planes. Since h, k, and l represent the lowest set of integers, Eq. (2.30) demands that $h_1/h = k_1/k = l_1/l = n$, n being an integer. In terms of n Eq. (2.30) becomes

$$2d(hkl) \sin \theta = n\lambda \qquad (2.31)$$

The foregoing analysis, originally due to von Laue, leads to the Bragg law of diffraction.

So far we have considered the condition for constructive interference of the diffracted waves and have not discussed the intensity of the diffracted waves. According to the classical theory, Eq. (1.2), when an electron is subjected to an electromagnetic wave of frequency ν, the electron will move about its equilibrium position in a simple harmonic motion with the same frequency. The electron, because of its accelerated motion, in turn emits radiation of frequency ν, contributing to the scattering of the beam. In the discussion so far, we have represented an atom as well as the electrons associated with the atom by a geometrical point. Since the distribution of electrons is finite in extent compared with the wavelength of an X-ray beam, we must consider the interference of waves scattered from different parts of the electron distribution. The ratio of the radiation amplitude scattered by the charge distribution in an atom to that scattered by a point electron is called the *atomic scattering factor*.

For simplicity, it is assumed that the electron distribution is spherically symmetrical and that the electron density in a volume element $r^2 \, dr \, d\Omega$ is given by $\varrho(r)r^2 \, dr \, d\Omega$, $d\Omega$ being the solid angle. In Fig. 2.26 the scattering plane ECF is the same as the one in Fig. 2.25, but the point electron at C is now replaced by an electron distribution $\varrho(r)$. According to Eq. (2.27), the phase difference between the rays scattered by this electron distribution and the point electron is given by

$$\phi = \left(\frac{2\pi}{\lambda}\right) (2 \sin \theta)(r \cos \vartheta) \qquad (2.32)$$

By superposition, the atomic scattering factor is

$$f = \int_0^\infty \int_0^\pi \varrho(r) \exp{(i\phi)}(r^2\, dr)(2\pi \sin\vartheta\, d\vartheta)$$

$$= \int_0^\infty 4\pi r^2 \varrho(r) \frac{\sin qr}{qr}\, dr \tag{2.33}$$

where $q = 4\pi(\sin\theta)/\lambda$. Note that for $\theta = 0$, $\phi = 0$ from Eq. (2.32), and thus Eq. (2.33) reduces to $f = \int_0^\infty 4\pi r^2 \varrho(r)\, dr = z$, the number of atomic electrons, as it should.

Besides the atomic scattering factor, there is another factor, called the *geometric structure factor S*, on which the intensity of a diffracted wave depends. Figure 2.27 shows the diffracted waves, indicated by lines, from various planes in crystals having cubic structures, the indices hkl being those of a simple cubic structure. Note that some of the lines from the simple cubic structure are missing in the diffraction patterns of other cubic structures such as the body-centered cubic structure. In discussing why these lines are missing, we shall use the representative units of Fig. 2.10 instead of the primitive cell as reference simply because all these structures have different primitive cells but the same basic cubic cell. If f_j is the atomic scattering factor of the jth atom which is situated at a distance $\mathbf{r} = u_j\mathbf{a} + v_j\mathbf{b} + w_j\mathbf{c}$ from the origin of the representative cell, then the ratio of the amplitude of the wave scattered by all atoms in the representative cell to that by a point electron situated at the origin is given by

$$F(hkl) = \sum_j f_j \exp{[2\pi i(u_j h + v_j k + w_j l)]} \tag{2.34}$$

Figure 2.26

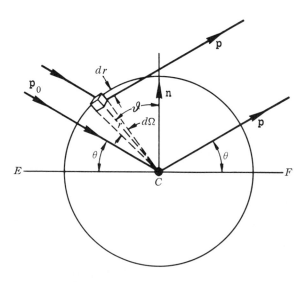

Scattering of an X-ray beam by an electron distribution.

which is generally called the structure factor.[3] The phase factor in the above equation is obtained by using Eqs. (2.22) and (2.27) together with the relations $\mathbf{n} = d\mathbf{r}^*$ and $2d \sin \theta = \lambda$.

When all atoms in the representative cell are identical, the atomic scattering factor f_j can be taken outside from the summation, and Eq. (2.34) becomes $F(hkl) = fS$, where the geometric structure factor S is

$$S = \sum_j \exp\left[2\pi i(u_j h + v_j k + w_j l)\right] \qquad (2.35)$$

In calculating S we must remember that the corner atoms are shared by eight cells and the face atoms by two cells. Thus, for a body-centered cubic cell, since the eight corner atoms, with uvw being 000, 100, 010, 001, 111, 110, 101, and 011, have the same phase factor, Eq. (2.35) becomes

$$S = 1 + \exp\left[\pi i(h + k + l)\right] \qquad (2.36)$$

The last term is due to the body-centered atom whose uvw is $\frac{1}{2}\frac{1}{2}\frac{1}{2}$. Again we should like to emphasize that the indices (hkl) refer to those of a simple cubic structure and so do uv and w. For planes with $h + k + l = \text{odd}$, the contribution from the body-centered atom exactly cancels out that from the corner atoms and hence $S = 0$, which explains why some lines are missing in Fig. 2.27. For a body-centered cubic cell containing two different atoms in a representative cell such as CsCl, Eq. (2.34) can be written as

$$F = f_A + f_B \exp\left[\pi i(h + k + l)\right] \qquad (2.37)$$

[3] The structure factor is used extensively in studying the structural changes in a ferroelectric crystal (Sec. 7.9) during a phase transition and in studying the magnetic ordering in an antiferromagnetic or in a ferrimagnetic crystal (Secs. 8.10 and 8.11).

Figure
2.27

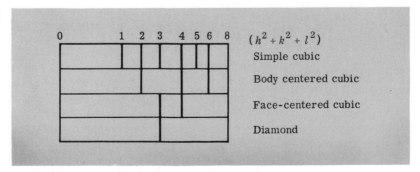

The X-ray diffraction from various cubic structures. The Miller indices (hkl) refer to the axes of a simple cubic structure. The existence of a diffracted wave from a given plane is indicated by a vertical line. (C. S. Barrett, "Structure of Metals," 2d ed., p. 136, McGraw-Hill Book Company, New York, 1952.)

When $h + k + l$ is odd, $F = f_A - f_B$, where A stands for Cs and B for Cl. Thus these lines are weak but no longer missing.

Finally, we shall briefly discuss experimental methods of X-ray diffraction. These are the powder method, the rotating-crystal method, and the Laue method. In the powder method illustrated in Fig. 2.28, a monochromatic X-ray beam is incident on a finely powdered specimen. Since a powder is a collection of many randomly oriented crystallites, the crystal planes which satisfy the Bragg law generate a cone that forms an angle θ with the original beam. If a photographic film is placed along a circle around the powder, diffracted rays intercepted by the film are marked by dark concentric rings. Knowing the distance AB and the radius AC, the diffraction angle θ is equal to $AB/4AC$ radians. According to Eqs. (2.25) and (2.31), planes with different indices will have different θ's and hence different values of AB. By a careful analysis of the photograph, the value of a, the lattice constant, can be determined. The rotating crystal method is used for measuring the interatomic distances in single crystals in an arrangement similar to Fig. 2.28 except that the crystal is rotating and that the film is cylindrically placed with respect to the axis of rotation. Suppose that the crystal is orientated with the c axis parallel to the axis of rotation. Then all planes with $l = 0$, that is, parallel to the c axis, will reflect in the horizontal plane. Other planes, being tilted, planes with $l = 1$ and $l = -1$ for example, will reflect in layers above and below the horizontal plane. The separation of these layers recorded on the film determines the spacing between crystal planes. In the Laue method, a white X-ray beam is incident on a stationary single crystal and the diffracted beam is recorded as darkened spots on a photographic film. Since the wavelength λ covers a wide range, many planes will satisfy the Bragg equation. The position of these spots on the Laue pattern will change, depending on the orientation of the crystal relative to the incident beam. Since known ways for indexing the Laue pattern are available, the method is most extensively used for

Figure 2.28

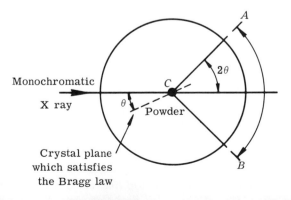

Monochromatic X ray

Powder

Crystal plane which satisfies the Bragg law

The powder method used in X-ray diffraction.

Figure
2.29

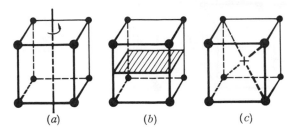

(a)   (b)   (c)

Symmetry operations in a crystal: (a) rotation through an axis, (b) mirror reflection in a plane, and (c) inversion through a point. The simple cubic structure is used for illustration.

study of crystal orientation. If a crystal is orientated in one of the principal directions, for example $\langle 100 \rangle$, $\langle 110 \rangle$, or $\langle 111 \rangle$ in a cubic crystal, the pattern will show the symmetry of the crystal with respect to that axis. For further discussions of X-ray technology, the reader is referred to the book by Guinier.[4]

2.7 SYMMETRY ELEMENTS

In Sec. 2.5, we stated that, starting with a primitive cell, we can generate the whole lattice by a translation vector **t** defined in Eq. (2.20). Besides translation, there are three other operations by which the periodic pattern of a lattice can be reproduced. They are (1) rotation through an axis, (2) mirror reflection in a plane, and (3) inversion through a point. These operations are illustrated in Fig. 2.29, using a primitive cubic cell as an example. If $\phi = 2\pi/n$ denotes the angle of rotation, obviously n must be an integer and the rotation axis is called an n-fold axis. However, there is a further restriction imposed by the periodicity of the lattice. Consider a two-dimensional lattice with a rotation axis through point P. If $\mathbf{d}_0$ represents the radius vector joining P with any lattice point A, after rotation point A takes a new position, one where point B originally was, and $\mathbf{d}_0$ becomes $\mathbf{d}$ as illustrated in Fig. 2.30. Since points A and B are lattice points, they must be separated by a translation vector

[4] A. Guinier, "X-ray Crystallographic Technology," Hilger and Watts Ltd., London, 1952.

Figure
2.30

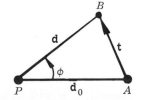

A two-dimensional lattice with a rotation axis through P. By definition of rotational symmetry, a lattice point A after rotation should fall upon another point B which is originally also a lattice point.

Figure 2.31

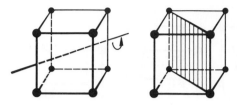

Diagrams showing the two-fold axis (parallel to the face diagonal) and the mirror plane perpendicular to the twofold axis in a simple cubic crystal.

t of Eq. (2.20), or, in other words, $\mathbf{d}_0 + u\mathbf{a} + v\mathbf{b} = \mathbf{d}$, where $\mathbf{a}$ and $\mathbf{b}$ are primitive vectors and u and v are integers. This restriction further limits n to the following values: 1, 2, 3, 4, or 6. Examples are: for $n = 6$, the c axis of a primitive hexagonal structure shown in Fig. 2.10; for $n = 4$, the rotation axis for a simple cubic structure shown in Fig. 2.29a; for $n = 3$, the body diagonal of a trigonal structure shown in Fig. 2.10; and for $n = 2$, an axis passing through the centers of two opposite rectangular faces of a primitive tetragonal structure shown in Fig. 2.10.

In addition to three fourfold axes, a simple cubic structure also possesses six twofold axes parallel to the six face diagonals and four threefold axes coincident with the four body diagonals, as illustrated in Figs. 2.31 and 2.32. A special symbol has been adopted to describe the symmetry elements in a crystal and its lattice type. For a sodium chloride crystal, it is $F(4/m)\bar{3}(2/m)$. The symbol X/m or $\frac{X}{m}$ means X-fold rotation axis with mirror plane perpendicular to it, while Xm means X-fold rotation axis with mirror plane or planes parallel to it. If a rotation followed by an inversion brings a crystal cell into self-coincidence, the crystal has a rotation-inversion axis denoted by $\bar{X}$. The symbols P, I, F, and C stand for primitive, body-centered, face-centered, and base-centered structures, respectively. At this point a brief comment about the trigonal crystal is in order. Figure 2.33 shows a special kind of hexagonal structure labeled R against the primitive hexagonal structure P. The hexagonal structure R has two extra atoms equally spaced along a body diagonal. The reader

Figure 2.32

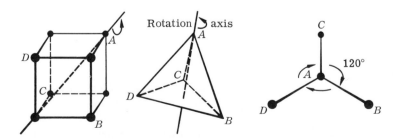

Diagrams showing the threefold axis (body diagonal) in a cubic crystal.

is asked to show that the primitive cell for such a structure belongs to the trigonal system of Fig. 2.10.

We should emphasize that crystals belonging to the same system do not necessarily have the same symmetry elements. For example, sodium chloride, diamond, and zinc blende structures are all cubic; yet the symmetry elements of diamond and zinc blende structures differ from each other and from those of sodium chloride structure. The combination of symmetry elements at a point is called a *point group*, and there are 32 point groups in the crystals shown in Table 2.4. We note that common symmetry elements exist among crystals classified under each system. For example, all crystals in the cubic system have a minimal symmetry of four threefold axes coincident with the four body diagonals ⟨111⟩. If the ⟨111⟩ axis of a cubic crystal is aligned parallel to an X-ray beam, the Laue pattern will show a threefold symmetry irrespective of whether the crystal has a sodium chloride, cesium chloride, calcium fluoride, diamond, zinc blende, or any other cubic structure. This explains why crystals are grouped into seven systems. Information about the symmetry of crystal systems is summarized in Table 2.5.

Now let us illustrate how to use the information contained in Table 2.4 by comparing the symmetry properties of simple cubic (class 29) and NaCl (class 32) structures. The existence of fourfold, threefold, and twofold symmetry axes in a simple cubic crystal is indicated by the symbol 432 in Table 2.4. A symmetry operation about any one of these axes, for example, a 90° rotation about the

Figure 2.33

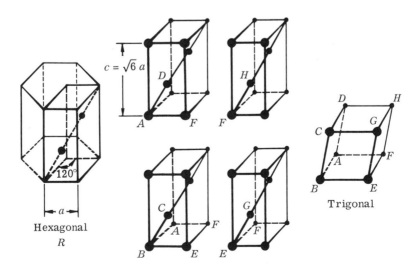

The hexagonal R structure which has two extra atoms equally spaced along a body diagonal. The atoms ABCDEFGH from four neighboring hexagonal cells form the primitive cell of the trigonal lattice.

fourfold axis, will bring a simple cubic structure into self-coincidence. The symmetry property of a NaCl structure can be best visualized by examining a small cubic cell (one-eighth of the structure shown in Fig. 2.11a) with Na and Cl atoms at alternate corners. To bring the NaCl structure into self-coincidence, a rotation about the symmetry axis must be followed by other operations. The notation $4/m$

TABLE 2.4 *The thirty-two crystal point groups*

System	Class no.	International symbol	Schoenflies symbol	Examples
Triclinic	1	1	C_1	Strontium ditartrate tetrahydrate
	2*	$\bar{1}$	$C_i = S_2$	FeAsS, P_2I_4, $CuSO_4$
Monoclinic	3	2	C_2	KOH, tartaric acid
	4	m	$C_s = C_{1h}$	$K_2S_4O_6$ (potassium tetrathionate)
	5*	$\dfrac{2}{m}$	C_{2h}	LiAs, ZrO_2, VO_2, CuF_2, AgO
Orthorhombic	6	$2\,2\,2$	$D_2 = V$	$Zn(OH)_2$, NH_2OH, $HgCl(CNS)$, Rochelle salt
	7	$m\,m\,2$	C_{2v}	NaCN, KCN, $HgBr_2$, HgI_2
	8*	$\dfrac{2}{m}\dfrac{2}{m}\dfrac{2}{m}$ (mmm)	$D_{2h} = V_h$	Ga, Cl_2, $CaCl_2$, InS, ZnSb, $BaSO_4$
Tetragonal	10	4	C_4	$PbMoO_4$
	9	$\bar{4}$	S_4	$LiNH_2$, $Ca_2Al_2SiO_7$
	13*	$\dfrac{4}{m}$	C_{4h}	NbO_2, PbFCl, CaHCl, PdS, $CaWO_4$
	12	$4\,2\,2$	D_4	$Np(\beta)$, TeO_2, NbP, $NiSO_4$
	14	$4\,m\,m$	C_{4v}	U (β), Pentaerythritol
	11	$\bar{4}\,2\,m$	$D_{2d} = V_d$	$ZnCl_2(\alpha)$, K_3PO_4
	15*	$\dfrac{4}{m}\dfrac{2}{m}\dfrac{2}{m}$ $(4/mmm)$	D_{4h}	B, white tin, NH_4CN, TiO_2, $ZrSiO_4$
Trigonal	16	3	C_3	Tl_2S, sodium periodate
	17*	$\bar{3}$	$C_{3i} = S_6$	$(Ca, Mg)CO_3$
	18	32	D_3	Se, Te, HgS, $SiO_2(\alpha)$
	19	$3\,m$	C_{3v}	NiS, MoS_2, NiSe, CNI
	20*	$3\dfrac{2}{m}$ $(\bar{3}\,m)$	D_{3d}	Sm, TiS, EuOF, CaI_2, $Ni(OH)_2$, $CaCO_3$
Hexagonal	23	6	C_6	Nephelite $(Na_6Al_8Si_9O_{34})$
	21	$\bar{6}$	C_{3h}	Li_2O_2
	25*	$\dfrac{6}{m}$	C_{6h}	$CaFCa_4(PO_4)_3$
	24	$6\,2\,2$	D_6	$SiO_2(\beta)$
	26	$6\,m\,m$	C_{6v}	ZnO, NiAs, CdI_2, AgI, ZnTe, GaN
	22	$\bar{6}\,m\,2$	D_{3h}	Fe_2P, Mn_2P, Cr_2As, FeS
	27*	$\dfrac{6}{m}\dfrac{2}{m}\dfrac{2}{m}$ $(6/mmm)$	D_{6h}	BN, GaSBe, Zn, MoS_2, H_2O(ice)
Cubic	28	23	T	BeF_2, $NiSbS$, CO, $NaClO_3$
	30*	$\dfrac{2}{m}\bar{3}$	T_h	Diamond, Ge, Si, ZnOS, Cu_2O, FeS_2
	29*	$4\,3\,2$	O	$Mn(\beta)$
	31	$\bar{4}\,3\,m$	T_d	$Mn(\alpha)$, NH_4Cl, ZnS, GaAs, $Zn(CN)_2$
	32*	$\dfrac{4}{m}\bar{3}\dfrac{2}{m}$ $(m3m)$	O_h	W, NaCl, NbO, CaF_2

* Possessing a center of inversion.

means a rotation of 90° about the fourfold axis followed by a mirror reflection through a plane perpendicular to the fourfold axis. The reader can easily verify that the symmetry operations represented by $\bar{3}$ and $2/m$ will reproduce the original pattern of the cell as does the $4/m$ operation.

Finally, we should point out that it is possible to combine a translation with a rotation or a reflection to produce a new symmetry operation. When a rotation (n-fold) is combined with a translation parallel to the rotation axis by a distance t/n which is a submultiple of the lattice translation t in the direction of the rotation axis, a *screw axis* is produced. When a reflection in a plane is combined with a translation parallel to the plane, a *glide plane* is formed. Therefore the possible symmetry elements include not only the point-group operations discussed earlier but also the screw-axis and glide-plane operations. These symmetry operations produce 230 basically different repetitive patterns called *space groups*. As mentioned earlier, most materials of technical importance have simple crystal structures. Furthermore, crystallography is a well-established field in itself, and a discussion of geometrical crystallography is outside the scope of this book. The reader is therefore referred to the loose-leaf compilation[5] by Wyckoff for crystal structures of materials and to the "International Tables[6] for X-ray Crystallography" by Henry and Lonsdale for a full description of the 230 space groups.

[5] R. W. G. Wyckoff, "Crystal Structures," Interscience Publishers, Inc., New York, 1948.

[6] N. F. M. Henry and K. Lonsdale, "International Tables for X-ray Crystallography," vol. 1, The Kynoch Press, Birmingham, 1952.

TABLE 2.5 *Crystal symmetry*

System	*Minimal symmetry*	*Maximal symmetry*
Triclinic	1 or $\bar{1}$	$\bar{1}$
Monoclinic	2 or m	$\dfrac{2}{m}$
Orthorhombic	222	$\dfrac{2}{m}\dfrac{2}{m}\dfrac{2}{m}$
Tetragonal	4	$\dfrac{4}{m}\dfrac{2}{m}\dfrac{2}{m}$
Cubic	4 threefold axes	$\dfrac{4}{m}\bar{3}\dfrac{2}{m}$
Hexagonal	6 or 3	$\dfrac{6}{m}\dfrac{2}{m}\dfrac{2}{m}$
Trigonal	3	$\bar{3}\dfrac{2}{m}$

2.8 CRYSTAL IMPERFECTIONS

In our discussion thus far, we have assumed an ideal lattice in which not only are the atoms arranged in a regular and repetitive pattern but they are also fixed in their position. In a real crystal, however, departure from an ideal structure can occur. First, the atoms are not stationary but vibrate about their equilibrium positions. It is this vibrational motion which constitutes the thermal property of a solid. Second, there are defects in the arrangement of atoms. The three principal kinds of defects are (1) point defects, such as interstitials, vacancies, and substitutional impurities, (2) line defects, commonly known as dislocations, and (3) plane defects, such as the grain boundary in a polycrystalline material. In the following material we shall give a brief general discussion of the various lattice imperfections and their effects on material properties.

Lattice vibration. In view of our previous discussion in Secs. 2.1, 2.3, and 2.4, the total energy of N atoms in a solid can be described as a sum of two opposite types of energy, the attractive and the repulsive energies. Both energies increase with decreasing interatomic distance R, but the repulsive energy dominates for small R, whereas the attractive energy dominates for large R. Hence an equilibrium situation is reached when the repulsive force exactly balances the attractive force. The situation is summarized in Fig. 2.34a. The position at which the total energy is a minimum deter-

Figure 2.34

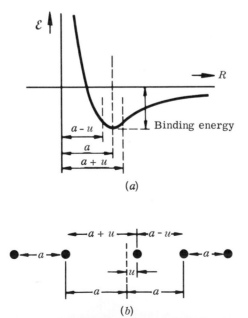

(a)

(b)

(a) Schematic representation of the energy of an atom in a solid as a function of interatomic distance and (b) a linear chain of atoms. If a certain atom is displaced from its equilibrium position by a distance u, the interatomic spacings between the atom and its nearest neighbors are changed to (a + u) and (a − u). Consequently, the energy of that atom also changes with u.

mines the lattice spacing of a solid. Consider a one-dimensional array of atoms in which one atom is displaced from its equilibrium position by a small displacement u as shown in Fig. 2.34b such that $u/a \ll 1$. If only the interaction between nearest neighbors is taken into consideration, the change in energy is given by

$$\Delta \mathcal{E} = \mathcal{E}(a + u) + \mathcal{E}(a - u) - 2\mathcal{E}(a) \tag{2.38}$$

An expansion of Eq. (2.38) into a Taylor series gives

$$\Delta \mathcal{E} = \frac{\partial^2 \mathcal{E}}{\partial u^2} u^2 = \gamma u^2 \quad \text{with} \quad \gamma = \frac{\partial^2 \mathcal{E}}{\partial u^2} \tag{2.39}$$

The following observations can be made about Eq. (2.39). The displaced atom experiences a force $f = -2\gamma u$ known as Hooke's law, and the equation describing the motion of the displaced atom is

$$M \frac{du^2}{dt^2} = -2\gamma u \tag{2.40}$$

where M is the mass of the atom. Note that Eq. (2.40) is the same as Eq. (1.42), the equation of motion for a harmonic oscillator with angular frequency of oscillation $\omega = \sqrt{2\gamma/M}$. If we assume that each atom oscillates independently of motions of other atoms, then according to the Bose-Einstein distribution law presented in Sec. 1.11 the total energy of N oscillators (or atoms) is

$$U = 3N \left[\frac{\hbar\omega}{2} + \frac{\hbar\omega}{\exp(\hbar\omega/kT) - 1} \right] \tag{2.41}$$

in which the term $\hbar\omega/2$ accounts for the zero-point energy and the factor 3 is for a three-dimensional lattice. A proof of Eq. (2.41) can be found in Prob. 1.24 of Chapter 1. As temperature increases, more oscillators are excited into higher energy states, and hence the energy U increases. The specific heat of a lattice is given by

$$C = \frac{dU}{dT} = 3Nk \left(\frac{\Theta_E}{T} \right)^2 \frac{\exp(\Theta_E/T)}{[\exp(\Theta_E/T) - 1]^2} \tag{2.42}$$

which was derived originally by Einstein. The symbol $\Theta_E = \hbar\omega/k$ is known as the Einstein characteristic temperature of a solid.

According to the classical theory, the average energy associated with each degree of freedom is kT ($kT/2$ each for kinetic and potential energies). Thus, for N three-dimensional harmonic oscillators, the total energy is $U = 3NkT$, which gives a lattice specific heat of $C = 3Nk$. For a mole of atoms, $N =$ Avogadro's number and C becomes $3R$, known as the Dulong and Petit value, where R is the gas constant. Note that at high temperatures such that $\Theta_E < T$, expansion of Eq. (2.42) gives the classical result. At low temperatures, however, Eq. (2.42) predicts a $T^{-2} \exp(-\Theta_E/T)$ temperature dependence, which differs considerably from the classical result.

Measurements of lattice specific heat in many solids at low temperatures confirm the negative exponential temperature dependence except at very low temperatures where the specific heat shows a T^3 dependence. Therefore vibrations of atoms in a lattice (*lattice vibrations*) behave in the same way as quantum mechanical harmonic oscillators whose energy is quantized in units of $\hbar\omega$ or $\varepsilon = (n + \frac{1}{2})\hbar\omega$, n being an integer. The word *phonon* is generally used to describe a quantum of energy $\hbar\omega$ associated with vibrational waves in a solid.

Let us estimate the frequency and amplitude of lattice vibration. For a simple cube, the force constant γ is related to the Young's modulus Y by $\gamma = aY$, a being the lattice constant. Taking $a = 2.5$ Å, $Y = 10^{11}$ newtons/m^2, and $M = 10^{-25}$ kg, we find (from $\omega = \sqrt{2\gamma/M}$) the frequency ν to be of the order of 3×10^{12} cps or in the far-infrared region. The amplitude of oscillation can be found from the average energy of a harmonic oscillator. At high temperatures $\gamma u^2 = kT$, giving a value for u of about 0.2 Å at room temperature. These values of ν and u fall in the range of values observed experimentally. Since u/a or $\Delta a/a$ is quite appreciable even at room temperature, we expect that lattice vibrations will affect material properties. This is indeed the case. For example, the mobility or mean free path of free carriers in pure covalent semiconductors such as Ge and Si is controlled by lattice vibrations. Furthermore, a phonon possesses momentum as well as energy, as does a photon. In electronic processes where the initial and final state of an electron may not have the same momentum, lattice vibrations may help these electrons to make the quantum mechanical transition. These processes will be discussed further in later chapters dealing with tunnel diodes and optical properties of solids.

Finally, we shall discuss how the Einstein theory can be improved. It is obvious from Fig. 2.34b that the atoms immediately to the left and to the right of the displaced atom also experience a force, for according to Newton's law the force is reciprocal. These atoms will also be set into motion, and so will the atoms next to these atoms. In other words, atoms are coupled to each other in a crystal. Consequently, the atoms will not oscillate with a single frequency but will have a band of frequencies. The situation is similar to coupled circuits. If N, the number of atoms involved, is large, the frequency band can be considered as continuous, and hence Eq. (2.41) can be replaced by

$$U = \int \frac{n(\omega)\hbar\omega \, d\omega}{\exp{(\hbar\omega/kT)} - 1} \tag{2.43}$$

where $n(\omega) \, d\omega$ represents the number of normal modes in the frequency range between ω and $\omega + d\omega$. Since the term representing the zero-point energy does not involve T, and hence does not contribute to specific heat, it is omitted in Eq. (2.43). The problem is now reduced to finding the frequency distribution function $n(\omega)$.

In the model considered by Debye, the frequency spectrum of a solid is obtained as if the solid were a continuous medium. The propagation of sound waves in such a medium can be described by an acoustic wave equation which in one-dimensional cases reads

$$\frac{\partial^2 u}{\partial x^2} = \frac{1}{v^2} \frac{\partial^2 u}{\partial t^2} \tag{2.44}$$

where v is the velocity of propagation of the sound wave. The solution of Eq. (2.44) is given by

$$u \sim \exp\left[i(\omega t - kx)\right] \tag{2.45}$$

with $k = \omega/v$. The number of normal modes in a cubic box of side L can be found in a manner similar to the one used for a radiation field in Sec. 1.11. The only difference is that for a sound wave, the wave can be either *longitudinal* or *transverse*. In the former, u is in the direction of propagation, that is, in the x direction, whereas in the latter, u can be in the y or z direction. Therefore the number of normal modes per unit volume is given by

$$n(\omega)\,d\omega = \frac{1}{2\pi^2} \sum k^2\,dk = \frac{\omega^2}{2\pi^2}\left(\frac{1}{v_l^3} + \frac{2}{v_t^3}\right)d\omega \tag{2.46}$$

The summation is for one longitudinal and two transverse modes of vibration, which are indicated by subscripts l and t, respectively.

Differentiating Eq. (2.43) with respect to T and using Eq. (2.46) for the value of $n(\omega)$, we find

$$C = 9Nk\left(\frac{T}{\Theta_D}\right)^3 \int_0^{x_m} \frac{e^x x^4\,dx}{(e^x - 1)^2} \tag{2.47}$$

where $x = \hbar\omega/kT$ is a dummy variable and the integration limit is such that

$$\int_0^{\omega_m} n(\omega)\,d\omega = 3N \qquad \text{and} \qquad x_m = \frac{\hbar\omega_m}{kT} = \frac{\Theta_D}{T} \tag{2.48}$$

The temperature Θ_D is known as the *Debye temperature*. Figure 2.35 shows the behavior of specific heat based on the classical Dulong and Petit, the Einstein, and the Debye models. At high temperatures, since $\hbar\omega/[\exp(\hbar\omega/kT) - 1]$ approaches kT in Eq. (2.43), the Debye model also gives the classical result, as does the Einstein model. At very low temperatures where $\Theta_D \gg T$, x_m can be replaced by infinity and thus Eq. (2.47) shows a T^3 temperature dependence, in agreement with experimental findings. The improvement of the Debye model is due to the inclusion of long wavelength (low frequency) phonons. Since these phonons have much lower energy, they are still excited even at very low temperatures and

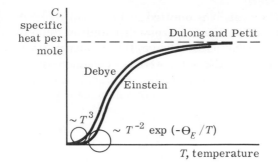

Figure 2.35

Comparison of the temperature variations of the specific heat of a solid based on different models.

hence contribute to the specific heat. In the Einstein model, only one kind of phonon with energy $\hbar\omega = k\Theta_E$ is considered. At very low temperatures, the occupancy of these phonons drops as $\exp(-\Theta_E/T)$ and consequently so does the specific heat.

Figure 2.36 shows the $n(\omega)$ curve and the ω against the k curve used in the Debye model with the cutoff value of ω_m given by $\omega_m = k\Theta_D/\hbar$. The representative values of the Debye temperature in degrees Kelvin (°K) are: for Ge, 366; for Si, 658; for Ag, 225; for NaCl, 308. A value of $\Theta_D = 300°K$ corresponds to a value of $\nu_m = 6 \times 10^{12}$ cps, which is in the same range as the frequency obtained from our earlier calculation. Therefore, in the Debye model, the frequency of the lattice wave extends from very low frequencies up to the far-infrared region. This is in general agreement with experimental observations. However, we should emphasize that the curves shown in Fig. 2.36 are based on a continuous medium and hence represent an approximation. The typical values for the velocity of elastic waves are about 2×10^3 m/sec and 4×10^3 m/sec for the transverse and longitudinal waves, respectively. Taking $v = 3 \times 10^3$ and $\nu = 6 \times 10^{12}$ cps, we find $\lambda = 5$ Å, which is about equal to the lattice spacing in solids. We can see

Figure 2.36

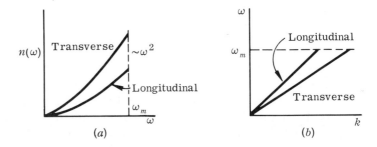

Schematic representations of (a) the function $n(\omega)$ as a function of ω and (b) the dispersion relation between ω and k. A common cutoff frequency ω_m for the longitudinal and transverse modes is assumed in the Debye model.

that the model based on a continuous medium is a rather poor approximation for high frequency (short wavelength) phonons. The ω against the k curve indeed deviates from being a straight line at high k values as we shall see in a more detailed discussion of phonon waves in a later section.

Vacancies and interstitials. From the discussion just presented, we see that the lattice vibration is a direct consequence of thermal energy associated with atoms. Figure 2.37a shows lattice defects of a different kind in an otherwise periodic structure. These lattice defects are known as *vacancies* and *interstitials*, representing, respectively, a vacant lattice site which is normally occupied and an atom occupying an interstitial space which is normally vacant. Figure 2.37b shows the creation of a vacant site by removing an atom from the interior to the surface of a crystal. During the process old bonds are broken and new bonds are created. It is clear from Fig. 2.37b that the number of broken bonds at the old site (in the interior) always exceeds that of bonds created at the new site (at the surface), and hence an energy ε_v is required to create a vacant site. Based on this consideration alone, it seems that vacancies would never be created. We must not forget, however, that associated with any order-disorder arrangement there is an entropy.

Consider the arrangement of n vacancies among N lattice sites. The total number of ways is $N!/n!(N-n)!$; thus according to statistical thermodynamics[7] the entropy S is defined by the Boltzmann relation $S = k \log W$ as

$$S = k \log \frac{N!}{n!(N-n)!} \tag{2.49}$$

[7] The reader may find the following references helpful. M. Born, "Atomic Physics," Appendix XXXV, p. 428, Hafner Publishing Co., New York, 1962. A. J. Dekker, "Solid State Physics," Appendices A and E, pp. 525 and 531, Prentice-Hall, Inc., Englewood Cliffs, N. J., 1957.

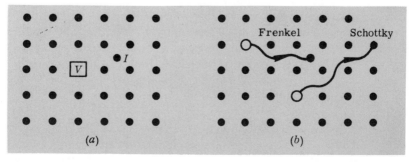

Illustrations of (a) *vacancies and interstitials, and* (b) *Frenkel and Schottky defects in a two-dimensional lattice.*

The increase in the free energy F (again defined in thermodynamics[7]) is

$$F = n\mathcal{E}_v - TS \tag{2.50}$$

Applying the Stirling approximation to Eq. (2.49) and minimizing F with respect to n in a manner outlined in Sec. 1.10, we find

$$\frac{n}{N - n} = \exp\left(\frac{-\mathcal{E}_v}{kT}\right) \quad \text{or} \quad n \simeq N \exp\left(\frac{-\mathcal{E}_v}{kT}\right) \quad \text{for } n \ll N \tag{2.51}$$

Typical values of $\mathcal{E}_v$ are about 1 ev; thus, at a temperature of $1000°K$, a value of n/N of about 10^{-5} is obtained.

In the above calculation we have neglected any change in the vibrational energy caused by vacancies. Using Eq. (1.129) in Eq. (1.127) and the Boltzmann definition of entropy, we find the entropy S' associated with $3N$ independent harmonic oscillators of frequency v to be

$$S' = 3Nk\left[\frac{ae^a}{e^a - 1} - \log(e^a - 1)\right] \tag{2.52}$$

where $a = h\nu/(kT)$ and $3N = \sum_s g_s$. Because of the lattice distortion produced by a vacancy, the oscillator frequency may shift from ν to ν'. The number of oscillators with frequency ν' is equal to $3nZ$, where Z is the coordination number and the factor 3 is the result of a three-dimensional lattice. The reader is asked to show that the inclusion of the change in S' in Eq. (2.50) leads to the following result (Probs. 2.16 and 2.17):

$$\frac{n}{N} \simeq \left(\frac{\nu}{\nu'}\right)^{3Z} \exp\left(\frac{-\mathcal{E}_v}{kT}\right) \quad \text{for } n \ll N \tag{2.53}$$

if $kT \gg h\nu$ and $h\nu'$, so that e^a and $e^a - 1$, respectively, can be replaced by 1 and a in Eq. (2.52).

In an ionic crystal it is often more favorable to form a pair of negative and positive ion vacancies so as to maintain electric charge neutrality. Since the numbers of cation and anion vacancies are equal, the entropy increase is just twice the amount given in Eq. (2.49). Following a procedure similar to the one used in deriving Eq. (2.51), we find

$$n \simeq N \exp\left(-\frac{\mathcal{E}_1}{2kT}\right) \tag{2.54}$$

where $\mathcal{E}_1$ is the energy required to form a vacancy pair. Both the single and pair vacancies are often called *Schottky defects*. Another type of lattice defect is caused by an interstitial atom. Because the interatomic distance is very much reduced by the interstitial atom, the repulsive force (Fig. 2.34) between this atom and its neighbors is very large. For example, in Al, the formation energies are, respectively, 0.75 ev and 3 ev for vacancies and interstitials. This large difference makes interstitials much less likely than vacancies in most crystals. However, an interstitial can be formed by transferring an atom from a lattice site to an interstitial position, leaving a vacancy behind. This type of lattice defect, known as the *Frenkel defect*, has a formation energy about equal to the sum of vacancy and interstitial energies. Because the defects are created in pairs, the activation energy is only one-half this sum. Note that a factor one-half also appears in the exponent in Eq. (2.54).

The number of ways of arranging n pairs of vacancies and interstitials among N lattice sites and N' interstitial positions is

$$W = \frac{N}{n!(N-n)!} \frac{N'!}{n!(N'-n)!} \tag{2.55}$$

If ε_2 is the formation energy of such a pair, then minimizing the free energy F with respect to n gives

$$\varepsilon_2 = kT \log \left[\frac{(N-n)(N'-n)}{n^2} \right] \quad \text{or} \quad n \simeq (NN')^{\frac{1}{2}} \exp \left(\frac{-\varepsilon_2}{2kT} \right) \tag{2.56}$$

for $n \ll N, N'$. One way of determining whether Schottky defects or Frenkel defects are the more common type is to measure the density of a crystal before and after these defects are introduced. In the production of Frenkel defects, atoms simply shift from a lattice site to an interstitial position, so the volume of the crystal is expected to remain unchanged. This is not the case with Schottky defects. One of the most common methods of producing lattice defects over and above the thermal equilibrium concentrations given by Eqs. (2.51), (2.54), and (2.56) is by quenching a crystal from a high temperature. Other methods include bombardments by high-energy particles and undergoing severe deformation by mechanical stress.

In addition to the forementioned types, lattice defects can also be produced by deviations from the stoichiometric composition. The most common and best known defects of this type are the *color centers* in alkali halides. If the excess alkali atoms take the normal positive ion positions, we find that there are not enough halogen atoms to fill in all the positions reserved for negative ions. Therefore negative ion vacancies are created. The vacancy, being surrounded by positive ions, attracts an electron. We may think of an electron moving about a negative ion vacancy qualitatively as behaving like an electron moving in a hydrogen atom. The electron bound to a negative ion vacancy is known as an F center. By the same token, if excess halogen atoms are present, positive ion vacancies are produced. The vacancy, being surrounded by negative ions, will repel an electron or, in the language of semiconductors, attract a hole. Figure 2.38 shows the ground and excited states of the F and V centers in KCl. Because of the large gap energy, pure alkali halide crystals are transparent to visible light. However, these crystals can be colored by the introduction of color centers which absorb light by exciting an electron from the valence band to the V center or by releasing an electron from the F center to the conduction band.

Dislocations. There are two types of dislocations, the edge and screw types illustrated in Fig. 2.39a and b. The distortion due to an *edge dislocation* may be considered as caused by the insertion of an extra plane of atoms into the upper half of the crystal. As shown in Fig. 2.39a, this half-plane is parallel to the yz plane and stops above the origin at a distance equal to one-half the lattice spacing. The *screw dislocation* may be imagined as being produced by cutting the crystal part way through and pushing the upper part one lattice spacing over. It is also obvious from Fig. 2.39 that both types of dislocations are due to shearing stresses whose direction is indicated by an arrow. Furthermore, we always can think of the crystal as consisting of slipped and unslipped regions. According to the alignment of atoms shown in Fig. 2.39a, the slip is in the x direction, and slip has occurred over the left half-plane. The boundary between the slipped and unslipped regions is the dislocation. Note that in the edge dislocation, the dislocation line (z axis) is perpendicular to the slip direction, while in the screw dislocation, the dislocation line (AB) is parallel to the slip direction (marked by an arrow).

In the early estimate of the elastic limit of perfect crystals, theoretical values were greater than the observed values by a factor of 10^2 or more. Referring to Fig. 2.40, in a model suggested by Frenkel, two neighboring atomic planes separated by a distance d are subject to a shear stress τ, and as a result the atoms in the upper and lower planes are displaced with respect to each other by a distance x. Since the lattice is periodic, so must be the shearing stress $\tau(x)$, and

Figure 2.38

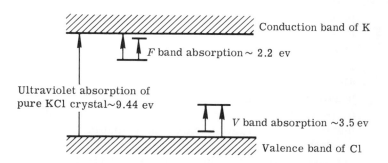

Energy levels created by F (an electron bound to a negative ion vacancy) and V (a hole bound to a positive ion vacancy) centers in KCl.

$\tau(x) = 0$ if $x = 0, a/2, a$, etc. The simplest periodic function satisfying these conditions is

$$\tau(x) = \tau_0 \sin\left(\frac{2\pi x}{a}\right) \tag{2.57}$$

For small x, the stress must be linearly proportional to the strain x through $\tau = Gx/d$, where G is the shear modulus and d is the spacing between two atomic planes. Thus, from Eq. (2.57), the maximum value of τ or the elastic limit is found to be $\tau_0 = Ga/(2\pi d)$. The difficulty with the model which predicts too high a value for τ_0 lies in the assumption that the atoms in the upper plane move simultaneously. The shear stress required to move atomic planes relative to each other can be greatly reduced if the movement is completed in several steps as suggested in Fig. 2.40b and c. Note that Fig. 2.40b is quite similar to Fig. 2.39a, the only difference being the way in which the lines are drawn to join the atoms. Actually, Fig. 2.39a serves as an intermediate step between Fig. 2.40b and c.

Figure
2.39

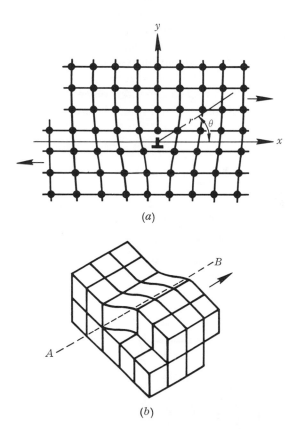

(a)

(b)

Illustrations of line defects: (a) edge and (b) screw dislocations.

Referring to Fig. 2.39, we see that the dislocation line marked by ⊥ represents atoms displaced half way or by a distance $a/2$ from their original alignment with atoms in the lower half of the crystal. The atoms on the right of the dislocation line are displaced by a distance less than $a/2$ and hence are still strongly attached to atoms in their old alignment. The atoms on the left of the dislocation line, however, are displaced more than $a/2$ and thus are in a new alignment with atoms which were apart from the atoms in the old alignment by a distance of one lattice spacing. Therefore the atoms along the dislocation line experience opposite pressures from atoms on opposite sides of the dislocation. To a first-order approximation, the forces on the dislocation line cancel, and the external stress required to move a dislocation will be very small. There are a number of ways to increase the mechanical strength of a material by impeding the motion of dislocation lines. A common example is the hardening of steel, in which tiny particles of iron carbide are formed to block the dislocation motion. We should also mention that tiny whiskers can be grown free of dislocations and that these whiskers have mechanical strengths close to the values predicted theoretically for perfect crystals. In summary, the elastic limit deduced from Eq. (2.57) applies only to dislocation-free or perfect crystals. In imperfect crystals, the elastic limit is very much reduced because the movement of atomic planes can take place in several steps through the movement of dislocation lines. The mechanical strength of a given

Figure 2.40

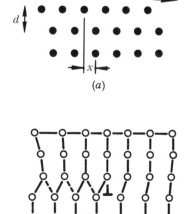

(a)

Schematic representation illustrating the movement of atomic planes. Diagram (a) assumes that all the planes in the upper half move simultaneously with respect to the planes in the lower half, while diagrams (b) and (c) assume that the movement takes place over several atomic planes and in several steps.

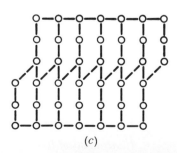

(b)   (c)

material depends on the relative easiness with which dislocation lines can move throughout the crystal.

Dislocations can be produced in a number of ways for experimental study, including plastic deformation of a crystal by bending or severe cold-work. The low-angle *grain boundary* between two crystals shown in Fig. 2.41 is another example; it consists of an array of edge dislocations. The density of dislocations, which is defined as the number of dislocation lines intersected by a plane of unit area in the crystal, can be measured by established methods. Among these is the etch pit technique. When a highly polished germanium or silicon surface is etched in an acid, conical pits are formed at places where dislocation lines intersect the surface. A precise measurement of the dislocation density is obtained by counting the number of etch pits. It is also possible to estimate dislocation densities from X-ray diffraction measurements. The observed spread of the angle θ used in the Bragg equation is a measure of the dislocation density.

Many of the effects associated with a dislocation line are produced directly or indirectly by the stress field. Referring to Fig. 2.39a, the normal stresses, τ_{rr} and $\tau_{\theta\theta}$ in the direction of r and θ, respectively, and the shear stress $\tau_{r\theta}$ are found to be

$$\tau_{rr} = \tau_{\theta\theta} = -\frac{Gb}{2\pi(1-\nu)}\frac{\sin\theta}{r}$$
$$\tau_{r\theta} = -\frac{Gb}{2\pi(1-\nu)}\frac{\cos\theta}{r} \tag{2.58}$$

where G is the shear modulus, ν is the Poisson ratio, and b is the magnitude of the slip. For most crystals, $\nu \simeq 0.3$; therefore, in the vicinity of the dislocation with $r \simeq b$, a stress of the order of $0.1G$ is expected. Under such a large stress field, impurity atoms may be attracted and precipitated along the dislocation line. In a semiconductor, the energy gap may be affected substantially by the stress field so that the electrical properties in regions near a disloca-

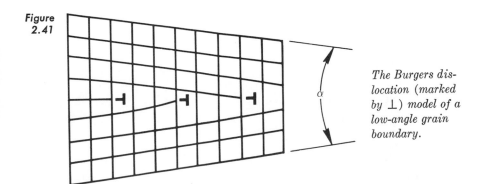

Figure 2.41

The Burgers dislocation (marked by ⊥) model of a low-angle grain boundary.

tion line differ from those in dislocation-free regions. For devices whose electrical characteristics are very sensitive to material properties, it is essential that the material have a very low dislocation density. Such is the case with semiconductor junction devices, especially near the junction region.

2.9 DIFFUSION AND ALLOYING PROCESSES

Many useful materials have foreign or impurity atoms purposely introduced into the material. Examples include Cr in Al_2O_3 used for laser action, Sb or P in Ge to make electrons available for conduction, and Cu and Cl in ZnS to act as luminescent centers. Diffusion and alloying processes are not only technologically important but also scientifically interesting for the mechanisms underlying these processes. Our present discussion is oriented toward the latter aspect. In trying to understand the diffusion process we must realize that besides vibrating about their equilibrium positions, atoms also possess a limited degree of freedom wandering through the lattice. This is implied in Fig. 2.37b. Had atoms not been able to move through the lattice, the concentrations of Schottky and Frenkel defects would depend on the history of a crystal but not on the temperature. The fact that there is a thermodynamical equilibrium concentration such as the one given by Eq. (2.51) tacitly implies that atoms can move about in the lattice to reach the equilibrium distribution.

Diffusion may be caused by three basic processes. These are (1) the migration of atoms through interstitial sites, (2) the interchange of positions between atoms and vacant lattice sites, and (3) the interchange of positions between two different neighboring atoms by rotation. Figure 2.42 shows the diffusion process by the interstitial mechanism. In moving from position 1 to position 2, the interstitial atom must squeeze through and push apart the atoms in the host lattice, and hence it experiences an energy barrier $\mathcal{E}_i$. It can be

Figure 2.42

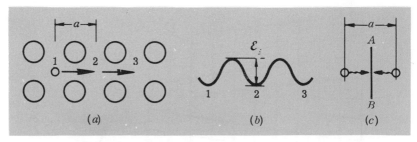

(a)   (b)   (c)

Diagrams illustrating (a) interstitial diffusion, (b) potential barrier experienced by the diffusing atom and (c) diffusion from left and right across the plane AB.

argued without formal proof that the fraction of time during which the interstitial atom has this requisite energy is only $\exp(-\varepsilon_i/kT)$. If ν is the atomic vibrational frequency with which an atom strikes the barrier, the probability p per unit time that the atom will go over the barrier is

$$p = \nu \exp\left(-\frac{\varepsilon_i}{kT}\right) \tag{2.59}$$

Consider now an imaginary plane which separates two interstitial positions, such as 1 and 2, on each side of the plane. The density of interstitial atoms per unit area on the right side is $C_2 = C(x + a/2)$, while the density on the left side is $C_1 = C(x - a/2)$. In terms of volume concentration N (atoms/m³), $C = Na$, a being the lattice constant. The net flow of atoms per unit area per second is

$$J = \frac{pC_1}{6} - \frac{pC_2}{6} = \frac{-pa}{6}\frac{\partial C}{\partial x} = \frac{-pa^2}{6}\frac{\partial N}{\partial x}$$
$$= -D\frac{\partial N}{\partial x} \qquad \text{atoms/m²-sec} \tag{2.60}$$

The quantity D, which has a dimension of m²/sec, is known as the *diffusion constant* and is given by

$$D = \frac{\nu a^2}{6}\exp\left(-\frac{\varepsilon_i}{kT}\right) \qquad \text{m²/sec} \tag{2.61}$$

The factor one-sixth accounts for the fact that out of six directions (to the right and left, forward and backward, upward and downward) only movement in one direction contributes to J.

Diffusion by the vacancy mechanism also involves an exponential temperature dependence. First of all, the concentration of vacancies is proportional to $\exp(-\varepsilon_v/kT)$ as given by Eq. (2.51). Secondly, even if the vacancy is at a nearest neighbor site, the diffusing atom still has to push other atoms apart to get to the vacant site. In other words, an activation energy of the same nature as ε_i is also involved. Therefore the total activation energy for this type of diffusion is $\varepsilon_i + \varepsilon_v$. The other mechanism, diffusion by rotating the positions of two atoms, requires a much higher activation energy than the interstitial and vacancy diffusion mechanisms according to theoretical calculations, and hence it is less likely to occur. In any event, the diffusion constant D is generally expressed as a function of temperature as

$$D = D_0 \exp\left(-\frac{\varepsilon}{kT}\right) \tag{2.62}$$

where ε is the *activation energy*. Selected values of D_0 and ε are given in Table 2.6.

In the experimental determination of D (for example, diffusion of element A in host lattice of element B) a layer of radioactive atoms (A) is electroplated or evaporated on a clean, polished surface of a given solid (B). The solid is then heated to and maintained at a temperature T for a period of time t. After the solid is allowed to return to normal temperature, the concentration of radioactive atoms is measured as a function of distance x. The reader is asked to show that conservation of particles leads to the following equation:

$$\frac{\partial N}{\partial t} = -\nabla \cdot \mathbf{J} = D \frac{\partial^2 N}{\partial x^2} \tag{2.63}$$

for a one-dimensional case. The solution of Eq. (2.63) depends on the initial (at a given time) and boundary (at a given x) conditions. For example, the following solutions

$$N(x,t) = \frac{A}{\sqrt{Dt}} \exp\left(-\frac{x^2}{4Dt}\right) \quad \text{and} \quad \frac{2B}{\sqrt{\pi}} \int_{x/2\sqrt{Dt}}^{\infty} \exp(-\xi^2)d\xi \tag{2.64}$$

actually correspond to two different physical situations (Probs. 2.22 and 2.23). Fitting Eq. (2.64) to experimental curves gives the value of D at a given T. Repeated measurements at different temperatures will result in a curve which can be further fitted to Eq. (2.62). The diffusion constant in semiconductors can be determined by a different and rather simple method. The impurity to be diffused into and that already existing in a semiconductor must be of opposite types, say p-type impurity into n-type semiconductor. Diffusion of the p-type impurity will convert part of the semiconductor from n-type to p-type if the concentration of the p-type impurity exceeds that of the n-type, thus forming a p-n junction. Knowing the original impurity concentration and the location of the junction, the diffusion constant can be calculated. In ionic crystals such as alkali and silver halides, electric conduction is entirely carried by ions. It can be shown that the ionic conductivity is given by

$$\sigma = \frac{N_0 e^2 D}{kT} = \left(\frac{N_0 e^2 a^2 \nu}{6kT}\right) \exp\left(-\frac{\varepsilon_i}{kT}\right) \tag{2.65}$$

which is based on the same physical model as Eq. (2.61), N_0 being the concentration of mobile ions of the type under consideration. A measurement of ionic conductivity at different temperatures thus

TABLE 2.6 *Selected values of D_0 and ε*

Material	As in Ge	Sb in Ge	In in Ge	C in Fe	Na in Na	Cu in Cu
D_0(m²/sec)	6.0×10^{-4}	1.2×10^{-3}	6.0×10^{-6}	2.0×10^{-5}	2.4×10^{-5}	2.0×10^{-5}
ε(ev)	2.5	2.5	2.5	0.87	0.45	2.1

determines the value of ε_i. The proof of Eq. (2.65) is left to the reader as a problem (see Prob. 2.25).

The next subject of our discussion is another metallurgical process during which an element A is dissolved in a solvent B; on cooling, the mixture forms a solid solution of A in B. There are two types of solid solutions, the interstitial and the substitutional. In the former, the dissolved atoms A, which are usually smaller than the solvent atoms B, occupy the interstitial spaces in the solvent structure. In most cases, however, the interstitial space is not large enough for the dissolved atom, causing an expansion of the solvent lattice and in turn a large local strain around the interstitial atom. Hence the percentage of foreign atoms which can be dissolved in an interstitial solution is small. An example of this type is austenite, which is an alloy of carbon in iron. Because interstitial atoms can serve to block the motion of dislocation lines, interstitial alloys usually have greater strength than pure metals. In a substitutional solid solution, the dissolved atoms A take positions in the solvent structure, which are normally occupied by the solvent atoms B. Since our present knowledge about the interstitial solution is rather incomplete, we shall concentrate our discussion on the substitutional solution.

The simplest type of substitutional solution is the one in which elements A and B not only have the same lattice structure but also form a complete solution within each other. Examples of such a system include the alloy of Au and Ag and that of Ge and Si. The other extreme is the case in which elements A and B have completely different structures and hence are insoluble in the solid phase. A diagram used to describe the state of a solid solution of a given composition at a given T is called the *phase diagram*. Figure 2.43 shows the phase diagrams of three representative systems. (1) The elements are mutually soluble in the solid phase and hence they form a single solid phase of homogeneous composition. (2) The elements are completely insoluble in each other in the solid state and the solid is a mixture of two solid structures. (3) The elements are partially soluble in each other. As shown in Fig. 2.43c, if the percentage of Cu and the temperature are such that they define a point inside a region marked α, there is only a single solid phase with a small amount of copper completely soluble in Ag. This is called the α phase. If the point happens to fall in the region marked solid $\alpha + \beta$, the solid actually is a mixture of two solid phases α and β, β being the phase with a small amount of Ag dissolved in Cu. In that case, the solid is no longer homogeneous.

Most impurities, such as Sb, P, Ga, and In, form a binary system with Ge or Si having phase diagrams similar to Fig. 2.43c. The fact that there exists a solid phase containing a small but finite percentage of impurities in these semiconductors is essential to the process of making semiconductor junctions by alloying. At this

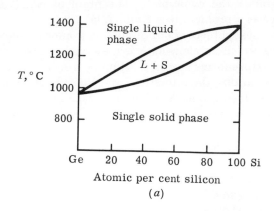

(a)

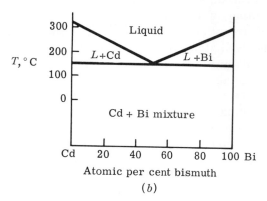

(b)

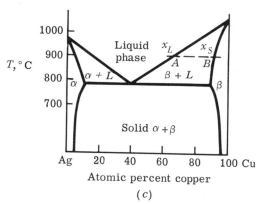

(c)

The phase diagrams of three representative systems for elements (a) completely soluble, (b) completely insoluble, and (c) partially soluble in each other.

point, we should mention another fact which is used in the purification of semiconductor materials. The distribution of impurities in the solid and liquid phases, x_S and x_L in Fig. 2.43c, defines a ratio $k = x_S/x_L$ known as the *distribution constant*. The values of k range from 10^{-1} for As and P in Si, 5×10^{-3} for Sb in Ge to 10^{-5} for Cu in Ge. Therefore, if a liquid zone is allowed to traverse from one end of a semiconductor bar to the other, the part which resolidifies becomes purer, whereas the liquid zone becomes richer in impurities. The process is known as *zone purification*. Through this process, Ge and Si with impurity concentration less than one part in 10^9 to 10^{10} can be obtained. Finally, we should add that systems of complete solubility are not limited to metal and semiconductor elements. Examples whose properties will be discussed later include Cr_2O_3 in Al_2O_3, GaSb in GaAs, and solid solutions of other III–V compounds.

In concluding this section we would like to present an elementary model to explain the behavior of solid solutions from an energy viewpoint. Let $\mathcal{E}_{AA}$, $\mathcal{E}_{BB}$ and $\mathcal{E}_{BA}$ represent the bond energy between a pair of nearest-neighbor atoms of the type AA, BB, and BA, respectively. If $2\mathcal{E}_{BA} > (\mathcal{E}_{AA} + \mathcal{E}_{BB})$, then two pairs of AB, AB bonds are energetically more favorable than pairs of AA and BB bonds, resulting in a complete mutual solubility. If $2\mathcal{E}_{BA} < (\mathcal{E}_{AA} + \mathcal{E}_{BB})$, mutual solubility happens only with the help from the TS term in free energy. For simplicity, we assume $\mathcal{E}_{AA} = \mathcal{E}_{BB} = \mathcal{E}_0$ and let $\mathcal{E} = 2(\mathcal{E}_0 - \mathcal{E}_{BA})$. For complete mutual solubility, the distribution of atoms must be random. If there are N_A A atoms and N_B B atoms for a total of $N = N_A + N_B$ atoms, the probability for an AA bond is x^2 and that for a BB bond is $(1 - x)^2$, x being equal to N_A/N. That leaves the probability for an AB bond to be $2x(1 - x)$. Thus the total binding energy of a lattice with coordination number Z is

$$\mathcal{E}_b = -\frac{ZN}{2} [x^2\mathcal{E}_0 + (1 - x)^2\mathcal{E}_0 + 2x(1 - x)\mathcal{E}_{BA}]$$

$$= -\frac{ZN}{2} [\mathcal{E}_0 - x(1 - x)\mathcal{E}] \tag{2.66}$$

The free energy of such a lattice is $F = \mathcal{E}_b - TS$, where

$$S = k \log \frac{N!}{N_A! N_B!} = -k[x \log x + (1 - x) \log (1 - x)]N \tag{2.67}$$

through the use of Stirling's approximation. For $\mathcal{E} > 0$, $\mathcal{E}_b$ favors the situation in which $x = 0$, meaning insolubility. However, this may be compensated by the TS term because a nonvanishing x would contribute to a positive S. Therefore, for a given x, there exists a critical temperature T_c above which the two elements form a single solid phase and below which they are a mixture of two solid phases.

To find the function $f(x,T_c)$ which relates T_c to x, we apply the condition that for complete solubility the free energy must be a minimum. The resultant curve generally has a shape similar to the curve shown in Fig. 2.43 except for the fact that at sufficiently high temperatures a liquid phase forms, and such a possibility has been ignored in our formulation of the problem. The reader may find more discussion on the subject in Prob. 2.26.

2.10   VIBRATIONAL MODES IN SOLIDS

In Sec. 2.8 we treated the vibration of each lattice atom as if it were independent of each other. Equation (2.40) is oversimplified on two accounts. First, the neighboring atoms are also displaced from their equilibrium positions and, second, the restoring force depends not on the absolute but on the relative displacement of an atom with respect to its environment. Figure 2.44a shows a linear chain consisting of similar atoms with mass M and spaced at a distance a apart from each other. If u_n denotes the displacement of the nth atom from its equilibrium position, then the force acting on the nth atom is given by

$$F_n = \gamma(u_{n+1} - u_n) - \gamma(u_n - u_{n-1}) \qquad (2.68)$$

where γ is the force constant, being the same as the γ of Eq. (2.39). For simplicity, only interactions between nearest neighbors are considered. Note that the force acting on the nth atom from the $(n-1)$th atom is opposite to that from the $(n+1)$th atom. If $u_n > u_{n-1}$, the force pulls the nth atom towards the $(n-1)$th atom.

Figure 2.44

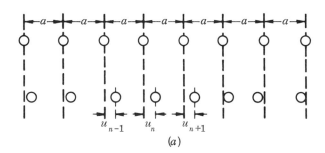

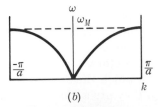

(a) A monatomic linear chain and (b) its vibrational dispersion (ω vs. k) curve.

From Eq. (2.68), the equation of motion of the nth atom is

$$M \frac{d^2 u_n}{dt^2} = \gamma(u_{n+1} + u_{n-1} - 2u_n)$$

(2.69)

Similar equations are obtained for other atoms.

It is obvious from Eq. (2.69) that the motion of an atom is coupled to that of other atoms in the linear chain. If there are N atoms in the chain, there will be N simultaneous differential equations. It seems highly desirable that a special technique be developed to solve these equations. First, we learn from the Debye model that the vibrational modes of a solid form an almost continuous spectrum from very low to far-infrared frequencies. The only approach to the problem is therefore by the method of Fourier analysis. For a particular Fourier component of lattice vibration having angular frequency ω and wavelength λ, the displacement must follow the wave motion or

$$u_n(\omega, k) = A \exp[i(\omega t + kna)]$$

(2.70)

where A is the amplitude of the Fourier component, na is the equilibrium position of the nth atom, and $k = 2\pi/\lambda$. For the $(n + 1)$th and $(n - 1)$th atom, u_{n+1} and u_{n-1} can be found from Eq. (2.70) by changing n to $n + 1$ and $n - 1$, respectively. Substituting Eq. (2.70) into Eq. (2.69) gives

$$-\omega^2 M = \gamma[\exp(ika) + \exp(-ika) - 2]$$

(2.71)

Combining the exponential terms in and taking the square root of Eq. (2.71), we find

$$\omega = \pm \left(\frac{4\gamma}{M}\right)^{1/2} \sin \frac{ka}{2}$$

(2.72)

which is plotted in Fig. 2.44b. For $ka = \pi$, ω reaches a maximum value of $\sqrt{4\gamma/M}$, which is $\sqrt{2}$ of the value found from Eq. (2.40). The factor $\sqrt{2}$ comes from the fact that the displacements of neighboring atoms are completely out of phase, or $u_{n-1} = u_{n+1} = -u_n$. For small values of ka or long wavelength phonons, ω/k is a constant equal to the velocity of elastic waves, and hence our previous analysis in Sec. 2.8 based on a continuous medium is correct only in regions where $ka < 1$.

For crystals consisting of two different kinds of atoms or having two or more atoms per primitive cell, the analysis just presented is no longer adequate. Consider the diatomic linear chain of Fig. 2.45a in which the even-numbered sites are occupied by atoms with mass M_2 and the odd-numbered sites by atoms with mass M_1. For such a diatomic chain, two equations of motion are needed:

$$M_2 \frac{d^2 u_{2n}}{dt^2} = \gamma(u_{2n+1} + u_{2n-1} - 2u_{2n})$$

$$M_1 \frac{d^2 u_{2n+1}}{dt^2} = \gamma(u_{2n+2} + u_{2n} - 2u_{2n+1})$$

(2.73)

We propose that the Fourier component of the lattice displacement be of the following form:

$$u_{2n} = A \exp [i(\omega t + k2na)]$$
$$u_{2n+1} = B \exp \{i[\omega t + k(2n + 1)a]\}$$

(2.74)

Substituting Eq. (2.74) into Eq. (2.73) gives

$$-\omega^2 M_2 A = 2B\gamma \cos ka - 2A\gamma$$
$$-\omega^2 M_1 B = 2A\gamma \cos ka - 2B\gamma$$

(2.75)

For a nontrivial solution, the determinant of coefficients must vanish or

$$\omega^2 = \gamma \left(\frac{1}{M_1} + \frac{1}{M_2}\right) \pm \gamma \sqrt{\left(\frac{1}{M_1} + \frac{1}{M_2}\right)^2 - \frac{4 \sin^2 ka}{M_1 M_2}}$$

(2.76)

Figure 2.46 shows a plot of ω as a function of k. The main feature is that the curve consists of two branches corresponding to the positive and negative signs in Eq. (2.76). The physical origin of these two branches lies not in the fact that two different masses are involved but in the fact that we allow two different modes of vibration in the proposed solution of the form given by Eq. (2.74). To understand why the two different branches correspond to two different modes of vibration, we consider the motion of two neighboring atoms. To make our point clear, we choose $ka = 0$. For the lower branch, $\omega = 0$; thus Eq. (2.75) leads to $A = B$. From Eq. (2.74) we see that the wave motions associated with odd-numbered and even-numbered atoms are in phase or, in other words, they move together with the same sense of direction as illustrated in Fig. 2.45b. For the upper branch of Fig. 2.46, at $ka = 0$, $\omega^2 =$

Figure 2.45

$$\overset{M_1}{\underset{\bullet}{}} \quad \overset{M_2}{\underset{\circ}{}} \quad \overset{M_1}{\underset{\bullet}{}} \quad \overset{M_2}{\underset{\circ}{}} \quad \overset{M_1}{\underset{\bullet}{}} \quad \overset{M_2}{\underset{\circ}{}} \quad \overset{M_1}{\underset{\bullet}{}} \quad \overset{M_2}{\underset{\circ}{}}$$

$$2n-1 \quad 2n \quad 2n+1 \quad 2n+2$$

(a)

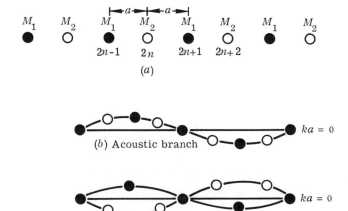

(b) Acoustic branch $\quad ka = 0$

(c) Optical branch $\quad ka = 0$

(a) A diatomic linear chain and (b), (c) the two different modes of vibration.

Figure 2.46

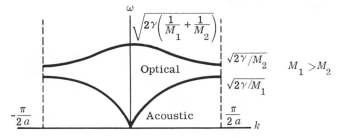

The frequency versus wave number (ω vs. k) curve for the two vibrational modes, acoustic (lower) and optical (upper) branches, of a diatomic linear lattice.

$2\gamma(M_1 + M_2)/M_1M_2$; thus Eq. (2.75) gives $M_2A + M_1B = 0$. Therefore odd-numbered and even-numbered atoms vibrate in an opposite sense as shown in Fig. 2.45c, resulting in a wider relative displacement of atoms and hence involving a larger energy. The upper and lower branches of the *dispersion curve* (ω against k curve) are generally known as the *optical* and *acoustic branches*, for the energies associated with these two modes of vibrations are, respectively, in the optical and acoustical range of the frequency spectrum.

The actual dispersion curves in Al and Si are shown in Fig. 2.47. These curves are obtained from measurements[8] of the diffuse scattering of monochromatic X rays and from the energy distribution of coherently scattered monoenergetic neutrons. In a three-dimensional lattice, the displacement of atoms can be either trans-

Figure 2.47

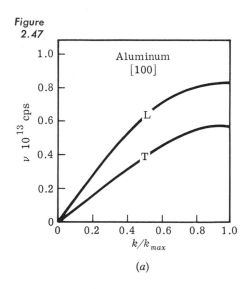

(a)

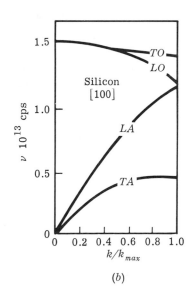

(b)

Experimental dispersion (ω vs. k) curves of lattice vibration in (a) Al and (b) Si.

verse or longitudinal to the direction of wave propagation, and these modes are marked by T and L in Fig. 2.47. The face-centered cubic lattice of Al has one atom per primitive cell, so only the acoustic branch exists, as in Fig. 2.44b. The diamond structure, on the other hand, consists of two interpenetrating face-centered cubic structures, and hence it contains two atoms per primitive cell, allowing two different modes of vibrations. This explains the appearance of the optical branch in Fig. 2.47b. We should point out that the dispersion curves depend on the direction of the **k** vector in the reciprocal lattice. For other directions we expect the curves to be different from those shown in Fig. 2.47. We should also mention that the density of states $n(\omega)$ which appears in the Debye model of lattice specific heat can also be obtained from X ray and neutron scattering measurements.[8] The most important part of our present discussion, however, is the simple concept that associated with a given proper mode of lattice vibration are a definite amount of energy represented by $\hbar\omega$ and a definite amount of momentum represented by $\hbar k$. We shall make use of these concepts in later discussions.

PROBLEMS

2.1 Find the Madelung constant for a one-dimensional lattice with alternate and evenly spaced negative and positive ions.

2.2 Consider a two-dimensional square lattice with alternate positive and negative ions. Choose any atom as the reference atom. Find the distance between this reference atom and its neighbors from the nearest neighbors up to the 9th nearest neighbors. Indicate also the charge types and the number of equivalent positions of those neighbors.

2.3 Consider the NaCl structure with a being the distance between two nearest neighbors. If the position of a given Na atom is chosen as the origin, the coordinates of other atoms can be arranged according to their distance from the origin as follows:

Atom type	Coordinates	Distance from origin	Number of equivalent positions
Na	000	0	1
Cl	$a00$	a	$6 = 2 \times 3$
Na	$aa0$	$\sqrt{2}\,a$	$12 = 4 \times 3$
. . .	. . . . . .	. . . . . .	. . . .
Na	$3a\,2a\,a$	$\sqrt{14}\,a$	$48 = 8 \times 3!$

[8] For the theory of these measurements and the analysis of the experiments, the reader is referred to the following articles. J. Laval, *J. Phys. Radium*, vol. 15, pp. 545 and 658, 1954; C. B. Walker, *Phys. Rev.*, vol. 103, p. 547, 1956; B. N. Brockhouse and P. K. Iyengar, *Phys. Rev.*, vol. 111, p. 747, 1958; B. N. Brockhouse, *Phys. Rev. Letters*, vol. 2, p. 256, 1959.

Complete the above table by filling in the information from $Na(a,a,0)$ to $Na(3a,2a,a)$. The number of equivalent positions is obtained by considering (1) negative as well as positive coordinates and (2) possible permutation of the coordinates.

2.4 The compressibility of a solid is defined as

$$K = \frac{-1}{V}\frac{\partial V}{\partial p}$$

where V is the volume of the crystal and p is the applied external pressure. Using the relation

$$d\mathcal{E} = -p\,dV$$

and the energy expression for an ionic crystal from Eqs. (2.8) and (2.9)

$$\mathcal{E} = N\left(\frac{B}{R^n} - \frac{Ae^2}{4\pi\epsilon_0 R}\right)$$

show that

$$\frac{1}{K} = \frac{(n-1)e^2 A}{72\pi\epsilon_0 R_0^4}$$

where A is the Madelung constant. In the above calculation the volume V can be taken as $2NR^3$ for a sodium chloride crystal, N being the total number of molecules.

2.5 Calculate the value of n for NaCl, using the values of 1.75 for the Madelung constant, 5.63 Å for the lattice constant and the experimental value of 3.3×10^{-12} cm²/dyne for compressibility. Using the calculated value of n, find the lattice energy $\mathcal{E}$ of the crystal. The experimental value of $\mathcal{E}$ is 180 kcal/mole or 7.8 ev/molecule.

2.6 Consult the loose-leaf compilation "Crystal Structures" by Wyckoff. Find the coordinates of the atoms in a unit cell of the following substances: FeS, ZnO, Se, Ga and CaF_2. Draw the unit cell and show the positions of a few representative atoms in the cell.

2.7 Find the distance between the nearest neighbors in terms of the dimension of the unit cell in the following structures: face-centered cubic, body-centered cubic, diamond and close-packed hexagonal structures. Fur-

Figure P2.7

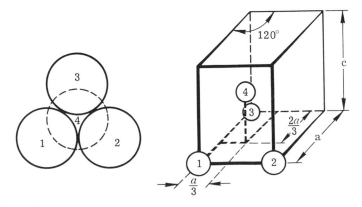

ther show that in the ideal hexagonal close-packed structure, the ratio of $c/a = \sqrt{8/3} = 1.633$. (Hint: Consider the atoms, 1, 2, 3, and 4 as being situated at the corners of a tetrahedron and making contacts with each other along the edge of the tetrahedron, as shown in Fig. P2.7.)

2.8 Refer to Fig. 1.18 and our discussion of thermionic emission in Sec. 1.12.

1. Write down the expressions equating the momentum components of an electron before and after escape.

2. Show that the density of electrons having sufficient energy to escape from the metal is approximately given by

$$\int dn = (2/h^3) \exp(-\mathcal{E}_w/kT) \int \exp[-(p_{x2}^2 + p_{y2}^2 + p_{z2}^2)/2m_0 kT] \, dp_{x2} \, dp_{y2} \, dp_{z2}$$

where p_{x2}, p_{y2}, and p_{z2} are, respectively, the values of the electron momentum along the x, y, and z directions after escape. What are the limits of integration in terms of the momentum before and after escape?

3. Show that an electron density n (electrons/m³) moving with a velocity v constitutes a current density $J = env$.

4. Using the results obtained in parts (3) and (2), write down an integral expression for the thermionic current density.

2.9 How many atoms belong to the unit cell of a face-centered cubic crystal? Using the three face diagonals of half length (a',b',c' of Fig. 2.20), construct a primitive cell and show the atoms in the primitive cell in relation to atoms in the original cubic structure. What is the volume of the primitive cell? [Hint: Volume = $|\mathbf{a}' \cdot (\mathbf{b}' \times \mathbf{c}')|$]

2.10 Repeat Prob. 2.9 for a body-centered cube. If the body-centered atom is taken as the origin, the three base vectors $\mathbf{a}'$, $\mathbf{b}'$, $\mathbf{c}'$ of the primitive cell end at $(\frac{1}{2},\frac{1}{2},-\frac{1}{2})$, $(-\frac{1}{2},\frac{1}{2},\frac{1}{2})$ and $(\frac{1}{2},\frac{1}{2},\frac{1}{2})$.

2.11 Show that any vector $\mathbf{r} = u\mathbf{a} + v\mathbf{b} + w\mathbf{c}$ lying in a plane (hkl) must satisfy the relation

$$uh + vk + wl = 0$$

Take the (110) plane in a cubic crystal as an example. Draw the directions of the following vectors: [001], [1$\bar{1}$0], [1$\bar{1}$1], and [110]. Identify the vector which is perpendicular to the plane and the vectors which are in the plane by using the above equation and Eq. (2.22).

2.12 Find a general expression of the structure factor of a sodium chloride crystal for X-ray diffraction by (hkl) planes. Show that the indices of allowed diffractions must be all odd or all even.

2.13 Find a general expression of the structure factor of a face-centered cubic crystal. Show that diffractions from (100) and (110) planes will be missing.

2.14 Imagine four collinear lattice points $FABC$ in a three-dimensional lattice separated by a translation vector $\mathbf{t}$ of Eq. (2.20). If the lattice has an n-fold rotational symmetry then a rotation about B of an angle $\phi = 2\pi/n$ should bring point C to another lattice point D. So should rotation about A

bring lattice point E to F. Since points D and E are lattice points and the line ED parallels the line $FABC$, the distance ED must be an integral multiple of t. From this relation, show that the allowed rotational symmetries have $n = 1, 2, 3, 4,$ or 6.

2.15 A further improvement of the Debye theory of specific heat was proposed by Born. Instead of using a maximum cutoff frequency ν_m common to both the longitudinal and transverse modes, Born suggested that it would be preferable to use a common cutoff wavelength λ_{min} such that

$$\lambda_{min} = (4\pi V/3N)^{1/3}$$

where N/V is the number of atoms per unit volume. Draw the ω versus k curve showing the difference between the two proposed models. Find Θ_D for the longitudinal and transverse modes for the Born model. Also modify Eqs. (2.47) and (2.48) accordingly. From Eq. (2.48), check the value of λ_{min} given above.

2.16 Show that the entropy S associated with $3N$ independent harmonic oscillator is given by Eq. (2.52) or

$$S = k \log W = 3Nk \left[\frac{ae^a}{e^a - 1} - \log (e^a - 1) \right]$$

where $a = h\nu/kT$ and W is defined in Eq. (1.127). Find the approximate expression for S which is valid at high temperatures such that $a < 1$.

It is assumed that the lattice vibrational frequency ν is changed to ν' if an atom has a vacancy as its nearest neighbor. For a lattice having $3n$ vacancies and coordination number Z, show that the entropy at high temperature is

$$S = 3k[N - nZ \log (\nu'/\nu)]$$

2.17 Using the result of Prob. 2.16, find the free energy of a lattice having $3n$ vacancies. By minimizing the free energy, show that the vacancy concentration is given by

$$\frac{n}{N - n} = \left(\frac{\nu}{\nu'} \right)^{3Z} \exp \left(-\frac{\mathcal{E}_v}{kT} \right)$$

2.18 Show that the ratio of the volume occupied by atoms to the total available volume for various structures is: (1) $\pi/6$ for simple cubic structure (2) $\pi\sqrt{3}/8$ for body-centered cubic structure (3) $\pi\sqrt{2}/6$ for face-centered cubic structure. The ratio is obtained by assuming that atoms are considered as spheres and that these spheres are in contact between nearest neighbors.

2.19 Consider a diamond structure as two interpenetrating face-centered cubic structures. The basic unit is made of four atoms situated at the corners of a tetrahedron and another atom at the center. The center atom makes contact with the corner atoms. Show that the ratio of the volume occupied by atoms to the total available volume in a diamond structure is $\pi\sqrt{3}/16$.

2.20 In a close-packed structure, there are two types of interstitial positions: one surrounded by four equidistant atoms is called a tetrahedral site and the other by six equidistant atoms is called an octahedral site. The atoms are considered as hard spheres making contact with each other. Find the maximum radius r of a sphere which can fit into the tetrahedral interstitial position. Express r in terms of R, the radius of the atom. Identify the position of the site in a hexagonal close-packed structure and in a cubic close-packed (face-centered) structure.

Figure
P2.20

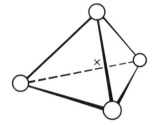

Figure
P2.21

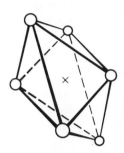

2.21 Repeat Prob. 2.20 for the octahedral interstitial position.

2.22 Calculate the rate of flow of particles into an elementary box with area A and length dx by diffusion. Considering the conservation of particles, show that

$$\frac{\partial N}{\partial t} = D \frac{\partial^2 N}{\partial x^2}$$

for one-dimensional diffusion. Verify that

$$N_1(x,t) = \left(\frac{A}{\sqrt{Dt}}\right) \exp\left(-\frac{x^2}{4Dt}\right)$$

is a solution to the partial differential equation and so is

$$N_2(x,t) = B \operatorname{erfc} \frac{x}{2\sqrt{Dt}} = \frac{2B}{\sqrt{\pi}} \int_{\frac{x}{2\sqrt{Dt}}}^{\infty} \exp\left(-\xi^2\right) d\xi$$

where erfc represents the complementary error function.

2.23 Consult any text which contains the table of error or probability integral (for example: Peirce, "A Short Table of Integrals"). Plot $N(x,t)$ as a function of x for three values of $Dt = 0.5$, 1 and 2. The purpose of the problem is to acquaint the reader with the general behavior of the two curves $N_1(x,t)$ and $N_2(x,t)$ as x and t vary. The first solution represents a physical situation in which a given finite amount of impurity is initially placed at $x = 0$ and let the impurity diffuse whereas the second solution represents a situation in which the concentration at $x = 0$ is maintained constant by constantly supplying impurities at $x = 0$. In the experiment discussed in the text, the second solution should be used.

2.24 A diffusion experiment is performed on a germanium slab originally doped with indium (p-type) to a concentration of $10^{14}/\text{cm}^3$, the diffusant being antimony (n-type). Given $D_0 = 1.2 \times 10^{-3}$ m²/sec and $\mathcal{E} = 2.5$ ev for Sb in Ge, find the diffusion constant at $T = 600°$K and $T = 900°$K. Using the value of D just obtained, calculate the depth of penetration of the n-region for $T = 600°$K and $t = 100$ hrs and for $T = 900°$K and $t = 10$ hrs if the concentration of Sb at surface or B in Eq. (2.64) has a value $B = 10^{18}$/cm³.

2.25 If an electric field E is applied to an ionic crystal, the barrier potential is modified by the electric field and so is the jumping probability p of Eq. (2.59). In other words, Eq. (2.60) should be written as $J = (p_1 C/6) - (p_2 C/6)$ where $p_1 = p(-a/2)$ and $p_2 = p(a/2)$. Show that the ionic current density is given by

$$J = (N_0 e^2 a^2 \nu/6kT)E \exp(-\mathcal{E}_i/kT)$$

In view of the relation $J = eN_0\mu E$, μ being defined as the mobility of ions, verify the Einstein relationship $\mu = eD/kT$.

2.26 Starting from Eqs. (2.66) and (2.67), show that

$$x/(1 - x) = \exp[-Z\mathcal{E}(1 - 2x)/2kT]$$

by minimizing the free energy. Sketch the x versus T curve by considering the regions where $x/(1 - x) < 1$ and where $(1 - x)/x < 1$ separately. Explain the phase diagram of Fig. 2.43c by comparing it with the x versus T curve just obtained. Your remarks should be directed toward explaining the physical state of the solution.

2.27 Compare the dispersion (or ω versus k) relation from Eq. (2.72) with the dispersion relation used in the Debye model by plotting them on the same graph. Comment on the validity of the Debye model regarding long and short wavelength lattice waves. Show that for long wavelength lattice waves, Eq. (2.72) predicts a longitudinal acoustic velocity of $v = \sqrt{Y/\rho}$ where Y is the elastic modulus and ρ the mass density of the medium.

REFERENCES

General

Azaroff, L. V., and J. J. Brophy: "Electronic Processes in Materials," McGraw-Hill Book Company, New York, 1963.

Cottrell, A. H.: "Theoretical Structural Metallurgy," St. Martin's, New York, 1957.

Dekker, A. J.: "Solid State Physics," Prentice-Hall, Inc., Englewood Cliffs, N.J., 1957.

Guy, A. G.: "Elements of Physical Metallurgy," Addison-Wesley, Reading, Mass., 1959.

Hannay, B. N. (ed.): "Semiconductors," Reinhold Publishing Corporation, New York, 1959.

Hume-Rothery, W.: "Atomic Theory for Students of Metallurgy," Institute of Metals, London, 1955.

Kittel, C.: "Introduction to Solid State Physics," John Wiley & Sons, Inc., New York, 1956.

Seitz, F.: "The Modern Theory of Solids," McGraw-Hill Book Company, New York, 1940.

Sproul, R. L.: "Modern Physics," John Wiley & Sons, Inc., New York, 1963.

Wert, C. A., and R. M. Thomson: "Physics of Solids," McGraw-Hill Book Company, New York, 1964.

Chemical bonds

Pauling, L.: "The Nature of the Chemical Bond," Cornell University Press, Ithaca, N.Y., 1960.

Slater, J. C.: "Quantum Theory of Molecules and Solids," McGraw-Hill Book Company, New York, 1963.

Crystallography and crystal structures

Buerger, M. J.: "Elementary Crystallography," John Wiley & Sons, Inc., New York, 1956.

Phillips, F. C.: "An Introduction to Crystallography," Longmans, Green & Co., Ltd., London, 1956.

Wyckoff, R. W. G.: "Crystal Structures," Interscience Publishers, New York, 1953.

Crystal imperfections

Cottrell, A. H.: "Dislocations and Plastic Flow in Crystals," Oxford University Press, Oxford, 1953.

Doremus, R. H., B. W. Roberts, and D. Turnbull (eds.): "Growth and Perfection of Crystals," John Wiley & Sons, Inc., New York, 1958.

Fisher, J. (ed.): "Dislocations and Mechanical Properties of Crystals," John Wiley & Sons, Inc., New York, 1957.

Read, W. T., Jr.: "Dislocations in Crystals," McGraw-Hill Book Company, New York, 1953.

Shockley, W. (ed.): "Imperfections in Nearly Perfect Crystals," John Wiley & Sons, Inc., New York, 1952.

Van Bueren, H. G.: "Imperfections in Crystals," North Holland Publishing Co., Amsterdam, 1960.

Diffusion

Le Claire, A. D.: Diffusion in Metals, "Progress in Metal Physics," Pergamon Press, London, vol. 1, 1949, and vol. 4, 1953.

Shewmon, P. G.: "Diffusion in Solids," McGraw-Hill Book Company, New York, 1963.

Phase diagrams

Hansen, M.: "Constitution of Binary Alloys," McGraw-Hill Book Company, New York, 1958.

Rhines, F. N.: "Phase Diagram in Metallurgy," McGraw-Hill Book Company, New York, 1956.

3 CONDUCTION MECHANISMS IN SEMICONDUCTORS

3.1 ELECTRIC CONDUCTION AND ENERGY BANDS

Consider a free electron gas with density n (electrons/m³) under the influence of an electric field E (volt/m). If τ is the mean free time of an electron between successive collisions either with other electrons or with positive ions, then the average *drift velocity* acquired by the electron in the direction of the field is given by

$$v = \frac{eE\tau}{m} = \mu E \tag{3.1}$$

and the conduction current density by

$$J = nev = ne\mu E = \sigma E \qquad \text{amp/m}^2 \tag{3.2}$$

where $\mu = e\tau/m$ (m²/volt-sec) is called the *mobility* of the electron and $\sigma = ne\mu$ [(Ω-m)⁻¹] is the conductivity of the free electron gas. Figure 3.1 shows the resistivity $\rho = 1/\sigma$ of some typical materials, ranging from 10^{-8} Ω-m for good conductors to 10^{13} Ω-m for good insulators. For such a wide range of conductivities, we must be careful in correctly interpreting the quantities involved in Eq. (3.2), especially the meaning of n.

In between the conductors and insulators lie materials generally referred to as *semiconductors*. Take silicon as an example and compare its electrical resistivity with that of sodium. Silicon has an electronic configuration $1s^2$, $2s^2$, $2p^6$, $3s^2$ and $3p^2$ while sodium has $1s^2$, $2s^2$, $2p^6$ and $3s^1$. The core electrons $1s^2$, $2s^2$ and $2p^6$ in both silicon

Figure 3.1

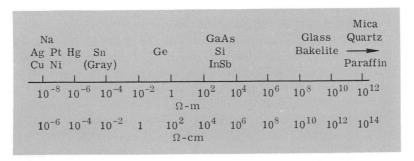

The range of resistivity observed at room temperature in different materials.

129

and sodium are tightly bound to the nucleus and hence they do not contribute to electrical conductivity. If we simply compare the number of valence electrons, for example 4 for silicon and 1 for sodium, we would conclude that silicon should be more conductive than sodium. This is contrary to fact.

The difference in the electrical behavior of a semiconductor and that of a conductor is best described in terms of an energy band diagram. When atoms are brought into proximity as in a crystal, the valence electrons of adjacent atoms interact very strongly. The situation is analogous to coupled resonant circuits. If two identical resonant circuits are weakly coupled, each circuit oscillates with the same frequency as an isolated resonant circuit. As the coupling becomes stronger, the resonant frequency is split into a pair of resonant frequencies and the two circuits oscillate as a whole unit.

The same applies to the atomic states. As the interaction between neighboring valence electrons becomes stronger, the quantized atomic energy levels begin to spread over a wider energy range, thus forming an energy band. The important consequences of the strong interaction are as follows. First, the quantum states no longer belong to a single atom but spread over the entire crystal. This justifies the use of Bloch wave functions in describing the motion of electrons in a solid, which we shall discuss later. Second, the number of quantum states in the energy band (analogous to the number of independent resonant frequencies) is the same as the number of atomic quantum states (analogous to the number of resonant circuits) from which the band is produced. These facts lead to the explanation of why sodium and silicon have a different electrical behavior.

Figure 3.2 shows the energy bands of carbon in the diamond structure. The core $1s^2$ electrons of carbon are tightly bound to the nucleus and hence unimportant in our present discussion. If there are N carbon atoms in a diamond crystal, the number of available quantum states associated with the $2s$ and $2p$ states is $2N$ and $6N$ respectively. At large atomic distances, the four valence electrons are divided such that the $2s$ band is completely filled and the $2p$ band only one-third filled. At smaller distances, however, the $2s$ and $2p$ states are mixed together as in coupled circuits and the total $8N$ quantum states are split equally between the lower and upper bands. The situation in silicon and germanium is qualitatively similar to that in diamond except that the $2s$ and $2p$ states should be changed to $3s$ and $3p$ and $4s$ and $4p$ respectively, and the energy scale is different. Therefore, at absolute zero temperature, the lower band of silicon is completely filled by the valence electrons and the upper band completely empty as shown in Fig. 3.3a. Sodium and all other alkali metals have a body-centered cubic structure. The $3s$ band of sodium has $2N$ available states while there are only N

Figure
3.2

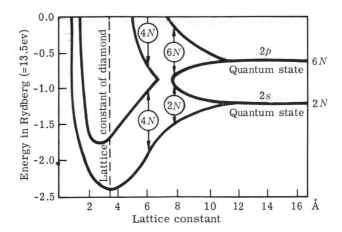

The energy-band diagram of diamond. (G. E. Kimball, J. Chem. Phys., vol. 3, p. 560, 1935.)

valence electrons in sodium; therefore, the band is only half filled, as Fig. 3.3*b* shows.

The difference in the electrical behavior of silicon and sodium is now evident. After acquiring energy from an applied electric field, the valence electrons in sodium can readily move into a vacant higher energy state and thus leave behind a new vacant state for

Figure
3.3

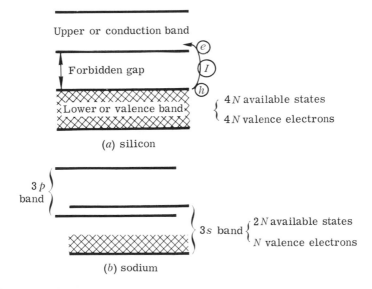

Comparison of the energy bands of (a) silicon and (b) sodium and the distribution of electrons in these bands.

other electrons to move into. Therefore, a simultaneous and continuous change of electronic state is possible, ensuring electric conduction. This applies to other materials with a partially filled band. In a semiconductor such as silicon, the lower band is completely filled at absolute zero temperature, and hence the valence electrons find themselves unable to move anywhere on account of the Pauli exclusion principle. In other words, if all the valence electrons in silicon are held in covalent bonds, no electric conduction is possible.

3.2 ELECTRIC CONDUCTION IN SEMICONDUCTORS

A semiconductor can be conductive only when both electrons and vacant states are made available in the same band. At absolute zero, there are electrons but no vacant states in the lower band while there are vacant states but no electrons in the upper band. Consequently, conduction is blocked in both bands. At a finite temperature, however, a certain number of valence electrons will appear in the upper band on account of thermal excitation, creating vacant states in the lower band. Conduction is now possible in both bands. The part of conductivity which arises on account of thermal excitation, process I in Fig. 3.3a, of electrons from the lower to the upper band, is called the *intrinsic conductivity* of a semiconductor. A semiconductor can also become conductive if certain impurity atoms are introduced.

For definiteness and clarity of presentation, we shall confine our present discussion to elemental semiconductors such as Ge and Si and leave compound semiconductors such as GaAs and CdS for later discussion in Sec. 3.10. The elements on the right side of silicon and germanium in the periodic table, such as phosphorus, arsenic, and antimony, have five valence electrons with an s^2p^3 configuration. If atoms of these elements are used to substitute a silicon or germanium atom, only four of the five valence electrons are needed to form the covalent bond with neighboring Si or Ge atoms, leaving one extra electron relatively free to move about in the crystal. The situation can be summarized as follows:

$$\text{Sb} \begin{pmatrix} \text{neutral with 5} \\ \text{valence electrons} \end{pmatrix} \rightleftharpoons \text{Sb}^+ \begin{pmatrix} \text{with 4 valence electrons} \\ \text{to complete the covalent bond} \end{pmatrix}$$
$$+ \ e \text{ (loosely attached to Sb}^+ \text{ atom)} \quad (3.3)$$

Figure 3.4 shows (a) the covalent structure of Ge with substitutional Sb impurity atoms and (b) the energy band diagram of Ge together with the energy levels of antimony in germanium. The upper and lower bands of a semiconductor shown in Figs. 3.3a and 3.4b are referred to as the *conduction* and *valence bands*, respectively, and the energy interval between the two bands is usually called the *energy gap.* The ionization energy which is the amount of energy required for the excess electron to break away from the Sb^+ atom is given by the energy $\mathcal{E}_c - \mathcal{E}_d$ in Fig. 3.4*b* where $\mathcal{E}_c$ and $\mathcal{E}_d$ are referred to as the energies of the conduction band edge and donor levels, respectively.

The word *donor* refers to those impurities which are capable of donating an excess electron to the conduction band. Playing a role just opposite to that of a donor are *acceptors* which have a valence of 3 with a configuration s^2p and hence must accept an electron from somewhere in order to complete the covalent bonds with neighboring atoms. A missing electron in the covalent bond is often called a *hole.* The elements on the left side of silicon and germanium in the periodic table, such as boron, aluminum, gallium and indium, are acceptor impurities. The role of an acceptor, using In as an exam-

Figure 3.4

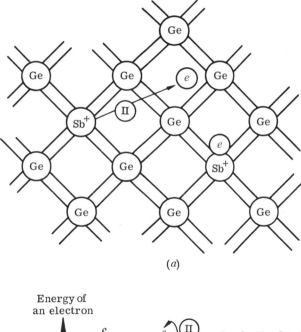

(*a*)

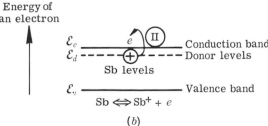

(*b*)

(*a*) *Substitution of germanium atoms by antimony atoms which act as a donor impurity and* (*b*) *energy levels created in the forbidden gap by antimony atoms.*

ple, can be stated in a way similar to Eq. (3.3) as follows:

$$\text{In} \begin{pmatrix} \text{neutral with} \\ \text{3 valence electrons} \end{pmatrix} \rightleftharpoons \text{In}^- \begin{pmatrix} \text{with 4 valence electrons to} \\ \text{complete the covalent bond} \end{pmatrix}$$
$$+ h \begin{pmatrix} \text{a missing electron or a hole} \\ \text{loosely attached to In}^- \text{ atom} \end{pmatrix} \quad (3.4)$$

and Fig. 3.5 is a pictorial illustration of Eq. (3.4) where the energy difference $\mathcal{E}_a - \mathcal{E}_v$ is the energy required for the missing electron to break away from the In$^-$ atom.

In sum, the donors possess an excess electron which upon excitation will jump into the conduction band and become available for conduction. Since there are many vacant states in the conduction band, all of the electrons that are ionized from the donor states will contribute to electric conduction. The acceptors, on the other hand, need an electron to complete the covalent bond, which may be

**Figure
3.5**

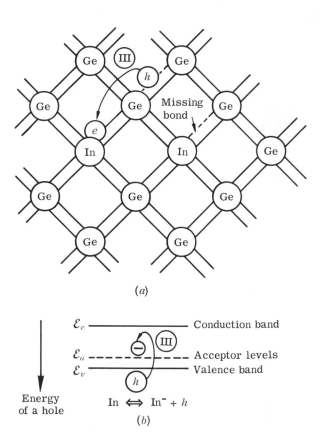

(a) *Substitution of germanium atoms by indium atoms which act as an acceptor impurity and (b) energy levels created in the forbidden gap by indium atoms.*

acquired from the valence band. Once having given the electron to the acceptor, the valence band has a vacant state which is now available to other electrons in the valence band. Since there are a large number of electrons in the valence band, all of the states made vacant by giving electrons to the acceptors will participate in the conduction process. These vacant states in the valence band are called holes. Therefore, electric conduction in the conduction band is by electrons while in the valence band it is by vacant states or holes. In the presence of both electrons and holes, Eq. (3.2) becomes

$$J = e(n\mu_n + p\mu_p)E = \sigma E \qquad (3.5)$$

where n and p are the concentration of electrons in the conduction band and that of holes in the valence band respectively and μ_n and μ_p are the respective mobilities of electrons and holes. A semiconductor is called *n-type* if the current carriers are predominantly electrons (negative charges) and it is called *p-type* if the current carriers are predominantly holes (missing electrons or positive charges).

It seems appropriate to make some pertinent comments about Eq. (3.5). First, we have assumed that the charges trapped by the donors and acceptors are immobile. At high impurity concentrations, the electron wave functions associated with the trapped charges may overlap, and hence the trapped charge may hop from one impurity atom to another, causing a current flow. The phenomenon known as *impurity band conduction* will be discussed later. Second, Eq. (3.5) is valid only for low electric field intensities. At high electric fields, extraordinary circumstances give rise to new phenomena. An electron may be excited directly from a state in the valence band to another state in the conduction band. This quantum-mechanical effect is known as the *tunnel effect*, and it will be discussed in Chapter 6. Nonlinear effects at high electric fields will also occur when the thermal contact between the current carriers and the host lattice can no longer handle the large amount of energy transfer required to keep the carriers in thermal equilibrium with the host lattice. This is known as the *hot carrier effect*, and it will be treated in Chapter 4. For our present discussion, however, we shall assume circumstances under which Eq. (3.5) is valid.

The principal quantities which determine the conductivity of a semiconductor are the concentrations of free carriers n and p in Eq. (3.5). These quantities depend not only on the concentrations N_d and N_a of donor and acceptor impurities but also on the temperature T which determines the degree of thermal excitation of carriers: processes I, II, and III in Figs. 3.3, 3.4, and 3.5, respectively. The resistivities of the various semiconductors Sn, Ge, Si, GaAs, and InSb given in Fig. 3.1 refer to the values obtainable in a pure crystal at room temperature. The values of the resistivity can change many orders of magnitude by changing either N_a, N_d, or T. The functional dependence of n and p on N_a, N_d, and T will be discussed in Secs. 3.4 and 3.5, pp. 140 and 146. The mobility factors μ_n and μ_p in Eq. (3.5) may not contribute as large a numerical factor as n and p in determining the value of σ but they are of fundamental importance. In a crystal, carriers are subject to an internal electric potential associated with the constituent nuclei; hence, in the mobility expression $\mu = e\tau/m$, the value of m should be replaced by an *effective mass* m^* which may differ from the free electron mass. Experimental determination of m^* will be discussed in a later section on cyclotron resonance. Free carriers also undergo many collision processes in a solid, such as scattering of electron motion by lattice vibration and by impurities. These scattering processes will be further discussed in Chapter 4 when we treat the subject of carrier transport in greater detail. The discussion given in this section serves merely as an introduction to what will follow so that the reader may acquire an overall perspective.

3.3 CONCEPTS OF MOBILITY AND DIFFUSION

The relationship between v and E given by Eq. (3.1) is conceptually much more involved than the simple expression suggests. First of all, electrons and holes in a semiconductor are in rapid motion because of their thermal energies. Using the mass of free electrons and $T = 300°\text{K}$, we have 10^5 m/sec for the order of the thermal velocity of an electron $(v = \sqrt{3kT/m})$. For germanium, the mobility for electrons $\mu_n = 3900$ cm²/volt-sec at room temperature. Hence, under an electric field $E = 10$ volts/cm, the average drift velocity is only 390 m/sec, much smaller than the thermal velocity. Figure 3.6 shows (a) the random or Brownian motion of an electron on account of its thermal energy (b) the drift motion of the electron due to the applied field and (c) the actual motion of the electron that is the combination of (a) and (b). Note that in Fig. 3.6a, the direction of motion changes randomly after each collision at various times t_2, t_3, etc. Furthermore, the time interval between successive collisions, that is, $t_4 - t_3$, $t_5 - t_4$, etc., changes from time to time, and the mean free time τ is the average time between successive collisions. Since the drift velocity is superimposed on a much larger random velocity, it does not make any sense to follow the motion of an individual electron. Instead, a meaningful average must be obtained. We can either average the motion of an electron over a period of time long compared with the mean free time or take an average of the motions of many electrons at any specific instant. Since the number of electrons involved in the averaging process is extremely large, the time average and the ensemble average should yield the same result.

We assume that the probability that an electron (or hole) makes a collision in a time interval dt is dt/τ, τ being a time constant. Let n_0 be the number of electrons which suffered a collision at time t_0 and assume that of these, n electrons have not suffered subsequent collisions during the time interval $t - t_0$; then

$$dn = -n\frac{dt}{\tau} \quad \text{or} \quad n = n_0 \exp\left(-\frac{t - t_0}{\tau}\right) \tag{3.6}$$

According to Eq. (3.6), there are $n \, dt/\tau$ electrons suffering collisions between t and $t + dt$ and for these electrons, the free time between collisions is $t - t_0$. Therefore, the mean free time for the n_0 electrons is given by

$$\langle t - t_0 \rangle_{\text{av}} = \frac{1}{n_0} \int_{t_0}^{\infty} (t - t_0) n \frac{dt}{\tau} = \tau \tag{3.7}$$

and is equal to the time constant used to describe the rate of collision processes.

Figure 3.6

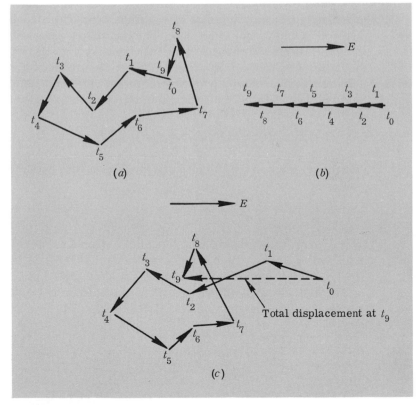

(a) (b)

(c)

Diagrams showing (a) the random motion of an electron, (b) the drift motion of an electron under an electric field, and (c) the superposition of these two motions.

The mean drift velocity can be obtained in a similar fashion. The velocity of an electron which suffered the last collision at t_0 is given by

$$v = v_0 - \frac{eE(t - t_0)}{m^*} \qquad (3.8)$$

where the first term represents the initial velocity after collision and the second term represents the velocity acquired under the applied field. Since the velocity v_0 after collision is completely random in its direction, the average of v_0 over many electrons is zero. To find the average value for v at time t, however, we need to know the distribution of electrons as a function of t_0.

Let w be the percentage of electrons that are uncollided at time t after a collision at t_0. According to Eq. (3.6)

$$w = \frac{n}{n_0} = \exp\left(-\frac{t - t_0}{\tau}\right) \qquad (3.9)$$

Note that the total number of electrons at any time t is always n_0. Therefore, we may restate the physical situation as follows. At a given time t, among the total n_0 electrons, some electrons had their last collision at t_{0_1} and some at t_{0_2}. The quantity w can also be interpreted as being the percentage of electrons at a given time t, that had their last collision at time t_0. Thus, the number of electrons having the last collision at a time interval between t_0 and $t_0 + dt_0$ and still uncollided at time t is given by

$$dn = n_0 \exp\left(-\frac{t - t_0}{\tau}\right) \frac{dt_0}{\tau} = \frac{n_0 w \, dt_0}{\tau} \qquad (3.10)$$

Therefore, the average velocity for the n_0 electrons can be found as follows:

$$\langle v \rangle = -\frac{eE}{n_0 m^*} \int_{-\infty}^{t} n_0 w(t - t_0) \frac{dt_0}{\tau} = -\frac{eE}{m^*}\tau = -\mu E \qquad (3.11)$$

The average drift velocity is independent of time as should be. The negative sign in Eq. (3.11) simply indicates that the direction of the drift velocity is opposite to that of E for electrons.

The above derivation can easily be extended to cases where the applied field is time-varying or $E = \text{Re } E_0 \exp(i\omega t)$. From the equation of motion, we have

$$v = \text{Re} - \frac{eE_0}{im^*\omega}[\exp(i\omega t) - \exp(i\omega t_0)] \qquad (3.12)$$

Carrying out a similar averaging process, we find

$$\langle v \rangle = \text{Re} - \frac{e}{m^*}\frac{\tau}{1 + i\omega\tau}E_0 \exp(i\omega t) \qquad (3.13)$$

where Re means taking the real part. Note that according to Eq. (3.13), the conductivity of a semiconductor becomes complex when $\omega\tau$ approaches unity. In germanium, τ is around 10^{-13} sec at room temperature; hence the relaxation or collision effect will not be noticeable until the frequency of the applied field exceeds that of millimeter waves. We can combine the results of Eqs. (3.11) and (3.13) into the following equation of motion for the average drift velocity $\langle v \rangle$ of electrons:

$$\frac{d\langle v \rangle}{dt} = -\frac{\langle v \rangle}{\tau} + \frac{F}{m^*} \tag{3.14}$$

where F is the force acting on free carriers. If the applied field is from a d-c source, $d\langle v \rangle/dt = 0$ and Eq. (3.14) reduces to Eq. (3.11). If the applied field is from a sinusoidally time-varying source, $d\langle v \rangle/dt = i\omega\langle v \rangle$, yielding the result of Eq. (3.13). Equation (3.14) is in a form similar to the general transport equation which will be treated in the following chapter.

In the derivation of the expression for electric current density, Eq. (3.5), nothing is said about the spatial variation of carrier concentrations, n and p. It is true that a local conductivity σ can still be defined, using Eq. (3.5), if n and p vary spatially. Our interest, however, is of a rather fundamental nature and the question is whether a nonconstant n and p in space will give rise to a new phenomenon. Consider the thermal motion of particles in a one-dimensional space as shown in Fig. 3.7 where v_{th} is the thermal velocity of particles and l is the mean free path. The number of particles moving across the boundary line A from left to right per unit time per unit area is $n_2 v_{th}/2$ while the corresponding number from right to left is $n_1 v_{th}/2$, n_2 and n_1 being the concentration of particles. The net rate of flow of particles per unit time per unit area is

$$S = v_{th}\frac{n_2 - n_1}{2} \tag{3.15}$$

Since only particles within one mean free path on each side of the boundary are able to move across A, the difference $n_2 - n_1$ in the above equation should be evaluated within such a distance, or

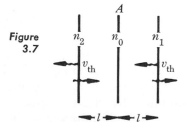

Figure 3.7

l = mean free path

v_{th} = thermal velocity

One-dimensional diffusion of free carriers.

$n_1 = n_0 + (\partial n/\partial x)l$ and $n_2 = n_0 - (\partial n/\partial x)l$ from Taylor's series expansion. Thus, Eq. (3.15) becomes

$$S = -v_{\text{th}} \frac{\partial n}{\partial x} l \qquad \text{particles/m}^2\text{-sec} \qquad (3.16)$$

If the particles under consideration are charged, then the thermal motion of these particles constitutes a current flow, the density of which is given by

$$J = -eS = ev_{\text{th}} l \frac{dn}{dx} = \frac{ekT\tau}{m^*} \frac{dn}{dx} = eD \frac{dn}{dx} \qquad \text{amp/m}^2 \quad (3.17)$$

where by definition $l = v_{\text{th}}\tau$ and $mv_{\text{th}}^2 = kT$. In a three-dimensional case, Eq. (3.17) still gives correctly the x component of the diffusion current density if v_{th} is replaced by $(v_{\text{th}})_x$. The current J defined in Eq. (3.17) is called the *diffusion current*. If a concentration gradient dn/dx exists, the natural tendency of random motion is to equalize the population of carriers if carriers are allowed to migrate freely. The coefficient D is generally called the *diffusion constant*. Note that according to Eq. (3.17), a definite relation exists between D and μ or $D = kT\tau/m^* = kT\mu/e$. This is known as the *Einstein relation*. A formal proof of the Einstein relation will be given in Chapter 5 when the drift and diffusion currents, Eqs. (3.5) and (3.17) respectively, will be derived from the Boltzmann transport equation. What we have not taken into consideration in the derivation of Eqs. (3.5) and (3.17) is the statistical distribution of carriers over different energy states. The Boltzmann transport equation not only takes the statistical nature of carriers into account but incorporates the time variation, such as that included in Eq. (3.14), of transport properties. The purpose of this section is to present a physical picture which leads to the derivation of Eqs. (3.5) and (3.17). With the physical picture in mind, the reader may feel more receptive to the more abstract approach in the formulation of the Boltzmann transport equation.

3.4 CONCENTRATIONS OF FREE CARRIERS AND IONIZED IMPURITIES IN NONDEGENERATE SEMICONDUCTORS

The concentrations of electrons and holes in a semiconductor, n and p, are determined by the distribution of electrons among the various energy states, the valence and conduction band states and the acceptor and donor impurity states. As discussed earlier in Sec. 3.1, electrons in a covalent semiconductor interact very strongly with other electrons and they are mutually exclusive so far as occupying a given energy state is concerned. The mutual exclusiveness commonly known as the Pauli exclusion principle leads naturally to the use of Fermi-Dirac statistics discussed in Sec. 1.12. Furthermore, it is assumed that the various energy states are in thermodynamic equilibrium, hence a common Fermi level is used for all the energy states. The derivation of carrier concentrations as functions of $\mathcal{E}_f$ is quite similar to that of Eq. (1.146) except for the following modifications. First, effective masses must be used instead of the free electron mass. Second, in a semiconductor, conduction and valence bands are separated by an energy gap so the energies of electrons in these two bands should be referred to their respective band edges. Figure 3.8 shows the relationship of the Fermi function $f(\mathcal{E})$ and the density of states function $g(\mathcal{E})$ with respect to the energy bands, the function $f(\mathcal{E})$ indicating the occupancy of a given state, while the function $g(\mathcal{E})\,d\mathcal{E}$, the density of available states within an energy interval $d\mathcal{E}$. Before we evaluate the concentrations of free carriers, the following points must be made clear.

First, for nondegenerate semiconductors, the Fermi level is somewhere in the forbidden gap. For electrons in the conduction band, the condition $(\mathcal{E} - \mathcal{E}_f)/kT > 1$ is satisfied; therefore, the Fermi function can be approximated by:

$$f_c(\mathcal{E}) \simeq \exp\left(\frac{\mathcal{E}_f - \mathcal{E}_c - \mathcal{E}_1}{kT}\right) \tag{3.18}$$

where $\mathcal{E}_1 = \mathcal{E} - \mathcal{E}_c$. For electrons in the valence band, the opposite, $(\mathcal{E}_f - \mathcal{E})/kT > 1$, is true. However, we are interested in the missing electrons in the valence band, or in other words the function $1 - f_v(\mathcal{E})$, which can be approximated by

$$1 - f_v(\mathcal{E}) = \frac{1}{1 + \exp\left[(\mathcal{E}_f - \mathcal{E})/kT\right]} \simeq \exp\left(\frac{\mathcal{E}_v - \mathcal{E}_f - \mathcal{E}_2}{kT}\right) \tag{3.19}$$

where $\mathcal{E}_2 = \mathcal{E}_v - \mathcal{E}$. Next, in view of the free electron theory dis-

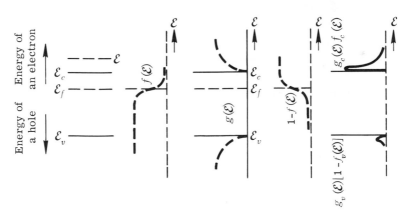

Figure 3.8

Plot of the Fermi function $f(\mathcal{E})$ with respect to the band edges. To calculate the electrons (occupied states) in the conduction band, $f(\mathcal{E})$ is used; to calculate the holes (empty states) in the valence band, the function $1 - f(\mathcal{E})$ is used.

cussed in Sec. 2.4, we shall assume that the electrons in the conduction band behave more or less like a free electron with an effective mass m_e, thus $\varepsilon_1 = p^2/2m_e$, and the electron concentration in the conduction band is given by

$$n = \int_{\varepsilon_c}^{\infty} f(\varepsilon)g(\varepsilon)\,d\varepsilon = \frac{8\pi}{h^3}\exp\left(-\frac{\varepsilon_c - \varepsilon_f}{kT}\right)\int_0^{\infty} p^2 \exp\left(-\frac{\varepsilon_1}{kT}\right)dp$$

$$= N_c \exp\left(-\frac{\varepsilon_c - \varepsilon_f}{kT}\right) \qquad \text{electrons/m}^3 \quad (3.20)$$

In obtaining Eq. (3.20), we have used the definite integral

$$2\int_0^{\infty} \exp\left(-\alpha u^2\right)du = (\pi/\alpha)^{1/2}$$

Since $\exp\left(-\varepsilon_1/kT\right)$ drops very fast with energy, the upper limit of integration can be taken as infinity. A quantity N_c defined as

$$N_c = 2\left(\frac{2\pi m_e kT}{h^2}\right)^{3/2} \qquad (3.21)$$

is generally referred to as the effective density of states in the conduction band, and it has a dimension of m^{-3}.

The calculation for the valence band is conceptually more difficult than that for the conduction band. For the time being, we shall take for granted that a hole behaves like a positive charge and hence the energy band diagram for holes is the one for electrons turned upside down. With this sign reversal, it is natural then to interpret ε_2 as the kinetic energy of a hole or $\varepsilon_2 = p^2/2m_h$ where m_h is the effective mass of a hole. We shall return to the concept of a hole with further clarifications in a later section. Once we accept this concept, we can readily obtain the following expression:

$$p = N_v \exp\left(-\frac{\varepsilon_f - \varepsilon_v}{kT}\right) \qquad \text{holes/m}^3 \qquad (3.22)$$

where N_v is the effective density of states in the valence band defined as

$$N_v = 2\left(\frac{2\pi m_h kT}{h^2}\right)^{3/2} \qquad (3.23)$$

and p is the concentration of holes or missing electrons in the valence band. If we take the mass of free electrons for m_e and m_h, then the values of N_c and N_v are 2.5×10^{19}/cm^3 at 300°K.

Next we shall discuss the distribution of electrons among the donor and acceptor states. The Fermi functions for these states are slightly different from those for the conduction and valence bands. Take the donor state as an example. Since four out of five valence electrons are tied in the covalent bond, we shall focus our attention on the extra valence electron. We shall use the word *occupied* or *unoccupied* to indicate whether the extra electron is with the donor atom or free to move about in the crystal. Therefore, the state which a donor impurity is in can be defined by the following equation:

Donor (electrically neutral, occupied by the extra electron) $\rightleftharpoons$ donor+ (positively charged, unoccupied) + a free electron   (3.24)

Equation (3.24) is similar to Eq. (3.3) except that now the emphasis is on the extra electron. Unlike the four electrons which form the covalent bond, the extra electron has a choice of two possible spin orientations. Let N_d be the concentration of donor impurities and n_d be the concentration of occupied donor states. The number of ways W_1 of arranging n_d electrons among N_d donor impurities is given by

$$W_1 = \frac{N_d!}{(N_d - n_d)!n_d!} 2^{n_d} \qquad (3.25)$$

The extra factor 2 to the power of n_d accounts for two possible spin orientations.

Let us take an extrinsic n-type semiconductor as an example. For simplicity, we shall pretend that acceptor and valence band states do not exist and consider the distribution of those extra electrons of the donor impurities only among the donor and conduction band states. We want to know how many of those extra electrons will still be attached to the donor states and how many will appear in the conduction band. The total number of ways W of arranging the N_d extra electrons is given by

$$W = W_1 W_2 \qquad (3.26)$$

where W_2 is the number of ways of arranging electrons in the conduction band

$$W_2 = \prod_s \frac{g_s!}{n_s!(g_s - n_s)!} \qquad (3.27)$$

which is identical to Eq. (1.136). Maximization[1] of W is subject to the following auxiliary conditions:

$$\sum_s n_s + n_d = N_d$$
$$\sum_s n_s \mathcal{E}_s + n_d \mathcal{E}_d = U \qquad (3.28)$$

---

[1]   Using the Boltzmann definition of entropy $S = k \log W$, the reader can show that the use of Lagrange's multipliers in Sec. 1.10 in the maximization of W is equivalent to the minimization of the free energy $F = U - TS$ of the system.

Following the same procedure as given in Sec. 1.12, we find, for maximum W,

$$\frac{n_d}{N_d} = \frac{1}{1 + \frac{1}{2} \exp(\beta + \alpha\mathcal{E}_d)} = f_d(\mathcal{E})$$

$$\frac{n_s}{g_s} = \frac{1}{1 + \exp(\beta + \alpha\mathcal{E})} = f_c(\mathcal{E}) \qquad (3.29)$$

According to our discussion in Sec. 1.12, $\alpha = 1/kT$ and $\beta = -\mathcal{E}_f/kT$ in Eq. (3.29).

The significant results from Eq. (3.29) are the following. First, a factor $\frac{1}{2}$ now appears in front of the exponential factor in $f_d(\mathcal{E})$ as a result of the twofold spin degeneracy. Second, the quantity β is the same in both Fermi functions, $f_d(\mathcal{E})$ for the donor states and $f_c(\mathcal{E})$ for the conduction band states. This justifies the use of a common Fermi level in Eqs. (3.18) and (3.19). Finally, taking donor states into consideration does not change the form of the Fermi function $f_c(\mathcal{E})$ for the conduction band states. In other words, we can treat the distribution of electrons among the acceptor and valence band states as if donor and conduction band states do not exist in deriving the Fermi functions, $f_a(\mathcal{E})$ and $f_v(\mathcal{E})$, for these states.

For the acceptor states, an additional electron is needed to complete the covalent bond. We shall again use the word *occupied* or *unoccupied* to mean whether the added electron is with the acceptor atom or not as illustrated by the following equation:

Acceptor (electrically neutral, unoccupied)
$+$ an added electron $\rightleftharpoons$ acceptor⁻
(negatively charged, occupied by the added electron)   (3.30)

Let N_a be the concentration of acceptor impurities and n_a be the concentration of occupied acceptor states. Then, the number of ways W_3 of arranging $N_a - n_a$ missing electrons among N_a acceptor impurities is given by

$$W_3 = \frac{N_a!}{(N_a - n_a)!n_a!} 2^{N_a - n_a} \qquad (3.31)$$

The factor 2 to the power of $N_a - n_a$ enters into Eq. (3.31) on account of two possible spin orientations for the missing electrons.[2]

---

[2] We should point out that in Eqs. (3.25) and (3.31), only spin degeneracy has been taken into account. This may be an oversimplification. In group IV elemental and III-V compound semiconductors, the conduction and valence bands are known to have s-like and p-like wave functions. The p-state is triply degenerate for $l = 1$. However. the degeneracy may be partly removed by the crystalline field and the spin-orbit coupling. Therefore, the degeneracy factor for the acceptor states can be 2, 4, or 6 in Eqs. (3.31) and (3.32), depending on the degeneracy of the valence band states. We take the degeneracy factor as 2 for acceptor states merely for the purpose of illustration.

The remaining n_a missing electrons of course are distributed among the valence band states. Maximization of the total number of ways $W = W_3 W_4$ of arranging N_a missing electrons among the acceptor impurities (W_3 ways) and the valence band states (W_4 ways) is achieved when

$$\frac{n_a}{N_a} = \frac{1}{1 + 2 \exp (\beta + \alpha \mathcal{E}_a)} = f_a(\mathcal{E}) \qquad (3.32)$$

for the acceptor states and

$$\frac{g_s - n_s}{g_s} = \frac{\exp (\beta + \alpha \mathcal{E})}{1 + \exp (\beta + \alpha \mathcal{E})} = 1 - f_v(\mathcal{E}) \qquad (3.33)$$

for the valence band states. In Eqs. (3.32) and (3.33), $\alpha = 1/kT$ and $\beta = -\mathcal{E}_f/kT$.

Now we can put everything together. The equation which ties everything together is the *charge neutrality equation*. Since electrons are merely redistributed among the various energy states in a semiconductor but not taken out of or put into the semiconductor, the crystal should remain electrically neutral. The equation which states this charge neutrality condition reads

$$n + N_a^- = p + N_d^+ \qquad (3.34)$$

On the left-hand side of Eq. (3.34), n represents the negative charge density due to conduction band electrons and N_a^- ($= n_a$) the charge density due to charged or occupied acceptor states. On the right-hand side, p is the positive charge density due to valence band holes and N_d^+ ($= N_d - n_d$) the charge density due to ionized or unoccupied donor states. Using Eqs. (3.20), (3.22), (3.29), and (3.32), we have

$$N_c \exp \left(-\frac{\mathcal{E}_c - \mathcal{E}_f}{kT} \right) + \frac{N_a}{1 + 2 \exp [(\mathcal{E}_a - \mathcal{E}_f)/kT]}$$

$$= N_v \exp \left(-\frac{\mathcal{E}_f - \mathcal{E}_v}{kT} \right) + \frac{N_d}{1 + 2 \exp [(\mathcal{E}_f - \mathcal{E}_d)/kT]} \qquad (3.35)$$

The above equation is a general, implicit expression which relates the Fermi level $\mathcal{E}_f$ to the impurity concentrations N_a and N_d at a given temperature T. Once $\mathcal{E}_f$ is known from Eq. (3.35), the free carrier concentrations n and p can be obtained from Eqs. (3.20) and (3.22). The actual evaluation of n and p will be discussed in the next section.

In conclusion, we should like to emphasize that the proper way of deriving the four Fermi functions, $f_d(\mathcal{E})$, $f_c(\mathcal{E})$, $f_a(\mathcal{E})$, and $f_v(\mathcal{E})$, is to consider the total number of ways $W = W_1 W_2 W_3 W_4$ of arranging the available electrons among the available donor, conduction band, acceptor, and valence band states. The forms of the various Fermi functions are the same irrespective of whether we take the product of $W_1W_2W_3W_4$ all at once or we treat each separately. The link which ties the four Fermi functions together is a common Fermi level. Therefore, the value of the Fermi function will be affected by the presence of other states even though the form is not. In other words, to find the form of the Fermi function for states under consideration we may ignore the other states, but in determining the distribution of electrons among the various energy states, we must include all energy states by requiring a common Fermi level for all these states. Finally, we should mention that in determining n and p, Eq. (3.34) or Eq. (3.35) should be used. Simplification of these equations is possible but bifurcation into two separate equations $n = N_d^+$ and $p = N_a^-$ is not permissible.

3.5 EVALUATION OF FERMI LEVEL AND CARRIER CONCENTRATIONS IN NONDEGENERATE SEMICONDUCTORS

Before discussing how the carrier concentrations n and p are evaluated in a given semiconductor, we note from Eqs. (3.20) and (3.22) that the product

$$ pn = N_cN_v \exp \left(-\frac{\mathcal{E}_c - \mathcal{E}_v}{kT} \right) = n_i^2 \qquad (3.36) $$

depends only on the intrinsic property of the semiconductor. In a pure (intrinsic) semiconductor where $N_a = N_d = 0$, Eqs. (3.34) and (3.36) lead to $n = p = n_i$. In other words, n_i gives the correct value of carrier concentrations in an intrinsic semiconductor; hence, n_i is generally referred to as the *intrinsic carrier concentration*. Let us now return to Eq. (3.35). We realize that in an impure semiconductor, a general solution for the Fermi level will be quite involved; therefore, it is desirable that approximations be made in Eq. (3.35). As will be seen shortly, the quantity n_i will be used as a guide in making such approximations.

Next, we shall examine the donor and acceptor states. To determine the extent to which these states are ionized, the following observations may be helpful. The ionization energies, $\mathcal{E}_c - \mathcal{E}_d$ and $\mathcal{E}_a - \mathcal{E}_v$, are around 0.01 ev in germanium and around 0.04 ev in silicon. For comparison, the thermal energy kT at room temperature is about 0.025 ev. Therefore, at room temperature and above, all the donor and acceptor impurities in germanium are almost completely ionized or in other words, $N_a^- \simeq N_a$ and $N_d^+ \simeq N_d$ in Eq. (3.34). At very low temperatures, however, the impurities are only partly ionized and hence the above approximations are no longer true. According to the value of n_i in relationship to those of N_a and N_d and also the degree of ionization of the impurity states, the behavior of n and p can be generally divided into three temperature regions: high, intermediate, and low.

Let us take an n-type sample in which $N_d \gg N_a$ as an example. In both the high and intermediate temperature regions, we may take $N_a^- = N_a$ and $N_d^+ = N_d$. Thus Eq. (3.34) becomes

$$n - p = N_d - N_a \tag{3.37}$$

Elimination of n or p from Eqs. (3.37) and (3.36) gives a quadratic equation in p or n, respectively, of which the solutions are

$$n = \frac{N_d - N_a}{2} + \sqrt{\left(\frac{N_d - N_a}{2}\right)^2 + n_i^2}$$

$$p = -\frac{N_d - N_a}{2} + \sqrt{\left(\frac{N_d - N_a}{2}\right)^2 + n_i^2} \tag{3.38}$$

If $n_i \gg (N_d - N_a)/2$, Eq. (3.38) reduces to

$$n \simeq n_i + \frac{N_d - N_a}{2}$$

$$p \simeq n_i - \frac{N_d - N_a}{2} \tag{3.39}$$

This corresponds to the high temperature case. If $n_i \ll (N_d - N_a)/2$, Eq. (3.38) can then be approximated by

$$n \simeq N_d - N_a + \frac{n_i^2}{N_d - N_a}$$

$$p \simeq \frac{n_i^2}{N_d - N_a} \tag{3.40}$$

Equation (3.40) corresponds to the intermediate temperature case.

The Fermi level in these temperature ranges can easily be obtained by relating n given by Eq. (3.39) or (3.40) to that of Eq. (3.20). In doing so, we shall neglect the second order terms in Eqs. (3.39) and (3.40). Furthermore, it is often convenient to refer the Fermi level to a reference energy level called the *intrinsic Fermi level*, $\mathcal{E}_{fi}$, which is defined by the following relation:

$$N_c \exp\left(-\frac{\mathcal{E}_c - \mathcal{E}_{fi}}{kT}\right) = N_v \exp\left(-\frac{\mathcal{E}_{fi} - \mathcal{E}_v}{kT}\right)$$

$$= \sqrt{N_c N_v} \exp\left(-\frac{\mathcal{E}_c - \mathcal{E}_v}{2kT}\right) \tag{3.41}$$

If $N_c = N_v$, $\mathcal{E}_{fi}$ will be exactly in the middle of the energy gap. In terms of $\mathcal{E}_{fi}$, we find

$$n = n_i \exp\left(-\frac{\mathcal{E}_{fi} - \mathcal{E}_f}{kT}\right)$$

$$p = n_i \exp\left(-\frac{\mathcal{E}_f - \mathcal{E}_{fi}}{kT}\right) \tag{3.42}$$

Obviously, in the high temperature region $\mathcal{E}_f \simeq \mathcal{E}_{fi}$. In the intermediate temperature range, $\mathcal{E}_f$ can be obtained by equating n in Eq. (3.40) to that in Eq. (3.42) or (3.20), thus

$$\mathcal{E}_f = \mathcal{E}_{fi} + \frac{\mathcal{E}_c - \mathcal{E}_v}{2} - \frac{kT}{2} \ln \frac{N_c N_v}{(N_d - N_a)^2}$$

$$= \mathcal{E}_c - kT \ln \frac{N_c}{N_d - N_a} \qquad (3.43)$$

Knowing $\mathcal{E}_f$, we can calculate from Eqs. (3.29) and (3.32) the error introduced by making $N_a^- = N_a$ and $N_d^+ = N_d$.

Now we shall discuss the low temperature case. Note that for a nondegenerate semiconductor, $N_d - N_a$ is smaller than N_c. According to Eq. (3.43), therefore, as the temperature decreases, the Fermi level of an n-type semiconductor moves toward the conduction band. As a consequence, the approximation $N_d^+ \simeq N_d$ becomes poorer and poorer, the value of p becomes smaller and smaller but the approximation $N_a^- \simeq N_a$ remains good. Thus, Eq. (3.35) can be approximated by

$$N_c \exp\left(-\frac{\mathcal{E}_c - \mathcal{E}_f}{kT}\right) + N_a = N_d \left[1 + 2 \exp\left(\frac{\mathcal{E}_f - \mathcal{E}_d}{kT}\right)\right]^{-1} \qquad (3.44)$$

It may be advisable to consider the simplest case first with $N_a = 0$. If the following substitutions are made: $\lambda = \exp(\mathcal{E}_f/kT)$, $\alpha = \exp(-\mathcal{E}_d/kT)$, $\beta = \exp(-\mathcal{E}_c/kT)$ and $b = N_d/N_c$; then Eq. (3.44) becomes

$$2\alpha\beta\lambda^2 + \beta\lambda - b = 0 \qquad (3.45)$$

The solution of Eq. (3.45) is

$$\lambda = \frac{1}{4\alpha}\left(-1 + \sqrt{1 + \frac{8b\alpha}{\beta}}\right) \qquad (3.46)$$

Depending on whether

$$\beta > 8b\alpha \qquad \text{or} \qquad \beta < 8b\alpha \qquad (3.47)$$

$$\lambda = \frac{b}{\beta} \qquad \text{or} \qquad \lambda = \left(\frac{b}{2\alpha\beta}\right)^{\frac{1}{2}} \qquad (3.48)$$

From Eq. (3.48), we have for $\beta > 8b\alpha$

$$n = N_c\beta\lambda = N_d \qquad (3.49)$$

and for $\beta < 8b\alpha$

$$n = N_c\beta\lambda = \sqrt{\frac{N_c N_d}{2}} \exp\left(-\frac{\mathcal{E}_c - \mathcal{E}_d}{2kT}\right) \qquad (3.50)$$

Equation (3.49) corresponds to the intermediate and Eq. (3.50) to the low temperature case.

It seems useful to summarize all the information accumulated from the above discussion for the case $N_a = 0$. The limits for which the different temperature ranges are defined can be obtained from the different inequalities [Eq. (3.47) for example] used in making the approximations, while the Fermi levels are obtained by equating n from Eq. (3.50) to that from Eq. (3.20). Table 3.1 summarizes all the relevant quantities for the behavior of a nondegenerate n-type semiconductor with $N_d < N_c$ and $N_a = 0$ as a function of temperature.

TABLE 3.1 *The variation of the carrier concentrations and the Fermi level in the three temperature regions for an n-type semiconductor with $N_a = 0$*

1. In the high temperature region

$$kT > \frac{\mathcal{E}_g}{\ln (N_c N_v / N_d^2)} = kT_1 \tag{3.51}$$

$$\mathcal{E}_f = \mathcal{E}_{fi} \qquad \text{and} \qquad n = p = n_i \tag{3.52}$$

2. In the intermediate temperature region

$$\frac{\mathcal{E}_g}{\ln (N_c N_v / N_d^2)} > kT > \frac{\mathcal{E}_c - \mathcal{E}_d}{\ln (N_c / 8 N_d)} \tag{3.53}$$

$$\mathcal{E}_f = \mathcal{E}_c - kT \ln \frac{N_c}{N_d} \tag{3.54}$$

$$n = N_d \qquad p = \frac{n_i^2}{N_d} \tag{3.55}$$

3. In the low temperature region

$$kT < \frac{\mathcal{E}_c - \mathcal{E}_d}{\ln (N_c / 8 N_d)} = kT_2 \tag{3.56}$$

$$\mathcal{E}_f = \frac{\mathcal{E}_c + \mathcal{E}_d}{2} - \frac{kT}{2} \ln (2 N_c / N_d) \tag{3.57}$$

$$n = \sqrt{\frac{N_c N_d}{2}} \exp \left(-\frac{\mathcal{E}_c - \mathcal{E}_d}{2kT} \right) \tag{3.58}$$

$$p = \sqrt{\frac{2 N_c}{N_d}} \, N_v \exp \left(-\frac{\mathcal{E}_c + \mathcal{E}_d - 2 \mathcal{E}_v}{2kT} \right) \tag{3.59}$$

The Fermi level $\mathcal{E}_f$ and the majority carrier concentration n are plotted as a function of temperature and reciprocal temperature respectively in Fig. 3.9. In the high temperature region $T > T_1$, intrinsic carriers due to thermal excitation dominate over carriers released from impurities and hence the region is generally referred to as the *intrinsic* region. In the intermediate and low temperature regions, majority carriers come from ionization of impurities, and hence these regions are called *extrinsic* regions. Furthermore, in the

Figure
3.9

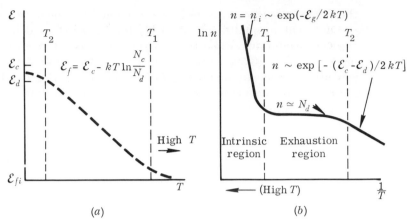

Curves showing the general dependences of (a) the Fermi level and (b) the majority carrier concentration n as functions of temperature in an n-type sample with $N_a = 0$.

intermediate temperature region, practically all impurities are ionized, and thus the region is sometimes called the *exhaustion* region.

Now we shall return to the case in which impurities of one kind, say donors, are partially compensated by impurities of the opposite kind, say acceptors, with $N_d > N_a$. In the high and intermediate temperature regions, we have already discussed the situation fully by Eqs. (3.38) through (3.44). However, we shall explain the physical process involved which leads to Eq. (3.40), and thus we may see what further approximations can be made to Eq. (3.44) in the low temperature region. Since p is by far the smallest quantity in Eq. (3.37), we shall ignore the valence band as we did in Eq. (3.44). The extra electrons released from the ionized donors are shared by the acceptor and conduction-band states as shown in Fig. 3.10a. Since the acceptor states are the lower energy states, they will be filled first. If the N_d donor impurities are completely ionized, of the N_d electrons released from donors, N_a electrons go to acceptor states, leaving $N_d - N_a$ electrons for the conduction band. This is exactly what Eq. (3.40) says. As the Fermi level approaches the donor energy level $\mathcal{E}_d$, fewer and fewer donor impurities are ionized. However, electrons must still be supplied to fill the acceptors. According to Eq. (3.29), if $\mathcal{E}_f = \mathcal{E}_d$, the density of ionized donor is equal to $N_d - n_d = N_d/3$. Thus, two quite different situations arise, depending on whether N_a is greater or smaller than $N_d/3$.

If $N_d > N_a > N_d/3$, then the Fermi level can never cross over the donor level because if it did, the ionized donor density would be smaller than the charged acceptor density, or in other words, $N_d^+ < N_a^-$. Since the donor impurities are the only source of electrons, the situation $N_d^+ < N_a^-$ is physically unrealizable as it is obvious from Fig. 3.10b. What happens in this case is that as the temperature decreases, the Fermi level is almost frozen close to but below $\mathcal{E}_d$ under the requirement $N_a^- < N_d^+$. The limit of the low temperature region is determined by the condition $n <$ or $> N_a$. For the former case, we can ignore n in Eq. (3.44); thus, we have

$$\mathcal{E}_f = \mathcal{E}_d - kT \ln \frac{2N_a}{N_d - N_a} \tag{3.60}$$

and from Eqs. (3.20) and (3.22)

$$n = \frac{N_c(N_d - N_a)}{2N_a} \exp\left(-\frac{\mathcal{E}_c - \mathcal{E}_d}{kT}\right) \tag{3.61}$$

$$p = \frac{2N_aN_v}{N_d - N_a} \exp\left(-\frac{\mathcal{E}_d - \mathcal{E}_v}{kT}\right) \tag{3.62}$$

Equations (3.60), (3.61), and (3.62) are valid if the temperature satisfies the following equation:

$$kT < \frac{\mathcal{E}_c - \mathcal{E}_d}{\ln\left[N_c(N_d - N_a)/2N_a^2\right]} = kT_2 \tag{3.63}$$

obtained from the condition $n < N_a$. Equations (3.60) and (3.63) should replace the corresponding quantities in Table 3.1 if $N_d > N_a > N_d/3$. The case with $N_a < N_d/3$ does not merit any further discussion because it is quite similar to the case with $N_a = 0$. The detailed analysis of such a case is left for the reader as a problem.

Note that the temperature dependence of the majority carrier concentration gives the activation energy of the dominant impurities. However, the question still remains as to whether the activation energy is equal to or only half the ionization energy of these impurities as the temperature dependences of Eqs. (3.50) and (3.61)

Figure 3.10

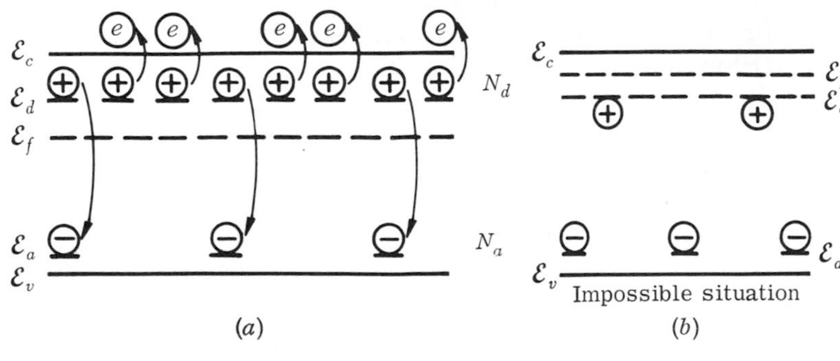

(a)　　　　(b)

Schematic illustration showing the distribution of electrons among available energy states in a compensated semiconductor (a semiconductor having both donor and acceptor impurities).

differ by a factor of 2 in the exponent. This is determined by the degree of compensation of impurities. By compensation of impurities we mean that the effect of one kind of impurity is partially canceled by the presence of the opposite kind. Because of compensation, a high resistivity does not necessarily mean a high degree of purity. A semiconductor can have a very high resistivity if it is highly compensated in such a way that the Fermi level is locked somewhere in the middle of the forbidden gap. This is possible with impurities which have energy levels deep in the forbidden gap. We should emphasize again that the carrier concentrations are determined by the Fermi level. The reader should be very careful about using a simple expression like Eq. (3.40) which is often seen in elementary textbooks on semiconductors to the exclusion of better expressions. If deep impurity levels exist in such large quantities that the condition $N_d > N_a > N_d/3$ or $N_a > N_d > N_a/3$ holds, then either Eq. (3.61) or the corresponding equation for the latter case may be needed. The reader should be especially cautious with large-gap semiconductors, such as Si, GaAs, and CdS, of the degree of compensation in such samples.

3.6 RESISTIVITY AND HALL MEASUREMENTS

In determining the electrical conductivity of a semiconductor, we must know two basic parameters, the concentrations of free carriers n and p and the mobilities of free carriers μ_n and μ_p. Now we shall discuss experiments which lead to the measurements of these parameters. To avoid spurious results due to nonuniform distribution of current and nonlinear behavior of the end contacts, the sample used in such measurements preferably has a shape shown in Fig. 3.11a. The broad areas on both ends are used for the current leads, the center arms for the Hall measurements and either pair of the outside arms on the same side for resistivity measurements. A simpler arrangement for resistivity measurement is the four-point-probe method shown in Fig. 3.11b. On account of nonuniform distribution of current, a correction factor must be used in finding the resistivity from the measured values of the total current I through the outer probes and the voltage V developed across the inner probes. Depending on the orientation of the probes with respect to the geometry of the sample, the pattern of current flow and hence the correction factor change. In a paper[3] by Valdes, the correction factor has been worked out as functions of l/s and t/s, the meaning of l, t, and s being self-evident from Fig. 3.11b. Since the resistivity measurement itself is straightforward, no further discussion is needed. Just as a reminder, it may be timely to point out that the measurement yields information on $\sigma = e\mu_n n + e\mu_p p$ of Eq. (3.5).

[3] L. B. Valdes, *Proc. IRE*, vol. 42, p. 420, 1954.

**Figure
3.11**

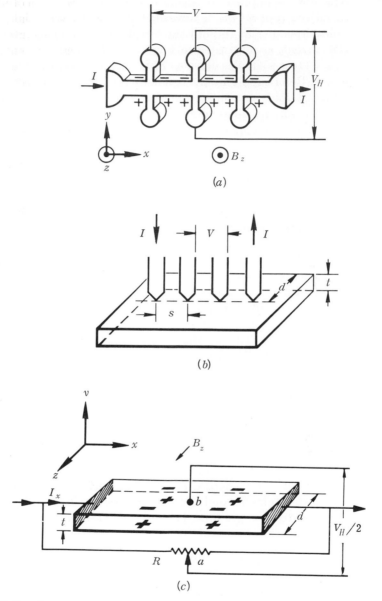

(a)

(b)

(c)

*Hall and resistivity measurements. The polarity of the Hall voltage is that for
a p-type semiconductor.*

When a magnetic field, B_z, perpendicular to the direction of current flow is applied to the sample of Fig. 3.11a, a voltage V_H, called the *Hall voltage*, is developed across the sample in the direction perpendicular to both the directions of the magnetic field and the current, as illustrated in Fig. 3.11a. The arrangement shown in Fig. 3.11c is a simpler version of Fig. 3.11a for Hall voltage measurements. Before applying the magnetic field, the potentiometer R is so adjusted that no voltage appears across terminals a and b. This ensures that terminals a and b are on the same equipotential surface. Under an applied magnetic field, the potentials of the top and bottom surfaces of the sample are shifted in opposite directions because of the Hall effect while the potential of terminal a remains the same. Consequently, the voltage developed across terminals a and b of Fig. 3.11c when a magnetic field is applied is only half the Hall voltage measured in the arrangement of Fig. 3.11a. We should mention that either arrangement, Fig. 3.11a or c, is simple enough if no further precaution against other effects is needed. Unfortunately, other effects do exist. For example, an additional voltage difference between different parts of the sample may arise if the temperature is not uniform over the sample. The effect known as the *thermoelectric effect* will be discussed in Chapter 4. In cases where the Hall voltage is small, not only should the undesirable effects be minimized but also experimental schemes should be worked out so that the Hall effect can be separated out from the other effects. For example, a reversal of the magnetic field changes the direction of the Hall voltage but not that of the ohmic drop.

Now we shall discuss the kind of information which can be obtained from the Hall voltage. Consider an extrinsic semiconductor in which one type of carrier dominates, say $p \gg n$. Under the influence of a magnetic field, as shown in Fig. 3.11a, the holes are driven towards the bottom surface of the specimen, leaving the bottom surface positively charged and the top surface negatively charged. Under open-circuit conditions, the electrostatic field, E_y, due to the accumulated charges must balance out the Lorentz force due to the magnetic field. This leads to the following equation:

$$eE_y = ev_xB_z \qquad \text{or} \qquad E_y = J_xB_z/pe \qquad (3.64)$$

where J_x is the current density. The ratio defined as

$$R_H = \frac{E_y}{J_xB_z} = \frac{1}{pe} \qquad (3.65)$$

is called the *Hall coefficient* or Hall constant and it is directly related to the free carrier concentration p. For an n-type extrinsic semiconductor, p should be replaced by n in Eq. (3.65) and the direction of the Hall field in Fig. 3.11 should be reversed. In terms of the experimentally measured quantities and the geometry of the sample of Fig. 3.11c, Eq. (3.65) becomes

$$R_H = \frac{V_H d}{BI} = \frac{1}{pe}$$ (3.66)

where V_H is the open-circuit Hall voltage and I is the total current. It is important to note that for semiconductors in the extrinsic region, the product of R_H and σ gives the mobility of the dominant carriers

$$R_H \sigma = \mu_p \text{ for } p\text{-type or } \mu_n \text{ for } n\text{-type}$$ (3.67)

Figures 3.12 and 3.13 show the measurements by Debye and Conwell of the resistivities and Hall coefficients of several germanium samples doped with various arsenic concentrations. In these samples, the acceptor concentration is much smaller than N_d; hence the equations in Table 3.1 apply. Take sample 51 as an example. The curve $ABCDE$ is divided into three regions: (1) the high temperature region AB, (2) the intermediate temperature region BC and (3) the low temperature region DE. Note that the low

Figure 3.12

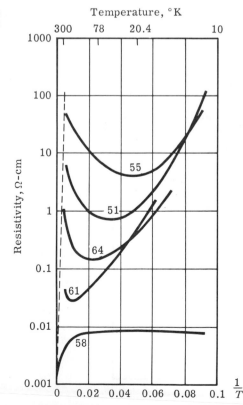

Plot of resistivity as a function of temperature in several germanium samples. (P. P. Debye and E. M. Conwell, Phys. Rev., vol. 93, p. 693, 1954.)

temperature region is expanded relative to the high and inter-mediate temperature regions on the $1/T$ scale, and actually we can see the gradual transition region CD from the intermediate to the low temperature case. The Hall coefficient in the region BC of Fig. 3.13 gives a value of $N_d = 1.4 \times 10^{14}/\text{cm}^3$. The slope of the high temperature part (AB) of the curve gives the thermal energy gap ($\mathcal{E}_c - \mathcal{E}_v$) which is 0.665 ev for germanium and the slope of the low temperature part (DE) of the curve gives the donor state ionization energy which is about 0.0125 ev for arsenic in germanium. Using Eq. (3.56) and a value of $N_c = 10^{19}/\text{cm}^3$ for Ge, we find the temperature of transition from the intermediate to low temperature case, $T_2 = 16°\text{K}$ which is marked on the curve $ABCDE$.

Now we shall make some brief comments about the other curves of Figs. 3.12 and 3.13. The curves from top to bottom are given with increasing impurity concentration. As N_d increases, the transition temperature T_2 moves toward higher temperatures in agreement with Eq. (3.56). A more accurate determination of the ionization energy $\mathcal{E}_c - \mathcal{E}_d$ is obtained by fitting curves of Fig. 3.13 to the more exact equation (3.44) instead of the approximate equation (3.50).

Figure 3.13

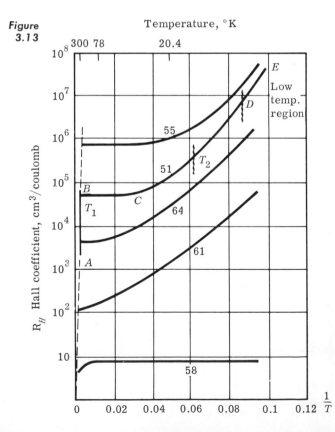

Plot of Hall coefficient as a function of temperature in several germanium samples. (P. P. Debye and E. M. Conwell, Phys. Rev., vol. 93, p. 693, 1954.)

In terms of n, Eq. (3.44) becomes

$$\frac{2(n + N_a)n}{N_d - N_a - n} = N_c \exp\left(-\frac{\mathcal{E}_c - \mathcal{E}_d}{kT}\right) \qquad (3.68)$$

after elimination of $\mathcal{E}_f$ through the use of Eq. (3.20). Figure 3.14 shows the values of $\mathcal{E}_c - \mathcal{E}_d$ obtained from a fit of Fig. 3.13 to Eq. (3.68) as a function of the impurity concentrations of the various Ge samples. As the impurity concentration increases beyond $10^{14}/\text{cm}^3$, the ionization energy begins to drop. The ionization energy finally goes to zero for $N_d > 2 \times 10^{17}/\text{cm}^3$. At such an impurity concentration, the donor energy level is no longer discrete but merges into the conduction band, signifying the formation of an impurity band[4] mentioned in Sec. 3.1. This is the reason why the Hall coefficient of sample 58 stays constant in the low temperature region. It is to be expected that impurity band conduction will become significant in sample 58 at low temperatures.

Figure 3.15 shows the mobility of electrons in the various germanium samples deduced from the curves in Figs. 3.12 and 3.13 through the use of Eq. (3.67). Since the theory concerning mobility will be treated in detail in Chapter 4, only the general features of Fig. 3.15 will be discussed here. For the purest sample, number 55, the mobility varies as $T^{-1.66}$ over quite a wide temperature range. The theoretical temperature dependence of mobility is $T^{-3/2}$ if the scattering of electron motion is due to lattice vibration. It is generally believed that in pure samples, lattice scattering dominates. As impurity concentration increases, however, not only the mobility drops but also it deviates further from the $T^{-3/2}$ dependence. The effect of impurity on mobility can be seen more clearly from the curve of Fig. 3.16 which is taken from Fig. 3.15 at 300°K and even a more drastic effect is expected at lower temperatures. We can sum-

---

[4] For a discussion of impurity band conduction, the reader is referred to the article by E. M. Conwell "Impurity Band Conduction in Germanium and Silicon," *Phys. Rev.*, vol. 103, p. 51, 1956.

Figure 3.14

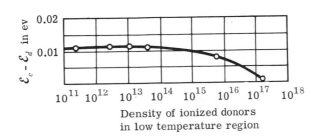

Density of ionized donors in low temperature region

Energy of donor impurities in Ge as a function of impurity concentration. (P. P. Debye and E. M. Conwell, Phys. Rev., vol. 93, p. 693, 1954.)

Figure 3.15

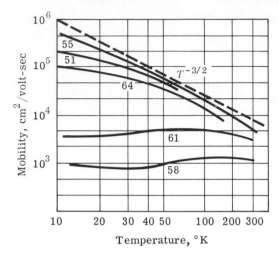

Mobility of electrons in Ge as a function of temperature. The different curves are for different impurity concentrations. (P. P. Debye and E. M. Conwell, Phys. Rev., vol. 93, p. 693, 1954.)

marize the picture as follows. Lattice scattering is the dominant mechanism in limiting the mobility of free carriers at high temperatures and low impurity concentrations. At low temperatures and high impurity concentrations, however, scattering by impurities may take over as the limiting mechanism. The same conclusion applies to the mobility of holes.

To complete the present discussion, we should add the following comments. The product of $R_H\sigma$ actually differs from the *drift mobility*[5] and the former is generally referred to as the *Hall mobility*.

[5] For a discussion of the measurements of drift mobilities, the reader is referred to Sec. 5.8 and the articles by M. B. Prince [*Phys. Rev.*, vol. 92, p. 681, 1953, and *Phys. Rev.*, vol. 93, p. 1204, 1954].

Figure 3.16

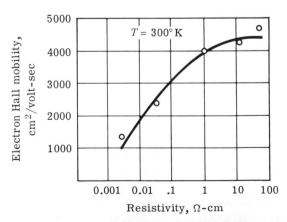

Plot of Hall mobility in germanium at 300°K as a function of resistivity. Since resistivity is inversely proportional to the impurity concentration, the plot shows implicitly the dependence of mobility on impurity concentration. (P. P. Debye and E. M. Conwell, Phys. Rev., vol. 93, p. 693, 1954.)

We can appreciate this subtle distinction only after a general discussion of the transport phenomena (see Chapter 4). The derivation of the Hall coefficient R_H can be generalized to include cases in which both types of carrier exist in comparable amount. The equation of motion, Eq. (3.14), for holes under cross electric and magnetic fields of Fig. 3.11a reads

$$\frac{dv_y}{dt} + \frac{v_y}{\tau} = \frac{e}{m_h}(E_y - v_x B_z) \tag{3.69}$$

where E_y is the Hall field. For d-c measurements $dv_y/dt = 0$; thus, Eq. (3.69) can be expressed in terms of mobility μ_p and the applied field E_x as

$$v_y = \mu_p(E_y - \mu_p E_x B_z) \tag{3.70}$$

In obtaining Eq. (3.70), v_x is taken to be $\mu_p E_x$ by neglecting the second order terms such as $v_y B_z$. For electrons, the corresponding equation is as follows:

$$v_y = -\mu_n(E_y + \mu_n E_x B_z) \tag{3.71}$$

Since the current in the y direction must be zero, that is

$$epv_{py} - env_{ny} = 0$$

substituting v_{py} from Eq. (3.70) and v_{ny} from Eq. (3.71) gives

$$E_y = \frac{\mu_p^2 p - \mu_n^2 n}{\mu_p p + \mu_n n} E_x B_z \tag{3.72}$$

and the corresponding Hall coefficient

$$R_H = \frac{E_y}{J_x B_z} = \frac{\mu_p^2 p - \mu_n^2 n}{e(\mu_p p + \mu_n n)^2} \tag{3.73}$$

The above equation reduces to Eq. (3.66) if $n = 0$. Note that since the mobilities for electrons and holes are different, it may be necessary to measure R_H over a temperature range in order to determine whether a given sample is n- or p-type. For example, $\mu_n > \mu_p$ in Ge, Si, and GaAs. It is possible that within a certain temperature range $\mu_p^2 p - \mu_n^2 n < 0$ and yet $p > n$. This happens when the sample is about to leave the intermediate and enter the high temperature region.

3.7 MOTION OF ELECTRONS IN A PERIODIC STRUCTURE

In fitting the temperature dependence of the Hall coefficient at low temperatures to Eq. (3.68), not only the ionization energy of donor or acceptor impurities can be found but also the effective density of states, N_c or N_v, in the conduction or valence band. The effective density of states is not simply a number but is related to a more fundamental quantity, the effective mass of free carriers, through Eqs. (3.21) and (3.23). The effective mass also enters in the expression of mobility. It seems highly desirable, therefore, that we should know something about the effective mass, its physical meaning, and its experimental determination. To appreciate the meaning of the effective mass in the simplest manner possible, we shall start our discussion with the Kronig-Penney model. Consider the motion of an electron through a one-dimensional, periodic, square-well potential as shown in Fig. 3.17. The time-independent Schrödinger equation reads:

$$\frac{\hbar^2}{2m} \frac{d^2\psi}{dx^2} + [\mathcal{E} - V(x)]\psi = 0 \tag{3.74}$$

where $\mathcal{E}$ is the energy of an electron, m the mass of a free electron, and $V(x)$ the potential energy of the electron given in Fig. 3.17. The question is whether an electron will be able to move to neighboring cells, say cells 3 and 5, if it starts out in one of the cells, say cell 1. Since $|\psi(x)|^2$ represents the probability of finding the electron with coordinate between x and $x + dx$, the question now becomes whether it is possible to find a nondecaying or propagating mode for $\psi(x)$.

In the region $0 < x < a$, $V = 0$; hence the solution of Eq. (3.74) takes the oscillatory form

$$\psi_1(x) = A \exp{(i\beta x)} + B \exp{(-i\beta x)} \tag{3.75}$$

where $\beta = \sqrt{2\mathcal{E}m}/\hbar$. In the region $0 > x > -b$, $V = V_0$ and $\mathcal{E} - V_0 < 0$; hence the solution takes the exponential form

$$\psi_2(x) = C \exp{(\alpha x)} + D \exp{(-\alpha x)} \tag{3.76}$$

where $\alpha = \sqrt{2m(V_0 - \mathcal{E})}/\hbar$. The coefficients A, B, C, and D are related to each other through the boundary conditions. Since the quantity $\mathcal{E} - V(x)$ is finite at the boundary, the derivative $d\psi/dx$

Figure 3.17

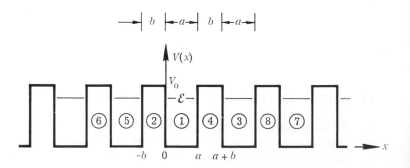

One-dimensional, periodic, potential $V(x)$ used in the Kronig-Penney model.

and the function ψ itself must be continuous at the boundary $x = 0$. In other words,

$$A + B = C + D$$
$$i\beta(A - B) = \alpha(C - D)$$

(3.77)

The question now arises as to what happens at the other boundary $x = a$ of region 1. Obviously Eq. (3.76) still satisfies the wave equation (3.74). If we consider the motion of the electron as a wave propagating through the periodic lattice potential $V(x)$, the solution $\psi_4(x)$ in region 4 should differ from $\psi_2(x)$ in region 2 by a phase factor $\exp [ik(a + b)]$. In other words, $\psi_4(x) = \psi_2(x) \exp [ik(a + b)]$. The boundary conditions at $x = a$ thus become

$$A \exp (i\beta a) + B \exp (-i\beta a)$$
$$= [C \exp (-\alpha b) + D \exp (\alpha b)] \exp [ik(a + b)]$$
$$i\beta[A \exp (i\beta a) - B \exp (-i\beta a)]$$
$$= \alpha[C \exp (-\alpha b) - D \exp (\alpha b)] \exp [ik(a + b)]$$

(3.78)

For a nontrivial solution, the determinant of the coefficients must be zero or

$$\frac{\alpha^2 - \beta^2}{2\alpha\beta} \sinh \alpha b \sin \beta a + \cosh \alpha b \cos \beta a = \cos k(a + b)$$

(3.79)

In order that we may derive some meaningful results from Eq. (3.79), the following assumptions are further made. We shall let $V_0 = \infty$ and $b = 0$ in such a way as to maintain a finite value of $\alpha^2 b = 2P/a$, P being a dimensionless quantity. Thus, Eq. (3.79) becomes

$$P \frac{\sin \beta a}{\beta a} + \cos \beta a = \cos ka$$

(3.80)

The left-hand side of Eq. (3.80) is plotted as a function of βa in Fig. 3.18 for $P = 3\pi/2$. Since $|\cos ka| \leq 1$, only those values of βa are allowed for which the left-hand side of Eq. (3.80) falls between $+1$ and -1. Within this range, which is marked with heavy lines in Fig. 3.18, for one value of βa there is a corresponding value of ka. Since β is directly related to the energy ε of the electron, a plot of ε against ka can be obtained from Fig. 3.18 and is shown in Fig. 3.19. The portions AB, CD, EF, and so on in Figs. 3.18 and 3.19 are the *allowed energy bands* because electrons having energies in these ranges can propagate from one cell to another quite freely. On the other hand, electrons having energies in the range BC, DE, and so on must have an imaginary k in order to satisfy Eq. (3.80); that means, the electron wave motion is highly attenuated, and hence these energy ranges are generally referred to as the *forbidden zones*. Further discussion of the Kronig-Penney model may be found at the end of this chapter in the problem section (Probs. 3.13 through 3.17).

Figure
3.18

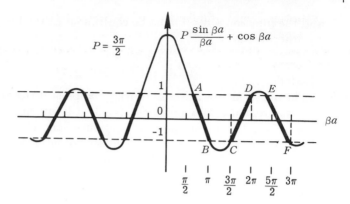

Plot of the function $P \dfrac{\sin \beta a}{\beta a} + \cos \beta a$ as a function of βa for $P = 3\pi/2$. Since $|\cos ka| \le 1$ in Eq. (3.80), only those values of βa are allowed for which the value of the function falls between $+1$ and -1. These values are indicated by the heavy lines in the figure.

A generalization of the above discussion will be made so that the results can be extended to cases of practical interest. First, in a solid, electrons are bound to the nuclei; therefore, as a first-order approximation, the lattice potential $V(x)$ and the wave function ψ in Eq. (3.74) should be replaced respectively by the atomic potential and the atomic wave functions of the electrons concerned, for example the $2s$ and $2p$ functions for diamond. However, the atomic potential is modified in a solid by the presence of neighboring atoms as shown in Fig. 3.20. As a refinement, the potential $V(x)$ in Eq. (3.74) experienced by the lth electron should be represented by the solid curve rather than the atomic potential (dashed curve). The

Figure
3.19

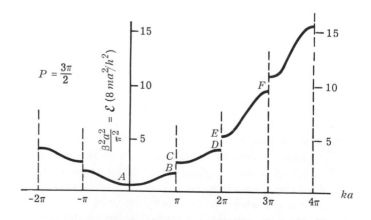

Plot of energy versus wave number for the Kronig-Penney model.

situation is quite similar to the one presented in our discussion of the Kronig-Penney model; or, stated more clearly, electrons can tunnel through the potential hill from one atom core to a neighboring core. Therefore, for electrons in a solid, the solution of Eq. (3.74) takes the following form

$$\psi(r,k) = u(r,k) \exp(i\mathbf{k} \cdot \mathbf{r}) \qquad (3.81)$$

where $u(r,k)$ is the atomic wave function and like the $\psi_1(x)$ of Eq. (3.75) has the same periodicity as the lattice potential. The factor $\exp i\mathbf{k} \cdot \mathbf{r}$ is the equivalent of $\exp[ik(a + b)]$ in Eq. (3.78). The wave function expressed in the form of Eq. (3.81) is known as the *Bloch wave function*. It can be proven mathematically that the solutions of the Schrödinger equation with a periodic lattice potential $V(r)$ can always be written in the above form with $u(r,k)$ having the same periodicity as the lattice potential.

3.8 CONCEPTS OF EFFECTIVE MASS

Now we shall complete the wave picture of electron motion in a periodic structure. The solution of the time-dependent Schrödinger equation, as discussed in Chapter 1, can be obtained from the time-independent solution as follows:

$$\psi(r,k,t) = \psi(r,k) \exp\left(-\frac{i\mathcal{E}t}{\hbar}\right) = u(r,k) \exp\left[i\left(\mathbf{k} \cdot \mathbf{r} - \frac{\mathcal{E}t}{\hbar}\right)\right] \qquad (3.82)$$

The Bloch wave, Eq. (3.82), is an extension of de Broglie's wave, Eq. (1.22), applied to an electron in a periodic lattice. Since the function $u(r,k)$ is the same from one lattice cell to another, the property of wave propagation is contained in the exponential factor $\exp[i(\mathbf{k} \cdot \mathbf{r} - \mathcal{E}t/\hbar)]$. This greatly simplifies the problem. We further note that the energy $\mathcal{E}$ is a function of the wave number k; therefore, Eq. (3.82) actually represents a wave packet. According to our

Figure 3.20

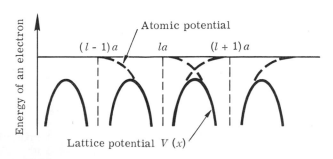

Atomic potential in a periodic lattice or the lattice potential.

discussion in Sec. 1.4, the group velocity of a wave packet containing electrons with energy $\mathcal{E}$ and wave vector $\mathbf{k}$ is given by

$$v_j = \frac{1}{\hbar} \frac{\partial \mathcal{E}}{\partial k_j} \tag{3.83}$$

where $j = x, y,$ or z. Equation (3.83) is the same as Eq. (1.27) and it is repeated here to refresh our memory. The expectation value of the momentum $\mathbf{p}$ associated with the wave packet is defined quantum mechanically as [Eq. (1.38) and Prob. (3.18)]:

$$\mathbf{p} = \frac{\hbar}{i} \int \psi^*(r,k) \,\text{grad}\, [\psi(r,k)] \, dx \, dy \, dz \tag{3.84}$$

It can be shown[6] through the use of the Schrödinger equation that Eq. (3.84) reduces to $\mathbf{p} = m\mathbf{v}$ where $\mathbf{v}$ is given by Eq. (3.83). As pointed out in Sec. 2.4, the definition of the group velocity serves as a link connecting wave and particle properties of an electron in the free electron theory of metals and semiconductors. We shall derive the effective mass (particle property) of an electron in a solid by use of Eq. (3.83).

If $\mathbf{F}$ is the force acting on electrons due to an applied field, the work done on an electron by the force must be equal to the gain in energy by the electron

$$\sum_j F_j v_j \, dt = d\mathcal{E} = \sum_j \frac{\partial \mathcal{E}}{\partial k_j} \, dk_j = \sum_j \hbar v_j \, dk_j \tag{3.85}$$

through the use of Eq. (3.83) for $\partial \mathcal{E}/\partial k_j$. Equating the coefficients of v_j in Eq. (3.85), we obtain $F_j \, dt = \hbar \, dk_j$ or vectorially the following equation:

$$\frac{d\mathbf{k}}{dt} = \frac{\mathbf{F}}{\hbar} \tag{3.86}$$

The equation of motion for the electron thus can be obtained by differentiating Eq. (3.83) with respect to t and substituting $F_i/\hbar$ for dk_i/dt through Eq. (3.86) as follows:

$$\frac{dv_j}{dt} = \frac{1}{\hbar} \sum_i \frac{\partial^2 \mathcal{E}}{\partial k_i \, \partial k_j} \frac{dk_i}{dt} = \frac{1}{\hbar^2} \sum_i \frac{\partial^2 \mathcal{E}}{\partial k_i \, \partial k_j} F_i \tag{3.87}$$

In general, the acceleration of the wave packet is related to the

[6] For such derivations, the reader is referred to E. Spenke, "Electronic Semiconductors," McGraw-Hill Book Company, New York, 1958, pp. 201–206; W. Shockley, "Electrons and Holes in Semiconductors," pp. 424–435, D. Van Nostrand Co., Inc., New York, 1950.

external field through an effective mass tensor

$$\frac{d}{dt}\begin{pmatrix} v_x \\ v_y \\ v_z \end{pmatrix} = \begin{pmatrix} m_{xx}^{-1} & m_{xy}^{-1} & m_{xz}^{-1} \\ m_{yx}^{-1} & m_{yy}^{-1} & m_{yz}^{-1} \\ m_{zx}^{-1} & m_{zy}^{-1} & m_{zz}^{-1} \end{pmatrix}\begin{pmatrix} F_x \\ F_y \\ F_z \end{pmatrix} \tag{3.88}$$

where m_{ij} is the element of the effective mass tensor for electrons in a crystal defined by the following expression:

$$m_{ij}^{-1} = \frac{1}{\hbar^2}\frac{\partial^2 \mathcal{E}}{\partial k_i\, \partial k_j} \tag{3.89}$$

In a semiconductor, only those electrons close to the bottom of the conduction band or the top of the valence band are of interest to us. We shall again use the Kronig-Penney model to make our viewpoints clear and take the point C in Fig. 3.19 as a specific example. Since Eq. (3.80) is an even function of k, for electrons having k close to that of C, their energy can be expressed as a quadratic function of k after Eq. (3.80) is expanded into a Taylor's series. For a three-dimensional lattice, the energy of a conduction band electron can be written as

$$\mathcal{E} = \mathcal{E}_c + \sum_i \frac{\hbar^2}{2m_{ii}}(k_i - k_{i0})^2 + \sum_{ij(i \neq j)} \frac{\hbar^2}{m_{ij}}(k_i - k_{i0})(k_j - k_{j0}) \tag{3.90}$$

where $i, j = x, y,$ or z and $i \neq j$. The k_{x0}, k_{y0}, and k_{z0} in Eq. (3.90) are values of k at the bottom of the conduction band; or in other words, $\mathcal{E}$ is a minimum at this point in k space. It turns out that for most semiconductors such as Ge, Si, and GaAs, the constant energy surface is either ellipsoidal (Ge, Si) or spherical (GaAs). In the case of an ellipsoidal energy surface, the axes of the ellipsoid in k space may not coincide with the directions of k_x, k_y and k_z. Expressed in terms of the axes, k_1, k_2 and k_3, of the ellipsoidal constant energy surface, Eq. (3.90) becomes

$$\mathcal{E} = \mathcal{E}_c + \frac{\hbar^2}{2m_1}(k_1 - k_{10})^2 + \frac{\hbar^2}{2m_2}(k_2 - k_{20})^2 + \frac{\hbar^2}{2m_3}(k_3 - k_{30})^2 \tag{3.91}$$

Equation (3.91) can be obtained from Eq. (3.90) through a rotation of the coordinate axes (Prob. 3.19). We emphasize that Eq. (3.91) comes from an expansion of a certain function relating $\mathcal{E}$ to k; hence the values of m_1, m_2, and m_3 can be considered as constants only within a limited range of k close to the bottom of the conduction band where most conduction band electrons are located. We further note that spherical energy surfaces are a special case of ellipsoidal energy surfaces with $m_1 = m_2 = m_3$.

Let us now discuss the valence band electrons. Referring to Fig. 3.19, we notice that the effective mass of electrons at point B is negative. From Eq. (3.80), we can see that the slope $\partial \mathcal{E}/\partial k$ is zero at both points B and C. Obviously, the curvatures of the $\mathcal{E}$ against k curve must have opposite signs near points B and C. At first, this may seem to be a very surprising result. We must not forget, however, that the effective mass of electrons in a solid differs from the free electron mass because of their interaction with the lattice potential. For electrons near the top of the valence band, the running wave solution given by Eq. (3.82) is no longer adequate. In order to maintain $\partial \mathcal{E}/\partial k$ or $v = 0$, a reflected wave must be added to the solution of Eq. (3.82).

As electrons near the top of the valence band acquire more energy from an external field, they move closer and closer to point B, and consequently these electrons will have a larger and larger component of the wave running opposite to the direction of the applied field. This gives a physical meaning to the negative effective mass. Note that the opposite is true for conduction band electrons. Further, it should be pointed out that according to Eq. (3.86), the total increment of momentum $\Delta \mathbf{p}_t$ during time Δt is equal to

$$\Delta \mathbf{p}_t = \mathbf{F} \, \Delta t = \hbar \, \Delta \mathbf{k} \qquad (3.92)$$

while the momentum $\Delta \mathbf{p}_e$ imparted to the electron during the same time interval is given by

$$\Delta \mathbf{p}_e = m \, \Delta \mathbf{v} = \frac{m}{m_{\text{eff}}} \mathbf{F} \, \Delta t = \frac{m}{m_{\text{eff}}} \hbar \, \Delta \mathbf{k} \qquad (3.93)$$

on account of Eq. (3.87). The difference between $\Delta \mathbf{p}_t$ and $\Delta \mathbf{p}_e$ is taken up by the lattice. Therefore, the fact that electrons at the top of the valence band have a negative effective mass does not violate any fundamental principle. For a more thorough discussion of the effective mass theory, the reader is referred to other texts which give a more detailed quantum mechanical treatment[7] of the subject.

The question still remains as to the manner in which valence band electrons contribute to current flow. First, let us divide the valence band into two portions, 1 and 2, with electron density n_1 and n_2. Since a full band does not contribute to any current

$$J = -en_1 v - en_2 v = -e \sum_1 v - e \sum_2 v = 0 \qquad (3.94)$$

In Eq. (3.94), n_1 and n_2 are represented by two summations which

[7] E. Spenke, "Electronic Semiconductors," McGraw-Hill Book Company, New York, 1958; W. Shockley, "Electrons and Holes in Semiconductors," D. Van Nostrand Co., New York, 1950; F. Seitz, "The Modern Theory of Solids," McGraw-Hill Book Company, New York, 1940; and A. H. Wilson, "The Theory of Metals," Cambridge University Press, Cambridge, 1935.

count the occupied states (electrons) per unit volume in the valence band.

When the valence band is only partially occupied, say portion 1 is occupied, the new current density J_1 can be found by counting the current contribution from the occupied states as follows:

$$J_1 = -en_1v = -e \sum_1 v \tag{3.95}$$

and J_1 is no longer zero. In view of Eq. (3.94), Eq. (3.95) becomes

$$J_1 = e \sum_2 v = en_2v = epv \tag{3.96}$$

Since portion 2 of the valence band is not occupied, n_2 in Eq. (3.96) represents the density of unoccupied states or the hole density p. Therefore, the transformation from Eq. (3.95) to Eq. (3.96) is significant physically in two respects. In Eq. (3.95), the summation 1 is over all occupied valence band states while in Eq. (3.96), the summation 2 is over all empty valence band states. Moreover, the sign associated with the charge changes from $-e$ to $+e$, meaning that an unoccupied state has the equivalent effect of a positive charge $+e$ so far as electric conduction is concerned. The velocity v in Eq. (3.96) is of course that of electrons and can be found from Eq. (3.87).

For simplicity, we assume that the effective mass is a scalar instead of tensor, thus we have

$$\mathbf{v} = \frac{-e\mathbf{E}}{m_{ev}} \tau = \frac{e\mathbf{E}}{m_h} \tau \tag{3.97}$$

where $\mathbf{E}$ is the applied electric field, τ the mean free time and m_{ev} the effective mass of electrons in the valence band. Since the unoccupied states are close to the top of the valence band, the quantity m_{ev} in Eq. (3.97) is negative according to the discussion in the preceding paragraph. However, the negative sign in m_{ev} cancels out that in $-e$, yielding the last equation of Eq. (3.97) in which $m_h = |m_{ev}|$ is a positive quantity and can be called the effective mass of a hole. Combining Eqs. (3.96) and (3.97), we have

$$J_1 = e \sum_2 v = e \sum_2 \frac{eE}{m_h} \tau = e\mu_p pE \tag{3.98}$$

This justifies the use of Eq. (3.5) which was arrived at from physical intuition.

In concluding this section, it may be useful to summarize the important information obtained from the above discussion. First, we shall compare the wave behavior of electrons in a solid with that of free electrons as listed in Table 3.2.

TABLE 3.2   *Comparison of the dynamic properties of electrons in a solid with those of free electrons*

FREE ELECTRONS	ELECTRONS IN A SOLID*
$\mathbf{p} = \hbar\mathbf{k}$	$p_i = \dfrac{m}{\hbar}\dfrac{\partial \mathcal{E}}{\partial k_i}$
$\mathbf{F} = \hbar\dot{\mathbf{k}} = \dot{\mathbf{p}}$	$\mathbf{F} = \hbar\dot{\mathbf{k}} = \dfrac{\overset{\leftrightarrow}{\mathbf{m}}_{\text{eff}}}{m}\dot{\mathbf{p}}$
$\mathbf{v} = \dfrac{\mathbf{p}}{m}$	$\mathbf{v} = \dfrac{\mathbf{p}}{m}$
$m\dot{\mathbf{v}} = \mathbf{F}$	$\overset{\leftrightarrow}{\mathbf{m}}_{\text{eff}}\dot{\mathbf{v}} = \mathbf{F}$

* The symbol $\leftrightarrow$ indicates a tensor.

Next, we shall list the essential properties of conduction band electrons and valence band holes (Table 3.3).

TABLE 3.3   *A summary of the essential properties of conduction band electrons and valence band holes (For simplicity, $\overset{\leftrightarrow}{\mathbf{m}}_{\text{eff}}$ is replaced by a scalar.)*

	Conduction band electrons	*Valence band holes*
Physical origin	Excess electrons	Missing electrons
Charge	$-e$	$+e$
Energy	$\mathcal{E} = \mathcal{E}_c + \dfrac{\hbar^2 k^2}{2m_e}$	$\mathcal{E} = \mathcal{E}_v + \dfrac{\hbar^2 k^2}{2m_h}$
Effective mass	$m_e = $ positive	$m_h = $ positive
Drift velocity	$v = \dfrac{F\tau}{m_e}$	$v = \dfrac{F\tau}{m_h}$

In the above table, the constant energy surfaces for both electrons and holes are assumed spherical and the values of k refer to those at the respective band edge. We also emphasize that in the conventional energy band diagram such as Figs. 3.8 and 3.10, the energy scale for electrons is upward while the energy scale for holes is downward. This is a direct result of the above discussion and hence justifies the steps taken in deriving Eq. (3.22).

3.9 THE CYCLOTRON RESONANCE EXPERIMENT

Theory. The cyclotron resonance experiment[8] provides the only direct determination of the effective masses in semiconductors. Consider the motion of a free electron in the presence of a d-c magnetic field B. It is well known that if there is no other force besides the Lorentz force, then the electron will perform a circular motion. If v is the velocity of the electron, the radius r of the circular motion must be such that $evB = mv^2/r$ or $\omega = v/r = eB/m$. The frequency $\nu = \omega/2\pi$ is known as the *cyclotron resonance frequency*. For $B = 10^3$ gauss $= 10^{-1}$ weber/m^2, $\nu = 2.8 \times 10^9$ cps, which falls in the microwave region. If an a-c electric field is applied in the plane perpendicular to the direction of the d-c magnetic field, resonant absorption of energy by the electron from the a-c field occurs when the frequency of the a-c field coincides with the cyclotron frequency.

If the effective mass in a semiconductor can be represented by a single constant m^*, then resonance absorption occurs only at a single frequency $\omega = eB/m^*$. It turns out that the situation is more complicated than what we have suggested. Actually the effective masses used in the expression of N_c and μ_n in Eqs. (3.21) and (3.11) may be numerically different. The effective mass used in Eq. (3.21) is called the *density-of-state effective mass* while that in Eq. (3.11) the *mobility effective mass*. This difference comes about because of the complicated band structure as we shall see later.

To start the discussion, we shall use the general energy expression of Eq. (3.91) for conduction band electrons. To derive the equation of motion from the energy equation, however, we must know the relationship between the k space and the coordinate space. In the previous chapter, it has been shown that the unit cell of a crystal can be defined in terms of three non-coplanar primitive vectors $\mathbf{a}'$, $\mathbf{b}'$ and $\mathbf{c}'$. In a similar manner, the wave vector $\mathbf{k}$ can be expressed in terms of a set of three vectors $\mathbf{a}^*$, $\mathbf{b}^*$ and $\mathbf{c}^*$ defined [Eq. (2.21)] as

$$\mathbf{a}^* = \frac{\mathbf{b}' \times \mathbf{c}'}{\mathbf{a}' \cdot (\mathbf{b}' \times \mathbf{c}')} \tag{3.99}$$

Similar expressions for $\mathbf{b}^*$ and $\mathbf{c}^*$ are obtained by rotating cyclically the vectors $\mathbf{a}'$ to $\mathbf{b}'$ to $\mathbf{c}'$ to $\mathbf{a}'$. The lattice structure which can be generated from $\mathbf{a}^*$, $\mathbf{b}^*$, and $\mathbf{c}^*$ is called the reciprocal lattice, a term generally used in X-ray diffraction studies. In terms of $\mathbf{a}^*$, $\mathbf{b}^*$, and $\mathbf{c}^*$,

$$\mathbf{k} = 2\pi(h_1\mathbf{a}^* + h_2\mathbf{b}^* + h_3\mathbf{c}^*) \tag{3.100}$$

[8] The free-electron theory of cyclotron resonance was first discussed by J. G. Dorfman (*Doklady. Nauk.*, U.S.S.R., vol. 81, p. 765, 1951) and R. B. Dingle (*Proc. Roy. Soc.*, London, A212, p. 38, 1952) and applied to non-spherical energy surfaces by W. Shockley (*Phys. Rev.*, vol. 90, p. 491, 1953).

The important criterion for the validity of this expression is that it preserves the periodicity of the function $\psi(r,k)$ defined by Eq. (3.81). To prove this, we perform a translation of the vector $\mathbf{r}$ such that

$$\mathbf{r}' = \mathbf{r} + l_1\mathbf{a}' + l_2\mathbf{b}' + l_3\mathbf{c}' \tag{3.101}$$

where l_1, l_2, and l_3 are integers. Noting that $\mathbf{a}' \cdot \mathbf{a}^* = 1$ but $\mathbf{b}' \cdot \mathbf{a}^* = 0$ and so on, the scalar product of $\mathbf{k}$ and $\mathbf{r}'$ gives

$$\mathbf{k} \cdot \mathbf{r}' = \mathbf{k} \cdot \mathbf{r} + 2\pi(h_1l_1 + h_2l_2 + h_3l_3) \tag{3.102}$$

It is obvious from Eq. (3.102) that $\psi(r',k) = \psi(r,k)$ if h_1l_1, h_2l_2, and h_3l_3 are integers.

In a cubic system, if the three primitive vectors $\mathbf{a}$, $\mathbf{b}$, and $\mathbf{c}$ are along the three principal axes x, y, and z (the three edges) of the cube and have a length a, then

$$\mathbf{a}^* = \mathbf{a}/a^2 \qquad \mathbf{b}^* = \mathbf{b}/a^2 \qquad \mathbf{c}^* = \mathbf{c}/a^2 \tag{3.103}$$

The important conclusion from Eq. (3.103) is that the directions of $\mathbf{k}_x$ (that is $\mathbf{a}^*$), $\mathbf{k}_y$, and $\mathbf{k}_z$ are the same as those of x, y, and z axes respectively. Returning to Eq. (3.91), we are now in a position to transform the energy equation in k space into an equation of motion in coordinate space. The constant energy surface represented by Eq. (3.91) is generally ellipsoidal in shape as shown in Fig. 3.21a and consequently it is called an ellipsoidal energy surface. When

Figure 3.21

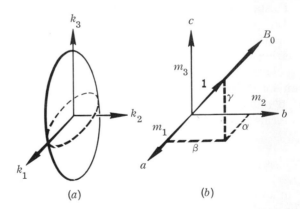

(a)   (b)

(a) *The constant energy surface in k space and (b) the three axes in coordinate space. In a cubic crystal, the axes in k space (reciprocal lattice space) are in the same direction as the axes in coordinate space through Eq. (3.103). The statement can be generalized to any other direction. For example, the [111] direction in k space coincides with the [111] direction in coordinate space. In (b) the coordinate axes $\mathbf{a}$, $\mathbf{b}$, and $\mathbf{c}$ are chosen to coincide with major axes of the ellipsoidal constant energy surface; hence, the three components of the mass tensor m_1, m_2, and m_3 can be used directly in the equation of motion, Eq. (3.105).*

$m_1 = m_2 = m_3$, the constant energy surface becomes spherical. Consider electron wave motion along the k_1 axis alone, that is $k_1 \neq 0$, $k_2 = k_3 = 0$, it is apparent from Eq. (3.91) that the effective mass for motion along the k_1 axis is m_1. Similarly, for motions along the k_2 and k_3 axes, the effective masses are respectively m_2 and m_3. Figure 3.21 shows how the effective masses along the axes in coordinate space can be derived from the constant energy surface in k space.

In view of the above discussion, Eq. (3.14) should be replaced by a more general equation of motion as follows:

$$\frac{d}{dt} (\overleftrightarrow{\mathbf{m}} \cdot \mathbf{v}) + \frac{\overleftrightarrow{\mathbf{m}} \cdot \mathbf{v}}{\tau} = -e(\mathbf{E} + \mathbf{v} \times \mathbf{B}) \qquad (3.104)$$

where $\mathbf{E}$ is the applied electric field, $\mathbf{B}$ the applied magnetic field, and $\overleftrightarrow{\mathbf{m}}$ the effective mass tensor. The case of interest here is one in which the magnetic field $\mathbf{B}$ is a d-c field having direction cosines α, β, γ as shown in Fig. 3.21b and the electric field $\mathbf{E}$ is a time-varying field with angular frequency ω. Thus, $\mathbf{v}$ will also vary with exp $i\omega t$. After canceling the time dependence on both sides of Eq. (3.104), we obtain

$$\begin{aligned}
i\omega' m_1 v_1 + eB\gamma v_2 - eB\beta v_3 &= -eE_1 \\
-eB\gamma v_1 + i\omega' m_2 v_2 + eB\alpha v_3 &= -eE_2 \\
eB\beta v_1 - eB\alpha v_2 + i\omega' m_3 v_3 &= -eE_3
\end{aligned} \qquad (3.105)$$

where $\omega' = (\omega\tau - i)/\tau$. The natural frequency of the system can be found by setting the determinant D of the coefficients of v_1, v_2, and v_3 equal to zero. Thus, we have

$$D = i\omega'[(eB)^2(m_1\alpha^2 + m_2\beta^2 + m_3\gamma^2) - \omega'^2 m_1 m_2 m_3] = 0 \quad (3.106)$$

or the cyclotron resonance frequency ω_c

$$\omega_c{}^2 = (eB)^2 \frac{m_1\alpha^2 + m_2\beta^2 + m_3\gamma^2}{m_1 m_2 m_3} \qquad (3.107)$$

by letting $\tau \to \infty$ in ω'. It should be pointed out that the subscripts 1, 2, and 3 in Eqs. (3.105) and (3.107) refer to the directions of the principal axes in k space and they may be different from the directions of the crystal axes. For a discussion of the mechanism by which electrons absorb energy from the a-c electric field, the reader is referred to the problem section at the end of this chapter (Probs. 2.20 and 2.21).

Experimental results on conduction band electrons in Ge and Si. It is also pointed out in the problem (Prob. 2.21) that the absorption curve shows a well-defined resonant frequency only for $\omega\tau > 1$. Since τ is very short, say of the order of 10^{-13} sec at room temperature, and increases with decreasing temperature, the cyclotron resonance

Figure
3.22

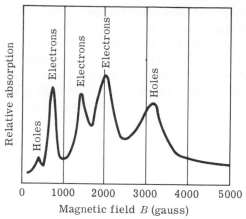

Measured at
24,000 mc/sec
in Ge with **B** in a
(110) plane at 60°
from a [001] axis

Cyclotron resonance absorption curve in Ge. (G. Dresselhaus, A. F. Kip, and C. Kittel, Phys. Rev., vol. 98, p. 368, 1955.)

experiment must be performed at very low temperatures (usually liquid He temperature) and very high frequencies (usually microwave frequencies). In applying Eq. (3.107) to experimental results, we shall use germanium as an example. Figure 3.22 shows the cyclotron resonance absorption curve of germanium near 24,000 Mc/sec and 4°K for a given orientation of **B**. To obtain a complete picture of the energy band structure near the conduction band edge, however, the resonance experiment must be carried out over different orientations of **B**. It happens in Ge that the conduction band has minima along the principal diagonals in k space, that means k[111] or its equivalents; therefore, the constant energy surfaces are ellipsoids along the principal diagonals in the reciprocal lattice as shown in Fig. 3.23a. The subscripts 1, 2, and 3 in Eq. (3.107) refer to the

Figure
3.23

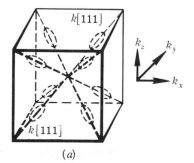

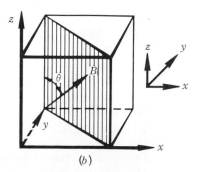

(a)   (b)

(a) *Ellipsoidal energy surfaces in k space for conduction electrons in Ge and* (b) *the direction of the magnetic induction field B with reference to the coordinate axes.*

principal axes of the ellipsoids and those axes are different for different ellipsoids.

Take the ellipsoid along the $k[111]$ direction as an example. One direction along which m_3 is defined can be chosen as the direction of $k[111]$ while the other two directions along which m_1 and m_2 are defined are in a plane perpendicular to the $k[111]$ direction in k space. It is found that the two effective masses, m_1 and m_2, thus defined are equal and are called the *transverse mass* m_t while the other effective mass m_3 is called the *longitudinal mass* m_l. In terms of m_t and m_l, Eq. (3.107) becomes

$$m^* = \frac{eB}{\omega_c} = \sqrt{\frac{m_l m_t^2}{m_l \gamma^2 + m_t(1 - \gamma^2)}} \tag{3.108}$$

where m^* is generally referred to as the cyclotron resonance effective mass.

If the applied magnetic field B is in the (110) crystal plane and makes an angle θ with the [001] crystal axis as shown in Fig. 3.23b, the direction cosine γ is equal to

$$\gamma = \frac{1}{\sqrt{3}}(1,1,1) \cdot \left(\frac{\sin\theta}{\sqrt{2}}, \frac{-\sin\theta}{\sqrt{2}}, \cos\theta\right) = \frac{\cos\theta}{\sqrt{3}} \tag{3.109}$$

Counting all the eight ellipsoids, we have all the possible values of γ listed below

$$\gamma = \pm\cos\theta/\sqrt{3} \qquad \text{(4 ellipsoids)}$$
$$\gamma = \pm(\cos\theta - \sqrt{2}\sin\theta)/\sqrt{3} \qquad \text{(2 ellipsoids)} \qquad \text{(3.110)[9]}$$
$$\gamma = \pm(\cos\theta + \sqrt{2}\sin\theta)/\sqrt{3} \qquad \text{(2 ellipsoids)}$$

experiment

Therefore, for each value of θ, there are three different values of m^* and consequently three different values of B for which the resonance condition of Eq. (3.107) is satisfied. This explains the appearance of three resonance peaks in Fig. 3.22. Knowing the value of ω_c and B from Fig. 3.22, the value of m^*/m can be found from Eq. (3.108). Figure 3.24 shows the plot of m^* as a function of θ. A fit of Eq. (3.108) to Fig. 3.24 gives the value of $m_l = 1.64m$ and $m_t = 0.082m$ where m is the free electron mass.

Figure 3.25a shows the cyclotron resonance effective masses for conduction band electrons in silicon. It is seen to be quite different from Fig. 3.24. The difference arises from the fact that the conduction band minima in silicon lie along the six equivalent $\langle 100 \rangle$ axes in k space and consequently there are six constant energy ellipsoids as shown in Fig. 3.25b. Similar to Ge, the constant energy surfaces in silicon can also be described in terms of two effective masses, the

[9] Because the conduction band minima are located on the boundary of the Brillouin zone (defined in Sec. 3.10), actually there are only four ellipsoids and the numbers of ellipsoids in parentheses should be divided by two.

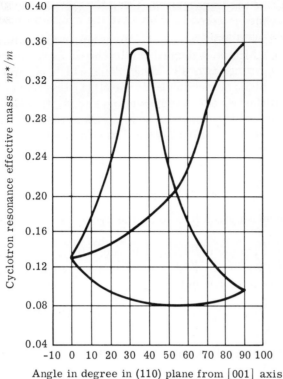

Figure 3.24

Cyclotron resonance effective mass m^*/m

Angle in degree in (110) plane from [001] axis

Plot of cyclotron resonance effective mass for conduction electrons in Ge as a function of the angle θ defined in Fig. 3.23b. (G. Dresselhaus, A. F. Kip, and C. Kittel, Phys. Rev., vol. 98, p. 368, 1955.)

Figure 3.25

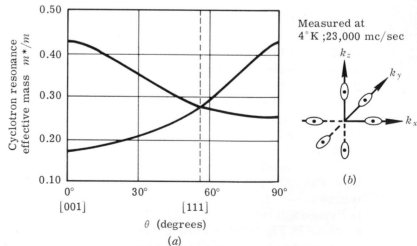

Cyclotron resonance effective mass m^*/m

θ (degrees)

[001]   [111]

(a)

Measured at 4°K ; 23,000 mc/sec

(b)

(a) Cyclotron resonance curves and (b) the constant energy surfaces for conduction electrons in Si. (G. Dresselhaus, A. F. Kip, and C. Kittel, Phys. Rev., vol. 98, p. 368, 1955.)

longitudinal m_l and the transverse m_t. Using the information given in Fig. 3.25, the reader is asked to show that the cyclotron resonance effective masses are given by

$$m^* = \left[\frac{2m_l m_t{}^2}{2m_t + (m_l - m_t)\sin^2\theta} \right]^{1/2} \quad \text{and}$$

$$\left[\frac{m_l m_t{}^2}{m_t + (m_l - m_t)\cos^2\theta} \right]^{1/2} \quad (3.111)$$

and that the values of m_l and m_t are $0.98m$ and $0.19m$ respectively.

Experimental results on valence band holes in Ge and Si. The behavior of valence band holes in both Ge and Si is quite different from that of conduction band electrons. First of all, the top of the valence band is situated at $k\langle 000 \rangle$ in k space. Second, the cyclotron resonance effective masses as shown in Fig. 3.26 for both Ge and Si do not show a very strong dependence upon θ, the angle which defines the orientation of the magnetic field with respect to the crystal axes as shown in Fig. 3.23b. Based on theoretical considerations, the energy of valence band electrons can be expressed in terms of the wave vector **k** as

$$\mathcal{E}(k) = -\frac{\hbar^2}{2m} \left\{ Ak^2 \pm [B^2 k^4 + C^2(k_x{}^2 k_y{}^2 + k_y{}^2 k_z{}^2 + k_z{}^2 k_x{}^2)]^{1/2} \right\}$$

$$(3.112)$$

What happens is that the energy for holes must be calculated to higher order approximations than that for electrons. The terms in brackets represent the result from such calculations. These terms must preserve the cubic symmetry of the crystal. The constants A, B, and C can be found by fitting the expression for m^* based on Eq. (3.112) to the curves in Fig. 3.26. The process of finding m^*, however, is quite involved, and the reader should consult the original articles[10] by Dresselhaus and by Lax for further details. The values of A, B, and C for Ge and Si are found to be

	A	B	C
Ge	13.1 (± 0.4)	8.3 (± 0.6)	12.5 (± 0.5)
Si	4.0 (± 0.1)	1.1 (± 0.4)	4.1 (± 0.4)

the numbers in the parentheses indicating possible experimental errors.

Figure 3.27 shows the constant energy surfaces of holes in the (110) plane of the k space. The terms involving C^2 in Eq. (3.112), which cause the variation of m^* with θ, are comparable in magnitude to the Ak^2 term. Hence, the anisotropic behavior of m^* (that means the variation of m^* as a function of θ) will be more pronounced in

[10] G. Dresselhaus, A. F. Kip and C. Kittel, *Phys. Rev.*, vol. 98, p. 368, 1955; R. N. Dexter, H. J. Zeiger, and B. Lax, *Phys. Rev.*, vol. 104, p. 637, 1956.

Figure
3.26

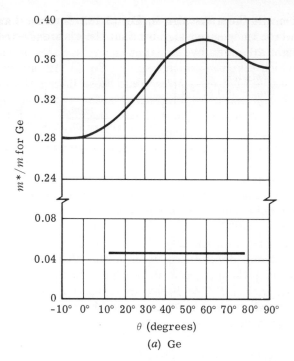

θ (degrees)

(a) Ge

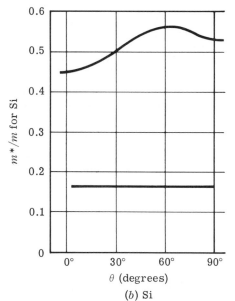

θ (degrees)

(b) Si

Cyclotron resonance curves for valence holes in (a) Ge and (b) Si. (G. Dresselhaus, A. F. Kip, and C. Kittel, Phys. Rev., vol. 98, p. 368, 1955.)

the case where the negative sign in Eq. (3.112) is taken than in the other case where the positive sign is taken. This is manifested in the curves of Fig. 3.26 and in the constant energy surfaces of Fig. 3.27. The holes which take the positive sign in Eq. (3.112) are generally referred to as the *light mass holes* while those taking the negative sign are the *heavy mass holes*. The energy surfaces near the top of the valence band in germanium and silicon, therefore, consist of two slightly distorted or warped spheres. For the heavy holes, the warped sphere protrudes in the $k[111]$ and its equivalent directions, while for the light holes the sphere is depressed along the $k[111]$ and its equivalent directions.

In sum, we have seen that the cyclotron resonance experiment reveals the detailed structure of the conduction and valence bands

Figure 3.27

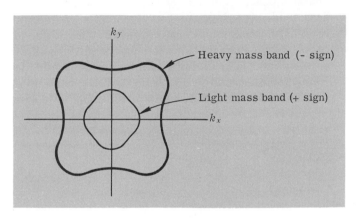

Projection of the constant energy surface for valence holes in the (001) plane of the k space.

near their respective band edges. It seems appropriate that a general discussion of the energy band structure in covalent semiconductors such as Ge, Si, and GaAs be given in connection with the cyclotron resonance experiment. This indeed will be the subject of our discussion in the following section. It is also important to see in what way the mobility and density-of-states expressions, Eqs. (3.11), (3.21), and (3.23), must be modified in view of the cyclotron resonance experiment. This will be done in a section devoted to the discussion of the mobility and density-of-states effective masses in Chapter 5.

3.10 SEMICONDUCTING MATERIALS

The word *semiconductor* is customarily used to refer to a specific class of material in which electric conduction is primarily carried by occupied states in the conduction band and empty states in the valence band. Furthermore, unlike metals, the concentrations of free electrons and holes can be controlled by the amount of impurities which are purposely introduced into a semiconductor. Common examples include: (1) elemental semiconductors such as silicon and germanium; (2) compound semiconductors formed by elements belonging to the third and fifth columns of the periodic table of the type IIIb-Vb such as GaAs, InSb, AlP, and GaP; (3) compound semiconductors formed by elements belonging to the second and sixth columns of the periodic table of the type IIb-VIb such as CdS, ZnS, CdTe, and ZnO. Table 3.4 shows some of the common properties of a selected number of elemental and compound semiconductors.

Chemical bond. The reader may have noticed that the gap energy has a tendency to be progressively larger in going from the elemental to the II-VI compound semiconductor. This is due to the increasing degree of ionicity in the chemical bond. The difference between covalent and ionic binding can be most vividly demonstrated in compound semiconductors. Take a III-V compound as an example. The IIIb element has an s^2p while the Vb element has an s^2p^3

TABLE 3.4* *Common properties of a selected number of semiconductors at 300°K*

Semi-con-ductor	Crystal structure†	Energy gap, ev	μ_n, cm²/volt-sec‡	μ_p, cm²/volt-sec‡	Static ϵ/ϵ_0	Index of refraction n^2
Si	D	1.11	1,350	480	11.7	11.7
Ge	D	0.67	3,900	1,800	16	16
GaAs	ZB	1.39	8,500	420	12.5	10.9
GaSb	ZB	0.67	4,000	1,400	15	14
InSb	ZB	0.17	77,000	750	17	15.7
CdTe	ZB	1.45	700	∼60	11	. . . . .
CdS	W, ZB	2.45(W)	200	. . . .	11.6	5.9
ZnO	W	3.30	200	. . . .	8.5	3.73

* The table is compiled from values quoted in the following references:
1. E. M. Conwell, Properties of Silicon and Germanium: II, *Proc. I.R.E.*, vol. 46, p. 1281, 1958.
2. T. S. Moss, "Optical Properties of Semiconductors," Academic Press Inc., New York, 1959.
3. C. Hilsum and A. C. Rose-Innes, "Semiconducting III-V Compounds," Pergamon Press, London, 1961.
4. N. B. Hannay (ed.), "Semiconductors," American Chemical Society Monograph Series, Reinhold Publishing Corporation, New York, 1959.
† D = diamond, ZB = Zinc blende, W = Wurtzite structures.
‡ Hall mobilities except in Ge and Si for which the values are drift mobilities. These values also represent the highest values reported thus far.

electron configuration outside a core of closed shells. For pure covalent binding which requires four valence electrons to form a tetrahedral sp^3-hybrid orbital as in germanium and silicon, we would expect that the IIIb element will capture an electron from the Vb element so that III$^-$ and V$^+$ ions are formed, each with four valence electrons. On the other hand, for pure ionic binding, the IIIb element will donate all its three electrons to the Vb element so that both the III^{3+} and V^{3-} ions can have closed-shell configurations as in the case of alkali halides. We see that the charge carried by each element in a compound semiconductor changes from one polarity to the other, depending on the nature of the chemical bond.

Based on the above discussion, the electron wave function ψ in a compound semiconductor can be written as a superposition of two idealized wave functions ψ_i and ψ_c

$$\psi = a_i\psi_i + a_c\psi_c \tag{3.113}$$

where ψ_i and ψ_c represent respectively the wave function for pure ionic and pure covalent binding and a_i and a_c are two constants. The degree of ionicity in a compound semiconductor depends on the ratio $|a_i/a_c|$ which in principle can be calculated theoretically by minimizing the energy of the state ψ defined in Eq. (3.113).

Information concerning charge transfer, however, can be obtained experimentally by measuring the dielectric constant of a compound semiconductor. As will be discussed in Chapter 7, the displacement of positive and negative charges in an ionic crystal gives rise to an electric polarization which contributes to part of the dielectric constant. According to an analysis[11] by Szigeti, the difference between the static and optical dielectric constant in a cubic crystal is given by

$$\frac{\epsilon}{\epsilon_0} - n^2 = \left(\frac{n^2 + 2}{3}\right)^2 \frac{(Sze)^2}{\epsilon_0} \frac{N}{\omega_t^2}\left(\frac{1}{M_1} + \frac{1}{M_2}\right) \tag{3.114}$$

where N is the number of ion pairs per unit volume, M_1 and M_2 are the masses of the ions, ω_t is the angular frequency of the transverse optical branch of the lattice vibration, z is the valence of the compound, and n is the optical index of refraction. In deriving Eq. (3.114), it is assumed that the effective charge of an ion is equal to Sze; hence the parameter S obtained in fitting Eq. (3.114) to the difference between the experimental values of ϵ/ϵ_0 and n^2 should be of value in estimating the ionic character of a compound.

In III-V compounds, the value[12] of S is found to vary from 0.10 in GaSb to 0.20 in GaP, taking $z = 3$ in Eq. (3.114). In II-VI compounds, the value of S is much higher as it is evident from Table 3.3

[11] B. Szigeti, *Trans. Faraday Soc.*, vol. 45, p. 155, 1949.
[12] G. Picus et al., *Jour. Phys. Chem. Solids*, vol. 8, p. 282, 1959; D. A. Kleinman and W. G. Spitzer, *Phys. Rev.*, vol. 118, p. 110, 1960.

that a large discrepancy exists between the static ϵ/ϵ_0 and the square of optical index of refraction n^2. The following values are quoted by Hutson in Hannay's book:[13] $S = 1$ for MgO, $S = 0.63$ for ZnO, and $S = 0.48$ for ZnS. Therefore, it is fair to say that the degree of ionicity increases markedly from III-V to II-VI compounds. This is consistent with the tendency of having a larger gap energy in II-VI compounds. We list in Table 3.5 the gap energy in various elemental and compound semiconductors. Since the gap energy is sensitive to interatomic distances, we choose for comparison materials with closest interatomic distances possible in the table.

Another measure of ionicity in a compound is the difference in the *electronegativity* of its constituent elements. By electronegativity, we mean the power of an atom to attract an electron to itself in a compound. In Table 3.6, we list the electronegativity values for the elements used in III-V and II-VI compounds. In comparing the static and optical dielectric constants in Eq. (3.114), only the magnitude of the charge transfer is determined, leaving the polarity of the charge still unspecified. However, it is clear from the electronegativity value listed in Table 3.6 that electrons will be transferred from the II*b* element to the VI*b* element in a II-VI compound and from the III*b* element to the V*b* element in a III-V compound, making II and III atoms positively charged. In other words, based on the argument of electronegativity, the polarity of the charge

[13] N. B. Hannay (ed.), *op. cit.*

TABLE 3.5 *Gap energy in a selected number of elemental and compound semiconductors**

Semiconductors	Si	AlP	ZnS	Ge	GaAs	InP	CdS
Interatomic distance, Å	2.34	2.36	2.34	2.44	2.44	2.54	2.52
Gap energy, ev	2.5 (1.11)	3.0	3.7	0.80 (0.665)	1.40	1.29	2.4

Semiconductors	InAs	AlSb	ZnTe	Gray-Sn	InSb	CdTe
Interatomic distance, Å	2.615	2.655	2.635	2.80	2.80	2.805
Gap energy, ev	0.36	1.62	2.0	0.17	0.36	1.6

* The values for Si and Ge are direct-gap energies measured across energy bands at $k[000]$, with the values of indirect-gap energies quoted in parentheses.

transfer is such that it tends to be in the direction of ionic rather than covalent binding.

Other comments concerning Table 3.6 are summarized as follows. The elements are so arranged that the degree of ionicity generally decreases from the top toward the bottom, from the left toward the right and from the upper left corner toward the lower right corner. According to Pauling, the percentage of ionic binding in a compound increases as the difference in electronegativity between its elements increases. Where the value of electronegativity is not sufficiently accurate in comparing the relative degree of ionicity in different compounds, data from measurements of the difference in the static and optical dielectric constants are used in the arrangement of Table 3.6. The percentage of ionic binding, based on Pauling's criterion, is estimated to be 22 per cent in CdS and ZnS and 9 per cent in InAs. However, these values are useful only in comparing the relative degree of ionicity in II-VI and III-V compounds. Theoretical analysis by Birman[14] and Browne[15] has led to a value close to 65 per cent ionic binding in ZnS, which agrees well with the value of 67 per cent concluded from an analysis of the piezo-electric constant by Saksena.[16]

The crystal structures of the various compounds are also shown in Table 3.6, together with the electronegativity values. The zinc blende structure may be considered as two interpenetrating cubic

[14] J. L. Birman, *Phys. Rev.*, vol. 109, p. 810, 1958.
[15] P. F. Browne, *J. Electronics*, vol. 2, p. 154, 1956.
[16] B. D. Saksena, *Phys. Rev.*, vol. 81, p. 1012, 1951.

TABLE 3.6 *Electronegativity values (according to the Pauling scale*) for elements used in III-V and II-VI compounds*

III \\ V	Al 1.5	In 1.6	Ga 1.6	II \\ VI	Mg 1.2	Cd 1.4	Zn 1.5	Be 1.5
N 3.0	W	W	W	O 3.5	NaCl	NaCl	W	W
P 2.1	ZB	ZB	ZB	S 2.5	NaCl	W ZB	W ZB	ZB
As 2.0	ZB	ZB	ZB	Se 2.3	NaCl	W ZB	W ZB	ZB
Sb 1.9	ZB	ZB	ZB	Te 2.1	W	ZB	ZB	ZB

W = Wurtzite structure   ZB = Zinc blende structure

* L. Pauling, "The Nature of the Chemical Bond," Cornell University Press, Ithaca, N.Y., 1948.

close-packed (face-centered) structures translated with respect to each other by ¼ of the body diagonal. The wurtzite structure (Fig. 3.28) may be considered as two interpenetrating hexagonal close-packed structures (Fig. 2.13) displaced with respect to each other by a distance $3c/8$ along the c axis. In these two structures, the nearest neighbors form identical tetrahedral units as shown in Figs. 2.4 and 3.28. Nine of the twelve second nearest neighbors also have identical locations, the other three (corresponding to atoms 10, 11, and 12 of Fig. 2.13) being rotated by an angle of 60 degrees in the zinc blende structure. Therefore, significant differences between the two structures come only in the third nearest neighbors. This is the reason why some compounds exist in both structures. However, a close examination shows that the wurtzite structure is more favorable for crystals with large charge differences between the two kinds of atoms. In other words, the general tendency is such that the wurtzite structure is more prone to have a higher degree of ionicity than the zinc blende structure.

Energy band structure. In Sec. 3.8, we have pointed out that the energies ε of conduction band electrons and valence band holes depend on the wave vector $\mathbf{k}$ associated with the wave motion of an electron. It is of theoretical interest as well as of practical importance to know the general behavior of the variation of ε with respect to $\mathbf{k}$. Figure 3.29 shows the two-dimensional plot of ε vs. $\mathbf{k}$ curve for

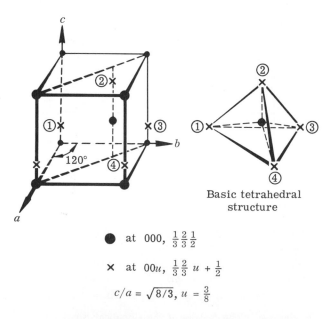

Figure 3.28

Basic tetrahedral structure

$\bullet$ at 000, $\frac{1}{3}\frac{2}{3}\frac{1}{2}$

$\times$ at $00u$, $\frac{1}{3}\frac{2}{3}\ u+\frac{1}{2}$

$c/a = \sqrt{8/3},\ u = \frac{3}{8}$

Ideal Wurtzite structure (hexagonal).

a selected number of elemental and compound semiconductors. Since the wave vector **k** is defined by three parameters k_x, k_y, and k_z, a two-dimensional plot requires either holding two parameters constant (for example k_y and k_z) or maintaining the direction of **k** constant. On account of the symmetry properties of a crystal, it is far

Figure 3.29

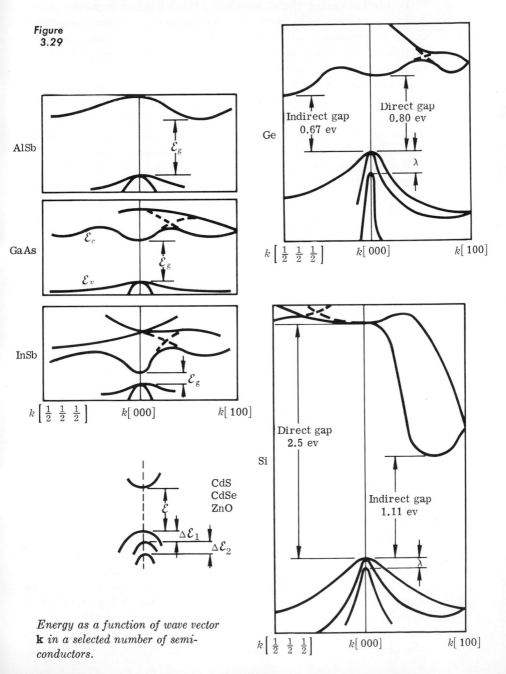

Energy as a function of wave vector **k** *in a selected number of semiconductors.*

more meaningful to express ε vs. $\mathbf{k}$ curve along a certain direction in k space (or the reciprocal lattice space). In Fig. 3.29, the two principal directions are chosen as [100] and [111], and the variations of ε along those two directions in k space are plotted respectively in the right half and the left half of the diagram.

Before explaining the difference in the ε against $\mathbf{k}$ curves shown in Fig. 3.29, we should like to discuss briefly the concept of *Brillouin zone*. As mentioned earlier in Sec. 3.1, the number of quantum states in the energy band based on the free-electron theory of metals should be the same as the number of atomic quantum states. This limits the size of the volume in k space. For example, in Eq. (3.20), the integration actually is carried out over a finite volume in k space (or p space). This volume is known as the Brillouin zone. In the following, we shall use the face-centered cubic structure as an example to illustrate how the Brillouin zone can be found.

Referring to Fig. 2.20b and our discussion in Sec. 2.5, the basis vectors of the primitive cell of a face-centered cubic structure are given by

$$\mathbf{a'} = \frac{\mathbf{a} + \mathbf{b}}{2} \qquad \mathbf{b'} = \frac{\mathbf{b} + \mathbf{c}}{2} \qquad \mathbf{c'} = \frac{\mathbf{c} + \mathbf{a}}{2} \qquad (3.115)$$

The basis vectors in the reciprocal lattice can be found from Eq. (3.99) as

$$\mathbf{a}^* = \frac{\mathbf{a} + \mathbf{b} - \mathbf{c}}{a^2} \qquad \mathbf{b}^* = \frac{-\mathbf{a} + \mathbf{b} + \mathbf{c}}{a^2} \qquad \mathbf{c}^* = \frac{\mathbf{a} - \mathbf{b} + \mathbf{c}}{a^2} \qquad (3.116)$$

Substituting Eq. (3.116) into Eq. (3.100) gives

$$\mathbf{k} = (2\pi/a^2)[(h_1 - h_2 + h_3)\mathbf{a} + (h_1 + h_2 - h_3)\mathbf{b} \\ + (-h_1 + h_2 + h_3)\mathbf{c}] \qquad (3.117)$$

Using the smallest sets of integers for h_1, h_2, and h_3, we find that the shortest nonzero $\mathbf{k}$'s are the eight vectors

$$(2\pi/a^2)(\pm\mathbf{a} \pm \mathbf{b} \pm \mathbf{c}) \qquad (3.118)$$

Figure 3.30 shows the Brillouin zone for a face-centered cubic crystal. The zone boundaries are determined by the eight vectors [Eq. (3.118)] at their midpoints but the six corners of the octahedron are truncated by the six planes normal to the six vectors $(2\pi/a^2)(\pm\mathbf{a})$, $(2\pi/a^2)(\pm\mathbf{b})$, and $(2\pi/a^2)(\pm\mathbf{c})$ at their midpoints. These six vectors are obtained by using the next-to-smallest sets of integers for h_1, h_2, and h_3 in Eq. (3.117); for example, $h_1 = h_2 = 1$ and $h_3 = 0$, and so on. The volume of the truncated octahedron shown in Fig. 3.30 is equal to $\frac{1}{2}(4\pi/a)^3$. Referring to Eq. (1.133) and expressing $\mathbf{k}$ in rectangular coordinates, we find the density of states per unit volume equal to $N/V = 4/a^3$, the number 4 being equal to the number of atoms per unit cell of a face-centered cubic crystal.

Therefore, the size of the Brillouin zone is so defined that the number of states calculated from Eq. (1.133) in a volume V of a given crystal is equal to the number of atomic states multiplied by the number of unit cells in volume V. This is one way of stating the reason for the truncation of the octahedron. The extreme values of k shown in Fig. 3.29 is expressed in units of $2\pi/a$; hence, the notations $k[100]$ and $k[\frac{1}{2} \ \frac{1}{2} \ \frac{1}{2}]$ are adopted. For a more detailed discussion of the Brillouin zones, the reader is referred to the books[17] by Seitz and by Kittel.

Now let us return to Fig. 3.29. A quick comparison of the curves for different semiconductors shows that there exist two principal types: the *direct-gap* and the *indirect-gap semiconductors*. In indirect gap semiconductors (Ge, Si, and AlSb), the minimum of the conduction band edge and the maximum of the valence band edge occur at different values of **k** while in direct-gap semiconductors (GaAs, InSb, CdSe, CdS, and ZnO), the band extrema have the same value of **k**. In an optical transition across the bands, quantum mechanical selection rules require that the wave number involved in such a transition be conserved. As will be discussed in later chapters, this essential difference between direct-gap and indirect-gap semiconductors will be manifested not only in the recombination process of excess carriers but in the optical absorption process also.

In Table 3.7, we list the gap energy and effective masses of a selected number of elemental and compound semiconductors. For indirect-gap semiconductors, two gap energies are given: the direct-gap energy which measures the energy difference across the bands

[17] F. Seitz, "The Theory of Modern Solids," McGraw-Hill Book Company, New York, 1940; C. Kittel, "Introduction to Solid State Physics," John Wiley & Sons, Inc., New York, 1956.

Figure 3.30

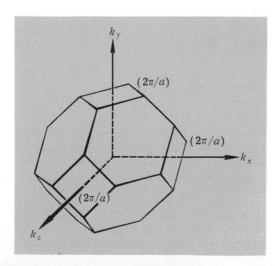

The first Brillouin zone in a face-centered cubic structure. The boundaries of the zone along the three principal axes are $2\pi\mathbf{a}/a^2$, $2\pi\mathbf{b}/a^2$, and $2\pi\mathbf{c}/a^2$, while those along the four body diagonals are (π/a^2) $(\pm \mathbf{a} \pm \mathbf{b} \pm \mathbf{c})$.

TABLE 3.7 *Band structures and effective masses of a selected number of elemental and compound semiconductors (Gap energies quoted in parentheses refer to values at 0°K.)*

	Band structure	Effective masses m^*/m_0	
		Conduction band	Valence band
Si	Indirect-gap $\mathcal{E}_g = 1.11$ ev $\mathcal{E}_{g0} = 2.50$ ev $\lambda = \ldots$	$m_l = 0.98m_0$ $m_t = 0.19m_0$ $m(k = 0) = \ldots$	0.50 0.16 $\ldots$
Ge	Indirect-gap $\mathcal{E}_g = 0.67$ ev $\mathcal{E}_{g0} = 0.80$ ev $\lambda = 0.28$ ev	$m_l = 1.58m_0$ $m_t = 0.082m_0$ $m(k = 0) = 0.037m_0$	0.30 0.04 $\ldots$
GaAs	Direct-gap $\mathcal{E}_g = 1.40$ ev $\lambda = 0.33$ ev	0.07	0.68 0.12 0.20
GaSb	Direct-gap $\mathcal{E}_g = (0.81$ ev) $\lambda = 0.86$ ev	0.047	0.23 0.06 $\ldots$
InAs	Direct-gap $\mathcal{E}_g = 0.36$ ev $\lambda = 0.43$ ev	0.024	0.41 0.025 0.083
InSb	Direct-gap $\mathcal{E}_g = 0.18$ ev(0.236) $\lambda = 0.90$ ev	0.0116	0.25 0.015 $\ldots$
CdS	Direct-gap $\mathcal{E}_g = 2.58$ ev $\Delta\mathcal{E}_1 = 0.016$ ev $\Delta\mathcal{E}_2 = 0.057$ ev	0.204	$m_\perp = 0.7m_0$ $m_\parallel = 5m_0$ ($\perp$ and $\parallel$ refer to the c axis)
ZnO	Direct-gap $\mathcal{E}_g = 3.35$ ev $\Delta\mathcal{E}_1 = 0.007$ ev $\Delta\mathcal{E}_2 = 0.038$ ev	0.38	1.8 (estimated)

REFERENCES: *Group IV semiconductor:* Dresselhaus, Kip, and Kittel, *Phys. Rev.*, vol. 98, p. 368, 1955; Lax, Zeiger, Dexter, and Rosenblum, *Phys. Rev.*, vol. 93, p. 1418, 1954; Zwerdling, Lax, Roth, and Button, *Phys. Rev.*, vol. 114, p. 80, 1959. *III-V compound:* H. Ehrenreich, *Jour. Appl. Phys.*, vol. 32, p. 2155, 1961; B. Lax et al., *Phys. Rev.*, vol. 122, p. 33, 1961. *II-VI compound:* J. J. Hopfield and D. G. Thomas, *Phys. Rev.*, vol. 122, p. 35, 1961; D. G. Thomas and J. J. Hopfield, *Phys. Rev.*, vol. 116, p. 573, 1959; R. E. Dietz, J. J. Hopfield, and D. G. Thomas, *Jour. Appl. Phys.*, vol. 32, p. 2282, 1961; A. R. Hutson, *Jour. Appl. Phys.*, vol. 32, p. 2287, 1961.

at $k[000]$ and the indirect-gap energy which is the energy difference between the conduction band minimum and valence band maximum. The indirect-gap energy ε_g is usually determined by a measurement of the temperature dependence of the Hall coefficient in the high temperature or intrinsic region, while the direct-gap energy ε_{g0} is directly measurable only in an optical absorption experiment in which a semiconductor absorbs a photon of energy $\hbar\omega = \varepsilon_{g0}$ by exciting an electron from the valence band to the conduction band at $k[000]$.

The energy expression, Eq. (3.91), for conduction band electrons can be applied both to electrons at conduction band minimum and to electrons at $k[000]$, yielding of course two separate sets of effective-mass values in an indirect-gap semiconductor. The effective masses m_t and m_l measured in a cyclotron resonance experiment refer to masses of electrons near the conduction band minimum. Since at ordinary temperatures, most conduction electrons occupy states near the conduction band minimum on account of energy considerations, the values of m_t and m_l quoted in Table 3.7 will determine the effective mass m^* in the mobility expression, Eq. (3.11), and in the density-of-states expression, Eq. (3.21). We shall work out the exact expression relating the effective masses in these expressions to m_t and m_l in Chapter 5.

If we apply Eq. (3.91) to conduction electrons at $k[000]$ in an indirect-gap semiconductor, we obtain not only a different constant term, that is ε_c, but also a different coefficient for the $(k - k_0)^2$ term, implying a different effective mass. It turns out that the constant energy surface near $k[000]$ is spherical and consequently only one value of $m(k = 0)$ is quoted in Table 3.7. Like the direct-gap energy ε_{g0}, this effective mass will come into the picture when we discuss the optical properties of an indirect-gap semiconductor.

The situation with respect to the conduction band is much simpler in direct-gap semiconductors such as GaAs. Since the conduction band minimum is located at $k[000]$ opposite the valence-band maximum, there is only one gap energy of importance. For example, the same energy $\varepsilon_g = 1.40$ ev is used for GaAs in the calculation of intrinsic carrier concentration n_i from Eq. (3.36) and in determining the threshold photon energy in an optical absorption process. Likewise, the value of the effective mass for conduction electrons is also unique in direct-gap semiconductors. The same value, for example $m^* = 0.07m_0$ in GaAs, is used in calculating the conductive property [Eqs. (3.11) and (3.21)] as well as the optical property of the semiconductor.

Let us now turn our attention to the valence band electrons. The electron wave function near $k[000]$ is s-like in the conduction band and p-like in the valence band. As a consequence, the valence band is triply degenerate at $k[000]$. On account of spin-orbit interaction, however, the degeneracy is partly removed, resulting in a split-off band with an energy lower than the other two bands by an amount given as λ in Table 3.7. Therefore, in general, there are three effective masses for valence band holes. Since the value of λ is fairly large compared to kT at ordinary temperatures, holes normally exist only in the two top valence bands. These two bands are referred to as the *heavy-hole* and *light-hole bands*. The values of m^* for these two bands can be determined from cyclotron resonance experiment by evaluating constants A, B, and C in Eq. (3.112). In general, the constant energy surface for holes is warped (Fig. 3.27), but for all practical purposes, the warped energy surface can be approximated by a spherical energy surface with an average effective mass. It is this average effective mass which is given in Table 3.7.

We also note from Fig. 3.29 that the valence band structures in direct-gap and indirect-gap semiconductors are qualitatively the same. Therefore, the difference between a direct-gap and indirect-gap semiconductor lies in the structure of the conduction band. It may be worthwhile to point out, however, that the II-VI compounds shown in Fig. 3.29 crystallize in the wurtzite form which is uniaxial. The uniaxial crystalline field further removes the degeneracy of the valence band at $k[000]$. The values of m^*/m_0 in CdS and ZnO are thus for the topmost valence band. Furthermore, the group theoretic analysis shows that based on the symmetry property of the conduction- and valence-band wave functions, both energy bands should have the general form

$$\mathcal{E} = A(k_x^2 + k_y^2) + Bk_z^2 \pm C(k_x^2 + k_y^2)^{1/2} \qquad (3.119)$$

where the c axis of the hexagonal structure is chosen to be in the z direction and the C term represents the splitting of the two spin states by the spin-orbit coupling and the crystalline field. Therefore, in general, the effective masses in hexagonal II-VI compounds are expected to be anisotropic, the degree of anisotropy depending on the ratio A/B. Studies of exciton structure by Hopfield[18] and by Wheeler and Dimmock[19] have shown conclusively that the conduction bands in CdS and CdSe are nearly spherical while the valence bands in these crystals are highly anisotropic. In cases where the ratio A/B is large, the hole mass parallel to the c axis can not be determined with accuracy from such studies.

Mobilities of free carriers. The difference between elemental and compound semiconductors is further reflected in the magnitude as well as the temperature dependence of their mobility. In group IV semiconductors, the mobility at high temperatures (say 200°K and

---

[18] J. J. Hopfield, *Jour. Appl. Phys.*, vol. 32, p. 2277, 1961.
[19] J. O. Dimmock and R. G. Wheeler, *Jour. Appl. Phys.*, vol. 32, p. 2271, 1961.

higher) is usually limited due to scattering of electron or hole motion by lattice vibration. Since the different scattering mechanisms will be discussed and the associated expressions for mobilities will be derived in the next chapter, we shall only quote those results which are pertinent to our present discussion. As discussed earlier in Sec. 2.10, p. 118, there are two principal modes of lattice vibration: the acoustic and optical modes. The limit of mobility due to acoustic modes of lattice vibration has the following principal dependence:

$$\mu_l = A T^{-3/2} m^{*-5/2} \varepsilon_1^{-2} \tag{3.120}$$

where ε_1 is the movement of the band edge per unit dilation caused by lattice vibration and A is a numerical constant which among other things contains the density of the crystal and the velocity of sound in the crystal. The reader is referred to Eq. (4.85) for a complete expression for μ_l.

In heavily doped semiconductors, scattering by ionized impurities plays an important role. The limit of mobility due to this process is given by Eq. (4.63) or

$$\mu_i = B T^{3/2} m^{*-1/2} N_i^{-1} \tag{3.121}$$

where N_i is the impurity concentration and B is another proportionality constant which varies very slowly with temperature as $\log (T/T_0)$. The lower value of the two predicted from Eqs. (3.120) and (3.121) limits the mobility of free carriers. From the functional dependences of μ_l and μ_i we see that impurity scattering is important in impure samples and at low temperatures. For elemental semiconductors Ge and Si, scattering by acoustic phonons and that by impurities are the principal mechanisms which limit the mobilities of electrons and holes. Except for fine details which may be attributed to other scattering mechanisms, both the magnitude and the temperature dependence of mobilities in Ge and Si can be explained on the basis of Eqs. (3.120) and (3.121).

In compound semiconductors, scattering of carriers by optical modes of lattice vibration plays a very important role. As pointed out in Sec. 2.10, p. 118, in the optical modes, the two neighboring atoms move in opposite directions relative to each other. Because of the ionic charge carried by each atom, their relative movement constitutes an electric polarization which in turn produces an electric field. Diffraction of the motion of free carriers by the field limits the mobility of carriers. According to an analysis by Howarth and Sondheimer,[20] the limit of mobility due to polar scattering is given by

$$\mu_{lo} = \frac{1}{2\alpha\omega_l} \frac{e}{m^*} \frac{8}{3\sqrt{\pi}} \frac{F(\Theta_l/T)[\exp(\Theta_l/T) - 1]}{(\Theta_l/T)^{1/2}} \tag{3.122}$$

[20] D. J. Howarth and E. H. Sondheimer, *Proc. Roy. Soc.*, London, vol. 70, p. 124, 1957.

where $\Theta_l = \hbar\omega_l/k$, $\hbar\omega_l$ being the energy associated with longitudinal optical phonons. In Eq. (3.122), α is a dimensionless interaction parameter defined as

$$\alpha = \frac{e^2}{4\pi\epsilon_0}\left(\frac{m^*}{2\hbar\omega_l}\right)^{\frac{1}{2}}\frac{1}{\hbar}\left(\frac{\epsilon_0}{\epsilon_\infty} - \frac{\epsilon_0}{\epsilon}\right) \tag{3.123}$$

where $\epsilon_\infty/\epsilon_0$ is the square of the optical index of refraction.

The polarization caused by relative displacement of ions not only is responsible for the interaction between the wave motion of electrons and the optical mode of lattice vibration but also contributes to the dielectric constant. Hence, it is expected that Eqs. (3.122) and (3.123) should be related. The function $F(\Theta_l/T)$ is a slowly varying function which decreases from unity at high temperatures to a minimum value 0.6 when $T = \Theta_l$ and then increases steadily with Θ_l/T, being equal to $(3/8)(\pi\Theta_l/T)^{\frac{1}{2}}$ at low temperatures. At temperatures $T < \Theta_l$, the main temperature dependence in Eq. (3.122) comes from the factor $[\exp(\Theta_l/T) - 1]$ which is the result of the Bose-Einstein factor for the occupancy number of optical phonons. The reader is referred to the book by Ziman[21] for a quantum mechanical derivation of Eq. (3.122) and also to an excellent review article by Hutson[22] for a discussion of the role of longitudinal and transverse optical modes of lattice vibration as well as for a physical picture of the electron-phonon interaction.

To estimate the value of μ_l in III-V compounds, we choose InAs and Ge for comparison. The effective mass m^* for Ge in Eq. (3.120) is the average of transverse and longitudinal masses $1/m^* = (2/m_t + 1/m_l)/3$ or $m^*/m_0 = 0.12$ while that for InAs is equal to $m^*/m_0 = 0.024$. Therefore, based on the ratio of the effective masses, Eq. (3.120) predicts a value of $\mu_l = 220{,}000$ cm²/volt-sec in InAs. This value is to be compared with the observed mobility of $33{,}000$ cm²/volt-sec and the calculated polar mobility μ_{lo} of $40{,}000$ cm²/volt-sec from Eq. (3.122). Similar comparisons can be made for other III-V compounds. The large difference in electron and hole mobilities in III-V compounds can also be explained by Eq. (3.122) as due to the difference in the effective masses. For example, the ratio of hole to electron effective mass in InAs is equal to 17, yielding a mobility ratio of 1/69 from Eq. (3.122). The observed ratio is 460/33,000 or 1/72. On the other hand, from Eq. (3.120) for acoustic lattice scattering, the ratio would be 1/1200, assuming a same deformation potential $\mathcal{E}_1$ for conduction and valence band edges. Based on the above calculations, it is generally believed that mobilities in III-V compounds are limited by polar

[21] J. M. Ziman, "Electrons and Phonons," Oxford University Press, Oxford, 1960.
[22] A. R. Hutson, Semiconducting Properties of Some Oxides and Sulfides, appeared in N. B. Hannay (ed.), "Semiconductors," Reinhold Publishing Corporation, New York, 1959.

scattering of Eq. (3.122) in pure samples. Ionized impurity scattering, however, will take over as the mobility limiting process in impure samples and at low temperatures.

Now let us discuss the scattering process which limits the mobilities in II-VI compounds. Among the II-VI compounds ZnO and CdS have been most extensively studied. It is known that both crystals are strongly piezoelectric. By *piezoelectric*, we mean that a polarization field (electric) is induced in a crystal by the application of a mechanical stress or vice versa. The reader is referred to Sec. 7.8 for a more detailed discussion of the piezoelectric effect. For our present discussion, the piezoelectric effect means that lattice vibration can now interact with electron motion through the electric field which accompanies the lattice wave. According to an analysis by Hutson,[23] the limit of mobility due to piezoelectric scattering is given by

$$\mu_{lp} = 1.44(m_0/m^*)^{3/2}(300/T)^{1/2}(K^2\epsilon_0/\epsilon)^{-1} \qquad (3.124)$$

where K is a dimensionless quantity known as the *electromechanical coupling constant*. Using $m^* = 0.38m_0$, $\epsilon/\epsilon_0 = 8.2$ and $K^2 = 0.074$ in ZnO, we find $\mu_{lp} = 640$ cm²/volt-sec at $T = 300°K$ from Eq. (3.124) as compared to the measured value of 200 cm²/volt-sec.

In summary, from the above discussion, we see that lattice vibrations can diffract the motion of free carriers in many different ways: (1) by modulating the band edges, (2) by inducing a polarization field caused by the relative movement of ionic charges, and (3) by inducing a polarization field through the piezoelectric effect. In elemental semiconductors, only the first mechanism exists. In compound semiconductors with zinc blende and wurtzite structures, all three mechanisms can exist simultaneously. From the measured values of mobilities in III-V compounds, it seems that the second mechanism is the dominating scattering mechanism.

The reader is cautioned against using the temperature dependence of mobility as the sole judge in determining the relative importance of different scattering processes. Ehrenreich[24] has shown that the temperature dependence in III-V compounds can be explained by a combination of polar scattering [Eq. (3.122)] and other scattering processes (for example, electron-hole scattering in InSb and ionized impurity scattering in GaAs). The most serious difficulty in comparing the theoretically calculated and experimentally measured values of mobility is the lack of controlled purity in compound semiconductors. A high resistivity sample does not necessarily mean a high purity sample because the sample may be highly compensated with donors and acceptors. The situation is even worse with the II-VI com-

[23]   A. R. Hutson, *Jour. Appl. Phys.*, vol. 32, p. 2287, 1961.
[24]   H. Ehrenreich, *Phys. Rev.*, vol. 120, p. 1951, 1960; *Jour. Phys. Chem. Solids*, vol. 9, p. 129, 1959.

pounds. Until recently, most measurements of the electrical properties of II-VI compounds were made on powder or polycrystalline samples, and p-type conductivity was never observed. Therefore, we still wait for a thorough analysis of the mobility data in II-VI compounds. However, judging from the experimental data in ZnO, it seems probable that piezoelectric scattering is dominant over acoustic and polar scattering. Since $\Theta_l = 920°K$ is relatively high in ZnO, polar scattering alone would predict around room temperature a much steeper temperature dependence of mobility than the temperature dependence experimentally observed. Furthermore, to explain the magnitude of the thermoelectric voltage in ZnO would require an unrealistically high deformation potential $\mathcal{E}_1$ if acoustic scattering were the dominant mechanism.

Gap energy states: impurity and exciton states. Figure 3.31 summarizes the information on energy levels created in the forbidden gap by impurity atoms in germanium and silicon. We shall start our discussion of impurity states with the group IIIb (acceptor) and group Vb (donor) impurities of which we have already given a qualitative description in Sec. 3.2. For these impurities the hydrogenic model works surprisingly well. We may think of a donor atom as consisting of a positively-charged core D^+ surrounded by a charge cloud with a negative charge e. In applying the Bohr model outlined in Sec. 1.3 to the donor atom, however, we must recognize the following fact. Because of the difference in the dielectric constant and the effective mass, the electron orbit in a semiconductor is much larger than that in a free atom and the ionization energy of a donor state should be considerably smaller than that of a hydrogen atom. According to Eq. (1.17), we have for the energy levels of a donor atom

$$\mathcal{E}_d(n) = \frac{-13.6(m^*/m_0)(\epsilon_0/\epsilon)^2}{n^2} \qquad \text{electron-volt} \qquad (3.125)$$

where n is the principal quantum number. A similar equation is obtained for the acceptor states.

The above equation is exact if the energy surface is spherical. To apply Eq. (3.125) to impurity states in Ge and Si, we must take an appropriate average of the effective masses. In estimating the ionization of donor states, we assume $1/m^* = (2/m_t + 1/m_l)/3$. Thus, we have: $\epsilon/\epsilon_0 = 15.8$ and $m^*/m_0 = 0.12$ in Ge, and $\epsilon/\epsilon_0 = 12$ and $m^*/m_0 = 0.27$ in Si. Substituting these values in Eq. (3.125) gives $\mathcal{E}_d = \mathcal{E}_d(\infty) - \mathcal{E}_d(1) = 0.65 \times 10^{-2}$ ev in Ge and 2.6×10^{-2} ev in Si. The calculated values of ionization energy are in fair agreement with the measured values. We should emphasize, however, that to calculate the ionization energy accurately, we must know the electron wave function, especially near the donor or acceptor ion core. Such information is available from electron spin resonance experiment.[25] The reader should consult the review article by Kohn[26] for the development of quantum theory for shallow donor and acceptor states.

Equation (3.125) should apply equally well to shallow donor and acceptor states in III-V compounds. The group IIb elements such as Zn, Cd, and Hg and the group VIb elements such as Se and Te will usually enter the lattice substitutionally, with the group II atoms replacing III atoms as acceptors and group VI atoms replacing V atoms as donors. Since electrons in III-V compounds have a smaller effective mass than holes, the ionization energy of donors is expected to be much smaller than that of acceptors. Furthermore, because purification is still a problem, most III-V compounds either have a high degree of compensation or have a high impurity concentration. As a consequence, impurity states are no longer discrete but form an impurity band which merges into either the conduction or valence band. Because of the high degree of compensation and the formation of impurity band, the measurement of Hall coefficient at low temperatures may not yield as much useful information as we expect from a sample of controlled purity as in the case of Ge and Si.

Let us now refer again to Fig. 3.31 for the deep-lying impurity states. These states not only have energy levels deep in the forbidden gap but may have several levels depending on their state of ionization. Take Zn as an example. Since Zn has a $4s^2$ configuration outside closed shells, it can accept a maximum of two electrons in Ge and Si. The two energy states (0.03 ev) and (0.09 ev) in Ge correspond to a singly ionized Zn^- and a doubly ionized Zn^{2-} respectively. In the same way, the transition elements Mn, Fe, Co, and Ni with two $4s$ electrons in the outer shell act as double acceptors. Electrons in the incomplete $3d$ shell apparently are unaffected. The other element worth mentioning is gold. Since it belongs to the first column of the periodic table, it can accept a maximum of three electrons. The three states from top to bottom in Fig. 3.31a correspond to Au^{3-}, Au^{2-}, and Au^-. In addition to these three levels, Au can also behave as a donor, indicating that it may lose its single valence electron to assume a noble gas configuration. The donor levels created by Mn and Fe in Si may indicate that these elements occupy interstitial positions in silicon. The reader is referred to the review article by Conwell[27] for further discussion of impurity states

[25] G. Feher, *Phys. Rev.*, vol. 103, p. 834, 1956; *Jour. Phys. Chem. Solids*, vol. 8, p. 486, 1958; G. W. Ludwig and H. H. Woodbury, Electron Spin Resonance in Semiconductors, in F. Seitz and D. Turnbull (eds.), "Solid State Physics," vol. 13, p. 223, Academic Press Inc., New York, 1962.

[26] W. Kohn, Shallow Impurity States in Si and Ge, in F. Seitz and D. Turnbull (eds.), "Solid State Physics," vol. 5, p. 258, Academic Press Inc., New York, 1957.

[27] E. M. Conwell, *Proc. IRE*, vol. 46, p. 1281, 1958.

Figure 3.31

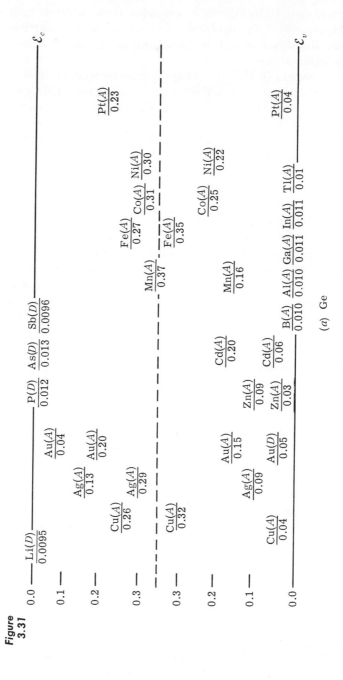

(a) Ge

$\mathcal{E}_c$

Li(D) $\overline{0.033}$ P(D) $\overline{0.044}$ As(D) $\overline{0.049}$ Sb(D) $\overline{0.039}$ Bi(D) $\overline{0.067}$

0.1 —
0.2 —
0.3 —
0.4 —
0.5 —

Au(A) $\overline{0.54}$ Zn(A) $\overline{0.55}$ Mn(D) $\overline{0.53}$ Fe(D) $\overline{0.55}$

- - - - - - - - - - - - - -

0.5 —   Cu(A) $\overline{0.49}$
0.4 —   Au(A) $\overline{0.35}$ Zn(A) $\overline{0.31}$
0.3 —   Cu(A) $\overline{0.24}$ Fe(D) $\overline{0.40}$
0.2 —
0.1 —

Zn,B (A) $\overline{0.126}$ Zn,Al (A) $\overline{0.092}$ $\overline{0.078}$ Zn,Ga (A) $\overline{0.083}$ B(A) $\overline{0.045}$ Al(A) $\overline{0.057}$ Ga(A) $\overline{0.065}$ In(A) $\overline{0.16}$ Tl(A) $\overline{0.26}$

$\mathcal{E}_v$

(b) Si

Energy levels of impurities in (a) Ge and (b) Si. (E. M. Conwell, Proc. IRE, vol. 46, p. 1281, 1958.)

in Ge and Si, and for references to the data presented in Fig. 3.31. We would like to mention, however, that these deep-lying states can act as recombination centers for excess carriers. Especially notorious is copper which diffuses very fast in Ge and Si. To obtain samples with long lifetime for excess carriers, it is essential that the concentration of deep-lying states should be kept as low as possible.

The role of other impurity states besides shallow donor and acceptor states has not been extensively studied in III-V compounds for the same reason mentioned earlier. It is expected, however, that a similar argument based on covalent bond (electrons needed to complete the tetrahedral bond) can be applied to impurities in III-V compounds. For example, there is evidence that Fe, Co, and Ni create deep-lying acceptor states in GaAs. However, any systematic study of deep-lying states must wait until we can control the purity of these compounds.

The situation with II-VI compounds is even more complicated than that with III-V compounds. According to the covalent bond model, we expect that the group I elements will serve as acceptors replacing the group II element while the group VII elements act as donors replacing the group VI element. To some extent, this covalent bond picture still applies. However, we also find that excess group II elements, such as excess Zn in ZnO and excess Cd in CdS and CdTe, act as donors. As a matter of fact, most insulating II-VI compounds can be made n-type by exposing the crystal to Zn or Cd vapor at high temperatures. The ionization energies for the donor levels seem to be in fair agreement with the value predicted from the hydrogenic model, for example, $\mathcal{E}_d = 0.05$ ev in ZnO and 0.02 ev in CdTe. The p-type conduction has been observed only in CdTe with an ionization energy which may vary from 0.2 ev to 0.4 ev. The level is tentatively assigned as being associated with copper. The dopants commonly used in CdS, CdTe, and CdSe are: Ag and Cu as acceptors and Ga and Cl as donors. However, a definite identification of the donor and acceptor species with the energy levels has not been achieved. Even in high resistivity materials, the concentrations of donors and acceptors are still high but they are nearly compensated. This makes a precise analysis of the Hall measurement very difficult.

Finally, we should like to discuss briefly energy states of another kind which exist in the forbidden gap. Because an electron and a hole have an attractive coulomb interaction, they together may form stable bound states called *exciton* states. Here again we may apply the hydrogenic model and obtain for the energy of the exciton states

$$\mathcal{E}_{\text{ex}} = -\frac{m_r e^4}{2\hbar^2 (4\pi\epsilon)^2} \frac{1}{n^2} \tag{3.126}$$

where n is an integer (the principal quantum number) indicating the various exciton states, m_r is the reduced mass related to the effective masses of electrons and holes by $m_r = m_e m_h/(m_e + m_h)$ and the energy $\mathcal{E}_{ex}$ is referred to the conduction band edge. The observation and interpretation of the exciton spectra[28] have greatly clarified the energy band structure in CdSe, CdS, and ZnO. Two conditions must be met, however, in order that bound exciton states may be observable. First, Eq. (3.126) is based on the assumption that screening due to other electrons and holes is negligible. Casella[29] has shown that bound states exist only if the *Debye screening length* L is much larger than the exciton Bohr radius a_{ex} with

$$L = (kT\epsilon/\bar{n}e^2)^{\frac{1}{2}} \qquad \text{and} \qquad a_{ex} = 4\pi\epsilon\hbar^2/(e^2 m_r) \qquad (3.127)$$

where $\bar{n}$ is the total density of conduction electrons and free holes. Second, since the energy $\mathcal{E}_{ex}$ is very small, of the order of 0.01 ev or so, the temperature must be very low so that an electron-hole pair will appear in one of the bound states instead of being in the ionized state (free electrons and holes). Based on these two considerations, conditions for studying exciton states are more favorably met in insulating materials and at low temperatures. Taking $\epsilon/\epsilon_0 = 16$ and $m_r/m_0 = 0.1$, we find $\mathcal{E}_{ex} = 5 \times 10^{-3}$ ev and $a_{ex} = 80$ Å. At $T = 3°K$, $\bar{n}$ must be smaller than $10^{14}/cm^3$ in order that $L > a_{ex}$.

PROBLEMS

3.1 The mobility of electrons in germanium is 3900 cm²/volt-sec at room temperature. Using the mass of electrons $m^* = 0.12\, m_0$ (the free electron mass), find the value of the relaxation time τ. At what frequency will the electric field and current be 45 degrees out of phase?

3.2 Using Eqs. (3.10) and (3.12), show that the average velocity of electrons is given by Eq. (3.13) and that Eq. (3.13) satisfies Eq. (3.14).

3.3 The mobility of electrons varies with temperature as $T^{-1.66}$ in germanium. Using the value of $\mu = 3900$ cm²/volt-sec at 300°K and the Einstein relationship $\mu = eD/kT$, find the diffusion constants of electrons in germanium at 300°K and 77°K (liquid nitrogen temperature). Also calculate the current density at the two temperatures caused by diffusion across the base region of an n-p-n transistor. The electron concentration at the emitter ($x = 0$) is $n = 10^{17}/cm^3$, and at the collector ($x = 10^{-3}$ cm) $n = 0$. For all practical purposes, the electron distribution can be assumed to vary linearly with x in the base region.

3.4 (a) The density of states in germanium is found to be $N_c = 1.05 \times 10^{19}/cm^3$ and $N_v = 0.51 \times 10^{19}/cm^3$ at 300°K. Find m_e and m_h. Calculate

[28] J. J. Hopfield, *Jour. Appl. Phys.*, vol. 32, p. 2277, 1961; Dietz, Hopfield, and Thomas, *Jour. Appl. Phys.*, vol. 32, p. 2282, 1961; J. O. Dimmock and R. G. Wheeler, *Jour. Appl. Phys.*, vol. 32, p. 2271, 1961.

[29] R. C. Casella, *Jour. Appl. Phys.*, vol. 34, p. 1703, 1963.

N_c and N_v at 77°K. Given $\mathcal{E}_g = 0.67$ and 0.70 ev at 300°K and 77°K, find the intrinsic carrier concentrations at these two temperatures. (b) The electron concentration is found to be $10^{17}/\text{cm}^3$ at 77°K. Assuming $N_a = 0$ and $\mathcal{E}_c - \mathcal{E}_d = 0.010$ ev, find the donor concentration N_d in the germanium sample.

3.5 Using the values of N_c and N_v given in Prob. 3.4 and the value of $\mathcal{E}_g = 0.67$ ev, find the electron and hole concentrations n and p at 300°K and 540°K in a germanium sample with $N_d = 5 \times 10^{15}/\text{cm}^3$ and $N_a = 2 \times 10^{15}/\text{cm}^3$. Using Eq. (3.43) and assuming $\mathcal{E}_c - \mathcal{E}_d = 0.01$ ev, estimate the temperature below which the donor states will no longer be completely ionized.

3.6 Using the information given in Table 3.1, find the values of T_1 and T_2 defining the three temperature regions for a sample with $\mathcal{E}_c - \mathcal{E}_d = 0.01$ ev, $\mathcal{E}_g = 0.67$ ev, $m_e = 0.55\ m_0$, $m_h = 0.35\ m_0$, $N_a = 0$ and $N_d = 10^{16}/\text{cm}^3$. Plot the value of the Fermi level as a function of temperature by choosing four strategically placed points on each side of T_1 and T_2.

3.7 Using the information given in Table 3.1 and the numerical values given in Prob. 3.6, find the electron concentration at 600°K, 300°K, and 15°K.

3.8 Refer to Figs. 3.12 and 3.13 for the Hall coefficient and resistivity data of various n-type germanium samples. It is found that for sample 55: $R_H = 10^6\ \text{cm}^3/\text{coulomb}$ at 20°K and $7 \times 10^5\ \text{cm}^3/\text{coulomb}$ between 30°K and 250°K; and $\rho = 4\ \Omega\text{-cm}$ at 20°K, 6 Ω-cm at 33°K, 10 Ω-cm at 50°K, 28 Ω-cm at 100°K, and 95 Ω-cm at 200°K. (a) Find the electron concentrations at 20°K, 50°K, and 100°K. Assuming $N_a = 0$, calculate N_d and the percentage of ionized donors at these temperatures. (b) Plot the values of electron mobility calculated from ρ and R_H as a function of temperature on a logarithmic scale and compare it with the $T^{-3/2}$ dependence. (c) Find the temperature dependence of τ and hence that of μ if it is assumed that the collision process is caused by lattice vibrations. It is known that the mean free path l of electrons for the process is inversely proportional to n_s of Eq. (1.129) where l is defined as $l = v_{\text{th}}\tau$, v_{th} being the thermal velocity of electrons.

3.9 The following data are found from Figs. 3.12 and 3.13:

R_H cm³/coulomb	8×10^5	6×10^4	6×10^3	3.5×10^2	9
ρ Ω-cm	10	1	0.2	0.07	0.008

for various germanium samples at 50°K. Calculate the electron mobilities in different samples. Examine the tendency of the variation of μ with respect to ρ and cite the possible cause for the observed variation in μ. If two different collision processes are present, it is necessary to add up the effects due to each process. Between the two possible choices $\mu = \mu_1 + \mu_2$ and $1/\mu = 1/\mu_1 + 1/\mu_2$, pick up the equation which you think is reasonable. Give reasons. How would you then treat the data presented above?

3.10 Derive Eq. (3.68) from Eq. (3.44) by expressing $\mathcal{E}_f$ in terms of n. Refer to sample 51 in Fig. 3.13. The values of R_H (cm³/coulomb) are:

5×10^4 at $50°K$, 1.8×10^5 at $20°K$, and 3×10^6 at $12.5°K$. Find the energy of the donor level with respect to the conduction band edge. The value of N_a is known to be $4.3 \times 10^{12}/cm^3$.

3.11 Take the ratio of the intrinsic carrier concentrations at two given temperatures. Find $n_i(T_2)/n_i(T_1)$ in Ge, Si and GaAs with $T_1 = 300°K$ and $T_2 = 310°K$, the gap energies being, respectively, 0.67 ev, 1.11 ev, and 1.42 ev at $300°K$. Also calculate the increase in the minority carrier concentrations (by how large a factor) in these materials per $10°$ increase in temperature around $300°K$. The impurity concentrations in these materials are such that $|N_d - N_a| \gg n_i$ around $300°K$, or, in other words, the semiconductors are in the exhaustion region.

3.12 Calculate the hole and electron concentrations in a 10 Ω-cm (at $300°K$) p-type germanium sample. The following data at $300°K$ are given: $n_i = 2.2 \times 10^{13}/cm^3$, $\mathcal{E}_g = 0.67$ ev, $\mu_n = 3900$ cm^2/volt-sec, and $\mu_p = 1900$ cm^2/volt-sec. For simplicity, it is assumed that the hole and electron mobilities have the same temperature dependence. Find the temperature at which the sample shows a zero Hall coefficient or $R_H = 0$.

3.13 Arrange the coefficients of A, B, C, and D in Eqs. (3.77) and (3.78) in a determinant. First form a new determinant by adding and subtracting the columns in the following manner: $I' = (I - II)/2i$, $II' = (I + II)/2$, $III' = (III + IV)/2$ and $IV' = (IV - III)/2$, the primed and unprimed roman numerals representing respectively the columns of the new and old determinants. Prove Eq. (3.79) by expanding the new determinant.

3.14 Plot $F(\beta a) = (P/\beta a) \sin \beta a + \cos \beta a$, Eq. (3.80), as a function of βa for $P = 3\pi/2$ as shown in Fig. 3.18. Then, for a given value of βa, find the corresponding value of ka. Note that for $|F(\beta a)| > 1$, k must be purely imaginary, corresponding to the forbidden gap. Plot $\beta^2 a^2/\pi^2$ as a function of ka in the forbidden gaps (BC and DE of Fig. 3.19) as well as in the allowed bands (AB, CD, and EF of Fig. 3.19).

3.15 Show that the group velocity of electrons at band edges (Points A, B, C, D, etc.) is zero by evaluating $\partial \beta/\partial k$ from Eq. (3.80). By expanding the energy expression $\mathcal{E}(k)$ at the band edge $k = k_0$,

$$\mathcal{E}(k) = \mathcal{E}(k_0) + A(k - k_0) + B(k - k_0)^2$$

show that $A = 0$ and $B = \hbar^2/2m^*$ where m^* is defined as $(m^*)^{-1} = \partial^2 \mathcal{E}/\partial(\hbar k)^2$.

3.16 Show that for $P = 3\pi/2$, the band edge at point B or $ka = \pi$, the value of βa is equal to π. In view of the fact that $\partial \beta/\partial k = 0$ or $\partial \mathcal{E}/\partial k = 0$ at point B, verify the following

$$\frac{\partial^2 \beta}{\partial k^2} = \frac{m}{\beta \hbar^2} \frac{\partial^2 \mathcal{E}}{\partial k^2} = -\frac{a \cos ka}{\dfrac{3\pi}{2} \dfrac{\cos \beta a}{\beta a} - \dfrac{3\pi}{2} \dfrac{\sin \beta a}{(\beta a)^2} - \sin \beta a}$$

Find the effective mass of electrons at the top of the band AB.

3.17 Repeat Prob. 3.16 for point C (with $\beta a = 3\pi/2$ and $ka = \pi$). Verify the same expression $\partial^2 \beta/\partial k^2$ and find the effective mass at the bottom of the band CD.

3.18   In the classical theory, the conservation of electric charge is written as

$$-\frac{\partial \varrho}{\partial t} = \nabla \cdot \mathbf{J}$$

where ϱ is the charge density and $\mathbf{J}$ the current density. The quantum mechanical charge density is given by

$$\varrho = e \int_{-\infty}^{\infty} \psi^* \psi \, dx \, dy \, dz$$

Following the procedure outlined in Sec. 1.5 [Eqs. (1.34) to (1.38)], show that the quantum-mechanical current density defined as

$$\mathbf{J} = e\mathbf{v} = e\langle \mathbf{p} \rangle / m = (e/m) \int_{-\infty}^{\infty} \psi^* \, \mathrm{grad} \, \psi \, dx \, dy \, dz$$

satisfies the charge conservation equation.

3.19   Show that a rotation of axes through an angle ϕ in a two-dimensional space brings a point (x,y) in the old coordinate system to the new coordinate system (x',y') with

$$x' = x \cos \phi + y \sin \phi$$
$$y' = y \cos \phi - x \sin \phi$$

Using the transformation, find the angle ϕ needed to eliminate the xy term in the expression $ax^2 + bxy + cy^2 = d$.

3.20   A cyclotron resonance experiment is performed on a semiconductor with spherical energy surfaces, that is $m_1 = m_2 = m_3 = m^*$. The d-c magnetic field is applied in the z direction with $B_z = B_0$ and the a-c electric field is in the xy plane with $E_x = E_1 \cos \omega t$ and $E_y = E_1 \sin \omega t$. Write down the equation of motion for v_x and v_y. Letting $v = v_x + iv_y$, show that the equation has a solution

$$v = -\frac{(eE_1 \tau / m) \exp (i\omega t)}{[1 + i(\omega - \omega_c)\tau]}$$

where $\omega_c = eB_0/m^*$. Find the expression for the a-c conductivity of the sample and discuss the meaning of the out-of-phase component.

3.21   Find the expressions for v_x and v_y in real notations in terms of $\sin \omega t$ and $\cos \omega t$ from v of the preceding problem. Find the expression of the power absorbed in the sample and plot the power absorption as a function of ω_c/ω (that means as a function of B_0 normalized with respect to $m^*\omega/e$) for two values of ω with $\omega\tau = 1$ and $\omega\tau = 5$. Discuss the sharpness of the resonance curve in relation to the value of $\omega\tau$.

3.22   Derive Eq. (3.111) for silicon. Find the angle θ at which the two resonance curves intersect. Identify the two curves with the corresponding equation of Eq. (3.111) and then obtain the values of m_t and m_l by examining the values of m^* at $\theta = 0°$ and $90°$.

3.23   By equating the experimental values of m^* at $\theta = 0°$ and $\theta = 90°$ of Fig. 3.24 to the values calculated from Eq. (3.108), determine the values of m_l and m_t. Using the values m_l and m_t just evaluated, construct the theoretical m^* versus θ curve for the case $\gamma = \cos \theta/\sqrt{3}$ and compare it with the experimental curve.

REFERENCES

Brooks, H.: Theory of the Electrical Properties of Germanium and Silicon, in L. Marton (ed.), "Advances in Electronics and Electron Physics," vol. 7, p. 87, Academic Press Inc., New York, 1940.

Conwell, E. M.: "Properties of Silicon and Germanium—I and II," *Proc. IRE*, vol. 40, p. 1327, 1952, and vol. 46, p. 1281, 1958.

"Conference on Semiconducting Compounds," in *Jour. Appl. Phys.*, vol. 32, October, 1961.

Hannay, N. B., ed.: "Semiconductors," Reinhold Publishing Corporation, New York, 1959.

Hilsum, C., and A. C. Rose-Innes: "Semiconducting III-V Compounds," Pergamon Press, New York, 1961.

Kittel, C.: "Introduction to Solid State Physics," John Wiley & Sons, Inc., New York, 1956.

Lax, B., and J. G. Mavroids: Cyclotron Resonance, in F. Seitz and D. Turnbull (eds.), "Solid State Physics," vol. 11, p. 261, Academic Press, Inc., New York, 1960.

Seitz, F.: "The Modern Theory of Solids," McGraw-Hill Book Company, New York, 1940.

Shockley, W.: "Electrons and Holes in Semiconductors," D. Van Nostrand Co., Inc., New York, 1950.

Spenke, E.: "Electronic Semiconductors" (translated by Jenny et al.), McGraw-Hill Book Company, New York, 1958.

4 TRANSPORT PHENOMENA IN SEMICONDUCTORS AND METALS

4.1 THE BOLTZMANN TRANSPORT EQUATION

We shall start our discussion with a brief review of what has been said in the preceding chapter about the conductive property of a semiconductor. Although Eq. (3.5) is based on a simple and rather naive model, it serves very nicely as a starting point from which we can introduce the concentrations of free carriers and discuss the concepts of effective masses. However, there is one important point which we introduced in Sec. 3.3 but have not discussed fully. In Eq. (3.14), the equation of motion is written in terms of an average velocity. The point in question is how this concept of taking a statistical average can be incorporated in the formulation of Eq. (3.5). Figure 3.6 may give us some insight into the problem. The effect of an applied field is to add a small component of drift velocity to an electron on top of a very large random velocity which the electron has after each collision. The reason why the random velocity does not result in any current lies in the fact that there are as many electrons moving in one direction as are moving in the opposite direction. Here, as we can see, the statistical distribution of electrons plays a decisive role in determining the macroscopic or group behavior of the system.

Let $f(\mathbf{r},\mathbf{v})$ be a distribution function in the six-dimensional space x, y, z, v_x, v_y and v_z called *phase space* such that the number of particles dN having position between $\mathbf{r}$ and $\mathbf{r} + d\mathbf{r}$ and velocity between $\mathbf{v}$ and $\mathbf{v} + d\mathbf{v}$ is given by

$$dN = f(\mathbf{r},\mathbf{v}) \, d\mathbf{r} \, d\mathbf{v} \qquad (4.1)$$

where $d\mathbf{r} = dx \, dy \, dz$ and $d\mathbf{v} = dv_x \, dv_y \, dv_z$ for brevity. The number of particles having a given position and velocity or the distribution function $f(\mathbf{r},\mathbf{v})$ may change with time on account of the following processes. First, particles may move from one position to another because of its velocity or may change from one velocity to another on account of the external applied fields. The process through which such changes of states occur is generally referred to as the *drift process*. Particles also change their states in phase space through collisions with other particles.

Consider an elementary volume $d\mathbf{r} \, d\mathbf{v}$ in the six-dimensional phase space in which the number of particles at a given time t is given by Eq. (4.1). On account of the drift process alone, the number of particles in this elementary volume must change. The particles originally in the box will move to another box in the phase space and will be replaced by particles coming from a box which has

$$\mathbf{r}' = \mathbf{r} - \mathbf{v}\,\Delta t \qquad \text{and} \qquad \mathbf{v}' = \mathbf{v} - \mathbf{a}\,\Delta t \qquad (4.2)$$

where $\mathbf{a}$ is the acceleration experienced by the particles. Thus, the change in $f(\mathbf{r},\mathbf{v})$ due to the drift process is given by

$$\Delta f_{\text{drift}} = f(\mathbf{r}',\mathbf{v}') - f(\mathbf{r},\mathbf{v})$$
$$= -\sum \frac{\partial f}{\partial x} v_x\,\Delta t - \sum \frac{\partial f}{\partial v_x} a_x\,\Delta t \qquad (4.3)$$

It is to be noted that changes in $f(\mathbf{r},\mathbf{v})$ caused by the drift process are smooth and continuous and hence the step taken in Eq. (4.3) by expanding $f(\mathbf{r}',\mathbf{v}')$ into a Taylor's series is valid.

The changes which particles suffer during a collision process are quite different in nature, as illustrated in Fig. 3.6. First, only the velocity of a particle is involved and second, the change in velocity is rather abrupt. For these reasons, the collision process must be treated differently. Let $S(\mathbf{v},\mathbf{v}')$ be the probability per unit time that a particle changes from a state having velocity between $\mathbf{v}$ and $\mathbf{v} + d\mathbf{v}$ to a state having velocity between $\mathbf{v}'$ and $\mathbf{v}' + d\mathbf{v}'$. The number of particles leaving the state $(\mathbf{r},\mathbf{v})$ per unit time to all other states is

$$\frac{(\Delta dN)_1}{\Delta t} = b\,d\mathbf{r}\,d\mathbf{v} = \int_{v'} f(\mathbf{r},\mathbf{v})S(\mathbf{v},\mathbf{v}')\,d\mathbf{v}'\,d\mathbf{v}\,d\mathbf{r} \qquad (4.4)$$

and the total number of particles coming to the state $(\mathbf{r},\mathbf{v})$ per unit time from all other states is

$$\frac{(\Delta dN)_2}{\Delta t} = c\,d\mathbf{r}\,d\mathbf{v} = \int_{v'} f(\mathbf{r}',\mathbf{v}')S(\mathbf{v}',\mathbf{v})\,d\mathbf{v}'\,d\mathbf{v}\,d\mathbf{r} \qquad (4.5)$$

where ΔdN represents the change in the number of particles in a volume element $d\mathbf{v}\,d\mathbf{r}$ in phase space. It should be pointed out that during the collision process, the position of the particle does not change and this fact can be incorporated into the formulation by replacing $S(\mathbf{v},\mathbf{v}')\,d\mathbf{v}'$ with $S(\mathbf{v},\mathbf{v}')\delta(\mathbf{r},\mathbf{r}')\,d\mathbf{v}'d\mathbf{r}'$ where $\delta(\mathbf{r},\mathbf{r}')$ is a Dirac delta function. The step $\int_{r'} \delta(\mathbf{r},\mathbf{r}')\,d\mathbf{r}' = 1$ is implicit in Eqs. (4.4) and (4.5). Combining Eqs. (4.4) and (4.5), and noting that $\Delta dN = (\Delta dN)_1 - (\Delta dN)_2 = \Delta f(\mathbf{r},\mathbf{v})\,d\mathbf{v}\,d\mathbf{r}$, we have the net rate of change in $f(\mathbf{r},\mathbf{v})$ due to the collision process as

$$\left.\frac{\partial f}{\partial t}\right|_{\text{collision}} = c - b \qquad (4.6)$$

by taking the limit $\Delta t \to 0$.

Under d-c conditions, the total time rate of change in $f(\mathbf{r},\mathbf{v})$ must be zero. From Eqs. (4.3) and (4.6), we have

$$\mathbf{v} \cdot \text{grad}_r f + \mathbf{a} \cdot \text{grad}_v f = c - b \qquad (4.7)$$

where the symbols grad_r and grad_v stand for derivatives with respect to coordinates x, y, and z and velocities v_x, v_y, and v_z, respectively. Equation (4.7) is generally known as the *Boltzmann transport equation* or Boltzmann equation of state defined in the classical $\mathbf{r}$ and $\mathbf{v}$ space.

Before discussing the solution of Eq. (4.7), we should clarify several important points which may have entered the reader's mind. In the quantum mechanical description of an electronic state, the distribution function is a function of $\mathbf{r}$ and wave vector $\mathbf{k}$, that is $f(\mathbf{r},\mathbf{k})$. It can be shown that the corresponding Boltzmann equation in the quantum space is

$$\frac{1}{\hbar} \text{grad}_k \, \mathcal{E} \cdot \text{grad}_r f + \frac{\mathbf{F}}{\hbar} \cdot \text{grad}_k f = c - b \qquad (4.8)$$

Equation (4.8) reduces to Eq. (4.7) if a correct interpretation is made. From Table 3.2, if $\mathcal{E} = \hbar^2 k^2 / 2m^*$ where m^* is the effective mass, then $\hbar^{-1} \text{grad}_k \, \mathcal{E} = \hbar \mathbf{k}/m^* = \mathbf{v}$. Furthermore,

$$\hbar^{-1}\mathbf{F} \cdot \text{grad}_k f = (m^* \mathbf{a} \cdot \text{grad}_v f) \frac{\partial v}{\partial \hbar k} = \mathbf{a} \cdot \text{grad}_x f$$

Therefore, for later applications, Eq. (4.7) can be applied to particles obeying quantum statistics with the understanding that $\mathbf{v} = \hbar \mathbf{k}/m^*$ and $\mathcal{E} = \hbar^2 k^2 / 2m^*$. The only restriction on Eq. (4.7) is that a d-c external field is assumed. Under a time-varying external disturbance, however, the total time rate of change in $f(\mathbf{r},\mathbf{v})$ may not be zero. The time-dependent Boltzmann transport equation reads

$$\frac{\partial f}{\partial t} = c - b - \mathbf{a} \cdot \text{grad}_v f - \mathbf{v} \cdot \text{grad}_r f \qquad (4.9)$$

It is worth mentioning that Eq. (4.9) bears some resemblance to Eq. (3.14) if a correct identification of the terms is made. The collision term in Eq. (4.9) can be viewed as a restoring force. Under thermal equilibrium, there are as many particles coming into as going out of a given state $(\mathbf{r},\mathbf{v})$, and hence $c - b = 0$. In the presence of an externally applied disturbance, the term $c - b$ is no longer zero and can be represented phenomenologically as

$$c - b = -(f - f_0)/\tau \qquad (4.10)$$

where f_0 is the equilibrium distribution function. After substituting Eq. (4.10) into Eq. (4.9), we see that the system will return to its equilibrium distribution f_0 with a characteristic time constant τ as soon as the external disturbance is removed. The way in which the system returns to equilibrium is through collisions. The same collision process which limits the drift velocity of particles in Eq. (3.14) is responsible for energy exchange between particles and hence the maintenance of internal equilibrium in the system as required by Eq. (4.9). It is not surprising, therefore, that the time constants τ in Eqs. (3.14) and (4.10) are the same.

A comparison of Eq. (4.9) with Eq. (3.14) shows that for every term in Eq. (4.9) except $\mathbf{v} \cdot \operatorname{grad}_r f$, there is a corresponding term in Eq. (3.14). The only difference is that Eq. (4.9) is expressed in terms of the distribution function f which is an implicit function of velocity instead of the drift velocity itself as in Eq. (3.14). The statistical nature of electrons is now incorporated right into the distribution function f. The term $\mathbf{v} \cdot \operatorname{grad}_r f$ gives rise to the diffusion phenomenon discussed briefly in Sec. 3.3. The existing parallelism between the Boltzmann transport equation and the discussion in Sec. 3.3 will become clear after the solution of Eq. (4.9) is found and applied to a specific problem.

Now we shall discuss the solution of the Boltzmann transport equation. For simplicity, we shall restrict our discussion to d-c conditions and to cases involving one-dimensional disturbances only. Thus, the Boltzmann equation reduces to

$$f - f_0 = -\tau \left(v_x \frac{\partial f}{\partial x} + a_x \frac{\partial f}{\partial v_x} \right) \tag{4.11}$$

The distribution function f_0 under a thermodynamic equilibrium condition is assumed to be known and our task is to find the new distribution function f under a nonequilibrium situation created by external disturbances such as an applied electric field or a temperature gradient. For small disturbances such that $(f - f_0)/f_0 \ll 1$, Eq. (4.11) can be solved by the standard method of successive approximation commonly known as iteration. To the first-order approximation, the function f on the right-hand side of Eq. (4.11) can be replaced by f_0 and hence

$$f = f_0 - \tau \left(v_x \frac{\partial f_0}{\partial x} + a_x \frac{\partial f_0}{\partial v_x} \right) \tag{4.12}$$

The above equation provides the starting point in our discussion of the group behavior of a system involving an enormous number of particles. Examples of such a system are electrons in a plasma, electrons in a metal, and electrons and holes in a semiconductor.

We shall illustrate the use of Eq. (4.12) by discussing the electrical and thermal properties of such a system. According to the definition of f given by Eq. (4.1), the electric current density J_x associated with the motion of electrons is equal to

$$J_x = -e\Sigma v_x = -e\iint f v_x \, d\mathbf{v} \qquad \text{amp/m}^2 \tag{4.13}$$

where $d\mathbf{v} = dv_x\, dv_y\, dv_z$ is the elementary volume in the velocity space. Since all known distribution functions (Boltzmann, Fermi-Dirac, and Bose-Einstein) are functions of energy and hence even functions of velocity, the integral $\iint f_0 v_x\, d\mathbf{v}$ must be zero. That means, under thermodynamic equilibrium (for which the distribution functions have been derived), no current is flowing in any arbitrary direction. Substituting Eq. (4.12) into Eq. (4.13) gives

$$J_x = e \iiint_{-\infty}^{\infty} \tau \left(v_x \frac{\partial f_0}{\partial x} + a_x \frac{\partial f_0}{\partial v_x}\right) v_x\, dv_x\, dv_y\, dv_z \qquad (4.14)$$

Since electrons in motion carry a kinetic energy, accompanying any electron flow there must be a corresponding energy flow. The thermal current density C_x which measures the rate of energy flow per unit area is defined as

$$C_x = \iiint_{-\infty}^{\infty} f v_x \mathcal{E}\, dv_x\, dv_y\, dv_z \qquad \text{watts/m}^2 \qquad (4.15)$$

where $\mathcal{E}$ is the kinetic energy of an electron. For free electrons with mass m, $\mathcal{E} = mv^2/2$. Using Eq. (4.12), we have

$$C_x = - \iiint_{-\infty}^{\infty} \tau \left(v_x \frac{\partial f_0}{\partial x} + a_x \frac{\partial f_0}{\partial v_x}\right) v_x \mathcal{E}\, dv_x\, dv_y\, dv_z \qquad (4.16)$$

Equations (4.14) and (4.16) are two basic equations from which the electrical and thermal properties of a system can be found.

4.2 ELECTRICAL AND THERMAL PROPERTIES OF AN ELECTRON GAS

Boltzmann or nondegenerate gas.   As an example, we shall consider electrons in an ionized gas. Since the density of electrons is relatively low in a gaseous medium, electrons are essentially noninteracting and hence the equilibrium distribution function f_0 is given by the Maxwell-Boltzmann distribution

$$f_0 = A \exp\left(-\frac{mv^2}{2kT}\right) \qquad (4.17)$$

where $A = n(m/2\pi kT)^{3/2}$, $v^2 = v_x^2 + v_y^2 + v_z^2$ and n is the density of electrons. To evaluate the expressions J_x and C_x, we assume that a constant field E_x and a uniform temperature gradient $\partial T/\partial x$ exist so that

$$\frac{\partial f_0}{\partial x} = \left(\frac{mv^2}{2kT} - \frac{3}{2}\right) \frac{f_0}{T} \frac{\partial T}{\partial x} \qquad (4.18)$$

$$a_x \frac{\partial f_0}{\partial v_x} = \frac{eE_x}{kT} v_x f_0 \qquad (4.19)$$

Note that Eqs. (4.17) to (4.19) and hence discussions to follow apply equally well to a nondegenerate semiconductor if the free-electron mass m is replaced by an effective mass m^* in Eqs. (4.17) to (4.19) and also in subsequent discussions.

With the help of the following integrals,

$$I_0 = \int_{-\infty}^{\infty} \exp\left(-\frac{mv_x^2\beta}{2}\right) dv_x = \left(\frac{2\pi}{m\beta}\right)^{\frac{1}{2}} \qquad (4.20)$$

$$I_j = \int_{-\infty}^{\infty} \left(\frac{mv_x^2}{2}\right)^j \exp\left(-\frac{mv_x^2\beta}{2}\right) dv_x$$

$$= (-1)^j \frac{d^j I_0}{d\beta^j} = (\tfrac{1}{2}) \cdots (\tfrac{1}{2} + j - 1) \frac{I_0}{\beta^j} \qquad (4.21)$$

and the use of Eqs. (4.17), (4.18), and (4.19), Eqs. (4.14) and (4.16), respectively, become

$$J_x = \frac{ne\tau}{m} \frac{\partial(kT)}{\partial x} + \frac{ne^2\tau}{m} E_x \qquad (4.22)$$

$$C_x = -\frac{5n(kT)\tau}{m} \frac{\partial(kT)}{\partial x} - \frac{5ne(kT)\tau}{2m} E_x \qquad (4.23)$$

In the above calculation, it is assumed that the mean free time τ is a constant for all electrons. If τ is a function of v, then the τ's in Eqs. (4.22) and (4.23) must be some kinds of a weighted average (see Prob. 4.2). Since the integrals in C_x and J_x involve different powers of v, the weighted average values of τ in Eqs. (4.22) and (4.23) will be different. The difference in the Hall and drift mobility mentioned in the preceding chapter arises also from the fact that different weighting factors are involved. We shall discuss the case in which τ is a function of v in a later section.

Now let us find the electrical and thermal conductivities of the electron gas. In a conductivity measurement, the ratio of J_x/E_x yields meaningful information only if the gas has a uniform property independent of x. In other words, $\partial T/\partial x$ in Eq. (4.22) must be equal to zero. Thus, we have

$$\sigma = ne^2\tau/m = e\mu n \qquad (4.24)$$

It is expected that Eq. (4.24) yields the same result as Eq. (3.2) for cases in which τ is independent of v. However, for most transport problems, it is more realistic to assume a constant mean free path than a constant mean free time τ. We shall have ample opportunities later in this chapter to discuss such cases in detail when we treat the different scattering processes in semiconductors.

In evaluating the thermal conductivity, J_x is set equal to zero and the value of E_x obtained from Eq. (4.22) is substituted back into Eq. (4.23). Thus, the thermal conductivity K is equal to

$$K = \frac{C_x}{\partial T/\partial x}\bigg|_{J_x=0} = \frac{5nk^2T\tau}{2m} \qquad (4.25)$$

Note that K is the electronic contribution to the thermal conductivity and it bears a general relationship to σ such that $K/\sigma T = 5k^2/2e^2$. The numerical factor in front, $5/2$ in the present case, depends on the specific assumption about τ but the dependence upon $(k/e)^2$ is rather general.

It may be instructive to retrace the mathematical steps and associate them with the physical processes taking place inside the system. When a temperature gradient exists, there is a constant rate of energy flow from the hot end toward the cold end. The energy is carried and transported by electrons through collisions with other electrons. Besides acting as energy carriers, electrons also experience a thermal pressure pushing them toward the cold end because electrons near the hot end have a larger kinetic energy or mean square velocity than those near the cold end. If electrical connections are made to the system, a current of density J_x flows even though $E_x = 0$. This is known as the *thermoelectric effect* and a current equal to J_x multiplied by the cross-sectional area is the short-circuit thermoelectric current. If no electrical connections are made to the system, no current flow is allowed. The condition $J_x = 0$ requires, however, that an electric field be set up inside the system to oppose the thermal force, thus creating an open-circuit voltage across the two open ends. This voltage is known as the thermoelectric voltage. The thermoelectric effects will be further discussed in a later section.

The points which we wish to bring into focus in the present discussion are the following. Since electrons are both carriers of energy and current, thermal effects naturally can give rise to electric effects and vice versa. The idea may seem strange at first that E_x can be different from zero even though no external voltage is applied. We must realize that there exist forces of many origins, thermal, diffusion, magnetic, and electric, and an interplay of the electric force with other forces may produce an internal electric field. The Hall voltage discussed in Sec. 3.6 is another example of an interplay, in that case, between the electric and magnetic forces. Later on, when we discuss semiconductor junction phenomena, we shall see that a built-in field exists in a *p-n* junction to counterbalance the diffusion force.

Fermi or degenerate gas. To acquaint the reader further with the statistical nature concerning the transport of carriers, we shall discuss the electrical and thermal conductivities of metals as examples involving quantum statistics. For simplicity, the energy surface is assumed spherical with an effective mass m^*. As stated in the preceding section, the classical Boltzmann equation, Eq. (4.7), may be applied to electrons in metals if an effective mass m^* is used instead of the free electron mass such that

$$\mathcal{E} = \frac{\hbar^2 k^2}{2m^*} \quad \text{and} \quad \mathbf{v} = \frac{\hbar \mathbf{k}}{m^*} \tag{4.26}$$

By the same token, Eqs. (4.14) and (4.16) are also valid. Given the definition of Eq. (4.1) and in view of Eq. (4.26), the distribution function f_0 in Eqs. (4.14) and (4.16) is equal to

$$dn = f_0 \, dv_x \, dv_y \, dv_z = 2(m^*/h)^3 F(\mathcal{E}) \, dv_x \, dv_y \, dv_z \tag{4.27}$$

where n is the density of electrons and $F(\mathcal{E})$ is the Fermi-Dirac distribution function

$$F(\mathcal{E}) = \frac{1}{1 + \exp\left[(\mathcal{E} - \mathcal{E}_f)/kT\right]} \tag{4.28}$$

The mathematical steps involved in the evaluation of J_x and C_x from Eqs. (4.14) and (4.16) are not difficult but rather tedious, therefore, it is necessary to make the bookkeeping as systematic as possible. First, for spherical energy surfaces, v_x^2 can be replaced by $v^2/3$ and the elementary volume in the velocity space by

$$dv_x \, dv_y \, dv_z = 4\pi v^2 \, dv = 2\pi (2/m^*)^{3/2} \mathcal{E}^{1/2} \, d\mathcal{E} \tag{4.29}$$

Second, the integrals in Eqs. (4.14) and (4.16) can be expressed in the form

$$A_p = \iiint\limits_{-\infty}^{\infty} f_0 \mathcal{E}^p \, dv_x \, dv_y \, dv_z = C \int_0^{\infty} \mathcal{E}^{q-1} F(\mathcal{E}) \, d\mathcal{E} \tag{4.30}$$

where $C = 4\pi (2m^*/h^2)^{3/2}$ and $q = p + \frac{3}{2}$. Integrating Eq. (4.30) by parts, we have

$$A_p = C \int_0^{\infty} \mathcal{E}^{q-1} F(\mathcal{E}) \, d\mathcal{E} = \frac{C}{q} \mathcal{E}^q F(\mathcal{E}) \Big|_0^{\infty} + \frac{C}{q} \int_0^{\infty} \mathcal{E}^q \left(-\frac{\partial F}{\partial \mathcal{E}} \right) d\mathcal{E} \tag{4.31}$$

The first term on the right-hand side is zero because $F(\mathcal{E})$ drops exponentially to zero as $\mathcal{E}$ goes to infinity and $\mathcal{E}^q$ goes to zero as $\mathcal{E}$ goes to zero.

The reason for transforming Eq. (4.30) to Eq. (4.31) is as follows. The function $(-dF/d\mathcal{E})$ has a very sharp peak at $\mathcal{E} = \mathcal{E}_f$ as shown in Fig. 4.1, and most contribution to the integral in Eq. (4.31) comes from the region where $\mathcal{E}$ is close to $\mathcal{E}_f$. Hence the function $\mathcal{E}^q$ can be expanded into a Taylor's series as

$$\mathcal{E}^q = (\mathcal{E}_f + \mathcal{E} - \mathcal{E}_f)^q$$
$$= \mathcal{E}_f^q \left[1 + q \frac{\mathcal{E} - \mathcal{E}_f}{\mathcal{E}_f} + \frac{q(q-1)}{2} \left(\frac{\mathcal{E} - \mathcal{E}_f}{\mathcal{E}_f} \right)^2 + \cdots \right] \tag{4.32}$$

If we now shift the origin to $\mathcal{E}_f$ by changing the variable from $\mathcal{E}$ to $\mathcal{E}' = \mathcal{E} - \mathcal{E}_f$, the new limits of integration in Eq. (4.31) are from $-\mathcal{E}_f$ to ∞ or approximately from $-\infty$ to ∞. Because $[-dF(\mathcal{E}')/d\mathcal{E}']$ is an even function of $\mathcal{E}'$, only the even terms in Eq. (4.32) contribute to the integral of Eq. (4.31). Using the definite integral

$$\int_{-\infty}^{\infty} \eta^2 \exp(\eta)\, d\eta/[1 + \exp(\eta)]^2 = \pi^2/3$$

we have

$$A_p = \iiint_{-\infty}^{\infty} f_0 \mathcal{E}^p \, dv_x \, dv_y \, dv_z = \frac{C}{q} \mathcal{E}_f^q \left[1 + \frac{q(q-1)\pi^2}{6} \left(\frac{kT}{\mathcal{E}_f} \right)^2 \right]$$

$$B_q = C \int_0^{\infty} \mathcal{E}^q \left(-\frac{dF}{d\mathcal{E}} \right) d\mathcal{E} = C\mathcal{E}_f^q \left[1 + \frac{q(q-1)\pi^2}{6} \left(\frac{kT}{\mathcal{E}_f} \right)^2 \right]$$

(4.33)

where $p = q - \frac{3}{2}$.

An immediate application of Eq. (4.33) is the establishment of a relationship between the density of electrons n and the Fermi level $\mathcal{E}_f$. At absolute zero temperature, realizing that $n = A_0$, Eq. (4.33) gives $n = (\frac{2}{3})C\mathcal{E}_f^{3/2}$ or

$$\mathcal{E}_{f0} = \left(\frac{3n}{2C} \right)^{2/3} = \frac{\hbar^2}{2m^*} (3\pi^2 n)^{2/3}$$

(4.34)

At a finite temperature T, since $(\mathcal{E} - \mathcal{E}_{f0})/\mathcal{E}_{f0} \ll 1$, the $\mathcal{E}_f$ in the correction term of Eq. (4.33) can be replaced by $\mathcal{E}_{f0}$, thus we have

$$\mathcal{E}_f = \mathcal{E}_{f0} \left[1 - \frac{\pi^2}{12} \left(\frac{kT}{\mathcal{E}_{f0}} \right)^2 \right]$$

(4.35)

Now we shall turn to the main task of finding J_x and C_x. In using

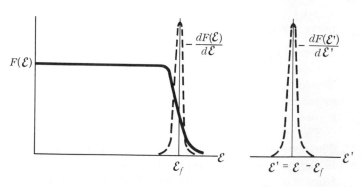

Figure 4.1

Diagram illustrating the behavior of the Fermi function $F(\mathcal{E})$ and its derivative $dF(\mathcal{E})/d\mathcal{E}$. Note that at $0°K$, $F(\mathcal{E})$ is a step function and hence $dF(\mathcal{E})/d\mathcal{E}$ behaves like a Dirac delta function $-\delta(\mathcal{E} - \mathcal{E}_f)$. At a finite temperature, the function $dF(\mathcal{E})/d\mathcal{E}$ is no longer a Dirac delta function but nevertheless it still shows a sharp peak around $\mathcal{E} = \mathcal{E}_f$.

Eqs. (4.14) and (4.16), we note that

$$a_x \frac{\partial f_0}{\partial v_x} = \frac{eE_x}{m^*} \left(-\frac{\partial F}{\partial \mathcal{E}} \right) m^* v_x 2 \left(\frac{m^*}{h} \right)^3 \tag{4.36}$$

$$v_x \frac{\partial f_0}{\partial x} = v_x \frac{\partial f_0}{\partial T} \frac{\partial T}{\partial x} = v_x \frac{\partial F}{\partial \mathcal{E}} (2kT) \left(\frac{m^*}{h} \right)^3 \frac{\partial T}{\partial x} \left[\frac{d}{dT} \left(\frac{\mathcal{E} - \mathcal{E}_f}{kT} \right) \right] \tag{4.37}$$

The term in the brackets of Eq. (4.37) can be split into two parts

$$\frac{d}{dT} \left(\frac{\mathcal{E} - \mathcal{E}_f}{kT} \right) = - \left[\frac{\mathcal{E}}{kT^2} + \frac{d}{dT} \left(\frac{\mathcal{E}_f}{kT} \right) \right] \tag{4.38}$$

according to the powers of $\mathcal{E}$. Assuming that the mean free path l is a constant and writing $\tau = l/v$ in Eqs. (4.14) and (4.16), we have

$$J_x = \frac{el}{3} \sqrt{\frac{2}{m^*}} \left\{ \frac{B_2}{T} \frac{\partial T}{\partial x} + B_1 \left[kT \frac{d}{dT} \left(\frac{\mathcal{E}_f}{kT} \right) \frac{\partial T}{\partial x} + eE_x \right] \right\} \tag{4.39}$$

$$C_x = \frac{l}{3} \sqrt{\frac{2}{m^*}} \left\{ \frac{B_3}{T} \frac{\partial T}{\partial x} + B_2 \left[kT \frac{d}{dT} \left(\frac{\mathcal{E}_f}{kT} \right) \frac{\partial T}{\partial x} + eE_x \right] \right\} \tag{4.40}$$

Let us first discuss the electric conductivity. Equation (4.39) for $\partial T/\partial x = 0$ gives

$$\sigma = \frac{J_x}{E_x} = \frac{e^2 l}{3} \sqrt{\frac{2}{m^*}} B_1 = \frac{e^2 l}{3} \sqrt{\frac{2}{m^*}} C\mathcal{E}_f \tag{4.41}$$

Noting that $n = (\frac{2}{3})C\mathcal{E}_f^{3/2}$, Eq. (4.41) becomes

$$\sigma = \frac{e^2 n}{m^*} \frac{l}{(2\mathcal{E}_f/m^*)^{1/2}} = \frac{e^2 n}{m^*} \tau(\mathcal{E}_f) \tag{4.42}$$

It is interesting to observe that Eq. (3.2), based on a very naive model, is still valid if τ is calculated at $\mathcal{E} = \mathcal{E}_f$, the top of the electron distribution, or from the expression $\tau = l/v$ with $m^* v^2/2 = \mathcal{E}_f$. This is quite different from the conclusion reached for the classical Boltzmann distribution, and this difference can be explained with the help of Fig. 4.2.

The solid curves of Fig. 4.2 represent (a) the Boltzmann and (b) the Fermi-Dirac distribution for a one-dimensional case, and the dashed curves represent the respective shifts in the distribution functions caused by the applied field as given by Eqs. (4.19) and (4.36) respectively for curves (a) and (b). The conduction current density J_x can be interpreted as a result of a slight redistribution of electrons in velocity space, removing electrons from $-v_x$ and adding electrons to $+v_x$ space. For the Boltzmann distribution, the effect is spread over the whole distribution, and hence τ must likewise be averaged over the entire distribution. For the Fermi-Dirac distribution, only electrons near the top of the distribution (that is with $\mathcal{E} \approx \mathcal{E}_f$) are affected as shown in Fig. 4.2b. This explains why $\tau(\mathcal{E}_f)$ is used in Eq. (4.42).

Figure 4.2

Maxwell–Boltzmann distribution

Fermi–Dirac distribution

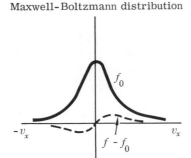
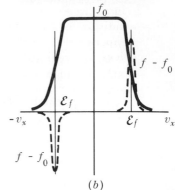

(a) (b)

Change in the distribution function in the presence of an applied field: (a) the Boltzmann distribution; (b) the Fermi-Dirac distribution.

The thermal conductivity K can be found from Eqs. (4.39) and (4.40) by eliminating E_x after setting $J_x = 0$. The result is

$$K = \frac{l}{3} \sqrt{\frac{2}{m^*}} \left(\frac{B_1 B_3 - B_2{}^2}{B_1 T} \right) \tag{4.43}$$

Substituting the values of B_1, B_2 and B_3 from Eq. (4.33) into (4.43) gives

$$K = \frac{l}{3} \sqrt{\frac{2}{m^*}} \frac{\pi^2 C \mathcal{E}_f}{3} k^2 T = \frac{\pi^2}{3} \frac{k^2}{m^*} n\tau(\mathcal{E}_f) \tag{4.44}$$

Again τ is evaluated at $\mathcal{E} = \mathcal{E}_f$. The ratio $K/\sigma T$ is equal to

$$\frac{K}{\sigma T} = \frac{\pi^2}{3} \left(\frac{k}{e} \right)^2 \tag{4.45}$$

and is known as the Wiedemann-Franz ratio. The theoretical value of 2.5×10^{-8} MKS unit is in good agreement with experiments for a number of metals as shown in Table 4.1.

TABLE 4.1   *Experimental values of $K/\sigma T$*

Metal	273°K	373°K
Al	2.19×10^{-8}	2.22×10^{-8}
Cu	2.23	2.34
Zn	2.22	2.34
Rh	2.56	2.54
Ag	2.31	2.37
Sn	2.52	2.49
Bi	3.31	2.89

It can be shown theoretically that the existence of a relaxation time τ in a degenerate electron gas (electrons in a metal) is a sufficient condition for the constancy of $K/\sigma T$. For impure metals and at low temperatures, the value of $K/\sigma T$ deviates quite considerably from the theoretical value. The reader is referred to a more advanced text[1] for further discussions. We should mention again that the thermal conductivity K that appears in Eqs. (4.44) and (4.45) is the contribution from electrons only and it does not include the contribution from lattice vibrations.

4.3 THE TEMPERATURE DEPENDENCE OF RESISTIVITY IN PURE METALS

From our discussion in the previous section, we see that electrical resistance arises on account of collision processes suffered by electrons, which are phenomenologically represented by a mean free time τ or a mean free path l in Eq. (4.42). In a perfect periodic crystal, the electron wave function is given by the Bloch wave function of Eq. (3.81). Because of the periodic nature of the wave function, the motion of electrons is unimpeded. Therefore, electrical resistance must come from deviations from a perfect lattice. As discussed in Sec. 2.8, lattice imperfections can be caused by lattice vibrations and structural defects such as dislocations, vacancies, interstitials, and impurity atoms. In the following, we shall present a qualitative discussion of the effect of lattice vibration on the resistivity of pure metals.

The resistivity of a pure metal is found by Grüneisen[2] to fit well over a wide temperature range the following empirical equation

$$\rho(T) = \frac{1}{\sigma(T)} = A T^5 \int_0^{\Theta/T} \frac{x^5 \, dx}{(e^x - 1)(1 - e^{-x})} = AG(T) \quad (4.46)$$

if the constants A and Θ are chosen properly. The value of Θ will in general be close to but slightly different from the Debye temperature Θ_D. At high temperatures, we recall that the average energy associated with a harmonic oscillator can be approximated by kT. As the temperature increases, the amplitude of lattice vibration increases, resulting in a shorter mean free path and in turn a higher electrical resistivity. In Eq. (4.46), the integral is proportional to $(\Theta/T)^4$ for $x \ll 1$; therefore, the linear relationship between $\rho(T)$ and T is indeed observed. At low temperatures such that $\Theta/T \gg 1$, the integral becomes a constant and thus the resistivity increases as T^5.

The physical origin of this T^5 law can be explained as follows.

[1] See for example: F. Seitz, "The Modern Theory of Solids," McGraw-Hill Book Company, New York, 1940; A. H. Wilson, "The Theory of Metals," Cambridge University Press, Cambridge, 1953.
[2] E. Grüneisen, *Ann. Physik*, vol. 16, p. 530, 1933.

The density of excited quanta of lattice vibrations, according to the Debye model, is equal to

$$N_{ex}(T) = \int_0^{\omega_m} \frac{n(\omega)\, d\omega}{\exp\,(\hbar\omega/kT) - 1} = B\left(\frac{T}{\Theta_D}\right)^3 \int_0^{x_m} \frac{x^2\, dx}{e^x - 1} \quad (4.47)$$

where B is a constant and $x = \hbar\omega/kT$. At low temperatures, as x approaches infinity, the density of excited modes of lattice vibration is proportional to T^3. Moreover, most contribution to Eq. (4.47) comes from phonons of low energy. These low-energy phonons are very ineffective in a collision process.

The Fermi level in a metal is of the order of several electron volts, 3.2 ev in Na for example; therefore, lattice phonons have insignificant energy in comparison with the kinetic energy of electrons at the top of the Fermi distribution. If we treat scattering of electron motion by phonons as a two-body collision problem, we find that the magnitude of the electron velocity is practically unchanged but its direction is deflected by an angle α such that $\sin\,(\alpha/2) = P/2p$ where P and p are the momenta of the phonon and electron respectively. As discussed in Sec. 2.8, the highest energy phonons have a wavelength of the order of lattice spacing a or a momentum value $P_0 \simeq \hbar/a$ which is also the magnitude of p. These high energy phonons are excited only at a temperature $T > \Theta_D$; hence, it is only for $T > \Theta_D$ a large deflection angle α results. Since the momentum and energy of a phonon are related by $k = \omega/v$ or $P = \hbar\omega/v$, v being the velocity of lattice wave, the momentum P at a temperature $T < \Theta_D$ is expected to be smaller than P_0 by a factor in the ratio of their energies or $P \simeq P_0(T/\Theta_D) = p(T/\Theta_D)$.

If an electron originally moves in the direction of an applied electric field, the loss of forward momentum during each collision process is given by

$$\Delta p = p(1 - \cos \alpha) \simeq p\frac{\alpha^2}{2} = \frac{p}{2}\left(\frac{T}{\Theta_D}\right)^2 \quad (4.48)$$

At temperatures $T > \Theta_D$, α can be fairly large; thus, a single collision can change completely the course of an electron. At temperatures $T < \Theta_D$, the number of collision processes required to reverse the direction of an electron motion is, according to Eq. (4.48), of the order of $(\Theta_D/T)^2$. Hence, the effectiveness of each collision process is reduced by a factor $(T/\Theta_D)^2$ at low temperatures. We must not forget that at the same time, the density of phonons decreases as temperature decreases according to a T^3 law. Therefore, the resistivity of a pure metal varies as T^5 at low temperatures. Finally, we should mention that $G(T)$ of Eq. (4.46) reduces to $\Theta^4 T/4$ for $T \gg \Theta$ and approaches $124T^5$ for $T \ll \Theta$. Therefore, it is possible to estimate the values of conductivity and mean free path of a metal at different temperatures from known values at a given temperature.

4.4 IMPURITY SCATTERING IN NONDEGENERATE SEMICONDUCTORS

The two most important mechanisms which limit the mobility of free carriers in a semiconductor are scattering by ionized impurities and scattering by lattice vibrations. Figure 4.3 illustrates the deflection of electron trajectories in the coulomb field of charged centers. In a solid, however, the classical picture of Fig. 4.3 must be modified. First, the electron motion is described by a wave of the form $\exp(i\mathbf{k} \cdot \mathbf{r})$ with a propagation vector $\mathbf{k}$. The change in the magnitude and direction of the velocity of an electron in the classical picture means a transition of the electron state from $\mathbf{k}$ to $\mathbf{k}'$ in the wave picture. The classical coulomb potential also needs modification. The perturbing potential of an impurity atom is usually taken to be

$$V = \frac{ze}{4\pi\epsilon r} \exp(-qr) \tag{4.49}$$

where ze is the effective charge of the impurity atom and ϵ the dielectric constant. The factor $\exp(-qr)$ takes into account the screening of the coulomb field by conduction-band electrons and valence-band holes.

Let us now refer to Eqs. (4.4) and (4.5). The quantum mechanical counterpart of the scattering probability $S(\mathbf{v},\mathbf{v}')$ is $S(\mathbf{k},\mathbf{k}')$ which can be calculated from perturbation theory as

$$
\begin{aligned}
S(\mathbf{k},\mathbf{k}') &= \frac{2\pi}{\hbar} \left| \int_0^\infty \exp(-i\mathbf{k} \cdot \mathbf{r}) \left(\frac{ze^2}{4\pi\epsilon r} e^{-qr} \right) \exp(i\mathbf{k}' \cdot \mathbf{r}) 4\pi r^2 \, dr \right|^2 \\
&= \frac{2\pi}{\hbar} \left\{ \frac{ze^2}{\epsilon(|\mathbf{k} - \mathbf{k}'|^2 + q^2)} \right\}^2
\end{aligned}
\tag{4.50}
$$

Because of the Pauli exclusion principle, a scattering process can take place only when the initial state is occupied and the final state is empty. Thus, the functions $f(\mathbf{r},\mathbf{v})$ and $f(\mathbf{r}',\mathbf{v}')$ in Eqs. (4.4) and (4.5) should be replaced by $f(\mathbf{k})[1 - F(\mathbf{k}')]$ and $f(\mathbf{k}')[1 - F(\mathbf{k})]$ respectively. Furthermore, we assume that during each scattering process, the exchange of energy is extremely small compared with the initial energy of an electron. That means, $|\mathbf{k}| = |\mathbf{k}'|$ in Eq. (4.50). The number of states in a volume V having momentum between p and $p + dp$ and the direction of p within a solid angle $d\Omega$ is given by $(2V/h^3)p^2 \, dp \, d\Omega$. In terms of energy, the density of state expression becomes

$$N(\mathcal{E}) \, d\mathcal{E} = (2/h^3)(2\mathcal{E}m^*)^{1/2}m^* \, d\Omega \, d\mathcal{E} \tag{4.51}$$

The quantity $N(\mathcal{E})$ multiplied by $f(\mathbf{k})[1 - F(\mathbf{k}')]$ or $f(\mathbf{k}')[1 - F(\mathbf{k})]$ represents the density of states involved in scattering, and it is to replace $f(\mathbf{r},\mathbf{v}) \, d\mathbf{v}'$ and $f(\mathbf{r}',\mathbf{v}') \, d\mathbf{v}'$, respectively, in Eqs. (4.4) and (4.5). In view of the above considerations and $f(\mathbf{k})F(\mathbf{k}') = f(\mathbf{k}')F(\mathbf{k})$, subtracting Eq. (4.4) from Eq. (4.5) gives

$$c - b = N_i \int [f(\mathbf{k}') - f(\mathbf{k})]S(\mathbf{k},\mathbf{k}') \frac{2m^*}{h^3} \sqrt{2\mathcal{E}m^*} \, d\Omega \qquad (4.52)$$

where the factor N_i is the concentration of ionized impurities.

In evaluating Eq. (4.52), we must know the change in the distribution function. It can be shown[3] that the following

$$f(\mathbf{k}') - f(\mathbf{k}) = [f(\mathbf{k}') - f_0(\mathbf{k}')] - [f(\mathbf{k}) - f_0(\mathbf{k})]$$
$$= (1 - \cos \theta)g(k) \qquad (4.53)$$

constitutes a self-consistent solution of the Boltzmann transport equation. Comparing Eq. (4.52) with Eq. (4.10), we define a mean free time $\tau(k)$

$$\frac{1}{\tau(k)} = \frac{c - b}{g(k)} = N_i \int (1 - \cos \theta)S(\mathbf{k},\mathbf{k}') \frac{2m^*}{h^3} \sqrt{2\mathcal{E}m^*} \, d\Omega \qquad (4.54)$$

where $d\Omega = 2\pi \sin \theta \, d\theta$ and θ is the scattering angle. Note that $\tau(k)$ depends only on the magnitude of k so it can also be designated as $\tau(\mathcal{E})$. Replacing $|\mathbf{k} - \mathbf{k}'|^2$ in Eq. (4.50) by $2p^2(1 - \cos \theta)/\hbar^2$ and letting

$$b = \frac{2p^2}{q^2\hbar^2} \qquad \text{and} \qquad x = 1 - \cos \theta$$

we find

$$\frac{1}{\tau(\mathcal{E})} = A \int_0^2 \frac{b^2 x \, dx}{(1 + bx)^2} = A \left[\log (1 + 2b) - \frac{2b}{1 + 2b} \right] \qquad (4.55)$$

where $A = 2\pi N_i(ze^2/4\pi\epsilon)^2(2\mathcal{E})^{-3/2}(m^*)^{-1/2}$.

In the classical analysis of the scattering problem, the scattering angle θ can be obtained from the electron trajectory and is found to be

$$\tan \left(\frac{\theta}{2} \right) = \frac{e^2}{4\pi\epsilon m^* v^2 a} \qquad (4.56)$$

where v is the velocity of an electron and the distance a is defined in Fig. 4.3. For electrons whose trajectories are within an annular ring of area $2\pi a \, da$, the deflections will be within a solid angle

$$d\Omega = 2\pi \sin \theta \, d\theta$$

Thus, a differential cross-section $\sigma(\theta)$ is defined as

$$\sigma(\theta) = \frac{2\pi a \, da}{2\pi \sin \theta \, d\theta} = \left(\frac{e^2}{8\pi\epsilon m^* v^2} \right)^2 \csc^4 \left(\frac{\theta}{2} \right) \qquad (4.57)$$

which is known as the Rutherford scattering law. If there are N_i

---

[3] See for example: F. Seitz, "The Modern Theory of Solids," McGraw-Hill Book Company, New York, 1940; A. H. Wilson, "The Theory of Metals," Cambridge University Press, Cambridge, 1953.

impurity atoms/cm³, each impurity atom has a region of influence which is chosen to be cubic. The edge a_1 of this cube is of course given by $a_1{}^3 = 1/N_i$. The mean free time τ_c between collisions is thus equal to $\tau_c = a_1/v$. However, not all collisions are equally effective. For small θ, the electron is hardly deflected at all, although it is still counted as a collision in τ_c. Therefore, a weighted average of some kind must be performed. The mean free time $\tau(\mathcal{E})$ of a carrier having velocity v is defined as

$$\frac{1}{\tau(\mathcal{E})} = \frac{v}{a_1} \frac{4}{\pi a_1{}^2} \int_0^{a_1/2} (1 - \cos\theta) 2\pi a\, da \qquad (4.58)$$

For $\theta = 0$, the electron trajectory is not deflected so the mean free time should be infinite as it is from Eq. (4.58). The integration in Eq. (4.58) is over a sphere having radius $a_1/2$ instead of a cube for simplicity. Substituting Eq. (4.57) into Eq. (4.58) gives

$$\frac{1}{\tau(\mathcal{E})} = \left(\frac{\pi}{2a_1}\right)\left(\frac{\mathcal{E}_1}{m^*}\right)^{1/2}\left(\frac{\mathcal{E}_1}{2\mathcal{E}}\right)^{3/2} \log\left[1 + \left(\frac{2\mathcal{E}}{\mathcal{E}_1}\right)^2\right] \qquad (4.59)^4$$

where $\mathcal{E}_1 = e^2/(2\pi\epsilon a_1)$.

The reader can easily show that the factor in front of the logarithm in Eq. (4.59) is equal to A in Eq. (4.55) if $z = 1$. Actually, the classical differential cross-section $\sigma(\theta)$ and the quantum-mechanical scattering probability $S(\mathbf{k},\mathbf{k}')$ should produce the same result if we let $q = 0$ in Eq. (4.50). However, difficulty arises because a pure coulomb potential gives a zero relaxation time. The introduction of a nonvanishing q (or physically the screening of the coulomb potential) prevents Eq. (4.54) from going to infinity. In the classical treatment, Eq. (4.58) is kept finite by confining the influence of the coulomb field to a box of volume $a_1{}^3$. Therefore, the origin for the

⁴ This formula was first introduced by E. M. Conwell and V. F. Weisskopf, *Phys. Rev.*, vol. 77, p. 388, 1950.

Figure 4.3

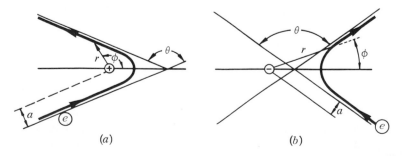

(a)   *(b)*

Deflection of the electron trajectory by ionized impurities: (a) ionized donors; (b) ionized acceptors.

apparent difference between Eqs. (4.55) and (4.59) lies not in the calculation itself but in the physical models which we take to make the coulomb potential behave properly.

Now let us return to Eq. (4.55). The screening length $1/q$ of the coulomb field has been calculated by Brooks[5] as

$$q^2 = \frac{4\pi e^2}{\epsilon kT} n \left(2 - \frac{n}{N_d}\right) \tag{4.60}$$

where n and N_d are, respectively, the ionized and total donor concentrations, the former being set equal to the concentration of free electrons. In the intermediate temperature range, donors are practically all ionized; thus, $n = N_d = N_i$, and the quantity $2b$ in Eq. (4.55) has a value

$$2b = \frac{8\pi\epsilon m^* kT}{e^2 h^2 n} \mathcal{E} \tag{4.61}$$

An analogous expression is obtained for p-type semiconductors. Taking $\mathcal{E} = 3kT = 0.075$ ev at room temperature, $n = 10^{14}/cm^3$ and m^* to be the free electron mass, the value of $2b$ is about 10^3. Since $2b$ is much larger than unity and $\log (2b)$ is a slowly varying function of b, the term in the bracket of Eq. (4.55) can be approximated by $\log (1 + 2b)$ with the value of b evaluated for $\mathcal{E} = 3kT$. The same is applied to the $\log [1 + (2\mathcal{E}/\mathcal{E}_1)^2]$ term in Eq. (4.59). Therefore, the mean free time $\tau(\mathcal{E})$ given by Eqs. (4.55) and (4.59) can be taken as proportional to $\mathcal{E}^{3/2}$ or v^3.

In averaging $\tau(\mathcal{E})$ over the velocity distribution, we note that if x is an odd integer

$$\langle v^x \rangle = \frac{\int_0^\infty v^x \exp(-m^* v^2/2kT) 4\pi v^2\, dv}{\int_0^\infty \exp(-m^* v^2/2kT) 4\pi v^2\, dv} = v_T^x \frac{2}{\sqrt{\pi}} \left(\frac{x+1}{2}\right)! \tag{4.62}$$

where $v_T^2 = 2kT/m^*$. With the help of Eqs. (4.21) and (4.62), we can evaluate the average value over the Boltzmann distribution of any function $g(v)$ which can be expanded into a power series in v. Thus, substituting Eq. (4.55) into Eq. (4.14) gives a value for the mobility μ_i limited by impurity scattering

$$\mu_i = \frac{e}{3kT} \langle v^2 \tau \rangle = \frac{2^{7/2}(kT)^{3/2}(4\pi\epsilon)^2}{\pi^{3/2} e^3 N_i \sqrt{m^*}} \frac{1}{\log\left[1 + \frac{6m^*(kT)^2\epsilon}{\pi e^2 \hbar^2 n}\right]} \tag{4.63a}$$

The reason for taking $\mathcal{E} = 3kT$ in the logarithm is because μ involves an integral of the form $\int \mathcal{E}^3 \exp(-\mathcal{E}/kT)\, d\mathcal{E}$ whose integrand has a maximum value at $\mathcal{E} = 3kT$. Substituting numerical constants

[5] H. Brooks," *Advan. Electron. Electron Phys.*," vol. 7, p. 87, 1955; *Phys. Rev.*, vol. 83, p. 879, 1951.

into Eq. (4.63a) gives

$$\mu_i = \frac{1.65 \times 10^{19}}{N_i} \left(\frac{T}{T_0}\right)^{3/2} \left(\frac{\epsilon}{\epsilon_0}\right)^2 \left(\frac{m}{m^*}\right)^{1/2}$$
$$\frac{1}{\log\left[1 + (10^{17}/n)(T/T_0)^2(\epsilon/\epsilon_0)\right]} \quad (4.63b)$$

where N_i and n are the ionized donor and free electron concentrations respectively, $T_0 = 300°K$, ϵ_0 is the dielectric constant of free space and m is the free electron mass. According to convention, μ is expressed in cm²/volt-sec and N_i and n are in cm⁻³. A similar expression is obtained for μ based on the classical model of Eq. (4.59), the only difference being that the logarithm term is given by $\log\left[1 + (3 \times 10^{11}/N_i^{2/3})(T\epsilon/T_0\epsilon_0)^2\right]$ instead.

In concluding this section, the following points are worth noting. The same weighting factor $(1 - \cos\theta)$ appears in Eqs. (4.54) and (4.58), which measures the relative change in the component of the electron velocity along the initial direction of motion. If a carrier moves in a certain direction, the current which it creates will not be lost immediately after the first collision but dissipated upon successive collisions. Note that the same factor $(1 - \cos\theta)$ enters into our discussion in the previous section, specifically Eq. (4.48), except that θ is replaced by α. Because an electron still retains part of its velocity after each collision, the relaxation time which appears in Eqs. (4.10) and (4.11) differs from the time between successive collisions. No such distinction is made in Sec. 3.3 since the relaxation time and the time between successive collisions would be the same in view of the assumption that the velocity of an electron is completely random after each collision.

The characteristic of impurity scattering is the $T^{3/2}$ dependence of μ. At higher temperatures, electrons on the average move faster and hence are less deflected by ionized impurities. Taking $N_i = n = 10^{18}$ cm⁻³, $T = 300°K$, $m^* = 0.13m$ and $\epsilon = 16\epsilon_0$ in germanium, we find $\mu_i = 1.5 \times 10^5$ cm²/volt-sec. Since μ for electrons in semiconductor-grade germanium is 3900 cm²/volt-sec at 300°K, there must be other processes, for example scattering by lattice vibrations, which limit the mobility of electrons in materials now available for semiconductor manufacture. According to Eq. (3.6), when two scattering mechanisms exist simultaneously,

$$dn = -n\left(\frac{dt}{\tau_1} + \frac{dt}{\tau_2}\right) = -n\frac{dt}{\tau} \quad \text{with} \quad \frac{1}{\tau} = \frac{1}{\tau_1} + \frac{1}{\tau_2} \quad (4.64)$$

where τ_1 and τ_2 are the relaxation times resulting separately from the two scattering mechanisms. If $\tau_1(\mathcal{E})$ and $\tau_2(\mathcal{E})$ have the same energy dependence, substitution of Eq. (4.64) into Eq. (4.14) obviously leads to

$$\frac{1}{\mu} = \frac{1}{\mu_1} + \frac{1}{\mu_2} \quad (4.65)$$

Unfortunately, $\tau_1(\mathcal{E})$ and $\tau_2(\mathcal{E})$, in general, depend differently on $\mathcal{E}$. When μ_1 and μ_2 are comparable in magnitude, Eq. (4.65) is rather inaccurate. However, it is still in common usage because of its simplicity.

4.5 LATTICE SCATTERING IN NONDEGENERATE SEMICONDUCTORS

The scattering of the motion of an electron by lattice vibrations is a wave phenomenon; therefore, it is conceptually difficult from a classical viewpoint. The following presentation is based on a physical model[6] originally due to Shockley and Bardeen. Figure 4.4a is a reproduction of the part of Fig. 3.2 relevant to our discussion, which

[6] W. Shockley and J. Bardeen, *Phys. Rev.*, vol. 77, p. 407, 1950; J. Bardeen and W. Shockley, *Phys. Rev.*, vol. 80, p. 72, 1950.

Figure 4.4

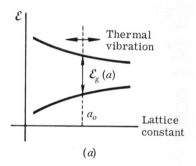

(a)

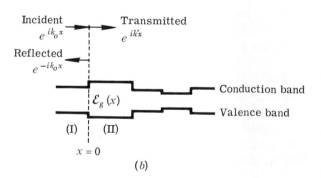

(b)

A physical model illustrating the effect of lattice vibration on the wave motion of electrons. The effect is due to modulation of lattice constant by lattice vibration. (a) Variation of the energy band structure as a function of lattice constant. (b) As atoms vibrate, there is a local variation of the lattice constant and consequently the energy band structure. This local variation in the energy band structure changes the propagation property of the electron, causing the scattering of electron wave motion.

shows the variation of the conduction and valence band edges as a function of the lattice constant. As lattice atoms vibrate about their equilibrium positions, their vibrational motion causes a local variation in the energies of the conduction and valence band edges, as shown in Fig. 4.4b. Since the energy of an electron is conserved, the gain in potential energy must be at the expense of the kinetic energy. Thus, from $\mathcal{E} = \mathcal{E}_c + \hbar^2 k_0^2/2m^*$, we have

$$-\delta\mathcal{E}_c = \hbar^2 k_0 \frac{\delta k_0}{m^*} \tag{4.66}$$

and a similar expression for valence band holes. For ease of discussion, we assume that the electron motion is one-dimensional and the change of $\mathcal{E}_c$ is abrupt.[7] At the boundary where $\delta\mathcal{E}_c$ occurs, the propagation constant of the wave changes, resulting in a reflected wave.

Referring to Fig. 4.4b, we have

$$\psi = \begin{cases} A(e^{ik_0x} + Re^{-ik_0x}) & \text{for } x < 0 \qquad (4.67a) \\ ATe^{ik'x} & \text{for } x > 0 \qquad (4.67b) \end{cases}$$

where ψ is the electron wave function, and R and T represent respectively the reflection and transmission coefficients. As pointed out in Sec. 3.7, the boundary condition for ψ is such that ψ and $\partial\psi/\partial x$ are continuous at $x = 0$, meaning

$$1 + R = T \qquad \text{and} \qquad k_0(1 - R) = k'T \tag{4.68}$$

Solving for R and using the result of Eq. (4.66), we obtain

$$|R|^2 = \left| \frac{k_0 - k'}{k_0 + k'} \right|^2 = \left(\frac{m^* \delta\mathcal{E}_c}{2\hbar^2 k_0^2} \right)^2 \tag{4.69}$$

In a three-dimensional lattice, the reflected wave may propagate in a different direction and hence a nonvanishing reflection coefficient means the scattering of electron wave motion.

The change in $\mathcal{E}_c$ can be expressed in terms of *lattice dilation* $\Delta = \delta V/V$ (V being the effective volume) as follows:

$$\delta\mathcal{E}_c = \mathcal{E}_{1c} \Delta \tag{4.70}$$

for small dilation Δ. The quantity $\mathcal{E}_{1c}$ is often referred to as the *deformation potential*. At high temperatures, the dilation of a solid can be estimated from its mechanical properties. If K is the compressibility of the solid $Kp = \delta V/V$, and the stored energy in the

---

[7] The word *abrupt* is used in the sense that the change in $\mathcal{E}_c$ occurs over a distance short compared with the wavelength associated with the wave motion of electrons.

lattice is given by $\frac{1}{2}p\,\delta V$ which on the average is equal to kT. In terms of K and $\mathcal{E}_{1c}$, Eq. (4.69) becomes

$$|R|^2 = \frac{m^{*2}\mathcal{E}_{1c}^2 KkT}{4\hbar^4 k_0^4 V} \tag{4.71}$$

In applying Eq. (4.71) to scattering problems, we must know the meaning of V. Remember that δV caused by lattice vibrations can be analyzed into Fourier components as running waves. If we take the lattice as a whole, part of the lattice will experience a negative δV and part a positive δV and the effect averages out. Therefore, the V in Eq. (4.71) must be the effective volume in which the lattice and electron wave motions interact strongly. It turns out that this region extends over $\lambda/4$ in each dimension where λ is the wavelength of the electron wave or $\lambda = 2\pi/k_0$. Since Eq. (4.71) has meaning only in this volume V, the mean free path l of an electron must also be referred to $\lambda/4$. Thus, we have

$$l = \frac{\lambda/4}{2|R|^2} = \left(\frac{\pi^3}{8c_{ll}K}\right)\frac{\pi\hbar^4 c_{ll}}{m^{*2}\mathcal{E}_{1c}^2 kT} \tag{4.72}$$

the factor 2 in $2|R|^2$ being on account of the fact that electrons suffer reflections at both the left and right sides of the cube. In Eq. (4.72), c_{ll} is the elastic constant. The factor in parentheses is the difference between the result based on Eq. (4.71) and the exact result which will be discussed shortly.

The quantum-mechanical treatment of the problem follows quite closely the steps outlined above. The complete Hamiltonian for a system of lattice atoms can be written as

$$\mathcal{H} = \sum_l \frac{P_l^2}{2M_l} + V(u_l) + \sum_i \frac{p_i^2}{2m_i} + \sum_i V_i(r_i, u_l) \tag{4.73}$$

where the first two terms represent the kinetic and potential energies associated with lattice atoms and they are the same as the terms in a much simplified form of Eq. (1.41) while the last two terms represent the kinetic and potential energies associated with electrons. Since the lattice atoms are no longer fixed in position, the potential energy of electrons, $V_i(r_i, u_l)$, depends on the displacement u_l of lattice atoms. The term V_i can be expanded in terms of u_l as follows:

$$V_i(r_i, u_l) = V_i(r_i) + \sum_l \mathbf{u}_l \cdot \text{grad } V_i \tag{4.74}$$

The last term in Eq. (4.74) is responsible for the interaction of phonon waves with electron waves. In a solid, as pointed out earlier in Sec. 2.10, it is necessary to express the lattice displacement vector

$\mathbf{u}_l$ in terms of its Fourier components[8]

$$\mathbf{u}_l = \frac{1}{\sqrt{N}} \sum_q \mathbf{X}_q e^{-i\mathbf{q}\cdot\mathbf{l}} \tag{4.75}$$

where $q = 2\pi/\lambda$, λ being the wavelength associated with a given mode of lattice vibration. The normalization constant N is equal to the number of lattice cells. The notation $\mathbf{l}$ is used for lattice coordinates to save confusion as $\mathbf{r}$ is used for electron coordinates. Since the Fourier component $\mathbf{X}_q$ now has a definite wave property, it is possible to relate $\mathbf{X}_q$ with the displacement vector of a harmonic oscillator discussed in Sec. 1.6.

In terms of the annihilation and creation operators A and B defined by Eqs. (1.55) and (1.66), Eq. (4.75) becomes

$$\mathbf{u}_l = -i \sum_q \left(\frac{\hbar}{2N\omega M}\right)^{\frac{1}{2}} e^{-i\mathbf{q}\cdot\mathbf{l}}(B_q - A_{-q})\frac{\mathbf{a}}{2} \tag{4.76}$$

where $\mathbf{a}$ is a unit vector in the direction of the displacement $\mathbf{u}_l$. If ψ and Φ represent respectively the lattice and electron wave functions, the quantity of interest is the matrix element $M(\mathbf{k},\mathbf{k}')$ connecting the electron state Φ_k to $\Phi_{k'}$ with the creation or annihilation of phonons. The process is known as *phonon scattering* of electron waves. The quantity $M(\mathbf{k},\mathbf{k}')$ is given quantum mechanically as

$$M(\mathbf{k},\mathbf{k}') = \int \psi_1^* \Phi_k^* \left(\sum_l \mathbf{u}_l \cdot \operatorname{grad} V_i\right) \psi_2 \Phi_{k'} \, d\sigma_l \, d\sigma_e \tag{4.77}$$

where $d\sigma_l$ and $d\sigma_e$ represent the elementary volumes in the lattice and electron space coordinates respectively. It can be shown by the Wigner-Seitz method that the term involving electron wave functions in Eq. (4.77) is equal to

$$\int \Phi_k^* (\mathbf{a} \cdot \operatorname{grad} V_i)\Phi_{k'} \, d\sigma_e = -i\mathbf{a} \cdot (\mathbf{k} - \mathbf{k}') \left(\frac{\delta\mathcal{E}_c}{\Delta}\right) \int \Phi_k^* \Phi_{k'} \, d\sigma_e \tag{4.78}$$

where Δ is the lattice dilation.

Since the electron wave function can be expressed in terms of Bloch wave functions, $\Phi_k^*(\mathbf{r} + \mathbf{l}) = \Phi_k^*(\mathbf{r}) \exp(-i\mathbf{k}\cdot\mathbf{l})$. Therefore, the integration in Eq. (4.78) over different lattice cells will have the same amplitude F but different phase factors. Combining this phase factor with $\exp(-i\mathbf{q}\cdot\mathbf{l})$ of Eq. (4.76), we have

$$\int_{\substack{\text{whole crystal}}} \Phi_k^* \exp(-i\mathbf{q}\cdot\mathbf{l})\Phi_{k'} \, d\sigma_e = \left[\sum_l \exp i(\mathbf{k}' - \mathbf{k} - \mathbf{q})\cdot\mathbf{l}\right] \int_{\substack{\text{cell}}} \Phi_k^*(\mathbf{r})\Phi_{k'}(\mathbf{r}) \, d\sigma_e \tag{4.79}$$

---

[8] For a thorough discussion, the reader is referred to J. M. Ziman, "Electrons and Phonons," Oxford University Press, Oxford, 1960.

The summation is over the whole crystal and it is zero unless

$$\mathbf{k}' = \mathbf{k} + \mathbf{q} \qquad (4.80)$$

This is known as the law of *conservation of momentum*.

We shall see in the next section that during a collision with the lattice, the magnitude of k hardly changes. Thus, we can replace the integral of Eq. (4.79) by $\int_{\text{cell}} |\Phi|^2 \, d\sigma_e$. For a one-electron wave function, Eq. (4.79) should have a value of unity if Eq. (4.80) is satisfied. Let us now turn to the lattice wave function. Accompanying grad V_i, there is $\mathbf{u}_l$. That means, the scattering of an electron by a lattice wave must be associated with either the creation or the annihilation of a phonon according to Eq. (4.76). Noting that the operational rules of A and B operators are given by Eq. (1.66), we find

$$|M(\mathbf{k},\mathbf{k}')|^2 = \frac{\hbar(2n_q + 1)}{4NM\omega} |\mathbf{a} \cdot (\mathbf{k} - \mathbf{k}')|^2 \left(\frac{\delta\mathcal{E}_c}{\Delta}\right)^2 \qquad (4.81)$$

where $n_q = [\exp(\hbar\omega/kT) - 1]^{-1}$. At high temperatures such that $\hbar\omega/kT < 1$, $n_q = kT/\hbar\omega$. In view of Eq. (4.80) and $NM = \rho V$, ρ and V being, respectively, the mass density and volume of the crystal, Eq. (4.81) becomes

$$|M(\mathbf{k},\mathbf{k}')|^2 = \frac{kT}{2\rho V} \left(\frac{q}{\omega}\right)^2 \mathcal{E}_{1c}^2 \qquad (4.82)$$

where $\mathcal{E}_{1c}$ is the deformation potential related to $\delta\mathcal{E}_c$ through Eq. (4.70).

The quantum-mechanical transition probability W per unit time that an electron be scattered from a $\mathbf{k}$ to $\mathbf{k}'$ state is given by

$$W = \frac{2\pi}{\hbar} \frac{|M(\mathbf{k},\mathbf{k}')|^2 N(\mathcal{E})}{2} \qquad (4.83)$$

where $N(\mathcal{E})$ is given by Eq. (4.51) and the factor $\frac{1}{2}$ is due to the fact that change of spin does not occur. We should point out that the scattering probability $S(\mathbf{k},\mathbf{k}')$ in general is related to the matrix element $M(\mathbf{k},\mathbf{k}')$ through the relation $S(\mathbf{k},\mathbf{k}') = (2\pi/\hbar)|M(\mathbf{k},\mathbf{k}')|^2$. Therefore, in Eq. (4.83), we follow the same procedure as we did in deriving Eq. (4.54). The reciprocal of the relaxation time τ is the average of W over $d\Omega$. Since $|M(\mathbf{k},\mathbf{k}')|^2$ is isotropic (that means not a function of θ), the weighting factor $(1 - \cos\theta)$ does not contribute anything new except that $\int(1 - \cos\theta) \, d\Omega = \int d\Omega = 4\pi$. Thus, using the relation $1/\tau(\mathcal{E}) = \int W \, d\Omega$, we have

$$\frac{1}{\tau(\mathcal{E})} = \frac{kT\mathcal{E}_{1c}^2 m^{*2}}{\pi\hbar^4 \rho v_s^2} \left(\frac{2\mathcal{E}}{m^*}\right)^{1/2} = \frac{v}{l} \qquad (4.84)$$

Note that the mean free path l defined as τv is a constant independent of the velocity of an electron. The value of l from Eq. (4.84) is exactly equal to that given in Eq. (4.72) without the factor in parentheses if we make use of the fact that the speed of sound wave $v_s = \omega/q = (c_{ll}/\rho)^{1/2}$. From Eq. (4.84), the electron mobility due to lattice scattering is

$$\mu_l = \frac{e}{3kT} \langle v^2\tau \rangle = \frac{2\sqrt{2\pi}}{3} \frac{\hbar^4 \rho v_s^2}{\varepsilon_{1c}^2 m^{*5/2}} \frac{1}{(kT)^{3/2}} \tag{4.85}$$

In germanium and silicon, the measured temperature dependence of electron and hole mobility deviates quite considerably from the $T^{-3/2}$ law. In the above analysis, we have considered only the long-wavelength acoustic phonons so that n_q can be replaced by $kT/\hbar\omega$ in Eq. (4.81). In a semiconductor having several equivalent conduction band minima, electrons may be scattered from one band minimum to another as well as scattered within the same band minimum. These processes are known as *intervalley* and *intravalley scattering* processes respectively. For intervalley scattering, a large change of electron momentum is involved, thus requiring the participation of short-wavelength phonons, both acoustic and optical. The contribution from the high-energy phonons to mobility varies much more rapidly than the $T^{-3/2}$ law. So far as the hole mobility is concerned, scattering of holes from the heavy to the light mass band is a possibility. The fact that only one type of hole has been observed in the Haynes-Shockley experiment to be discussed later is a strong indication that interband scattering is important. A complete theory[9] of free-carrier mobility should include all possible scattering processes to allow the participation of low- and high-energy phonons. Finally we should emphasize that the quantum-mechanical derivation of Eq. (4.85) presented above is rather sketchy. The step that is especially significant is expressed in Eq. (4.78). For lack of space, we are compelled to leave out all the steps and their justifications leading to Eq. (4.78). For a detailed treatment of the electron-phonon scattering problem, the reader is referred to the books by Ziman and by Shockley.[10]

4.6 HIGH-FIELD MOBILITY AND HOT ELECTRON EFFECT

In the previous derivation of the mobility of free carriers, a tacit assumption is implied that the applied electric field is small so that the free carriers are in thermodynamical equilibrium with the host

[9] For a more extensive discussion, the reader is referred to H. Brooks, Electrical Properties of Germanium and Silicon, in *Advan. Electron. Electron Phys.*, vol. 7, p. 87, 1955.
[10] J. M. Ziman, "Electrons and Phonons," Oxford University Press, Oxford, 1960; W. Shockley, "Electrons and Holes in Semiconductors," D. Van Nostrand Co., Inc., New York, 1950.

lattice. More specifically, the temperature T in the mobility expression is taken without question as the lattice temperature. Returning to Eq. (3.5) or

$$J = e[n\mu_n(T) + p\mu_p(T)]E \qquad (4.86)$$

we see that a linear relationship exists between the current density J and the electric field E as long as T is not a function of E. Figure 4.5 shows the power flow diagram in which the electron system receives power from the electric field and dissipates power to the lattice. By electron system, we mean that electrons in the conduction band and holes in the valence band form a thermodynamic equilibrium system such that the distribution of electrons among the available energy states can be described by statistical laws in terms of a common temperature T_e called the *electron temperature*. From thermodynamics, we know that for power to flow from the electron system to the lattice, the electron temperature T_e must exceed the lattice temperature T_l. When the electric field is small, the difference between T_e and T_l is insignificant and hence Ohm's law is obeyed. Since T_e is expected to increase with increasing E, deviations from Ohm's law will occur at sufficiently high electric fields.

Experimental observations of deviations from Ohm's law were first reported by Ryder[11] in n-type germanium. Figure 4.6 shows the more recent result obtained by Prior,[12] using a more refined experimental technique. To explain the experimental result on a quantitative basis, we must examine the power flow between the electron system and the lattice. In the following, we shall present a treatment of the subject discussed in a paper by Shockley.[13] First, let us estimate the amount of energy exchange in a collision process between electrons and acoustic phonons. If ε_1 and p_1 and ε_2 and p_2 represent respectively the energy and momentum of an electron before and

[11] E. J. Ryder and W. Shockley, *Phys. Rev.*, vol. 81, p. 139, 1951; E. J. Ryder, *Phys. Rev.*, vol. 90, p. 766, 1953.

[12] A. C. Prior, *Jour. Phys. Chem. Solids*, vol. 12, p. 175, 1960.

[13] W. Shockley, *Bell System Tech. Jour.*, vol. 30, p. 4, October, 1951.

Figure 4.5

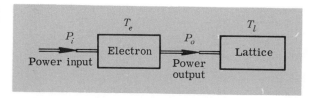

Power flow between two heat reservoirs: (1) the electron system containing conduction-band electrons and valence-band holes; (2) the host lattice.

after the collision, conservation of energy and momentum requires

$$\mathcal{E}_2 - \mathcal{E}_1 = \frac{p_2{}^2}{2m^*} - \frac{p_1{}^2}{2m^*} = \pm \hbar\omega = \pm \hbar cq \qquad (4.87a)$$

$$\mathbf{p}_2 - \mathbf{p}_1 = \pm \hbar\mathbf{q} \qquad (4.87b)$$

where $\hbar\omega$ and $\hbar\mathbf{q}$ are the energy and momentum of the phonon involved. The positive and negative signs in Eq. (4.87) correspond respectively to the phonon absorption and emission processes. For acoustic phonons, $\omega = cq$, c being the velocity of a phonon; therefore, maximum exchange of energy occurs when q is a maximum or in other words $\mathbf{p}_2 = -\mathbf{p}_1$. Substituting $\hbar q = p_2 + p_1$ into the energy equation of Eq. (4.87), we find

$$p_2 - p_1 = \pm(\omega/q)2m^* = \pm 2m^*c$$
$$\mathcal{E}_2 - \mathcal{E}_1 = \pm \hbar qc = \pm c(p_1 + p_2) \qquad (4.88)$$

The average thermal velocity v_T of an electron is around 10^7 cm/sec at room temperature (taking $\mathcal{E}_1 = kT/2$ and m^* equal to the free electron mass), while the value of c is around 5×10^5 cm/sec. Therefore, the values of $\Delta p/p$ and $\Delta\mathcal{E}/\mathcal{E}$ for electrons, being equal to $2c/v_T$ and $4c/v_T$ from Eq. (4.88), are about 0.1 and 0.2 respectively. This justifies our earlier assumption that $p_1 = p_2$ or $k_1 = k_2$ in the derivation of mobility in the low-field region. In dealing with energy exchange in the high-field region, however, this difference in p_1 and p_2 can no longer be ignored. What is even more important is that we must treat the phonon absorption and emission processes separately. Referring to Eq. (4.81) in conjunction with Eq. (1.66), we note that instead of having $(2n_q + 1)$ lumped together, the matrix element

Figure 4.6

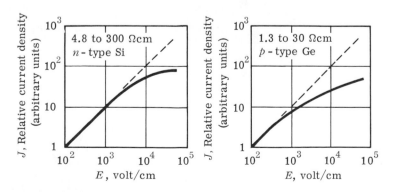

The relation between current density and electric field intensity observed in silicon and germanium. Note the nonlinear behavior at high fields. (A. C. Prior, Jour. Phys. Chem. Solids, vol. 12, p. 175, 1960.)

M_a for the phonon absorption process is proportional to

$$n_q = \frac{1}{\exp{(\hbar\omega/kT_l)} - 1} \simeq \frac{kT_l}{\hbar\omega} - \frac{1}{2} \qquad (4.89)$$

while that M_e for the phonon emission process is proportional to

$$n_q + 1 = \frac{\exp{(\hbar\omega/kT_l)}}{\exp{(\hbar\omega/kT_l)} - 1} \simeq \frac{kT_l}{\hbar\omega} + \frac{1}{2} \qquad (4.90)$$

Therefore, in terms of M given by Eq. (4.82),

$$M_a{}^2 = M^2 \frac{\hbar\omega}{2kT_l} n_q \qquad \text{and} \qquad M_e{}^2 = M^2 \frac{\hbar\omega}{2kT_l} (n_q + 1) \qquad (4.91)$$

Next, comes the question about the density of the end states $N(\varepsilon)/2$ in Eq. (4.83). In a volume V,

$$\tfrac{1}{2}N(\varepsilon_2)\, d\varepsilon_2 = \tfrac{1}{2}N(p_2)\, dp_2 = \frac{V}{h^3} 2\pi m^* p_2\, d\varepsilon_2 \sin\theta\, d\theta \qquad (4.92)$$

where θ is the scattering angle. Equation (4.92) is exactly the same as Eq. (4.50) except that now care must be taken in integrating Eq. (4.92) over phase space. Consideration of momentum conservation as illustrated in Fig. 4.7 gives

$$q^2 = k_1{}^2 + k_2{}^2 - 2k_1k_2 \cos\theta \qquad \text{or} \qquad q\, dq = k_1k_2 \sin\theta\, d\theta \qquad (4.93)$$

Figure 4.7

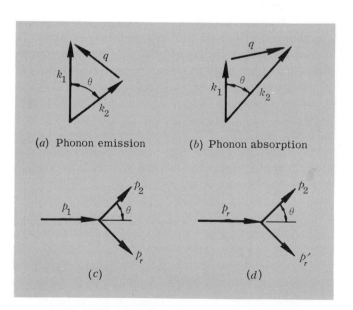

(a) Phonon emission   (b) Phonon absorption

(c)   (d)

Consideration of momentum conservation in a scattering process involving (a) phonon emission and (b) phonon absorption. The classical counterparts of these two processes are illustrated in (c) and (d).

for fixed initial and final states with k_1 and k_2. We can change $d\theta$ into dq in Eq. (4.92) and integrate Eq. (4.92) over dq instead of $d\theta$. In doing so, we must remember the energy conservation law $\delta(\mathcal{E}_1 - \mathcal{E}_2 - \hbar cq)$, meaning that in integrating Eq. (4.92),

$$\int \frac{V}{h^3} 2\pi m^* p_2 \, d\mathcal{E}_2 \frac{q \, dq}{k_1 k_2} = \frac{V}{h^2} \frac{m^*}{k_1} q \, dq = n \qquad (4.94)$$

on account of the Dirac delta function $\delta(\mathcal{E}_1 - \mathcal{E}_2 - \hbar cq)$. The symbol n is used merely to represent $(Vm^*/h^2 k_1)q \, dq$. Combining Eq. (4.94) with Eq. (4.91), we find that the transition rates, W_a and W_e respectively for the phonon absorption and emission processes, are given by

$$W_a = \frac{2\pi}{\hbar} M_a^2 n_a \qquad \text{and} \qquad W_e = \frac{2\pi}{\hbar} M_e^2 n_e \qquad (4.95)$$

where n_a and n_e are obtained by letting q equal to q_a and q_e in Eq. (4.94).

The average gain in the energy of an electron $\langle \delta\mathcal{E} \rangle$ during a collision process is given by the energy transfer during a phonon absorption or emission process multiplied by the probability of such an occurrence. Analytically, we have

$$\delta\mathcal{E} = \hbar\omega_a \frac{W_a}{W_a + W_e} - \hbar\omega_e \frac{W_e}{W_a + W_e} \qquad (4.96)$$

After we cancel the common factors which appear in both the denominator and numerator, Eq. (4.96) becomes

$$\delta\mathcal{E} = \frac{\hbar\omega_a n_q q_a \, dq_a - \hbar\omega_e (n_q + 1) q_e \, dq_e}{n_q q_a \, dq_a + (n_q + 1) q_e \, dq_e} \qquad (4.97)$$

where n_q and $n_q + 1$ are given, respectively, by Eqs. (4.89) and (4.90). In Eq. (4.97), $\omega_a = cq_a$ and $\omega_e = cq_e$. Since $\Delta p/p$ and $\Delta\mathcal{E}/\mathcal{E}$ are shown to be small in our earlier discussion, we can approximate $k_2 = k_1$ in Eq. (4.93) and thus obtain

$$q \simeq q_0 = 2k_1 \sin (\theta/2) \qquad (4.98)$$

This approximation, however, is good only for the summation terms in Eq. (4.97). For the difference terms, we must substitute into Eq. (4.93) the value of k_2 obtained from the energy equation of Eq. (4.87), and then make the approximation. The results are as follows (Prob. 4.8):

$$
\begin{aligned}
q_a &= q_0 \left(1 + \frac{m^* c q_0}{2\hbar k_1^2}\right) = 2k_1 \sin\left(\frac{\theta}{2}\right)\left[1 + \left(\frac{m^* c}{\hbar k_1}\right)\sin\left(\frac{\theta}{2}\right)\right] \\
q_e &= q_0 \left(1 - \frac{m^* c q_0}{2\hbar k_1^2}\right) = 2k_1 \sin\left(\frac{\theta}{2}\right)\left[1 - \left(\frac{m^* c}{\hbar k_1}\right)\sin\left(\frac{\theta}{2}\right)\right]
\end{aligned}
\qquad (4.99)
$$

Substituting Eqs. (4.98), (4.99), (4.89), and (4.90) into Eq. (4.97) and averaging $\sin^2 (\theta/2)$ over $d\Omega = 2\pi \sin \theta \, d\theta$, we find

$$\langle \delta\varepsilon \rangle = 4c^2 \left(m^* - \frac{\hbar^2 k_1^2}{4kT_l} \right) \tag{4.100}$$

A brief explanation of the physical origins of these terms is now in order. The quantum-mechanical phonon emission and absorption processes illustrated in Fig. 4.7a and b can be thought of classically as collision processes between two hard spheres. The fact that the mass M associated with phonons is larger than that m associated with electrons can be seen from the consideration of average energy that $mv^2 = Mc^2 = kT$. Since $c < v$, $M \gg m$. First, we consider the collision process involving an electron with initial velocity v_1 and a mass M initially at rest as illustrated in Fig. 4.7c. Applying the laws of conservation of energy and momentum gives

$$\left(1 + \frac{m}{M} \right)^2 p_r^2 = 2p_1^2 \left(1 + \frac{m}{M} \sin^2 \theta \right)$$
$$- \sqrt{4p_1^4 \cos^2 \theta \left[1 - \left(\frac{m}{M} \right)^2 \sin^2 \theta \right]} \tag{4.101}$$

where p_r is the momentum of the phonon after collision. Since $M \gg m$, the energy loss by an electron can be approximated as follows:

$$\delta\varepsilon = \frac{-p_r^2}{2M} \simeq \frac{-c^2 p_1^2}{kT_l} (1 - \cos \theta) \tag{4.102}$$

by neglecting terms with m/M and using $Mc^2 = kT_l$. Equation (4.102), after being averaged over $d\Omega$, is exactly equal to the term with minus sign in Eq. (4.100) since the average of $(1 - \cos \theta)$ over $d\Omega$ is one. Opposite to the process of Fig. 4.7c is the process of Fig. 4.7d in which a mass M with an initial velocity c collides with an electron initially at rest. For $M \gg m$, the scattered electron moves in the same direction as the mass M or $\theta = 0$. Equating the energy and momentum before and after collision, we find that the energy gain of an electron is given by $\delta\varepsilon = 2m^* c^2$. There is a difference of a factor of two between the exact and the approximate results.

The amount of energy acquired by an electron from the applied field E is equal to

$$\left(\frac{d\varepsilon}{dt} \right)_{\text{field}} = \text{force} \times \text{velocity} = eEv_d = e\mu E^2 \tag{4.103}$$

where v_d is the drift velocity of electrons. The time rate of energy loss is given by the energy loss per collision divided by the mean free time τ or

$$\left(\frac{d\mathcal{E}}{dt}\right)_{phonon} = \frac{\langle\delta\mathcal{E}\rangle}{\tau} = 4c^2\left(m^* - \frac{m^{*2}v_1{}^2}{4kT_l}\right)\frac{v_1}{l} \qquad (4.104)$$

where l is the mean free path. Averaging Eq.(4.104) over the velocity distribution and using Eq. (4.62), we find

$$\left(\frac{d\mathcal{E}}{dt}\right)_{phonon} = \frac{8c^2}{\sqrt{\pi}}\left(m^* - \frac{m^{*2}v_e{}^2}{2kT_l}\right)\frac{v_e}{l} \qquad (4.105)$$

Since the electron temperature T_e differs from the lattice temperature T_l, we must distinguish between v_e and v_T with

$$v_e = \left(\frac{2kT_e}{m^*}\right)^{\frac{1}{2}} \qquad \text{and} \qquad v_T = \left(\frac{2kT_l}{m^*}\right)^{\frac{1}{2}} \qquad (4.106)$$

According to Eq. (4.85), the mobility μ in Eq. (4.103) can be written as

$$\mu = \frac{e}{3kT_e}\langle v^2\tau\rangle = \frac{e}{3kT_e}\langle vl\rangle = \frac{2el}{3\sqrt{\pi}}\frac{v_e}{kT_e} = \frac{4el}{3\sqrt{\pi}\,m^*v_e} \qquad (4.107)$$

which in turn results in the following equalities

$$\mu = \frac{v_T}{v_e}\mu_0 \qquad \text{and} \qquad \frac{l}{v_e} = \frac{3\sqrt{\pi}\,m^*}{4e}\frac{v_T}{v_e}\mu_0 \qquad (4.108)$$

where μ_0 is the electron mobility at the lattice temperature.

Substituting Eq. (4.108) into Eqs. (4.103) and (4.105) and equating the rate of energy loss to that of energy gain, we find

$$\left(\frac{v_e}{v_T}\right)^4 - \left(\frac{v_e}{v_T}\right)^2 = \frac{3\pi}{32}\left(\frac{\mu_0 E}{c}\right)^2 \qquad (4.109)$$

At low electric fields such that $\mu_0 E \ll c$, the solution of Eq. (4.109) leads to

$$\mu = \mu_0\frac{v_T}{v_e} \simeq \mu_0\left[1 - \frac{3\pi}{64}\left(\frac{\mu_0 E}{c}\right)^2\right] \qquad (4.110)$$

When $\mu_0 E \gg c$, the mobility varies as

$$\mu = 1.36\,\mu_0\left(\frac{c}{\mu_0 E}\right)^{\frac{1}{2}} \qquad \text{or} \qquad v_d = 1.36\,(c\mu_0 E)^{\frac{1}{2}} \qquad (4.111)$$

causing the deviation from Ohm's law. The reader is asked to show that at $\mu_0 E = 1.85c$, the mean square velocity v_e of an electron has an energy corresponding to $1.62T_l$. For $T_l = 300°\text{K}$, $T_e = 492°\text{K}$.

The optical phonons in germanium and silicon have a minimum energy corresponding to a frequency ν of 7×10^{12} and 12×10^{12} cps (Fig. 2.47b) or an equivalent temperature of $335°\text{K}$ and $570°\text{K}$ respectively. Therefore, in very high field regions, electrons will have sufficient energy to excite optical phonons. If the mean time τ_{opt} between collisions with optical modes of vibration is sufficiently short so that the acoustic scattering process can be neglected, then we have

$$\left(\frac{d\mathcal{E}}{dt}\right)_{\text{field}} = \frac{e^2 E^2 \tau_{\text{opt}}}{m^*} = \left(\frac{d\mathcal{E}}{dt}\right)_{\text{opt}} = \frac{h\nu_{\text{opt}}}{\tau_{\text{opt}}} \tag{4.112}$$

From Eq. (4.112), the drift velocity is given by

$$v_d = eE\tau_{\text{opt}}/m^* = (h\nu_{\text{opt}}/m^*)^{1/2} \tag{4.113}$$

Therefore, if the optical scattering process were so efficient that electrons were able to dissipate its energy by exciting optical phonons, the drift velocity of electrons would approach a constant independent of E.

Let us compare the main features concerning the two scattering processes with acoustic (A) and optical (O) modes of vibration. First, the amount of energy transferred in a collision process is vastly different, being equal to $(2c/v_T)kT$ and $h\nu_{\text{opt}}$ for the A and O processes respectively. Second, according to our discussion in the previous section, the reflection coefficient $|R|^2$ is proportional to the energy of a vibrational mode. This is why the factor kT appears in Eq. (4.72). Therefore, based on energy considerations, we expect the ratio of the mean free path $l_{\text{opt}}/l_{\text{ac}} \simeq (kT/h\nu_{\text{opt}})$. However, this ratio favorable to the O process is compensated by the fact that in order for an electron to participate in the O process, it must acquire the threshold energy $h\nu_{\text{opt}}$. If v_1 and v_2 are the initial and final velocity of an electron before and after each collision,

$$v_2{}^2 = v_1{}^2 - 2h\nu_{\text{opt}}/m^* \tag{4.114}$$

Since the limiting process is to allow v_2 after each collision to build up gradually to a value which exceeds $(2h\nu_{\text{opt}}/m^*)^{1/2}$, the mean free time τ_{opt} can be estimated as follows:

$$\tau_{\text{opt}} = \frac{l_{\text{opt}}}{v_2} = \tau_{\text{ac}}\left(\frac{v_1}{v_2}\right)\left(\frac{l_{\text{opt}}}{l_{\text{ac}}}\right) = \tau_{\text{ac}}\left(\frac{v_1}{v_2}\right)\left(\frac{kT}{h\nu_{\text{opt}}}\right) \tag{4.115}$$

The factor (v_1/v_2) can tip the balance in favor of the A process. Finally, we should point out that at ordinary lattice temperatures, the optical modes are hardly excited. This explains why only the optical phonon emission process is considered here.

In view of Eq. (4.64), Eq. (4.103) in the presence of both the A and O processes becomes

$$\left(\frac{d\mathcal{E}}{dt}\right)_{\text{field}} = \frac{e^2 E^2 \tau_{\text{opt}}\tau_{\text{ac}}}{m^*(\tau_{\text{opt}} + \tau_{\text{ac}})} \tag{4.116}$$

while Eq. (4.104) takes the form

$$\left(\frac{d\mathcal{E}}{dt}\right)_{\text{phonon}} = 4m^*c^2\left(\frac{v_1}{l_{\text{ac}}}\right)\left[1 - \left(\frac{v_1}{v_T}\right)^2\right] - \frac{h\nu_{\text{opt}}}{\tau_{\text{opt}}} \qquad (4.117)$$

Note that the above equations are valid only when $m^*v_1^2 > 2h\nu_{\text{opt}}$. The general behavior of the current density J as a function of E is illustrated in Fig. 4.8. Region a and lower part of region b where no participation of optical phonons is possible can be described by Eqs. (4.110) and (4.111) respectively. In region c, the last term in Eq. (4.117) is dominating, therefore, the drift velocity of electrons is almost constant as given by Eq. (4.113). For still higher E, the velocity v_1 in Eq. (4.117) becomes so large that the middle term in Eq. (4.117) takes over because of its cubic dependence on v_1. This is the reason why the current density J rises again in region d. Comparing Fig. 4.8 and the experimental curves shown in Fig. 4.7, we see that the flat velocity region predicted by Eq. (4.113) has not

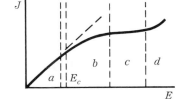

Figure
4.8

General behavior of the current density as a function of applied electric field as expected from theoretical considerations given in the text.

been clearly confirmed by experiments. The result is not surprising in view of our discussion in conjunction with Eq. (4.115). For further discussions of high-field mobility, the reader is referred to the article by Conwell.[14]

4.7 THERMOELECTRIC EFFECTS

Since electrons in a solid are carriers of energy as well as electric charges, thermal effects naturally can produce electrical effects and vice versa. When a current flows across a junction between two different materials, heat is generated or absorbed at the junction, depending on the direction of the current. This is known as the *Peltier effect*. Another commonly known thermoelectric phenomenon is the *Seebeck effect*. When two contacts to a metal or semiconductor are kept at different temperatures, a potential difference develops across the contacts. The various thermoelectric effects are not only related to each other but based on the same energy principle. We are all familiar with the law of conservation of

[14]   E. M. Conwell, *Jour. Phys. Chem. Solids*, vol. 8, p. 234, 1959.

electric charges, which can be put into an analytical form as

$$\frac{\partial \varrho}{\partial t} = -\nabla \cdot \mathbf{J} \tag{4.118}$$

where ϱ is the charge density and $\mathbf{J}$ the current density. A similar equation regarding the flow of thermal energy exists;

$$\frac{\partial H}{\partial t} = -\nabla \cdot \mathbf{C} \tag{4.119}$$

where H is the density of heat energy in joules/m³ and $\mathbf{C}$ the thermal current density in joules/m²-sec.

In writing down the above two equations, we must not forget the assumption invoked in their derivation: there are no electron sinks and sources as well as no heat sinks and sources in the medium. In the next chapter, we shall see how Eq. (4.118) should be modified by the carrier recombination and generation processes. The fact that Eq. (4.119) is incomplete is quite obvious because it does not even include the heat generated in the medium due to ohmic losses. In the presence of J_x and E_x, Eq. (4.119) should read

$$\frac{\partial H}{\partial t} = J_x E_x - \frac{\partial C_x}{\partial x} \tag{4.120}$$

for a one-dimensional flow. We can express E_x and C_x in terms of the current density J_x and the temperature gradient $\partial T/\partial x$ through the use of Eqs. (4.39) and (4.40), and thus obtain

$$\frac{\partial H}{\partial t} = \frac{J_x^2}{\sigma} + \frac{J_x}{e} T \left[\frac{d}{dT} \left(\frac{B_2}{B_1 T} - \frac{\mathcal{E}_f}{T} \right) \right] \frac{dT}{dx} - \frac{d}{dx} \left(K \frac{dT}{dx} \right) \tag{4.121}$$

where σ and K are, respectively, the electrical and thermal conductivities given by Eqs. (4.41) and (4.43). The first and last terms in Eq. (4.121) represent the heat generated by ohmic losses and the heat lost by thermal conduction while the middle term, being proportional to the product of J_x and (dT/dx), is responsible for the various thermoelectric effects.

Let us consider a metal-semiconductor contact as shown in Fig. 4.9. For convenience, we take the semiconductor to be p-type and further suppose that a current of holes flows across the contact from

Figure 4.9

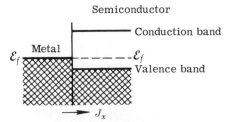

Semiconductor

Conduction band

Metal

$\mathcal{E}_f$ —————— $\mathcal{E}_f$

Valence band

J_x

↑ Energy of an electron

↓ Energy of a hole

Energy-band diagram at a metal-semiconductor contact.

the metal to the semiconductor. The ratio of the heat generated per second at the junction to the current density is known as the *Peltier coefficient* given by

$$\pi_{m \to s} = \int_m^s \frac{T}{e} \frac{d}{dx} \left(\frac{B_2}{B_1 T} - \frac{\mathcal{E}_f}{T} \right) dx$$

$$= \frac{T}{e} \left[\left(\frac{B_2}{B_1 T} - \frac{\mathcal{E}_f}{T} \right)_s - \left(\frac{B_2}{B_1 T} - \frac{\mathcal{E}_f}{T} \right)_m \right] \tag{4.122}$$

The subscripts m and s stand for metal and semiconductor respectivley and the limits of integration are from a point just inside the metal to another just inside the semiconductor across the junction.

To consider the Seebeck effect, let us take a p-type semiconductor whose ends are connected to the same metal but maintained at two different temperatures as shown in Fig. 4.10. The open-circuit voltage developed around the metal-semiconductor loop is given by

$$V_s = \frac{1}{e} \oint \left[\frac{B_2}{B_1 T} + T \frac{d}{dT} \left(\frac{\mathcal{E}_f}{T} \right) \right] \frac{dT}{dx} dx = \frac{1}{e} \oint \left(\frac{B_2}{B_1 T} - \frac{\mathcal{E}_f}{T} \right) \frac{dT}{dx} dx$$

$$\tag{4.123}$$

This is known as the *Seebeck voltage*. In deriving Eq. (4.123), we first set $J_x = 0$ in Eq. (4.39) and solve for E_x. Next, we use the fact that $\oint d\mathcal{E}_f = 0$. In view of Eq. (4.122), Eq. (4.123) becomes

$$V_s = \int_T^{T+\Delta T} \frac{\pi_{m \to s}}{T} dT = (Q_m - Q_s) \Delta T \tag{4.124}$$

for a small temperature difference ΔT. The quantity Q is often known as the *thermoelectric power* of a given substance and it is expressed in volts/degree Kelvin. From Eq. (4.124), we have

$$Q = \frac{1}{e} \left(\frac{\mathcal{E}_f}{T} - \frac{B_2}{B_1 T} \right) \tag{4.125}$$

Note that the sign in Eq. (4.124) depends on the chosen direction of

Figure 4.10

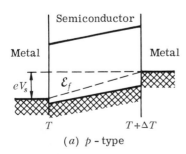

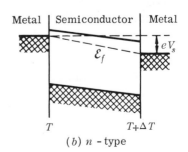

(a) p-type   (b) n-type

A Seebeck voltage V_s develops inside a semiconductor when its two ends are maintained at two different temperatures and in contact with a metal.

integration around the semiconductor-metal loop. We choose the direction to be positive if the loop integral goes from the hot end to the cold end in the metal.

Let us compare the magnitude of Q in metals with that in semiconductors. For metals, Eq. (4.33) is applicable and it leads to

$$\frac{B_2}{B_1} = \mathcal{E}_f \left[1 + \frac{\pi^2}{3} \left(\frac{kT}{\mathcal{E}_f} \right)^2 \right] \qquad \text{or} \qquad Q = -\frac{\pi^2}{3} \frac{k^2 T}{e \mathcal{E}_f} \qquad (4.126)$$

For nondegenerate semiconductors, the definitions of B_2 and B_1 are still given by Eq. (4.33) but their values must be evaluated from Eq. (4.62) after the function $kT(\partial F/\partial \mathcal{E})$ is approximated by $\exp[-(\mathcal{E} - \mathcal{E}_f)/kT]$. Thus, for constant mean free path, we have

$$\frac{B_2}{B_1} = \frac{m^* \langle v^3 \rangle}{2 \langle v \rangle} = 2kT \qquad (4.127)$$

In order to obtain Q from Eq. (4.125), we must know the reference energy level. In Eqs. (3.18) and (3.19), we let the Fermi-Dirac distribution be approximated by the Boltzmann distribution with the notation that $\mathcal{E}_1 = \mathcal{E} - \mathcal{E}_c$ and $\mathcal{E}_2 = \mathcal{E}_v - \mathcal{E}$, respectively, for conduction band electrons and valence band holes. It is obvious that the band edges are used as the reference energy levels. Therefore, the energy $\mathcal{E}_f$ in Eq. (4.125) stands for $\mathcal{E}_f - \mathcal{E}_v$ in p-type and $\mathcal{E}_f - \mathcal{E}_c$ in n-type semiconductors. Furthermore, the quantity B_2/B_1 which is equal to $\langle \mathcal{E}^2 \tau \rangle / \langle \mathcal{E} \tau \rangle$ is a weighted average of some sort of the kinetic energy of electrons or holes, τ being the relaxation time. For nondegenerate semiconductors, $\mathcal{E}$ and $\mathcal{E}_f$ lie on the opposite sides of the respective band edge. Since the terms have opposite signs in Eq. (4.125), contributions from B_2/B_1 and $\mathcal{E}_f$ should be additive, yielding

$$Q_n = -\frac{(\mathcal{E}_c - \mathcal{E}_f) + 2kT}{eT} \qquad \text{and} \qquad Q_p = \frac{(\mathcal{E}_f - \mathcal{E}_v) + 2kT}{eT}$$

$$(4.128)$$

for n- and p-type semiconductors, respectively. For a near intrinsic semiconductor,

$$Q = \frac{\mu_p p Q_p + \mu_n n Q_n}{\mu_p p + \mu_n n} \qquad (4.129)$$

Since Q_n and Q_p have opposite signs, a decrease in Q is expected when a semiconductor moves from the extrinsic region toward the intrinsic region.

From the above discussion, we see that the thermoelectric powers Q in metals and degenerate semiconductors are small because the terms B_2/B_1 and $\mathcal{E}_f$ almost cancel out. Take sodium as an example. The observed value of Q/T is -2.8×10^{-8} volt/$(°K)^2$ while the calculated value is -0.8×10^{-8} volt/$(°K)^2$ with $\mathcal{E}_f = 3.12$ ev in Na.

Note that in the derivation of Eq. (4.39), we assumed a constant mean free path. If the mean free path depends on energy, an additional term appears in Eq. (4.126) and this could be a source of error in our calculated value. In any case, the value of Q in alkali metals is of the order of 10 μv/$°$K at room temperature which is to be compared with a value of several hundred μv/$°$K in typical semiconductors. Therefore, in Eqs. (4.122) and (4.124), the contributions to π and V_s from the metal can be neglected.

Figure 4.11 shows the measured value of Q in germanium for three different impurity concentrations. For an acceptor concentration $N_a = 5.7 \times 10^{15}$/cm^3 and $N_v = 5.4 \times 10^{18}$/cm^3, the value of $\mathcal{E}_f - \mathcal{E}_v$ is $7kT$, resulting in a value of $Q = 750$ μv/$°$K at 300$°$K in agreement with the measured value. At high impurity concentrations and low temperatures, impurity scattering plays an important role. Using the definition of mean free path $l = v\tau$ in Eq. (4.64), we find

$$\frac{\mathcal{E}^{1/2}}{l} = \frac{1}{A\mathcal{E}^{3/2}} + \frac{\mathcal{E}^{1/2}}{l_0} \qquad \text{or} \qquad l = \frac{l_0 A \mathcal{E}^2}{l_0 + A\mathcal{E}^2} \qquad (4.130)$$

where the $\mathcal{E}^{-3/2}$ energy dependence is obtained from Eq. (4.59) for impurity scattering, A being a proportionality constant, while the $\mathcal{E}^{1/2}$ term is derived from Eq. (4.84) for lattice scattering, l_0 being the mean free path due to lattice scattering alone. In the region where Eq. (4.130) must be used, Eq. (4.128) should likewise be modified to fit the experimental curves of Fig. 4.11.

In the following, we shall present a simple physical interpretation of the quantity Q so that we can see how it will be affected. According

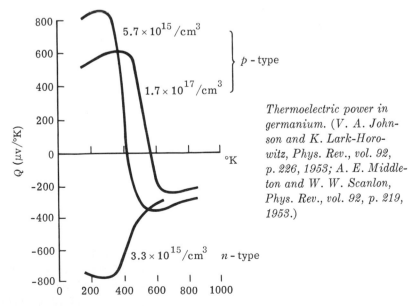

Thermoelectric power in germanium. (V. A. Johnson and K. Lark-Horowitz, Phys. Rev., vol. 92, p. 226, 1953; A. E. Middleton and W. W. Scanlon, Phys. Rev., vol. 92, p. 219, 1953.)

to Eqs. (3.2) and (3.17), the electron current density is

$$J_e = e\mu_n nE + e\frac{\partial}{\partial x}(v_{th}ln) \tag{4.131}$$

where the first term is caused by the applied field and the second term by diffusion. We note that in the derivation of Eq. (3.17), a uniform temperature throughout the sample is assumed. If a temperature gradient exists, v_{th}, being equal to $(kT/m^*)^{1/2}$, also varies with x and hence it should be differentiated as in Eq. (4.131). Under the open-circuit condition of Fig. 4.10, Eq. (4.131) yields

$$E = -\frac{1}{\mu_n n}\left(v_{th}l\frac{dn}{dx} + n\frac{dv_{th}l}{dx}\right) = -E_1 - E_2 \tag{4.132}$$

Assuming that

$$l = A\mathcal{E}^r \tag{4.133}$$

and noting that $\mu_n = el/(m^* v_{th})$, we find

$$E_1 = \frac{v_{th}l}{\mu_n}\frac{d(\log n)}{dx} = \frac{kT}{e}\frac{d(\log n)}{dx} \tag{4.134}$$

$$E_2 = \frac{1}{\mu_n}\frac{d(v_{th}l)}{dx} = \frac{r + \frac{1}{2}}{e}\frac{d(2\mathcal{E})}{dx} = \frac{r + \frac{1}{2}}{e}\frac{dkT}{dx} \tag{4.135}$$

The last equation is obtained in two steps: first by taking $v_{th} = (2\mathcal{E}/m^*)^{1/2}$ and second by realizing that the average value of $\mathcal{E}$ is $kT/2$. The reader is asked to derive E_1 and E_2 directly from Eq. (4.14) by setting $J_x = 0$ and solving for E_x. Equation (4.134) is obtained by considering the variation of n with respect to x in Eq. (4.14), whereas Eq. (4.135) is obtained by using $\partial f_0/\partial x$ from Eq. (4.18).

Let us refer to Fig. 4.10b. The Seebeck voltage V_s actually consists of two parts resulting from (1) the change in the Fermi level and (2) the existence of an internal electric field E. The change in $\mathcal{E}_f$ can be found from Eqs. (3.20) and (3.21) as

$$-\frac{d(\mathcal{E}_c - \mathcal{E}_f)}{dT} = kT\frac{d(\log n)}{dT} - \frac{3k}{2} - \frac{\mathcal{E}_c - \mathcal{E}_f}{T} \tag{4.136}$$

while the internal electric field is given by $E_1 + E_2$. Thus, we have

$$V_s = \frac{\Delta(\mathcal{E}_c - \mathcal{E}_f)}{e} - (E_1 + E_2)\Delta x = -\frac{k}{e}\left(\frac{\mathcal{E}_c - \mathcal{E}_f}{kT} + r + 2\right)\Delta T \tag{4.137}$$

Equation (4.137) reduces to Q_n of (4.128) if $r = 0$. From the above analysis, we see that the thermoelectric power depends on the mechanism of scattering. The value of r in Eq. (4.137) changes from zero for lattice scattering to two for impurity scattering. A fit of Eq. (4.137) to the curves of Fig. 4.11, therefore, determines the value of r and hence the relative importance of different scattering mechanisms in different temperature regions.[15]

In concluding this section, we shall discuss briefly principles governing the design of a practical thermoelectric generator. As shown in Fig. 4.12, two semiconducting elements (1 and 2) are joined at the ends by two heat reservoirs which are made of metal blocks (3, 4, and 5) maintained at two different temperatures. For the thermoelectric voltages, V_1 and V_2, to be additive, the two semiconductors must be of opposite types. The electric power delivered to the load of resistance R is

$$W_1 = I^2R = Q^2(T_1 - T_2)^2R/(R + r)^2 \qquad (4.138)$$

where $Q = |Q_1| + |Q_2|$ and $r = r_1 + r_2$, r_1 and r_2 being the resistances of the two semiconducting arms. The heat power input to the circuit can be calculated as follows. The heat loss due to conduction is given by

$$W_2 = K(T_1 - T_0) \qquad (4.139)$$

where $K = K_1 + K_2$ is the thermal conductance expressed in joules/deg-sec. From Eq. (4.139), it is necessary to deduct the power returned to the hot junction through joule heating. If it is assumed that I^2r is divided equally between the hot and cold junctions, the net loss is equal to $W_2 - I^2r/2$. Another heat loss mechanism is the Peltier heat. Referring to Fig. 4.9, a current flow from metal to p-type semiconductor cools the junction because an energy ($\mathcal{E}_f -$

---

[15] At low temperatures, a new effect known as the phonon-drag effect plays an important role. The effect arises because the electrical current upsets appreciably the isotropic phonon distribution. For further discussion of the effect, the reader is referred to: H. P. R. Frederiske, *Phys. Rev.*, vol. 92, p. 248, 1953; T. H. Geballe and G. W. Hull, *Phys. Rev.*, vol. 98, p. 940, 1955; C. Herring, *Phys. Rev.*, vol. 92, p. 857, 1953, and Role of Low Energy Phonons in Thermoelectricity and Thermal Conduction in M. Schön and H. Welker (eds.), "Halbleiter und Phosphore," Braunschweig, Freidrich Vieweg & Sohn, 1958.

Figure 4.12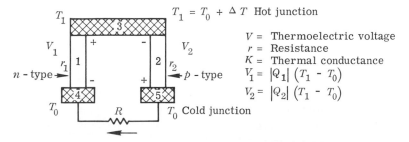

A practical thermoelectric generator. For thermoelectric voltages to be additive, the two semiconductor arms must be of opposite types: one n-type and the other p-type.

$\mathcal{E}_v + 2kT$) is taken away from the metal for each hole migrating across the junction. This can be deduced quantitatively from Eqs. (4.122), (4.125) and (4.128). The situation with the n-type semiconductor is just the opposite. Therefore, we see from Fig. 4.12 that the Peltier effect cools the hot junction. Since the effect is contrary to what is desired, it constitutes a loss given by

$$W_3 = QT_1I \tag{4.140}$$

In view of the above discussion, the efficiency η of a thermoelectric generator is equal to

$$\eta = \frac{W_1}{W_2 + W_3 - I^2r/2}$$
$$= \frac{T_1 - T_0}{T_1} \frac{m/(m+1)}{1 + (m+1)/(zT_1) - (T_1 - T_0)/2(m+1)T_1} \tag{4.141}$$

where $m = R/r$ and $z = Q^2/(Kr)$. Since the thermoelectric effect, unlike joule heating, is a reversible process, the efficiency of such a device resembles that of a Carnot engine. It is clear then that a large temperature difference $T_1 - T_0$ is needed for high efficiency. The quantity z can be called the figure of merit for a thermoelectric element. Note that the product of Kr is independent of the physical dimensions and thus z can be expressed in terms of thermal and electrical conductivities as

$$z = \frac{Q^2\sigma}{K_{el} + K_{ph}} \tag{4.142}$$

where K_{el} and K_{ph}, respectively, are the electronic and lattice contributions to thermal conductivity and K_{el} is related to σ through Eqs. (4.24) and (4.25).

To maximize z, we first must choose materials with low K_{ph}. For typical thermoelectric materials such as PbSe and PbTe, K_{ph} is of the order of 10 to 20×10^{-3} watts/deg-cm. The value can be further reduced by either introducing impurities into or forming solid solutions of PbSe and PbTe. The next step is to choose the right resistivity. From Fig. 4.11, we see that Q reaches a maximum value when a semiconductor is in the extrinsic region but not too far from the intrinsic region. Therefore, a high Q means a relatively high resistivity which in turn results in a high ohmic loss. It is obvious that a compromise between a high Q and a low ohmic loss must be reached. Realizing that $K_{el} = A\sigma$ where A is a proportionality constant, Eq. (4.142) can be maximized with respect to σ and such a maximizing process leads to the following results

$$\log \sigma = \log (e\mu_p N_v) - 2K_{el}/K_{ph}$$
$$Q = 2(k/e)(1 + K_{el}/K_{ph}) \tag{4.143}$$

Knowing K_{ph}, μ_p, and N_v, the values of K_{el} and σ can be found from Eqs. (4.143) and (4.45).

To estimate the value of z, K_{el} is first assumed zero. Taking $m^* = m_0$ and $\mu_p = 500$ cm²/volt-sec, we find $N_v = 2.5 \times 10^{19}$/cm³ and $\sigma = 2000$ (ohm-cm)$^{-1}$ at 300°K. Obviously, the semiconductor is so heavily doped that it is almost degenerate. Substitution of σ into Eq. (4.45) gives a value of $K_{el} = 1.5 \times 10^{-2}$ watts/deg-cm. Assuming that $K_{ph} = 2 \times 10^{-2}$ watts/deg-cm, we obtain from Eq. (4.143) a value of $Q = 3 \times 10^{-4}$ volt/deg and from Eq. (4.142) a value of $z = 1.7 \times 10^{-3}$. These performance characteristics are considered to be quite respectable by present standards. For $T_1 = 600$°K, $T_0 = 300$°K and $m = 1$, we find from Eq. (4.141) that the efficiency is about 5 per cent which is close to an efficiency of 3 per cent found in practical thermoelectric generators. At present, the efficiency is limited not only by the Peltier heat but also by the upper limit of temperature attained in practice. As long as K_{ph} is the controlling mechanism for thermal conductivity, the value of Q and hence that of z are held pretty much to the values calculated above. The highest attainable temperature is determined not only by the melting point of a given thermoelectric material but also by the condition that the material stays in the extrinsic region. In the actual design of a thermoelectric generator, the doping concentration in the semiconductor must vary from the cold end to the hot end so that Eq. (4.143) is always satisfied. Therefore, the value of Q calculated at 300°K should be repeated for other temperatures. Such a calculation, however, requires the knowledge how μ_p and K_{ph} change with temperature. Finally, we may add that the reverse process for thermoelectric generation of electricity is thermoelectric cooling. This is attained by applying a voltage of such a polarity as to reverse the direction of current flow in Fig. 4.12. Heat thus will be constantly taken away from the cold junction and added to the hot junction.

4.8 GALVANOMAGNETIC EFFECTS

In Sec. 3.6, we have seen that Hall measurement yields valuable information about the concentration of free carriers. The purpose of our present discussion is to refine our treatment of the effect of a magnetic field upon the conductive property of a semiconductor so that we may not only obtain a more accurate result but also derive more information about the semiconductor from such a measurement. In the presence of electric and magnetic fields, an electron experiences an acceleration $\mathbf{a}$:

$$\mathbf{a} = -\frac{e}{m^*}(\mathbf{E} + \mathbf{v} \times \mathbf{B}) \qquad (4.144)$$

For simplicity, $\mathbf{E}$ is taken to be in the xy plane and $\mathbf{B}$ to be in the z direction. Furthermore, a uniform temperature is assumed so the Boltzmann transport equation reduces to

$$f - f_0 = -\tau \sum_{x,y,z} a_x \frac{\partial f}{\partial v_x} \tag{4.145}$$

Our experience with the Hall voltage tells us that if $\mathbf{B}$ is in the z direction, it is the velocity in the xy plane which enters into the equation of motion. Thus, intuitively we feel that the solution of Eq. (4.145) should take the following form:

$$f = f_0 + v_x g_1 + v_y g_2 \tag{4.146}$$

Substituting f from Eq. (4.146) into Eq. (4.145) and equating the coefficients of v_x and v_y, we find

$$\begin{aligned}
g_1 + \left(\frac{e\tau B}{m^*}\right) g_2 &= e\tau E_x \frac{\partial f_0}{\partial \varepsilon} \\
-\left(\frac{e\tau B}{m^*}\right) g_1 + g_2 &= e\tau E_y \frac{\partial f_0}{\partial \varepsilon}
\end{aligned} \tag{4.147}$$

where $\varepsilon = m^*(v_x{}^2 + v_y{}^2 + v_z{}^2)/2$. In deriving Eq. (4.147), terms involving the products of g and E as well as the derivatives of g are regarded as negligible. Equation (4.147) has a solution of the form

$$\begin{aligned}
g_1 &= e\tau \left(\frac{\partial f_0}{\partial \varepsilon}\right) \frac{E_x - sE_y}{1 + s^2} \\
g_2 &= e\tau \left(\frac{\partial f_0}{\partial \varepsilon}\right) \frac{sE_x + E_y}{1 + s^2}
\end{aligned} \tag{4.148}$$

with $s = eB\tau/m^*$. For nondegenerate semiconductors, $\partial f_0/\partial \varepsilon = -f_0/kT$. Substitution of Eqs. (4.148) and (4.146) into Eq. (4.13) gives

$$J_x = \frac{e^2}{kT} (a_1 E_x - a_2 E_y) \tag{4.149a}$$

$$J_y = \frac{e^2}{kT} (a_2 E_x + a_1 E_y) \tag{4.149b}$$

where the quantities a_1 and a_2 are given by

$$a_1 = \frac{4\pi A}{3} \int_0^\infty \frac{\tau}{1 + s^2} v^4 \exp\left(-\frac{m^* v^2}{2kT}\right) dv$$

and

$$a_2 = \frac{4\pi A}{3} \cdot \int_0^\infty \frac{\tau s}{1 + s^2} v^4 \exp\left(-\frac{m^* v^2}{2kT}\right) dv \tag{4.150}$$

In Eq. (4.150), $A = n(m^*/2\pi kT)^{3/2}$ is the same as defined in Eq. (4.17), n is the electron concentration, and $v = (v_x{}^2 + v_y{}^2 + v_z{}^2)^{1/2}$.

The Hall coefficient can easily be found from Eq. (4.149b) by setting $J_y = 0$ as we did in Sec. 3.6. Thus, we find

$$J_x = \frac{e^2}{kT} \frac{a_1{}^2 + a_2{}^2}{a_1} E_x$$

$$R_H = \frac{E_y}{J_x B} = \frac{kT}{e^2} \frac{a_2}{a_1{}^2 + a_2{}^2} \frac{1}{B}$$

(4.151)

If it is assumed that τ is independent of velocity, the factors involving s and hence B can be taken outside the integral. Thus, the quantity $a_1{}^2 + a_2{}^2$ will have a $(1 + s^2)^{-1}$ dependence and so will the quantity a_1. That means, the conductivity $\sigma = J_x/E_x$ is independent of the magnetic field B. What happens is that the Hall field E_y in Eq. (4.149) exactly cancels the effect of B on $a_1 E_x$ so the magnetoresistive effect vanishes. On the other hand, if τ is a function of velocity as we have shown in Secs. 4.5 and 4.6, we expect that σ will be a function of B. This is known as the *magnetoresistive effect*.

To evaluate R_H, we may assume that the applied field B is small such that $(1 + s^2)$ may be approximated by one in Eq. (4.150). In the region where lattice scattering is dominant, $\tau = l/v$ where the mean free path l is independent of v and the quantities a_1 and a_2 can be evaluated through the use of Eqs. (4.20), (4.21), and (4.62). Substitution of the calculated values of a_1 and a_2 into Eq. (4.151) gives

$$R_H = 3\pi/8ne \qquad (4.152)$$

Equation (4.152) is the equivalent of Eq. (3.65) except for a numerical factor which depends on the mechanism of scattering or the velocity dependence of τ.

Let us now estimate the effect of the s^2 term in Eq. (4.150). As an estimate, the value of $s = eB\tau/m^*$ can be approximated by $s = \mu B$, μ being the mobility of free carriers. For $B = 1$ weber/m² and $\mu = 4000$ cm²/volt-sec, $s = 0.4$. Therefore, for $B\mu < 0.5$, we can expand $(1 + s^2)^{-1}$ as $1 - s^2$ in Eq. (4.150), thus obtaining

$$a_1 = \frac{8\pi A}{3} l \left(\frac{kT}{m^*}\right)^2 \left[1 - \frac{e^2 B^2 l^2}{(2kT)m^*}\right]$$

$$a_2 = \frac{A}{3} \frac{eB}{m^*} l^2 \left(\frac{2\pi kT}{m^*}\right)^{3/2} \left[1 - \frac{2e^2 B^2 l^2}{kTm^*}\right]$$

(4.153)

Keeping terms to the first order in B^2 in Eq. (4.151), we find

$$\sigma = \sigma_0 \left[1 - \left(\frac{1}{2} - \frac{\pi}{8}\right) \frac{e^2 B^2 l^2}{m^* kT}\right]$$

$$R_H = R_{HO} \left[1 - \left(1 + \frac{\pi}{8}\right) \frac{e^2 B^2 l^2}{m^* kT}\right]$$

(4.154)

where $\sigma_0 = en\mu$, $\mu = 4el/(3\sqrt{2\pi m^* kT})$, and R_{HO} is given by Eq.

(4.152). It is interesting to note that for small B, Eq. (4.154) leads to

$$\frac{\sigma_0 - \sigma}{\sigma_0} = \left(1 - \frac{\pi}{4}\right)\left(\frac{3\sqrt{\pi}}{4} B\mu\right)^2 = 0.38 B^2 \mu^2 \qquad (4.155)$$

The above equation is based on the assumption that l is a constant independent of v.

Several comments about the magnetoresistive effect are now in order. If lattice scattering is dominant, the integrals a_1 and a_2 can be evaluated exactly. The values of $(\sigma_0 - \sigma)/\sigma_0$ and R_H/R_{HO} are plotted in Fig. 4.13 in terms of a dimensionless quantity $3\sqrt{\pi}\,B\mu/4$. For large values of B,

$$\frac{\sigma_0 - \sigma}{\sigma_0} \Rightarrow 0.132 \qquad \text{and} \qquad \frac{R_H}{R_{HO}} \Rightarrow \frac{8}{3\pi} = 0.849 \qquad (4.156)$$

It is evident from Fig. 4.13 that the quadratic dependence of Eq. (4.154) is good only for $(B\mu)^2 < 0.1$. Furthermore, Eq. (4.154) is based on the assumption that $\mathbf{B}$ is perpendicular to $\mathbf{J}$. This is called the transverse magnetoresistive effect. For $\mathbf{E}$ and $\mathbf{B}$ to be in any arbitrary directions, the solution of the Boltzmann transport equation is found to be[16]

$$f = f_0 + e\tau\left(\frac{\partial f_0}{\partial \mathcal{E}}\right)\frac{\mathbf{E}\cdot\mathbf{v} + \dfrac{e\tau}{m^*}\mathbf{E}\cdot(\mathbf{v}\times\mathbf{B})}{1 + s^2} \qquad (4.157)$$

---

[16] See, for example, C. Kittel, "Elementary Statistical Mechanics," pp. 201–204, John Wiley & Sons, Inc., New York, 1958.

Figure 4.13

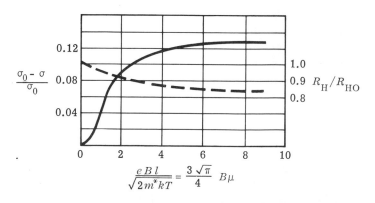

Theoretical curves showing the magnetoresistive effect (solid curve) and Hall coefficient (dashed curve) as a function of the magnetic flux density B expressed in terms of a dimensionless quantity Bμ. The curves are based on the assumptions that the mean free path l of carriers is a constant independent of the velocity of the carrier and that the energy surface is spherical.

If **B** is parallel to the applied **E**, the scalar product of **E** and ($\mathbf{v} \times \mathbf{B}$) is zero. Therefore, no longitudinal galvanomagnetic effect is expected as should be the case from our elementary discussion in Sec. 3.6.

The first experimental investigation of the magnetoresistive effect in Ge was made by Pearson and Suhl.[17] Figure 4.14 shows the variation of $\Delta\rho/(\rho B^2)$ as a function of temperature in an n-type sample of resistivity 11.5 Ω-cm measured at 0.4 weber/m². The fact that the sample has a high resistivity ensures that the mobility of electrons is limited by lattice scattering at moderate temperatures. Hence we can use the results just obtained in analyzing experimental findings. It is expected from Eq. (4.155) that $\Delta\rho/(\rho B^2)$ should follow the temperature variation of μ^2 or a T^{-3} dependence. This is indeed the case for temperatures above 150°K. Taking $\mu = 0.39$m²/volt-sec, a value of 0.06 is predicted from Eq. (4.155), which is in good agreement with the measured value of 0.08 at room temperature. What is interesting, however, is the magnitude of the longitudinal magnetoresistive effect which is not only there but even bigger than the transverse effect. The complexity of the effect is further manifested in Figs. 4.15 and 4.16 which show respectively the measured values of $\Delta\rho/\rho$ and $\Delta\rho/(\rho B^2)$ at high and low B.

At very high B, Fig. 4.15 shows that the measured value of $\Delta\rho/\rho$ in n-type Ge far exceeds the saturation value of 0.132 predicted from Fig. 4.13. It turns out that for electrons, it is necessary to

[17] G. L. Pearson and H. Suhl, *Phys. Rev.*, vol. 83, p. 768, 1951.

Figure 4.14

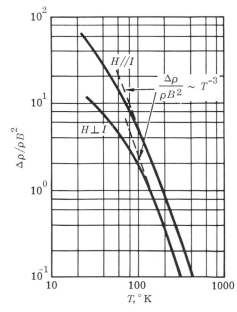

Magnetoresistive effect in Ge plotted as a function of temperature at $B = 0.4$ weber/m². The Ge sample is n-type and has a resistivity of 11.5 Ω-cm. (G. L. Pearson and H. Suhl, Phys. Rev., vol. 83, p. 768, 1951.)

Figure
4.15

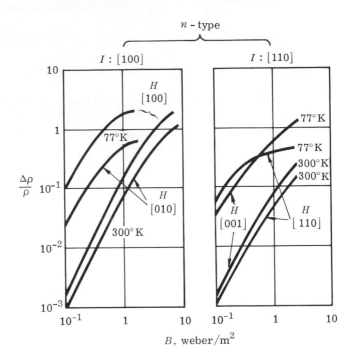

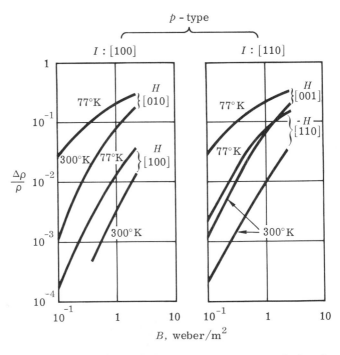

Magnetoresistive effect in Ge *plotted as a function of magnetic flux density B. The different curves in each graph are taken at different temperatures for a fixed orientation of* **B** *with respect to* **I**. (*G. L. Pearson and H. Suhl, Phys. Rev., vol. 83, p. 768, 1951.*)

considcr not only the existence of several conduction band minima but also the nonspherical energy surfaces in Ge and Si. However, the calculations involved are so complicated that it is not possible to obtain a detailed comparison of the theoretical and experimental results. To make the calculation tractable, Gold and Roth[18] assumed a constant τ instead of a constant l and obtained a reasonable agree-

[18] L. Gold and L. M. Roth, *Phys. Rev.*, vol. 103, p. 61, 1956. Actually, B. Abeles and S. Meiboom first demonstrated (*Phys. Rev.*, vol. 95, p. 31, 1954) that the ellipsoidal energy bands were responsible for the anisotropy of magnetoresistance in germanium. The Abeles-Meiboom formulation takes into account the dependence of τ on energy.

Figure 4.16

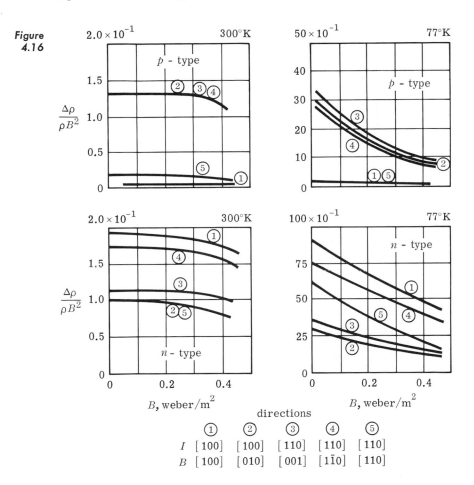

Magnetoresistive effect in Ge plotted as a function of magnetic flux density B. The different curves in each graph are for different orientations of B with respect to I at a given temperature. (G. L. Pearson and H. Suhl, Phys. Rev., vol. 83, p. 768, 1951.)

ment between theory and experiment. It is, therefore, the complexity of the energy-band structure which complicates the magnetoresistive effect. For the p-type sample, the measured value of longitudinal effect is much smaller than that of transverse effect in qualitative agreement with theory, but the calculated value of μ from Eq. (4.155) differs quite considerably from the value of μ obtained from other measurements. Calculations by Willardson, Harman, and Beer[19] show that the discrepancy can be fully accounted for if the light hole band is included.

4.9 ELECTRON-PHONON INTERACTION AND ULTRASONIC AMPLIFICATION IN SEMIMETALS AND PIEZOELECTRIC SEMICONDUCTORS

Amplification of ultrasonic waves has recently been reported by Hutson, McFee, and White[20] in CdS crystals when the drift velocity of the optically excited electrons exceeds the velocity of sound by the application of a strong d-c drift field (Fig. 4.17). In a piezoelectric crystal, the mechanical stress T and strain S are coupled to the electric field E and displacement D through the following relations:

$$T = cS - dE \tag{4.158a}$$

$$D = dS + \epsilon E \tag{4.158b}$$

where c is the elastic constant, ϵ the dielectric constant and d the

[19] R. K. Willardson, T. C. Harman, and A. C. Beer, *Phys. Rev.*, vol. 96, p. 1512, 1954. For further references, see B. Lax and J. G. Mavroides, *Phys. Rev.*, vol. 100, p. 1650, 1955.

[20] A. R Hutson, J. H. McFee, and D. L. White, *Phys. Rev. Letters*, vol. 7, p. 237 1961.

Figure 4.17

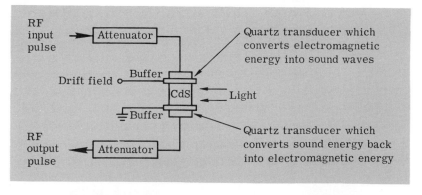

Schematic diagram showing the experimental setup for ultrasonic amplification in CdS. Since the experiment is sensitive to carrier concentrations present in the sample, light beam is used as a controllable source for exciting carriers in CdS.

piezoelectric constant. Normally, the quantities c, ϵ, and d are tensors. For simplicity, we assume that the ultrasonic wave propagates in the x direction and there is only one nonzero piezoelectric constant which couples the field in the x direction to the ultrasonic wave. Under these assumptions, we can replace the tensor quantities by scalars and treat Eq. (4.158) as a one-dimensional problem in which T, S, D, and E depend on the x coordinate only. The reader is referred to Sec. 7.8 for a more general and extensive treatment of the piezoelectric effect.

Referring to Eq. (7.89) which relates the acceleration to the stress, we find

$$\rho \frac{\partial^2 \xi}{\partial t^2} = \frac{\partial T}{\partial x} = c \frac{\partial^2 \xi}{\partial x^2} - d \frac{\partial E}{\partial x} \qquad (4.159)$$

where ξ is the displacement of the medium (Fig. 7.14c) and the strain is defined as $S = \partial \xi / \partial x$ by Eq. (7.77). If $d = 0$, Eq. (4.159) reduces to the ordinary elastic wave equation, Eq. (2.44), with $\xi = u$ and $v = \sqrt{c/\rho}$, ρ being the mass density of the medium. In a piezoelectric crystal, the electric field which accompanies the lattice wave through Eq. (4.158b) interacts with the motion of free carriers. If the drift velocity of electrons is much smaller than the sound velocity (low drift-field case), this interaction can only bring about the scattering of electron motion known as piezoelectric scattering [Eq. (3.124)]. In other words, in the low drift-field case, the electrons on the one hand and the lattice vibrations on the other can be treated as independent systems and the piezoelectric coupling as a small perturbation. When the drift velocity of electrons exceeds the sound velocity (high drift-field case), the electrons are strongly coupled to the lattice waves through the space-charge waves set up by the motion of electrons. In the following, we shall establish the relation between the space-charge wave and the piezoelectric field E.

Since the effect is associated with majority carriers (electrons in n-type CdS), we shall consider the equations of motion for electrons only. According to Eqs. (3.2) and (3.17), the electron current density is given by

$$J = e\mu_n n(E_0 + E) + eD_n \frac{\partial n}{\partial x} \qquad (4.160)$$

where E_0 is the d-c part (drift field) and E the a-c part (piezoelectric field) of the electric field. The electron concentration $n = n_0 + fn_s$ can also be divided into two parts: the spatial-independent part n_0 and the part fn_s due to space-charge waves. The fraction f accounts for a division of the space charge between the conduction band and the trap states which are known to exist in the forbidden gap of CdS. The part n_0 neutralizes the charges carried by ionized impurity states and holes in the valence band; hence the space-charge part n_s

is related to D through Poisson's equation

$$\frac{\partial D}{\partial x} = \varrho = -en_s \tag{4.161a}$$

and to the current density J through the continuity equation

$$\frac{\partial J}{\partial x} = -\frac{\partial \varrho}{\partial t} = e\frac{\partial n_s}{\partial t} \tag{4.161b}$$

where ϱ is the space-charge density. Differentiating Eq. (4.161a) with respect to t and eliminating $\partial n_s/\partial t$ and J from Eqs. (4.161b) and (4.160), we find

$$-\frac{\partial^2 D}{\partial x\,\partial t} = efD_n\frac{\partial^2 n_s}{\partial x^2} + e\mu_n E_0 f\frac{\partial n_s}{\partial x} + e\mu_n n_0\frac{\partial E}{\partial x} \tag{4.162}$$

Since we are interested only in the terms varying at the fundamental frequency of acoustic vibration, that is, terms linearly dependent on E and n_s (the a-c part), the cross-product of $n_s E$ is discarded in the above equation.

Assuming a wave dependence $\exp[i(kx - \omega t)]$ for T, S, D, E, ξ, and n_s, we find, respectively, from Eqs. (4.161a) and (4.158b):

$$ikD = -en_s \quad\text{and}\quad D = i\,dk\xi + \epsilon E \tag{4.163}$$

Substituting the same dependence $\exp[i(kx - \omega t)]$ in Eq. (4.162) and eliminating D and n_s by Eq. (4.163), we obtain the following relation between E and ξ:

$$-\frac{ikd}{\epsilon}\xi = E\left(1 + \frac{i\sigma}{\epsilon}\frac{1}{\omega + fk\mu_n E_0 + ifD_n k^2}\right) \tag{4.164}$$

The other equation relating ξ to E is obtained from Eq. (4.159) as follows

$$i\,dkE = (\rho\omega^2 - ck^2)\xi \tag{4.165}$$

The propagation constant $k = \beta + i\alpha$ can be obtained by eliminating E and ξ from Eqs. (4.164) and (4.165). The procedure of finding k is greatly simplified if we realize the following. For $d = 0$ (no piezoelectric coupling), $k = k_0 = \omega\sqrt{\rho/c}$. With piezoelectric coupling, it is expected that the major change in k will be in the attenuation constant α but not in the phase constant β. Consequently, we replace k by k_0 on the right-hand side of Eq. (4.164) and thus obtain the following equation for k

$$k^{-2} = k_0^{-2}\left[1 + \frac{K^2(\gamma + j\omega/\omega_D)}{\gamma + j(\omega/\omega_D + \omega_c/\omega)}\right] \tag{4.166}$$

where $\gamma = 1 + fv_d/v_s$, $\omega_D = v_s^2/(fD_n)$ and $\omega_c = \sigma/\epsilon$, v_s being the ordinary velocity of sound, v_d the drift velocity of electrons, and ω_c the reciprocal of the dielectric relaxation time. The parameter

$K^2 = d^2/\epsilon c$ is often referred to as the electromechanical coupling constant. Strictly speaking, $d^2/\epsilon c = K^2/(1 - K^2)$. Since $K^2 \ll 1$, K can be approximated by $d/\sqrt{\epsilon c}$. Refer to Eq. (7.98) for the definition of the electromechanical coupling constant.

Taking the imaginary part of k after expanding the term in the brackets of Eq. (4.166), we find

$$\alpha = \frac{K^2}{2} \frac{\omega_c}{v_s \gamma} \left[1 + \frac{\omega_c^2}{\gamma^2 \omega^2} \left(1 + \frac{\omega^2}{\omega_c \omega_D} \right)^2 \right]^{-1} \tag{4.167}$$

The reader is asked to obtain from the real part of k, the expression for the velocity of sound, $v = \omega/\beta$, which in a piezoelectric crystal will be different from $v_s = \sqrt{c/\rho}$. Figure 4.18 shows the variation of α and $v^2 - v_s^2$ as a function of the drift velocity of electrons. We see that amplification of ultrasonic and space-charge waves is possible if $\gamma = 1 + f v_d/v_s < 0$. The quantity γ comes from the terms $\omega + f k \mu_n E_0$ in Eq. (4.164), which in turn originate from $\partial^2 D/\partial x\, \partial t$ and $e \mu_n E_0 f\, \partial n_s/\partial x$ in Eq. (4.162). The physical process for the buildup of the space-charge wave is now clear and can be understood as follows.

The term $\partial^2 D/\partial x\, \partial t = \partial \varrho/\partial t$ represents the depletion of the space charge as a result of the time variation of ϱ. The space charge ϱ can be replenished through the drift term $e f \mu_n n_s E_0$. For $\gamma = 0$, these two terms exactly balance out and hence $\alpha = 0$. When the term $e f \mu_n E_0\, \partial n_s/\partial x$ exceeds $\partial \varrho/\partial t$, the space-charge wave starts to build up in its amplitude, resulting in the amplification of the ultrasonic

Figure 4.18

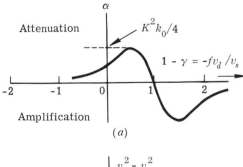

(a)

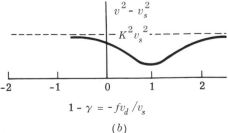

(b)

Schematic representation of (a) the attenuation (or amplification) constant and (b) the velocity of propagation $v^2 - v_s^2$ for the sound wave as functions of the drift velocity v_d of carriers.

wave through piezoelectric coupling. Note that for $\gamma < 0$, it is necessary for v_d and v_s to have opposite signs. The difference in sign is consistent with our explanation that one term represents the depletion while the other represents the replenishing of the space charge.

The above simple physical picture can be further refined by taking the diffusion term $\partial^2 n_s / \partial x^2$ and the dielectric relaxation term $\partial E / \partial x$ of Eq. (4.162) into consideration. We note from Eqs. (4.164) and (4.165) that if $\sigma = 0$, k will always be real and hence no amplification of ultrasonic wave is possible in an ideal insulator. This explains why optically excited carriers are needed in the experimental setup of Fig. 4.17. The phase relation between E and ξ can be changed by varying the d-c drift field E_0. For high drift fields such that $f\mu_n E_0 = fv_d \gg v_s$ and $\epsilon fkv_d \gg \sigma$, the phase angles of E and ξ again will be related to each other by $\pi/2$ as in the case $\sigma = 0$. This explains why the amplification factor α drops at high values of v_d. An optimum condition is reached when ω_c or σ/ϵ is about equal to $\gamma\omega$. The diffusion term always works against the bunching of space charge and hence it is not surprising to see that the term involving ω_D in Eq. (4.167) effectively cuts down the amplification constant irrespective of the value of γ. The ratio $\omega^2/(\omega_c\omega_D)$ can be written as

$$\frac{\omega^2}{\omega_c\omega_D} = \frac{\omega^2\epsilon fD_n}{v_s^2 e\mu_n n} = f\left(\frac{2\pi L}{\lambda}\right)^2 \tag{4.168}$$

where L is the Debye screening length given by Eq. (3.127) and λ is the wavelength of the acoustic vibration. Electron debunching through diffusion will become important when the sound wavelength is smaller than the Debye screening length.

In the experimental setup (Fig. 4.17) of Hutson, McFee, and White, the transducer is a Y-cut quartz which converts a 1-μsec pulse of rf signal (15 Mc/sec) into a shear wave. The sound wave then propagates through the buffer into CdS in which the sound wave is amplified. The second transducer converts the amplified sound wave back into an rf signal which can be displayed on an oscilloscope. In later experiments, both longitudinal (compressional) and shear sound waves have been employed with the direction of polarization parallel to the c axis. We should also point out that spontaneous oscillation of sound waves should be possible if both ends of a piezoelectric crystal are made perfectly reflecting to the sound wave. Referring to Fig. 4.18a, we see that the reflected wave with v_d in the direction of v_s will have a positive γ and hence it will be attenuated. However, the attenuation curve is non-symmetrical with respect to v_d/v_s. Therefore, for $|v_d| > |v_s|$, the acoustic wave still will experience a net amplification in making a round trip through the piezoelectric crystal. For further discussions of elastic wave propagation and experimental observation of ultrasonic amplification in piezoelectric crystals, the reader is referred to a series of articles by Hutson, White, and McFee.[21]

The possibility of achieving ultrasonic amplification in semimetals is discussed in articles by Hopfield[22] and independently by Dumke and Haering.[23] By *semimetals*, we mean those elements whose conduction and valence bands overlap slightly. Such is the case with Bi, As, Sb, Se, Te, and some of their alloys with transition metals. Had it not been for the slight overlap in the energy bands, the valence band would be completely filled and the conduction band completely empty at the absolute zero temperature. Because of the slight overlap, a small number of electrons go from the valence band into the conduction band, creating an equal number of holes and electrons respectively in the two bands. In a Hall voltage or thermoelectric measurement, the type of carrier with a higher value of mobility (or τ/m^*) dominates. This explains why Bi shows a negative (electron) Hall coefficient while As and Sb show positive (hole) Hall coefficients.

As pointed out by Dumke and Haering in the aforementioned paper, the electron-phonon interaction in nonpiezoelectric crystals is through the deformation potential which is much weaker than the piezoelectric coupling. However, the large carrier concentration obtainable in a semimetal should be able to compensate for the relatively small coupling constant and still result in an appreciable interaction. In the scheme proposed by Dumke and Haering, a semimetal is placed in cross electric field E_0 (in the positive x direction) and magnetic flux density B (in the negative z direction). Under the circumstances, Eqs. (4.149a) and (4.149b) should apply (except for the sign of B). Assuming that τ is independent of velocity, we obtain from Eq. (4.150) the following:

$$a_1 = \frac{kT\tau}{m^*} \frac{n}{1+s^2} \quad \text{and} \quad a_2 = \frac{kT\tau}{m^*} \frac{sn}{1+s^2} \qquad (4.169)$$

where $s = eB\tau/m^*$ and n is the concentration of free electrons. Therefore, under a high magnetic flux density such that $s \gg 1$, Eq. (4.149a) reduces to

$$J_{ny} = -en\frac{E_0}{B} + e\mu_n n E_y \qquad (4.170)$$

where $\mu_n = e\tau/(m^*s^2)$. The negative sign appears in Eq. (4.170) on account of the change in the direction and hence the sign of B.

[21] A. R. Hutson and D. L. White, *Jour. Appl. Phys.*, vol. 33, p. 40, 1962; A. R. Hutson, J. H. McFee, and D. L. White, *Phys. Rev. Letters*, vol. 7, p. 237, 1961; D. L. White, *Jour. Appl. Phys.*, vol. 33, p. 2547, 1962; J. H. McFee, *Jour. Appl. Phys.*, vol. 34, p. 1548, 1963.

[22] J. J. Hopfield, *Phys. Rev. Letters*, vol. 8, p. 311, 1962.

[23] W. P. Dumke and R. R. Haering, *Phys. Rev.*, vol. 126, p. 1974, 1962.

Strictly speaking, Eqs. (4.169) and (4.170) apply only to non-degenerate semiconductors. In the case of semimetals, $\partial f_0/\partial \mathcal{E} = 2(m^*/h)^3 \, \partial F/\partial \mathcal{E}$ should be used in Eq. (4.148) where F is the Fermi-Dirac distribution and $\partial F/\partial \mathcal{E}$ can be approximated by a Dirac delta function at $\mathcal{E} = \mathcal{E}_f$. Using Eqs. (4.148) and (4.14) and following a similar procedure outlined in Sec. 4.2, the reader can easily verify that Eq. (4.170) applies equally well to metals and degenerate semiconductors if the relaxation time τ is evaluated at the Fermi level $\mathcal{E}_f$ or $\tau = \tau(\mathcal{E}_f)$.

Since any bunching of free carriers as a result of electron-phonon interaction brings about a concentration gradient in the carrier density, we must add a diffusion term to Eq. (4.170), obtaining the following equation for electrons

$$J_{ny} = -en(E_0/B) + eD_n(\partial n/\partial y) + e\mu_n n E_{yn} \qquad (4.171a)$$

and a similar equation for holes

$$J_{py} = ep(E_0/B) - eD_p(\partial p/\partial y) + e\mu_p p E_{yp} \qquad (4.171b)$$

where $\mu_n = m_e^*/(\tau_n B^2)$ and $\mu_p = m_h^*/(\tau_p B^2)$ with the values of τ_n and τ_p evaluated at Fermi level $\mathcal{E}_f$. The reader is asked to show that in metals and degenerate semiconductors, the diffusion constant D is related to the mobility μ by

$$eD = (2/3)\zeta\mu \qquad (4.172)$$

where ζ is the Fermi level referred to the respective band edges or $\zeta = \mathcal{E}_f - \mathcal{E}_c$ for electrons and $\zeta = \mathcal{E}_v - \mathcal{E}_f$ for holes. Equation (4.172) should be compared to the Einstein relationship $eD = \mu kT$ for non-degenerate semiconductors.

The electric fields, E_{yp} and E_{yn}, in Eq. (4.171) actually consist of two parts: one part, E_s, due to the space charge created by the electron-phonon interaction and the other due to the deformation potential of Eq. (4.70). As discussed in Sec. 4.5, lattice vibration modulates the band edges which affect the potential energy of electrons and holes. Let δ be the amplitude of the strain wave propagating in the y direction with a phase factor $\exp[i(ky - \omega t)]$. Then, according to Eq. (4.70), the conduction and valence band edges can be written respectively as

$$\mathcal{E}_c = \mathcal{E}_{c0} + \mathcal{E}_{c1}\delta \exp[i(ky - \omega t)] \qquad (4.173a)$$

$$\mathcal{E}_v = \mathcal{E}_{v0} + \mathcal{E}_{v1}\delta \exp[i(ky - \omega t)] \qquad (4.173b)$$

where $\mathcal{E}_{c1}$ and $\mathcal{E}_{v1}$ are the deformation potentials. Note that $\mathcal{E}_c$ and $\mathcal{E}_v$ represent the potential energy of an electron in the conduction and valence bands respectively. Thus, the effective electric fields acting on electrons and holes are respectively

$$E_n = d\mathcal{E}_c/e\,dy = (i\mathcal{E}_{c1}\,\delta k/e)\exp[i(ky - \omega t)] \qquad (4.174a)$$

$$E_p = d\mathcal{E}_v/e\,dy = (i\mathcal{E}_{v1}\,\delta k/e)\exp[i(ky - \omega t)] \qquad (4.174b)$$

and the electric fields in Eq. (4.171) are given by $E_{yn} = E_s + E_n$ and $E_{yp} = E_s + E_p$.

As will be shown in the next chapter, the deviation from space-charge neutrality is exceedingly small in a high conductivity material. Applying the charge-neutrality condition to Eq. (4.171) means that the ratio $(p - n)/(p + n) \ll 1$. In other words, we may equate n to p in Eq. (4.171) and also write the continuity equations as

$$e\frac{\partial n}{\partial t} = \frac{\partial J_{ny}}{\partial y} \quad \text{and} \quad e\frac{\partial n}{\partial t} = -\frac{\partial J_{py}}{\partial y} \qquad (4.175)$$

Multiplying the first equation by μ_p and the second equation by μ_n and adding them, we obtain

$$\frac{\partial n}{\partial t} = -\frac{E_0}{B}\frac{\partial n}{\partial y} + D\frac{\partial^2 n}{\partial y^2} + \mu(E_n - E_p)\frac{\partial n}{\partial y} + n\mu\frac{\partial}{\partial y}(E_n - E_p) \qquad (4.176)$$

where $D = (\mu_p D_n + \mu_n D_p)/(\mu_n + \mu_p)$ and $\mu^{-1} = \mu_n^{-1} + \mu_p^{-1}$. In solving the above equation, we replace n by n_0 (the equilibrium electron concentration) in the last term and ignore the third term in comparison with the first and last terms on the right-hand side of Eq. (4.176). Using the dependence $\exp[i(ky - \omega t)]$ for n and $(E_n - E_p)$, we find

$$n = n_0 + \frac{ikn_0\mu(E_n - E_p)}{Dk^2 + i[(E_0 k/B) - \omega]} \qquad (4.177)$$

The average energy transfer per unit volume $\langle dW/dt \rangle$ from the electronic system to the lattice is given by

$$\left\langle \frac{dW}{dt} \right\rangle = \text{Re}\left[\frac{J_{ny}E_n^* + J_{py}E_p^*}{2}\right] \qquad (4.178)$$

where the symbol Re means the real part and the asterisk means the complex conjugate. Substituting the values of J_{ny} and J_{py} obtained from Eqs. (4.177) and (4.175) into Eq. (4.178), we obtain

$$\left\langle \frac{dW}{dt} \right\rangle = \frac{en_0\mu(\mathcal{E}_{c1} - \mathcal{E}_{r1})^2\,\delta^2 k^2}{2e^2}\frac{v_s[(E_0/B) - v_s]}{(Dk)^2 + [(E_0/B) - v_s]^2} \qquad (4.179)$$

The attenuation constant α of the acoustic wave is equal to $\alpha = -(1/v_s W)\langle dW/dt \rangle$ where $W = cS^2/2 = \rho v_s{}^2\,\delta^2/2$, ρ being the mass density of the crystal. Thus, we have

$$\alpha = -\frac{n_0 u(\mathcal{E}_{c1} - \mathcal{E}_{v1})^2 k^2}{e\rho v_s{}^2}\frac{(E_0/B) - v_s}{(Dk)^2 + [(E_0/B) - v_s]^2} \qquad (4.180)$$

Amplification of ultrasonic wave is possible if $E_0/B > v_s$. The reader is referred to the article[24] by Toxen and Tansal for a detailed account of the experiment by which ultrasonic amplification was observed in bismuth.

Shortly after the experimental observation of ultrasonic amplification in CdS, the effect of electron-phonon interaction has further manifested itself in a group of interesting experiments. Figure 4.19 shows the current-voltage characteristics observed[25] by Smith in piezoelectric semiconductors (CdS and GaAs) and by Esaki in bismuth in the presence of a strong magnetic field applied in a direction perpendicular to the drift field. Let us discuss the case of piezoelectric semiconductors first. In Eq. (4.160), the d-c part of the current density J_0 actually consists of two terms:

$$J_0 = e\mu_n(n_0 E_0 + f\langle n_s E\rangle_{av}) \tag{4.181}$$

the first term being the ordinary drift current density while the second term represents a new component known as the *acousto-electric current* which arises from electron-phonon interaction. From Eqs. (4.163) and (4.164), the reader can easily show

$$n_s = -k\sigma E/(e\omega\gamma) = -\sigma E/(ev_s\gamma) \tag{4.182}$$

by neglecting the diffusion term $ifD_n k^2$ in Eq. (4.164) and approximating k by ω/v_s. Taking the time average of $n_s E$, we find

$$J_0 = e\mu_n n_0 E_0 \left(1 - \frac{f\mu_n|E|^2}{2v_s E_0 \gamma}\right) \tag{4.183}$$

In the region for ultrasonic amplification, $\gamma < 0$ and the velocity v_s of the acoustic wave is in a direction opposite to that of the drift field E_0; or in other words, the acoustoelectric current is of such a

[24] A. M. Toxen and S. Tansal, *Phys. Rev. Letters*, vol. 10, p. 481, 1963.
[25] R. W. Smith, *Phys. Rev. Letters*, vol. 9, p. 87, 1962; L. Esaki, *Phys. Rev. Letters*, vol. 8, p. 4, 1962.

Figure 4.19

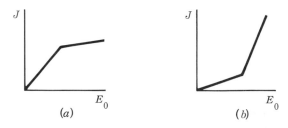

(a)   *(b)*

Current density against drift field curve in (a) piezoelectric semiconductors and (b) bismuth. The kink occurs in (a) when μE_0 is greater than the sound velocity and in (b) when the drift velocity of carriers in the direction perpendicular to $\mathbf{E}_0$ and $\mathbf{B}$ exceeds the sound velocity.

polarity that it opposes the ordinary drift (ohmic) current. When the value of E_0 exceeds the threshold required for amplification, the amplitude of the acoustic wave and hence that of E become sufficiently large to make the acoustoelectric current comparable in magnitude to the ohmic current. The kink appears in the J-E_0 characteristic of Fig. 4.19a as a direct consequence of a partial cancellation of the ohmic current by the acoustoelectric current.

Let us now discuss the case with semimetals. In the presence of a strong magnetic induction field B (in the negative z direction), the current densities in the x direction (J_{nx} for electrons and J_{px} for holes) are given by

$$J_{nx} = e\mu_n n E_0 + en E_{yn}/B \qquad (4.184a)$$

$$J_{px} = e\mu_p p E_0 - ep E_{yp}/B \qquad (4.184b)$$

The above set of equations is obtained by using in Eq. (4.149a) the values of a_1 and a_2 from Eq. (4.169). The acoustoelectric current density J_a is thus equal to

$$J_a = (e/B)\langle nE_{yn} - pE_{yp}\rangle_{\mathrm{av}} = (e/B)\langle n(E_n - E_p)\rangle_{\mathrm{av}} \qquad (4.185)$$

Substituting Eq. (4.177) into Eq. (4.185) gives

$$J_a = \frac{e\mu n_0}{2B} \frac{(E_0/B) - v_s}{(Dk)^2 + [(E_0/B) - v_s]^2} |E_n - E_p|^2 \qquad (4.186)$$

Note that the acoustoelectric current is now in the same direction as the ohmic current if $E_0/B > v_s$. This is in agreement with Fig. 4.19b.

In concluding this section, we should mention that current oscillations have been observed in the experiments of Smith and Esaki when the drift velocity of free carriers exceeds the sound velocity (that is, for $\mu E_0 > v_s$ in Fig. 4.19a and $E_0/B > v_s$ in Fig. 4.19b). As pointed out by Hutson,[26] nonlinear loss mechanisms may bring about a relaxation type of oscillation in the envelope of the drift current and the period of oscillation should correspond to the time required for the sound wave to make a round trip between the ends of the sample. The model given above which links the non-ohmic behavior with the amplification of ultrasonic waves has been confirmed through the experimental work of McFee.[27] Finally, we should also point out that acoustoelectric effects have been reported in GaAs by Beale and Pomerantz[28] and in Ge by Pomerantz.[29] For a general discussion of the acoustoelectric effect in semiconductors and semimetals, the reader is referred to articles by Weinreich[30] and by Eckstein.[31]

[26] A. R. Hutson, *Phys. Rev. Letters*, vol. 9, p. 296, 1962.
[27] J. H. McFee, *Jour. Appl. Phys.*, vol. 34, p. 1548, 1963.
[28] J. R. A. Beale and M. Pomerantz, *Phys. Rev. Letters*, vol. 13, p. 198, 1964.
[29] M. Pomerantz, *Phys. Rev. Letters*, vol. 13, p. 308, 1964.
[30] G. Weinreich, T. M. Sanders, and G. H. White, *Phys. Rev.*, vol. 114, p. 33, 1959.
[31] S. G. Eckstein, *Jour. Appl. Phys.*, vol. 35, p. 2702, 1964.

PROBLEMS

4.1 Using the definition of $f_0(\mathbf{r},\mathbf{v})$ given by Eq. (4.1) and realizing that $dv_x\, dv_y\, dv_z = 4\pi v^2\, dv$ in polar coordinates, show that in Eq. (4.17), $A = n(m^*/2\pi kT)^{3/2}$ and that Eq. (4.17) is a restatement of Eq. (1.125). Calculate the average kinetic energy associated with electrons in a nondegenerate gas.

4.2 Show that if the relaxation time τ is a function of velocity v, then the τ's in Eq. (4.22) should be replaced by two average τ's such that

$$\langle \tau \rangle_1 = \frac{m}{kT}\left(\frac{m}{2kT}\langle v_x{}^2 v^2 \tau \rangle - \tfrac{3}{2}\langle v_x{}^2 \tau \rangle\right) \qquad \langle \tau \rangle_2 = \frac{m}{kT}\langle v_x{}^2 \tau \rangle$$

where the symbol $\langle\,\rangle$ represents an average over the distribution $f_0(\mathbf{r},\mathbf{v})4\pi v^2\, dv$. Given $\tau = Bv^3$, verify that $\langle \tau \rangle_1 = (20/\sqrt{\pi})(Bv_T{}^3)$ and $\langle \tau \rangle_2 = (8/\sqrt{\pi})(Bv_T{}^3)$ where $v_T = \sqrt{2kT/m}$.

4.3 With the help of Eq. (4.33), find the average kinetic energy $\langle \mathcal{E}_0 \rangle$ of electrons in a metal at 0°K. Show that at a finite temperature, the Fermi level is given by Eq. (4.35) and the average kinetic energy by

$$\langle \mathcal{E}(T) \rangle = \langle \mathcal{E}_0 \rangle \left[1 + \frac{5\pi^2}{12}\left(\frac{kT}{\mathcal{E}_{f0}}\right)^2 \right]$$

4.4 The electrical conductivity σ of Au has a value 5×10^5 $(\Omega\text{-cm})^{-1}$ at room temperature. Using $\Theta = 175°K$, calculate the value of σ at $T = 50°K$ from Grüneisen's empirical formula. The mean free path at room temperature is estimated to be around 10^{-5} cm in good conductors such as Au. Find the mean free path of electrons in Au at 50°K.

4.5 The electron concentration in Au is known to be 5.9×10^{22} cm^{-3}. Using the free electron theory of metals and taking m^* for Au as the free electron mass, find the Fermi level and the mean free path of electrons at room temperature. Refer to the conductivity value given in Prob. 4.4.

4.6 Check the dimensions of $S(\mathbf{k},\mathbf{k}')$ and $c - b$ in Eqs. (4.50) and (4.52) respectively. Explain what happens to the factor $d\mathcal{E}$ in Eq. (4.51). Draw the contour of constant energy in a two-dimensional k space and show the initial and final states of an electron during transition.

4.7 Find an expression relating the mean free path to the mobility of electrons in the region where the impurity scattering is dominant. For simplicity use $\tau_i = Bv^3$ where B is a proportionality constant.

4.8 From Eq. (4.93) and the energy conservation relation

$$\frac{\hbar^2 k_1{}^2}{2m^*} = \pm\hbar cq + \frac{\hbar^2 k_2{}^2}{2m^*}$$

(upper sign for emission and lower sign for absorption), verify Eq. (4.99). The result can be obtained by letting $k_1 = k$ and $k_2 = k + \Delta k$ where Δk is assumed to be much smaller than k so that the second order terms containing $(\Delta k)^2$ can be neglected in the energy and momentum conservation equations.

4.9 From Eq. (4.97), show that $\delta\mathcal{E} = 8c^2[m^* - \hbar^2k_1{}^2/(4kT_l)]\sin^2(\theta/2)$.

4.10 A critical electric field E_c is defined such that

$$E_c = (32/3\pi)^{1/2}(c/\mu_0)$$

Taking $c = 5 \times 10^5$ cm/sec and $\mu_0 = 3700$ cm²/volt-sec, calculate E_c. Find the mobility and the electron temperature at $E = E_c$.

4.11 When a wire carrying a current is placed in a uniform temperature gradient, heat will be produced. This is known as the *Thomson heat*. The rate of heat absorbed $(\partial H/\partial t)$ per $J_x(dT/dx)$ is called the Thomson coefficient σ_T. Show that

$$\sigma_T = \frac{T}{e}\frac{d}{dT}\left(\frac{\mathcal{E}_f}{T} - \frac{B_2}{B_1 T}\right)$$

Find a general expression for σ_T in terms of $\langle v^x\tau\rangle$. Apply the result to a metal.

4.12 Given $N_a = 1.7 \times 10^{17}$/cm³ and $N_v = 5.4 \times 10^{18}$/cm² (at room temperature), plot the thermoelectric power Q as a function of temperature between 200°K and 400°K. Compare the calculated curve with the curve shown in Fig. 4.11.

4.13 Show that the product of $Q^2\sigma$ can be expressed as

$$Q^2\sigma = \left(\frac{k}{e}\right)^2[A^2 - 2A\log\sigma + (\log\sigma)^2]\sigma$$

where $A = 2 + \log(2\mu_pN_v)$ for a p-type semiconductor. Show also that $\sigma = e\mu_pN_v$ and $Q = 2k/e$ for maximum $Q^2\sigma$.

4.14 Show that for maximum z of Eq. (4.142)

$$\log\sigma = \log(e\mu_pN_v) - 2K_{el}/K_{ph} \qquad Q = 2(k/e)(1 + K_{el}/K_{ph})$$

by noting $K_{el} = A\sigma$ where A is a constant.

4.15 Show that if τ is independent of velocity, Eq. (4.149) reduces to the ordinary equation of motion, Eq. (3.69).

4.16 Verify Eqs. (4.152) and (4.153). State the conditions under which these two equations are valid.

4.17 Outline the procedure for the calculation of the Hall coefficient in the region where processes of impurity and lattice scattering are equally important.

4.18 Following the discussion in Sec. 4.5, show that the reciprocal of the relaxation time for an electron of wave vector $\mathbf{k}$ due to one-phonon scattering process is

$$\frac{1}{\tau} = \frac{m^*k}{4\pi^2\hbar^3}\int_0^\pi\int_0^{2\pi}|M(\mathbf{k},\mathbf{k}')|^2(1 - \cos\theta)\sin\theta\,d\theta\,d\phi$$

where θ is the angle between $\mathbf{k}$ and $\mathbf{k}'$, the wave vectors of the electron before and after scattering. Also verify that the matrix element given in

Eq. (4.84) can be written as

$$|M(\mathbf{k,k'})|^2 = \frac{kT}{2cV}\,\mathcal{E}_{1c}{}^2$$

where $\mathcal{E}_{1c}$ is the deformation potential, c is the elastic constant and k is the Boltzmann constant.

4.19 Using Eq. (4.164) and the definition of S in relation to ξ, show that the electric potential accompanying a lattice wave is given by

$$\phi = \frac{id}{\epsilon q}S \qquad \text{joule/coulomb}$$

where d is the piezoelectric constant, S is the strain and q is the wave number of the lattice wave. Comment on the assumptions made in the derivation of the above equation.

The potential ϕ contributes to the scattering of electrons in a piezoelectric crystal. In the treatment of piezoelectric scattering, we note that the quantity $e\phi/S$ corresponds to $\mathcal{E}_{1c}$ of Eq. (4.70) in lattice scattering. Using the result obtained in the preceding problem and following the same procedure outlined in Sec. 4.5, show that

$$\frac{1}{\tau} = \frac{e^2 m^{*\frac{1}{2}}}{2^{\frac{3}{2}}\pi\hbar^2}\frac{(K^2)_{\mathrm{av}}}{\epsilon}\frac{kT}{\mathcal{E}^{\frac{1}{2}}}$$

where $K = \sqrt{d^2/(\epsilon c)}$ is the electromechanical coupling constant.

From the expression for $1/\tau$, verify the mass and temperature dependence of the mobility due to piezoelectric scattering expressed in Eq. (3.124).

4.20 Derive Eq. (4.171) and verify Eq. (4.172) for semimetals.

4.21 Read the articles by G. Weinreich, T. M. Sanders, and G. H. White [*Phys. Rev.*, vol. 114, p. 33, 1959] and S. G. Eckstein [*J. Appl. Phys.*, vol. 35, p. 2702, 1964]. Describe briefly the acoustoelectric effects expected in semiconductors and semimetals. Discuss in particular how the acoustoelectric current is defined and outline the procedure by which the current may be evaluated.

REFERENCES

Bardeen, J.: "Electrical Conductivity of Metals," *Jour. Appl. Phys.*, vol. 11, p. 88, 1940.

Beer, A. C.: Galvanomagnetic Effects in Semiconductors, *Solid State Phys.*, Supplementary monographs No. 4, 1963.

Brooks, H.: Theory of the Electrical Properties of Germanium and Silicon, *Advan. Electron. Electron Phys.*, vol. 7, p.87, 1955.

Herring, C.: "Theory of the Thermoelectric Power of Semiconductors," *Phys. Rev.*, vol. 96, p. 1163, 1954.

————: "Transport Properties of a Many-Valley Semiconductor," *Bell System Tech. Jour.*, vol. 34, p. 237, 1955.

Ioffe, A. F.: "Semiconductor Thermoelements and Thermoelectric Cooling," translated by Infosearch Limited, London, 1957.

Kittel, C.: "Introduction to Statistical Physics," John Wiley & Sons, Inc., New York, 1958.

Seitz, F.: "The Modern Theory of Solids," McGraw-Hill Book Company, New York, 1940.

Shockley, W.: "Electrons and Holes in Semiconductors," D. Van Nostrand Co., New York, 1950.

————: "Hot Electrons in Germanium and Ohm's Law," *Bell System Tech. Jour.*, vol. 30, p. 999, 1951.

Wilson, A. H.: "The Theory of Metals," Cambridge University Press, Cambridge, 1953.

Ziman, J. M.: "Electrons and Phonons," Oxford University Press, Oxford, 1960.

5 TRANSPORT AND RECOMBINATION OF EXCESS CARRIERS

5.1 DRIFT AND DIFFUSION CURRENTS IN NONDEGENERATE SEMICONDUCTORS

The transport phenomena discussed in the previous chapter are associated with carriers normally present in a semiconductor. However, in many semiconductor devices, we must deal with the transport of *excess carriers*, that is, carriers whose concentration exceeds the thermal equilibrium concentration. As a starting point, we shall develop the current expression under circumstances which are typical of a semiconductor junction device. Consider a situation in which the temperature is maintained uniform but the free-carrier concentrations n and p, for reasons to be explained later, are nonuniform; that is, $\partial n/\partial x \neq 0$ and $\partial p/\partial x \neq 0$. With an applied field E_x in the x direction, the current density J_x, using the result of the Boltzmann transport equation, can be written as

$$J_x = e \iiint_{-\infty}^{\infty} \tau \left(v_x \frac{\partial f_0}{\partial x} + a_x \frac{\partial f_0}{\partial v_x} \right) v_x \, dv_x \, dv_y \, dv_z \tag{5.1}$$

Equation (5.1) is identical to Eq. (4.14) and is repeated here for easy reference.

For ease of discussion, the following assumptions are made: (1) only electrons are considered; (2) the relaxation time τ is a constant independent of v; (3) the energy surface is spherical so that Eqs. (4.27) and (4.29) are still applicable. However, for a nondegenerate semiconductor, the Fermi-Dirac function $F_c(\varepsilon)$ for conduction-band electrons can be approximated by Eq. (3.18) or

$$F_c(\varepsilon) \simeq \exp \left(\frac{\varepsilon_f - \varepsilon_c - \varepsilon}{kT} \right) \tag{5.2}$$

where $\varepsilon = \hbar^2 k^2 / 2m^*$ is the kinetic energy of an electron. Noting that $f_0 = 2(m^*/h)^3 F_c(\varepsilon)$, we have

$$a_x \frac{\partial f_0}{\partial v_x} = \frac{eE_x}{m^*} \left[\frac{F_c(\varepsilon)}{kT} \right] m^* v_x 2 \left(\frac{m^*}{h} \right)^3 \tag{5.3}$$

It is obvious that Eq. (5.1) involves a common integral of the following type:

$$I = \iiint_{-\infty}^{\infty} m^* v_x^2 f_0 \, dv_x \, dv_y \, dv_z \tag{5.4}$$

Using Eqs. (4.20), (4.21) and (3.20), we find

$$I = 2kT(2\pi m^* kT/h^2)^{3/2} \exp\left(-\frac{\mathcal{E}_c - \mathcal{E}_f}{kT}\right) = nkT \tag{5.5}$$

Substituting Eq. (5.3) into Eq. (5.1), we have

$$J_x = e\tau \left(\frac{\partial}{\partial x} \int_0^\infty v_x^2 f_0 4\pi v^2 \, dv + \frac{eE_x}{kT} \int_0^\infty v_x^2 f_0 4\pi v^2 \, dv\right) \tag{5.6}$$

Note that the carrier concentration n is nonuniform because the quantity $(\mathcal{E}_c - \mathcal{E}_f)$ but not the temperature is a function of x. Therefore, the differentiation with respect to x can be taken outside the integral in Eq. (5.6). The reader can verify our contention by assuming a nonvanishing $\partial(\mathcal{E}_c - \mathcal{E}_f)/\partial x$. Substituting Eqs. (5.4) and (5.5) into Eq. (5.6) gives

$$J_x = \frac{e\tau kT}{m^*} \frac{\partial n}{\partial x} + \frac{e^2 \tau n}{m^*} E_x = eD_n \frac{\partial n}{\partial x} + e\mu_n n E_x \tag{5.7}$$

The field induced term $e\mu_n n E_x$ is generally referred to as the *drift current* while the term $eD_n \, \partial n/\partial x$ caused by a concentration gradient $\partial n/\partial x$ is usually called the *diffusion current*. It is interesting to note that a definite relationship exists between the diffusion constant D_n and the mobility μ_n as follows:

$$D_n = \frac{\tau kT}{m^*} = \frac{kT}{e} \mu_n \tag{5.8}$$

A similar expression is obtained for holes. The relationship is known as the Einstein relation. If the relaxation time τ is a function of the energy or velocity of an electron, τ should be kept inside the integral in Eq. (5.6). However, the same integral appears in the expressions of the diffusion and drift currents. In other words, the value of τ will be weighted equally in D_n and μ_n. Therefore, the Einstein relation is true in nondegenerate semiconductors irrespective of any specific assumption about τ.

5.2   THE DENSITY-OF-STATE AND MOBILITY EFFECTIVE MASSES

Equation (5.7) is based on a single effective mass, or, therefore, on a spherical energy surface. In direct-gap semiconductors such as GaAs, GaSb, and many other III-V compounds where the conduction-band minimum is located at $k[000]$, the constant energy surface for conduction electrons is spherical and hence Eq. (5.7) is directly applicable. For conduction electrons in indirect-gap semiconductors (Ge and Si) and valence-band holes, however, the situations are not so straightforward. In view of the discussion in Sec. 3.10, it seems appropriate to discuss how Eq. (5.7) should be modified for ellipsoidal energy surfaces. Consider the action upon electrons of an electric field E which has components $u_1'E$, $u_2'E$, and $u_3'E$ along the **a**, **b**, and **c** axes as shown in Fig. 5.1, u_1', u_2', and u_3' being the direction cosines of a unit vector **u** in the direction of the applied field. The directions of **a**, **b**, and **c** are so chosen that they coincide with the axes of the ellipsoid and hence

$$\mathcal{E} = \frac{\hbar^2}{2}\left(\frac{k_1{}^2}{m_1} + \frac{k_2{}^2}{m_2} + \frac{k_3{}^2}{m_3}\right) = \tfrac{1}{2}(m_1v_1{}^2 + m_2v_2{}^2 + m_3v_3{}^2) \quad (5.9)$$

It is understood that the k's in Eq. (5.9) are increments in k_1, k_2, and k_3 referring to the values of k_{10}, k_{20}, and k_{30} at one of the conduction-band minima. Therefore, Eq. (5.9) is the same as Eq. (3.91).

Refer to Eq. (4.12) from which Eq. (5.1) and hence Eq. (5.7) are derived. For simplicity, we shall consider the effect of an electric field alone. Since the electric field now has three components, Eq. (4.12) should be replaced by the following equation:

$$f = f_0 + \tau eE\left(\frac{u_1'}{m_1}\frac{\partial f_0}{\partial v_1} + \frac{u_2'}{m_2}\frac{\partial f_0}{\partial v_2} + \frac{u_3'}{m_3}\frac{\partial f_0}{\partial v_3}\right)$$

$$= f_0 + \tau eE(u_1'v_1 + u_2'v_2 + u_3'v_3)\frac{\partial f_0}{\partial \mathcal{E}} \quad (5.10)$$

Figure 5.1

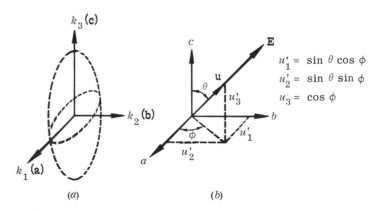

(a) (b)

$$u_1' = \sin\theta\cos\phi$$
$$u_2' = \sin\theta\sin\phi$$
$$u_3' = \cos\phi$$

The orientation of the electric field **E** with respect to the axes (**a**, **b**, and **c**) of the constant-energy ellipsoid.

where $\mathbf{v} - \hbar^{-1} \operatorname{grad}_k \mathcal{E}$. The drift current density is given by

$$J_a = -e \iiint_{-\infty}^{\infty} \tau e E(u_1' v_1 + u_2' v_2 + u_3' v_3) v_1 \frac{\partial f_0}{\partial \mathcal{E}} \, dv_1 \, dv_2 \, dv_3 \quad (5.11)$$

Since $\partial f_0 / \partial \mathcal{E}$ is an even function in v_1, v_2, and v_3, the only surviving term is the term involving $v_1{}^2$. Furthermore, from Eq. (5.2), $\partial f_0 / \partial \mathcal{E} = -f_0 / kT$, thus Eq. (5.11) becomes

$$J_a = \frac{e^2 \tau E u_1'}{m_1 kT} \iiint_{-\infty}^{\infty} m_1 v_1{}^2 f_0 \, dv_1 \, dv_2 \, dv_3 \quad (5.12)$$

Similar expressions are obtained for J_b and J_c.

To evaluate Eq. (5.12), we must find a new expression similar to Eq. (4.27) which defines f_0. For an ellipsoidal energy surface,

$$dn' = \frac{2F(\mathcal{E})}{(2\pi)^3} \, dk_1 \, dk_2 \, dk_3 = 2 \frac{m_1 m_2 m_3}{h^3} F(\mathcal{E}) \, dv_1 \, dv_2 \, dv_3 \quad (5.13)$$

Thus, $f_0 = 2(m_1 m_2 m_3 / h^3) F(\mathcal{E})$ in Eq. (5.12). Using Eqs. (4.20) and (4.21) and $F(\mathcal{E})$ given by Eq. (5.2), we find

$$J_a = \frac{e^2 \tau E u_1'}{m_1} n' \quad (5.14)$$

$$n' = 2 \left(\frac{2\pi kT}{h^2} \right)^{3/2} (m_1 m_2 m_3)^{1/2} \exp \left(-\frac{\mathcal{E}_c - \mathcal{E}_f}{kT} \right) \quad (5.15)$$

The above equations are the results due to one ellipsoid only. For N ellipsoids, the free carrier density n' will be multiplied N times. If we define a *density-of-state effective mass* m_e such that

$$m_e{}^3 = N^2 m_1 m_2 m_3 = N^2 m_t{}^2 m_l \quad (5.16)$$

m_t and m_l, respectively, being the transverse and longitudinal masses, then the total free-carrier density n can be written as

$$n = n'N = 2 \left(\frac{2\pi m_e kT}{h^2} \right)^{3/2} \exp \left(-\frac{\mathcal{E}_c - \mathcal{E}_f}{kT} \right) \quad (5.17)$$

Equation (5.17) is the familiar expression shown in most books on semiconductors. However, the use of Eq. (5.17) does not necessarily imply a spherical energy surface. For ellipsoidal energy surfaces, the m_e in Eq. (5.17) is related to m_t and m_l by Eq. (5.16).

Now let us turn to Eq. (5.14) and discuss the effect on the current density J of many ellipsoidal energy surfaces. The fact that different ellipsoids may be oriented in different directions (Fig. 3.23) makes the computation rather complicated. To generalize Eq. (5.14), we note that u_1' can be written as $(\mathbf{a} \cdot \mathbf{u})$ where vectors $\mathbf{a}$ and $\mathbf{u}$ may refer to the crystal axes x, y, and z. After adding contributions from J_b and J_c to Eq. (5.14), we find the total current density equal to

$$\mathbf{J} = e^2 \tau E n' \left[\mathbf{a} \, \frac{(\mathbf{a} \cdot \mathbf{u})}{m_1} + \mathbf{b} \, \frac{(\mathbf{b} \cdot \mathbf{u})}{m_2} + \mathbf{c} \, \frac{(\mathbf{c} \cdot \mathbf{u})}{m_3} \right] \tag{5.18}$$

The component of $\mathbf{J}$ in the direction of the electric field is equal to

$$J_u = e^2 \tau E n' \left[\frac{(\mathbf{a} \cdot \mathbf{u})^2}{m_1} + \frac{(\mathbf{b} \cdot \mathbf{u})^2}{m_2} + \frac{(\mathbf{c} \cdot \mathbf{u})^2}{m_3} \right] \tag{5.19}$$

where $\mathbf{a}$, $\mathbf{b}$, and $\mathbf{c}$ are unit vectors along the a, b, and c axes respectively. For curiosity, we shall also compute the current density J_w in a direction perpendicular to the electric field. For definiteness, take $\mathbf{w} = \mathbf{a} \times \mathbf{u}/|\mathbf{a} \times \mathbf{u}|$. Thus, we have

$$J_w = \mathbf{J} \cdot \mathbf{w} = e^2 \tau E n' [(\mathbf{b} \cdot \mathbf{u})(\mathbf{c} \cdot \mathbf{u})] \, \frac{\dfrac{1}{m_3} - \dfrac{1}{m_2}}{|\mathbf{a} \times \mathbf{u}|} \tag{5.20}$$

It is interesting to note that if only one ellipsoid is taken into consideration, the current density J_w in a direction perpendicular to the electric field is, in general, not zero. This is so because the distribution of electrons is shifted in the velocity space along the v_w axis in a manner similar to the distribution shown in Fig. 4.2a. The reader may then ask why such an effect has not been observed in germanium and silicon which have ellipsoidal energy surfaces. The answer lies in the cubic symmetry of the diamond structure.

In germanium, as discussed in Sec. 3.8, the conduction-band minima are at the edge of the Brillouin zone along the four principal diagonals $k[111]$, $k[\bar{1}\bar{1}1]$, $k[\bar{1}1\bar{1}]$, and $k[1\bar{1}\bar{1}]$ of the reciprocal lattice. We should take further note that the ellipsoidal energy surfaces with major axes located along the $k[111]$ and $k[\bar{1}\bar{1}\bar{1}]$ are merely a continuation of each other in the k space as shown in Fig. 5.2a. The same applies to the energy surfaces with major axes located along the other pair of axes, $k[\bar{1}\bar{1}1]$ and $k[11\bar{1}]$ for example. Therefore, only four energy surfaces should be counted, or, $N = 4$ in Eq. (5.16). Furthermore, in order that the four diagonals may preserve the tetragonal symmetry of the diamond lattice, either the set $k[111]$, $k[\bar{1}\bar{1}1]$, $k[\bar{1}1\bar{1}]$, and $k[1\bar{1}\bar{1}]$ or the equivalent set $k[\bar{1}\bar{1}\bar{1}]$, $k[11\bar{1}]$, $k[1\bar{1}1]$, and $k[\bar{1}11]$ must be chosen. Here the former set is chosen. The situation is illustrated in Fig. 5.2. Our next problem is to apply the above information to Eq. (5.19). If we let $m_3 = m_l$ and $m_1 = m_2 = m_t$, then Eq. (5.19) becomes

$$J_u = e^2 \tau E n' \sum \left[\frac{1}{m_t} + \left(\frac{1}{m_l} - \frac{1}{m_t} \right) (\mathbf{u} \cdot \mathbf{c})^2 \right] \tag{5.21}$$

Note that $(\mathbf{u} \cdot \mathbf{c})$ is the direction cosine between the electric field and the major axis of the constant energy ellipsoid, and in the case of germanium, $\mathbf{c}$ is a unit vector along one of the four principal diagonals.

Since Eq. (5.21) is expressed in a form unaffected by a rotation of the coordinate system, we shall use the crystal axes as the reference coordinates in which the unit vector $\mathbf{u}$ has three components (u_1, u_2, u_3). Referring to the coordinate axes, x, y, and z, the four principal diagonals of the four ellipsoids are

$$\mathbf{c_1}: \frac{1}{\sqrt{3}}(1,1,1) \qquad \mathbf{c_3}: \frac{1}{\sqrt{3}}(\bar{1},1,\bar{1})$$
$$\mathbf{c_2}: \frac{1}{\sqrt{3}}(\bar{1},\bar{1},1) \qquad \mathbf{c_4}: \frac{1}{\sqrt{3}}(1,\bar{1},\bar{1}) \tag{5.22}$$

Thus, we find

$$\sum_{1,2,3,4} (\mathbf{u} \cdot \mathbf{c_1})^2 = \tfrac{4}{3}(u_1{}^2 + u_2{}^2 + u_3{}^2) = \tfrac{4}{3} \tag{5.23}$$

Summing Eq. (5.21) over the four ellipsoids gives

$$J_u = \frac{e^2 \tau E}{3}\left(\frac{2}{m_t} + \frac{1}{m_l}\right)(4n') = \frac{e^2 \tau E n}{m_e^*} \tag{5.24}$$

Figure 5.2

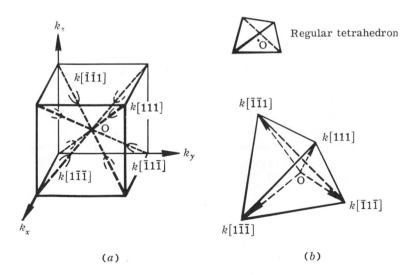

(a) The eight constant-energy ellipsoids have their major axes lying along the four principal (body) diagonals in germanium. Since the conduction-band minima are located at the edge of the Brillouin zone, the two ellipsoids along the same principal diagonal (for example, the two ellipsoids along $k[111]$ and $k[\bar{1}\bar{1}\bar{1}]$ indicated by solid and dotted contours) are merely a continuation of each other. (b) The chosen set of ellipsoidal axes preserves the tetrahedral symmetry of the diamond structure.

where m_e^* is defined as the *mobility effective mass*

$$\frac{1}{m_e^*} = \frac{1}{3}\left(\frac{2}{m_t} + \frac{1}{m_l}\right) \tag{5.25}$$

Note that the density-of-state effective mass defined in Eq. (5.16) is different from the mobility effective mass, but they are both functions of m_l and m_t, the two fundamental masses used in defining the constant energy surfaces.

The question whether the total current density **J** has components in directions perpendicular to the electric field is of considerable importance and remains unanswered. If we choose the two unit vectors **a** and **b** such that

$$\mathbf{a} = \frac{\mathbf{c} \times \mathbf{u}}{\sqrt{1 - (\mathbf{c} \cdot \mathbf{u})^2}}$$

then

$$\mathbf{b} = \frac{\mathbf{c} \times (\mathbf{c} \times \mathbf{u})}{\sqrt{1 - (\mathbf{c} \cdot \mathbf{u})^2}} = \frac{\mathbf{u} - \mathbf{c}(\mathbf{c} \cdot \mathbf{u})}{\sqrt{1 - (\mathbf{c} \cdot \mathbf{u})^2}} \tag{5.26}$$

where **c** is a unit vector along one of the principal diagonals. According to Eq. (5.18) with $m_3 = m_l$ and $m_1 = m_2 = m_t$, we have

$$J_c = e^2 \tau E n' \frac{(\mathbf{c} \cdot \mathbf{u})}{m_l} \qquad J_a = 0 \qquad J_b = e^2 \tau E n' \frac{\sqrt{1 - (\mathbf{c} \cdot \mathbf{u})^2}}{m_t} \tag{5.27}$$

The total current density along the x axis is given by

$$J_x = e^2 \tau E n' \sum \left[\frac{(\mathbf{c} \cdot \mathbf{u})(\mathbf{c} \cdot \mathbf{x})}{m_l} + \frac{(\mathbf{b} \cdot \mathbf{x})\sqrt{1 - (\mathbf{c} \cdot \mathbf{u})^2}}{m_t}\right] \tag{5.28}$$

The above summation is taken over the four ellipsoids. The reader is asked to show that Eq. (5.28) reduces to $J_x = u_1 J_u$. Similar expressions are also obtained for $J_y (= u_2 J_u)$ and $J_z (= u_3 J_u)$. This is equivalent to saying that the total current flows in the direction of the applied field. The mutual cancellation of the currents in a direction perpendicular to the electric field is a direct consequence of the symmetry of the four ellipsoids.

Now we shall discuss the transport properties of valence-band holes. First, there exist three valence bands as discussed in Sec. 3.8. However, at ordinary temperatures, only the top two bands, referred to as the light and heavy hole bands, are occupied by holes and hence only these bands need be considered. The warped energy surfaces shown in Fig. 3.27 can be approximated by spherical energy surfaces with an average effective mass. From Fig. 3.26, the average effective masses for light and heavy holes are $m_1 = 0.04m_0$ and $m_2 = 0.35m_0$ in Ge and $m_1 = 0.17m_0$ and $m_2 = 0.51m_0$ in Si, respectively. Since the total hole concentration is a sum of the concentrations of light and heavy holes, the density-of-state effective mass used in Eq. (3.23) is given by

$$m_h^{3/2} = m_1^{3/2} + m_2^{3/2} \tag{5.29}$$

In the expression of electrical conductivity, $\sigma = e\mu_p p$ for a p-type sample, μ_p is the effective conductivity mobility

$$\mu_p = \frac{\mu_{p1}p_1 + \mu_{p2}p_2}{p_1 + p_2} \tag{5.30}$$

where the subscripts 1 and 2 refer to the light and heavy holes respectively. Cyclotron resonance experiments show that the light and heavy holes have approximately the same relaxation time τ as evidenced from the widths of the resonance curves. Assuming $\tau_1 = \tau_2$ in Eq. (5.30), a mobility effective mass can be defined as

$$\frac{1}{m_h^*} = \frac{m_1^{1/2} + m_2^{1/2}}{m_1^{3/2} + m_2^{3/2}} \tag{5.31}$$

Furthermore, because the energy expression for holes, Eq. (3.112), preserves the cubic symmetry of the crystal, the hole conduction current also lies in the direction of the applied field as does the electron conduction current.

In concluding this section, we would like to emphasize that for conduction-band electrons in indirect-gap semiconductors, only two fundamental masses exist, m_t and m_l. However, under different circumstances, different effective masses are defined and can be expressed in terms of m_t and m_l. Those discussed thus far include the cyclotron resonance, the density-of-states, and the mobility effective masses. The reader may wonder whether a fourth mass, the diffusion effective mass, is needed. It can be shown that the effective mass which appears in the diffusion coefficient D is the same as the mobility effective mass. So far as the conductive properties of conduction-band electrons are concerned, the effective masses defined in Eqs. (5.8), (5.16), and (5.25) are sufficient. For valence band holes, the corresponding equations are Eqs. (5.8), (5.29), and (5.31). There is, however, an important difference between these equations. The two masses, m_t and m_l, appear in Eqs. (5.16) and (5.25) on account of the ellipsoidal energy surfaces, while the two masses, m_1 and m_2, enter into Eqs. (5.29) and (5.31) because of the existence of two different valence bands.

5.3 THE THERMAL EQUILIBRIUM CONDITION

In the previous sections we have derived the current expression, Eq. (5.7), and expressed the various parameters involved in terms of the energy-band structure of a semiconductor. In order that Eq. (5.7) can be put to good use, a general discussion of the condition for thermal equilibrium is needed. Let us consider a simple situation which arises from joining together two uniform semiconductors of the same material but of opposite types as shown in Fig. 5.3a. Clearly, there are more electrons on the n side and more holes on the p side; therefore, there exists a region in which $\partial n/\partial x$ and $\partial p/\partial x$ are nonzero. The question naturally arises as to whether there will be any current flow if there is no external voltage applied. The answer hinges on what happens in the boundary region where $\partial n/\partial x$ and $\partial p/\partial x$ are nonzero. A statement is often made that in a system under thermal equilibrium, the Fermi level $\mathcal{E}_f$ stays constant throughout the entire system, creating a situation shown in Fig. 5.3b. In proving this statement, two approaches will be used, not only to show the physical significance of the Fermi level, but also to indicate the possible use of Fermi level under a nonequilibrium situation.

Consider electrons having energies $\mathcal{E}_2$ and $\mathcal{E}_1$ respectively bordering on the n side and p side of the junction as shown in Fig. 5.3a. For simplicity, a spherical energy surface with an effective mass m^* is assumed. The number of electrons moving across the boundary in a time interval dt from the p side to the n side in an area $dy\ dz$ is given by

$$\left\{ 2\left(\frac{m^*}{h}\right)^3 \frac{dv_{x1}\,dv_{y1}\,dv_{z1}}{1+\exp\left[(\mathcal{E}_1-\mathcal{E}_{fp})/kT\right]} \right\} v_{x1}\,dt\,dy\,dz \qquad (5.32)$$

the term in the braces representing the density of occupied elec-

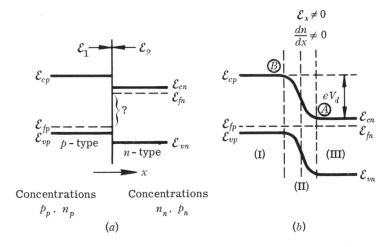

Figure 5.3

(a) *The situation arises from joining an n-type and a p-type semiconductor together. What happens near the boundary region is a matter of special interest. (b) Under thermal equilibrium, the Fermi levels on both sides of the junction stay the same. The change in the electrostatic potential in the transition region is such that the drift and diffusion terms in Eq. (5.7) exactly balance out each other.*

trons in the conduction band. A similar expression is obtained for the corresponding number of electrons moving from the n side to the p side. Detailed balancing requires that these two numbers must be equal under thermal equilibrium. That means,

$$\frac{v_{x1} \, dv_{x1} \, dv_{y1} \, dv_{z1}}{1 + \exp\left[(\mathcal{E}_1 - \mathcal{E}_{fp})/kT\right]} = \frac{v_{x2} \, dv_{x2} \, dv_{y2} \, dv_{z2}}{1 + \exp\left[(\mathcal{E}_2 - \mathcal{E}_{fn})/kT\right]} \tag{5.33}$$

Furthermore, the energy of an electron must be conserved in crossing the boundary, $\mathcal{E}_1 = \mathcal{E}_2$, or

$$\mathcal{E}_{cp} + \frac{m^*}{2}\left(v_{x1}^2 + v_{y1}^2 + v_{z1}^2\right) = \mathcal{E}_{cn} + \frac{m^*}{2}\left(v_{x2}^2 + v_{y2}^2 + v_{z2}^2\right) \tag{5.34}$$

Due to the fact that the potential energy $\mathcal{E}_{cp}$ or $\mathcal{E}_{cn}$ of an electron remains the same in the y and z directions, $v_{y1} = v_{y2}$ and $v_{z1} = v_{z2}$. Thus, we have

$$v_{x1} \, dv_{x1} \, dv_{y1} \, dv_{z1} = v_{x2} \, dv_{x2} \, dv_{y2} \, dv_{z2} \tag{5.35}$$

In view of Eq. (5.34), substituting Eq. (5.35) into Eq. (5.33) yields

$$\mathcal{E}_{fp} = \mathcal{E}_{fn} \tag{5.36}$$

The situation is illustrated in Fig. 5.3b.

Now let us examine whether Eq. (5.36) is consistent with the macroscopic condition for thermal equilibrium, that is, $J = 0$. It is obvious from Fig. 5.3b that in order to make $\mathcal{E}_{fp} = \mathcal{E}_{fn}$, a difference in the potential energy, $eV_d = \mathcal{E}_{cp} - \mathcal{E}_{cn}$, results, creating a transition region in which there exists an internal electric field E_x. In the meantime, the diffusion term in Eq. (5.7) is also nonzero in the transition region because $\mathcal{E}_c - \mathcal{E}_f$ and hence n vary with x. Under thermal equilibrium, $J_x = 0$, thus, Eq. (5.7) yields

$$-eE_x \, dx/kT = dn/n \tag{5.37}$$

For ease of discussion, it is assumed that the transition region ends abruptly at points A and B in Fig. 5.3b, or in other words, $E_x = 0$ for $x < x_B$ and $x > x_A$. Justification for the use of such an idealized model will be given in the next chapter where a more detailed treatment of a p-n junction will be made. Integrating Eq. (5.37) from A to B gives

$$\frac{n_{p0}}{n_{n0}} = \exp\left(\int_A^B \frac{dn}{n}\right) = \exp\left(-\frac{eV_d}{kT}\right) \tag{5.38}$$

where n_{p0} and n_{n0} are the equilibrium concentrations of electrons in the bulk p and n regions respectively. The potential difference $V_d = (\mathcal{E}_{cp} - \mathcal{E}_{cn})/e$ is generally referred to as the *built-in* or *diffusion potential* in a p-n junction. Substituting Eq. (3.20) into Eq. (5.38) gives

$$(\mathcal{E}_{cp} - \mathcal{E}_{fp}) - (\mathcal{E}_{cn} - \mathcal{E}_{fn}) = eV_d \tag{5.39}$$

or $\mathcal{E}_{fp} = \mathcal{E}_{fn}$, the same as Eq. (5.36). Therefore, the condition $\mathcal{E}_{fp} = \mathcal{E}_{fn}$ under thermal equilibrium can be obtained from either microscopic detailed balancing or macroscopic current. The reader is asked to show that an equation similar to Eq. (5.38) and leading to the same result of Eq. (5.39) is obtained for valence-band holes.

To generalize the results obtained from the above discussion, the following observations are made. The free energy F of a composite system is given by

$$dF = p\,dV - S\,dT + \sum_i \phi_i\,dn_i \tag{5.40}$$

where p, V, S, and T stand for pressure, volume, entropy, and temperature respectively while ϕ_i and n_i are the electrochemical potential and concentration of particles in the subsystem i. It can be shown that the Fermi level used mostly by physicists and electrical engineers and the electrochemical potential used mostly by chemists and metallurgists are the same. Therefore, under constant T and V, the change in the free energy in transferring dn electrons from the subsystem i to j is equal to

$$dF = (\mathcal{E}_{fi} - \mathcal{E}_{fj})\,dn \tag{5.41}$$

Under thermal equilibrium, the free energy is a minimum, and Eq. (5.41) requires that $\mathcal{E}_{fi} = \mathcal{E}_{fj}$. The result expressed in Eq. (5.36), therefore, is general regardless of any specific model used.

5.4 THE TREATMENT OF NONEQUILIBRIUM SITUATIONS

From discussions in the preceding section, it is clear that for a nonvanishing J_x, the Fermi levels $\mathcal{E}_{fn}$ and $\mathcal{E}_{fp}$ must be different. This corresponds to the situation when an external voltage is applied to the p-n junction as shown in Fig. 5.4. If it is assumed that the differ-

Figure 5.4

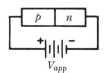

When a voltage V_a is applied across a semiconductor junction, the Fermi levels are separated by eV_a. The quasi-Fermi levels $\mathfrak{F}_n$ and $\mathfrak{F}_p$ are introduced here to describe the minority carrier concentrations under a nonequilibrium situation.

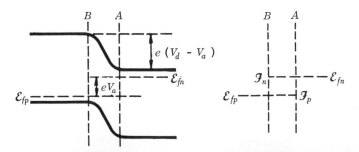

ence in the potential energy $\mathcal{E}_{cp} - \mathcal{E}_{cn}$ now becomes $e(V_d - V_a)$ and that an expression similar to Eq. (5.38) is still valid, then the minority carrier concentration on the p side under the applied voltage is given by

$$n_p = n_n \exp\left[-\frac{e(V_d - V_a)}{kT}\right] = n_{p0} \exp\left(\frac{eV_a}{kT}\right) \qquad (5.42)$$

In the above equation, the majority carrier concentration is of course unchanged, or in other words, $n_n = n_{n0}$. We shall further justify the use of Eq. (5.42) or rather the assumptions leading to Eq. (5.42) in the next chapter when we discuss the p-n junction more thoroughly. Accepting Eq. (5.42), we find that an excess minority carrier concentration $n_1 = n_p - n_{p0}$ is created near the junction such that

$$n_1 = n_p - n_{p0} = n_{p0}\left[\exp\left(\frac{eV_a}{kT}\right) - 1\right] \qquad (5.43)$$

This phenomenon is known as *minority carrier injection*.

There are other means which can create a nonequilibrium situation or an excess carrier concentration. These include the generation of electron-hole pairs by light and by energetic charged particles. Many semiconductor devices among which transistors and photodetectors are good examples have the useful information carried by excess carriers. In other words, an incoming signal creates an excess carrier concentration which in turn causes an output voltage or current. We shall devote the remainder of this section to a discussion of the use of the Fermi level and the charge-neutrality condition under a nonequilibrium situation.

When excess carriers are present, the distribution of electrons among the conduction-band states can still be described by a Fermi function $f_c(\mathcal{E})$:

$$f_c(\mathcal{E}) = \frac{1}{1 + \exp\left[(\mathcal{E} - \mathfrak{F}_n)/kT\right]} \qquad (5.44)$$

such that the concentration of conduction-band electrons is given by

$$n = n_0 + n_1 = \int_{\mathcal{E}_c}^{\infty} N_c(\mathcal{E}) f_c(\mathcal{E}) \, d\mathcal{E} = N_c \exp\left(-\frac{\mathcal{E}_c - \mathfrak{F}_n}{kT}\right) \qquad (5.45)$$

where $N_c(\mathcal{E}) \, d\mathcal{E}$ is the density of conduction-band states in an energy interval $d\mathcal{E}$ and $N_c = 2(2\pi m^* kT/h^2)^{3/2}$. Similar expressions are obtained for valence-band states. The quantity $\mathfrak{F}_n$ in Eq. (5.44) is called the *quasi-Fermi level* for electrons. The use of quasi-Fermi levels is illustrated in the following examples. If it is assumed that in a p-n junction, $\mathfrak{F}_n$ runs from $\mathcal{E}_{fn}$ on the n side straight to the p side as shown in Fig. 5.4, then

$$n_p = N_c \exp\left(-\frac{\mathcal{E}_c - \mathfrak{F}_n}{kT}\right) = n_{p0} \exp\left(\frac{eV_a}{kT}\right) \qquad (5.42a)$$

yielding the same result as Eq. (5.42). As will be shown in the next chapter, the current density J_x of Eq. (5.7) can also be expressed in terms of $\mathfrak{F}_n$.

The concept of quasi-Fermi levels is extremely useful when we discuss the distribution of electrons among various energy states. Suppose that there exists in the forbidden gap a certain kind of available energy states called *traps* which can be in either of two possible states, the occupied and the unoccupied state. The ratio of occupied trap states to the total trap states can be written as

$$\frac{N_t \text{ (occupied)}}{N_t} = f_t(\mathcal{E}_t) = \frac{1}{1 + \exp\left[(\mathcal{E}_t - \mathfrak{F}_t)/kT\right]} \tag{5.46}$$

where $\mathcal{E}_t$ is the trap energy and $\mathfrak{F}_t$ the quasi-Fermi level for traps. The fact that two different quasi-Fermi levels $\mathfrak{F}_n$ and $\mathfrak{F}_t$ are used in Eqs. (5.44) and (5.46) simply means that the distribution of excess electrons between the conduction-band and trap states will not be the same as the thermal-equilibrium distribution of electrons. Quasi-Fermi levels are introduced mainly to simplify calculations under a nonequilibrium situation as the discussion in the next section will testify.

Now let us discuss the charge-neutrality condition. The current density $\mathbf{J}$, the charge density ϱ, and the electric field $\mathbf{E}$ are related to each other through the following equations:

$$\mathbf{\nabla} \cdot \mathbf{E} = \varrho/\epsilon \tag{5.47}$$

$$\mathbf{J} = \sigma\mathbf{E} \tag{5.48}^1$$

$$\mathbf{\nabla} \cdot \mathbf{J} = -\frac{\partial \varrho}{\partial t} \tag{5.49}$$

Substituting Eqs. (5.47) and (5.48) into Eq. (5.49) gives

$$\frac{\partial \varrho}{\partial t} = -\frac{\sigma}{\epsilon}\varrho \tag{5.50}$$

The solution of Eq. (5.50) says that if there exists an unbalanced charge density ϱ, the charge will disappear with a time constant ϵ/σ which is generally called the *dielectric relaxation time*. Take $\epsilon = 16\epsilon_0 = 1.4 \times 10^{-10}$ farad/m, $\sigma = 1$ (Ω-cm)$^{-1}$, $\epsilon/\sigma = 1.4 \times 10^{-12}$ sec. The values of ϵ and σ are chosen for a typical semiconductor such as germanium. Therefore, unless we are dealing with transient

---

[1] Here we want to consider the effect produced by any unbalanced amount of opposite charges alone; hence, the diffusion term is not included in Eq. (5.48). The transient behavior of excess carriers governed by diffusion processes will be discussed in Sec. 5.7. As the calculations (given below and in Sec. 5.7) indicate, the characteristic time involved in a diffusion process is much longer than the dielectric relaxation time. In other words, the current density created by the space-charge field E is normally much larger by several orders of magnitude than the diffusion current density. This justifies the use of Eq. (5.48) in analyzing the effect produced by space charges.

phenomena of time duration shorter than 10^{-12} sec or treating insulators with σ many orders of magnitude smaller than 1 $(\Omega\text{-cm})^{-1}$, the charge-neutrality condition $\rho = 0$ can always be assumed.[2]

From the charge-neutrality condition, we have under thermal equilibrium

$$n_0 + N_a^- + N_t^- = p_0 + N_d^+ \qquad (5.51)$$

and under a nonequilibrium situation

$$n_0 + n_1 + N_a^- + N_t^- + n_t = p_0 + p_1 + N_d^+ \qquad (5.52)$$

where N_t^- represents the thermal-equilibrium density of negatively charged trap states, n_1 and p_1 are the excess electron and hole concentrations respectively, and n_t is the concentration of excess electrons captured by trap states. In Eqs. (5.51) and (5.52) it is assumed that the donor and acceptor states are almost completely ionized and hence not affected. Subtracting Eq. (5.51) from Eq. (5.52) gives

$$n_1 + n_t = p_1 \qquad (5.53)$$

A similar expression $n_1 = p_1 + p_t$ applies to semiconductors having hole traps. In using Eq. (5.53), we must distinguish two types of traps.[3] A *temporary storage trap* captures only one type of carrier, and hence n_t or p_t is nonzero. A *recombination trap* acts as an intermediary for the recombination of excess carriers by simultaneously capturing an electron from the conduction band and a hole from the valence band. Hence, for materials possessing recombination centers only, $n_1 + n_t = p_1 + p_t$ and $n_t = p_t$, yielding

$$n_1 = p_1 \qquad (5.54)$$

Of course, Eq. (5.54) also applies to a situation in which no traps whatsoever are involved. A more detailed examination of the validity of Eq. (5.54) will be given in a later section.

5.5 CARRIER RECOMBINATION PROCESSES

There are two basic processes by which electrons and holes may recombine with each other. In the first process, electrons in the con-

---

[2] The charge-neutrality condition does not apply, of course, to the space-charge region of a semiconductor junction such as the one shown as region AB in Fig. 5.3b. In semi-insulating materials such as high-resistivity CdS, the charge-neutrality condition can also be violated for a time short compared with the dielectric relaxation time.

[3] The definition adopted here for temporary storage traps and recombination centers is highly idealized for ease of discussion. In materials such as CdS and other II-VI compounds where traps play a very important part, the demarcation line between the two species may not be so clear-cut; hence, the treatment given below should be modified accordingly. For a discussion of the situation in materials literally infested with traps, the reader is referred to R. H. Bube, "Photoconductivity of Solids," John Wiley & Sons, Inc., New York, 1960.

duction band simply makes a direct transition to the valence band. In the second process, electrons and holes recombine through some intermediary states known as recombination centers. These recombination centers are usually lattice imperfections of some sort. These two processes are illustrated in Fig. 5.5. It is clear that the first process is an inherent property of the semiconductor itself while the second process depends very much on the nature of the lattice imperfections. Therefore, we call the lifetimes resulting from the first and second recombination processes respectively the intrinsic and extrinsic lifetime.

The intrinsic recombination process. We shall discuss the intrinsic recombination process first. Let $\langle P \rangle$ be an average probability per unit time that an electron in the conduction band makes a transition to a vacant state in the valence band. Since the recombination process involves both the occupied states in the conduction band and the vacant states in the valence band, the total rate of recombination is given by

$$R = \langle P \rangle np \tag{5.55}$$

as for any bimolecular process. However, we must not forget that there is a constant generation of carriers. Under thermal equilibrium, the generation rate G must be equal to the recombination rate R, that is $G = \langle P \rangle n_0 p_0$ where n_0 and p_0 are respectively the thermal equilibrium electron and hole concentrations. The net rate of recombination, U, is equal to

$$U = R - G = \langle P \rangle (np - n_0 p_0) \tag{5.56}$$

If n_1 and p_1 are the excess carrier concentrations such that $n = n_0 + n_1$ and $p = p_0 + p_1$, then the charge-neutrality condition

Figure 5.5

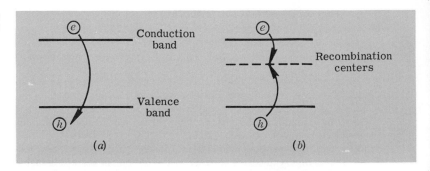

Schematic illustration showing two basic recombination processes: (a) the intrinsic process which is an inherent property of a given semiconductor, and (b) the extrinsic process which depends on the nature of the recombination centers created by certain kinds of lattice imperfections.

requires that $n_1 = p_1$. Under the condition that n_1, $p_1 \ll (n_0 + p_0)$, Eq. (5.56) becomes

$$\frac{dn_1}{dt} = -U = -\langle P \rangle (p_0 + n_0) n_1 \qquad (5.57)$$

In the macroscopic description of the recombination process, a *lifetime* τ is often so defined that

$$\frac{dn_1}{dt} = -\frac{n_1}{\tau} \qquad (5.58)^4$$

A comparison of Eq. (5.58) with Eq. (5.57) gives

$$\tau = \frac{1}{\langle P \rangle (n_0 + p_0)} \qquad (5.59)$$

Since a detailed discussion on the quantum-mechanical treatment of the intrinsic recombination process will be made in Chapter 11, only a brief outline of the process will be presented. First, a distinction must be made between a direct-gap and an indirect-gap semiconductor. As shown in Fig. 5.6a and b, the recombination processes involve an emission of electromagnetic radiation. However, the light emission process is subject to the quantum-mechanical selection rule that the wave vectors of electrons in their initial and final states must be conserved. In a direct-gap semiconductor, conduction-band electrons can make direct quantum-mechanical transitions to vacant valence-band states, resulting in an emission of light shown as process II in Fig. 5.6a. In an indirect-gap semiconductor, because

---

4 To save confusion, we use τ to denote the lifetime of excess carriers as distinct from τ used previously for the relaxation time of free carriers.

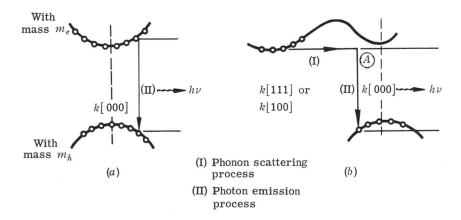

(I) Phonon scattering process

(II) Photon emission process

Figure 5.6 *Diagram showing the intrinsic recombination process (a) in a direct-gap semiconductor and (b) in an indirect-gap semiconductor.*

the conduction-band minima and valence-band maxima are located at different values of k in the reciprocal lattice, electrons must be first scattered by lattice vibrations to an intermediate state A having a proper value of k (process I) and then make a transition from A to a vacant valence-band state (process II). In view of the above discussion, the recombination processes presented in Fig. 5.6a and b are respectively called the direct and indirect radiative recombination processes.

Note that if n_1 or p_1 is comparable to or greater than $(n_0 + p_0)$, τ decreases with increasing n_1 and p_1 according to Eqs. (5.56) and (5.58). The decrease, however, can not continue indefinitely. The average probability $\langle P \rangle$ in Eq. (5.56) can be expressed in terms of a quantum-mechanically calculable quantity P_{cv} as

$$\langle P \rangle = \frac{1}{p_0 n_0} \sum P_{cv} f_c(\mathcal{E})[1 - f_v(\mathcal{E})] \qquad (5.60)$$

where P_{cv} is the transition probability per unit time per conduction- and valence-band state, $f_c(\mathcal{E})$ and $f_v(\mathcal{E})$ are the respective Fermi functions and the summation is over all possible conduction- and valence-band states. An ultimate situation is reached if for every occupied conduction-band state, there is a corresponding, vacant valence-band state. This corresponds to the situation in which n and p in Eq. (5.56) are so large that both the conduction and valence bands become degenerate. Under such circumstances, the lifetime τ reaches an ultimate lower limit. For the direct radiative recombination process of Fig. 5.6a, this lower limit is called the *spontaneous lifetime* and is of the order of 10^{-10} sec in typical semiconductors. This value is shorter than a typical value of 10^{-8} sec for atomic transitions on account of the differences in dielectric constants and effective masses for electrons in the crystal lattice and in the isolated atom.

Now let us make some estimate of the intrinsic lifetime of excess carriers. In direct-gap semiconductors, if we take the degenerate carrier concentrations as $N_c = N_v = 10^{18}/\text{cm}^3$, then at $(n_0 + p_0) = 10^{14}/\text{cm}^3$, the value of τ is estimated from Eq. (5.59) as $10^{-10} N_c/(n_0 + p_0)$ or 10^{-6} sec. The situation in indirect-gap semiconductors is quite different. Because the indirect radiative recombination process is a two-step process, the probability of its occurrence is greatly reduced. Using the optical absorption data, Van Roosbroeck and Shockley[5] and later Dumke[6] have concluded that the value of τ is of the order of 0.5 sec in intrinsic germanium. Similar estimates have also been made for silicon. Obviously, the radiative recombination process does not play a very effective role in the decay of excess carriers in indirect-gap semiconductors.

[5] W. Van Roosbroeck and W. Shockley, *Phys. Rev.*, vol. 94, p. 1558, 1954.
[6] W. P. Dumke, *Phys. Rev.*, vol. 105, p. 139, 1957.

The extrinsic recombination process. Next, we shall discuss the recombination process through lattice imperfections, mostly due to impurities having energy levels deep in the forbidden gap. The model is originally due to Shockley and Read[7] and independently proposed by Hall.[8] As shown in Fig. 5.7, a recombination center or trap can be in either of two possible states, the occupied or the unoccupied. In its unoccupied state, it is ready to receive an electron from the conduction band (process 1) or the valence band (process 4). In its occupied state, it can give its electron to either the conduction band (process 2) or the valence band (process 3). For process 1, the rate of capturing an electron is proportional to the concentration of the unoccupied trap states and also to that of the occupied conduction-band states, hence the rate can be written as

$$c_n(\mathcal{E})N_t(1 - f_t)f_c(\mathcal{E})N_c(\mathcal{E})\, d\mathcal{E} \tag{5.61}$$

where N_t is the concentration of recombination traps, f_t and $f_c(\mathcal{E})$ are, respectively, the Fermi functions of the traps and the conduction-band states, $N_c(\mathcal{E})\, d\mathcal{E}$ is the density of conduction band states within an energy interval $d\mathcal{E}$, and $c_n(\mathcal{E})$ is a proportionality constant. A similar expression is obtained for process 2.

The net rate of capturing electrons from the conduction band is, therefore, equal to

$$dU_{cn} = N_t N_c(\mathcal{E})[c_n(\mathcal{E})(1 - f_t)f_c - e_n(\mathcal{E})(1 - f_c)f_t]\, d\mathcal{E} \tag{5.62}$$

In the above equation, the second term in the bracket comes from

[7] W. Shockley and W. T. Read, *Phys. Rev.*, vol. 87, p. 835, 1952.
[8] R. N. Hall, *Phys. Rev.*, vol. 83, p. 228, 1951, and vol. 87, p. 387, 1952.

Figure 5.7

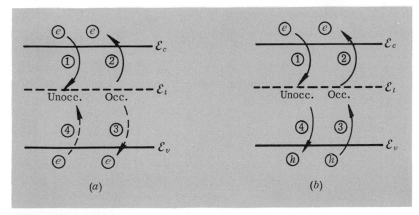

Extrinsic recombination processes through traps as proposed by Shockley and Read and independently by Hall: (a) in terms of conduction-band electrons and valence-band electrons and (b) in terms of conduction-band electrons and valence-band holes.

process 2 or the emission process for which $e_n(\mathcal{E})$ is the proportionality constant. Under thermal equilibrium, $dU_{cn} = 0$ as required by detailed balancing. Thus, Eq. (5.62) together with Eqs. (5.44) and (5.46) gives

$$\frac{e_n}{c_n} = \left(\frac{f_c}{1 - f_c}\right)_0 \left(\frac{1 - f_t}{f_t}\right)_0 = \exp\left(-\frac{\mathcal{E} - \mathcal{E}_t}{kT}\right) \qquad (5.63)$$

by noting that $(\mathfrak{F}_n)_0 = (\mathfrak{F}_t)_0 = \mathcal{E}_f$ under the thermal-equilibrium situation as indicated by the subscript 0.

When excess carriers are present, dU_{cn} of Eq. (5.62) is no longer zero. This is a direct consequence of the fact that the quasi-Fermi levels $\mathfrak{F}_n$ and $\mathfrak{F}_t$ in Eqs. (5.44) and (5.46) are different under a non-equilibrium situation. Substituting the value of e_n from Eq. (5.63) and the value of $(1 - f_c)$ from Eq. (5.44) into Eq. (5.62), we have

$$U_{cn} = \left[N_t(1 - f_t) - N_t f_t \exp\left(-\frac{\mathfrak{F}_n - \mathcal{E}_t}{kT}\right)\right] \int_{\mathcal{E}_c}^{\infty} c_n(\mathcal{E}) N_c(\mathcal{E}) f_c(\mathcal{E}) \, d\mathcal{E}$$

$$(5.64)$$

The above equation can be greatly simplified if the following substitutions are made:

$$nC_n = \int_{\mathcal{E}_c}^{\infty} c_n(\mathcal{E}) N_c(\mathcal{E}) f_c(\mathcal{E}) d\mathcal{E} \qquad (5.65)$$

$$n' = n \exp\left(-\frac{\mathfrak{F}_n - \mathcal{E}_t}{kT}\right) = N_c \exp\left(-\frac{\mathcal{E}_c - \mathcal{E}_t}{kT}\right) \qquad (5.66)$$

The quantity C_n can be interpreted as the average of $c_n(\mathcal{E})$ over the conduction-band states while n' is the concentration of conduction-band electrons that would exist if the Fermi level coincided with the trap energy $\mathcal{E}_t$. In terms of n' and C_n, Eq. (5.64) becomes

$$U_{cn} = C_n N_t (1 - f_t) n - C_n N_t f_t n' \qquad (5.67)$$

An entirely similar expression is obtained by considering hole capture by the trap, processes 3 and 4 of Fig. 5.7b, as follows:

$$U_{cp} = C_p N_t f_t p - C_p N_t (1 - f_t) p' \qquad (5.68)$$

where the quantity p' is defined as

$$p' = p \exp\left(-\frac{\mathcal{E}_t - \mathfrak{F}_p}{kT}\right) = N_v \exp\left(-\frac{\mathcal{E}_t - \mathcal{E}_v}{kT}\right) \qquad (5.69)$$

In the steady state, the net rate of capture of electrons must be equal to that of holes. Equating $U_{cn} = U_{cp}$ gives

$$f_t = \frac{C_n n + C_p p'}{(p + p')C_p + (n + n')C_n} \qquad (5.70)$$

Substituting Eq. (5.70) back into either Eq. (5.67) or Eq. (5.68), we obtain

$$U = U_{cn} = U_{cp} = \frac{N_t C_n C_p (pn - p'n')}{C_p(p + p') + C_n(n + n')} \tag{5.71}$$

Following the same procedure as outlined by Eqs. (5.56), (5.57), and (5.59), and noting that $p'n' = p_0 n_0$, we find

$$\tau = \frac{\tau_n(p_0 + p') + \tau_p(n_0 + n')}{p_0 + n_0} \tag{5.72}$$

under the condition that the excess carrier concentrations n_1 and p_1 are much smaller than the equilibrium carrier concentrations n_0 and p_0. In the above equation, $\tau_n = 1/(C_n N_t)$ and $\tau_p = 1/(C_p N_t)$.

In Fig. 5.8, the general behavior of $\log \tau$ is plotted as a function of the thermal-equilibrium Fermi level $\mathcal{E}_f$ for a given trap density N_t and trap energy $\mathcal{E}_t$. Since p_0 and n_0 are functions of $\mathcal{E}_f$ and so is the resistivity of a given sample, Fig. 5.8 is in effect a plot of $\log \tau$ against resistivity. When $\mathcal{E}_f$ is very close to $\mathcal{E}_v$, the sample is almost degenerate p-type and p_0 is the dominating term in Eq. (5.72). Thus, we have $\tau = \tau_n = 1/(C_n N_t)$. Physically, the sample is flooded with holes so the lifetime is controlled by the ability of the trap to capture electrons from the conduction band. The opposite is true when $\mathcal{E}_f$ is close to $\mathcal{E}_c$. If $\mathcal{E}_f$ gradually moves away from $\mathcal{E}_v$, soon the term $\tau_p n'$ catches up with $\tau_n p_0$ and hence τ can be approximated by $\tau_p n'/p_0$. The rise of τ in the center portion of Fig. 5.8 is due to the scarcity of free carriers n_0 and p_0. Figure 5.8 also shows that $C_n \gg C_p$. In other

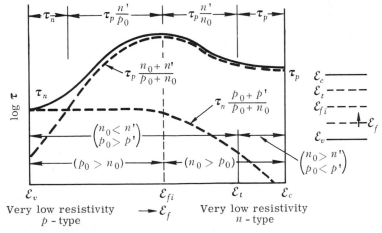

Figure 5.8

Variation of the lifetime τ of excess carriers as a function of the Fermi level $\mathcal{E}_f$ according to Eq. (5.72). The trap energy is taken to be in the upper half of the forbidden gap and C_n is assumed to be much larger than C_p.

words, it is much easier for traps to capture electrons than holes. Therefore, in the middle range, the lifetime is limited by the ability of traps to capture holes as indicated by the $\tau_p n'$ dependence of τ in Fig. 5.8.

Not only the general shape of the curve shown in Fig. 5.8 has been borne out by experiments but also the temperature dependence of τ has been verified. From the activation energy of τ, the energy of the recombination centers can be found and in turn the kind of impurities can be identified. Of the impurities investigated, copper and nickel are notoriously known to be lifetime shorteners in germanium. Not only do they readily capture electrons and holes but these metals also diffuse rapidly and have fairly large solid solubilities in germanium. By careful preparation and purification, germanium samples with τ of the order of several milliseconds can be obtained. However, the value is still several orders of magnitude shorter than the indirect radiative recombination lifetime. Therefore, it is safe to say that in indirect-gap semiconductors, the impurity-caused or extrinsic recombination process controls the lifetime of free carriers.

We should mention that the extrinsic recombination process generally does not result in any appreciable emission of electromagnetic radiation in contrast to the radiative recombination process discussed earlier. It is the radiative recombination process which is utilized in the production of coherent electromagnetic radiation in a semiconductor laser to be discussed in Chapter 12. The reader may ask whether the nonradiative recombination process will compete with and thus impede the radiative recombination process. The answer may be stated as follows. In a semiconductor laser, not only heavily doped materials are used but also heavy injection is needed. Under such circumstances, the direct radiative recombination lifetime almost approaches the spontaneous radiative lifetime. Fortunately, the nonradiative recombination lifetime as given by Eq. (5.72) becomes equal to $(\tau_n + \tau_p)$ as the excess carrier concentrations n_1 and p_1 approach infinity. For impurities not to impede laser action, therefore, N_t must be kept low such that $(\tau_n + \tau_p)$ should be longer than the spontaneous radiative lifetime.

5.6 THE CONTINUITY EQUATION AND THE TIME-DEPENDENT DIFFUSION EQUATION

In many applications of semiconductor devices and measurements of transport properties in semiconductors, electric pulses and time-varying signals are used. It is the purpose of the present discussion to develop a set of general equations which are capable of describing the behavior of free carriers under time-varying as well as d-c (constant) excitations. Since most of our past discussions started with electrons, this time we shall speak first of holes. The hole-

Figure
5.9

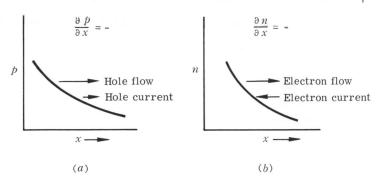

$$\frac{\partial p}{\partial x} = -$$

$$\frac{\partial n}{\partial x} = -$$

p

Hole flow
Hole current

n

Electron flow
Electron current

$x \longrightarrow$

$x \longrightarrow$

(a)

(b)

Diagrams showing that (a) a negative hole-concentration gradient constitutes a positive hole current but (b) a negative electron-concentration gradient constitutes a negative electron current.

current density, similar to Eq. (5.7), is given by

$$J_h = -eD_p \frac{\partial p}{\partial x} + e\mu_p p E_x \tag{5.73}$$

The negative sign in front of the diffusion term is due to the fact that a negative $\partial p/\partial x$ constitutes a positive hole current. A comparison with the electron diffusion current is made in Fig. 5.9.

The equation which relates the time variation of charges with currents is the *continuity equation*. Consider a one-dimensional flow of current as shown in Fig. 5.10. The net charge carried by holes entering a box of length dx and cross-sectional area A in a time interval dt is given by

$$A[J_h(x) -- J_h(x + dx)] \, dt = - \frac{\partial J_h}{\partial x} \, dx \, A \, dt \tag{5.74}$$

Therefore, the net increase in the number of holes inside the box is equal to that supplied by the current J_h minus that lost through recombination, or mathematically

$$\frac{\partial p}{\partial t} = - \frac{1}{e} \frac{\partial J_h}{\partial x} - \frac{p_1}{\tau} \tag{5.75}$$

In the above equation, p_1 is the concentration of excess holes and

Figure
5.10

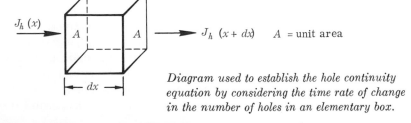

$J_h(x)$

A

A

$J_h(x + dx)$ A = unit area

dx

Diagram used to establish the hole continuity equation by considering the time rate of change in the number of holes in an elementary box.

$p = p_0 + p_1$, p_0 being the equilibrium hole concentration. Equation (5.75) is known as the continuity equation. In cases where no generation or recombination of carriers is considered and where no excess carriers are involved, the second term on the right-hand side of Eq. (5.75) is zero. In this latter form, the continuity equation appears in many other places, for example in the electromagnetic theory. In semiconductors, however, the recombination term must be included whenever excess carriers are involved.

Substituting Eq. (5.73) into (5.75) gives

$$\frac{\partial p}{\partial t} = D_p \frac{\partial^2 p}{\partial x^2} - \frac{\partial}{\partial x}(\mu_p p E_x) - \frac{p_1}{\tau} \tag{5.76}$$

A similar equation is also obtained for electrons, which reads

$$\frac{\partial n}{\partial t} = D_n \frac{\partial^2 n}{\partial x^2} + \frac{\partial}{\partial x}(\mu_n n E_x) - \frac{n_1}{\tau} \tag{5.77}$$

Equations (5.76) and (5.77) are the time-dependent diffusion equations for holes and electrons respectively. To help simplify the above two equations, the following observations are made. For most applications, the thermal-equilibrium concentrations n_0 and p_0 are independent of x and t; hence n and p can be replaced by n_1 and p_1 respectively in terms involving differentiations with respect to x or t. Thus, Eqs. (5.76) and (5.77), respectively, become

$$\frac{\partial p_1}{\partial t} = D_p \frac{\partial^2 p_1}{\partial x^2} - \mu_p p \frac{\partial E_x}{\partial x} - \mu_p E_x \frac{\partial p_1}{\partial x} - \frac{p_1}{\tau} \tag{5.78}$$

$$\frac{\partial n_1}{\partial t} = D_n \frac{\partial^2 n_1}{\partial x^2} + \mu_n n \frac{\partial E_x}{\partial x} + \mu_n E_x \frac{\partial n_1}{\partial x} - \frac{n_1}{\tau} \tag{5.79}$$

From the above two equations and Poisson's equation

$$\frac{\partial E_x}{\partial x} = \frac{e(p_1 - n_1)}{\epsilon} \tag{5.80}$$

we hope to obtain a partial differential equation in n_1 (or p_1) by eliminating p_1 (or n_1) and E_x.

It is instructive as well as helpful that we reexamine the charge-neutrality condition. For the sake of generality, we let

$$p_1 = n_2 + n_3 = n_2(1 + \delta)$$
$$n_1 = n_2 - n_3 = n_2(1 - \delta) \tag{5.81}$$

In Eq. (5.81), n_2 and n_3 may be called respectively the charge-neutrality part and the anti-charge-neutrality part of the excess carrier concentration. If the charge-neutrality condition is observed, $n_3 = 0$ and $\delta = 0$. Therefore, the quantity δ can be interpreted as a parameter which measures the departure from strict observance of the charge-neutrality condition. In terms of δ, the second terms in

Eqs. (5.78) and (5.79), respectively, become

$$\mu_p p \frac{\partial E_x}{\partial x} = \frac{e\mu_p p}{\epsilon} \frac{2\delta}{1+\delta} p_1 \tag{5.82a}$$

$$\mu_n n \frac{\partial E_x}{\partial x} = \frac{e\mu_n n}{\epsilon} \frac{2\delta}{1-\delta} n_1 \tag{5.82b}$$

For an extrinsic sample, say n-type, the ratio $(e\mu_n n/\epsilon)$ can be approximated by σ/ϵ. As pointed out in Sec. 5.4, the value of σ/ϵ is of the order of 10^{12} sec^{-1} in a 1 Ω-cm Ge sample. For comparison, the value of $1/\tau$ as pointed out in the previous section ranges from 10^7 sec^{-1} in low grade materials to 10^3 sec^{-1} in high quality materials. The argument again demands that the charge-neutrality condition should be observed or that the quantity δ in Eq. (5.81) should be extremely small. Otherwise, the term represented in Eq. (5.82) would completely dominate over the other terms in Eqs. (5.78) and (5.79), and hence no interesting phenomena would be expected from diffusion of minority carriers.

According to the above discussion, p_1 and n_1 in Eqs. (5.78) and (5.79) can be replaced by n_2 since δ is expected to be small. However, the term involving $(\partial E_x/\partial x)$ should still be retained because a very small deviation from the charge-neutrality condition makes the term comparable in magnitude to other terms in Eqs. (5.78) and (5.79). Eliminating the $(\partial E_x/\partial x)$ term from these two equations gives

$$\frac{\partial n_2}{\partial t} = D' \frac{\partial^2 n_2}{\partial x^2} + \mu' E_x \frac{\partial n_2}{\partial x} - \frac{n_2}{\tau} \tag{5.83}$$

In the above equation, the quantities D' and μ' are called the effective diffusion constant and mobility of excess carriers and are related to D_n, D_p, μ_n, and μ_p as follows:

$$D' = \frac{D_n D_p (n+p)}{D_p p + D_n n} \tag{5.84}$$

$$\mu' = \frac{\mu_n \mu_p (p-n)}{\mu_p p + \mu_n n} \tag{5.85}$$

Since D' and μ' are functions of n_1 and p_1, Eq. (5.83) is nonlinear. However, if the thermal-equilibrium majority-carrier concentration is much larger than both the excess carrier concentration and the thermal-equilibrium minority-carrier concentration, for example, $p_0 \gg n_0$, n_1, and p_1, then Eqs. (5.84) and (5.85) can be approximated by $D' = D_n$ and $\mu' = \mu_n$. Therefore, for extrinsic materials under small-injection conditions, Eq. (5.83) becomes linearized and the coefficients D' and μ' respectively take the values of the diffusion constant and mobility of minority carriers.

Since $n_2 \gg n_3$, Eq. (5.83) is sufficient in predicting the movement of excess carriers. We may obtain a measure of the degree of violation of the charge-neutrality condition or a measure of n_3/n_2 by subtracting Eq. (5.79) from Eq. (5.78):

$$\frac{\partial n_3}{\partial t} + e(\mu_n n + \mu_p p)\frac{n_3}{\epsilon} + \frac{n_3}{\tau} = (D_p - D_n)\frac{\partial^2 n_2}{\partial x^2}$$

$$- (\mu_n + \mu_p)E_x\frac{\partial n_2}{\partial x} \quad (5.86)$$

Knowing the solution of n_2 from Eq. (5.83), the value of n_3 can be calculated from Eq. (5.86). Again it can be seen that on the left-hand side of Eq. (5.86), the second term will be the dominating term for pulses of nanosecond or longer duration. Furthermore, the value of terms on the right-hand side of Eq. (5.86) is of the order of $\partial n_2/\partial t$ or n_2/τ, whichever is larger, as seen from Eq. (5.83). Therefore, the ratio of n_3/n_2 is of the order of t_1/t_2 where $t_1 (= \epsilon/\sigma)$ is the dielectric relaxation time and t_2 is the pulse duration or the lifetime, whichever is smaller. Since the charge-neutrality condition plays such an important role in determining the solution of the time-dependent diffusion equation, a reexamination of its validity from a different viewpoint is instructive even though it leads to the same conclusion.

5.7 DISCUSSION OF THE SOLUTION OF THE TIME-DEPENDENT DIFFUSION EQUATION

Since the process of finding the steady-state solution of the diffusion equation is rather straightforward, the task is left to the reader as a problem. Here we shall devote our discussion to transient solutions. The most important solution of the time-dependent diffusion equation is the response to a unit pulse. For definiteness, we shall treat an extrinsic n-type semiconductor under small-injection conditions. The one-dimensional diffusion equation takes the following form:

$$\frac{\partial p_1}{\partial t} = D_p\frac{\partial^2 p_1}{\partial x^2} - \mu_p E_x\frac{\partial p_1}{\partial x} - \frac{p_1}{\tau} + f(x,t) \quad (5.87)$$

where $f(x,t)$ is the source function which represents the injection of excess carriers. By using a moving coordinate system such that

$$x_1 = x - \mu_p E_x t \quad \text{and} \quad t_1 = t \quad (5.88)$$

and letting $\qquad p_1 = p' \exp\left(-\frac{t}{\tau}\right) \qquad (5.89)$

Eq. (5.87) can be transformed into a standard form which is well-known in the theory of heat conduction:

$$\frac{\partial p'}{\partial t_1} = D_p\frac{\partial^2 p'}{\partial x_1{}^2} + f(x,t) \quad (5.90)$$

For a unit impulse function, $f(x,t) = \delta(x_1)\,\delta(t_1)$ in Eq. (5.90) where $\delta(x_1)$ and $\delta(t_1)$ are the Dirac delta functions. A simple way of solving Eq. (5.90) is through Laplace transformation. Let

$$\bar{p} = \mathcal{L}p' = \int_0^\infty \exp\,(-st_1)p'\,dt_1 \tag{5.91}$$

After taking the Laplace transform, Eq. (5.90) becomes

$$D_p \frac{\partial^2 \bar{p}}{\partial x_1{}^2} = s\bar{p} - p_1(t=0) - \delta(x_1) \tag{5.92}$$

If Eq. (5.92) is integrated with respect to x, we find

$$D_p \frac{\partial \bar{p}}{\partial x_1}\bigg|_{x_1=0+} - D_p \frac{\partial \bar{p}}{\partial x_1}\bigg|_{x_1=0-} = -\int_{0-}^{0+} \delta(x_1)\,dx_1 = -1 \tag{5.93}$$

The symbols $0+$ and $0-$ represent, respectively, points just right and just left of the source. In other words, because of the source, there is a discontinuity in the derivative as given by Eq. (5.93). For the present discussion, we shall take the simplest case of no initial excess carriers or $p_1(t=0) = 0$, then the solution of Eq. (5.92) can be written as

$$\bar{p} = B \exp\,(-\alpha x_1) + C \exp\,(\alpha x_1) \tag{5.94}$$

where $\alpha = (s/D_p)^{1/2}$.

If the sample is infinite in extent, $\bar{p}$ must be zero at $x_1 = \pm\infty$. That means,

$$\bar{p} = B \exp\,(-\alpha x_1) \qquad \text{for } x_1 > 0 \tag{5.95}$$
$$\bar{p} = C \exp\,(\alpha x_1) \qquad \text{for } x_1 < 0 \tag{5.96}$$

Applying Eq. (5.93) to the solution at $x_1 = 0$ and using the boundary condition $\bar{p}\,|_{0+} = \bar{p}\,|_{0-}$, we find $B = C = (2\alpha D_p)^{-1}$. Taking the inverse Laplace transform of $\bar{p}$ and expressing $\mathcal{L}^{-1}\bar{p}$ in terms of the original dependent and independent variables p_1, x, and t, we obtain, except for the $\exp\,(-t/\tau)$ term, the famous Gaussian distribution

$$p_1(x,t) = \frac{\exp\,(-t/\tau)}{(4\pi D_p t)^{1/2}} \exp\left[-\frac{(x - \mu_p E_x t)^2}{4D_p t}\right] \tag{5.97}$$

As pointed out earlier, Eq. (5.83) and also Eq. (5.87) are based on a one-dimensional flow of excess carriers; therefore, Eq. (5.97) applies only to a long sample having a uniform distribution of excess carriers over the transverse plane. We also should mention that a factor N/A should appear in front on the right-hand side of Eq. (5.97) if the source injects N holes over a cross-sectional area A at $x = 0$ and $t = 0$ into the sample.

Another case of importance is one in which the source at $x = 0$ is a step function in time. In other words, the concentration of excess carriers, $p_1(x,t)$, is everywhere zero for $t < 0$ and becomes unity at $x = 0$ for $t > 0$. For simplicity, it is assumed that the electric field is negligibly small and the lifetime is infinitely long so the second and third terms on the right-hand side of Eq. (5.87) can be ignored. Noting that the Laplace transform of unity is $1/s$, the solution of $\bar{p}$ becomes

$$\bar{p} = \begin{cases} \dfrac{\exp\,(-\alpha x)}{s} & \text{for } x > 0 \\[2mm] \dfrac{\exp\,(\alpha x)}{s} & \text{for } x < 0 \end{cases} \tag{5.98}$$

The inverse Laplace transform of $\bar{p}$ gives

$$p_1(x,t) = \operatorname{erfc}\frac{x}{2\sqrt{Dt}} = \frac{2}{\sqrt{\pi}}\int_{\frac{x}{2\sqrt{Dt}}}^{\infty} \exp\,(-\xi^2)\,d\xi \tag{5.99}$$

An obvious extension of the above solution is to the case where the source function at $x = 0$ is a general function $\phi(t)$ of time instead of being a step function. To use the method of superposition, the time scale is divided into small intervals of $\Delta\lambda$ each as shown in Fig. 5.11a, then the response to the function $\phi(\lambda)$ of Fig. 5.11b can be considered as the difference of the responses to the source functions of Fig. 5.11c and d. Therefore, the response to the source function of Fig. 5.11b is given by $\Delta u(x,t,\lambda)$:

$$\begin{aligned} \Delta u(x,t,\lambda) &= \phi(\lambda)\left[\operatorname{erfc}\frac{x}{2\sqrt{D(t-\lambda)}} - \operatorname{erfc}\frac{x}{2\sqrt{D(t-\lambda-\Delta\lambda)}}\right] \\[2mm] &= \phi(\lambda)\,\Delta\lambda\,\frac{\partial}{\partial\lambda}\operatorname{erfc}\frac{x}{2\sqrt{D(t-\lambda)}} \end{aligned} \tag{5.100}$$

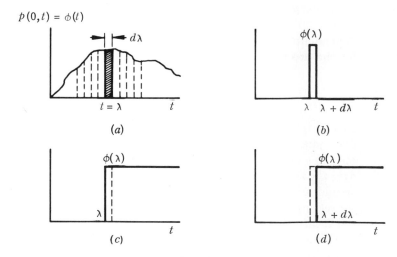

(a)

(b)

Figure 5.11

(c)

(d)

Diagrams showing how the response to an arbitrary excitation can be obtained from the response to a step function by applying the method of superposition.

The total response $p_1(x,t)$ is equal to

$$p_1(x,t) = \int_0^t [du(x,t,\lambda)/d\lambda]\, d\lambda$$

$$= \int_0^t \frac{x}{2\sqrt{\pi D(t-\lambda)^3}} \exp\left[-\frac{x^2}{4D(t-\lambda)}\right] \phi(\lambda)\, d\lambda \quad (5.101)$$

We should point out that the equation of heat conduction has been treated thoroughly by Carslaw and Jaeger.[9] The time-dependent diffusion equation differs from the equation of heat conduction in that the concentration of excess carriers appears as the dependent variable in the former while the temperature appears in the latter. The continuity equation in the charge density in the former corresponds to the continuity equation in the heat-energy flux in the latter. We also realize that the examples given above are very much idealized. For example, the assumption that the concentration of excess carriers is uniform over the transverse plane may not be a valid one if excess carriers recombine at the surface. The reader is referred to the book by Carslaw and Jaeger for treatments of cases involving different boundary conditions as well as for discussions of the equation in different coordinate systems.

So far as the motion of excess carriers in semiconductors is concerned, solutions of the types given by Eqs. (5.97), (5.99), and (5.101) are important for two reasons. The geometry of a semiconductor junction device is such that a rigorous three-dimensional solution of the problem is tedious and long. Therefore, simplifying assumptions must be made in order to make the solution physically interpretable and mathematically manageable. The prototype solutions presented in this section offer a good starting point for the analysis of real problems. The prototype solution is known as the Green's function, which is unique for a definite set of initial and boundary conditions, that is, a definite set of values for $p_1(x,0)$ and $p_1(x_0,t)$. In a more complicated situation, the actual solution can be expressed in terms of a particular prototype solution which satisfies a well-defined set of initial and boundary conditions. Both Eqs. (5.97) and (5.99) are prototype solutions. In deriving Eq. (5.101) from Eq. (5.99) by the method of superposition, we have further demonstrated the use of the prototype solution [Eq. (5.99)] as a starting point to obtain the solution [Eq. (5.101)] for a more complicated problem. If we use the Laplace transform, Eq. (5.101) can be obtained from Eq. (5.99) by the use of the convolution theorem. The use of Eqs. (5.97), (5.99), and (5.101) will be demonstrated in the next chapter when we discuss the transient behavior of a p-n diode.

[9] H. S. Carslaw and J. C. Jaeger, "Conduction of Heat in Solids," Oxford University Press, Oxford, 1959.

5.8 THE HAYNES-SHOCKLEY EXPERIMENT

In the previous sections, we have discussed the mechanisms governing the flow as well as the decay of excess carriers and have worked out mathematical expressions describing the motion of excess carriers. In the experiment which we are about to describe, we shall demonstrate the actual operation of these mechanisms in accordance with theory. Consider the experimental setup of Fig. 5.12a in which the contacts A and B serve to inject and collect respectively the minority carriers, holes in the present case, and the voltages V_1 and V_2 are used respectively to create a uniform electric field along the sample and to bias the collector contact. With proper modifications, the experimental setup should work for either n-type or p-type samples although an n-type sample is chosen in Fig. 5.12a for the purpose of illustration. In order that the expressions worked out in the previous section can be directly applied to the experiment, the sample must have the shape of a long bar with a uniform cross-section.

The experiment can be described as follows. At time $t = 0$, holes are injected into the bulk sample in the form of a pulse of very short duration as shown by the top trace of Fig. 5.12b. After hole injection occurs, two things happen. Within a time of the order of the dielectric relaxation time an excess electron concentration neutralizes the

Figure 5.12

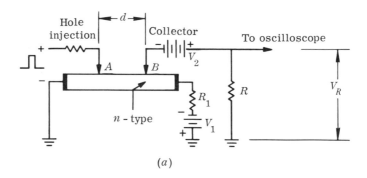

(a)

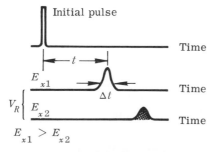

(b)

Haynes-Shockley experiment: (a) the experimental setup; (b) the experimental observation of the arrival of the injected carriers at the collector for different values of the drift field.

minority carrier concentration in the injection region. The increase in the carrier concentration immediately reduces the resistance of the sample, thus creating a voltage across the resistor R_1. However, this voltage is not important to the experiment for reasons that will become clear shortly. The motions of the excess carriers thus created, both minority and majority, are governed by Eqs. (5.78) and (5.79). As pointed out in the discussion following these two equations, the charge-neutrality condition demands that the excess majority carriers move with excess minority carriers. This conclusion may seem incorrect at first because majority and minority carriers carry opposite electric charges. The explanation lies in the fact that excess carriers behave differently from thermal-equilibrium carriers.

We shall recapitulate the physical situation as follows. As indicated by Eq. (5.81), the excess carrier concentrations actually consist of a charge-neutrality part n_2 and an anti-charge-neutrality part n_3. The anti-part n_3 creates a variation in the local electric field as governed by Eq. (5.82). The effect due to $\partial E_x/\partial x$ on n_2 is so large (Probs. 5.11 and 5.12) that it overcomes the influence of E_x in Eq. (5.79). This explains why the excess majority carriers can move in the same direction as the excess minority carriers in so far as the charge-neutrality part of excess carriers is concerned. Since $n_2 \gg n_3$, we are interested in the movement of n_2 but not that of n_3. For the movement of n_2, the excess carriers can be considered as a charge cloud consisting of both minority and majority carriers and this charge cloud moves as a whole. The motion of this charge cloud is dictated by Eq. (5.83) and in the presence of an electric field, the charge cloud drifts in the same direction as minority carriers. With the polarity of the voltage V_1 indicated in Fig. 5.12, the charge cloud drifts toward the right.

The voltage developed across R_1 will last as long as the excess carriers; therefore, it does not give much useful information except that the signal decays as excess carriers recombine. The situation is quite different with the voltage developed across another resistor R. A voltage V_R appears across R only when excess holes in moving down the sample are collected at the contact B, thus creating a current flow through R. The lower two traces of Fig. 5.12b show V_R in relation to the initial pulse on the time scale for two different values of E_x. The time lapse between the initiation of the injection pulse and the arrival of excess carriers at the contact B is a measure of the drift velocity of excess carriers. Simultaneously with the drift motion, excess carriers are also dispersing because of diffusion. This explains why the pulse shown in the lower traces is not as sharp as the initial pulse shown in the top trace. In the following discussion, we shall analyze the experiment in accordance with Eq. (5.97). To satisfy the condition on which Eq. (5.97) is based, the injection contact must be far away from both ends of the sample. The aim of the experiment is to measure μ_p, D_p, and τ in Eq. (5.97). The experiment is generally known as the Haynes-Shockley experiment[10] because of their original work. The setup shown in Fig. 5.12a using pulse techniques is an improved version of the original experiment.

According to Eq. (5.97), the maximum of the hole density occurs at a time t_0 which is approximately given by

$$d = x_B - x_A = \mu_p E_x t_0 \tag{5.102}$$

That means, the maximum moves along the bar with a drift velocity $\mu_p E_x = d/t_0$. Knowing t_0, d, and E_x, the value of μ_p can be calculated from Eq. (5.102). The exact equation for the time of arrival of the maximum is obtained by differentiating $\log p_1(x,t)$ of Eq. (5.97) with respect to t and setting the differential equal to zero:

$$-\frac{1}{\tau_p} - \frac{1}{2t} + \mu_p E_x \frac{d - \mu_p E_x t}{2D_p t} + \frac{(d - \mu_p E_x t)^2}{4D_p t^2} = 0 \tag{5.103}$$

Letting $t = t_0 + t_1$ and realizing that $t_1 < t_0$, we set t equal to t_0 except in the term involving $(d - \mu_p E_x t)$, and further, we neglect the last term in Eq. (5.103). Thus, we obtain

$$t = t_0 - \frac{kT}{eE_x} \frac{1}{\mu_p E_x} \left(1 + \frac{2t_0}{\tau_p}\right) \tag{5.104}$$

where t is the first-order solution for the time of arrival of the maximum.

The diffusion constant D can be measured from the width of the Gaussian distribution function. The signal V_R drops to $1/e$ of its peak value approximately when

$$(d - \mu_p E_x t)^2 = 4D_p t \tag{5.105}$$

If t' and t'' are the two values of time which satisfy Eq. (5.105) and $\Delta t = t' - t''$, then the value of D_p can be approximated by

$$D_p = (\mu_p E_x)^2 \frac{(\Delta t)^2}{16t_0} \tag{5.106}$$

Again, the approximation is reasonably good only when the factor $\exp(-t/\tau_p)$ does not change appreciably over the range Δt. Another factor which may affect the measurement of D_p is the physical size of the collector contact. Theoretically, Eq. (5.105) is true only for a single point. For a collection of collector points, Eq. (5.97) should be summed over a range of x. In other words, the spread in the distance between contacts A and B effectively increases the apparent spread Δt of the signal. To check whether the experiment gives

[10] J. R. Haynes and W. Shockley, *Phys. Rev.*, vol. 81, p. 835, 1951.

genuine results in the mobility and diffusion-constant measurements, the experiment should be repeated for different values of E_x.

It is also possible to obtain the value of the lifetime of excess carriers from the Haynes-Shockley experiment. Note that the Gaussian distribution or Eq. (5.97) has the following property:

$$\int_{-\infty}^{\infty} p_1(x,t) \exp\left(\frac{t}{\tau}\right) dx = \int_{-\infty}^{\infty} \frac{\exp\left[-(x - \mu_p E_x t)^2/4D_p t\right]}{(4\pi D_p t)^{1/2}} dx = 1$$

(5.107)

and the quantity unity is just the amount of carriers initially injected into the sample. Physically, Eq. (5.107) means that had it not been for the recombination process, the amount of excess carriers initially injected into the sample would remain constant. If η is the collector efficiency defined as the ratio of holes collected by the contact B to the total number of holes passing through a transverse plane at B, then for N holes injected into the sample, $N \exp(-t_0/\tau)$ holes will survive the recombination process and of these, $\eta N \exp(-t_0/\tau)$ will be collected by the contact B where t_0 is the approximate drift time between contacts A and B. The area S under the curves of Fig. 5.12b is, of course, directly proportional to the total holes collected or

$$S = \int_{\text{pulse}} V_R \, dt = A\eta N \exp\left(-\frac{t_0}{\tau}\right) = A\eta N \exp\left(-\frac{d}{\mu_p E_x \tau}\right) \quad (5.108)$$

A plot of log S against $(d/\mu_p E_x)$ gives a straight line of slope $1/\tau$.

In concluding this section, we would like to stress the elegance of the Haynes-Shockley experiment. In a single setup, it gives a vivid demonstration of the drift, diffusion, and recombination processes of excess carriers. So far as the actual determination of μ, D, and τ is concerned, however, only the mobility measurement is simple and straightforward enough to yield accurate results. The thing to be careful about is that the electric field should be high enough so that the value of t_0 obtained from Eq. (5.102) is a good approximation. To avoid heating, the voltage V_1 can also be in the form of a pulse. For the measurement of D, extreme care must be taken to ensure that no erroneous effects exist to increase the apparent width of the pulse. The difficulty lies in the fact that it is hard to tell whether the pulse is genuinely Gaussian merely by looking at it. So far as the measurement of τ is concerned, there are other methods which give a direct reading of τ and hence may be preferred over the Haynes-Shockley experiment.

The values of drift mobility measured in germanium by Prince[11] and in silicon by Ludwig and Watters[12] are given in Table 5.1.

[11]   M. B. Prince, *Phys. Rev.*, vol. 92, p. 681, 1953, and vol. 93, p. 1204, 1954.
[12]   G. W. Ludwig and R. L. Watters, *Phys. Rev.*, vol. 101, p. 1699, 1956.

TABLE 5.1   *Drift mobilities in cm²/volt-sec*

Ge	Si
$\mu_n = 3900 \pm 100$ $\mu_p = 1900 \pm 50$ } at 300°K	$\mu_n = 1350 \pm 100$ $\mu_p = 480 \pm 15$ } at 300°K

Since the measurements were made on high-purity samples, for resistivities higher than 10 Ω-cm in germanium and ranging from 19 to 170 Ω-cm in silicon, the mobilities given in the above table are lattice mobilities. In other words the relaxation time τ in the expression of mobility $\mu = e\tau/m^*$ is due to scattering by lattice vibrations. Another point worth mentioning concerns valence-band holes. Since there are heavy-mass and light-mass holes, conceivably two separate pulses with different drift velocities would be observed. This is not found experimentally. The explanation lies in the fact that holes make rapid transitions from the light- to heavy-mass band and vice versa. According to an analysis by Rittner,[13] if the interband transition time is very short compared with the lifetime τ and the drift time t_0 but long compared with the relaxation time τ, then both types of holes propagate and broaden as a single pulse.

5.9   SURFACE STATES AND SURFACE RECOMBINATION VELOCITY

In the previous discussions, we have treated the flow of excess carriers as a one-dimensional problem by assuming that nothing extraordinary happens at the surface of a semiconductor. This oversimplification is a necessary step to the derivation of physically meaningful results from the diffusion equation without getting deeply involved in the mathematics. However, the assumption is not a realistic one. The manufacture of high-performance semiconductor devices with stable electrical characteristics requires not only a careful surface preparation but also a stable ambient for the device. We shall present a physical description of a semiconductor surface followed by a discussion of what may happen at the surface. Examples will be given in the next section to illustrate how surface effects can be taken into account.

It is generally agreed that chemically-etched semiconductor surfaces such as those of germanium or silicon are covered with thin layers of oxides. Because of the mismatch of the crystallographic structures and dimensions at the semiconductor-oxide interface boundary, imperfections in the structure are likely to exist and to create energy levels within the forbidden gap. The extra energy levels created at a semiconductor surface are commonly called *surface*

[13]   E. S. Rittner, *Phys. Rev.*, vol. 101, p. 1291, 1956.

states.[14] There are two different kinds of surface states. The energy states created at the semiconductor-oxide interface boundary are generally called the fast surface states. Since these states are in intimate electrical contacts with the bulk semiconductor, they can reach a state of equilibrium with the bulk within a relatively short period of time (of the order of nanoseconds or less). There is another kind of surface states characteristically different from the fast surface states. These states may be attributed to either chemisorbed ambient ions[15] or imperfections in the oxide. Carrier transfer from such states to the bulk semiconductor has either to overcome a high potential barrier presented by the large energy gap of the oxide or to tunnel through the oxide layer; therefore, such a charge-transfer process involves a long time constant, typically of the order of seconds or longer.

The existence of two types of surface states, *fast* and *slow*, is illustrated schematically in Fig. 5.13. The following is a brief description of the effects which these states will produce on the electrical properties of a semiconductor. First of all, the slow states have a much larger density than the fast states and these slow states generally carry either positive or negative electric charges, depending on the nature of these states, in a way similar to donor and acceptor states. To keep the surface electrically neutral, the semiconductor near the surface has to supply an equal amount of opposite electric charges. As a result, the carrier concentration near the surface is different from that inside the bulk semiconductor. The situation is

14   The existence of surface states was first postulated by J. Bardeen, *Phys. Rev.*, vol. 71, p. 717, 1947.

15   The effect of ambient gas on semiconductor surface was clearly demonstrated in the experiment of W. H. Brattain and J. Bardeen, *Bell System Tech. J.*, vol. 32, p. 1, 1953.

Figure 5.13

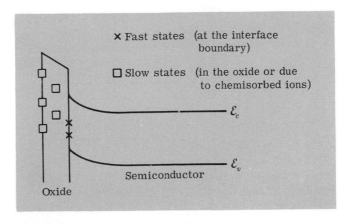

Surface states which exist near the surface of a semiconductor.

illustrated in Fig. 5.14. Since the Fermi-level $\mathcal{E}_f$ stays constant under thermal equilibrium, the electrostatic potential has to change in order to accommodate the change in the carrier concentrations near the surface. To repel electrons and attract holes toward the surface, the electrostatic potential for electrons must bend up near the surface. Under the circumstance, region A near the surface will be positively charged as shown in Fig. 5.14a, and these positive charges are to neutralize the negative charges of the slow states.

Because of the slow states, the carrier concentrations at the surface not only can change but may change so drastically that the surface conductivity may be of the opposite type to the bulk. In other words, if the bulk semiconductor is n-type with $n_0 > p_0$, the hole and electron concentrations p_s and n_s at the semiconductor-oxide interface may have the inequality sign reversed or $p_s > n_s$. The situation is illustrated in Fig. 5.14a in which an *inversion layer* is established with p-type conductivity near the surface and n-type conductivity in the bulk. Another situation may also arise with $n_s > p_s$ and $n_s > n_0$. The transition region from the bulk to the surface, which

Figure 5.14

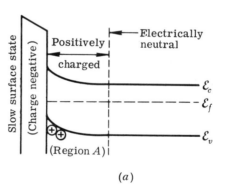

(a)

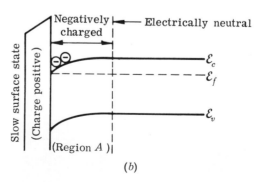

(b)

Potential barrier created at the surface of a semiconductor due to (a) negative charges and (b) positive charges carried by the slow surface states.

retains the same conductivity type but has accumulated more free carriers, is given the name *accumulation layer*. The words inversion and accumulation indicate the state of surface carriers with reference to that of bulk carriers. In summarizing the above discussion, we conclude that the slow states determine the conductivity type of a semiconductor surface. However, a discussion of the effects associated with the surface conductivity will be postponed until the next chapter where surface effects on semiconductor devices will be discussed.

Now we shall turn our attention to the fast states. The reader who has some experience with semiconductor surfaces will have observed that a mechanically roughened surface such as a sandblasted surface lowers the effective lifetime of excess carriers drastically. Undoubtedly, the interface fast surface states play an important role in carrier-recombination processes. Now we shall apply the Shockley-Read model of recombination through traps to the surface recombination process. Figure 5.15a, like Fig. 5.14, shows the energy-band diagram of a typical semiconductor surface and the relevant energy levels. The quantity ϕ_s is the difference in electron volts between the Fermi level $\mathcal{E}_f$ and the intrinsic Fermi level $\mathcal{E}_{fi}$ at the surface. We should point out that the intrinsic Fermi level, defined in Eq. (3.41), is introduced only for operational convenience, and is unrelated to the Fermi level $\mathcal{E}_f$. More specifically, in Fig. 5.15a, the Fermi level must stay constant under thermal equilibrium while the intrinsic Fermi level must change with distance near the surface in order to satisfy Eq. (3.42).

In applying the result obtained in Sec. 5.5 or specifically Eq. (5.71)

Figure 5.15

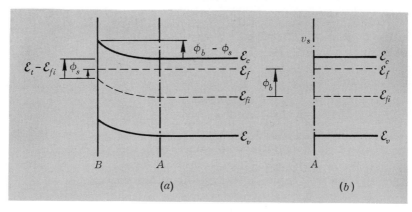

Schematic diagrams indicating how (a) the effect of the fast surface states can be incorporated into (b) the mathematical formulation by a parameter v_s called the surface recombination velocity. The arrows (in diagram a) indicate the directions in which the quantities ϕ_s and $\phi_b - \phi_s$ are measured as positive quantities.

to the surface recombination process, the following points must be made clear. First, the bulk carrier concentrations p and n in Eq. (5.71) must be replaced by the carrier concentrations p_s and n_s defined at the interface boundary. Second, a quasiequilibrium situation is assumed to prevail between electrons at the surface and electrons in the bulk and also between holes at the surface and holes in the bulk; and hence the following relations exist:

$$n_s = n \exp\left(-\frac{\phi_b - \phi_s}{kT}\right) \qquad p_s = p \exp\left(\frac{\phi_b - \phi_s}{kT}\right) \qquad (5.109)$$

on account of the surface barrier $\phi_b - \phi_s$. In terms of p_s and n_s, Eq. (5.71) or the net rate of recombination becomes

$$U = \frac{N_t C_n C_p (p_s n_s - p' n')}{C_p(p_s + p') + C_n(n_s + n')} \qquad (5.110)$$

where n' and p' according to the definition given in Sec. 5.5 are, respectively, given by

$$n' = N_c \exp\left(-\frac{\mathcal{E}_c - \mathcal{E}_t}{kT}\right) = n_i \exp\left(\frac{\mathcal{E}_t - \mathcal{E}_{fi}}{kT}\right)$$
$$p' = N_v \exp\left(-\frac{\mathcal{E}_t - \mathcal{E}_v}{kT}\right) = n_i \exp\left(-\frac{\mathcal{E}_t - \mathcal{E}_{fi}}{kT}\right) \qquad (5.111)$$

and $\mathcal{E}_t$ is the energy of the fast states.

Because of the surface recombination process, the flow of excess carriers along a bar sample discussed in Sec. 5.7 is no longer a one-dimensional problem. In other words, the boundary conditions at the surface must be incorporated into the mathematical formulation. Figure 5.15b is a reproduction of Fig. 5.15a except that the surface region between AB is removed and replaced mathematically by a parameter v_s defined through the relationship

$$U = v_s n_1 = v_s p_1 \qquad (5.112)$$

where $n_1 = p_1$ is the concentration of excess carriers at the boundary A in Fig. 5.15b. Note that N_t in Eq. (5.110) is the density of surface traps and hence it has a dimension of cm^{-2} while N_t in Eq. (5.71) is of cm^{-3}. That means, the quantity U in Eq. (5.112) has a dimension of cm^{-2} sec^{-1} instead of cm^{-3} sec^{-1} in Eq. (5.71). Therefore, the parameter v_s has a dimension of velocity or cm/sec and for that reason, it is called the *surface recombination velocity*.

Noting that from Eq. (5.109), $p_s n_s = pn = (p_0 + n_1)(n_0 + n_1)$ and that from Eq. (5.111), $p'n' = n_i{}^2 = p_0 n_0$, we have

$$p_s n_s - p'n' = (p_0 + n_0 + n_1)n_1 \qquad (5.113)$$

For small injection, the concentrations p_s and n_s can be approximated by their respective equilibrium concentrations p_{s0} and n_{s0}:

$$p_{s0} = N_v \exp\left(-\frac{\mathcal{E}_{fi} + \phi_s - \mathcal{E}_v}{kT}\right) = n_i \exp\left(-\frac{\phi_s}{kT}\right)$$

$$n_{s0} = N_c \exp\left(-\frac{\mathcal{E}_c - \mathcal{E}_{fi} - \phi_s}{kT}\right) = n_i \exp\left(\frac{\phi_s}{kT}\right)$$

(5.114)

Substituting Eqs. (5.111), (5.113), and (5.114) into Eq. (5.110) gives

$$v_s = \frac{U}{n_1}$$

$$= \frac{N_t C(p_0 + n_0)/2n_t}{\cosh\left[(\mathcal{E}_t - \mathcal{E}_{fi} - \phi_0)/kT\right] + \cosh\left[(\phi_s - \phi_0)/kT\right]} \quad \text{cm/sec}$$

(5.115)

where the quantity $\phi_0 = (kT/2) \ln (C_p/C_n)$ is introduced simply to give Eq. (5.115) a neater form and $C = (C_p C_n)^{1/2}$. The important feature of Eq. (5.115) is the dependence of v_s upon ϕ_s. As the ambient condition changes, so does ϕ_s as evidenced by Fig. 5.14. This explains why a stable ambient condition is essential for the stability of a semiconductor device.

Let us now apply Eq. (5.112) to the surface of a semiconductor. In Fig. 5.16a, a bar sample of uniform cross section is shown. Because of the surface recombination process, the distribution of excess carriers is no longer uniform over the cross section but decreases toward the surface as shown in Fig. 5.16b. The rate of diffusion of excess carriers toward the surface is given by $D_p\ \partial p/\partial y$ and $D_p\ \partial p/\partial z$ and this rate must be equal to the rate of recombination at the surface, or mathematically

$$D_p \frac{\partial p_1}{\partial y} = \pm v_s p_1 \quad \text{at } y = \begin{cases} 0 \\ b \end{cases}$$

$$D_p \frac{\partial p_1}{\partial z} = \pm v_s p_1 \quad \text{at } z = \begin{cases} 0 \\ c \end{cases}$$

(5.116)

Figure 5.16

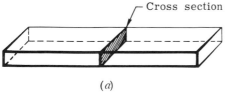

(a)

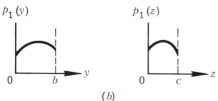

(b)

Treatment of the flow of excess carriers along a bar sample as a three-dimensional problem. Because of surface recombination, the distribution of excess carriers over the cross section is no longer uniform.

In finding the solution of the diffusion equation, Eq. (5.87), appropriate boundary conditions similar to those presented in Eq. (5.116) must be imposed. In Eq. (5.116), it is implied that holes are the minority carriers. The actual solution of the diffusion equation together with the boundary conditions of Eq. (5.116) will be discussed in the next section.

5.10 PHOTOCONDUCTIVE METHOD OF MEASURING LIFETIME AND SURFACE RECOMBINATION VELOCITY

Theory. The purpose of this section is twofold: to discuss the effect of the boundary conditions upon the solution of the diffusion equation and to introduce the photoconductive experiment. Since a thorough discussion of the latter requires a complete understanding of the former, we shall discuss the solution of the diffusion equation first. Consider a thin semiconductor slab of thickness b and width c as shown in Fig. 5.17. The top surface of the semiconductor is uniformly illuminated with a nonpenetrating light. By *nonpenetrating*, we mean a light having energy greater than the gap energy of the semiconductor. The absorption coefficient of such a light is around 100 cm^{-1} or more in indirect-gap semiconductors and around 1000 cm^{-1} or more in direct-gap semiconductors. Therefore, it is reasonable to assume that the light is completely absorbed at the surface, creating excess carriers.

If the rate of generating excess carriers is K, then the boundary conditions at the illuminated and unilluminated surfaces are respectively given by

$$D_p \frac{\partial p_1}{\partial y} = K - v_s p_1 \qquad \text{at } y = b$$

$$D_p \frac{\partial p_1}{\partial y} = v_s p_1 \qquad \text{at } y = 0$$

(5.117)

Now we must discuss the diffusion equation itself. Equation (5.83) can be generalized into a three-dimensional diffusion equation as follows:

$$\frac{\partial p_1}{\partial t} = D_p \left(\frac{\partial^2 p_1}{\partial x^2} + \frac{\partial^2 p_1}{\partial y^2} + \frac{\partial^2 p_1}{\partial z^2} \right) - \mu_p E_x \frac{\partial p_1}{\partial x} - \frac{p_1}{\tau}$$

(5.118)

It is understood from Eqs. (5.117) and (5.118) that the semiconductor under discussion is n-type and the applied electric field is in the x direction. To simplify the solution of Eq. (5.118), we shall make the following assumptions. Since the light is uniformly illuminated in the x-z plane, the $\partial^2 p_1/\partial x^2$ and $\partial^2 p_1 /\partial z^2$ terms in Eq. (5.118) can be taken to be zero. Furthermore, the applied electric field is so small that the term $\mu_p E_x \, \partial p_1/\partial x$ is negligible in comparison with other terms. Thus, Eq. (5.118) becomes

$$\frac{\partial p_1}{\partial t} = D_p \frac{\partial^2 p_1}{\partial y^2} - \frac{p_1}{\tau} \tag{5.119}$$

We shall reexamine the assumptions just made when we discuss the photoconductive experiment at a later time.

Taking the Laplace transform of Eq. (5.119), we find

$$\frac{\partial^2 \bar{p}}{\partial y^2} - \lambda^2 \bar{p} = 0 \tag{5.120}$$

where $\lambda^2 = (1 + s\tau)/D_p\tau$ and

$$\bar{p} = \int_0^\infty p_1 \exp{(-st)}\, dt \tag{5.121}$$

The boundary conditions, Eq. (5.117), after the Laplace transformation, become

$$D_p \frac{\partial \bar{p}}{\partial y} = -v_s\bar{p} + \frac{K}{s} \qquad \text{at } y = b$$
$$D_p \frac{\partial \bar{p}}{\partial y} = v_s\bar{p} \qquad \text{at } y = 0 \tag{5.122}$$

The solution of Eq. (5.120) is given by

$$\bar{p} = A \cosh \lambda y + B \sinh \lambda y \tag{5.123}$$

Applying the boundary conditions expressed in Eq. (5.122) to Eq. (5.123) gives

$$A = \frac{D_p\lambda K}{s[(D_p^2\lambda^2 + v_s^2)\sinh \lambda b + 2v_sD_p\lambda \cosh \lambda b]}$$
$$B = \frac{v_s K}{s[(D_p^2\lambda^2 + v_s^2)\sinh \lambda b + 2v_sD_p\lambda \cosh \lambda b]} \tag{5.124}$$

Figure 5.17

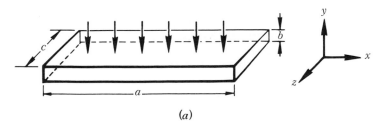

(a)

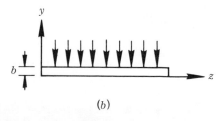

(b)

A thin semiconductor slab of cross-section bc is uniformly illuminated with a nonpenetrating light.

Substituting Eq. (5.124) into Eq. (5.123), we have

$$\bar{p} = \frac{K[D_p\lambda \cosh \lambda y + v_s \sinh \lambda y]}{s[(D_p^2\lambda^2 + v_s^2) \sinh \lambda b + 2v_s D_p\lambda \cosh \lambda b]} \tag{5.125}$$

The inverse Laplace transform of Eq. (5.125),

$$p_1 = \frac{1}{2\pi i} \int_{\epsilon - i\infty}^{\epsilon + i\infty} \bar{p} \exp (st) \, ds \tag{5.126}$$

can be found by the standard method of contour integration and the use of Cauchy's residue theorem. It may be instructive to treat the steady-state solution first. The integrand of Eq. (5.126) has a single pole at $s = 0$; therefore, the steady-state solution of p_1 is given by

$$(p_1)_0 = \frac{K[\sinh (y/L) + \beta \cosh (y/L)]}{v_s[(1 + \beta^2) \sinh (b/L) + 2\beta \cosh (b/L)]} \tag{5.127}$$

where $\beta = D_p/(Lv_s)$ and $L = (D_p\tau)^{1/2}$. The steady-state change in the conductance of the sample is equal to

$$\begin{aligned}(\Delta G)_0 &= \int_0^b \int_0^c \frac{e(\mu_p p_1 + \mu_n n_1) \, dy \, dz}{a} \\ &= \frac{CKL[\cosh (b/L) - 1 + \beta \sinh (b/L)]}{av_s[(1 + \beta^2) \sinh (b/L) + 2\beta \cosh (b/L)]}\end{aligned} \tag{5.128}$$

where $C = ce(\mu_n + \mu_p)$. Depending on the values of b/L and β, Eq. (5.128) can be simplified to give physically interpretable results.

For $b/L > 1$, $\cosh (b/L) \simeq \sinh (b/L) \gg 1$, thus Eq. (5.128) becomes

$$(\Delta G)_0 = \frac{CK}{a} \tau_{\text{eff}} \tag{5.129}$$

and the quantity τ_{eff} is defined as

$$\frac{1}{\tau_{\text{eff}}} = \frac{1}{\tau} + \frac{v_s}{L} \tag{5.130}$$

For $b/L < 1$, an expansion of the terms in Eq. (5.128) gives

$$(\Delta G)_0 = \frac{bCK(\beta + b/2L)}{av_s[(1 + \beta^2)(b/L) + 2\beta]} \tag{5.131}$$

There exist two possibilities: $\beta > 1$ and $\beta < 1$. For $\beta > 1$, we have $\beta \gg b/2L$ and $\beta^2 \gg 1$; hence Eq. (5.131) can be approximated by

$$(\Delta G)_0 = \frac{CK}{a} \tau_{\text{eff}} \tag{5.132}$$

and τ_{eff} is equal to

$$\frac{1}{\tau_{\text{eff}}} = \frac{1}{\tau} + \frac{2v_s}{b} \tag{5.133}$$

For $\beta < 1$, Eq. (5.131) can still be approximated by Eq. (5.132) but

$$\frac{1}{\tau_{\text{eff}}} = \frac{2v_s}{b} \tag{5.134}$$

It turns out that the combination of $\beta < 1$ and $b/L < 1$ means $\beta b/L = b/v_s\tau \ll 1$ and Eq. (5.133) automatically reduces to Eq. (5.134). Therefore, only Eqs. (5.130) and (5.133) need be considered.

Let us now discuss the transient part of Eq. (5.126). Equation (5.125) has poles along the imaginary axis $\lambda = i\xi$ such that

$$\tanh \lambda b = -\frac{2v_s D_p \lambda}{v_s^2 + D_p^2 \lambda^2} \qquad \text{or} \qquad \tan\left(\frac{\xi b}{2}\right) = \frac{v_s}{D_p \xi} \tag{5.135}$$

Figure 5.18 shows the graphic solution of Eq. (5.135). For small values of v_s such that $v_s b/D_p \ll 1$, the first root of Eq. (5.135) is approximately equal to

$$\xi_1 = \pm \left(\frac{2v_s}{D_p b}\right)^{1/2} \tag{5.136}$$

and the subsequent roots can be approximated by

$$\xi_n = \pm \frac{(n-1)\pi}{b} \qquad n = 2, 3, 4, \ldots \tag{5.137}$$

Following the standard procedure of contour integration and the use of Cauchy's residue theorem, we find the transient part of p_1

$$p_1(t) = \frac{2K}{b} \sum_{n-1}^{\infty} \exp\left(-s_n t\right)$$

$$\frac{\alpha_n(\alpha_n \cos \xi_n y + \gamma \sin \xi_n y)}{s_n[(\alpha_n^2 - \gamma^2 - 2\gamma) \cos \alpha_n + 2(\gamma + 1)\alpha_n \sin \alpha_n]} \tag{5.138}$$

Figure 5.18

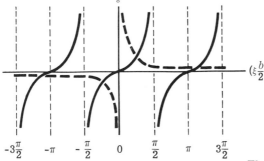

$\zeta = \tan\left(\xi\frac{b}{2}\right)$ (Solid curve)

$\zeta = v_s / (D_p \xi)$ (Dotted curve)

The graphic solution of Eq. (5.135) which is given by the intersection of the solid curve $\zeta = \tan(\xi b/2)$ with the dotted curve $\zeta = v_s/(D_p\xi)$.

where $\alpha_n = \xi_n b$, $\gamma = bv_s/D_p$, and

$$s_n = D_p\xi_n{}^2 + \tau^{-1} \tag{5.139}$$

To get the change in conductance ΔG, Eq. (5.138) must be integrated over dy and dz in the same way as in Eq. (5.128). Since the exact expression for ΔG is rather complicated, the following observation is made. First, ΔG can be written as

$$\Delta G = \sum_{n=1}^{\infty} \Delta G_n \exp\left(-s_n t\right) \tag{5.140}$$

where for large n, the ratio of ΔG_{n+1} to ΔG_n can be approximated by

$$\frac{\Delta G_{n+1}}{\Delta G_n} \simeq \frac{s_n \alpha_n}{s_{n+1}\alpha_{n+1}} \simeq \frac{\xi_n{}^3}{\xi_{n+1}^3} \simeq \left(\frac{n-1}{n}\right)^3 \tag{5.141}$$

Equation (5.141) says not only that the series in Eq. (5.140) converges rapidly but also that most of the contribution to ΔG comes from terms with small n. Now let us consider the effect of the decay time constant $1/s_n$. Substituting Eqs. (5.136) and (5.137) into Eq. (5.139) gives

$$s_1 = \frac{1}{\tau} + \frac{2v_s}{b}$$

$$s_n = \frac{1}{\tau} + D_p(n-1)^2 \frac{\pi^2}{b^2} \qquad \text{for } n = 2, 3, 4, \ldots \tag{5.142}$$

Therefore, the term ΔG_n with a larger index n not only has a smaller initial value but also decays much faster because of a smaller time constant. A combination of these two factors makes the following approximation of Eq. (5.140) permissible:

$$\Delta G \simeq \Delta G_1 \exp\left(-\frac{t}{\tau_{\text{eff}}}\right) \qquad \text{for } t > \tau_{\text{eff}} \tag{5.143}$$

where $1/\tau_{\text{eff}} = s_1 = 1/\tau + 2v_s/b$.

The experiment. Figure 5.19a shows the experimental setup in a photoconductive measurement. In an actual experiment, only part of the sample needs to be illuminated. The excess carriers are expected to extend over a diffusion length $(D\tau)^{1/2}$ beyond the illuminated region on both sides. If the length of the illuminated region is much larger than $(D\tau)^{1/2}$, the end effects will be negligibly small. It is also important that the applied electric field along the sample should be small so that excess carriers will recombine before they drift out of the sample. In other words, the unilluminated portion should be of sufficient length a' such that $a'/(\mu_p E_x)$ should be much greater than τ. If these two conditions are met, then the results obtained from the previous analyses can be directly applied to the experiment of Fig. 5.19a. In Fig. 5.19b, the upper trace represents the waveform of the light source while the lower trace is that of photoconductance which is measured by the voltage V_R developed across the resistor R.

Since the steady-state change of conductance, $(\Delta G)_0$ of Eq. (5.132), is proportional to v_s according to Eq. (5.133), the amplitude of the voltage V_R is a measure of the change in v_s. Figure 5.20 shows the experimental variation of $(\Delta G)_0$ as a function of ϕ_s/kT observed by Wang and Wallis.[16] According to Eq. (5.115), v_s has a maximum at $\phi_s = \phi_0$. This is manifested by the minimum in $(\Delta G)_0$ of Fig. 5.20. The experiment was performed on a germanium slab 0.04 cm thick, with a resistivity of 35 Ω-cm, n-type, and having a bulk lifetime of 800 microseconds. The following values were used in fitting the theoretical curve from Eqs. (5.132), (5.133), and (5.115) to the experimental points: $(v_s)_{max} = 55$ cm/sec, $\cosh\left[(\mathcal{E}_t - \mathcal{E}_{fi} - \phi_0)/kT\right] = 8$,

[16] S. Wang and G. Wallis, *Phys. Rev.*, vol. 105, p. 1459, 1957.

Figure 5.19

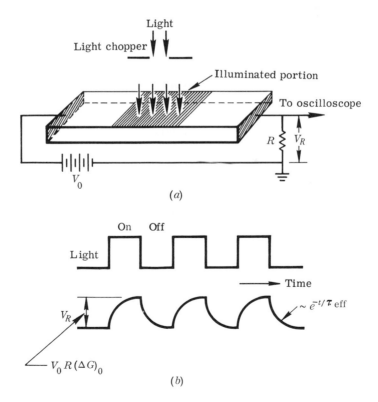

(a)

(b)

The photoconductive experiment for measuring the lifetime of excess carriers: (a) the experimental setup and (b) the experimental observation of the voltage V_R developed across a resistor R due to conductance changes induced in the semiconductor by the light.

and $C_p/C_n = 9$. For a well-etched surface, a value of v_s as low as 30 cm/sec may be obtained in Ge and between 200 to 500 cm/sec in Si.

The interesting part of the photoconductive experiment is the decay of the photoconductance. The quantity τ_{eff} used in Eq. (5.143) actually has a broader meaning attached to it than Eq. (5.143) suggests. The reader is asked to show (Prob. 5.27) that in the Haynes-Shockley experiment, the excess carriers drifting down the bar sample will decay with a time constant τ_{eff}:

$$\frac{1}{\tau_{\text{eff}}} = \frac{1}{\tau} + \frac{2v_s}{b} + \frac{2v_s}{c} \tag{5.144}$$

for small v_s if recombination at all four surfaces is taken into account. Equation (5.144) reduces to Eq. (5.133) when $c \gg b$. The point which we want to make is the following. The photoconductive experiment differs from the Haynes-Shockley experiment in the manner in which excess carriers are initially excited and distributed along the sample. However, after the light is turned off, the distribution of excess carriers in the transverse plane in the photoconductive experiment will readjust itself and approach that in the Haynes-Shockley experiment. This explains why after a certain period of time, the photoconductance decays according to Eq. (5.143) with a time constant τ_{eff} given by Eq. (5.144), which is the same as the decay constant in the Haynes-Shockley experiment. Therefore, Eq. (5.144) is a general expression for the effective lifetime of excess carriers drifting down a bar sample of rectangular cross section. If the surface recombination velocity is very large, then the solution of Eq. (5.135) is $\xi_1 b/2 = \pi/2$ and Eq. (5.144) must be replaced by

$$\frac{1}{\tau_{\text{eff}}} = \frac{1}{\tau} + D_p \pi^2 \left(\frac{1}{b^2} + \frac{1}{c^2} \right) \tag{5.145}$$

Note that for a semiconductor sample with well-etched surfaces, $v_s = 30$ cm/sec for example, and of fair thickness, say 0.1 cm, the

Figure 5.20

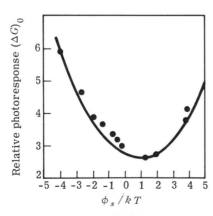

Experimental curve showing the variation of photoconductance $(\Delta G)_0$ versus the surface potential expressed in units of kT. (S. Wang and G. Wallis, Phys. Rev., vol. 105, p. 1459, 1957.)

photoconductive decay gives a direct measurement of τ up to 500 μsec. Furthermore, the method of exciting excess carriers is not limited to photoexcitation. For example, in the Haynes-Shockley experiment where excess carriers are created by injection, the voltage developed across R_1 will have a transient part which decays like the lower trace of Fig. 5.19b. The theory developed in this section for the decay of the excess conductance is independent of the means by which excess carriers are excited and for that reason, the experiment probably should be called the conductive decay instead of the photoconductive decay experiment. The conductive-decay method of measuring bulk lifetime can be used to study the effects of lattice imperfections on τ; Fig. 5.21 shows the measurements reported by Curtis et al.[17] It is believed that irradiations by fast neutrons and gamma rays produce interstitials with energy levels at 0.20 ev below the conduction band. These energy levels act as recombination centers. It can also be seen from Fig. 5.21 that in certain regions, the density of such recombination centers increases and hence the lifetime decreases proportionately with the amount of exposure.

To verify the Shockley-Read-Hall model of recombination, we should report measurements of lifetime as a function of temperature. Figure 5.22 shows the measurement reported by Wertheim[18] in 7 Ω-cm, n-type silicon by electron bombardments. According to our discussion in Sec. 5.5, for an n-type semiconductor, the lifetime varies as

$$\tau = \tau_p + \tau_p \frac{n'}{n_0} \qquad \text{or} \qquad \tau = \tau_p + \tau_n \frac{p'}{n_0} \qquad (5.146)$$

[17] Curtis, Cleland, Crawford, and Pigg, *Jour. Appl. Phys.*, vol. 28, p. 1161, 1957.
[18] G. K. Wertheim, *Phys. Rev.*, vol. 105, p. 1730, 1957.

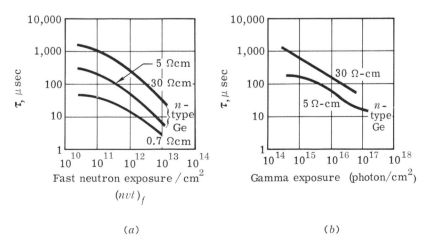

Figure 5.21 *Experimental curves showing the effects of bombardments (a) by fast neutrons and (b) by gamma rays on the lifetime of excess carriers in Ge. (Curtis, Cleland, Crawford, and Pigg, Jour. Appl. Phys., vol. 28, p. 1161, 1957.)*

depending on whether $n' \gg p'$ or $p' \gg n'$. Therefore, if the trap energy is near the conduction band, τ should vary as $\exp\left[-(\mathcal{E}_c - \mathcal{E}_t)/kT\right]$ and if it is near the valence band, as $\exp\left[-(\mathcal{E}_t - \mathcal{E}_v)/kT\right]$. In either case, the exponential variation with temperature is borne out by the experiment.

Besides vacancies and interstitials, impurities can also act as recombination centers. Measurements of lifetime as a function of temperature have been reported in germanium doped with copper. However, the picture there is complicated by the fact that copper impurities also act as temporary storage traps at low temperatures. The reader is referred to the article by Shulman and Wyluda[19] for further details.

Figure 5.22

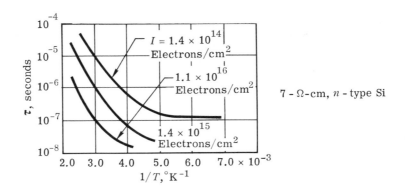

$7\text{-}\Omega\text{-cm}, \, n\text{-type Si}$

Experimental curves showing the effects of electron bombardment on the lifetime of excess carriers in Si. The lifetime is measured as a function of temperature for different intensities of electron bombardment. (G. K. Wertheim, Phys. Rev., vol. 105, p. 1730, 1957.)

PROBLEMS

5.1 Derive the equation corresponding to Eq. (5.7) for holes.

5.2 Prove the Einstein relation [Eq. (5.8)] for the general case where the relaxation time τ is a function of velocity $\tau = g(v)$. Consulting Sec. 4.2, show that Eq. (5.8) is no longer true for degenerate semiconductors.

5.3 Calculate the density-of-state and mobility effective masses of electrons and holes in germanium.

5.4 Calculate the density-of-state and mobility effective masses of electrons and holes in silicon.

5.5 Prove the statement $J_x = u_1 J_u$ by using J_x from Eq. (5.28) and J_u from Eq. (5.21) and by showing that the summation $\Sigma(\mathbf{c} \cdot \mathbf{u})(\mathbf{c} \cdot \mathbf{x})$ over the four ellipsoids is equal to $4u_1/3$.

[19] R. G. Shulman and B. J. Wyluda, *Phys. Rev.*, vol. 102, p. 1455, 1956.

5.6 Derive Eq. (5.36) by applying the method of detailed balancing to valence-band holes. Also find the ratio of the hole concentrations p_{n0}/p_{p0} following the same approach leading to Eq. (5.38).

5.7 A germanium diode is made of materials with resistivities $\rho = 1\ \Omega$-cm p-type and $\rho = 0.1\ \Omega$-cm n-type. Using the values of μ_n, μ_p, N_c, and N_v given in Probs. 3.4 and 3.12, find the built-in voltage V_d.

5.8 Refer to our discussion in Sec. 3.5 or to Fig. 3.9 about the change of the Fermi level as a function of temperature in an n-type semiconductor. If it is assumed that the trap energy $\mathcal{E}_t$ lies halfway between $\mathcal{E}_d$ and $\mathcal{E}_{fi}$, mark on Fig. 3.9 the temperature T_3 at which $n_0 = n'$ in Eq. (5.72). Discuss the temperature dependence of lifetime in the intermediate temperature region $T_2 < T < T_1$.

5.9 Repeat the preceding problem for the case in which the trap energy lies half-way between $\mathcal{E}_{fi}$ and $\mathcal{E}_v$. Assuming that $N_c = N_v$, mark on Fig. 3.9 the temperature T_3 at which $n_0 = p'$ in Eq. (5.72). Discuss the temperature dependence of lifetime for $T_2 < T < T_1$, assuming that τ_n and τ_p are of the same order of magnitude. The temperatures T_1 and T_2 are defined in Sec. 3.5.

5.10 It is desirable to determine the activation energy of the lifetime τ so that the recombination centers can be identified. In certain temperature regions, we find that τ decreases as $\exp(-\mathcal{E}/kT)$, indicating that either the term $\tau_n p'$ or $\tau_n n'$ dominates. Discuss the conditions (regarding temperature and trap energy) under which the said exponential dependence may be expected. How is $\mathcal{E}$ related to the trap energy under each circumstance? By comparing the conditions for n-type and p-type samples, comment on the possibility of establishing whether $\mathcal{E}_t$ lies closer to the conduction or valence band in cases where τ_n and τ_p differ by several orders of magnitude.

5.11 The charge-neutrality condition imposes an important restriction on the movement of excess carriers. Suppose that for some unknown reasons, the excess carriers were separated from each other as shown in the accompanying figure. Plot the electric field as a function of x by solving Poisson's equation. For simplicity, the problem is taken to be one-dimensional and the carrier concentrations are such that $n_0 \gg n_1 = p_1 \gg p_0$ with the subscripts 0 and 1 indicating respectively the thermal-equilibrium and excess concentrations. Discuss the movements of electrons and holes in regions ab and cd as a result of the dipolar field alone (exclusive of the diffusion and ordinary drift currents). Show by qualitative arguments that the center

Figure P5.11

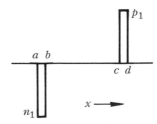

of gravity of the excess-charge distribution moves rapidly toward the initial position of p_1.

5.12 Consider an n-type germanium sample with $n_0 = 10^{16}/cm^3$ and $p_0 = 6 \times 10^{10}/cm^3$. Refer to the previous problem and calculate the electric field as a result of the polarization of opposite charges with densities $n_1 = p_1 = 10^{13}/cm^3$. Also calculate the rate with which region ab is depleted of electrons by taking the difference of electron current densities at points a and b or $J_a - J_b$. Repeat the calculation for holes in region cd. Comment on the movements of n_1 and p_1. Also estimate the time required to complete the process and compare your estimate with the dielectric relaxation time ϵ/σ.

5.13 Show that the d-c steady-state solution of Eq. (5.83) is given by

$$n_2 = A \exp\left(\frac{x}{L_1}\right) + B \exp\left(-\frac{x}{L_2}\right)$$

Express L_1 and L_2 in terms of D, μ, and E. One case of practical importance corresponds to the situation prevailing in the base region of a transistor. Assuming that $E = 0$ and that at $x = 0$ (the emitter), $n_2 = n_2(0)$ and that at $x = W$ (the collector), $n_2 = 0$, verify that

$$n_2 = \frac{n_2(0)[\sinh (W - x)/L]}{\sinh(W/L)}$$

where $L = \sqrt{D\tau}$.

5.14 Verify that Eq. (5.97) satisfies the diffusion equation, Eq. (5.87), for $t > 0$. Show that at $t = 0$, $p_1(x,t)$ of Eq. (5.97) behaves like a delta function as it should. In other words, at $t = 0$, $p_1(x,t) = \infty$ at $x = 0$ and zero elsewhere, and $\int_{-\infty}^{\infty} p_1(x,0)\, dx = 1$.

5.15 Given: $\tau = 10^{-5}$ sec, $\mu_p = 1900$ cm^2/volt-sec, and $D_p = 50$ cm^2/sec in a typical germanium sample. Plot $p_1(x,t)$ of Eq. (5.97) as a function of x with $E = 5$ volts/cm for $t = 10^{-6}$, 2×10^{-6}, 5×10^{-6}, and 10^{-5} sec. Compare the areas under the curve at $t = 10^{-6}$ and 10^{-5} sec and verify that they are in the ratio of $\exp[-(t_1 - t_2)/\tau]$.

5.16 Equation (5.97) is the solution of the time-dependent diffusion equation for a unit impulse input at $x = 0$. Suppose that in Eq. (5.92), instead of having a unit impulse $\delta(x_1)$, the initial excess carrier distribution $p_1 (t = 0)$ is nonzero. If $p_1(t = 0)$ is given by a Gaussian distribution

$$p_1(t = 0) = \frac{1}{\sqrt{\pi}\,\Delta} \exp\left(-\frac{x^2}{\Delta^2}\right)$$

use physical reasoning to show that the solution $p_1(x,t)$ to the diffusion equation is also a Gaussian function. Express $p_1(x,t)$ in terms of τ, μ_p, D_p, E, and Δ.

5.17 Obtain Eq. (5.104) from Eq. (5.103). Using Eq. (5.104), show that Eq. (5.102) is a good approximation if during the time of drift (1) the dispersion caused by diffusion is smaller than the drift distance and (2) the effect on the shape of the pulse due to recombination is insignificant within the dispersion caused by diffusion. Using the numerical values given in Prob. 5.15, calculate the value of E_x needed to make Eq. (5.102) a valid approximation for $d = 0.5$ cm.

5.18 Refer to Fig. 5.15a. Show that (a) the charge density $\varrho(x)$ near the surface is given by

$$\varrho(x) = 2en_i[\sinh(\phi_b/kT) - \sinh(\phi(x)/kT)]$$

(b) the potential $\phi(x)$ near the surface satisfies the equation

$$\left(\frac{\partial\phi(x)}{\partial x}\right)^2 = \frac{4e^2n_ikT}{\epsilon}[\cosh(\phi/kT) - (\phi/kT)\sinh(\phi_b/kT) + A]$$

where A is a constant, resulting from the integration of Poisson's equation.

5.19 Apply boundary conditions to find the constant A of Prob. 5.18. Using Gauss's law, find the relation between the surface charge density in the slow states and the surface potential ϕ_s.

5.20 Find expressions in integral form relating the total free carrier concentrations $\int p(x)\, dx$ and $\int n(x)\, dx$ near the surface to the surface potential ϕ by converting the variable of integration from x to ϕ. The integral expressions in general cannot be evaluated explicitly. Consider the case in which the surface potential is such that the charge densities contributed by charged donors and acceptors are much smaller than either the electron or the hole charge density in the space-charge region near the surface. Evaluate the integral expression explicitly for such cases.

5.21 Using the expressions given in Prob. 5.18, find the total charge (coulomb/area of surface) and the surface potential ϕ as a function of x in the space-charge region near the surface for an intrinsic sample with $\phi_b = 0$.

5.22 Take $C_p/C_n = 9$ and $\mathcal{E}_t - \mathcal{E}_{fi} - \phi_0 = 0.068$ ev in Eq. (5.115). Plot $v_s/(v_s)_{max}$ as a function of ϕ_s for ϕ_s between $\pm 6kT$ and compare the curve with Fig. 5.20. Explain the difference between the two curves and locate the energy levels of the recombination centers.

5.23 Refer to Eq. (5.115) and the expression for $\varrho(x)$ in Prob. 5.18. Discuss qualitatively the behaviors of the total conductance G (including conductance due to free carriers in region AB of Fig. 5.15) and the surface recombination velocity in a germanium sample (1 Ω-cm, n-type). Sketch the variation of G and v_s as functions of ϕ_s. The information about the sample (1 Ω-cm, n-type) is given so that you may use G at $\phi_s = \phi_b$ as a reference conductance for comparison and the exact value of ϕ_b is not required. Indicate the locations of ϕ_s for minimum G and for maximum v_s (in both cases $C_p > C_n$ and $C_p < C_n$). Also mark the line for ϕ_s across which the majority carriers in the region AB of Fig. 5.15 change from n- to p-type.

5.24 Derive Eq. (5.125) from Eq. (5.123) by filling in the missing steps.

5.25 Obtain the approximate results of Eqs. (5.130) and (5.133) from Eq. (5.128). Explain physically why the results are different.

5.26 Refer to Eqs. (5.135) and (5.138). Show that for $n = 1$, the excess carrier distribution over the cross section can be represented by a cosine function

$$p_1 = p_{10} \cos \left[\xi_1 \left(y - \frac{b}{2} \right) \right]$$

as shown in the accompanying figure. Also verify that for small values of v_s, $\xi_1 = [2v_s/(D_p b)]^{1/2}$ and for large values of v_s, $\xi_1 b = \pi$.

Figure P5.26

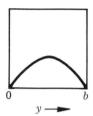

5.27 Consider the distribution of excess carriers given in Prob. 5.26. Calculate the average diffusion velocity $\bar{v}_d$ of excess carriers, which is defined as

$$e \bar{p}_1 \bar{v}_d = e D_p \frac{\partial p_1}{\partial y}$$

where $\bar{p}_1$ is the spatial average of p_1. Refer to Eqs. (5.144) and (5.145). Show that the effective lifetime due to surface recombination velocity is equal to the average transit time $(\bar{t} = b/2\bar{v}_d)$ for excess carriers to reach the surface.

REFERENCES

Brooks, H.: Theory of the Electrical Properties of Germanium and Silicon, in *Advan. Electron. Electron Phys.*, vol. 7, p. 87, 1955.

Bube, R. H.: "Photoconductivity of Solids," John Wiley & Sons, Inc., New York, 1957.

Carslaw, H. S., and J. C. Jaeger, "Conduction of Heat in Solids," Oxford University Press, Oxford, 1959.

Hannay, N. B. (ed.): "Semiconductors," Reinhold Publishing Corp., New York, 1959.

Kingston, R. H. (ed.): "Semiconductor Surface Physics," University of Pennsylvania Press, Philadelphia, 1957.

Moll, J. L.: "Physics of Semiconductors," McGraw-Hill Book Company, New York, 1964.

Shockley, W.: "Electrons and Holes in Semiconductors," D. Van Nostrand Company, Inc., New York, 1950.

Spenke, E.: "Electronic Semiconductors," McGraw-Hill Book Company, New York, 1958.

6 SEMICONDUCTOR JUNCTION DEVICES

6.1 p-n JUNCTION DIODES

As the heading of the chapter suggests, our discussion in this chapter will be centered around semiconductor devices. Its purpose is two-fold: to develop a working theory and to consider various applications of semiconductor devices. It is most fitting that we start our discussion with a simple *p-n* junction. Not only will the theory developed for a junction diode be found to underlie many other semiconductor devices such as transistors, but we can also continue almost without interruption from where we left off in the previous chapter, with especial reference to Sec. 5.4. As we recall, we have introduced the concept of quasi-Fermi levels for diodes under nonequilibrium conditions such as applied bias. The assumption made in Sec. 5.4 is that the quasi-Fermi levels for minority carriers, $\mathfrak{F}_n$ in the *p* region and $\mathfrak{F}_p$ in the *n* region, remain constant in the transition region AB as shown in Fig. 5.4. What happens to the quasi-Fermi levels actually depends very much upon the properties of the transition region itself, especially its width. Therefore, it is highly desirable that we study the transition region in some detail.

Space-charge region and junction capacitance. First, let us consider a *p-n* junction under thermal equilibrium as shown in Fig. 6.1 and define a net charge-density function $\varrho(x)$ such that

$$\varrho(x) = e[p + N_d^+ - n - N_a^-] \tag{6.1}$$

For simplicity, it is assumed that the donor and acceptor impurities are completely ionized and that $N_a^- - N_d^+ = N_1$ in the *p* region and $N_d^+ - N_a^- = N_2$ in the *n* region. Outside the transition region AB, the charge-neutrality condition prevails, or $\varrho(x) = 0$ for $x > x_B$ and for $x < -x_A$. In the transition region $x_B > x > -x_A$, however, the charge-neutrality condition is no longer in effect. At a point x in the *n* region, the electron (majority carrier) concentration is given by

$$n = N_c \exp\left(-\frac{\mathcal{E}_c - \mathcal{E}_f}{kT}\right) = n_{n0} \exp\left(-\frac{e\phi}{kT}\right) \tag{6.2}$$

where ϕ is the negative of the electrostatic potential and, as seen in Fig. 6.1, it is a positive quantity. The quantity n_{n0} represents the equilibrium electron concentration in the n region outside the transition region and n_{n0} is equal to N_2 if the hole (minority carrier) concentration is negligibly small compared to n_{n0}. Under such circumstances, Eq. (6.1) for $x > 0$ becomes

$$\varrho(x) = eN_2 \left[1 - \exp\left(-\frac{e\phi}{kT} \right) \right] \qquad (6.3)$$

The profile of the electrostatic potential V is easily obtainable through the use of Poisson's equation

$$\frac{\partial^2 V}{\partial x^2} = -\frac{\partial^2 \phi}{\partial x^2} = -\frac{\varrho(x)}{\epsilon} \qquad (6.4)$$

in conjunction with Eq. (6.3), ϵ being the dielectric constant of a given semiconductor. The region in which $\varrho(x)$ is nonzero is generally referred to as the *space-charge region*, hitherto called the transition region. To facilitate solving Eq. (6.4), we take note of the following: The value of n given by Eq. (6.2) drops very sharply as soon as $e\phi$ reaches a value of the order of kT. Since the value of kT at room temperature is around 0.025 ev which is at least an order of magnitude smaller than eV_d, V_d being the difference in the electrostatic potential between the n and p regions, the net charge density $\varrho(x)$ takes the value of eN_2 practically for the whole length of the transition region OB. As a first-order approximation, Poisson's equation becomes

$$\frac{\partial^2 V}{\partial x^2} = -\frac{eN_2}{\epsilon} \qquad \text{for } 0 < x < x_B \qquad (6.5)$$

If the impurity concentration N_2 is uniform, that is, independent of x in the n region, integrating Eq. (6.5) gives the following results:

$$\frac{\partial V}{\partial x} = -\frac{eN_2}{\epsilon}(x - x_B)$$

$$V = -\frac{eN_2}{2\epsilon}x(x - 2x_B) \qquad (6.6)$$

**Figure
6.1**

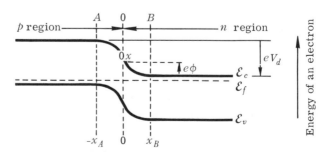

Energy-band diagram of a semiconductor p-n junction under thermal equilibrium. The difference in the electrostatic potential V_d is generally referred to as the diffusion or built-in potential.

Applying similar arguments to the p region, we have

$$\frac{\partial V}{\partial x} = \frac{eN_1}{\epsilon}(x + x_A)$$

$$\text{for } -x_A < x < 0 \qquad (6.7)$$

$$V = \frac{eN_1}{2\epsilon} x(x + 2x_A)$$

For further clarification of the steps taken in arriving at Eqs. (6.6) and (6.7), we refer to Fig. 6.2. Briefly, the space-charge region is assumed to end abruptly at $x = x_B$ and $x = -x_A$. Therefore, the value of the electric field $E_x = -\partial V/\partial x$ is taken to be zero at $x = x_B$ and $x = -x_A$. Furthermore, the electrostatic potential at the boundary between the n and p regions is used as the reference potential, that is $V = 0$ at $x = 0$. Equating the values of E_x at $x = 0$ gives

$$eN_2 x_B = eN_1 x_A \qquad (6.8)$$

Equation (6.8) says that the total positive charges stored in the space-charge region on the n side must equal the corresponding negative charges stored on the p side. Setting $V = -V_A$ at $x = -x_A$, $V = V_B$ at $x = x_B$, and $V_d = V_A + V_B$, we find

$$V_d = \frac{e}{2\epsilon}(N_1 x_A{}^2 + N_2 x_B{}^2) \qquad (6.9)$$

Eliminating x_A or x_B from Eqs. (6.8) and (6.9) gives

$$x_A = \left(\frac{2\epsilon V_d N_2}{eN_1}\frac{1}{N_1 + N_2}\right)^{1/2}$$

$$\qquad (6.10)$$

$$x_B = \left(\frac{2\epsilon V_d N_1}{eN_2}\frac{1}{N_1 + N_2}\right)^{1/2}$$

and the total width of the transition region W:

$$W = x_A + x_B = \left(\frac{2\epsilon V_d}{e}\frac{N_1 + N_2}{N_1 N_2}\right)^{1/2} \qquad (6.11)$$

The above analysis can be extended to a nonequilibrium situation

Figure 6.2

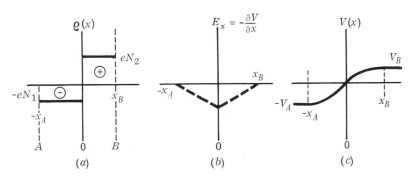

Diagrams showing the variations of (a) charge density, (b) electric field, and (c) electrostatic potential in the space-charge region AB of Fig. 6.1.

in which a voltage is applied to the p-n junction. The situation is illustrated in Fig. 6.3. The only modification is that the difference in the electrostatic potential now changes to $V_d \mp V_a$ under an applied voltage V_a, the negative and positive signs corresponding respectively to the forward- and reverse-bias conditions. As long as $e(V_d - V_a)$ is an order of magnitude larger than kT, the approximations used in deriving Eqs. (6.8) to (6.11) are still valid. From Eq. (6.8), the charge Q stored in a junction of an area A is given by

$$\frac{Q}{A} = eN_2 x_B = eN_1 x_A = \left(\frac{2e\epsilon N_1 N_2}{N_1 + N_2}\right)^{1/2} (V_d \mp V_a)^{1/2}$$

$$= W \frac{eN_1 N_2}{N_1 + N_2} \tag{6.12}$$

Figure 6.3

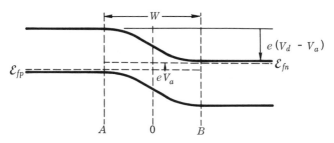

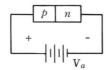

(a) Forward - bias

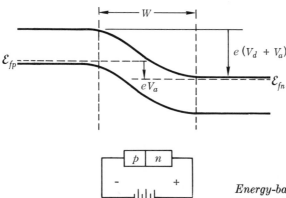

(b) Reverse - bias

Energy-band diagrams of a semiconductor junction under biased conditions.

Therefore, the capacitance C_T associated with the stored charge Q in the transition region is equal to

$$C_T = \left| \frac{dQ}{dV_a} \right| = \frac{A}{2} \left(\frac{2e\epsilon N_1 N_2}{N_1 + N_2} \right)^{1/2} (V_d \mp V_a)^{-1/2}$$

$$= \frac{AW}{2} \frac{eN_1 N_2}{N_1 + N_2} \frac{1}{V_d \mp V_a} = \frac{A\epsilon}{W} \tag{6.13}$$

For an alloyed diode, the alloyed side usually has an impurity concentration far greater than the base side; therefore, the expression $N_1 N_2/(N_1 + N_2)$ in Eqs. (6.12) and (6.13) reduces to N_1 or N_2, whichever is smaller. It is also obvious from Eq. (6.10) and Fig. 6.2 that the space-charge region extends farther into the region with lower impurity concentration. To estimate the width of the space-charge region and the magnitude of junction capacitance, we take germanium with a base impurity concentration 10^{15} cm^{-3}. At a junction potential $V_d - V_a = 0.4$ volt, W is found to be 0.8×10^{-4} cm and C/A to be 2×10^{-8} farad/cm^2.

Minority carrier injection and ideal diode characteristics. Before we analyze the characteristics of a junction diode, we must re-examine the validity of the assumption that the quasi-Fermi levels stay constant in the space-charge region. From Eqs. (5.7) and (5.45), the current density due to conduction-band electrons can be expressed in terms of the quasi-Fermi level $\mathfrak{F}_n$ for electrons as

$$J_n = e\mu_n nE + eD_n \frac{\partial n}{\partial x} = \mu_n n \frac{\partial \mathfrak{F}_n}{\partial x} \tag{6.14}$$

realizing that $\partial \mathcal{E}_c/\partial x = -e\, \partial V/\partial x = eE$. Under steady-state conditions, it is required by the continuity equation

$$\frac{\partial n}{\partial t} = \frac{1}{e} \frac{\partial J_n}{\partial x} - \frac{n_1}{\tau} \tag{6.15}$$

that if recombination of excess carriers in the space-charge region is negligible, then J_n should be independent of x. Let us now consider the situation outside the space-charge region. As we shall see later, the excess minority-carrier concentration will eventually drop to zero either through the recombination process or through the action of the base contact. In other words, $\mathfrak{F}_n$ moves gradually toward $\mathcal{E}_{fp}$ in the p region and $\mathfrak{F}_n$ practically equals $\mathcal{E}_{fp}$ within a distance d determined by the diffusion length $\sqrt{D_n \tau}$ or the proximity of the base contact, whichever is smaller. The situation is illustrated in Fig. 6.4. Since the current density J_n must be continuous at point A, the change in the quasi-Fermi level in the space-charge region can be estimated from Eq. (6.14) as follows

$$\Delta \mathfrak{F}_n \Big|_B^A \simeq \Delta \mathfrak{F}_n \Big|_A^C \frac{\Delta x_{AB}}{\Delta x_{AC}} \frac{n_{AC}}{n_{AB}} \tag{6.16}$$

Since the ratio, n_{AC}/n_{AB}, of the average electron concentration in region AC to that in region AB is much less than 1,

$$\frac{\Delta \mathfrak{F}_n \Big|_B^A}{\Delta \mathfrak{F}_n \Big|_A^C} \leq \frac{\Delta x_{AB}}{\Delta x_{AC}} \tag{6.17}$$

The width of the space-charge region was estimated previously to be of the order of 10^{-4} cm while $\sqrt{D_n \tau} = 10^{-2}$ cm with $D_n = 100$ cm^2/sec and $\tau = 10^{-6}$ sec. It is obvious that almost all of the drop in $\mathfrak{F}_n$ occurs outside the space-charge region.

In summary, if recombination of excess carriers in the space-charge region is negligible and if the width of the space-charge region is very thin compared with the diffusion length, then the assumption is valid that the quasi-Fermi levels remain approximately constant in the space-charge region. Under such circumstances, the minority-carrier concentrations become

$$n_p = n_{p0} \exp\left(\frac{eV_a}{kT}\right) \qquad \text{at } x = -x_A \tag{6.18a}$$

$$p_n = p_{n0} \exp\left(\frac{eV_a}{kT}\right) \qquad \text{at } x = x_B \tag{6.18b}$$

where n_{p0} and p_{n0} represent the equilibrium minority-carrier concentrations, electron and hole, in the p and n regions, respectively.

Figure 6.4

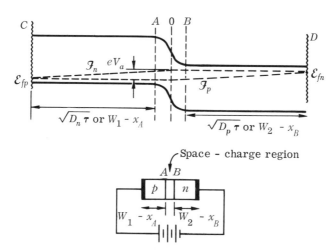

Diagram showing the variation of quasi-Fermi levels $\mathfrak{F}_n$ and $\mathfrak{F}_p$ in a semiconductor junction. The quasi-Fermi levels are for minority carriers. When the excess carrier concentrations decrease to zero, the quasi-Fermi levels should approach the respective Fermi levels ($\mathfrak{F}_n$ to $\mathcal{E}_{fp}$ and $\mathfrak{F}_p$ to $\mathcal{E}_{fn}$), indicating the return to thermal equilibrium.

We again refer to Fig. 6.3. For a forward bias, V_a is positive in Eq. (6.18) and the minority-carrier concentrations given by Eq. (6.18) exceed the corresponding equilibrium minority-carrier concentrations. Here we have excess minority carriers in the respective p and n regions. This is called *minority-carrier injection*. On the other hand, for a reverse bias, the minority-carrier concentrations fall below their respective equilibrium concentrations. In this case, we have *minority-carrier extraction* or a deficiency of minority carriers.

To find the current flow through a junction diode we must refer to Eqs. (5.78) and (5.79) or Eq. (5.83). Since almost all the applied voltage V_a appears across the space-charge region, the drift term is unimportant outside the space-charge region. Thus, under steady-state d-c conditions, the diffusion equation applicable to Fig. 6.4 reads

$$D_p \frac{\partial^2 p_1}{\partial x^2} - \frac{p_1}{\tau} = 0 \qquad \text{for } x > x_B \tag{6.19}$$

A similar expression is obtained for excess electrons in the region $x < -x_A$. The general solution of Eq. (6.19) is given by

$$p_1 = A \exp\left(-\frac{x}{L}\right) + B \exp\left(\frac{x}{L}\right) \tag{6.20}$$

where $L = \sqrt{D_p \tau}$. The above equation is subject to the boundary condition that at the base contact $x = W_2$, $p_1 = 0$, and at the boundary of the space-charge region $x = x_B$, $p_{10} = p_n - p_{n0}$ where p_n is given by Eq. (6.18). Two different approximations may be taken, depending on whether W_2 is greater or smaller than L. The reader is asked to show that the general solution of Eq. (6.20) can be approximated by

$$p_1 = p_{10} \exp\left(-\frac{x - x_B}{L}\right) \tag{6.21}$$

if $(W_2 - x_B) \gg L$ in a wide-base diode and

$$p_1 = \frac{p_{10}(W_2 - x)}{W_2 - x_B} \tag{6.22}$$

if $(W_2 - x_B) \ll L$ in a narrow-base diode. Note that Eq. (6.22) satisfies Eq. (6.19) if the p_1/τ term in Eq. (6.19) is neglected. By a narrow-base diode, we therefore mean that recombination of excess carriers is negligible in the base region. From Eq. (6.21) or (6.22), the current density due to diffusion of holes is equal to

$$J_p = -eD_p \frac{\partial p_1}{\partial x}\bigg|_{x=x_B} = \frac{eD_p}{d} p_{n0}\left[\exp\left(\frac{eV_a}{kT}\right) - 1\right] \tag{6.23}$$

In the above equation, the distance d is given by either L or $W_2 - x_B$, whichever is smaller. According to Eq. (6.23), for large bias voltage in the reverse direction such that $eV_a/kT < -1$, the current approaches a constant value commonly referred to as the *reverse saturation current*.

Now we should clarify the steps taken in securing the boundary condition at the base contact and in obtaining the current expression J_p of Eq. (6.23). Theoretically, a rectifying junction results between two materials in contact if excess (or deficit) carriers exist at the junction. This usually happens at a metal-semiconductor contact. However, the degree of rectification is relative. A contact is said to be ohmic if no appreciable excess (or deficit) carriers may exist at the contact, or $n_1 = p_1 = 0$. This boundary condition is used in evaluating the coefficients A and B in Eq. (6.20). The other point concerns the composition of diode current. If the following continuity equations are subtracted from each other,

$$\frac{\partial p_1}{\partial t} = -\frac{1}{e}\frac{\partial J_p}{\partial x} - \frac{p_1}{\tau} \tag{6.24a}$$

$$\frac{\partial n_1}{\partial t} = \frac{1}{e}\frac{\partial J_n}{\partial x} - \frac{n_1}{\tau} \tag{6.24b}$$

we see that the current density $J_2 = J_n + J_p$ is independent of x in view of the fact that $n_1 = p_1$. It may be instructive to examine the nature of the current as derived from Eqs. (6.21) and (6.22).

Note that the current density J_2 is due to the injection of holes into the n region. The reason that we evaluated J_p at $x = x_B$ is because at $x = x_B$, the current is constituted by hole diffusion alone, or in other words, $J_2 = J_p$ at $x = x_B$. As the injected holes diffuse from B toward D in Fig. 6.4, they recombine with electrons. This explains why p_1 decreases with x in Eq. (6.21), but it is not the whole story. The n

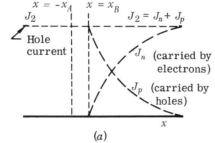
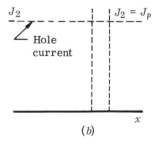

Figure 6.5

Diagrams showing the composition of the current created by injection of holes into and subsequent diffusion of holes in the n region. The current in the p region is carried by holes (the majority carrier). In the n region, however, two distinct cases exist. Figure 6.5a represents the case in which practically all injected holes disappear because of the recombination process before they reach the base contact, while Fig. 6.5b represents the opposite case in which very little recombination takes place in the base region.

region must supply electrons during the recombination process, creating a movement of electrons toward the left and thus an electron current. In other words, in Eq. (6.24), because n_1/τ is nonzero so is $\partial J_n/\partial x$. Figure 6.5a shows the composition of the current created by diffusion of holes as a function of x for cases where recombination processes play an important role. The situation is quite different for cases where recombination of excess carriers is relatively insignificant. Note that Eq. (6.22) gives a constant hole current as shown in Fig. 6.5b in contrast to Fig. 6.5a. The current J_2 in the p region is of course carried by holes in both cases. We should also point out that an expression J_1 similar to Eq. (6.23) is obtained for electron injection into the p region, and the total current density J

Figure 6.6

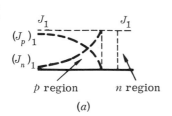

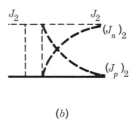

(a) (b)

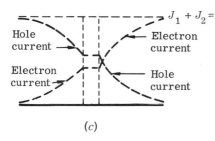

(c)

Diagrams showing the composition of total current (Fig. 6.6c) which is the sum of the current due to hole injection (Fig. 6.6b) from the p into the n region and that due to electron injection (Fig. 6.6a) from the n into the p region.

is a sum of J_1 and J_2. Figure 6.6 shows the composition of the total current as a function of distance when both electron and hole injections are considered.

6.2 CARRIER STORAGE AND TRANSIENT RESPONSE

Small signal a-c equivalent circuit and diffusion capacitance. To find the steady-state response of a junction diode to a sinusoidally time-varying signal, we can write the time-dependent diffusion equation as follows

$$\frac{\partial p_1}{\partial t} = i\omega p_1 = D_p \frac{\partial^2 p_1}{\partial x^2} - \frac{p_1}{\tau} \tag{6.25}$$

where ω is the angular frequency of the time-varying voltage. If v_0 and v_1 are, respectively, the d-c and a-c part of the applied voltage, then the excess hole concentration can also be separated into a d-c and an a-c part in the following manner:

$$p_1 = p_{n0} \left[\exp \frac{e(v_0 + v_1)}{kT} - 1 \right] = (p_1)_{d\text{-c}} + (p_1)_{a\text{-c}}$$

$$= p_{n0} \left[\exp \left(\frac{ev_0}{kT} \right) - 1 \right] + \left(\frac{ev_1}{kT} \right) p_{n0} \exp \left(\frac{ev_0}{kT} \right) \qquad (6.26)$$

In the above equation, it is assumed that the a-c applied voltage is small such that $ev_1/kT < 1$ and hence $\exp(ev_1/kT)$ can be approximated by $1 + ev_1/kT$. The solution of Eq. (6.25) is quite similar to that of Eq. (6.19), the d-c solution, except that τ should be replaced by $\tau/(1 + i\omega\tau)$ and that at $x = x_B$, $p_1 = (p_1)_{a\text{-c}}$ of Eq. (6.26). Thus, the a-c part of the excess hole concentration obeys the following equation:

$$(p_1)_{a\text{-c}} = \frac{ev_1}{kT} p_{n0} \exp \left(\frac{ev_0}{kT} \right) \frac{\sinh (W_2 - x)/L'}{\sinh (W_2 - x_B)/L'} \qquad (6.27)$$

where $L' = \sqrt{D_p\tau/1 + i\omega\tau}$.

Now we shall work out the current expression for the case $(W_2 - x_B) \ll L$. It is assumed that the p region is much more heavily doped than the n region so that the effect due to diffusion of holes predominates. Differentiating Eq. (6.27) with respect to x and expanding the hyperbolic functions into a power series, we find the alternating current in a junction of area A

$$i_1 = \frac{ev_1}{kT} p_{n0} \exp \left(\frac{ev_0}{kT} \right) \left(\frac{eD_p}{W_2 - x_B} \right) \left[1 + \frac{i\omega}{3D_p} (W_2 - x_B)^2 \right] A \qquad (6.28)$$

Obviously, the alternating current has a conductive and a capacitive component. A comparison of Eq. (6.28) with Eq. (6.23) shows that if the d-c bias voltage is relatively large so that $ev_0/kT > 1$, then the equivalent conductance and capacitance G and C_D of a junction diode can be approximated, respectively, as follows:

$$G = \frac{ei_0}{kT} \qquad \text{and} \qquad C_D = \frac{ei_0}{kT} \frac{(W_2 - x_B)^2}{3D_p} \qquad (6.29)$$

where i_0 is the direct current flowing through the diode. Figure 6.7 shows the small signal a-c equivalent circuit for a junction diode.

Note that in Fig. 6.6a, the current density J_1 in the n region is constituted by majority carriers and hence it must be supported by an electric field. The same is true for the current density J_2 in the p region. The voltage necessary to support J_1 and J_2 outside the diffusion region is represented by the voltage drop developed across R in Fig. 6.7, R being generally called the *spreading resistance* of a junction diode. The additional capacitance C_T in Fig. 6.7 is the transi-

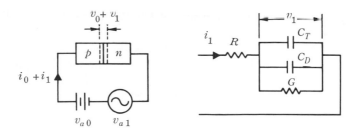

Figure
6.7

A-C equivalent circuit of a junction diode. To get the junction voltage v_1, the voltage drop across the spreading resistance must be subtracted from the applied voltage v_{a1}.

tion-region capacitance given by Eq. (6.13) due to charges stored in the space-charge region. The capacitance C_D given by Eq. (6.29) is generally referred to as the *diffusion capacitance* because it is associated with the diffusion of excess carriers. Figure 6.8 shows the variation of C_T and C_D as functions of the d-c bias voltage v_0. In the forward direction, C_D increases much faster with increasing forward bias than C_T. In the reverse direction, C_T decreases much more slowly with increasing reverse bias than C_D. Therefore, C_D usually dominates over C_T for forward-biased diodes and C_T over C_D for reverse-biased diodes.

Switching response of junction diodes. The equivalent circuit of Fig. 6.7 is useful only for small variations of the applied voltage. As shown in Fig. 6.8, the values of C_D and C_T, especially C_D, change drastically with the applied bias voltage v_0. Obviously, for switching diodes in which the mode of operation swings from v_0 being positive to negative or vice versa, an entirely different approach other than the small-signal, equivalent-circuit analysis must be undertaken.

Figure
6.8

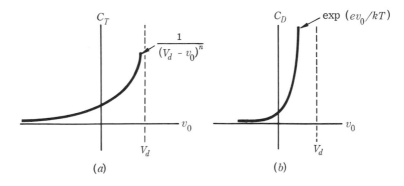

Diagrams showing the dependence of the junction capacitance C_T and that of the diffusion capacitance C_D upon the d-c voltage v_0 applied across the junction.

Consider the arrangement of Fig. 6.9 in which the applied voltage in a diode circuit is suddenly switched from a positive to a negative direction at time $t = 0$. The question is how does the current respond to this voltage change. To answer this question we must know the minority-carrier distribution in the base region. For ease of discussion, it is again assumed that the p side of the junction is much more heavily doped than the n side so only the injection of holes into the n region need be considered.

First, we realize that any change of hole distribution in the n region must be accomplished through the process of diffusion, and hence any redistribution of holes is a time-consuming process. In changing from the steady-state distribution in the forward direction to that in the reverse direction, the excess hole concentration right at the junction must pass through zero. Depending on whether the excess hole concentration p_1 at the edge of the space-charge region is greater or smaller than zero, the transient behavior of a junction diode can be divided into two phases: I and II, as shown in Fig. 6.10. During the phase I operation, the hole concentration p_n at the junction is larger than the equilibrium hole concentration p_{n0}. According to Eq. (6.18), the junction voltage is still biased in the forward direction although the applied voltage is already in the reverse direction. Since the voltage drop across the junction cannot be large in the forward direction, the current is mainly determined by the external resistance r or $I_r = V_r/r$. During the phase II operation, the hole concentration at the junction goes down below the equilibrium hole concentration and hence the junction becomes reversely biased. The junction voltage builds up gradually from zero to the full reverse applied voltage V_r, except for the small drop across the resistance r, and the current decreases gradually from V_r/r to I_s, the reverse saturation current of the diode.

The following mathematical analysis of the transient behavior of the hole distribution is given for a wide-base diode in which W, the

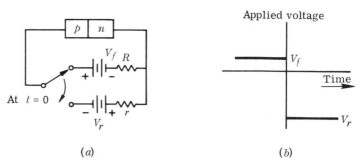

(a)   (b)

Diagrams showing that the applied voltage in a diode circuit is suddenly switched from positive V_f (forward direction) to negative V_r (reverse direction).

basewidth, is much larger than $L (= \sqrt{D_p\tau})$, the hole diffusion length. The time-dependent diffusion equation

$$\frac{\partial p_1}{\partial t} = D_p \frac{\partial^2 p_1}{\partial x^2} - \frac{p_1}{\tau} \qquad (6.30)$$

for excess hole concentration p_1 takes the form

$$D_p \frac{\partial^2 \bar{p}}{\partial x^2} = -p'(0) + s\bar{p} \qquad (6.31)$$

after the Laplace transform $\bar{p} = \mathcal{L}p'$ and the substitution $p_1 = p'$ exp $(-t/\tau)$. In Eq. (6.31), $p'(0)$ represents the initial distribution of excess holes. For the phase I operation,

$$p'(0) = p_{1f} \exp\left(-\frac{x}{L}\right) = p_{n0}\left[\exp\left(\frac{ev_{f0}}{kT}\right) - 1\right] \exp\left(-\frac{x}{L}\right) \quad (6.32)$$

where p_{1f} is the excess hole concentration at the edge of the space-charge region under a forward bias v_{f0} applied across the junction. It is also understood that the origin for x in Eq. (6.32) is taken right at the edge of the space-charge region. The general solution of Eq. (6.31) is given by

$$\bar{p} = B \exp(-\beta x) + \frac{p'(0)}{s - \alpha} \qquad (6.33)$$

where $\beta = \sqrt{s/D_p}$ and $\alpha = 1/\tau$.

The solution presented above is subject to the following boundary condition which relates the concentration gradient of p_1 at $x = 0$

Figure 6.10

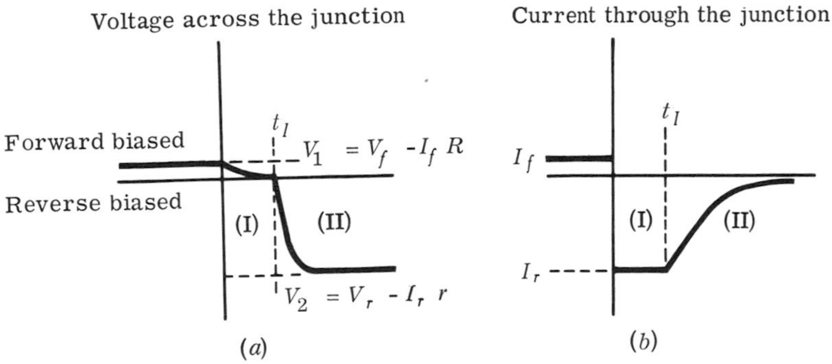

Diagrams showing (a) the voltage across and (b) the current through a junction diode as functions of time during the switching process. Note that the junction voltage remains in the forward-biased condition for a certain length of time after the applied voltage switches its direction. The operating condition of the diode is divided into two phases, depending on whether the junction voltage is forward-biased (phase I) or reverse-biased (phase II).

to the current flow or

$$\frac{-I_r}{eA} = -D_p \frac{\partial p_1}{\partial x}\bigg|_{x=0} = -D_p \exp\left(-\frac{t}{\tau}\right) \frac{\partial p'}{\partial x}\bigg|_{x=0}$$

In terms of $\bar{p}$, the above equation becomes

$$D_p \frac{\partial \bar{p}}{\partial x}\bigg|_{x=0} = \frac{I_r}{eA} \mathcal{L} \exp\left(\frac{t}{\tau}\right) = \frac{I_r}{eA(s-\alpha)} \tag{6.34}$$

where A is the cross-sectional area of the junction and $\mathcal{L}$ stands for Laplace transform. Applying the boundary condition of Eq. (6.34) to Eq. (6.33) yields

$$\bar{p} = \frac{L}{eAD_p} \frac{I_f}{s-\alpha} \exp\left(-\frac{x}{L}\right) - \frac{I_r + I_f \exp(-\beta x)}{eAD_p} \frac{\exp(-\beta x)}{\beta(s-\alpha)} \tag{6.35}$$

where $I_f = eAD_p p_{1f}/L$ is the forward junction current and $L = \sqrt{D_p \tau}$. Inverse Laplace transform of Eq. (6.35) results in the following:

$$p_1 = \frac{L}{2eAD_p} \left\{ 2I_f \exp(-v) - (I_r + I_f)\left[\exp(-v) \operatorname{erfc}\left(\frac{v}{2u} - u\right) \right.\right.$$
$$\left.\left. - \exp(v) \operatorname{erfc}\left(\frac{v}{2u} + u\right) \right] \right\} \tag{6.36}$$

In the above equation, erfc is the complementary error function and $v = x/L$ and $u = \sqrt{t/\tau}$. Changing the variable of integration in the erfc from $\xi = (v/2u) \pm u$ to u, Eq. (6.36) is further simplified to

$$p_1 = p_{1f} \exp\left(-\frac{x}{L}\right) - \frac{(I_r + I_f)LK(v,u)}{eAD_p} \tag{6.37}$$

with the understanding that

$$K(v,u) = \frac{2}{\sqrt{\pi}} \int_0^u \exp\left(-u^2 - \frac{v^2}{2u^2}\right) du \tag{6.38}$$

Figure 6.11 shows the progressive change with time of the excess hole distribution in the base region (that is, the n region) after the applied voltage is switched from a positive to a negative polarity. The phase I operation ends when the excess hole concentration at the junction drops to zero. Thus, letting $p_1 = 0$ at $t = t_1$ and $x = 0$ in Eq. (6.38) and noting that $K(0,u) = \operatorname{erf}(u)$, we obtain

$$\operatorname{erf}\left(\sqrt{\frac{t_1}{\tau}}\right) = \frac{I_f}{I_f + I_r} \tag{6.39}$$

The error function erf (u) can be approximated by $2u/\sqrt{\pi}$ if $u < 1$. Thus for $I_f/I_r < 0.5$, the time t_1 for phase I operation can be calculated with reasonable accuracy from $t_1 = 0.8 \, \tau(I_f/I_r)^2$.

After $t > t_1$, phase II operation begins. In Eq. (6.31), the initial excess hole distribution $p'(0)$ for phase II operation must be set

equal to the excess hole distribution in the phase I operation at
$t = t_I$. We immediately see that it would be very difficult to find the
particular solution of Eq. (6.31) if p_1 evaluated at $t = t_I$ from Eq.
(6.36) is taken as $p'(0)$ in Eq. (6.31). At this point, a reasonable
approximation must be made.

Here we follow the approach taken by Kingston[1] in his analysis
of the switching response of a junction diode. The initial excess hole
distribution is chosen to be the same as that in phase I; that is, at
$t = 0$, $p'(0)$ is given by Eq. (6.32). However, the boundary condition
at $x = 0$ is that for a reverse-biased diode or

$$\bar{p}(= -p_{n0}) \simeq 0 \qquad \text{at } x = 0 \tag{6.40}$$

The solution of Eq. (6.31) now reads

$$\bar{p} = \frac{p_{1f}[\exp(-x/L) - \exp(-\beta x)]}{s - \alpha} \tag{6.41}$$

Inverse Laplace transform of $\bar{p}$ gives

$$p_1 = \frac{LI_f}{2eAD_p}\left[2\exp(-v) - \exp(-v)\,\text{erfc}\left(\frac{v}{2u} - u\right)\right.$$
$$\left. - \exp(v)\,\text{erfc}\left(\frac{v}{2u} + u\right)\right] \tag{6.42}$$

For $u(= \sqrt{t/\tau}) = \infty$, $\text{erfc}(u) = 0$ and $\text{erfc}(-u) = 2$. Thus, at

[1]   R. H. Kingston, *Proc. IRE*, vol. 42, p. 829, 1954.

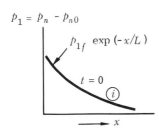

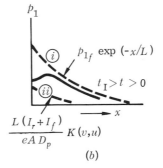

**Figure
6.11**

(a)

(b)

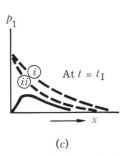

(c)

*Diagrams showing the progressive change with
time of the excess hole distribution in the base
region: (a) at $t = 0$, the initial distribution; (c)
at $t = t_I$, the distribution at the end of the phase
I operation; and (b) at $0 < t < t_I$, the intermediate distribution. According to Eq. (6.37), p_1
consists of two terms which are denoted by (i), the
first term, and (ii), the second term.*

Figure 6.12

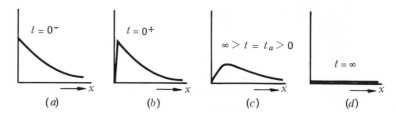

(a)   (b)   (c)   (d)

Diagrams showing the progressive change with time of the excess hole distribution in the base region. The distribution is obtained by assuming that the initial distribution of p_1 is given by Eq. (6.32) and that the boundary condition is given by Eq. (6.40). Note that the use of Eq. (6.40) forces an abrupt change of the slope (that means an infinite current) at $t = 0$.

$t = \infty$, p_1 is zero everywhere. That means, the excess hole distribution decreases gradually toward zero in the phase II operation in the manner shown in Fig. 6.12. However, our main interest here is not the time variation of p_1 but rather the behavior of the diode current as a function of time. Differentiating Eq. (6.42) with respect to x gives

$$I = -eAD_p \frac{\partial p_1}{\partial x}\bigg|_{x=0} = -I_f\left[\text{erf}\,(u) + \frac{\exp\,(-u)}{\sqrt{\pi u}} - 1\right] \quad (6.43)$$

Figure 6.13a shows the general shape of the diode current I in the second phase. The current goes to infinity at $t = 0$ because of the artificial boundary condition imposed on the solution. The condition that p_1 at $x = 0$ is nonzero for $t < 0$ and zero for $t > 0$ requires an infinite slope for p_1 at $x = 0$ and hence an infinite current. This is illustrated in Fig. 6.12a and b. Apparently, the solution p_1 of Eq. (6.42) near $t = 0$ is far from being the actual excess hole distribution.

Figure 6.13

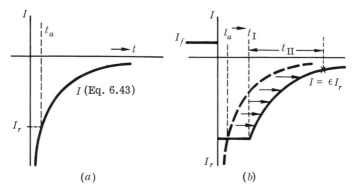

(a)   (b)

(a) The diode current calculated from the distribution of Fig. 6.12. (b) The diagram illustrates how the time axis of (a) must be shifted so that the curve can be used to approximate the actual diode current in the phase II operation.

Note, however, that at a time $t = t_a$ in Fig. 6.13a, the current I has the same value I_r as the current in the phase I operation. That means, the excess hole distribution given by Eq. (6.42) at $t = t_a$ has the same value and the same slope at $x = 0$ as the actual excess hole distribution given by Eq. (6.36) at $t = t_I$. Therefore, we can reasonably expect that for $t \geq t_a$, the approximate solution will be close to the actual solution. It has been shown by Kingston[2] and by Lax and Neustadter[3] that this is indeed the case. To obtain a complete solution of the current, we match the current I (phase I) $= I$ (phase II) at $t = t_I$ by shifting I (phase II) in time scale from t_a to t_I as illustrated in Fig. 6.13b. It should be pointed out that I approaches zero instead of I_s, the reverse saturation current of a diode, as t goes to infinity in Eq. (6.43). In our calculation, the reverse saturation current is taken to be negligibly small on account of the approximation ($\bar{p} \simeq 0$) made in Eq. (6.40).

The recovery time t_r of a diode can be defined as the time required for the current to drop to a certain percentage of I_r, say to ϵI_r where $\epsilon < 1$ is a preset value. From Fig. 6.13b, $t_r = t_I + t_{II}$. In Fig. 6.14a, $u = \sqrt{t_I/\tau}$ is plotted as a function of I_r/I_f according to Eq. (6.39), and hence the value of t_I can be read off from Fig. 6.14a for a chosen value of I_r/I_f in a switching circuit. In Fig. 6.14b, the behavior of I/I_f is plotted as a function of u. The time lapse for the quantity I/I_f to drop from I_r/I_f to $\epsilon I_r/I_f$ is the value for t_{II} as is obvious from Fig. 6.14b. As will be discussed shortly, the above analysis can easily be extended to narrow-base diodes where $W < L$. The only difference is that u will be normalized by a different time instead of the lifetime τ ($u = \sqrt{t/\tau}$ in the wide-base diode case). Therefore, Fig. 6.14 may be considered as universal curves in estimating the recovery time of a junction diode.

In Fig. 6.15, we compare the excess hole distribution in a wide-base diode (Fig. 6.15a) with that in a narrow-base diode. In the wide-base diode, our attention should be concentrated on the region within a diffusion length of the junction because it is the diffusion of minority carriers but not the drift of majority carriers which causes the time lag between the applied voltage and the diode current. The diffusion capacitance calculated earlier and the recovery time discussed above originate from the same physical process, that is the diffusion of minority carriers across the region where excess minority carriers are stored. A *diffusion velocity* v_d can be defined as

$$v_d = \frac{I}{Aep_1} = \frac{D_p}{p_1}\frac{\partial p_1}{\partial x} \qquad (6.44)$$

For the distribution of Fig. 6.15a, $v_d = D_p/L$. Hence, the time

[2] *Ibid.*
[3] B. Lax and S. F. Neustadter, *Jour. Appl. Phys.*, vol. 25, p. 1148, 1954.

**Figure
6.14**

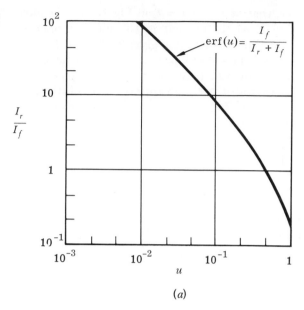

$$\mathrm{erf}(u)= \frac{I_f}{I_r + I_f}$$

(a)

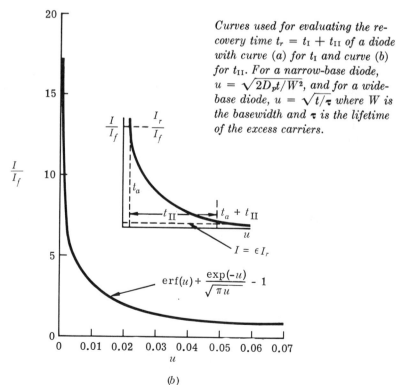

$$\mathrm{erf}(u) + \frac{\exp(-u)}{\sqrt{\pi u}} - 1$$

$I = \epsilon I_r$

Curves used for evaluating the recovery time $t_r = t_{\mathrm{I}} + t_{\mathrm{II}}$ of a diode with curve (a) for t_{I} and curve (b) for t_{II}. For a narrow-base diode, $u = \sqrt{2D_p t/W^2}$, and for a wide-base diode, $u = \sqrt{t/\tau}$ where W is the basewidth and τ is the lifetime of the excess carriers.

(b)

required for diffusion across the region where excess holes are stored is given by $t = L/v_d = \tau$ which is used as the time basis for normalization in the case of wide-base diodes. For the distribution of Fig. 6.15b, $v_d = 2D_p/W$ if p_1 in Eq. (6.44) is taken as the average value of the excess hole concentration in the base region. The diffusion time in this case is equal to $t = W/v_d = W^2/2D_p$. The transit times, τ and $W^2/2D_p$, can be considered as the characteristic response times of wide-base and narrow-base diodes respectively. Therefore, we expect that curves in Fig. 6.14 are still useful in estimating the recovery time of a narrow-base diode provided that the new u is made equal to $\sqrt{2D_p t/W^2}$. A complete and more accurate analysis of the narrow-base diode problem will be left for the reader as a problem.

Figure 6.15

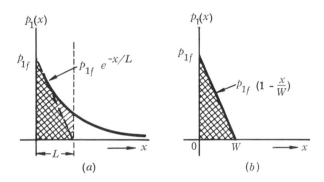

Comparison of the distributions of excess carriers in (a) *wide-base and* (b) *narrow-base diodes. Cross-hatched areas represent the region where most excess minority carriers are stored. The characteristic response time of a diode is defined as the average transit time for excess carriers to diffuse across the charge-storage region.*

The reader is also referred to articles by Lax and Neustadter[4] and by Gossick[5] for further discussions of the transient behavior of junction diodes.

6.3 JUNCTION TRANSISTORS[6]

Under d-c operating conditions. A transistor can be described as two diodes connected back to back, one operated in the forward and the other in the reverse direction. Refer to the p-n-p structure of Fig.

[4] *Ibid.*
[5] B. R. Gossick, *Jour. Appl. Phys.*, vol. 26, p. 1356, 1955, and vol. 27, p. 905, 1956.
[6] To follow the history of the development of transistors, the reader may find it rewarding to read the following series of articles by Bardeen, Brattain, and Shockley: *Phys. Rev.*, vol. 74, pp. 230–232, 1948; *Phys. Rev.*, vol. 75, p. 1208, 1949; *Bell System Tech. J.*, vol. 28, p. 435, 1949.

6.16a and its energy-band diagram in Fig. 6.16b. Holes and electrons are injected respectively into the n and p regions of the forward-biased diode on the left. The focus of our attention is on holes for a p-n-p transistor. Holes after being injected into the middle n region will continue to move across the n region into the p region on the right, which, because of the reverse bias, represents the lower energy region for holes. The middle region is generally referred to as the *base region* while the forward-biased junction is called the *emitter* and the reverse-biased junction is called the *collector*. From the qualitative description just presented, the currents through the emitter, base, and collector terminals under the d-c bias condition of Fig. 6.16a can be written respectively as

$$I_e = I_{en} + I_{ep}$$
$$I_c = \beta I_{ep} + I_{c0} = \beta\gamma I_e + I_{c0} = \alpha I_e + I_{c0} \qquad (6.45)$$
$$I_b = (1 - \alpha)I_e - I_{c0}$$

In Eq. (6.45), the term I_{c0} represents the reverse saturation current of the collector junction. The *current-amplification factor* α is a product of the *emitter efficiency* γ and the *transport efficiency* β. Of the two components I_{en} and I_{ep} of the emitter current, only one component, that is the hole current I_{ep} in a p-n-p transistor, will appear in the collector and hence become useful. The ratio of the

Figure 6.16

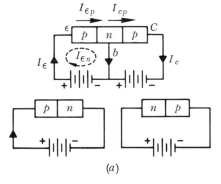

(a)

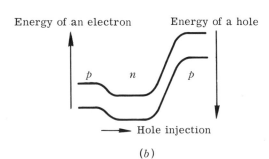

Energy of an electron   Energy of a hole

Hole injection

(b)

Diagrams showing (a) the structure of a transistor which can be considered as two diodes connected back to back and (b) the energy-band diagram for a p-n-p transistor. Note that holes will continue to move into the p-region on the right after being injected into the n-region because of energy considerations.

useful component to the total emitter current is defined as the emitter efficiency or $\gamma = I_{ep}/I_e$. Second, on account of recombination processes in the base region, part of the injected minority carriers will be lost during their journey from the emitter to the collector. The transport efficiency is defined as the percentage of injected carriers having survived the recombination process. Obviously, the quantity α is less than unity and the reader may wonder how transistors can be used as amplifiers. The answer lies in the fact that the emitter circuit is a low-impedance circuit while the collector circuit is a high-impedance circuit. Even though the current amplification is less than unity, power amplification is possible on account of the large ratio between the output and input voltages. For a quantitative analysis of the amplifying action of a transistor, we must treat the diffusion of minority carriers in the base region under a-c operating conditions. Since I_{c0} is a d-c current, it will not appear in the a-c equivalent circuit. However, the various parameters β, γ, and α, and the concepts of the physical processes associated with them are still useful in developing a dynamic relationship between I_e and I_c.

A-C small signal equivalent circuit. In the following discussion, we shall again assume a $p\text{-}n\text{-}p$ structure, similar to that of Fig. 6.16a but under simultaneous d-c and a-c bias conditions. For $n\text{-}p\text{-}n$ transistors, the discussion will be the same except that the role of holes and electrons will be interchanged. Figure 6.17a shows the geometry of an alloyed transistor in which the emitter and collector

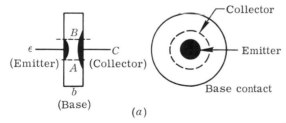

(a)

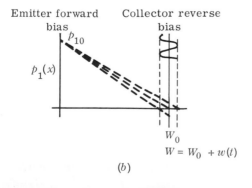

(b)

Diagrams showing (a) the geometry of an alloyed transistor and (b) the distribution of excess carriers (holes in the present case) in the base region. Note that because of the a-c variation of the collector voltage, the boundary of the collector junction and hence the width of the base region also vary with time.

junctions are made by alloying processes. The presence of the base lead and hence the base current makes the problem seem difficult. Our immediate task, then, is to see if the diffusion of injected minority carriers across the base region can be treated as a one-dimensional problem under justifiable assumptions. First, the quantity α for reasonably good transistors is close to 1, say 0.99, thus the base current I_b is going to be much smaller than the emitter or collector current. Second, the base contact in a practical transistor is far away from the region AB in Fig. 6.17 where the injected carriers are concentrated. This means, the presence of the base contact will not affect significantly the distribution of injected carriers in region AB. In addition, it is assumed that the surface of the transistor is well etched so that the surface recombination velocity is low and consequently almost all injected carriers will diffuse toward the collector rather than toward the surface. Under these assumptions, it is a reasonably good approximation to treat the diffusion of injected carriers in the base region as a one-dimensional problem.

Figure 6.17b shows the one-dimensional model of the base region in which the boundary condition at the emitter junction ($x = 0$) is given by $p_1 = p_{n0} [\exp (ev_e/kT) - 1]$ while that at the collector junction ($x = W$) by $p_1 = p_{n0} [\exp (ev_c/kT) - 1]$. Since the collector is negatively biased and the emitter positively biased, $|p_1(x = 0)| \gg |p_1(x = W)|$ and thus it is a reasonable approximation that $p_1(x = W)$ can be set equal to zero. The problem is now reduced to the same form as the diode problem discussed in Sec. 6.2 except for the following consideration: In a transistor, there is an a-c voltage appearing across the collector junction. Since the collector is reverse-biased, the a-c voltage will not affect significantly the condition that $p_1(x = W) = 0$ but it will modulate the width of the space-charge region and thereby the width of the base region. This is called the *basewidth modulation effect* and is due to collector junction bias. The width of the emitter space-charge region is much smaller than that of the collector on account of the forward-bias condition, hence the effect of the emitter bias on the basewidth can be neglected.

Let $W = W_0 + w(t)$ where W_0 and $w(t)$ denote, respectively, the d-c and time-varying parts of W. Since $p_1(x = W) = 0$, $p_1(x = W_0)$ is now a function of time. This is illustrated in Fig. 6.17b with dotted lines indicating the moving boundary of the collector junction. As a first-order approximation,

$$p_1(x = W_0) = p_1(x = W) - \frac{\partial p_1}{\partial x} w(t) \tag{6.46}$$

To find the a-c value of p_1 at $x = W_0$, we substitute the d-c distribution of p_1 into $\partial p_1/\partial x$ in the above equation. Furthermore, for a practical transistor, the basewidth is much smaller than the diffusion length so that Eq. (6.22) applies. Therefore, the boundary condition

at $x = W_0$ becomes

$$(p_1)_{\text{a-c}} = \frac{p_{10}}{W_0} w(t) \qquad (6.47)$$

by noting that $W_2 - x_B$ in Eq. (6.22) is presently equal to W_0. The other boundary condition at $x = 0$, the emitter junction, is $(p_1)_{\text{a-c}} = (ev_{\epsilon 1}/kT)p_{10}$ as given by Eq. (6.26) where $p_{10} = p_{n0} \exp (ev_{\epsilon 0}/kT)$, $v_{\epsilon 0}$ and $v_{\epsilon 1}$ being, respectively, the d-c and a-c parts of the emitter bias.

The solution of the time-dependent diffusion equation subject to the above boundary conditions is given by

$$(p_1)_{\text{a-c}} = p_{10} \left[\frac{ev_{\epsilon 1}}{kT} \frac{\sinh (W_0 - x)/L'}{\sinh (W_0/L')} + \frac{w(t)}{W_0} \frac{\sinh (x/L')}{\sinh (W_0/L')} \right] \qquad (6.48)$$

where $L' = \sqrt{D_p \tau / (1 + i\omega\tau)}$ as in the diode case. From Eq. (6.48), the a-c emitter and collector current respectively can be evaluated as follows:

$$i_\epsilon = -AeD_p \frac{\partial p_1}{\partial x}\bigg|_{x=0}$$

$$= \frac{AeD_p}{L'} (p_{10}) \left[\frac{ev_{\epsilon 1}}{kT} \coth \frac{W_0}{L'} - \frac{w(t)}{W_0} \operatorname{csch} \frac{W_0}{L'} \right]$$

$$i_c = -AeD_p \frac{\partial p_1}{\partial x}\bigg|_{x=W_0} \qquad (6.49)$$

$$= \frac{AeD_p}{L'} (p_{10}) \left[\frac{ev_{\epsilon 1}}{kT} \operatorname{csch} \frac{W_0}{L'} - \frac{w(t)}{W_0} \coth \frac{W_0}{L'} \right]$$

where A is the cross-sectional area of the junction. The current expressions can easily be written in terms of input, output, and transfer admittances as

$$i_1 = Y_{11}v_1 + Y_{12}v_2$$
$$i_2 = Y_{21}v_1 + Y_{22}v_2 \qquad (6.50)$$

with the emitter circuit as the input circuit and the collector circuit the output circuit. The equivalent circuit representing Eq. (6.49) is shown in Fig. 6.18 and the evaluation of Y_{11}, Y_{12}, Y_{21}, and Y_{22} is given below.

Figure 6.18

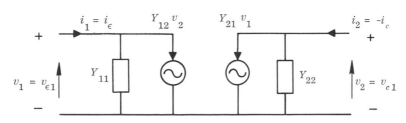

Circuit representation of Eq. (6.49).

Let $Y_{\epsilon 0} = \dfrac{eI_\epsilon}{kT}$ be the d-c admittance of the emitter junction, I_ϵ being the d-c emitter current. Expanding the hyperbolic functions in a Taylor's series, neglecting the high-order terms in $W_0^2/D_p\tau$, and recognizing that $I_\epsilon = AeD_p p_{10}/W_0$, we find

$$Y_{11} = Y_{\epsilon 0} \frac{1 + i\omega/\omega_\alpha}{1 + i\omega/(3\omega_\alpha)}$$

$$Y_{21} = -Y_{\epsilon 0} \frac{\beta_0}{1 + i\omega/(3\omega_\alpha)}$$

(6.51)

In the above equation, $\omega_\alpha = 2D_p/W_0^2$ and $\beta_0 = [1 + W_0^2/(2D_p\tau)]^{-1}$. We shall discuss the physical meaning of these quantities presently. The current, due to excess holes diffusing across the base region, suffers both amplitude and phase changes, the amplitude change being due to recombination processes and the phase change to the transit time in the base region. Refer to the excess hole distribution for a narrow-base transistor. The transit time across the base region is given by $W_0^2/2D_p$ as discussed at the end of Sec. 6.2. Therefore, it is no wonder that Y_{11} is advanced in phase with respect to Y_{21} by a factor $1 + i\omega/\omega_\alpha$ since Y_{11} comes into the expression of i_ϵ while Y_{21} comes into that of i_c. It is expected that the rate of decay of injected holes is proportional to $\exp(-t/\tau)$, t being the transit time of holes. For $t < \tau$, $\exp(-t/\tau)$ can be approximated by $(1 + t/\tau)^{-1} = \beta_0$. This explains why β_0 appears in the expression of Y_{21}.

The dependence of the emitter and collector currents upon the collector voltage can be found as follows. The space-charge region of the collector junction extends from the fixed junction boundary (fixed by metallurgical process) into the n and p regions. However, only the part extending into the n region in a p-n-p transistor affects the width of the base region. Therefore, using x_B from Eq. (6.10) and replacing V_d by $V_d + v_{c0}$, we find

$$w(t) = + \left| \frac{\partial x_B}{\partial v_{c0}} \right| v_{c1}$$

(6.52)

where v_{c0} and v_{c1} are, respectively, the d-c and a-c parts of the collector bias. The sign in Eq. (6.52) is in accordance with the polarity of v_{c1} defined in Fig. 6.18. Substituting Eq. (6.52) into Eq. (6.49) and comparing Eq. (6.49) with Eq. (6.50), we obtain

$$Y_{22} = Y_{c0} \frac{1 + i\omega/\omega_\alpha}{1 + i\omega/(3\omega_\alpha)}$$

$$Y_{12} = -Y_{c0} \frac{\beta_0}{1 + i\omega/(3\omega_\alpha)}$$

(6.53)

where $Y_{c0} = (I_\epsilon/W_0)|\partial x_B/\partial v_{c0}|$. Note that the phase factor $(1 + i\omega/\omega_\alpha)$ and the amplitude factor β_0 in Y_{22} and Y_{12} are indicative of the fact that the effect of v_{c1} upon i_c and i_e starts out at the collector junction.

To complete our discussion of the circuit representation of a transistor, we shall add the following comments: First, the space-charge region capacitance discussed in Sec. 6.1 must be added to the circuit of Fig. 6.18 as well as the base resistance from the base contact to the diffusion region AB of Fig. 6.17. Furthermore, there are other versions of circuit representation. Figure 6.19 shows one version (with the base terminal common to both input and output circuits) of the complete equivalent circuit of a transistor at low frequencies. For $\omega < \omega_a$, all admittances Y_{11}, Y_{12}, Y_{21}, and Y_{22} become real. In Fig. 6.19, $r_\epsilon = 1/Y_{\epsilon 0}$ and $r_c = 1/Y_{c0}$. It is also to be noted that the current generator in the collector circuit is now dependent upon the emitter current instead of the emitter voltage. The factor γ in α again accounts for the fact that the emitter current has a current component which does not appear in Eq. (6.49), namely the component due to diffusion of electrons into the p region of the emitter junction. The reader is referred to articles by Early[7] for further discussions of the base-width modulation effect and by Zawels[8] for various circuit representations of junction transistors.

Finally, we should point out that in analyzing the transient behavior of transistor circuits a compromise must be made between simplicity and accuracy of the analysis. The large-signal theory developed in Sec. 6.2 (also applicable to transistors with proper modifications) and the small-signal a-c equivalent circuit developed above are often too cumbersome for circuit engineers in their routine

[7] J. M. Early, *Bell System Tech. J.*, vol. 32, p. 1271, 1953, and vol. 32, p. 1305, 1953.
[8] J. Zawels, *Jour. Appl. Phys.*, vol. 25, p. 976, 1954.

Figure 6.19

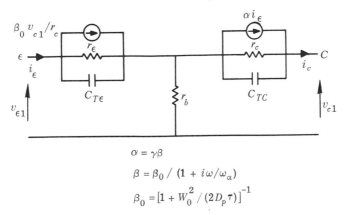

$$\alpha = \gamma\beta$$
$$\beta = \beta_0 / (1 + i\omega/\omega_\alpha)$$
$$\beta_0 = [1 + W_0^2 / (2D_p\tau)]^{-1}$$

Low-frequency equivalent circuit of a transistor in common base connection with the base terminal common to both input and output circuits.

design. Many approximation methods have been suggested for this purpose. Among these, the charge-control method[9] useful for small-signal analysis and the approach taken by Ebers and Moll[10] should be mentioned, and the reader is referred to the cited references for a general discussion on the subject.

Design considerations. For the following discussion, let us consider the equivalent circuit of Fig. 6.20, a simplified version of Fig. 6.19. Since $r_c \gg r_\epsilon$ and $C_{TC} > C_{T\epsilon}$, the frequency response of the circuit is determined by the time constant $r_c C_{TC}$ of the collector circuit and not by $r_\epsilon C_{T\epsilon}$. This is the reason why $C_{T\epsilon}$ is omitted in Fig. 6.20. It can easily be shown from circuit analysis that the effect of the feedback current generator $\beta_0 v_{c1}/r_c$ can be incorporated into the base-emitter circuit by changing r_b to r_b'. Suppose that the transistor is connected on the input side to a voltage generator with an input impedance R_g and on the output side to a load with load resistance R_L. If η is defined as the ratio of the output power ($i_c^2 R_L$) to the maximum power available ($v_g^2/4R_g$), then under matched input and output conditions, the maximum available gain at very low frequencies is approximately equal to $\eta_{\max} = \alpha^2 r_c/r_b'$. The reader is asked to check the result.

At higher frequencies, not only does the shunting effect of C_{TC} become noticeable, but the transit-time effect also comes into play. As a rough approximation without a detailed analysis of the high-frequency equivalent circuit, $\eta_{\max}$ can be inferred from the low-frequency result as

$$\eta = \frac{\alpha_0^2}{(1 + i\omega/\omega_\alpha)^2} \frac{r_c}{r_b'(1 + i\omega/\omega_c)} \tag{6.54}$$

[9] See, for example: Discussion in (1) *SEEC Notes* 2, "PEM: Physical Electronics and Circuit Models of Transistors," chap. 10, by P. E. Gray and co-authors, John Wiley & Sons, Inc., New York, 1964; (2) J. L. Moll, "Physics of Semiconductors," chap. 8, McGraw-Hill Book Company, New York, 1964; and (3) J. J. Sparkes, A Study of the Charge Control Parameters of Transistors, *Proc. IRE*, vol. 48, p. 1696, 1960.

[10] J. J. Ebers and J. L. Moll, *Proc. IRE*, vol. 42, p. 1761, 1954; J. L. Moll, *Proc. IRE*, vol. 42, p. 1173, 1954.

Figure 6.20

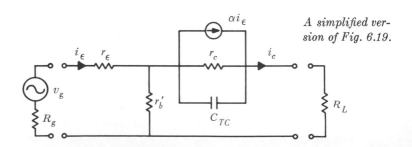

A simplified version of Fig. 6.19.

where $\omega_c = 1/(r_c C_{TC})$ and $\omega_\alpha = 2D/W_0{}^2$. As ω approaches ω_c or ω_α, $|\eta|$ decreases. For a high-frequency transistor, the basewidth W_0 must be small and the collector junction capacitance C_{TC} must be low. Usually the collector capacitance poses a more serious limitation, or in other words, $\omega_c < \omega_\alpha$. Under such circumstances, the gain-bandwidth product of a transistor is given by

$$GB = \eta_0 \omega_c = \frac{\alpha_0{}^2}{r_b' C_{TC}} \qquad (6.55)$$

It is obvious that r_b' must be small for high gain.

There are other considerations besides the gain and frequency response of a transistor. The current I_{c0}, being a d-c current, contributes to the heat dissipation in the collector junction. It is also known that a junction breaks down if the junction voltage approaches a certain limit in the reverse direction. A detailed discussion of the breakdown phenomenon will be given in a later section. For the present discussion, it suffices to say that a high breakdown voltage requires a low doping concentration N_a or N_d. Another effect which concerns the power handling capacity of a transistor is the base-resistance bias effect. To minimize the effect, it is essential to reduce the base current and the base resistance. A low base current requires a high emitter efficiency. Our task is to consider all these effects and to choose our design parameters such as the basewidth and the doping concentrations accordingly.

Unfortunately, the requirements for meeting the above considerations are often conflicting. Consider, for example, the basewidth. Keeping other conditions the same, a narrower basewidth means a higher β and a higher ω_α but in the meantime a higher r_b which in turn results in a lower gain-bandwidth product and lower power handling capacity. Now let us discuss the choice of the base resistivity. A lower resistivity means a higher gain-bandwidth product but at the same time it results in a lower emitter efficiency, a lower breakdown voltage and a higher collector capacitance. The choice of the design parameters of a transistor necessarily involves compromises to suit best a set of particular needs. For example, for a power transistor, the power handling capacity is more important than the frequency response. It is also possible that some of the conflicts in requirements can be resolved or minimized by using advanced technology and better design techniques.

First, let us discuss the emitter junction. According to Eq. (6.23), the ratio of the current due to hole diffusion to that due to electron diffusion is proportional to p_{n0}/n_{p0} or inversely proportional to n_n/p_p and σ_n/σ_p, σ_n and σ_p being respectively the conductivity of the n and p regions. In a transistor, the emitter efficiency γ depends on σ_e/σ_b in a similar way, the higher the ratio of emitter to base conductivity, the higher the emitter efficiency. Therefore, we should seek those impurities with high solubilities as dopants in the emitter region. Now, we shall turn our discussion to the collector junction. Equation (6.23) also tells us that the reverse saturation current I_{c0} decreases with decreasing p_{n0} and n_{p0}, or in other words, I_{c0} decreases with increasing σ_n and σ_p. However, as mentioned earlier, a higher doping concentration means a lower breakdown voltage and a higher collector capacitance. The use of an n-i-p structure for the collector junction has been advanced by Early[11] and has proved very successful. The intrinsic region sandwiched between the n and p regions not only increases the separation between the charges stored in the n and p regions but also takes up most of the voltage drop across the junction. Therefore, a p-i-n junction has a lower collector capacitance and higher breakdown voltage than a corresponding p-n junction without the intermediate intrinsic region.

6.4 DIFFUSED (DRIFT) TRANSISTORS

In the previous discussion, we have assumed a uniform distribution of impurities in the base region. This is the case with alloyed or grown transistors. In the alloying process, a small pellet of metal dopant, say In, is placed on top of a semiconductor slab, say n-type germanium doped with Sb, and the combination is then placed in an oven heated to a temperature above the eutectic temperature. A molten zone of In-Ge mixture forms of which the composition is determined by the Ge-In phase diagram. If the temperature is lowered toward the eutectic point, part of the molten zone begins to solidify. The recrystallized region which contains mostly Ge is deposited on the parent Ge slab which acts as a seed for recrystallization. Since In has a finite solid solubility in Ge, the recrystallized Ge is converted into p-type, creating an abrupt p-n junction. If a similar process is carried out on the other side of the wafer, a p-n-p structure is formed. The impurity distribution in the base region is not disturbed during the alloying process and hence can be assumed uniform in an alloyed transistor.

In the grown transistor, the doping concentration is controlled by the growth rate of the solid which is pulled from a melt containing both donors and acceptors. For each impurity, there exists a definite ratio C_S/C_L called the segregation constant K, C_L and C_S being, respectively, the solubility of the given impurity in the liquid and solid phase. Since K is usually smaller than 1, the impurities rejected from the solid phase momentarily will pile up in the liquid phase at the solid-liquid interface. The smaller the K value, the larger will be the pileup. If the pulling speed or rate of solidification is suddenly increased, the rejected impurities may not have enough time to be dispersed uniformly throughout the liquid zone and thus they are

[11]   J. M. Early, *Bell System Tech. J.*, vol. 32, p. 517, 1954.

incorporated in the solid with a concentration higher than the equilibrium C_S. For impurities with a larger K, a smaller amount of impurities will be rejected from the solidifying zone and consequently a less drastic change of the doping concentration results if the pulling speed suddenly changes. Figure 6.21 shows the concentrations of acceptor and donor impurities at intervals of slow and rapid pull in a crystal originally doped with both impurities, K for acceptor impurities being greater than that for donor impurities. Since the pulling speed cannot change instantaneously, the change of impurity concentrations in the transition region must be gradual, resulting in a graded junction. Furthermore, because $K < 1$, as the solid phase advances, the liquid zone becomes richer in impurities. This explains why the impurity concentrations have a positive slope toward the right in Fig. 6.21. However, the increase is so slight that the doping concentration in the base region can be considered essentially constant.

The situation is quite different with diffused transistors. One of the most important tasks in the fabrication of transistors is the accurate control of the basewidth. For high-frequency units, say $\omega_\alpha = 10^8$ rad/sec, we have $W_0 = 10^{-3}$ cm for $D = 50$ cm^2/sec. Obviously an accurate control of such a thin basewidth with ordinary mechanical means is out of the question. It is made possible, however, through the diffusion techniques. If a semiconductor slab originally doped with an acceptor concentration N_a is placed in an oven in an inert atmosphere containing a partial pressure of a donor impurity, then part of the semiconductor near the surface is converted into n-type through the diffusion of donor impurities into the solid. The concentration of donor impurity, C_x, at a distance x below the surface is given by

$$C_x = C_0 \operatorname{erfc}\left(\frac{x}{2\sqrt{Dt}}\right) \tag{6.56}$$

The point at which the C_x-x curve intercepts the N_a line defines the boundary of a p-n junction as shown in Fig. 6.22a. In Eq. (6.56), C_0 is the donor concentration at surface which is determined by the

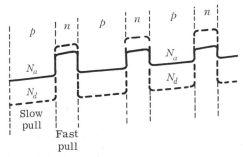

Acceptor and donor concentrations at intervals of slow and rapid pull in a rate-grown process. The donor which has a smaller segregation constant shows a much larger change in the concentration than the acceptor as the pulling speed changes.

partial vapor pressure of the diffusant, D the diffusion constant of the diffusant, and t the duration of diffusion. The diffusion of impurities, unlike the diffusion of electrons and holes, usually requires that impurity atoms in changing position must overcome a potential barrier ε. Therefore, as given by Eq. (2.62), D in Eq. (6.56) varies exponentially with temperature T as

$$ D = D_0 \exp\left(-\frac{\varepsilon}{kT}\right) \tag{6.57} $$

By accurately controlling the duration and temperature of diffusion, the penetration depth d of Fig. 6.22a can be held to a value within reasonable limits.

For lack of space, it is impossible to discuss in great detail metallurgical processes which are important in semiconductor technology. The reader is referred to the following references for further discussions on (1) purification of semiconductors: W. G. Pfann, "Zone-Melting," John Wiley & Sons, Inc., New York, 1958; (2) alloying: J. S. Saby and W. C. Dunlap, Jr., *Phys. Rev.*, vol. 90, p. 630, 1953; (3) rate-grown junction: R. N. Hall, *Phys. Rev.*, vol. 88, p. 139, 1952; (4) diffusion in Ge: W. C. Dunlap, Jr., *Phys. Rev.*, vol. 94, p. 1531, 1954; and (5) diffusion in Si: C. S. Fuller and J. A. Ditzenberger, *Jour. Appl. Phys.*, vol. 27, p. 544, 1956. The "Transistor Technology" series published by D. Van Nostrand Co., which contain a wealth of technical information, should be especially noted.

The transistor structure shown in Fig. 6.22b can be obtained (1) through a double-diffusion process of donors and acceptors at two different temperatures or (2) through a post-alloy diffusion process in which both donors and acceptors are first alloyed into the semiconductor and then diffuse at different rates inside the semiconductor.

Figure 6.22

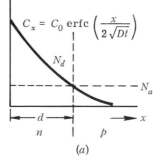

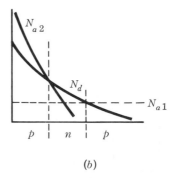

(a) (b)

(a) *Formation of a p-n junction by diffusing donor impurities into a p-type substrate. The depth d of the junction is controlled by the duration (t) and temperature (T and hence the value of D) of diffusion. (b) Formation of a p-n-p transistor by a double diffusion process.*

It is by the very nature of the diffusion process that the impurity concentration changes a great deal in the base region. In the following, we shall discuss the possible effect of such a nonuniform distribution upon the performance of a transistor. As our discussion in the previous section shows, the three important factors which limit the performance of a high-frequency transistor are: (1) the transit time, (2) the collector capacitance, and (3) the base resistance; hence our discussion will be centered around these three quantities. For definiteness we take a p-n-p transistor and for ease of discussion we shall approximate the complementary error function of Eq. (6.56) by an exponential function or

$$N_d = N(x) = N_0 \exp\left(-\frac{x}{l}\right)$$
(6.58)

where l is a characteristic distance. Assuming that the donors are fully ionized, we have

$$n = N_c \exp\left(-\frac{\mathcal{E}_c - \mathcal{E}_f}{kT}\right) = N_d = N_0 \exp\left(-\frac{x}{l}\right)$$
(6.59)

Since $\mathcal{E}_f$ under thermal equilibrium stays constant throughout the base region, $\mathcal{E}_c$ must change with x, as shown in Fig. 6.23b, creating an electric field E_x.

If V represents the electrostatic potential experienced by electrons, then the electric field E_x is given by

$$E_x = -\frac{\partial V}{\partial x} = \frac{1}{e}\frac{\partial \mathcal{E}_c}{\partial x} = \frac{-kT}{e}\frac{1}{N}\frac{dN}{dx} = \frac{kT}{el}$$
(6.60)

From Fig. 6.23 we see that the diffusion force on electrons is toward the right; thus the electric force must be toward the left to counterbalance the diffusion force. This is consistent with the sign in Eq.

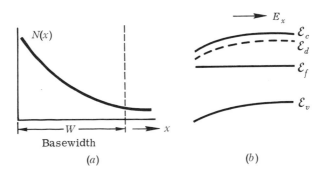

Figure 6.23

(a)

(b)

(a) Variation of the impurity concentration in the base region which in turn results in (b) the change of the electrostatic potential (that is, $\mathcal{E}_c$ and $\mathcal{E}_v$) with distance.

(6.60). Under an electric field, the expression for the hole current density becomes

$$J_p = e\mu_p p_1 E_x - eD_p \frac{\partial p_1}{\partial x} = -eD_p \left(\frac{p_1}{N} \frac{dN}{dx} + \frac{dp_1}{dx} \right) \qquad (6.61)$$

where p_1 is the excess hole density in the base region. If it is assumed that recombination of excess carriers is negligible in the base region, J_p must be independent of x as required by the continuity equation. Thus, we have

$$\frac{dp_1}{dx} + \frac{p_1}{N} \frac{dN}{dx} = -\frac{J_p}{eD_p} \qquad (6.62)$$

The solution of Eq. (6.62) is given by

$$p_1(x) = \frac{J_p}{eD_p N(x)} \left[\int_x^W N(x) \, dx \right] \qquad (6.63)$$

in view of the boundary condition that at the collector junction $x = W$, $p_1 = 0$.

The total charge Q stored in the base region of a cross-sectional area A is equal to $Q = A \int_0^W e p_1(x) \, dx$. It seems reasonable that the transit time t can be defined as the time required to move Q charges across the base region to constitute a total current $J_p A$. Thus, we have

$$t = \frac{Q}{AJ_p} = \frac{\int_0^W e p_1 \, dx}{J_p} \qquad (6.64)$$

For a uniform distribution of impurities, $N(x) = $ const, thus Eq. (6.63) reduces to $p_1(x) = p_{10}(1 - x/W)$ and Eq. (6.64) reduces to $t = W^2/2D_p$, realizing that $J_p = eD_p p_{10}/W$ where p_{10} is the excess hole concentration at the emitter junction. The results are exactly the same as those derived previously for the case of a narrow-base diode. This is what we expect in view of an earlier assumption ($\tau = \infty$) leading to Eq. (6.62). Substituting Eq. (6.58) into Eq. (6.63) and Eq. (6.64), we find

$$p_1(x) = \frac{lJ_p}{eD_p} \left[1 - \exp\left(-\frac{W - x}{l} \right) \right] \qquad (6.65)$$

$$t = \frac{l}{D_p} \left[W - l + l \exp\left(-\frac{W}{l} \right) \right] \qquad (6.66)$$

Now let us compare the performance of transistors having impurity distributions: (A) uniform, (B) decreasing toward the collector junction, and (C) increasing toward the collector junction as shown in Fig. 6.24a. In making the comparison, it is necessary to impose common constraints on the parameters. For our present discussion, we choose the following as the common constraints: (1) the same impurity concentration N_0 at the emitter junction, (2) the same emitter current density J_p, and (3) the same base resistance r_b. Except for a geometric factor, the base resistance is proportional to R which has a dimension of resistance and is defined as

$$\frac{1}{R} = e\mu_n \int_0^W N(x)\, dx = e\mu_n N l \left[1 - \exp\left(-\frac{W}{l}\right) \right] \qquad (6.67)$$

To maintain the same base resistance, we must have

$$W_0 = l\left[1 - \exp\left(-\frac{W_1}{l}\right) \right] = l\left[\exp\left(\frac{W_2}{l}\right) - 1 \right] \qquad (6.68)$$

Note that the expressions derived for case B can easily be adopted for case C by changing l to $-l$. Using Eqs. (6.65) and (6.68), we also find that for same J_p, the excess hole concentrations at the emitter junction $p_1(0)$ are the same for all three cases.

We further see from Eq. (6.61) that the slope $\partial p_1/\partial x$ at the collector junction must be the same for all three cases since $p_1 = 0$ at the collector junction. In view of the above discussion, the hole distribution in the base region takes the general shape shown in Fig. 6.24b. For case B, the electric field is aiding the hole current and thus, for the same J_p, case B requires a lower $\partial p_1/\partial x$ at the emitter junction. The opposite is true for case C. It appears on the surface that the electric field in case B may cut down the transit time and hence improve the high-frequency response of a transistor. Actually, the situation depends very much on the constraints which we impose, as we shall see presently. For $W_0/l < 1$, Eq. (6.68) reduces to $W_1 = W_0(1 + W_0/2l + W_0^2/3l^2)$. From Eq. (6.66), we find $t = l(W_1 - W_0)/D_p = (W_0^2/2D_p)(1 + 2W_0/2l)$. This result shows that

Figure 6.24

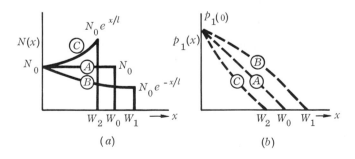

(a) (b)

(a) Three impurity distributions under consideration. For the same base resistance, the basewidths W_0, W_1, and W_2 for cases A, B, and C are such that $W_1 > W_0 > W_2$. (b) Excess hole distributions in the base region. Same r_b demands the same $p_1(x)$ at the emitter junction for all three cases through Eqs. (6.65) and (6.68). Same emitter current demands the same slope at the collector junction through Eq. (6.61).

even though the electric field is aiding, the transit time in case B is longer than that in case A because of a larger basewidth. Among the three cases, case C has the shortest transit time. The situation is entirely different if the constraint requiring the same base resistance is changed to that requiring the same basewidth. The reader is asked to work out the latter case as a problem. We also should add that the three cases under discussion have the same emitter efficiency but case B has the lowest collector capacitance if the doping concentrations in both the emitter and collector are the same for all three cases. It is obvious that we should take all these factors into consideration in evaluating the effect of the built-in electric field. As a historical note, a transistor having a nonvanishing electric field in the base region due to a nonuniform impurity distribution is generally referred to as a drift transistor.[12] The small-signal frequency response of a drift transistor has been studied by Varnerin[13] and by Moll and Ross[14] while the switching response of such a device has been analyzed by Muto and Wang.[15]

6.5 CONSIDERATIONS AT HIGH INJECTION LEVELS

The current-amplification factor of a transistor is not a constant but is a function of the emitter current. In Fig. 6.25, the quantity $1/(1 - \alpha)$ or I_ϵ/I_b is shown as a function of I_ϵ for a typical power transistor. The value of $1/(1 - \alpha)$ may reach a maximum value around 150 and drop to around 30 at high emitter currents. To explain the effect, we shall examine what constitutes the base current. As mentioned earlier in Sec. 6.3, of the two components (hole and electron) of I_ϵ, only one component goes to the collector while the other constitutes a part of the base current which will be denoted as I_{b1}. According to the definition of emitter efficiency γ, $I_{b1} = (1 - \gamma)I_\epsilon$. Another source of base current comes from recombination of injected carriers in the base region. Referring back to Eq. (6.45) as

[12]   H. Kromer, "Zur Theorie des Diffusion—und des Drift Transistors," *Arch. elek. Ubertragung*, vol. 8, pp. 223–228, 363–369, 499–504, 1954.

[13]   L. J. Varnerin, *Proc. IRE*, vol. 47, p. 523, 1959.

[14]   J. L. Moll and I. M. Ross, *Proc. IRE*, vol. 44, p. 72, 1958.

[15]   S. Y. Muto and S. Wang, *IRE Trans. on Electron Devices*, vol. ED-6, p. 183, 1962.

Figure 6.25

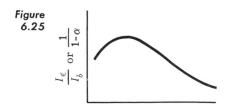

Curve showing the variation of $1/(1 - \alpha)$ or I_ϵ/I_b as a function of the emitter current in a power transistor.

an example, $I_{b2} = I_{ep} - I_{cp}$ which can be approximated by $I_e W_0^2/(2D\tau)$ for a narrow-base transistor. The third component of base current, which has not been discussed previously, is due to recombination of injected carriers at the surface of the base region.

The most effective region for surface recombination is near the emitter where the concentration of injected carriers is high. However, injected carriers are drawn toward the collector as well as toward the surface, the former process being the result of the collector bias, the latter, that of recombination. Since these two processes are directly competing, the effective region of surface recombination will be limited to a radius $r_e + W_0$ where W_0 and r_e are, respectively, the basewidth and the radius of the emitter. If v_s is the surface recombination velocity and p_{10} the excess hole concentration at the emitter junction,

$$I_{b3} = ev_s p_{10}[\pi(r_e + W_0)^2 - \pi r_e^2] = \frac{2v_s W_0^2 I_e}{Dr_e} \qquad (6.69)$$

for $W_0 \ll r_e$. Summing up all the three components, we have

$$I_b = \left[(1 - \gamma) + \frac{W_0^2}{2D\tau} + \frac{2v_s W_0^2}{Dr_e}\right] I_e \qquad (6.70)$$

In the discussion presented so far, we have always assumed that the excess minority-carrier concentration is smaller than the equilibrium majority-carrier concentration. In a power transistor, this is no longer true. In the following, we shall discuss possible effects at high injection levels upon the three components of the base current.

First, at high injection levels, an electric field E is set up in the base region in such a way as to oppose the majority-carrier flow and to aid the minority-carrier flow. It is understood in the present discussion that the impurity distribution in the base region is uniform so the electric field has an origin different from the one discussed in the previous section. Consider a p-n-p transistor. The electron and hole current densities in the base region are respectively given by

$$J_n = e\mu_n n E_x + eD_n \frac{\partial n}{\partial x} \qquad (6.71)$$

$$J_p = e\mu_p p E_x - eD_p \frac{\partial p}{\partial x} \qquad (6.72)$$

As stated in the previous chapter, the charge-neutrality condition requires that $n_1 = p_1$, n_1 and p_1 being respectively the excess majority- and minority-carrier concentrations. The question then is raised as to what happens to J_n in view of a nonvanishing n_1. The answer lies in the fact that the drift term and the diffusion term exactly cancel out in J_n. Solving for E_x from Eq. (6.71) by setting

$J_n = 0$ and substituting E_x into Eq. (6.72), we find

$$J_p = -eD_p \left(1 + \frac{p}{n}\right) \frac{\partial p_1}{\partial x} = -eD \frac{\partial p_1}{\partial x} \qquad (6.73)$$

If a parameter Z is defined as the ratio of excess minority-carrier concentration to the equilibrium majority-carrier concentration or $Z = p_1/n_0$, then Eq. (6.73) can be expressed in terms of an effective diffusion constant $D = D_p(1 + 2Z)/(1 + Z)$. The new effective diffusion constant should be used throughout Eq. (6.70).

The lifetime of free carriers also changes with the excess carrier concentrations. According to the Shockley-Read model of recombination or Eq. (5.71), the lifetime will vary with Z as $\tau_0(1 + aZ)/(1 + Z)$ where τ_0 is the lifetime at low injection levels and a is a constant which according to the data of Burton et al.[16] has a value 1.8 in 4 Ω-cm, n-type Ge and a value 0.9 in 4 Ω-cm, p-type Ge. Let us now look at the emitter efficiency. The component of emitter current in a p-n-p transistor due to electron diffusion into the emitter is proportional to n_p which is in turn related to n_n through $n_p = n_n \exp\left[-e(V_d - V_a)/kT\right]$ where V_d and V_a are, respectively, the diffusion and applied voltage across the emitter junction. The voltage dependence will cancel out in $I_{\epsilon n}/I_{\epsilon p}$ but the change in n_n must be taken into account. At high injection levels, $n_n = n_0 + p_1 = n_0(1 + Z)$ where n_0 is the equilibrium majority-carrier concentration. Summarizing all the effects discussed so far, the base current or Eq. (6.70) at high injection levels becomes

$$I_b = \left[(1 - \gamma_0) \frac{(1 + Z)^2}{1 + 2Z} + \frac{W_0^2}{2D_0\tau_0} \frac{1 + Z}{1 + 2Z} \frac{1 + Z}{1 + aZ}\right.$$
$$\left. + \frac{2v_s W_0^2}{D_0 r_\epsilon} \frac{1 + Z}{1 + 2Z} g(I_\epsilon)\right] I_\epsilon \qquad (6.74)$$

The subscript 0 in γ_0, τ_0 and D_0 refers to the respective values found at low injection levels.

The extra factor $g(I_\epsilon)$ in Eq. (6.74) takes care of another effect known as the *base-resistance bias effect*[17] briefly mentioned in Sec. 6.3. Referring to the geometry of a typical transistor shown in Fig. 6.17a, we see that the base current will establish a potential difference as we move from the center toward the periphery of the emitter junction. Since the emitter side is heavily doped, it can be considered as an equipotential surface. The potential difference created by the base current makes the center region of the emitter junction less forwardly biased than the region near the periphery. That means the excess carriers will be heavily concentrated near the periphery as will be the current. Going back to Eq. (6.69), we note that p_{10} refers

[16] J. A. Burton et al., *Jour. Phys. Chem.*, vol. 57, p. 853, 1953.
[17] R. N. Fletcher, *Proc. IRE*, vol. 43, p. 551, 1955, and vol. 43, p. 1669, 1955.

to the concentration of excess holes at places, namely near the periphery of the emitter junction, where surface recombination takes place. Because of a nonuniform distribution of excess carriers, the second step in Eq. (6.69) breaks down at high injection levels. The factor $g(I_\epsilon)$ in Eq. (6.74) relates the concentration of excess holes at the periphery to the emitter current as follows:

$$I_\epsilon = \frac{eD}{W} \int_0^{r_\epsilon} p_{10}(r) 2\pi r \, dr = \frac{eD}{W} p_{10}(r_\epsilon) \frac{\pi r_\epsilon^2}{g(I_\epsilon)} \qquad (6.75)$$

Obviously, the value of $g(I_\epsilon)$ reduces to unity for a uniform distribution of injected carriers and can be considerably greater than one at high emitter currents. The reader will find the actual evaluation of $g(I_\epsilon)$ and the dependence of $1/(1 - \alpha)$ upon I_ϵ in Prob. 6.19.

Now, we shall give a qualitative explanation of Fig. 6.25. The initial rise of $1/(1 - \alpha)$ is primarily due to an increase in the effective diffusion constant while the falloff is due to an increased surface recombination of excess carriers. The point at which the falloff begins depends upon how soon the base-resistance bias effect becomes important. Of course, the resistivity of the base region is a factor. However, the most effective way of reducing the effect is to reduce the base current itself. In Eq. (6.70), the $(1 - \gamma)$ term is usually the dominating term. One way, widely used and proved successful, to improve the falloff effect is to increase the conductivity or the doping concentration of the emitter. In a p-n-p transistor, it is found that gallium helps to increase the solubility of indium in germanium and hence in such transistors, the falloff in $1/(1 - \alpha)$ occurs at a higher emitter current than in similar transistors without gallium doping in the emitter. For a detailed analysis of the falloff effect, the reader is referred to a paper by Wang and Wu.[18]

6.6 SEMICONDUCTOR DIODES AS NONLINEAR CAPACITORS IN PARAMETRIC AMPLIFIERS

A *parametric amplifier* is a device which uses nonlinear reactance elements to convert input energy of a given frequency to output energy at a different frequency. To illustrate the principle of parametric amplification,[19] we consider an idealized tank circuit consisting of a lossless capacitor of capacitance C and a lossless inductor of inductance L. If there is initial energy stored in the tank circuit, the stored energy will oscillate back and forth from the inductor where it is in the form of magnetic energy ($LI^2/2$) to the capacitor where it is in the form of electrostatic energy ($Q^2/2C$), I and Q being the current through the inductor and the charge stored in the capaci-

[18] S. Wang and T. T. Wu, *IRE Trans. on Electron Devices*, vol. 6, p. 162, 1959.
[19] For a general discussion, see E. D. Reed, *IRE Trans. on Electron Devices*, vol. 6, p. 216, 1959; A. Nergaard, *RCA Review*, vol. 20, p. 3, 1959.

tor respectively. In Fig. 6.26a, the charge Q is shown as a function of time, oscillating with an angular frequency $\omega = 1/\sqrt{LC}$. The amplitude of oscillation will remain constant as long as no energy is further added to or taken away from the tank circuit. There are two obvious ways to change the energy stored in a capacitor: by changing either Q or C. In a diode parametric amplifier, the energy is added by modulating the capacitance C.

Figure 6.26

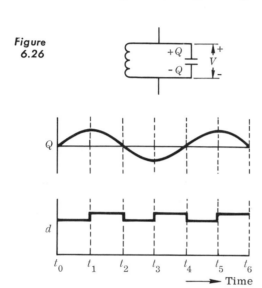

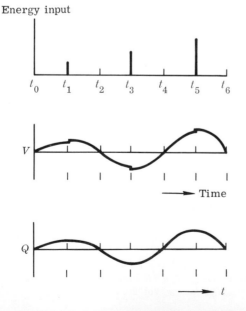

(a) *Oscillation of the stored charge in a lossless tank circuit if the values of the circuit elements are held constant.*

(b) *To illustrate the principle of parametric amplification, we pull the plates of the capacitor apart when Q is maximum (at t_1, t_3, etc.) and push the plates back to their initial separation when Q is zero (at t_2, t_4, etc.).*

(c) *Energy is pumped into the tank circuit at t_1, t_3, etc., in the form of work done in pulling the plates apart but no energy is taken away from the circuit at t_2, t_4, etc., since $Q = 0$.*

(d) *Variation of V as a function of time.*

(e) *Variation of Q as a function of time.*

Suppose that the plates of the capacitor are pulled apart when Q reaches a maximum and are pushed back to the initial separation when Q is zero. Figure 6.26b shows the separation d as a function of time. For our present qualitative description, it is assumed that the modulation in d is small; therefore, the change in natural frequency of the tank circuit will be small and any effects produced by the frequency shift will not be considered. Note that energy is pumped into the tank circuit at t_1, t_3, t_5, . . . , in the form of work done against coulombic attraction in pulling the plates apart but no energy is taken out from the tank circuit at t_2, t_4, t_6, . . . , as illustrated in Fig. 6.26c. The voltage V developed across the capacitor jumps abruptly at t_1, t_3, t_5, . . . , because of the change in d, and the amplitude of oscillation in Q increases every half-cycle because of the energy increase in the tank circuit. This simple example illustrates that amplification of a signal is possible if the value of the reactance can be varied in a proper phase relationship with the signal.

A semiconductor diode is a natural candidate for the role required of a time-varying reactance element in a parametric amplifier. As discussed in Sec. 6.1, the transition region capacitance C_T is nonlinear and can be controlled by an applied voltage. If an a-c voltage v_1 is superimposed on the d-c bias voltage v_0 of a semiconductor diode, then according to Eq. (6.13), C_T can be separated into a d-c component C_0 and an a-c component C_1 as follows:

$$C_T = C_0 + C_1 = K(V_d + v_0 + v_1)^{-m}$$
$$= K(V_d + v_0)^{-m}\left(1 - \frac{mv_1}{V_d + v_0}\right) \tag{6.76}$$

In deriving the above equation, it is assumed that $v_1 < V_d + v_0$ and that the diode is biased in the reverse direction. It is also understood that $m = \frac{1}{2}$ for abrupt junctions and $m = \frac{1}{3}$ for graded junctions and that K is the constant in C_T in front of the voltage dependence. Suppose that the a-c voltage actually consists of two parts: one being the signal to be amplified and the other being the voltage, usually referred to as the *pump*, to control the transition-region capacitance C_T. Thus,

$$v_1 = v_s \sin (\omega_s t + \theta) + v_p \sin \omega_p t \tag{6.77}$$

the subscripts s and p referring to the signal and pump voltage, respectively.

The a-c current going through the diode is given by

$$i_1 = \frac{dQ}{dt} = \frac{dQ}{dv}\frac{dv_1}{dt} = C_0\left(1 - \frac{mv_1}{V_d + v_0}\right)\left(\frac{dv_1}{dt}\right) \tag{6.78}$$

If we substitute Eq. (6.77) into Eq. (6.78), we see that nothing interesting happens to the current component at the signal frequency unless $\omega_p = 2\omega_s$. In a parametric amplifier, actually three a-c voltages are needed, one at the signal frequency ω_s, another at the pump frequency ω_p and the third at the idle frequency ω_i. The frequencies are such that $\omega_i + \omega_s = \omega_p$ so that mixing of the idle voltage with the pump voltage by the nonlinear element can produce a component at the signal frequency. The case $\omega_i = \omega_s$, $\omega_p = 2\omega_s$ is generally referred to as the degenerate mode of operation and is the case which we shall consider presently. Substituting Eq. (6.77) into Eq. (6.78), we find the current at ω_s:

$$i_1(\omega_s) = \omega_s C_0 v_s \cos(\omega_s t + \theta) + \frac{m v_p C_0}{V_d + v_0} \frac{\omega_p - \omega_s}{2} v_s \sin(\omega_s t - \theta) \quad (6.79)$$

Therefore, the input signal sees an admittance Y:

$$Y = j\omega_s C_0 + \frac{m v_p \omega_s C_0}{2(V_d + v_0)} \exp(-j2\theta) \quad (6.80)$$

The first term simply represents the capacitive reactance of the junction diode and there is nothing unusual about it. The second term is the result of parametric coupling of the pump to the signal voltage through the nonlinear C_T. Energy transfer from the pump to the signal is possible if $i_1(\omega_s)$ sees a negative conductance which reaches a maximum value when $\theta = \pi/2$. Figure 6.27a shows the phase relation between the pump and the signal voltage for $\theta = \pi/2$. The variation in the junction capacitance C_T due to the pump voltage is plotted in Fig. 6.27b together with the time variation of d taken

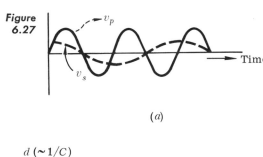

(a)

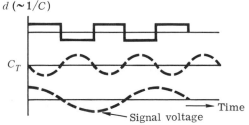

(b)

(a) Phase relationship between the pump and signal voltage for $\theta = \pi/2$ in Eq. (6.77). (b) Comparison of the phase relations between the a-c capacitance and the signal voltage in two cases: top curve for a square wave of the pump, middle curve for a sinusoidal variation of the pump, and bottom curve for the signal voltage.

from Fig. 6.26b. Note that the same phase relationship exists between the capacitance and the signal voltage in both cases.

After this introductory discussion, we can extend our analysis to the case in which three a-c voltages, the signal, the pump, and the idle, are coupled together through nonlinear capacitors. For appreciable amplification with practical devices, many semiconductor diodes in some form of cascade arrangement are needed. Furthermore, to prevent feedback from the output to the input circuit due to reflections at the output end, isolation between the input and output signal becomes necessary. For these reasons, a practical parametric amplifier is generally of the traveling-wave type. Consider a transmission line loaded with semiconductor diodes at regular intervals. For ease of discussion, however, the loading shall be taken as uniformly distributed along the whole length of the transmission line, and as a result, the capacitance per unit length of the line can be written as $c = c_0 + fv$ where c_0 is the constant part of the combined distributed capacitances of the line and the diodes while fv represents the time-varying part of the distributed capacitance of the diodes. Under such circumstances, the transmission line equation reads

$$\frac{\partial i}{\partial z} = -(c_0 + fv)\frac{\partial v}{\partial t} \quad \text{and} \quad \frac{\partial v}{\partial z} = -l\frac{\partial i}{\partial t} \tag{6.81}$$

where l is the distributed inductance of the line. In the above equation, the current i and the voltage v actually consist of three components at the signal, the pump, and the idle frequencies. In this section the symbol j is used to mean $j = \sqrt{-1}$ because the symbol i is used for current.

In solving Eq. (6.81), we shall use complex notations. For the signal voltage, we propose a solution of the following form:

$$v_s + v_s^* = a_s \exp\left[j(\omega_s t - k_s z)\right] + \text{complex conjugate} \tag{6.82}$$

and similar expressions for v_p and v_i. Note that the product of fv_p and $\partial v_i^*/\partial t$ produces a function varying with $\exp(j\omega_s t)$ if $\omega_p = \omega_s + \omega_i$ and so does the product of fv_i^* and $\partial v_p/\partial t$. Substituting Eq. (6.82) and similar expressions into Eq. (6.81) and collecting terms varying with $\exp(j\omega_s t)$, we find

$$\frac{\partial i_s}{\partial z} = -\left(c_0\frac{\partial v_s}{\partial t} + fv_p\frac{\partial v_i^*}{\partial t} + fv_i^*\frac{\partial v_p}{\partial t}\right)$$
$$\frac{\partial v_s}{\partial z} = -l\frac{\partial i_s}{\partial t} \tag{6.83}$$

In reference to Eq. (6.80), it was pointed out that for constant energy transfer from the pump to the signal voltage, a constant phase ($\theta = \pi/2$) must be maintained. The same is true with traveling-wave parametric amplifiers. A constant phase relationship must exist between v_s and $v_p v_i^*$ along the whole length of the transmission line, or in other words, $k_p = k_s + k_i$. Otherwise, if the phase relation is not satisfied, energy is converted from the pump into the signal frequency along certain sections of the transmission line where the phase angle is right, but it is returned back to the pump frequency along other sections of the line where the phase angle is wrong.

The conservation of energy (or angular frequency) and that of momentum (or wave vector)

$$\omega_p = \omega_s + \omega_i \qquad \text{and} \qquad \mathbf{k}_p = \mathbf{k}_s + \mathbf{k}_i \qquad (6.84)$$

are two basic requirements[20] concerning the interaction of different waves. These two conservation laws will be referred to in later applications, for example in the harmonic generation of laser beams and in the scattering of a light beam by phonon waves. Eliminating i_s from Eq. (6.83), using the explicit forms of v_p and v_i from Eq. (6.82) in Eq. (6.83), and applying Eq. (6.84) to Eq. (6.83), we obtain

$$\frac{\partial^2 v_s}{\partial z^2} = -\omega_s^2 \left(lc_0 + lf \frac{a_p a_i^*}{a_s} \right) v_s \qquad (6.85)$$

and similar expressions for v_s^*, v_p, v_p^*, v_i, and v_i^*.

Our next step is to find the propagation constant for v_s. Without the pump and idle voltages, the signal voltage will propagate with a constant phase velocity along a lossless transmission line, or in other words, $k_s^2 = \omega_s^2 lc_0$ and a_s is a constant in Eq. (6.82). Because of the term $lfa_p a_i^*/a_s$ in Eq. (6.85), energy is constantly fed to the wave at signal frequency and, as a result, the amplitude of the wave will grow. It is expected, however, that the rate of change in a_s and a_s^* with distance will be slow so that $\partial a_s/\partial z$ is much smaller than $k_s a_s$. Substituting Eq. (6.82) into Eq. (6.85) and neglecting terms involving $\partial^2 a_s/\partial z^2$, we find

$$-2jk_s \frac{\partial a_s}{\partial z} = -\omega_s^2 (lf)(a_p a_i^*) \qquad (6.86)$$

A similar expression

$$2jk_i \frac{\partial a_i^*}{\partial z} = -\omega_i^2 (lf)(a_p^* a_s) \qquad (6.87)$$

is obtained for v_i^*.

In a parametric amplifier, the pump voltage is usually of sufficient strength so that its amplitude can be assumed essentially constant along the transmission line even though energy is constantly extracted from it. Eliminating a_i^* from Eqs. (6.86) and (6.87) gives

$$4k_i k_s \frac{\partial^2 a_s}{\partial z^2} = \omega_i^2 \omega_s^2 (lf)^2 |a_p|^2 a_s \qquad (6.88)$$

---

[20] See, for example, discussion by P. K. Tien, *Jour. Appl. Phys.*, vol. 29, p. 1347, 1958.

The above equation says that the signal voltage grows with an amplification constant $\alpha = (f/2c_0)|a_p|(k_ik_s)^{1/2}$. It should be pointed out that if the pump voltage is a traveling wave propagating in the positive z direction, only the forward signal wave will be able to maintain a constant phase with respect to the pump wave as required by Eq. (6.84). The reflected signal wave, even if it does exist, will be attenuated by transmission line losses as it propagates along the line. Since only the forward signal wave is amplified, the problem of isolation mentioned earlier is greatly minimized. We further note that parametric amplification of the signal voltage is possible only when an idle voltage exists along with v_s and v_p. Therefore, the amplitude of the idle voltage also increases in the forward direction of propagation.

The reader is asked to show that if the transmission line is terminated with a load equal to its characteristic impedance, the powers delivered to the load at the signal and idle frequencies are proportional to their respective frequencies or $P_s/P_i = \omega_s/\omega_i$. In a parametric amplifier using lossless nonlinear reactance elements, a general relationship known as the *Manley-Rowe relation*[21] exists between powers delivered and absorbed at different frequencies. The relation is given as follows

$$\sum_m \frac{nP_{mn}}{m\omega_1 + n\omega_0} = 0 \tag{6.89}$$

where ω_0 and ω_1 are frequencies of two interacting voltages, m and n are two integers, and P_{mn} is the power delivered or absorbed at a frequency $\omega = m\omega_1 + n\omega_0$. Take $\omega_0 = \omega_s$, $\omega_1 = \omega_i$, and $\omega_p = \omega_0 + \omega_1$. For $n = 1$, two values of m, 1 and zero, are allowed, hence Eq. (6.89) gives $(P_p/\omega_p) + (P_s/\omega_s) = 0$. That means, if power is delivered to the signal voltage with P_s being positive, power must be supplied by the pump voltage with P_p being negative.

In concluding this section, we should mention that parametric amplification is also possible with nonlinear inductive elements using magnetic materials.[22] It will be shown in Chapter 10 that various nonlinear interactions exist in a ferromagnetic or ferrimagnetic substance in the presence of d-c and a-c magnetic fields. The present discussion, however, is limited to parametric excitation in semiconductor diodes. Design considerations concerning the diode include the following: First, the variation of C_T with voltage is larger in the forward than in the reverse direction. However, the conductance G of a diode is larger and hence the quality factor of the diode ($Q = \omega C_T/G$) is smaller in the forward than in the reverse direction. Therefore, the choice of the d-c bias is pretty much a com-

[21] J. M. Manley and H. E. Rowe, *Proc. IRE*, vol. 44, p. 904, 1956, and vol. 47, p. 2115, 1959.
[22] P. K. Tien and H. Suhl, *Proc. IRE*, vol. 46, p. 700, 1958.

promise between a high quality factor and a high coefficient for parametric coupling. The d-c operating point also determines the maximum excursion allowed for the pump voltage. These design considerations will be further illustrated with examples in the problem set.

6.7 IMPACT IONIZATION AND AVALANCHE BREAKDOWN

In Sec. 6.1, we have derived the I-V characteristic of an ideal diode in which the current in the reverse direction approaches a constant value known as the reverse saturation current. In a practical diode, however, we usually find that the diode current rises slowly as bias is increased in the reverse direction until the voltage reaches near a critical value at which the current increases very rapidly. In Fig. 6.28, the I-V characteristics of ideal and practical diodes are drawn for comparison. The slow rise of current in region I of Fig. 6.28b may occur for several reasons. In a narrow-base diode, the space-charge region extends a fair distance into the base region. As the voltage increases, the width of the space-charge region increases, reducing the effective basewidth and hence raising the reverse saturation current. Surface effects can also cause a slow rise of reverse current. As pointed out in Sec. 5.9, an inversion layer may exist on a semiconductor surface, creating a surface p-n junction[23] in parallel with the bulk p-n junction. As the voltage increases, the effective area of the inversion layer increases, thereby raising the reverse current. The rapid rise of current in region II of Fig. 6.28b is different in nature from the slow rise in region I, and the phenomenon is generally known as *breakdown* in a junction diode.

There are two types of breakdown which can cause a sudden rise of current: namely, direct *tunneling* of electrons across the forbidden

[23] W. L. Brown, *Phys. Rev.*, vol. 91, p. 518, 1953; H. Christensen, *Proc. IRE*, vol. 42, p. 1371, 1954; A. L. McWhorter and R. H. Kingston, *Proc. IRE*, vol. 42, p. 1376, 1954.

Figure 6.28

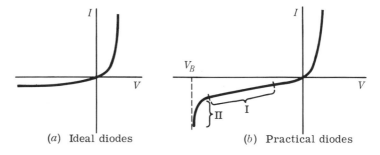

(a) Ideal diodes   (b) Practical diodes

Comparison of the I-V characteristics of (a) ideal diodes and (b) practical diodes.

gap and creation of electron-hole pairs through *impact ionization* by high-energy carriers. Both processes require a very high electric field to exist in the space-charge region of a reverse-biased diode. However, a rough demarcation line for these two processes can be drawn. For Ge and Si diodes, if the critical voltage V_B, generally referred to as the *breakdown voltage*, is above 8 volts, impact-ionization process dominates; and if V_B is below 4 volts, tunneling process is in control. The reason that the impact-ionization process results in a higher breakdown voltage is because the voltage required to produce a high electric field is larger in wider junctions. Since the tunneling process will be treated in a subsequent section, our present discussion will be limited to the impact-ionization process.

The reader who has had some familiarity with the Townsend discharge in gases[24] will find that a similar process takes place inside a semiconductor junction. In a gaseous discharge, electron and positive-ion pairs are created in an inelastic collision process between gas molecules and energetic electrons having energies greater than the ionization energy of the gas molecule. In the same way, electron-hole pairs can be created in an inelastic collision process[25] involving energetic electrons or holes. For simplicity, we shall consider a hypothetical case of two parallel-plate electrodes of distance W apart, one emitting electrons and the other emitting holes at a rate of n_0 electrons and p_0 holes per sec per m² respectively. In an elementary distance dx, the number of electron-hole pairs created per sec per m² by electrons and holes, respectively, can be written as

$$dn = dp = \beta_1 n\, dx \qquad \text{and} \qquad dn = dp = \beta_2 p\, dx \qquad (6.90)$$

where β_1 and β_2 are the ionization coefficients for the two processes. We shall postpone a discussion of the nature of β_1 and β_2 until we have completed our analysis of the breakdown mechanism.

To simplify our discussion, we assume that only electron emission takes place from electrode A but no hole emission from electrode B. The situation is illustrated in Fig. 6.29, and our attention will be focused on what is going on within the elementary distance dx. Because of the ionization process the number of electrons arriving per sec per m² from the left is equal to $n_0 + n_1$ where n_1 is the number of electrons generated per sec per m² in the space between A and C of Fig. 6.29. Since electron-hole pairs are also created in the space between B and D, holes are coming from the right into dx at a rate, say p_2 holes per sec per m². Therefore, according to Eq. (6.90), we have

$$dn_1 = \beta_1(n_0 + n_1)\, dx + \beta_2 p_2\, dx \qquad (6.91)$$

[24] See, for example, L. B. Loeb, "Basic Processes of Gaseous Electronics," University of California Press, Berkeley, 1955.

[25] For early work, read K. G. McKay, *Phys. Rev.*, vol. 94, p. 877, 1954. For theoretical analysis, refer to P. A. Wolff, *Phys. Rev.*, vol. 95, p. 1415, 1954.

Under steady-state, electrons must arrive at electrode B at a rate equal to the total rate by which electrons leave electrode A and electrons are created within the space $ACDB$. In other words, if n_f is the number of electrons arriving at B per sec per m², then $n_f = n_0 + n_1 + n_2$ with $n_2 = p_2$. Eliminating n_2 from Eq. (6.91) gives

$$\frac{dn_1}{dx} = (\beta_1 - \beta_2)(n_0 + n_1) + n_f\beta_2 \qquad (6.92)$$

Applying the boundary conditions that at $x = 0$, $n_1 = 0$ and at $x = W$, $n_1 + n_0 = n_f$ to the solution (Prob. 6.27) of Eq. (6.92), we find

$$M_e = \frac{n_f}{n_0} = \frac{1}{1 - \int_0^W \beta_1 \left\{ \exp\left[- \int_0^x (\beta_1 - \beta_2)\, dx' \right] \right\} dx} \qquad (6.93)$$

where M is generally referred to as the *multiplication factor*.

In applying Eq. (6.93) to a junction diode, it is obvious that the electrodes A and B of Fig. 6.29 correspond respectively to the boundaries of the space-charge region on the p and n sides. The symbols n_0 and p_0 multiplied by the electronic charge e and the area A of the junction actually are equal to the component of the reverse saturation current in a junction diode due to electrons and holes respectively. Because of the impact-ionization processes, however, the electron current will not be proportional to n_0 but to n_f. Therefore, the multiplication factor M_e defined by Eq. (6.93) is the ratio of the diode current to the reverse saturation current for diodes with a heavily doped n region. For diodes with a heavily doped p region, the reverse saturation current is mostly due to hole diffusion in the

Figure 6.29

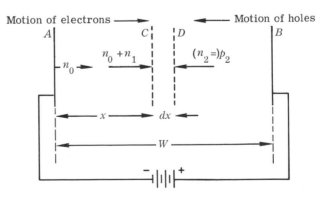

Electrons are emitted from electrode A at a rate of n_0 electrons per sec per m², and no holes are emitted from electrode B. Because of the ionization process, electron-hole pairs are created in regions AC and BD. Electrons created in region AC will move in a direction toward C while holes created in region BD will move in a direction toward D.

n region and hence an expression similar to Eq. (6.93) but with β_1 and β_2 interchanged must be used.

It may be instructive to discuss the case with $\beta_1 = \beta_2$. Equation (6.93) says that M_e goes to infinity if $\int_0^W \beta_1 \, dx = 1$. This means that an electron in going through the space-charge region from A to B creates one electron-hole pair and the hole thus generated in turn creates another electron-hole pair on its way from B to A. As the process goes on, the carriers inside the region AB multiply themselves and the current increases indefinitely. This type of breakdown is called the *avalanche breakdown*. It is worthwhile to point out that feedback mechanisms are needed in building up any chain reaction. A common example is the feedback loop used in an oscillator circuit. Hence, in an avalanche breakdown, both the β_1 and β_2 processes are needed, with one process acting as a feedback loop for the other.

Now we shall discuss the nature of the ionization coefficients β_1 and β_2. In an ionization process, only those carriers can participate which have energies larger than the minimum energy ε_i required to create an electron-hole pair.[26] To acquire this energy, a carrier must travel in an electric field E a distance d such that $eEd = \varepsilon_i$. If l is the mean free path of carriers, according to Eq. (3.9), the percentage of carriers uncollided at $x = d$ after a collision at $x = 0$ is equal to exp $(-d/l)$. If it is assumed that a carrier after acquiring the minimum energy ε_i soon takes part in an impact-ionization process, the number of electron-hole pair production is given by $\Delta n = n \exp(-d/l)$. According to the definition of β_1 and β_2 given by Eq. (6.90), $\Delta n = n\beta_1 d$; therefore,

$$\beta_1 = A E \exp\left(-\frac{B}{E}\right) \tag{6.94}$$

where $A = e/\varepsilon_i$ and $B = \varepsilon_i/(el)$ and a similar expression is obtained for β_2. Although Eq. (6.94) is based on a very much simplified model, it does provide a qualitative description of the behavior of β_1 or β_2 with respect to the electric field.

There are many ways for the experimental determination of the multiplication factors M_e and M_h. One obvious way is to measure the current of a diode in the reverse direction as a function of the applied voltage in the breakdown region. To avoid heating effects, pulse techniques must be employed. However, the method still suffers from the fact that there usually exists a region of slow rise of the reverse current, shown as region I in Fig. 6.28. This makes the value of the reverse saturation current ill-defined. A more accurate way of determining β_1 and β_2 is to measure the photoresponse of a diode as a function of the reverse voltage. The carriers generated by photons are multiplied in the space-charge region and so is the photocurrent.

[26]   A. G. Chynoweth and K. G. McKay, *Phys. Rev.*, vol. 108, p. 29, 1957.

Therefore, the multiplication factor is simply the ratio of the photo-current in the breakdown region to the normal photocurrent.

To calculate the ionization coefficient from the multiplication factor, however, is rather involved. First, Eq. (6.93) involves both β_1 and β_2. Furthermore, the electric field changes and so do β_1 and β_2 in the space-charge region. Hence, a brief discussion of the problem seems necessary. Figure 6.30 shows the ionization coefficients β_1 and β_2 reported by Miller[27] from measurements on alloyed Ge p-n junctions. Although β_1 and β_2 change very rapidly with E, the ratio of β_2/β_1 stays reasonably constant. Under such circumstances, Eq. (6.93) becomes

$$\frac{1}{c-1} \exp\left[\int_0^W (c-1)\beta_1 \, dx \right] = \frac{1}{c-1} + 1 - \frac{1}{M_e} \qquad (6.95)$$

where $c = \beta_2/\beta_1$ is a constant.

In integrating the expression $\int_0^W \beta_1 \, dx$, we note the following. Within the limited range of E in which β_1 or the impact-ionization process is appreciable, β_1 can be approximated by $\beta_1 = aE^n$. According to Eq. (6.6), the field in an alloyed or step junction is given by $E = E_{max}(x - x_B)/x_B$. Quoting the result from Prob. 6.28, we have $\int_0^W \beta_1 \, dx = aE_{max}^n W/(n+1)$. Thus, in principle, the left-hand side of Eq. (6.95) can be expressed in terms of a and E_{max} as $f(a, E_{max})$.

[27] S. L. Miller, *Phys. Rev.*, vol. 99, p. 1234, 1955. For related work in Si, see A. G. Chynoweth, *Phys. Rev.*, vol. 109, p. 1537, 1958.

Figure 6.30

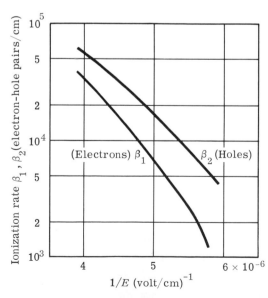

(Electrons) β_1 β_2 (Holes)

Ionization rate β_1, β_2 (electron-hole pairs/cm)

$1/E$ (volt/cm)$^{-1}$

Ionization coefficients for electrons and holes in germanium. (S. L. Miller, Phys. Rev., vol. 99, p. 1234, 1955.)

Therefore, the slope of the multiplication-factor curve gives the following:

$$\frac{d}{dV}\left(1 - \frac{1}{M_e}\right) = \frac{df}{dE_{max}} \frac{dE_{max}}{dV} \tag{6.96}$$

Since E_{max} is a known function of the bias voltage V, the constant a in the expression of β_1 can be determined from Eq. (6.96). The reader is asked to carry out the detailed calculation necessary for the determination of β_1 from the M_e curve in Prob. 6.28.

To check whether our proposed method of attack is valid or not, we must ascertain from experiments that the constants n and c indeed exist. From Eq. (6.95), breakdown occurs when

$$\beta_1(E_{max})W = \frac{(n+1)(\log c)}{c - 1} \tag{6.97}$$

Since E_{max} and W are known functions of the bias voltage V and the impurity concentrations, Eq. (6.97) predicts a relation between V_B, the breakdown voltage, and the doping concentrations as follows:

$$V_B = K\left(\frac{N_1 N_2}{N_1 + N_2}\right)^m \tag{6.98}$$

where $m = (1-n)/(1+n)$, N_1 and N_2 are the doping concentrations defined in Sec. 6.1, and K is a proportionality constant. The above dependence is indeed verified by experiments. In Ge, the value of m is found to be -0.725 according to Miller[28] and hence n has a value around 6.25. To aid in selecting doping concentrations for a given V_B, the following information also reported by Miller is useful. For $N_1 = 10^{16}/cm^3$ and $N_1 \ll N_2$, $V_B = 26$ volts. For other doping concentrations, V_B can be calculated from Eq. (6.98).

The determination of the other constant $c(= \beta_2/\beta_1)$ involves a procedure of trial and error. First, a value of c is assumed. Then, the value of β_1 can be found from Eq. (6.96) and the value of β_2 from a similar equation for β_2. The procedure repeats itself until a self-consistent value of c is found. Obviously, information concerning M_e and M_h must be obtained from separate measurements on diodes with heavily doped n regions and those with heavily doped p regions respectively. The data shown in Fig. 6.30 are derived from an analysis of the results from such measurements.

Finally, we should point out that from the slope of the curves shown in Fig. 6.30 or the quantity $B = \mathcal{E}_i/(el)$, we may get some insight into the physical processes governing the impact ionization process. For example, the slope of the β_1 versus $1/E$ curve of Fig. 6.30 implies a value of $B = 1.4 \times 10^6$ volt/cm. If the ionization energy $\mathcal{E}_i$ is taken as the gap energy of germanium, a value of 200 Å is obtained. Actually, $\mathcal{E}_i$ must be larger than the gap energy because

[28] *Ibid.*

in a collision process, momentum must also be conserved. This could reduce the value of l to around 100 Å in germanium. The mean free path l can also be calculated from the mobility data. As an estimate without taking the proper average over the Boltzmann distribution, its value is given by $l = m^*\mu v_{th}/e$ or around 1000 Å. The large difference in l points out that the scattering process affecting the ionization process is quite different from that governing the mobility of carriers.

Actually, these two mean free paths refer to scattering of carriers by different modes, acoustic and optical, of lattice vibration. As we recall our discussion in Sec. 2.10, acoustic modes in general are characterized by a relatively small energy; therefore, they are largely responsible for the scattering of carriers at low fields. Optical modes, on the other hand, have a fairly high threshold energy around 0.03 ev in Ge and 0.05 ev in Si. Under high fields, carriers get rid of their energy most effectively by exciting optical phonons. The mean free path around 100Å should be taken as that for collisions with optical phonons. Such information concerning l_{opt} will be of value in analyzing the hot-electron effect discussed in Sec. 4.6.

We should point out that Eq. (6.94) is based on a very much simplified model. It is apparent that when a carrier has acquired an energy greater than $\mathcal{E}_i$, the carrier can lose its energy in either optical phonon generation or impact ionization. For a proper treatment of the ionization process, these two processes must be considered as being in parallel operation. Furthermore, multiple scattering processes must also be included. Since the threshold energy of optical phonons is much smaller than $\mathcal{E}_i$, energetic carriers still have sufficient energy for impact ionization even after several scattering processes involving the generation of optical phonons. The reader is referred to articles by Shockley[29] and by Moll and coauthors[30] for further discussions on the role of optical phonons, on the determination of $\mathcal{E}_i$ as well as on the multiple scattering processes.

6.8 AVALANCHE TRANSISTORS AND *pnpn* STRUCTURES

In this section, we shall discuss devices which make use of the impact ionization process. The V-I characteristics[31] of avalanche transistors and *pnpn* structures possess a negative-resistance region separating two stable regions; therefore, these devices are useful for bistable operations. First, let us consider a *p-n-p* transistor operated as a

[29] W. Shockley, *Solid-state Electronics*, vol. 2, p. 35, 1961.

[30] J. L. Moll and R. Van Overstraeten, *Solid-state Electronics*, vol. 6, p. 147, 1963; J. L. Moll and N. I. Meyer, *Solid-state Electronics*, vol. 3, p. 155, 1961.

[31] For a general discussion, see S. L. Miller and J. J. Ebers, *Bell System Tech. J.*, vol. 34, p. 883, 1955; J. J. Ebers, *Proc. IRE*, vol. 40, p. 1361, 1952; J. L. Moll *et al.*, *Proc. IRE*, vol. 44, p. 1174, 1956.

two-terminal device shown in Fig. 6.31a. It may be instructive that the two extreme cases with $R = 0$ and $R = \infty$ be discussed first. For $R = 0$, the emitter junction is short-circuited and hence the transistor operates as a diode. The V-I characteristic for such an operation is shown as curve A in Fig. 6.31b. The current I increases very rapidly as the voltage V approaches the breakdown voltage V_B of the collector junction. For $R = \infty$, the base circuit is open-circuited. Thus, according to Eq. (6.45), equating $I_b = 0$ gives

$$I = I_\epsilon = I_c = \frac{I_{c0}}{1 - \alpha} \qquad (6.99)$$

In this mode of operation, the emitter is biased slightly in the forward direction even though most of the applied voltage is across the collector junction. This explains why the current I is greater than I_{c0} by a factor of $1/(1 - \alpha)$. Equation (6.99) is true, however, only at low voltages. When impact ionization becomes significant, both components of the collector current (the current injected by the emitter and the collector reverse saturation current) will be multiplied in the collector junction. That means I_{c0} and αI_ϵ in Eq. (6.45) should be replaced by MI_{c0} and αMI_ϵ, respectively, or Eq. (6.99) becomes

$$I = \frac{MI_{c0}}{1 - \alpha M} \qquad (6.100)$$

where M is the multiplication factor. For simplicity, we have assumed that $M_e = M_h = M$ in Eq. (6.100).

From Eq. (6.100), it is obvious that the breakdown in an open-base transistor will occur much sooner than that in a diode of similar doping profiles because of the internal amplification α in a transistor. If the multiplication curve is approximated by an empirical law of the following form:

$$\frac{1}{M} = 1 - \left(\frac{V}{V_B}\right)^m \qquad (6.101)$$

Figure 6.31

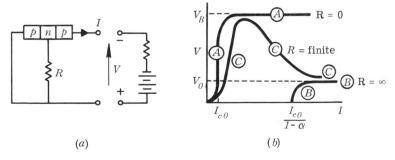

(a) (b)

(a) A p-n-p transistor operated as a two-terminal device and (b) its V-I characteristic.

then breakdown occurs when $1 - \alpha M = 0$ or

$$V_0 = V_B(1 - \alpha)^{1/m} \qquad (6.102)$$

The value of m in Eq. (6.101) is found to be around 3 in p-n-p and around 6 in n-p-n Ge transistors. If $\alpha = 0.95$, $V_0 = 0.37\ V_B$ in p-n-p transistors. In Fig. 6.31b, the V-I curve for mode B operation with $R = \infty$ is shown together with that for mode A operation with $R = 0$. The question naturally arises as to what happens to the V-I curve if the value of R is finite. We will probably have a better perspective on the problem if our discussion is based on the relative magnitudes of R and the resistance of the emitter junction.

The voltage V_R developed across the resistance R is

$$V_R = I_b R = [I_{c0} - (1 - \alpha)I_e]R \qquad (6.103)$$

Note that the emitter current I_e must be limited to a value smaller than $I_{c0}/(1 - \alpha)$ by Eq. (6.103) in order to keep the emitter junction forward-biased. This means that if impact ionization is absent, self-biasing keeps the resistance of the emitter junction high. The situation changes drastically, however, when ionization becomes appreciable. Under such circumstances, Eq. (6.103) should be replaced by

$$V_R = [M I_{c0} + (\alpha M - 1)I_e]R \qquad (6.104)$$

If $\alpha M > 1$, any increase in I_e raises the forward bias on the emitter junction which results in further increase in I_e. Obviously, the resistance R_e of the emitter junction will be very low in this case. If the value of R is so chosen that it lies between the extreme values of R_e obtained with M equal to zero and M different from zero, both modes of operation are possible: mode A for $R < R_e$ and mode B for $R > R_e$. The V-I characteristic for a finite R, therefore, follows the curve C of Fig. 6.31b. The point at which the transistor operates is of course determined by the load resistance.

After a qualitative description of the V-I characteristic, it seems appropriate that a quantitative analysis should follow. The question is how soon does the negative resistance region begin. From Eq. (6.45), the incremental current dI is given by

$$dI = dI_c = \alpha M\, dI_e + (I_{c0} + \alpha I_e)\, dM \qquad (6.105)$$

Since $1/R_e = dI_e/dV_R = eI_e/kT$, differentiation of Eq. (6.104) results in the following

$$dI_e[R_e - R(\alpha M - 1)] = R(I_{c0} + \alpha I_e)\, dM \qquad (6.106)$$

Eliminating dI_e from Eqs. (6.105) and (6.106) gives

$$\frac{dI}{dV} = \frac{(I_{c0} + \alpha I_e)(R + R_e)}{R_e - R(\alpha M - 1)} \frac{dM}{dV} \qquad (6.107)$$

Figure 6.32

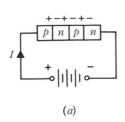

(a)

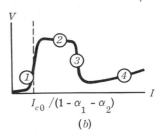

$$I_{c0}/(1 - \alpha_1 - \alpha_2)$$

(b)

(a) A pnpn structure and (b) its V-I characteristic.

The condition for dI/dV to be negative is that

$$R > \frac{R_\epsilon}{\alpha M - 1} \qquad \text{or} \qquad eRI_\epsilon \frac{\alpha M - 1}{kT} > 1 \qquad (6.108)$$

The occurrence of the negative resistance region may be explained as follows. Once $\alpha M - 1 > 0$, any further increase in I_ϵ reduces R_ϵ. That means, less V is required in $(\alpha M - 1)$ in order to satisfy Eq. (6.108). We should also point out that if I_{c0} in Eq. (6.107), I_0 in the expression $I_\epsilon = I_0 \exp (eV_R/kT)$, and m in Eq. (6.101) are known, the curve C of Fig. 6.31b can be computed. However, we shall not burden the reader with such a process because it is rather cumbersome and does not lend any further physical insight into the problem.

Let us now turn to the *pnpn* structure. If a bias voltage V is applied with a polarity as shown in Fig. 6.32a, a V-I characteristic illustrated in Fig. 6.32b is observed. At low voltages, the *pnpn* structure can be considered as two transistors, one *p-n-p* and the other *n-p-n*, with a common middle region as shown in Fig. 6.33a. Note that the middle *n-p* junction is reverse-biased and the two outer *p-n* junctions are forward-biased. Figure 6.33b shows an artificial cut of the *pnpn* structure so that the currents associated with either the *p-n-p* or the *n-p-n* section of the structure can be treated in the same manner as they are treated in a transistor. For example, I_{B1} is the collector current of the *p-n-p* transistor and hence it is equal to

$$I_{B1} = \alpha_1 I_{\epsilon 1} + I_{c01} \qquad (6.109)$$

Figure 6.33

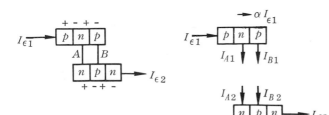

(a)

(b)

A pnpn structure can be considered as two transistors with a common middle region. Note that the middle junction is reverse-biased while the two outer junctions are forward-biased.

On the other hand, I_{B2} corresponds to the base current in the n-p-n transistor,

$$I_{B2} = (1 - \alpha_2)I_{e2} - I_{c02} \qquad (6.110)$$

Since no electrical connection is made to the middle p and n regions, $I_{B1} = I_{B2}$ and $I_{e1} = I_{e2} = I$. Equating Eq. (6.109) with Eq. (6.110) gives

$$I = \frac{I_{c0}}{1 - (\alpha_1 + \alpha_2)} \qquad (6.111)$$

where $I_{c0} = I_{c01} + I_{c02}$ is the reverse saturation current of the middle n-p junction.

We should point out that Eq. (6.111) is the same as Eq. (6.99) except that α is replaced by $(\alpha_1 + \alpha_2)$ on account of two transistors being connected in tandem. Therefore, it is expected that the $pnpn$ structure will break down when $\alpha_1 M_h + \alpha_2 M_e = 1$. However, something happens in the breakdown region. When the current increases, the values of α_1 and α_2 also increase for the reasons discussed in Sec. 6.5. In other words, for larger currents, less M_h and M_e are needed and hence less voltage is required to maintain the breakdown condition $\alpha_1 M_h + \alpha_2 M_e = 1$. This explains why the V-I characteristic shows a negative resistance in region 2 of Fig. 6.32b. For sufficiently large currents, the values of α_1 and α_2 may become sufficiently large so that $\alpha_1 + \alpha_2$ may be greater than unity even without multiplication in the middle n-p junction. When this happens, the middle junction suddenly switches from a reverse- to a forward-biased condition.

The physical process responsible for the switching is as follows. For simplicity, we assume that I_{c0} is negligibly small. If I is the current flowing through the $pnpn$ structure, the rates of hole injection from the left into the n region and of electron injection from the right into the p region of the middle junction are respectively $\alpha_1 I/e$ and $\alpha_2 I/e$. Since the current going through the middle junction is also I, only I/e injected carriers out of a total of $(\alpha_1 + \alpha_2)I/e$ carriers will be able to pass through the middle junction per sec. The rest of the injected carriers must pile up near the boundary of the middle junction, holes in the n region and electrons in the p region. The piling up of these carriers reduces the rate of diffusion of holes in the n region by reducing $\partial p/\partial x$. Similar statements apply to electron diffusion in the p region. However, this is not the whole story. As discussed in Sec. 6.2 on the switching response of a junction diode, the voltage across a junction is determined by the concentrations of minority carriers at the junction. The piling up of carriers at the middle junction causes the minority-carrier concentrations to exceed their respective equilibrium concentrations, thus changing the bias condition of the middle junction. The change occurs as soon as $(\alpha_1 + \alpha_2)$ becomes greater than unity. This explains why the slope of the V-I curve in region 3 of Fig. 6.32b is rather steep.

In region 4 of Fig. 6.32b, all three junctions are positively biased. If V_1, V_m, and V_2 are respectively the bias voltages on the left, middle, and right junctions, the current flow in the middle junction is given by

$$I_m = \alpha_1 I_{s1}(e^x - 1) - I_{sm}(e^y - 1) + \alpha_2 I_{s2}(e^z - 1) \qquad (6.112)$$

where $x = eV_1/kT$, $y = eV_m/kT$, and $z = eV_2/kT$. It is understood that I_{s1}, I_{sm}, and I_{s2} are the reverse saturation currents of the left, middle, and right junctions. Referring to Eq. (6.23), they represent the respective coefficients in front of the voltage dependence. Equation (6.112) is obtained by considering separately the actions of each bias alone, for example that of V_1 alone with $V_2 = V_m = 0$, and superimposing the current contributions. The first term in Eq. (6.112) is the result of V_1 with the middle junction acting as a collector having a zero bias voltage. Expressions for the currents through the left and right junctions, I_1 and I_2, are obtained in the same way. From these expressions, the voltage V in region 4 of Fig. 6.32b can be solved in terms of the current I, realizing that $I = I_1 = I_m = I_2$ and $V = V_1 - V_m + V_2$. If it is assumed that all the α's are equal and the two components (hole and electron) of I_{sm} are equal, then we find

$$V = \frac{kT}{e} \log \left[\left(\frac{I_{sm}}{I_{s1}I_{s2}} \right) \frac{(IA_1 + I_{s1})(IA_2 + I_{s2})}{(IA_m + I_{sm})} \right] \qquad (6.113)$$

where $A_1 = A_2 = (2 - \alpha)/2(1 - \alpha^2)$ and $A_m = (2\alpha - 1)/(1 - \alpha^2)$.

Region 2 and region 4 of Fig. 6.32 represent respectively the off- and on-mode of operation of the *pnpn* structure. In the off-region, $\alpha_1 + \alpha_2 < 1$. In the on-region, $\alpha_1 + \alpha_2 > 1$. This requirement is also implicit in Eq. (6.113). Since I is greater than I_{s1}, I_{s2}, and I_{sm}, the condition for the existence of region 4 or for a real V is that $2\alpha = \alpha_1 + \alpha_2 > 1$. Although the avalanche breakdown does not play a role in region 4, it is responsible for bringing about the transition from the off-mode into the on-mode of operation. Impact ionization is instrumental in increasing the current which in turn raises the value of $\alpha_1 + \alpha_2$.

In comparing the performances of a *pnpn* structure and an avalanche transistor, we note that the latter has a smaller off-mode current but a much larger sustaining voltage for the on-mode operation. Since all three junctions are operated in the forward direction, the sustaining voltage for a *pnpn* structure is exceedingly low, of the order of kT/e. The difficulty with the *pnpn* structure is that the initial value of $\alpha_1 + \alpha_2$ should be less than unity. Without the off-region, a *pnpn* structure would simply appear as an electric short-circuit. The other important consideration is the switching speed of these devices.[32] Since recombination of excess carriers in the breakdown region is extremely fast, time response of these devices is limited by capacitance effects and recovery of the junction itself.

6.9 TUNNEL DIODES

Theory of tunneling. Most of us are familiar with processes of thermionic, photoelectric, and field emission in vacuum tubes. The basic requirement common to all the processes is to get electrons over a potential barrier at the surface of a cathode. For comparison, the three processes are illustrated in Fig. 6.34. In thermionic emission, the cathode temperature is raised to increase substantially the number of electrons in the tail of the Fermi distribution having energies greater than the work function of the cathode. In photoelectric emission, the energy is supplied by the impinging photons so that excited electrons may have enough energy to escape from the cathode. The physical mechanisms for the release of electrons in these two processes are quite clear from the classical viewpoint based on energy considerations. The emission of electrons by the application of a high electric field, however, is quite different in nature from the other two processes. In the process shown in Fig. 6.34c, electrons penetrate the barrier by quantum-mechanical tunneling instead of climbing over it as in the processes shown in Fig. 6.34 a and b. The tunneling process in a semiconductor diode is quite similar to, but more complex in nature than, the field emission in vacuum tubes. This is the reason why field emission is discussed first.

The field-emission current derived by Fowler and Nordheim for a triangular barrier depends on the electric field E as

$$I = AE^2 \exp\left(-\frac{B}{E}\right) \qquad (6.114)$$

[32] For a discussion of the a-c response of a p-n junction operated near breakdown, refer to J. L. Moll *et al.*, *Phys. Rev.*, vol. 110, p. 612, 1958.

Figure 6.34

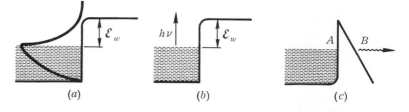

(a) (b) (c)

Processes of (a) thermionic, (b) photoelectric, and (c) field emission in vacuum tubes.

where A is a constant involving the work function $\mathcal{E}_w$ and B is given by

$$B = \frac{2(2\mathcal{E}_w)^{3/2}m^{1/2}}{3e\hbar} \tag{6.115}$$

Equation (6.114) which agrees well with experimental results has special significance in our present discussion because of its exponential dependence on $1/E$. If we recall our treatment of electron wave motion involving a potential well in Sec. 3.7, the solution to the wave equation becomes exponentially decaying in the region where the energy $\mathcal{E}$ of an electron is smaller than the potential barrier V_0. More precisely,

$$\psi = C \exp\left(-\frac{\sqrt{2m(V_0 - \mathcal{E})}\, x}{\hbar}\right) \tag{6.116}$$

where C is a normalization constant. In order that Eq. (6.116) may be applicable to field emission problems, the following observations are made. First, referring to the triangular potential barrier of Fig. 6.34c, $V_0 - \mathcal{E} = \mathcal{E}_w - eEx$ for electrons at the top of the Fermi distribution; hence, the exponent in Eq. (6.116) should be replaced by an integral. According to the physical meaning of ψ, a transmission probability T can be defined as follows

$$T = \frac{|\psi_B|^2}{|\psi_A|^2} = \exp\left(-\frac{2\int_0^{x_B}\sqrt{2m(\mathcal{E}_w - eEx)}\, dx}{\hbar}\right) \tag{6.117}$$

where $x_B = \mathcal{E}_w/(eE)$. It is easy to show that T is identical to $\exp[-(B/E)]$ of Eq. (6.114). Therefore, the limiting process in tunneling is the probability with which an electron may be able to transmit through a potential barrier when it hits one.

Some modifications must be made, if we are to apply the above analysis to the tunneling process in a semiconductor. In the presence of a high electric field, the electron energy diagram is tilted in a manner shown in Fig. 6.35a. Analogous to the field emission, electrons arriving at point A located at the top of the valence band have a nonvanishing transmission probability T for their appearance at point B in the bottom of the conduction band. In calculating T, however, it is not at all obvious what form of potential barrier an electron experiences in the region AB. If it is assumed that an electron immediately after disappearing from the valence band behaves like a conduction-band electron, then the conservation of energy gives the following relation

$$eEx + \mathcal{E}_v - \frac{\hbar^2 k_v^2}{2m_v} = \mathcal{E}_c + \frac{\hbar^2 k_c^2}{2m_c} \tag{6.118}$$

where the term eEx represents the energy acquired from the field. If it is further assumed that no scattering mechanism is involved in the tunneling process, another restriction is imposed by the conservation of wave vectors or $\mathbf{k}_c = \mathbf{k}_v = \mathbf{k}$. Using this relation in Eq. (6.118) and realizing that $\mathcal{E}_c - \mathcal{E}_v = \mathcal{E}_g$, we have

$$k = \frac{i \sqrt{2m_r(\mathcal{E}_g - eEx)}}{\hbar} \tag{6.119}$$

where the mass m_r, generally referred to as the *reduced mass*, is related to the effective masses m_c and m_v of conduction- and valence-band electrons by

$$m_r = \frac{m_c m_v}{m_c + m_v} \tag{6.120}$$

Note that Eq. (6.119) will result in a transmission probability T similar to that given by Eq. (6.117), and hence the term $\mathcal{E}_g - eEx$ in Eq. (6.119) may be considered by analogy as the potential barrier experienced by electrons in region AB. The question arises as to whether the value of k obtained from Eq. (6.118) is compatible with the energy-band structure or not. As discussed in Sec. 3.10, semiconductors can be classified according to the energy-band structure into two types: direct-gap and indirect-gap. In indirect-gap semiconductors such as Ge and Si, the conduction-band minima and valence-band maxima are located at different values of k in k space; hence, momentum will not be conserved in a direct tunneling process across the bands. However, besides the conduction-band minima, there exists in Ge and Si a $k[000]$ valley directly opposite to the valence-band maxima in k space. Tunneling between the $k[000]$ valley and the valence-band maxima will conserve momentum. Our discussion will be limited to tunneling processes which conserve momentum, or in other words, tunneling processes (1) across the

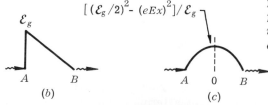

Figure 6.35

(a) *The energy-band diagram of a semiconductor in the presence of a high electric field.* (b) *The potential barrier experienced by an electron if it is considered to be a conduction-band electron after leaving the valence band.* (c) *Modified form of the potential barrier which is compatible with the band theory requiring $k = 0$ at the band edges.*

bands in a direct-gap semiconductor and (2) between the $k[000]$ valley and the valence band in an indirect-gap semiconductor.

In view of the above discussion, the value of k is zero at points A and B. Obviously, Eq. (6.119) does not have the right property at point A; therefore the potential barrier of Fig. 6.35b must be modified. The simplest form of potential barrier $\mathcal{E}_b$ which satisfies the boundary conditions at points A and B is shown in Fig. 6.35c, with

$$\mathcal{E}_b = \frac{(\mathcal{E}_g/2)^2 - (eEx)^2}{\mathcal{E}_g} \tag{6.121}$$

where the origin $x = 0$ is chosen midway between A and B. Note that the condition $\mathcal{E}_b = 0$ at x_B and $-x_A$ gives $x_A + x_B = \mathcal{E}_g/(eE)$ in agreement with the energy diagram of Fig. 6.35a. Replacing $\mathcal{E}_g - eEx$ in Eq. (6.118) by $\mathcal{E}_b$ of Eq. (6.121), we find

$$k_x{}^2 = -k_\perp^2 - 2m_r \frac{(\mathcal{E}_g/2)^2 - (eEx)^2}{(\hbar^2 \mathcal{E}_g)} \tag{6.122}$$

where $k_\perp^2 = k_y{}^2 + k_z{}^2$.

The separation of k^2 into $k_x{}^2$ and $k_\perp^2$ in Eq. (6.122) is necessary because while the kinetic energy of an electron involves k^2, the transmission probability T is determined by the wave number k_x in the direction of tunneling only. Therefore, for electrons having momentum perpendicular to the direction of tunneling, the tunneling points are determined from Eq. (6.122) by equating $k_x = 0$. These points are shown as C and D in Fig. 6.35a. Likewise, in calculating T, the limits of integration are from $-x_C$ to x_D. If we let $u = 2eEx/(4\mathcal{E}_\perp \mathcal{E}_g + \mathcal{E}_g{}^2)^{1/2}$ and $\mathcal{E}_\perp = \hbar^2 k_\perp^2/(2m_r)$, then

$$-2 \int_{-x_C}^{x_D} \mathrm{Im}\,(k_x)\,dx = -\frac{m_r{}^{1/2}(4\mathcal{E}_\perp \mathcal{E}_g + \mathcal{E}_g{}^2)}{\sqrt{2}\,e\mathcal{E}_g{}^{1/2}E\hbar} \int_{-1}^{1} (1 - u^2)^{1/2}\,du$$

where $\mathrm{Im}\,(k_x)$ stands for the imaginary part of k_x. From the above equation, the transmission probability T is found to be

$$T = \exp\left(-\frac{\pi m_r{}^{1/2}\mathcal{E}_g{}^{3/2}}{2\sqrt{2}\,e\hbar E}\right) \exp\left(-\frac{\sqrt{2}\,\pi m_r{}^{1/2}\mathcal{E}_\perp \mathcal{E}_g{}^{1/2}}{e\hbar E}\right) \tag{6.123}$$

Equation (6.123) agrees with the quantum-mechanical expression derived by Kane[33] except for a numerical factor $\pi^2/9$ in front. A comparison of Eq. (6.123) with Eqs. (6.114) and (6.115) shows that the exponents have a similar dependence on quantities like m, $\mathcal{E}$, and E but a slightly different numerical factor. The reader may wonder whether the above calculation is worth presenting if the difference in the results is small. The answer is in the affirmative

[33] E. O. Kane, *Jour. Phys. Chem. Solids*, vol. 12, p. 181, 1959, and *Jour. Appl. Phys.*, vol. 32, p. 83, 1961.

because the calculation requires a subtle consideration of the band theory.

As discussed in Sec. 3.7, allowed energy bands differ from energy states in the forbidden gap in that for the former, the value of k is real while for the latter, k is imaginary. In the complex k space, the conduction and valence bands actually come together at a certain point designated as κ_b on the imaginary axis. In a tunneling process, we may think of an electron as moving along the imaginary k axis and making a smooth transition from one band to the other at κ_b as illustrated in Fig. 6.36. The function given by Eq. (6.121) not only provides a smooth transition between the two bands but also has the desirable properties at band edges. That is the reason why Eq. (6.121) is picked as representing the potential barrier.

The reader also may question the validity of using the Bloch wave function as solution to the Schrödinger wave equation in the presence of a very high electric field since a tacit approval of such usage is implicit in the definition of T given by Eq. (6.117). We realize that even a field strength of 10^6 volts/cm produces only a potential difference of 3×10^{-2} volt in a distance of one lattice constant which is taken to be 3×10^{-8} cm. This potential difference is to be compared with a variation in the lattice potential of the order of 10 volts resulting from the atomic cores within one lattice constant. Therefore, the use of Bloch wave function constitutes a good approximation. As a historical note, both Eqs. (6.121) and (6.117) were used in substantially the same form by Zener[34] in his original analysis of the tunneling phenomenon.

Tunneling in semiconductor junctions. Before applying the above analysis to a tunneling device, we shall present a qualitative description of the important features of a tunnel diode. First, let us consider a junction made of a nondegenerate, direct-gap semiconductor of which the energy band diagram is shown in Fig. 6.37.

[34] C. Zener, *Proc. Roy. Soc.*, London, vol. 145, p. 523, 1934.

Figure
6.36

Energy-band diagram in the complex k space of a direct-gap semiconductor.

Tunneling is possible in the space-charge region of a p-n junction only when there are occupied states (initial states) on one side and empty states (final states) on the other side of the junction. Furthermore, leaving the consideration of momentum conservation for a later discussion, the initial and final states must have the same energy. It is obvious from Fig. 6.37a that the condition for tunneling is met when the diode is biased in the reverse direction, the arrow across the gap in the junction region being indicative of the tunneling process. Applying the same reasoning to Fig. 6.37b, we find that tunneling is not possible because the final states would end up in the forbidden gap. Therefore, in a nondegenerate semiconductor junction, if the junction electric field is high enough, tunneling may contribute significantly to current flow in the reverse direction but add nothing to the forward current. The breakdown in narrow junctions with breakdown voltages below 4 volts is caused primarily by tunneling and such a breakdown is generally referred to as the *Zener breakdown*.

The situation in the forward direction is quite different, however, for diodes made of degenerate semiconductors in which the Fermi level lies above the conduction-band edge in n-type materials and below the valence-band edge in p-type materials. When a small forward bias V_a is applied, electrons in the conduction band on the n side having energies between $\mathcal{E}_{fn}$ and $\mathcal{E}_{fn} - eV_a$ will be able to tunnel to vacant states in the valence band on the p side having energies between $\mathcal{E}_{fp}$ and $\mathcal{E}_{fp} + eV_a$. The situation is illustrated in Fig. 6.38a. As V_a further increases, the number of available states for tunneling increases and so does the tunneling current. However, a situation of maximum available states is reached when $\mathcal{E}_{fn}$ is about to rise above $\mathcal{E}_v$ on the p side or $\mathcal{E}_{fp}$ to go below $\mathcal{E}_c$ on the n side, depending upon which comes first. This corresponds to the situation

Figure 6.37

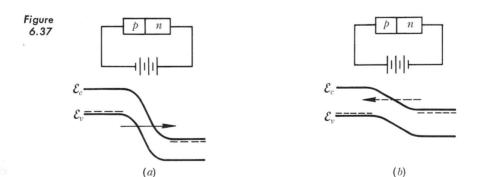

Energy-band diagram and tunneling process in a nondegenerate junction made of direct-gap semiconductors. Tunneling is possible in a reverse-biased junction (case a) but not possible in a forward-biased junction (case b).

shown in Fig. 6.38*b*. Finally, upon still further increase of V_a, the conduction-band edge on the n side rises above the valence-band edge on the p side. When this happens, tunneling is no longer possible as is apparent from Fig. 6.38*c*. Therefore, the I-V characteristic in the tunneling region follows the solid curve of Fig. 6.38*d*. Diodes having such characteristics are known as *Esaki diodes*[35] named after the discoverer of the phenomenon in a forward-biased, degenerate junction. The dotted curve in Fig. 6.38*d* represents the ordinary diffusion current due to minority-carrier injection and it is much smaller than the tunnel current in the tunneling region. We shall return to the I-V characteristic of a tunnel diode and discuss its fine details after a quantitative analysis of the tunnel current.

The number of electrons striking the junction per sec per unit area of the junction is given by the product of the density of occupied states and the velocity of these electrons. Thus the incident current density associated with such a motion is equal to

$$dJ_i = ev_x(2/h^3)f(\mathcal{E})\,dp_x\,dp_y\,dp_z \qquad (6.124)$$

By the definition of the transmission coefficient T, the tunnel current dJ_t is related to the incident current precisely by a factor T. However, we note that the transmission coefficient T of Eq. (6.123) is derived by assuming a completely empty conduction band. The transmission coefficient will be reduced by a factor $[1 - f(\mathcal{E}_2)]$ for a partially filled conduction band. The energies $\mathcal{E}_1$ and $\mathcal{E}_2$ in the Fermi function represent the energy of the initial and final states respectively. Therefore, the tunneling current density from the valence band into the conduction band across a semiconductor junction is

[35]   L. Esaki, *Phys. Rev.*, vol. 109, p. 603, 1958.

Figure 6.38

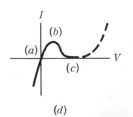

(a)   (b)   (c)

(*a*) *Energy-band diagram in a degenerate junction.* (*b*) *Forward tunneling is possible when the occupied conduction-band states on the n side fall within the energy range of the empty valence-band states on the p side.* (*c*) *Forward tunneling ceases when the conduction-band edge on the n side rises above the valence-band edge on the p side.* (*d*) *Typical I-V characteristic of a tunnel diode.*

given by

$$J_t(v \to c) = \int e \frac{4\pi}{h^3} Tf(\mathcal{E}_1)[1 - f(\mathcal{E}_2)]v_x \, dp_x p_\perp \, dp_\perp \qquad (6.125)$$

A similar expression is obtained for the current tunneling in the opposite direction. The net tunneling current density is the difference between the two expressions or

$$J_t = \int e \frac{4\pi}{h^3} T[f(\mathcal{E}_1) - f(\mathcal{E}_2)]m_r \, d\mathcal{E}_x \, d\mathcal{E}_\perp \qquad (6.126)$$

where $\mathcal{E}_\perp = p_\perp^2/2m_r$ and $\mathcal{E}_x = p_x^2/2m_r$.

We shall now apply Eq. (6.126) to a junction diode. Since Eq. (6.126) is very complicated for a general analysis, reasonable simplifying steps must be taken. First, the temperature is assumed low enough so that the Fermi functions can be taken as step functions. Second, the field in the junction region will be replaced by a uniform average field of strength $E = (V_a + V_d)/W$, the meanings of V_a, V_d, and W being shown in Fig. 6.3 and defined in Eqs. (6.11) and (6.12). Furthermore, the effective masses m_c and m_v are assumed equal. In the following, we shall discuss fully the tunnel current in the reverse direction only. The various symbols to be used in the discussion are defined in Fig. 6.39a.

As the applied voltage V_a changes, the Fermi levels $\mathcal{E}_{fp}$ and $\mathcal{E}_{fn}$ move with respect to each other. Consequently, the difference $f(\mathcal{E}_1) - f(\mathcal{E}_2)$ takes the value unity for an energy range eV_a between $\mathcal{E}_{fn}$ and $\mathcal{E}_{fp}$ and is zero outside the energy range as shown in Fig. 6.40. The transmission probability T of Eq. (6.123) can be written as a product of T_1 and T_2, T_2 being the part dependent on $\mathcal{E}_\perp$. Since E in Eq. (6.123) is now replaced by $(V_a + V_d)/W$, T_1 can be taken out-

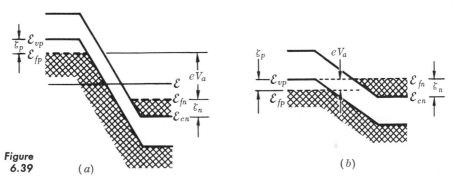

Figure 6.39 (a) (b)

Energy-band diagrams of a degenerate semiconductor junction: (a) biased in the reverse direction and (b) biased in the forward direction. For computational purposes, the electric field in the junction region is taken to be constant.

side the integral in Eq. (6.126). Therefore, in Eq. (6.126), we are left with an integral of the following type:

$$D = \int (\Delta f) \exp\left(-\frac{\mathcal{E}_\perp}{\bar{\mathcal{E}}_\perp}\right) d\mathcal{E}_x \, d\mathcal{E}_\perp \tag{6.127}$$

where $\Delta f = f(\mathcal{E}_1) - f(\mathcal{E}_2)$ is given in Fig. 6.40b and

$$\bar{\mathcal{E}}_\perp = \frac{e\hbar E}{\sqrt{2\pi m_r{}^{\frac{1}{2}}\mathcal{E}_g{}^{\frac{1}{2}}}} \tag{6.128}$$

To evaluate the integral of Eq. (6.127), we must consider the following energy-momentum relations:

$$\mathcal{E}_1 = \mathcal{E}_{vp} - \mathcal{E} = \frac{\hbar^2(k_\perp^2 + k_x{}^2)}{2m_v} \tag{6.129}$$

$$\mathcal{E}_2 = \mathcal{E} - \mathcal{E}_{cn} = \frac{\hbar^2(k_\perp^2 + k_x{}^2)}{2m_c} \tag{6.130}$$

Since $k_\perp$ before and after tunneling must be equal, the maximum value of $\mathcal{E}$ allowed in Eq. (6.127) is determined by the maximum value allowed by both Eqs. (6.129) and (6.130). For example, if $\mathcal{E} = \mathcal{E}_{fp}$, $(\mathcal{E}_\perp)_{\max} = \zeta_p$ from Eq. (6.129) and $(\mathcal{E}_\perp)_{\max} = eV_a + \zeta_n > \zeta_p$ from Eq. (6.130). Obviously, the maximum allowable value for $\mathcal{E}_\perp$ should be the lower value of the two. Therefore, for $\mathcal{E}_{vp} > \mathcal{E} > \mathcal{E}_{cn} + (eV_a + \zeta_n + \zeta_p)/2$, Eq. (6.129) should be used and for $\mathcal{E}_{cn} + (eV_a + \zeta_n + \zeta_p)/2 > \mathcal{E} > \mathcal{E}_{cn}$, Eq. (6.130) should be used. In other words, the integral D is split into two parts integrated over the respective ranges 1 and 2 as follows:

$$D = \int_1 \bar{\mathcal{E}}_\perp\{1 - \exp\left[-(\mathcal{E}_{vp} - \mathcal{E} - \mathcal{E}_x)/\bar{\mathcal{E}}_\perp\right]\} \, d\mathcal{E}_x$$
$$+ \int_2 \bar{\mathcal{E}}_\perp\{1 - \exp\left[-(\mathcal{E} - \mathcal{E}_{cn} - \mathcal{E}_x)/\bar{\mathcal{E}}_\perp\right]\} \, d\mathcal{E}_x \tag{6.131}$$

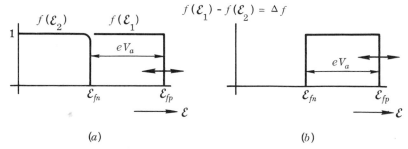

Figure 6.40

(a)   (b)

(a) The plot of respective Fermi functions as a function of energy on both sides of the junction. (b) The difference in the Fermi functions. It is nonzero when there is an applied voltage V_a.

Since the value of $\mathcal{E}_\perp = \mathcal{E}_{vp} - \mathcal{E} - \mathcal{E}_x$ runs between a minimum value of ζ_p to a maximum value of $(eV_a + \zeta_n + \zeta_p)/2$ in range 1 and between ζ_n and $(eV_a + \zeta_n + \zeta_p)/2$ in range 2, integrating Eq. (6.131) gives

$$\frac{D}{\bar{\mathcal{E}}_\perp} = eV_a + \bar{\mathcal{E}}_\perp \left[\exp\left(-\frac{\zeta_p}{\bar{\mathcal{E}}_\perp}\right) + \exp\left(-\frac{\zeta_n}{\bar{\mathcal{E}}_\perp}\right) \right.$$
$$\left. - 2\exp\left(-\frac{eV_a + \zeta_n + \zeta_p}{2\bar{\mathcal{E}}_\perp}\right) \right] \qquad (6.132)$$

Let us now examine the relative importance of the terms in Eq. (6.132). Taking $N_1 = N_2 = 10^{19}/\text{cm}^3$, $\epsilon = 16\epsilon_0$, and $V_d + V_a = 1.5$ volts in Eq. (6.11), we find $W = 2.5 \times 10^{-6}$ cm and an average field $E = 6 \times 10^5$ volts/cm. Further, if $m_c = 0.037m$, $m_v = 0.20m$, and $n = p = 10^{19}/\text{cm}^3$, Eq. (4.34) gives $\zeta_p = 0.07$ ev and $\zeta_n = 0.37$ ev. A value of $\bar{\mathcal{E}}_\perp = 0.045$ ev is obtained from Eq. (6.128) by using $\mathcal{E}_g = 0.8$ ev and $m_r = 0.031m$. From these calculations, it is obvious that for $eV_a > 1$ ev in the reverse direction, the term involving $\bar{\mathcal{E}}_\perp$ in Eq. (6.132) is much smaller than eV_a. Hence, the tunnel current density or Eq. (6.126) can be approximated by the following expression

$$J_t = \frac{4\pi e m_r}{h^3} (\bar{\mathcal{E}}_\perp)(eV_a) \exp\left(-\frac{\pi m_r^{1/2} \mathcal{E}_g^{3/2}}{2\sqrt{2}e\hbar E}\right) \qquad (6.133)$$

For a reverse bias of 1 volt, Eq. (6.133) gives a current density of $7 \times 10^6 \times e^{-8}$ or 2.3×10^3 amp/cm^2 which is very large indeed.

A physical interpretation of Eq. (6.133) is as follows. Because of the factor $\exp(-\mathcal{E}_\perp/\bar{\mathcal{E}}_\perp)$, only those states having $\mathcal{E}_\perp$ smaller than $\bar{\mathcal{E}}_\perp$ contribute appreciably to the tunneling current. Other states having larger $\mathcal{E}_\perp$ will see a much longer effective length of tunneling region ($x_C + x_D$ in Fig. 6.35a as compared to $x_A + x_B$ for $\mathcal{E}_\perp = 0$) and hence a much smaller T. The number of states participating in the tunneling process is further limited by the condition that the initial state must be occupied and the final state must be empty. The factor eV_a in Eq. (6.133) is a result of such a consideration as it is evident from the difference in the Fermi functions plotted in Fig. 6.40b.

Even though the number of tunneling states is limited by various considerations, the numerical factor except V_a in front of the exponential factor in Eq. (6.133) is still very large, being of the order of 7×10^6 amp/cm^2-volt. The only factor which limits the current is the exponential factor. Note that the function $\exp(-B/E)$ increases much less rapidly with E in the region $E > 0.1B$ than in the region $E < 0.1B$. If an arbitrary limit for breakdown in the reverse direction is set at $E = 0.1B$, then a value of the breakdown field $E = 5 \times 10^5$ volt/cm is obtained from Eq. (6.133) for the case just considered. The calculated value of E can be considered as typical in semiconductors such as Ge, Si, and GaAs, even though the exact breakdown field varies from material to material, depending on m_r and $\mathcal{E}_g$.

It should also be pointed out that although the ionization coefficient β in the avalanche breakdown has a dependence upon the electric field similar to that of the transmission probability T, the width of the space-charge region plays a different role in these two processes. According to Eq. (6.97), the product of (βW) determines the condition for avalanche breakdown. The wider the region in which carriers multiply each other, the smaller is the required value of β. On the other hand, for tunneling, the width of the space-charge region comes into the picture only in determining the field strength in the junction. Therefore, assuming that the dependences of β and T upon E are the same, the avalanche process is favored in a wider junction. Since a wider junction requires a larger applied voltage for the same E, the avalanche breakdown is the dominating process in junctions with higher breakdown voltages. The value of 4 to 8 volts mentioned earlier as the demarcation line between the two processes in Ge and Si is obtained from experiments.

The situation in the forward direction is similar to but more complicated than that in the reverse direction. Assuming that only states having $\mathcal{E}_\perp$ smaller than $\bar{\mathcal{E}}_\perp$ contribute significantly to tunneling, we can rewrite Eq. (6.127) as $D = \bar{\mathcal{E}}_\perp \int (\Delta f)\, d\mathcal{E}_x$ and hence the number of states in tunneling is determined by the Fermi functions Δf. As shown in Fig. 6.41, the quantity D first increases with the applied voltage V_a until the empty zone on the p side is completely within the occupied zone on the n side or vice versa, depending upon whether $\zeta_p < \zeta_n$ or $\zeta_p > \zeta_n$. This is obvious from Fig. 6.39b. For the present case, $\zeta_p < \zeta_n$. Upon further increase of eV_a, the number of available states for tunneling is limited by the empty states on the p side. Therefore, the value of D remains constant until $\mathcal{E}_{fp}$ begins to fall below $\mathcal{E}_{cn}$. When this happens, the value of D decreases linearly

Figure 6.41

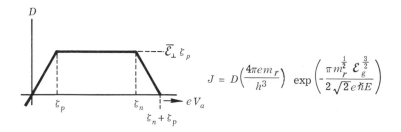

$$J = D\left(\frac{4\pi e m_r}{h^3}\right) \exp\left(-\frac{\pi m_r^{\frac{1}{2}} \mathcal{E}_g^{\frac{3}{2}}}{2\sqrt{2}\, e\hbar E}\right)$$

The variation of the density-of-state function D as a function of the applied forward voltage. The quantity D which has a dimension of energy² actually is proportional to the density of states participating effectively in the tunneling process.

with eV_a. Eventually, the empty zone on the p side and the occupied zone on the n side go completely out of alignment if $\mathcal{E}_{vp}$ falls below $\mathcal{E}_{cn}$. This explains why the value of D goes to zero at $eV_a = \zeta_n + \zeta_p$ in Fig. 6.41. For the case just presented, $\zeta_n + \zeta_p = 0.44$ ev which is still much larger than $\bar{\mathcal{E}}_\perp$. Therefore, if we actually carry out an integration of $\exp(-\mathcal{E}_\perp/\bar{\mathcal{E}}_\perp)$ over $\mathcal{E}_\perp$ similar to that in Eq. (6.131), an expression similar to Eq. (6.132) is obtained in which the term involving $\bar{\mathcal{E}}_\perp$ will be much smaller than eV_a for most of eV_a in the forward direction. In other words, our earlier assumption that only those states having $\mathcal{E}$ smaller than $\bar{\mathcal{E}}_\perp$ contribute significantly to tunneling is still justified.

The actual I-V characteristic of a tunnel diode looks more like the rounded curve shown in Fig. 6.38d than the trapezoidal D-V_a curve of Fig. 6.41. First, the factor T_1 is also a function of V_a through E. Further, the Fermi functions Δf are not exactly step functions at finite temperatures. Finally, in the region before diffusion current sets in and after the tunneling current begins to fall, there is an additional current component known as excess current. At very high concentrations, impurity states are no longer discrete but form a band. Tunneling through impurity bands or other gap states is a definite possibility. All these factors tend to round off the trapezoidal curve of Fig. 6.41. We should also point out that contrary to ordinary diffusion current, tunneling current is much smaller in the forward direction than in the reverse direction on account of a smaller E. At $eV_a = \zeta_n$ in Fig. 6.41, $V_a + V_d = \mathcal{E}_g + \zeta_p = 0.87$ ev, reducing E to 4.6×10^5 volts/cm. That means, T_1 drops from a previous value of $\exp(-8)$ to $\exp(-10.5)$, or by a factor of 12. So far as the applications of tunnel diodes are concerned, the forward characteristic is much more interesting than the reverse characteristic because of the negative-resistance region. Needless to say, tunnel diodes can be used in amplifier and oscillator circuits. Since no minority-carrier storage is involved, tunnel diodes are most suitable for fast switching and pulse shaping.

Tunneling and energy-band structure. In the above discussion, we have considered the tunneling process which involves transitions between conduction- and valence-band states having the same wave number **k**. In an indirect-gap semiconductor where the band extrema occur at different **k**, tunneling is possible only under certain circumstances. Take germanium as an example. As discussed in Sec. 3.10, the conduction-band minima occur at edge of the Brillouin zone along the four principal diagonals in the reciprocal lattice, that is $k[111]$ and its equivalent axes. Therefore, direct tunneling between the conduction- and valence-band extrema is not possible because of the difference in wave numbers associated with the extrema. However, above the $k[111]$ minima, there exists a potential valley at $k[000]$ as illustrated in Fig. 3.29. If the Fermi level on the p side is

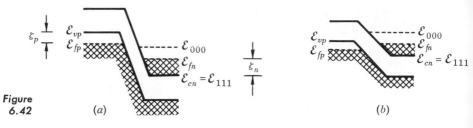

Figure
6.42

(a)   (b)

Energy-band diagram of Ge which is an indirect-gap semiconductor. (a) Direct tunneling is possible in the reverse direction when the applied voltage is large enough such that $\mathcal{E}_{fp}$ is above $\mathcal{E}_{000}$ of the $k[000]$ valley. (b) Since the $k[000]$ valley normally is not occupied even in degenerate n-type Ge, direct tunneling is not possible in the forward direction.

raised by a reverse voltage such that $\mathcal{E}_{fp} > \mathcal{E}_{000}$, tunneling of electrons from the valence band into the $k[000]$ valley is now possible. This is shown in Fig. 6.42a. If V_a is the reverse voltage, direct tunneling begins when

$$eV_a = \mathcal{E}_{000} - \mathcal{E}_{111} - \zeta_n \qquad (6.134)$$

Since the density-of-states effective mass is fairly large for electrons in the conduction-band minima, the value of ζ_n is small compared with $\mathcal{E}_{000} - \mathcal{E}_{111}$ even at high doping concentrations. This also means that the $k[000]$ valley will not be occupied even in the n region of an Esaki diode. Therefore, direct tunneling is not possible in the forward direction as is evident from Fig. 6.42b.

Besides the direct tunneling process, however, there is also the indirect tunneling process assisted by impurity or phonon scattering. As shown in Fig. 6.43, the tunneling states extend into the forbidden gap along the imaginary axis as in Fig. 6.36 but they do not meet at a

Figure
6.43

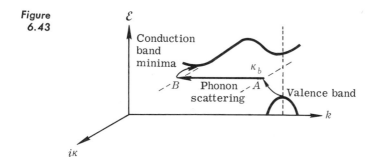

Energy-band diagram in the complex k space of an indirect-gap semiconductor. Since the conduction-band minimum and valence-band maximum will not meet along the imaginary k axis, electrons must be scattered by phonons or impurities in a tunneling process to conserve momentum.

common **k**. The difference in momentum between the two states is supplied by phonons or impurities in a scattering process as indicated by an arrow AB in Fig. 6.43. The expressions for the transmission probability T and current density for the phonon-assisted indirect tunneling have been worked out by Keldysh[36] and Kane[37] but the derivation of these expressions is too involved for a simple and yet meaningful presentation here. In brief, the indirect tunneling current density involving phonon emission can be written in the following form:

$$J_{ind} = C_i D_i (N_S + 1) \exp(-\beta) \qquad \text{with } \beta = \frac{4(2m_{rx})^{1/2}(\mathcal{E}_g \pm \hbar\omega)^{3/2}}{e\hbar E}$$

(6.135)

In the above equation, the upper $(+)$ and lower $(-)$ signs are respectively for n to p (forward bias) and p to n (reverse bias) tunneling. The quantity $\hbar\omega$ is the energy of the phonon involved in the scattering process and N_S is the phonon occupation number. For tunneling process involving the absorption of a phonon, $(N_S + 1)$ should be replaced by N_S and the upper and lower signs in β should be interchanged.

The quantity D_i in Eq. (6.135) represents the effective density-of-states participating in indirect tunneling and is similar in nature to D of Eq. (6.132). However, now the restriction on momentum conservation is removed and a different result is expected. The mass m_{rx} in β is the reduced mass of valence-band holes and conduction-band electrons along the direction of tunneling. Needless to say, m_c to be used in Eq. (6.120) is the effective mass of electrons in the conduction-band minima for indirect tunneling while m_c is the effective mass of $k[000]$ valley electrons for direct tunneling. The quantity C_i in Eq. (6.135) can be calculated quantum-mechanically. Since indirect tunneling is a two-step process, it involves a second-order perturbation calculation. Among other things, the matrix element which connects the two states A and B in Fig. 6.43 is involved in C_i. This matrix element measures the effectiveness of phonon scattering and its value can be inferred from other experiments. The reader should refer to Kane's article for an explicit expression for C_i. However, for our present discussion, it is worthwhile to compare the magnitudes of J_{ind} from Eq. (6.135) and J_t from Eq. (6.133). It turns out that in the reverse direction, D_i can also be approximated by $eV_a\bar{\mathcal{E}}_\perp$. If N_S is taken to be zero as it is at low temperatures, the numerical factor except V_a in front of the $\exp(-\beta)$ in Eq. (6.135) is around 600 amp/cm²-volt in germanium. This value is to be compared with 7×10^6 amp/cm²-volt from a previous calculation for direct tunnel-

[36] L. V. Keldysh, *Zh. Eksperim. i Theor. Fiz.*, vol. 34, p. 962, 1958 (translation in *Soviet Phys.—JETP*, vol. 34, p. 665, 1958).

[37] E. O. Kane, *Jour. Appl. Phys.*, vol. 32, p. 83, 1961.

ing. Therefore, a very large change in the magnitude of the current by a factor of around 10^4 is anticipated in going from indirect to direct tunneling.

Figure 6.44a shows the I-V characteristic of a germanium tunnel diode in the reverse direction, reported by Tiemann and Fritzsche.[38] The sharp rise in the current around 140 mvolts is indicative of the onset of direct tunneling. From Eq. (6.134), with $\zeta_n = 0.02$ ev, the separation between the conduction-band minima and the $k[000]$ valley minimum is found to be 0.16 ev in agreement with optical measurements on degenerate germanium. Figure 6.44b shows the current in the indirect tunneling region on a much expanded scale. As pointed out earlier, indirect tunneling is assisted by phonons or impurities. However, there is a definite amount of energy (Figs. 2.46 and 2.47) associated with phonons of a certain wave number needed for momentum conservation. When the bias voltage is large enough to permit the emission of such phonons, indirect tunneling sets in. The current increases abruptly again when a higher energy phonon of the same wave number can be emitted. The places at which the discontinuities in the slope of the I-V curve occur give the energy of

[38] J. J. Tiemann and H. Fritzsche, *Phys. Rev.*, vol. 132, p. 2506, 1963.

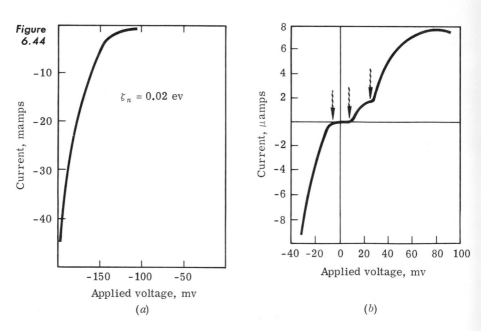

Figure 6.44

$\zeta_n = 0.02$ ev

(a) *The I-V characteristic of a germanium tunnel diode which shows the onset of direct tunneling around -0.14 volt. (b) Curve a on a much expanded scale near the origin to show the onset of indirect tunneling. (J. J. Tiemann and H. Fritzsche, Phys. Rev., vol. 132, p. 2506, 1963.)*

these phonons. The energies 8 and 28 mvolts obtained from Fig. 6.44b can be identified as associated with transverse and longitudinal acoustic phonons. In concluding, we should point out that conservation of wave number is also observed in an optical transition across the bands. Therefore, these phonons will enter the picture again when we discuss optical absorption measurements in Chapter 11.

PROBLEMS

6.1 A p-n germanium junction is made of an alloyed p region with a hole concentration of $2 \times 10^{19}/cm^3$ and a base n region of $1 \ \Omega$-cm resistivity. Given: $N_c = 1.05 \times 10^{19}/cm^3$ and the Fermi level on the p side can be taken as equal to $\mathcal{E}_v$. Calculate the built-in voltage V_d. Also compute the width W of the space-charge region and the junction capacitance per unit area (C_T/A) at zero applied bias voltage. What is the ratio of the extension of the space-charge region into the n and p regions?

6.2 Calculate the junction capacitance per unit area (C_T/A) in a germanium diode with $N_a = 2 \times 10^{19}/cm^3$ on the p side and $N_d = 10^{15}/cm^3$ on the n side for (a) a forward bias of 0.2 volt, (b) no applied bias voltage, and (c) a reverse bias of 10 volts. Also find the value and the location of maximum electric field in the junction for parts (a), (b), and (c). Derive an expression relating the maximum electric field to the impurity concentrations and the applied voltage.

6.3 Consider a graded p-n junction in which the impurity concentrations in the transition region vary linearly with distance away from the junction. In other words, $N_1 = ax$ for $x < 0$ and $N_2 = bx$ for $x > 0$ in Eq. (6.5). Plot $\varrho(x)$ and $E(x)$ as a function of x. Show that $ax_A{}^2 = bx_B{}^2$ on account of equal stored charges on opposite sides of the junction. Also verify the following expressions for the width W of the space-charge region and the junction capacitance per unit area C_T/A:

$$W = \left[\frac{3\epsilon V(\sqrt{a} + \sqrt{b})^2}{eab} \right]^{1/3} \quad \text{and} \quad \frac{C_T}{A} = \frac{2\epsilon}{W}$$

where $V = V_d \mp V_a$. Check the dimensions of these expressions.

6.4 Consider a p-i-n diode which has a middle region made of intrinsic material sandwiched between p and n materials. Draw diagrams (similar to Fig. 6.2) showing $\varrho(x)$, $E(x)$, and $V(x)$ as a function of x in the transition region. Verify the following expression

$$V = V_d \mp V_a = \frac{Qd}{\epsilon} + \frac{(N_1 + N_2)Q^2}{2e\epsilon N_1 N_2}$$

where Q is the charge stored per unit area of the junction and d is the width of the intrinsic region. The other symbols have the same meaning as the corresponding ones in Eq. (6.13). Show that the junction capacitance is equivalent to two capacitors in series or

$$\frac{1}{C_T} = \frac{d + W}{\epsilon A}$$

6.5 Show that the exact solution of Eq. (6.19) is

$$p_1 = p_{10} \frac{\sinh (W_2 - x)/L}{\sinh (W_2 - x_B)/L}$$

by evaluating coefficients A and B of Eq. (6.20) from the boundary conditions: $p_1 = 0$ at $x = W_2$ and $p_1 = p_{10}$ at $x = x_B$. Verify the two approximate expressions, Eqs. (6.21) and (6.22), under the appropriate conditions specified in the text.

6.6 Write down the expression corresponding to Eq. (6.23) due to diffusion of electrons in the p region. Apply these two equations to an alloyed p-n junction in which the p side is the much more heavily doped side. Show that the current is predominantly determined by hole diffusion into the n region. Discuss the temperature dependence of the reverse saturation current (current under large reverse bias) for the case $L > W_2 - x_B$ and for the case $L < W_2 - x_B$.

6.7 Equation (6.23) is based on the assumption that generation or recombination of excess carriers is unimportant in the space-charge region. This assumption is no longer true in silicon diodes and in germanium diodes at low temperatures biased in the reverse direction. Refer to articles by Pell for further details (E. M. Pell, *Jour. Appl. Phys.*, vol. 26, p. 658, 1955, and vol. 27, p. 768, 1956). Apply Eq. (5.71) to the space-charge region of a diode and find the net rate of generation of excess carriers. Assuming that carriers are swept out of the space-charge region as soon as they are generated, derive the current expression. Show that this component of current has a temperature dependence as p'/C_n or n'/C_p.

6.8 Consider the switching response of a narrow-base diode in which $W \ll L$ so that the p_1/τ term in Eq. (6.30) can be neglected. Show that the Laplace transform of $p_1(\bar{p} = \mathcal{L}p_1)$ is given by

$$eAD_p\bar{p} = \frac{1}{s}\left[I_f(W - x) - \frac{(I_r + I_f)\sinh(\beta W - \beta x)}{\beta \cosh(\beta W)} \right]$$

where $\beta = \sqrt{s/D_p}$. The above equation is for the constant current phase (phase I) and it corresponds to Eq. (6.35) for the wide-base diode. Find the inverse Laplace transform of $\bar{p}$, that is p_1, which can be expressed in a series $\sum_n A_n \cos(n\pi x/W) \exp(-\alpha_n t)$. What is α_n?

6.9 Show that $\bar{p}$ in a narrow-base diode, corresponding to Eq. (6.41), is given by

$$eAD_p\bar{p} = \frac{I_f}{s}\left[(W - x) - \frac{W\sinh(\beta W - \beta x)}{\sinh(\beta W)} \right]$$

Find p_1 by taking the inverse Laplace transform of $\bar{p}$. Show that the current decays as $2I_f \sum_n (-1)^n \exp(-n^2\pi^2 D_p t/W^2)$ where n is an integer.

6.10 In the exact treatment of diode switching by B. Lax and S. F. Neustadter (*Jour. Appl. Phys.*, vol. 25, p. 1148, 1954), the excess hole concentration is given by

$$p'(x,t) = p_1 \exp\left(\frac{t}{\tau}\right)$$

$$= \frac{1}{2\sqrt{\pi Dt}} \int_0^\infty f(x') \left\{ \exp\left[-\frac{(x-x')^2}{4Dt}\right] - \exp\left[-\frac{(x+x')^2}{4Dt}\right] \right\} dx'$$

$$+ \frac{x}{2\sqrt{\pi D}} \int_0^t \frac{g(\lambda)}{(t-\lambda)^{3/2}} \exp\left[-\frac{x^2}{4D(t-\lambda)}\right] d\lambda$$

It is assumed that the solution satisfies the initial condition $p'(x,0) = f(x)$ and the boundary condition $p'(0,t) = g(t)$. Referring to our discussion in Sec. 5.7, explain how the above expression can be obtained from Eqs. (5.97) and (5.99) by the methods of image and superposition.

6.11 Apply the expression given in the previous problem to a diode. How do the functions $f(x')$ and $g(t)$ relate to the excess hole distribution and the junction voltage in the diode? In what region (with regard to x) is the expression applicable? Read the original article by B. Lax and S. F. Neustadter, and compare the essential results of their calculation with those of Kingston's.

6.12 Consider a p-n-p transistor operated under d-c conditions. Taking

$$(p_1)_{\text{d-c}} = A \sinh\frac{W-x}{L} + B \sinh\frac{x}{L}$$

as the solution of the diffusion equation, find the constants A and B in terms of the equilibrium carrier concentrations in the base region and the bias voltages on the emitter and collector junctions. Calculate the hole components of the emitter and collector currents. Assuming that the electron components of currents in Eq. (6.45) are the same as those calculated for diodes, find the analytic expressions for β, γ, and I_{c0} in Eq. (6.45).

6.13 Refer to the discussion presented in the previous problem. Show that the emitter efficiency γ in a p-n-p transistor is given by

$$\gamma^{-1} = 1 + \frac{\sigma_b}{\sigma_\epsilon}\frac{L_{pb}}{L_{n\epsilon}} \tanh\frac{W}{L_{pb}}$$

where σ and L represent conductivity and diffusion length respectively and the subscripts b and ϵ refer to the base and emitter regions respectively. For a transistor with $W = 5 \times 10^{-3}$ cm, $\tau = 10^{-4}$ sec in the base region and 10^{-7} sec in the emitter region, $\rho_b = 5$ Ω-cm and $\rho_\epsilon = 0.001$ Ω-cm, calculate the values of γ and β_0 (the transport efficiency). At what frequency does the phase factor in the various admittances become appreciable?

6.14 The ratio of the emitter resistance to the collector resistance is sometimes called the basewidth modulation factor η. Show that

$$\eta = \frac{kT}{eW}\frac{\partial x_B}{\partial v_{c0}}$$

Estimate the value of η for $v_{c0} = -10$ volts, assuming that $\rho_b = 5$ Ω-cm, $W = 5 \times 10^{-3}$ cm, and the collector side is heavily doped through alloying.

Find the ratio of the time constants of the emitter and collector circuits $(r_e C_{Te}/r_c C_{TC})$ if $V_d = 0.40$ volt and $V_{\epsilon 0} = 0.10$ volt. The area and the doping concentration of the emitter junction are taken to be the same as those of the collector junction. With the polarity of the voltage and the direction of the current chosen as those given in Fig. 6.18, compare the relative phase of the components of i_e due to v_{e1} and v_{c1} and hence determine the sign in front of $w(t)$ in Eq. (6.49) for a p-n-p and an n-p-n transistor.

6.15 (a) Express r_b' of Fig. 6.20 in terms of r_b and β_0/r_c of Fig. 6.19. (b) Show that matched conditions at the input and output terminals (Fig. 6.20) require

$$R_g = r_1\left[1 - \frac{(r_b' + \alpha r_c)r_b'}{r_1 r_2}\right]^{1/2} \qquad R_L = r_2\left[1 - \frac{(r_b' + \alpha r_c)r_b'}{r_1 r_2}\right]^{1/2}$$

(c) Verify that under matched conditions

$$\eta = \left[\frac{\alpha r_c + r_b'}{\sqrt{r_1 r_2} + \sqrt{r_1 r_2 - r_b'(r_b' + \alpha r_c)}}\right]^2$$

where $r_1 = r_b' + r_e$ and $r_2 = r_b' + r_c$. Since $r_c \gg r_b' \gg r_e$ and $\alpha \simeq 1$, the above equation reduces to $\eta \simeq \alpha^2 r_c/r_b'$.

6.16 (a) Using the Einstein relation and assuming that all donors are ionized, derive Eq. (6.61). (b) Starting from Eq. (6.61) explain why the different distributions in Fig. 6.24b have different slopes at the emitter. (c) Referring again to Fig. 6.24b in conjunction with Eq. (6.64), show from the viewpoint of the amount of charge stored in the base region that case C has the shortest and case B the longest transit time. (d) Assuming the same impurity concentrations in the emitter and collector, compare the effects of the three distributions on E_{max} in the space-charge region, on the emitter efficiency, on the junction capacitances, and on the basewidth modulation effect.

6.17 Plot $p_1(x)$ from Eq. (6.65) as a function of x, assuming the same basewidth and the same current density for the three cases. Compare the excess hole concentrations at the emitter and then the transit times. Comment on the effects of the three distributions upon E_{max} in the space-charge region, the emitter efficiency, the junction capacitances, and the basewidth modulation effect.

6.18 (a) Show that the ratio $I_{\epsilon n}/I_{\epsilon p}$ in a p-n-p transistor can be approximated at high injection levels by

$$\frac{I_{\epsilon n}}{I_{\epsilon p}} = \frac{D_n L_p}{D_p L_n}\frac{n_{n0} + p_1}{p_{p0}}\tanh\frac{W_0}{L_p}$$

assuming that the p side is much more heavily doped. In the above equation, which diffusion constant, D_n or D_p, is affected by high injection? (b) Show that for a narrow-base transistor ($W_0 \ll L_p$), the parameter Z can be evaluated from emitter current I_ϵ and base conductivity σ_n as follows:

$$Z\frac{1 + 2Z}{1 + Z} = \frac{\mu_n W_0 I_\epsilon}{D_{p0}\pi r_\epsilon^2 \sigma_n}$$

(c) Given: $W_0 = 2 \times 10^{-3}$ cm, $\mu_n = 3900$ cm²/volt-sec, $D_{p0} = 50$ cm²/sec, $\sigma_n = 1$ (Ω-cm)$^{-1}$, and $r_e = 0.1$ cm. Find the value of the emitter current at which $p_1 = n_{n0}$ in the base region. (d) Given: $D_n = 100$ cm²/sec, $\tau_0 = 10^{-7}$ sec and $\sigma_p = 500$ (Ω-cm)$^{-1}$ in the emitter region, $\tau_0 = 10^{-4}$ sec in the base region, and $v_s = 50$ cm/sec. Find the relative magnitudes of I_{b1}, I_{b2} and I_{b3} in Eq. (6.70).

6.19 By analyzing the base-resistance-bias effect upon current flow in the base region, R. N. Fletcher (*Proc. IRE*, vol. 43, p. 551, 1955) has shown that the current density varies laterally according to

$$J(y) = \frac{J(r_e)}{[1 + (r_e - y)/y_0]^2}$$

where $y_0^{-2} = (e\rho/2WkT)(1 - \alpha)J(r_e)$ has a dimension of cm^{-2}, $J(r_e)$ being the current density at the periphery of the emitter and y being the lateral distance measured from the center of the emitter. Show that averaging the above expression in a one-dimensional case gives

$$g(I_e) = 1 + \frac{r_e}{y_0} = 1 + \sqrt{Hg(I_e)}$$

where $H = (e\rho/2WkT)(1 - \alpha)I_e$. Plot $g(I_e)$ as a function of H.

6.20 (a) From Eq. (5.71), show that in a semiconductor with $n \gg n'$ or $p' \gg p$, τ is independent of the injection level or $a = 1$ in Eq. (6.74). (b) Given: $1 - \gamma_0 = 10^{-3}$, $W_0^2/2D_0\tau_0 = 4 \times 10^{-4}$, $2v_sW_0^2/D_0r_e = 10^{-4}$, and the value of $g(I_e)$ from the expression derived in the previous problem. Plot I_e/I_b from Eq. (6.74) as a function of I_e, assuming that the value of $(1 - \alpha)$ which relates I_e to H is taken to be the value at low injection levels. Use the numerical values given in Prob. 6.18.

6.21 From Eqs. (6.78) and (6.79), find the admittance seen by the pump voltage. Show that power is conserved or $P_p + P_s = 0$ for any phase angle θ. Obtain the ratio P_p/P_s from the Manley-Rowe relationship. Explain the apparent inconsistency if there is any.

6.22 (a) For a semiconductor alloyed diode with $N_d = 10^{16}$/cm³ and $W = 2 \times 10^{-3}$ cm, show that at a forward bias of V_f volts, the Q of the diode is equal to

$$Q = \frac{3.6 \times 10^{-6}\omega}{\sqrt{V_d - V_f \exp{(eV_f/kT)}}}$$

If V_d is assumed to be 0.46 volt, find the maximum value of V_f allowed to give Q a value of 1000 at a frequency 5×10^9 cycles/sec. (b) In view of the above discussion, the maximum swing of the pump voltage v_p is limited to $v_0 + V_f$ to ensure a high Q, where v_0 is the reverse bias on the diode and V_f is the value calculated above. Show that for maximum negative conductance, the value of v_0 should be so chosen that

$$v_0 = \frac{V_d - (m + 1)V_f}{m}$$

where the parameter m is defined in Eq. (6.76).

6.23 Suppose that in Eq. (6.77) three voltages, the pump, the signal, and the idle, exist:

$$v_1 = v_s \sin (\omega_s t + \theta_s) + v_i \sin (\omega_i t + \theta_i) + v_p \sin (\omega_p t + \theta_p)$$

Show that with $\omega_p = \omega_s + \omega_i$, the admittance seen by the signal voltage is equal to

$$Y = j\omega_s C_0 + \frac{m v_p v_i \omega_s C_0}{2(V_d + v_0) v_s} \exp [j(\theta_p - \theta_i - \theta_s)]$$

Compare the phase relationship for maximum power transfer concluded from the above equation with that expressed in Eq. (6.84).

6.24 Show that the amplitudes of the signal and idle voltage waves are in the ratio $|a_s/a_i|^2 = k_s/k_i$ for the wave traveling in the positive z direction. Assume that the transmission line is terminated at its characteristic impedance. Obtain the power relation $P_s/P_i = \omega_s/\omega_i$ and check the above relation against the conclusion drawn from Eq. (6.89).

6.25 Consider a lumped transmission line of which the nth section is shown in the accompanying figure. Assume that the two node voltages are

Figure
P6.25

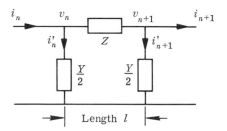

related through $v_{n+1} = v_n \exp (-j\beta l) = \Sigma a_s \exp j[\omega_s t - \beta_s (n + 1)l]$ where the summation is over the signal, idle, and pump voltages. If $Y = j\omega(C_0 + c_1 v)$, express i_n and i_{n+1} in terms of a_s, a_i, a_p, and their complex conjugates. By equating the signal component of i_n to i_{n+1} through $i_{n+1} = i_n \exp (-j\beta_s l)$ show that

$$4 \sin^2 \left(\frac{\beta_s l}{2}\right) = \omega_s^2 \left(LC_0 + \frac{c_1 a_i^* a_p}{a_s}\right)$$

A similar expression is obtained for $\beta_i l$. By letting $\beta_s l = \beta_{s0} l + j\alpha$ and $\beta_i l = \beta_{i0} l + j\alpha$, show that

$$\alpha = \frac{\beta_{s0} \beta_{i0} |a_p| l}{\sqrt{\sin \beta_{s0} l \sin \beta_{i0} l}} \left(\frac{c_1}{2C_0}\right)$$

6.26 Read A. L. McWhorter and R. H. Kingston, Channels and Excess Reverse Current in Grown Germanium p-n Junction Diodes, *Proc. IRE*, vol. 42, p. 1376, 1954. Discuss the experimental evidence establishing the existence of surface channel and its effect on the excess junction current.

Comment also on the variation of the effective length of surface channel with junction voltage.

6.27 Using $f(x) = \exp\left[-\int_0^x (\beta_1 - \beta_2)\, dx'' \right]$ as an integrating factor, show that Eq. (6.92) can be integrated to give

$$(n_0 + n_1)f(x) = n_f \int_0^x \beta_2 f(x')\, dx' + A$$

where A is an integration constant. Also apply the boundary conditions to the above equation to get

$$n_f \left[f(W) - \int_0^W \beta_2 f(x)\, dx \right] = n_0$$

The above equation is identical to Eq. (6.93) by noting that

$$\int_0^W (\beta_1 - \beta_2)f(x)\, dx = f(0) - f(W) = 1 - f(W)$$

6.28 Show that for an alloyed junction

$$\int_0^W \beta_1\, dx = aE_{\max}^n \frac{W}{n+1}$$

Also verify that $\beta_1(E_{\max}) = aE_{\max}^n$ is related to the slope of the M_e curve by

$$\frac{1}{c-1}\frac{d}{dv_0}\left(1 - \frac{1}{M_e} \right) = \left(\frac{1}{c-1} + 1 - \frac{1}{M_e} \right)\frac{W}{2}\frac{\beta_1(E_{\max})}{V_d + v_0}$$

where v_0 is the applied reverse voltage.

6.29 From the dependence of $E_{\max}^n W$ upon the impurity concentrations N_1 and N_2 and the applied voltage v_0, prove the dependence of V_B upon N_1 and N_2 as given by Eq. (6.98). Given $m = -0.725$ in Ge, plot V_B as a function of N_1, assuming that $N_2 \gg N_1$. Choose the value for the base resistivity of a transistor if the collector bias is 20 volts.

6.30 Assume $c = 4$ and $n = 6.25$ in Eq. (6.97). Find the value of W needed for breakdown at $1/E_{\max} = 4 \times 10^{-6}$ cm/volt. From $E_{\max}$ and W, calculate V_B and N_i (the smaller of N_1 and N_2). Repeat the calculation for $1/E_{\max} = 5 \times 10^{-6}$ cm/volt. Construct the V_B versus N_i curve from the two calculated points and compare your calculated values with the values predicted from Eq. (6.98).

6.31 Read the article by Kidd et al., RCA Review, vol. 16, p. 16, 1955. Discuss the general features of V-I characteristic of a transistor with controlled base current and explain the reason for the existence of a negative resistance region.

6.32 Write down the expressions corresponding to Eq. (6.112) for the current through the two outer junctions. Be sure that you count the right amount of contributions to I_{sm} from hole and electron injection in these expressions. Prove Eq. (6.113) under the assumptions stated in the text.

REFERENCES

Biodi, F. J. (ed.): "Transistor Technology," vol. 2, D. Van Nostrand Co., Inc., New York, 1958.

Chang, K. K. N.: "Parametric and Tunnel Diodes," Prentice-Hall, Inc., Englewood Cliffs, N.J., 1964.

Jonscher, A. K.: Physics of Semiconductor Switching Devices, in "Progress in Semiconductors," vol. 6, p. 143, Heywood & Co., Ltd., London, 1962.

————: "Principle of Semiconductor Device Operation," G. Bell & Sons, Ltd., London, 1960.

Middlebrook, R. D.: "An Introduction to Junction Transistor Theory," John Wiley & Sons, Inc., New York, 1957.

Moll, J. L.: "Physics of Semiconductors," McGraw-Hill Book Company, New York, 1958.

Shive, J. N.: "The Properties, Physics and Design of Semiconductor Devices," D. Van Nostrand Co., Inc., New York, 1959.

Shockley, W.: "Electrons and Holes in Semiconductors," D. Van Nostrand Co., Inc., New York, 1950.

7 DIELECTRIC AND FERROELECTRIC MATERIALS

7.1 POLARIZATION AND DIPOLE MOMENT

In the macroscopic description of the dielectric properties of a substance, the principal parameter is ϵ, the *dielectric constant*[1] (farad/m). For example, this parameter enters in the calculation of the capacitance of a condenser and it also appears in Maxwell's equations. However, the value of ϵ used in these two cases may be drastically different even though we are referring to the same material. Whenever the macroscopic parameter ϵ is used, it is necessary to specify the frequency and temperature range within which the material is operated and also the orientation of the applied electric field with respect to the crystal axes. To explain these and other aspects of ϵ, a model based on the atomic structure of a material is needed. In this chapter, only the simple classical models will be discussed. They are useful because some of the most significant results derived from them are essentially correct. Furthermore, they serve to illustrate the basic ideas leading to more sophisticated calculations based on quantum mechanics.

A basic concept from an atomic viewpoint is that of *electric dipole moment*. For a system of charged particles, the dipole moment is defined as

$$\mathbf{p} = \sum_i q_i \mathbf{r}_i \qquad \text{coul-m} \tag{7.1}$$

where q_i is the charge associated with the ith particle and $\mathbf{r}_i$ is the displacement of the particle with respect to the origin. In an electrically neutral system (which we assume in the following discussions), $\mathbf{p}$ as defined by Eq. (7.1) is independent of the origin of the coordinate system. If we suppose that the origin is shifted by a vector $\mathbf{r}_0$, then the dipole moment in the new coordinate system becomes $\mathbf{p}' = \sum_i q_i(\mathbf{r}_i + \mathbf{r}_0) = \sum_i q_i \mathbf{r}_i + \mathbf{r}_0 \sum_i q_i = \mathbf{p}$.

---

[1] In common with many other authors we have the problem of symbols and names for parameters which have been somewhat controversial during the years when the MKS system of units was becoming established. We find it simplest to regard *permittivity* and *dielectric constant* as synonymous and to use *relative dielectric constant* for the dimensionless number ϵ/ϵ_0.

The macroscopic counterpart of p is the *electric polarization* $\mathbf{P}$ which appears in the auxiliary Maxwell's equations as follows

$$\mathbf{D} = \epsilon_0\mathbf{E} + \mathbf{P} \tag{7.2}$$

where $\mathbf{D}$ (coul/m^2) is the electric flux density or electric displacement, $\mathbf{E}$ (volt/m) the electric field and ϵ_0 (farad/m) the dielectric constant of free space. The macroscopic $\mathbf{P}$ is the volume average of the dipole moment $\mathbf{p}$. In the following, a simple proof of this fact based on a specific physical model is presented.

The specific example to be considered is the electronic contribution to electric polarization, which arises from the displacement of electrons in an atom relative to the nucleus in the presence of an electric field. Suppose that a single electron is bound harmonically to a nucleus and x is the relative displacement under a static field E_0. Then we have[2]

$$\gamma x = -eE_0 \tag{7.3}$$

where γ is the force constant. The dipole moment associated with each atom is

$$p = -ex \tag{7.4}$$

and the potential energy stored in each atom is

$$\frac{\gamma x^2}{2} = \frac{E_0 p}{2} \tag{7.5}$$

If N is the number of atoms per unit volume, the stored-energy density is

$$W = \frac{E_0 N p}{2} \tag{7.6}$$

The stored-energy density can also be calculated macroscopically and is given by

$$W = \int \mathbf{E} \cdot d\mathbf{P} \tag{7.7}$$

For the system under consideration, $\mathbf{P}$ is proportional to $\mathbf{E}$ as is implied by Eqs. (7.3) and (7.4), thus Eq. (7.7) becomes

$$W = \frac{E_0 P}{2} \tag{7.8}$$

Comparing Eqs. (7.6) and (7.8) and generalizing the result, we find

$$\mathbf{P} = N\mathbf{p} = \frac{\sum\limits_i q_i \mathbf{r}_i}{V} \tag{7.9}$$

---

[2] For ease of discussion, we treat the problem as one-dimensional. Therefore, in the equations to follow, we use the magnitudes of the electric field, the displacement, and the dipole moment.

In other words, the macroscopic polarization $\mathbf{P}$ (coul/m^3) is a statistical average of the atomic dipole moment $\mathbf{p}$ (coul-m) over a large volume V.

There are other contributions to electric polarization besides the electronic polarization. The ionic contribution comes from the displacement of positive ions relative to the negative ions in an ionic crystal. In substances possessing permanent electric dipoles, a contribution called orientational polarization arises as a result of reorientation of the dipoles in the presence of an electric field. The simple example given above concerns only the electronic polarization under a static field. The dynamic behavior of the various kinds of polarization is the subject of discussion in the following sections.

7.2 ELECTRONIC POLARIZABILITY

The force equation under static conditions, Eq. (7.3) in the previous section, can easily be extended to cases where the electric field varies sinusoidally with time. The modified force equation under a time-varying field reads

$$m \frac{d^2x}{dt^2} = -\beta \frac{dx}{dt} - \gamma x - eE_1 \exp(i\omega t) \qquad (7.10)$$

which is similar to the equation describing the time variation of electric charges in a resonant circuit. First, the system has a natural angular frequency $\omega_0 = \sqrt{\gamma/m}$, m being the electron mass. Thus, by analogy to an electric circuit, the dielectric is expected to show a strong absorption when ω is close to ω_0. Second, there must be some loss or damping mechanisms to limit the amplitude of the forced oscillation. According to electromagnetic theory of radiation, an oscillating dipole radiates energy; thus, radiation can be one of the damping mechanisms. The term $\beta(dx/dt)$, like resistance in electric circuit, represents an equivalent retarding force caused by damping.

It may be instructive to make an estimate of the parameters in Eq. (7.10) based on classical mechanics so that the results predicted from the theory can be compared with experimental values. To make calculations possible, however, drastic assumptions must be made. For simplicity, consider the hydrogen atom as an example. First, the electron-charge distribution is assumed to be of uniform density confined within the Bohr radius a_1. In the presence of an electric field, the whole electron distribution is shifted by a distance x relative to the position of the nucleus, x being much smaller than a_1. The coulombic force which is responsible for the restoring force in Eq. (7.10) is equal to

$$e \left(\frac{x}{a_1} \right)^3 \frac{e}{4\pi\epsilon_0 x^2} = \frac{e^2}{4\pi\epsilon_0 a_1{}^3} x = \gamma x$$

Hence, the resonant angular frequency ω_0 is found to be

$$\omega_0 = \sqrt{\frac{\gamma}{m}} = \left(\frac{e^2}{4\pi\epsilon_0 a_1{}^3 m}\right)^{1/2} \tag{7.11}$$

The static *electronic polarizability* α_e is defined as the ratio of the electronic dipole moment to the electric field. According to Eqs. (7.3) and (7.4), we have

$$\alpha_e = \frac{p}{E} = \frac{e^2}{\gamma} = 4\pi\epsilon_0 a_1{}^3 \tag{7.12}$$

The correct value of α_e for a normal hydrogen atom based on quantum mechanical calculations[3] is $\alpha_e = 18\pi\epsilon_0 a_1{}^3$. Using the value of $a_1 = 0.53 \times 10^{-10}$m from Eq. (1.16) in Eqs. (7.11) and (7.12), we find $\omega_0 = 4.1 \times 10^{16}$ rad/sec and $\alpha_e = 1.7 \times 10^{-41}$ farad-m^2. For gases, the electric field E in Eq. (7.2) can be taken simply as the applied field; thus, we can relate $\mathbf{D}$ to $\mathbf{E}$ by a parameter called the dielectric constant ϵ. According to Eq. (7.2), ϵ is related to α_e by the following equation

$$\epsilon = \frac{D}{E} = \epsilon_0 + N\alpha_e \tag{7.13}$$

The value of N is 5×10^{25}/m^3 for atomic hydrogen gas under normal pressure and temperature, and this gives a value of $(\epsilon - \epsilon_0)/\epsilon_0$ of the order of 10^{-4}. The values of ω_0 and $(\epsilon - \epsilon_0)/\epsilon_0$ predicted from the classical theory agree well with experimental findings.

Notice that the value of ω_0 given above is in the ultraviolet region. As frequency increases, radiation damping becomes more and more important. Even though there may be other damping mechanisms such as collisions, the calculation below, based on radiation damping alone, may serve to set a lower limit for the value of β in Eq. (7.10). Classically, an oscillating dipole radiates energy at an average rate (Prob. 1.5):

$$\text{Power radiated} = \sqrt{\frac{\mu}{\epsilon}} \frac{\pi}{3} \left(\frac{e\omega x_0}{\lambda}\right)^2 \tag{7.14}$$

in a medium with permeability μ and dielectric constant ϵ. In Eq. (7.14), x_0 is the amplitude and λ is the wavelength of the oscillation. In Eq. (7.10), the term $\beta(dx/dt)$ represents an effective retarding force; therefore, the product $\beta(dx/dt) \cdot (dx/dt)$ is equal to the instantaneous power loss due to damping mechanisms. For harmonic oscillations, the average value of $\beta(dx/dt)^2$ is $\beta x_0{}^2\omega^2/2$ and thus,

[3] See for example, J. H. Van Vleck, "The Theory of Electric and Magnetic Susceptibilities," pp. 203–205, Oxford University Press, Oxford, 1932; N. F. Mott and I. N. Sneddon, "Wave Mechanics and Its Applications," pp. 166–168, Clarendon Press, Oxford, 1948.

from Eq. (7.14), we have

$$\beta = \sqrt{\frac{\mu}{\epsilon}} \frac{2\pi}{3} \frac{e^2}{\lambda^2} = \frac{\mu e^2 \omega^2}{6\pi c} \qquad (7.15)$$

where $c(= 1/\sqrt{\mu\epsilon})$ is the velocity of light in the medium.
The steady-state solution of Eq. (7.10) reads

$$x = -\frac{eE_1/m}{\omega_0^2 - \omega^2 + i\omega\beta/m} \exp(i\omega t) \qquad (7.16)$$

and following Eq. (7.12), the a-c electronic polarizability is

$$\alpha_e = \frac{e^2/m}{\omega_0^2 - \omega^2 + i\omega\beta/m} = \alpha_e' - i\alpha_e'' \qquad (7.17)$$

where α_e' and α_e'' are readily obtained as

$$\alpha_e' = \frac{e^2(\omega_0^2 - \omega^2)/m}{(\omega_0^2 - \omega^2)^2 + (\omega\beta/m)^2} \qquad (7.17a)$$

$$\alpha_e'' = \frac{e^2\omega\beta/m^2}{(\omega_0^2 - \omega^2)^2 + (\omega\beta/m)^2} \qquad (7.17b)$$

Note that α_e is, in general, a complex number. The current density associated with electronic polarization is given by

$$J = -\Sigma e\dot{x} = E_1 N(i\omega\alpha_e' + \omega\alpha_e'') \exp(i\omega t) \qquad (7.18)$$

where N is the number of electronic dipoles per unit volume. Since the part of the current associated with α_e'' is in phase with the field, it constitutes a power loss. In Fig. 7.1, α_e' and α_e'' are plotted as functions of frequency. In the region $\omega_0 + \Delta\omega > \omega > \omega_0 - \Delta\omega$, α_e'' is appreciable; therefore this is referred to as the absorption region. The width of the absorption region is usually expressed in units of reciprocal time and it is a measure of the effectiveness of various

Figure 7.1

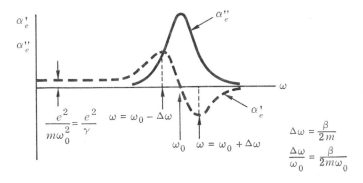

Illustration showing the variation of α_e' and α_e'', the real and imaginary part of the electronic polarizability, as functions of frequency.

damping mechanisms. For example, at $\omega = 4 \times 10^{16}$ rad/sec, Eq. (7.15) gives a value of $\beta/2\pi m \approx 10^9$ sec^{-1}. In solids, the width of the absorption curve is usually much broader than this value, indicating that other mechanisms such as collisions must be more effective causes for damping than radiation.

Although the calculation of γ based on a classical model should not be taken too seriously, the phenomenological equation of motion, Eq. (7.10), is still quite useful in explaining the shape of the absorption and dispersion curves and in correlating experimental results. For example, an elimination of γ from Eqs. (7.11) and (7.12) yields

$$\alpha_e = \frac{e^2}{m\omega_0{}^2} \tag{7.19}$$

For most ionic and molecular crystals, $\omega_0 \approx 6 \times 10^{15}$ rad/sec and $N \approx 2 \times 10^{28}$ atoms/m³. Thus, in Eq. (7.19), if the value of m is taken as the free-electron mass, we have $\epsilon/\epsilon_0 \approx 2$, a value which is of correct order of magnitude in the optical region. It should be pointed out that in solids, internal electric fields arising from the polarization itself can be appreciable. Under such circumstances, it is incorrect to use Eq. (7.13) for an exact calculation of ϵ. The effect caused by the internal field will be discussed in Sec. 7.5. The calculation given above can only serve as an order-of-magnitude estimate to show that even though we do not know how to calculate γ, useful information can still be derived from Eq. (7.10).

7.3 IONIC POLARIZABILITY

In an ionic crystal, such as KBr, LiF, and so on, relative displacement of positive ions with respect to negative ions in the presence of an electric field gives rise to additional polarization. Since ionic polarization involves the motion of ions which are heavier than electrons by a factor of more than a thousand, the resonant frequency of the ionic system is not in the ultraviolet but in the infrared region, as indicated by the mass ratio. Although the shapes of the absorption and dispersion curves due to ionic polarization are quite similar to those due to electronic polarization except for the frequency shift, it is worthwhile to derive the force equation governing the relative motion of ions.

In setting up Eq. (7.10), we considered an atom to be essentially isolated. The assumption is a valid one because the intra-atomic force is much stronger than the interatomic force. For ionic polarization, however, motion of neighboring ions is coupled together through interatomic forces. Consider a linear diatomic chain having masses M_1 located at odd-numbered lattice points x_{2n-1}, x_{2n+1}, . . . , and M_2 located at even-numbered lattice points x_{2n}, x_{2n+2}, . . . , as shown in Fig. 7.2.

If only interatomic forces between nearest neighbors are considered, the force equations for the $(2n-1)$th and $2n$th atoms, respectively, read [Eq. (2.73)]

$$M_1 \frac{d^2 x_{2n-1}}{dt^2} = \gamma[(x_{2n} - x_{2n-1}) - (x_{2n} - x_{2n-1})_0]$$
$$-\gamma[(x_{2n-1} - x_{2n-2}) - (x_{2n-1} - x_{2n-2})_0] \quad (7.20a)$$

$$M_2 \frac{d^2 x_{2n}}{dt^2} = \gamma[(x_{2n+1} - x_{2n}) - (x_{2n+1} - x_{2n})_0]$$
$$-\gamma[(x_{2n} - x_{2n-1}) - (x_{2n} - x_{2n-1})_0] \quad (7.20b)$$

where γ is the interatomic force constant which is different from the intra-atomic force constant in Eq. (7.3), and the subscript 0 denotes an equilibrium situation. Equation (7.20) resembles a transmission-line equation based on a lumped-circuit viewpoint and has wavelike solutions. Thus, let

$$x_{2n-1} = u \exp[ik(2n-1)a] \qquad x_{2n} = v \exp[ik(2n)a] \quad (7.21)$$

where u and v are periodic functions of time and the exponential is a phase factor. From Eq. (7.21), it is obvious that the phase of the elastic wave changes by an amount $2ka$ in going from one atom to the nearest atom of the same kind where k is the wave number associated with the lattice wave. Substituting Eq. (7.21) into Eq. (7.20) and realizing that $a = (x_{2n} - x_{2n-1})_0 = (x_{2n-1} - x_{2n-2})_0$, etc., we find

$$M_1 \frac{d^2 u}{dt^2} = 2\gamma(v \cos ka - u)$$
$$M_2 \frac{d^2 v}{dt^2} = 2\gamma(u \cos ka - v) \quad (7.22)$$

Note that Eq. (7.22) is exactly the same as Eq. (2.75) with $A = v$ and $B = u$. The equation is rederived here for refreshing our memory as well as for easy reference.

To study the effects of an electric field upon the motion of ions, certain simplifications of Eq. (7.22) must be made. For waves in the visible and near infrared regions, λ is of the order of 5×10^{-7} m

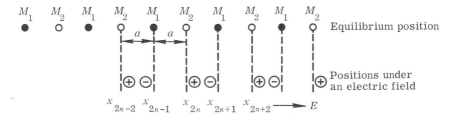

Figure 7.2 *Movement of positive and negative ions in an ionic crystal under the influence of an electric field.*

which is considerably larger than the interatomic distance, $a \approx 3 \times 10^{-10}$ m. Furthermore, for a strong interaction between the electric field and the motion of ionic charges, the phase constant of the electric field should match that of the lattice wave. In other words, the value of k in Eq. (7.22) may be taken as $2\pi/\lambda$ and hence the value of ka is much smaller than unity for the frequency range being considered. Therefore, under an electric field, the force equation reads

$$M_1 \frac{d^2u}{dt^2} = -2\gamma(u - v) - eE_1 \exp{(i\omega t)}$$

$$M_2 \frac{d^2v}{dt^2} = -2\gamma(v - u) + eE_1 \exp{(i\omega t)}$$

(7.23)

supposing that M_1 carries a negative charge and M_2 a positive charge. Solving for relative displacement of positive and negative ions, we have

$$u - v = \frac{e/M}{\omega_0^2 - \omega^2} E_1 \exp{(i\omega t)}$$

(7.24)

where $M = M_1 M_2 / (M_1 + M_2)$ and $\omega_0 = \sqrt{2\gamma/M}$. If damping mechanisms are included in Eq. (7.24), the general expression for *ionic polarizability* is

$$\alpha_i = \frac{e^2/M}{(\omega_0^2 - \omega^2) + i\omega\beta/M}$$

(7.25)

Equation (7.25) has the same form as Eq. (7.17) except that ω_0 is now in the infrared region and that different damping mechanisms, i.e., different values of β, are associated with ionic polarization. Therefore, Eq. (7.25) will not be further discussed to avoid undue repetition.

7.4 ORIENTATIONAL POLARIZABILITY

Molecules can be divided into two classes, *polar* and *nonpolar*, according to whether they do or do not possess an electric dipole moment. The most commonly known polar molecule is water. In water, the H—O—H bond is not straight but forms an angle of 105 degrees. Since hydrogen and oxygen atoms in water are positively and negatively charged respectively, water molecules have a permanent electric dipole moment pointing in the direction which bisects the bond angle. Without an electric field, the dipoles are randomly oriented in all directions with equal probability and hence the resultant polarization is zero. In the presence of an electric field, a reorientation of dipoles takes place, giving rise to a non-vanishing polarization often referred to as *orientational polarization*.

For gases and liquids (for example, dilute solutions of polar molecules in a nonpolar liquid), molecules and hence dipoles associated with polar molecules are relatively free to change their orientations. The only force that prevents permanent dipoles from complete alignment with the electric field is that due to thermal agitation. Consider an elementary dipole with dipole moment p_0 placed in an electric field E. The torque acting on the dipole is

$$T = p_0 E \sin \theta \tag{7.26}$$

and the potential energy of the dipole is

$$W = \int_{\pi/2}^{\theta} T \, d\theta = -p_0 E \cos \theta \tag{7.27}$$

where θ is the angle between p_0 and E. Let n be the number of molecules whose dipole moment lies in an element of solid angle $d\Omega (= \sin \theta \, d\theta \, d\phi)$. Then, according to the Boltzmann distribution law,

$$n = A \exp \left(-\frac{W}{kT} \right) \tag{7.28}$$

where A is a proportionality constant which can be determined from the total number of molecules N through the following relation

$$N = A \int \exp \left(-\frac{W}{kT} \right) d\Omega \tag{7.29}$$

Because of the Boltzmann factor, there are more dipoles pointing in the direction of E than dipoles with other orientations; and consequently there is a net dipole moment in the direction of the electric field. The average dipole moment in the direction of E is given by

$$\bar{p}_0 = \int_0^{\pi} \int_0^{2\pi} p_0 \cos \theta \, \frac{n \, d\Omega}{N} = p_0 \left[\coth \left(\frac{p_0 E}{kT} \right) - \left(\frac{kT}{p_0 E} \right) \right] \tag{7.30}$$

The function inside the brackets is known as the *Langevin function* and it is represented by $\mathcal{L}(x)$ where $x = p_0 E / kT$ for the present case (cf. Fig. 8.12). Let us now estimate the order of magnitude of x. For p_0, take e as the electronic charge and r as the interatomic distance 10^{-10} m, then we have $p_0 \approx 10^{-29}$ coul-m. For $E = 10^5$ volts/m, $p_0 E \approx 10^{-24}$ joule which is to be compared with the room temperature value of $kT \approx 4 \times 10^{-21}$ joule. Thus, in the ordinary range of field strength and temperature, $x \approx \frac{1}{4000} \ll 1$. Under such circumstances, the Langevin function, after being expanded into a Taylor's series, can be approximated by $x/3$; and the *orientational polarizability* is

$$\alpha_0 = \frac{\bar{p}_0}{E} = \frac{p_0^2}{3kT} \tag{7.31}$$

In solids, because of interactions with neighbors, molecules no longer have the freedom of orienting in any arbitrary direction. Instead, they have a limited number of stable orientations, depending on the symmetry of the molecule and the bond between molecules. In the following, we shall discuss the model originally proposed by Fröhlich[4] as an example. Consider a single particle with electric charge e having two equilibrium positions A and B separated by a distance $2b$. This charged particle naturally corresponds to a positive ion in a polar molecule. In the presence of an electric field, there is an energy difference $W_A - W_B$ because different potential energies are associated with positions A and B, respectively.

$$W_A - W_B = 2ebE \cos \theta \qquad (7.32)$$

where θ is the angle which $\mathbf{E}$ makes with the direction of AB. Let n_A and n_B be the number of particles situated in positions A and B respectively. According to the Boltzmann distribution law,

$$\frac{n_B}{n_A} = \exp \left(\frac{2ebE \cos \theta}{kT} \right) \qquad (7.33)$$

If the midpoint between A and B is taken as the origin, the dipole moment associated with position A in the $\mathbf{E}$ direction is $-eb \cos \theta$ while that associated with B is $eb \cos \theta$. Therefore, the average dipole moment per particle is

$$\bar{p}_0 = eb \cos \theta \frac{n_B - n_A}{n_B + n_A} = eb \cos \theta \tanh \left(\frac{ebE \cos \theta}{kT} \right) \qquad (7.34)$$

and the orientational polarizability

$$\alpha_0 = \frac{(eb \cos \theta)^2}{kT} \qquad \text{for} \; \frac{ebE \cos \theta}{kT} \ll 1 \qquad (7.35)$$

Although Eqs. (7.31) and (7.35) are quite similar in form, the physical processes involved are different. In the case of gases and liquids, the effect of an electric field upon dipoles can be pictured as turning the dipoles by the field. In the case of solids, however, molecules and hence electric dipoles associated with them no longer have the rotational freedom. Instead, the dipoles may jump from one equilibrium position to another. In the presence of an electric field, the two positions will be occupied with different probabilities according to Eq. (7.33). This difference in the rotational freedom of molecules is further manifested most remarkably in the dynamic behavior of polar substances.

To explain the frequency dependence of orientational polarizability, we need to introduce the concept of relaxation time. Any reorientation of dipoles cannot occur instantaneously in response to changes in electric field. The rate at which a system of dipoles approaches the new equilibrium distribution is measured by a relaxation time τ. The phenomenological equation describing the dynamic

[4] H. Fröhlich, "Theory of Dielectrics," pp. 17–19, Clarendon Press, Oxford, 1949.

behavior of $\bar{p}_0$ in the presence of a changing electric field is give by

$$\tau \frac{d\bar{p}_0}{dt} = -\bar{p}_0 + \alpha_0 E_1 \exp(i\omega t) \tag{7.36}$$

Here, we shall not present the original derivation of Eq. (7.36) due to Debye.[5] However, we can readily see that Eq. (7.36) is of correct form. First, if ω approaches zero, Eq. (7.36) predicts that $\bar{p}_0 = \alpha_0 E$ in agreement with Eq. (7.31). Furthermore, if the electric field is removed, the system returns to the equilibrium value zero with a characteristic time constant τ in accordance with our previous definition of τ.

The steady-state solution of Eq. (7.36) reads

$$\bar{p}_0 = \frac{\alpha_0 E_1}{1 + i\omega\tau} \exp(i\omega t) \tag{7.37}$$

yielding the a-c orientational polarizability

$$\alpha_d = \alpha_d' - i\alpha_d'' = \frac{\alpha_0}{1 + i\omega\tau} \tag{7.38}$$

where $\qquad \alpha_d' = \dfrac{\alpha_0}{1 + \omega^2\tau^2} \qquad$ and $\qquad \alpha_d'' = \dfrac{\omega\tau\alpha_0}{1 + \omega^2\tau^2} \qquad$ (7.39)

The real and imaginary parts of α_d are plotted in Fig. 7.3 as functions of frequency. The reason that α_d drops rapidly for $\omega > 1/\tau$ is because the dipoles no longer can follow the rapid field variations. The value of τ can be obtained from a best fit to Eq. (7.38) by measuring α_d as a function of ω.

In liquids, the relaxation time τ is related to the viscosity η through the following expression, originally due to Debye:

$$\tau = \frac{4\pi a^3 \eta}{kT} \tag{7.40}$$

where a is the radius of the polar molecule which is assumed spherical. Equation (7.40) is only approximate because the polar molecule may not be spherical and furthermore the macroscopic η may be different from the microscopic η which governs the rotation of polar molecules. To make an order-of-magnitude estimate of τ, we take

[5] P. Debye, "Polar Molecules," chap. 5, Dover Publications, New York, 1945.

Figure 7.3

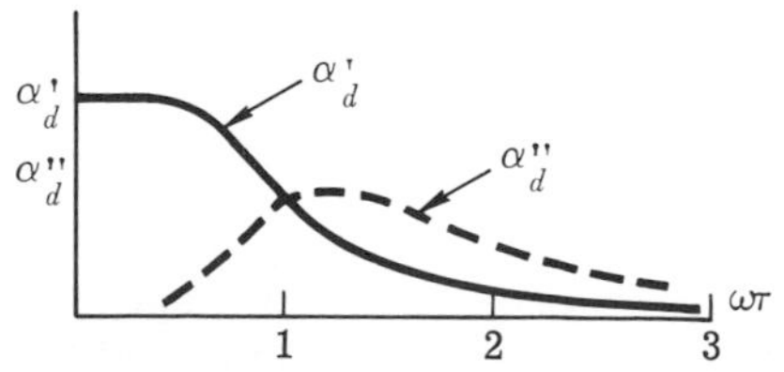

Illustration showing the variation of α_d' and α_d'', the real and imaginary part of the orientational polarizability, as functions of frequency.

$a = 1.4 \times 10^{-8}$ cm, $\eta = 0.01$ poise at $T = 300°$K, and find $\tau = 10^{-11}$ sec. The experimental results on pure polar liquids, such as alcohol and water, show a dispersion region ranging from 1 cm to 1 m wavelength, corresponding to a value of τ from 5×10^{-12} to 5×10^{-10} sec.

The relaxation times in solids are usually several orders of magnitude longer than in liquids. As pointed out earlier in this section, a reorientation of dipoles in solids involves the movement of an ion from one equilibrium position, A, to another, B. Since positions A and B represent positions of minimum energy, the ion in jumping from A to B must overcome a potential barrier, say of energy ϕ. The probability per unit time of making such a jump is proportional to $\exp(-\phi/kT)$. The relaxation time can be interpreted as the average time required or the reciprocal of the probability for the jump. Hence, τ is expected to have the following temperature dependence

$$\tau = C \exp\left(\frac{\phi}{kT}\right) \tag{7.41}$$

where C is a constant. For ice, as an example, the experimental value of τ can be best fitted by taking $C = 5 \times 10^{-16}$ sec and $\phi = 9 \times 10^{-20}$ joule. From Eq. (7.41), the values of τ are 6×10^{-5} sec and 1.2×10^{-2} sec at $-10°$C and $-50°$C respectively. Since the magnitude of τ is determined by the ease with which a polar molecule can rotate inside a solid, any structural change of the solid which prohibits the rotational freedom of polar molecules should appear in the polarization measurement, as α_d should drop practically to zero. This has been a subject of many fruitful investigations. For detailed information on the subject, the reader should refer to a specialized text[6] on dielectric properties and structure of materials. In Sec. 7.7, we shall give simple examples to illustrate the connection between the dielectric behavior and structure of a substance.

7.5 INTERNAL FIELDS

In the previous discussions, we have purposely avoided the use of the dielectric constant ϵ because the relationship between ϵ and α is by no means straightforward. In deriving the polarizability, it is tacitly understood that the electric field E is the total field acting on an atom, ion, or dipole, and hence it includes the externally applied field and the field due to the dipoles. The dipole field at atom site i caused by the dipole moment of all other atoms is given by (Prob. 7.10)

$$\mathbf{E}_i = \sum_{j \neq i} \frac{3(\mathbf{p}_j \cdot \mathbf{r}_{ij})\mathbf{r}_{ij} - r_{ij}^2 \mathbf{p}_j}{4\pi\epsilon_0 r_{ij}^5} \tag{7.42}$$

[6] See, for example, R. J. W. Le Fèvre, "Dipole Moments," Methuen & Co., Ltd., London, 1948; C. P. Smyth, "Dielectric Behavior and Structure," McGraw-Hill Book Company, New York, 1955.

where $\mathbf{p}_j$ is the dipole moment of atom j and $\mathbf{r}_{ij}$ is a radius vector pointing from atom j toward i. Since the summation in Eq. (7.42) is over the whole crystal, it will be very difficult to carry out the summation. Special steps must be taken to evaluate $\mathbf{E}_i$.

Consider a dielectric material placed in an electric field $\mathbf{E}_0$ applied externally as shown in Fig. 7.4. Imagine that a microscopic spherical cavity is cut out of the specimen, the sphere of radius a being very small compared with the physical dimensions of the specimen but very large compared with atomic dimensions. First, we shall separate the contribution to $\mathbf{E}_i$ due to dipoles inside the cavity and call it $\mathbf{E}_1$. For the remaining contribution, since the summation is over a large volume, we can replace the microscopic dipole moment $\mathbf{p}_j$ by a macroscopic polarization $\mathbf{P}$. From classical electrostatics,[7] a nonvanishing $\mathbf{P}$ is equivalent to a volume charge density $\varrho = -\operatorname{div} \mathbf{P}$ and a surface-charge density $\varrho_s = (\mathbf{n} \cdot \mathbf{P})$ where $\mathbf{n}$ is an outward normal to the surface. For uniform $\mathbf{P}$, $\varrho = 0$. Physically, the dipole field is now divided into three parts: (1) the aforementioned $\mathbf{E}_1$, (2) the field $\mathbf{E}_2$ due to charges induced on the surface of the specimen, and (3) the field $\mathbf{E}_3$ due to charges induced on the surface of the

[7] See, for example, R. W. P. King, "Electromagnetic Engineering," vol. 1: *Fundamentals*, pp. 24–36, McGraw-Hill Book Company, New York, 1945; J. D. Jackson, "Classical Electrodynamics," pp. 103–112, John Wiley & Sons, Inc., New York, 1962.

Figure 7.4

(a)

The local field acting on a dipole consists of (1) $\mathbf{E}_0$, the external applied field, and (2) $\mathbf{E}_i$, the internal field contributed by dipoles themselves. To facilitate the evaluation of $\mathbf{E}_i$ in Eq. (7.42), the internal field is further decomposed into three parts: (1) $\mathbf{E}_1$, the field due to dipoles inside the fictitious cavity; (2) $\mathbf{E}_2$, the field due to surface charges induced on the surface of the specimen; and (3) $\mathbf{E}_3$, the field due to surface charges induced on the surface of the fictitious cavity.

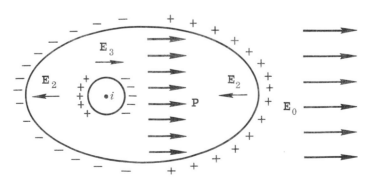

(b)

cavity. Therefore, the total local field acting on the ith atom is given by

$$\mathbf{E}_{loc} = \mathbf{E}_0 + \mathbf{E}_1 + \mathbf{E}_2 + \mathbf{E}_3 \tag{7.43}$$

The situation is schematically shown in Fig. 7.4.

The dipole field is divided in the above manner for two reasons. First, the physical meanings of $\mathbf{E}_1$, $\mathbf{E}_2$, and $\mathbf{E}_3$ are now clear, and second, the values of E_1, E_2, and E_3 can be calculated under the circumstance that polarization inside the dielectric is uniform. The argument may appear to be circular since we must know $\mathbf{E}_{loc}$ to evaluate $\mathbf{P}$. It turns out that the polarization inside a specimen is uniform if the specimen is of homogeneous composition and has the shape of a general ellipsoid. Let us first discuss $\mathbf{E}_1$. Take the z axis as the direction of the dipole moment $\mathbf{p}$. In Eq. (7.42), $\mathbf{p}_j \cdot \mathbf{r}_{ij} = pz_{ij}$. For crystals of cubic symmetry, $\Sigma 3z_{ij}^2 = \Sigma r_{ij}^2$, $\Sigma z_{ij}x_{ij} = 0$ and $\Sigma z_{ij}y_{ij} = 0$; therefore, E_1 is identically zero. For other simple lattices, the value of E_1 can be calculated and is nonzero. However, to limit the scope of our discussion, we shall assume that $\mathbf{E}_1 = 0$. The field $\mathbf{E}_3$ due to the surface charges induced on the fictitious cavity is usually referred to as the *Lorentz field*. Consider an elementary circular ring having a polar angle θ with the z axis as shown in Fig. Fig. 7.4a. The total charge on the ring is equal to

$$dQ = P \cos \theta (2\pi a \sin \theta)(a \, d\theta)$$

The electric field at the center of the cavity is, by symmetry, along the z axis and equal to

$$E_3 = \int_0^\pi \frac{P \cos \theta (2\pi a \sin \theta)(a \, d\theta)}{4\pi\epsilon_0 a^2} \cos \theta = \frac{P}{3\epsilon_0} \tag{7.44}$$

It should be pointed out that the calculations given here are based on the assumption that the condition for a uniform polarization is satisfied.

The field $\mathbf{E}_2$ due to charges induced on the surface of the specimen is called the *depolarization field* because it is in the opposite direction to $\mathbf{P}$. For a homogeneous ellipsoidal specimen with a uniform polarization $\mathbf{P}$, having components $\mathbf{P}_x$, $\mathbf{P}_y$, and $\mathbf{P}_z$,

$$\mathbf{E}_2 = -N_x\mathbf{P}_x - N_y\mathbf{P}_y - N_z\mathbf{P}_z \tag{7.45}$$

where N_x, N_y, and N_z are called *depolarization factors*. For illustration, take a rectangular slab, thickness $2b$, as shown in Fig. 7.5. Due to P_z, that is, the polarization charges on the top and bottom surfaces, an electric field E_z arises where

$$E_z = \int \frac{-2P_z(2\pi b \tan \theta)(b \sec^2 \theta \, d\theta)}{4\pi\epsilon_0 b^2 \sec^2 \theta} \cos \theta \tag{7.46}$$

The fields due to P_x and P_y are similar to Eq. (7.46) with subscripts changed to x and y respectively. For a thin slab or disc infinitely extended in x and y directions, the integration limits are: For E_z, θ extends from 0 to $\pi/2$ and for E_x and E_y, from 0 to an infinitesimal angle δ. Comparing the result obtained from Eq. (7.46) with Eq. (7.45), we have

$$N_x = N_y = 0 \qquad \text{and} \qquad N_z = \frac{1}{\epsilon_0} \qquad (7.47)$$

For a spherical sample, the calculation is exactly the same as in Eq. (7.44) for E_3, and hence $N_x = N_y = N_z = 1/3\epsilon_0$. For an infinitely long, circular cylinder, $N_x = N_y = 1/2\epsilon_0$ and $N_z = 0$ where the z direction is taken along the axis of the cylinder. The proof is left for the reader as a problem (7.12). Note that for all three cases, $N_x + N_y + N_z = 1/\epsilon_0$. This general property of the depolarization factor is true for any ellipsoidal specimen[8] as long as x, y, and z are the three principal axes of the ellipsoid. With this general rule in mind, the values of N_x, N_y, and N_z can easily be obtained through reasoning. For example, in the case of a long cylinder, $N_z = 0$ in view of the fact that the polarization charges are infinitely far away in the z direction and then $N_x = N_y = 1/2\epsilon_0$ by virtue of symmetry. The three geometrical shapes presented in the above discussion are commonly used in dielectric and magnetic experiments. The require-

[8] For a general calculation of the depolarizing factors in an ellipsoidal specimen, the reader is referred to a similar calculation for demagnetizing factors by J. A. Osborn, *Phys. Rev.*, vol. 67, p. 351, 1945, and by E. C. Stoner, *Phil. Mag.*, vol. 36, p. 803, 1945.

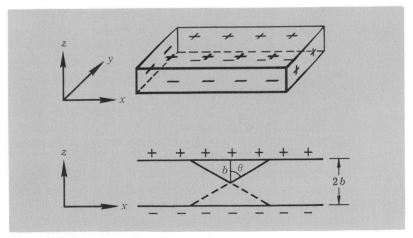

A rectangular dielectric slab with opposite charges induced on the two opposite surfaces of the slab. The example is used to illustrate how the depolarizing field E_2 of Eq. (7.45) can be evaluated.

ment that a sample is infinitely long or infinitely thin is met in practice if one dimension is larger or smaller by a factor of 5 or more than the other two dimensions.

7.6 POLARIZABILITY AND DIELECTRIC CONSTANT

In the macroscopic description of the dielectric properties of a given material, the dielectric constant ϵ appears in the relationship between $\mathbf{D}$ and $\mathbf{E}$ as follows:

$$\mathbf{D} = \epsilon \mathbf{E} \tag{7.48}$$

$\mathbf{E}$ being the electric field existing in a dielectric medium. Equation (7.48) is an alternate but equivalent expression of Eq. (7.2) describing the same physical system. For the present discussion, we only consider the field-induced polarization as given in Secs. 7.2 to 7.4; therefore, $\mathbf{P}$ in Eq. (7.2) is linearly proportional to and in the direction of $\mathbf{E}$, and consequently ϵ in Eq. (7.48) is a constant and scalar. Anisotropic and nonlinear effects in dielectrics will be discussed later.

In the previous section, the local field $\mathbf{E}_{loc}$ acting on an atom was divided into four parts: $\mathbf{E}_0$, $\mathbf{E}_1$, $\mathbf{E}_2$, and $\mathbf{E}_3$. The reason for doing so was to give a clear picture of the physical origins of these fields. However, in practice, it is much more convenient to combine $\mathbf{E}_0$ and $\mathbf{E}_2$, so $\mathbf{E} = \mathbf{E}_0 + \mathbf{E}_2$. The field $\mathbf{E}$ is the field actually existing inside the dielectric and it is the same $\mathbf{E}$ as given in Eqs. (7.2) and (7.48). A simple proof of this statement is given below for a specific example. Consider a thin dielectric slab placed in a uniform electric field as shown in Fig. 7.6 with $\mathbf{E}_0$ perpendicular to the face of the dielectric. According to Eq. (7.2) and the electric-flux continuity equation,

$$\mathbf{D} = \epsilon_0 \mathbf{E} + \mathbf{P} = \epsilon_0 \mathbf{E}_0 \tag{7.49}$$

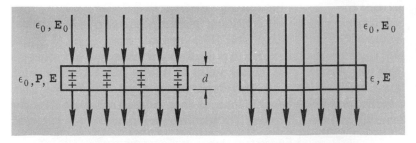

A thin dielectric slab placed in a uniform electric field normal to the surfaces of the slab. The example is used to show that the field existing inside the dielectric is equal to the external applied field and the depolarizing field. Note that a dielectric can be considered either as a medium with a dielectric constant ϵ_0 and a polarization $\mathbf{P}$ or as a medium with a dielectric constant ϵ.

Using the information given in Eqs. (7.45) and (7.47), we find

$$\mathbf{E} = \mathbf{E}_0 - \mathbf{P}/\epsilon_0 = \mathbf{E}_0 + \mathbf{E}_2 \tag{7.50}$$

where $\mathbf{E}$ is the electric field existing in the dielectric. If the dielectric slab is used between two parallel plates of a condenser of separation d, then $E = V/d$ where V is the voltage applied across the condenser.

In terms of $\mathbf{E}$, the local field can be expressed as

$$\mathbf{E}_{loc} = \mathbf{E} + \mathbf{E}_3 + \mathbf{E}_1 = \mathbf{E} + \frac{\mathbf{P}}{3\epsilon_0} \tag{7.51}$$

The polarization $\mathbf{P}$ is proportional to $\mathbf{E}$ through the equation

$$\mathbf{P} = N\alpha\mathbf{E}_{loc} \tag{7.52}$$

where α is the total polarizability including α_e, α_i, and α_d. After elimination of $\mathbf{E}_{loc}$ from Eqs. (7.51) and (7.52), we obtain

$$\mathbf{P} = \frac{3\epsilon_0 N\alpha}{3\epsilon_0 - N\alpha} \mathbf{E} \tag{7.53}$$

Substituting Eq. (7.53) into Eq. (7.2) and comparing Eq. (7.2) with Eq. (7.48), we find

$$\epsilon = \epsilon_0 \left(1 + \frac{3N\alpha}{3\epsilon_0 - N\alpha} \right) \tag{7.54}$$

which can be easily rearranged to the more familiar *Clausius-Mossotti* equation,

$$\frac{\epsilon - \epsilon_0}{\epsilon + 2\epsilon_0} = \frac{N\alpha}{3\epsilon_0} \tag{7.55}$$

We should point out that Eq. (7.55) applies only to materials in which electrons are spatially well localized, and further, that the polarization is due to elastic displacement of charges (that is, for nonpolar contributions to α from α_e and α_i). In a highly polarizable electron system with $\epsilon > 10\epsilon_0$, the electron wave function is no longer localized but spreads over many interatomic distances. As a consequence, the local-field correction is small, and Eq. (7.13) should be used instead. Such is the case with metals and covalent semiconductors. Caution should also be exercised in applying Eq. (7.55) to polar substances.

If the value of α_0 from Eq. (7.31) is substituted into Eq. (7.54), we see that the dielectric constant would become infinitely large at a critical temperature T_c:

$$T_c = \frac{Np_0^2}{9k\epsilon_0} \tag{7.56}$$

Take again p_0 about 10^{-29} coul-m and $N \simeq 2 \times 10^{28}$ molecules/m³, then we find $T_c \approx 1000°K$. This is contrary to the known properties of polar liquids and solids. Therefore, it seems highly desirable to retrace our steps in deriving the Clausius-Mossotti equation. The main assumption, that **P** is parallel to **E**, should be reexamined. For nonpolar substances, the induced dipole moments, according to the equation of motion, are in the direction of the applied field and hence the assumption is valid. For polar substances, however, the permanent dipoles are randomly oriented in space. It is true that the resultant macroscopic polarization **P** is in the direction of **E**, but so far as the local field is concerned, the focus of our attention is on the atomic **p** and not on the macroscopic **P**. In the following, the treatment of polar substances originally due to Onsager[9] is presented.

A departure from the previous treatment is that the local field to be calculated is not an average local field but the field acting on a specific dipole with a particular orientation. Hence, the radius of the fictitious cavity is taken as the radius of the polar molecule. The contribution to the local field can now be divided into two parts (Fig. 7.7): (1) the field $\mathbf{E}_4$ due to the applied field **E** and (2) the field $\mathbf{E}_5$, called the reaction field, due to polarization charges induced by the dipole itself. Considering the fact that the cavity is immersed in a dielectric medium of dielectric constant ϵ, the field $\mathbf{E}_4$ induced inside the cavity by an external field **E** is given by

$$\mathbf{E}_4 = \mathbf{E}\,\frac{3\epsilon}{2\epsilon + \epsilon_0} \tag{7.57}$$

The reaction field $\mathbf{E}_5$, induced by the dipole **p**, is related to **p** by

$$\mathbf{E}_5 = \frac{\mathbf{p}}{4\pi a^3}\,\frac{2(\epsilon - \epsilon_0)}{\epsilon_0(2\epsilon + \epsilon_0)} \tag{7.58}$$

according to the analysis by Onsager (Prob. 7.13). In Eq. (7.58), a is the radius of the fictitious cavity.

To help understand the nature of the reaction field, the following

---

[9] L. Onsager, *Jour. Am. Chem. Soc.*, vol. 58, p. 1486, 1936.

Figure
7.7

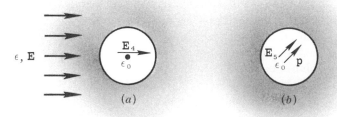

Onsager's model of calculating the local field. The radius a of the fictitious cavity is taken to be the radius of the polar molecule. The local field $\mathbf{E}_{loc}$ is decomposed into two parts: (a) $\mathbf{E}_4$, because the cavity is immersed in a dielectric medium of dielectric constant ϵ with a uniform field $\mathbf{E}$ and (b) $\mathbf{E}_5$, due to polarization charges induced by the molecule due to its own presence.

observation may be instructive. If the assumption is made that $\mathbf{p}$ is parallel to $\mathbf{E}$, then according to Eqs. (7.9), (7.2), and (7.48)

$$\frac{3\mathbf{p}}{4\pi a^3} = \mathbf{P} = (\epsilon - \epsilon_0)\mathbf{E} \tag{7.59}$$

Substituting Eq. (7.59) into Eq. (7.58) and adding Eqs. (7.57) and (7.58), we find

$$\mathbf{E}_{\text{loc}} = \mathbf{E}_4 + \mathbf{E}_5 = \mathbf{E}\,\frac{\epsilon + 2\epsilon_0}{3\epsilon_0} = \mathbf{E} + \frac{\mathbf{P}}{3\epsilon_0} \tag{7.60}$$

the same as Eq. (7.51) which is used in the derivation of the Clausius-Mossotti equation. However, the difference between Eq. (7.60) on the one hand and Eqs. (7.57) and (7.58) on the other hand is significant. In a polar substance, the part of the local field acting on a polar molecule due to the molecule itself depends on the orientation of the molecule and it does not necessarily lie in the direction of the applied field. Under the circumstance, the contribution to the local field $\mathbf{E}_{\text{loc}}$ due to $\mathbf{p}$ and that due to $\mathbf{E}$ should be treated separately.

Now we shall calculate the macroscopic $\mathbf{P}$ based upon Eqs. (7.57) and (7.58). The total dipole moment of a polar molecule can be expressed as

$$\mathbf{p} = \mathbf{p}_0 + \alpha_{ei}\mathbf{E}_{\text{loc}} \tag{7.61}$$

where $\mathbf{p}_0$ represents the permanent dipole moment while $\alpha_{ei}\mathbf{E}_{\text{loc}}$ with $\alpha_{ei} = \alpha_e + \alpha_i$ represents the induced dipole moment due to elastic displacement of charges. Substituting Eqs. (7.57) and (7.58) into Eq. (7.61), we find

$$\mathbf{p} = c\mathbf{p}_0 + d\mathbf{E} \tag{7.62}$$

where
$$\frac{1}{c} = 1 - \frac{2(\epsilon - \epsilon_0)}{4\pi\epsilon_0(2\epsilon + \epsilon_0)}\frac{\alpha_{ei}}{a^3} \tag{7.63}$$

$$d = \alpha_{ei}\left(\frac{3\epsilon}{2\epsilon + \epsilon_0}\right)c \tag{7.64}$$

Since $\mathbf{p}_0$ can be oriented in any arbitrary direction with respect to the applied field, Eq. (7.62) must be averaged over the Boltzmann distribution. The potential energy associated with the permanent dipole part of $\mathbf{p}$ is given by

$$W = -c\mathbf{p}_0 \cdot \mathbf{E}_4 = -c\,\frac{3\epsilon}{2\epsilon + \epsilon_0}\,p_0 E \cos\theta \tag{7.65}$$

In deriving the above equation, we note that the self-energy or the $\mathbf{p}_0 \cdot \mathbf{E}_5$ term does not depend on θ and hence it need not be included in W for the averaging process. Following the same averaging process as was used in deriving Eq. (7.31), we find for polarization

$$\mathbf{P} = N\mathbf{p} = N\left(\frac{c^2 p_0^2}{3kT}\frac{3\epsilon}{2\epsilon + \epsilon_0} + d\right)\mathbf{E} \tag{7.66}$$

and for the dielectric constant

$$\epsilon - \epsilon_0 = \frac{P}{E} = N \left(\frac{c^2 p_0{}^2}{3kT} \frac{3\epsilon}{2\epsilon + \epsilon_0} + d \right) \tag{7.67}$$

If ϵ_∞ is the dielectric constant at high frequencies such that only the elastic displacements of charges contribute to polarization, then

$$\frac{\epsilon_\infty - \epsilon_0}{\epsilon_\infty + 2\epsilon_0} = \frac{N\alpha_{ei}}{3\epsilon_0} \tag{7.68}$$

Substitution of Eq. (7.68) into Eqs. (7.63) and (7.64) yields

$$\frac{1}{c} = \frac{3\epsilon_0(2\epsilon + \epsilon_\infty)}{(2\epsilon + \epsilon_0)(\epsilon_\infty + 2\epsilon_0)} \tag{7.69}$$

$$d = \frac{3\epsilon}{N} \frac{\epsilon_\infty - \epsilon_0}{2\epsilon + \epsilon_\infty} \tag{7.70}$$

A test of the Onsager equation (7.67) can be made by measuring ϵ and ϵ_∞ of a polar molecule and comparing the value of p_0 from Eq. (7.67) with the measured value of p_0 while the polar molecule is in a gaseous state. Experimental results on many polar liquids show a general agreement of these two values. For example, a value of 1.23×10^{-23} coul-m for p_0 is obtained in liquid nitromethane from Onsager's equation, which agrees well with the value of 1.17×10^{-23} coul-m in the gaseous nitromethane. Finally, we should like to point out the apparent difference between Eqs. (7.55) and (7.67). In the Onsager equation, the polarization arising from a reorientation of permanent dipoles is no longer lumped together with that due to an elastic displacement of charges as in the Clausius-Mossotti equation. The difference is again a manifestation of the fact that the local fields produced by permanent and induced dipoles are oriented differently and hence they should be treated separately.

7.7 MOLECULAR STRUCTURE AND DIELECTRIC PROPERTIES OF MATERIALS

The division of polarization into three types, electronic, ionic, and orientational, is of a very general nature. We shall start our discussion with gases mainly because we need not be concerned with the effects of internal fields and hence can use Eq. (7.13) for ease of discussion. A generalization of Eq. (7.13) to include all three kinds of polarization requires that α_e be substituted for by the total polarizability α where

$$\alpha = \alpha_e + \alpha_i + \alpha_d \tag{7.71}$$

The behavior of $\alpha = \alpha' - i\alpha''$ as a function of frequency has a general shape schematically illustrated in Fig. 7.8. Significant changes in α occur only when one component of polarization can no longer respond to fast changes in the applied field. For α_d, the change usually occurs in the radio or microwave region; for α_i, in the infra-red region; and for α_e, in the visible or ultraviolet region. Since these changes occur at very different frequencies, the contribution of each component to α can easily be separated out by measuring the polarizability at three different frequencies ω_1, ω_2, and ω_3 as shown in Fig. 7.8.[10] According to the frequency variation of α, dielectric materials can be divided into three classes: (1) substances showing α_e only, (2) substances having both α_e and α_i, and (3) substances having all three components. Which type a given material falls into is mainly determined by its molecular structure and the nature of the chemical bonds.

The first class (those substances with electronic contribution only) contains all materials consisting of a single type of atom. Common examples are germanium, nitrogen (in solid, liquid, or gaseous state), etc. By reason of symmetry, these materials cannot possibly possess a permanent dipole moment or contain dipolar groups of atoms as in the case of ionic crystals. For materials in this group, the polarizability and hence the dielectric constant stay nearly constant up to the ultraviolet region. In other words, the static dielectric constant ϵ_s can be related to the optical index of refraction n through $\epsilon_s = n^2 \epsilon_0$ where ϵ_0 is the dielectric constant of free space.

The second class (those substances with both electronic and ionic contributions to $\mathbf{P}$) contains materials whose molecular structure

[10] In a complex substance, there may be several resonances in the optical and infrared regions. The single resonance shown in the figure is for illustration only.

Figure 7.8

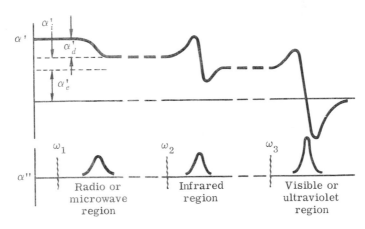

Schematic representation showing the frequency dependence of the polarizability α in a substance for which all three contributions to polarizability, α_e, α_i, and α_d, exist.

possesses certain degrees of symmetry so that the total dipole moment is zero though the molecule may consist of dipolar groups of atoms. A simple example is CO_2. The three atoms are arranged in a linear chain with the carbon atom $(+)$ halfway between the two oxygen atoms $(-)$, resulting in a cancellation of dipole moments. Other examples are methane CH_4 and carbon tetrachloride CCl_4 with the four hydrogen or chlorine atoms at the corners and the carbon atom at the center of a regular tetrahedron. Of course, ionic crystals of alkali halides are also representatives of this group. The compounds of groups II-VI and groups III-V are, in general, much less ionic than the alkali halides, meaning a much smaller dispersion in the infrared region. However, even in SiC which is a IV-IV compound, the dielectric constant increases from $6.7\epsilon_0$ in the visible region to $10\epsilon_0$ in the low frequency region, indicating a moderate degree of ionicity. The reader is referred to Eq. (3.114) and Sec. 3.10 for a discussion of ionicity of a crystal.

All materials made of dipolar molecules belong to the third group. Common examples are H_2O, CH_3Cl, NH_3, etc. In contrast to CO_2, water has a bond angle of $105°$ and hence possesses a dipole moment. The replacement of a hydrogen atom in CH_4 by a chlorine atom destroys the symmetry of the tetrahedral structure, creating a net dipole moment. The dipole moment of a polar gas can be determined from a measurement of the temperature dependence of the static dielectric constant. Such a plot is shown in Fig. 7.9 for several gases. As we recall, the polarizability $\alpha_i + \alpha_e$ due to elastic displacement of charges is independent of temperature while the orientational polarizability α_0 is inversely proportional to the temperature. It is obvious from Fig. 7.9 that both CCl_4 and CH_4 are nonpolar, but the other three gases are polar. Furthermore, from an extrapolation of the straight lines to $1/T = 0$, we obtain the quantity $\alpha_i + \alpha_e$ for the various gases.

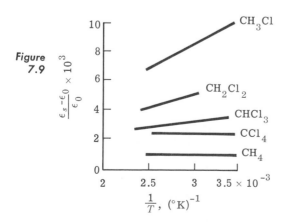

Figure 7.9

$$\epsilon_s - \epsilon_0 = N(\alpha_e + \alpha_i + \alpha_d)$$

Measurement of the temperature dependence of the static dielectric constants in several polar gases.

In contrast to ionic crystals which have one ion at each lattice point, the polar molecules, in forming a solid, have a whole molecule at each lattice point. In most molecular solids, the turning of a dipole in the presence of an applied field means a rotation of the whole molecule. In Fig. 7.10, the dielectric constant of hydrogen sulfide is plotted as a function of temperature measured at a frequency of 5 kc. The sudden drop in ϵ at around 100°K is a direct consequence of a disorder-order transition. In the ordered state, the dipoles are all frozen in and hence the dielectric constant becomes temperature independent.

One of the most important and interesting polar substances is H_2O. The dielectric constant of ice increases first with decreasing temperature from a value at the freezing point not far from that of water, indicating that molecules are relatively free to rotate. However, upon continued decrease in temperature, the dielectric constant measured at low frequencies decreases rapidly toward the high-frequency value due to α_e and α_i only. Two theories have been advanced to explain the dielectric behavior of ice. The obvious one is that the drop is due to limited freedom of rotation. The other suggested mechanism involves the transfer of a proton. The bond between neighboring water molecules, called the hydrogen bond, is supposed to be directed along the O—H bond of one molecule toward the oxygen ion of the other molecule and can be represented as O—H$\cdots$O where the dotted line indicates the hydrogen bond. Since the dipole moment of an H_2O molecule depends on the direction of the H—O—H bond angle, any transfer of a proton from one molecule to the other means a change from O—H$\cdots$O to O$\cdots$H—O and hence a change in the direction of the dipole moment as schematically illustrated in Fig. 7.11. In jumping from one oxygen atom to another, the proton has to overcome a potential-energy barrier; therefore, the relaxation time is expected to increase very rapidly with decreasing temperature as discussed in Sec. 7.4.

The examples given above serve to show that valuable information about the structural symmetry and rotational freedom of molecules

Figure 7.10

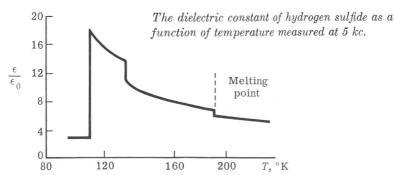

The dielectric constant of hydrogen sulfide as a function of temperature measured at 5 kc.

can be obtained from measurements of the dielectric constant. For an extensive discussion on the subject, the reader should consult a specialized book[11] devoted entirely to dielectric properties. We shall also leave, for discussion in the subsequent sections, a large class of crystalline solids which have anisotropic dielectric properties and solids which possess frozen-in permanent dipoles. What is left for this section is the dielectric loss of a given substance. Since absorption in the optical region will be discussed in Chapter 11 dealing with optical properties of solids, our present discussion will be confined to losses at low frequencies, that is, losses associated with reorientation of dipoles.

From Eqs. (7.54) and (7.67), it is evident that if the orientational polarizability α_d is given by Eq. (7.38), the dielectric constant is in general a complex number. For gases, the relationship between ϵ and α is simple, and hence, according to Eq. (7.13), the dielectric constant can be related neatly to the relaxation time τ as follows:

$$\epsilon = \epsilon' - i\epsilon'' = \epsilon_0 + N\alpha_{ei} + \frac{N\alpha_d}{1 + i\omega\tau}$$
$$= \epsilon_0 + N\alpha_{ei} + \frac{N\alpha_0}{1 + \omega^2\tau^2} - i\,\frac{N\alpha_0\omega\tau}{1 + \omega^2\tau^2} \qquad (7.72)$$

For liquids and solids, the relationship between ϵ and α can be de-

[11] See, for example, C. P. Smyth, "Dielectric Behavior and Structure," McGraw-Hill Book Company. New York, 1955; H. Fröhlich, "Theory of Dielectrics," Clarendon Press, Oxford, 1949.

Figure 7.11

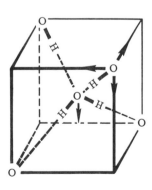 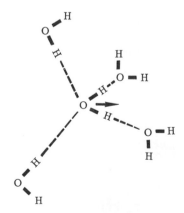

An H_2O molecule in ice with its four neighbors. The heavy lines indicate the bond within the molecule while the dashed lines indicate the hydrogen bond between neighboring molecules. The arrows indicate the direction of the dipole moment associated with each molecule. Note that the dipole associated with the corner molecule may have three possible orientations depending on the location of the other hydrogen atom.

duced from either Eq. (7.55) or Eq. (7.67), but it is so complicated that only qualitative arguments will be presented here. The general shape of ϵ' and ϵ'' according to Eq. (7.72) is plotted in Fig. 7.12 where $\epsilon_\infty = \epsilon_0 + N\alpha_{ei}$ represents the dielectric constant at high frequencies while $\epsilon_s = \epsilon_0 + N\alpha_{ei} + N\alpha_0$ is the static dielectric constant. Now a brief discussion of the meaning of a complex dielectric constant is in order.

Consider a parallel-plate condenser having a medium of dielectric constant ϵ sandwiched in between. The displacement current density, due to a sinusoidally time-varying field $E = E_1 \exp (i\omega t)$, is given by

$$J = \frac{\partial D}{\partial t} = (\omega\epsilon'' + i\omega\epsilon')E_1 \exp (i\omega t) \tag{7.73}$$

It is clear from Eq. (7.73) that the effect of ϵ'' is equivalent to that of a finite conductivity σ. In other words, since the current density J now has an in-phase component with the field E, the dielectric is lossy. The average power loss per unit volume of the condenser is $\omega\epsilon''E_1^2/2$ while the maximum stored-energy density is $\epsilon'E_1^2/2$. Therefore, the quality factor Q associated with the dielectric is given by

$$Q = \frac{\text{maximum stored energy}}{\text{average power loss}} \times \omega = \frac{\epsilon'}{\epsilon''} = \frac{1}{\tan \delta} \tag{7.74}$$

The angle δ is usually referred to as the *loss angle*. In Table 7.1, the values of ϵ'/ϵ_0 and $\tan \delta$ at three different frequencies are given for several insulating materials.

To illustrate the point presented earlier in this section, take NaCl as an example. At 25°C, the values of ϵ'/ϵ_0 stay constant in the frequency range from 10^2 to 10^{10} cps, indicating no dipolar contribution. This conclusion also agrees with the fact that the loss angle in this region is very small indeed. At 85°C, however, a noticeable dipolar contribution appears. A similar situation exists for rutile. An

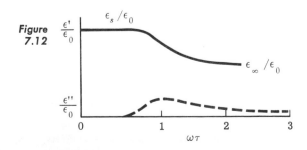

Figure 7.12

The frequency dependence of dielectric constant as predicted from Eq. (7.72). The curve only covers the frequency region where the change in the dielectric constant is caused solely by a reorientation of permanent dipoles. The imaginary part of ϵ, that is, ϵ'', contributes to dielectric loss.

TABLE 7.1   *Dielectric constant and loss-angle values in the low frequency region*

Material	Frequency = 10^2 cps		10^6 cps		10^{10} cps	
	ϵ'/ϵ_0	$(\tan \delta)$ $\times 10^4$	ϵ'/ϵ_0	$(\tan \delta)$ $\times 10^4$	ϵ'/ϵ_0	$(\tan \delta)$ $\times 10^4$
Sodium chloride						
(25°C)	5.90	<1	5.90	<2	5.90	<5
(85°C)	6.35	170	5.98	<2	5.97	<4
Titanium oxide (rutile) ⊥ to optical axis						
(25°C)	87.3	110	85.8	2		
Fused quartz						
(25°C)	3.78	8.5	3.78	2	3.78	1
Polystyrene						
(25°C)	2.56	0.5	2.56	0.7	2.54	4.3

excellent compilation of dielectric data can be found in the book edited by Von Hippel.[12]

7.8   PIEZOELECTRIC EFFECT

The word *piezo* is derived from a Greek word meaning to press, and hence *piezoelectricity* literally means pressure electricity. The converse piezoelectric effect refers to mechanical distortions created by the application of an electric field. Piezoelectricity is distinguished from electrostriction in that the former is characterized by a one-to-one correspondence between the direct and converse effect. In other words, for piezoelectricity, a reversal of the electric field reverses the sign of the resulting strain while for electrostriction, the strain is an even function of the applied field. In general, electrostrictive effects are much smaller than piezoelectric effects except in certain ferroelectric crystals. In the following, our discussion will be confined to piezoelectric effects.

The best known crystal giving a piezoelectric effect is quartz. It is worthwhile to see qualitatively what happens inside a quartz crystal when a stress is applied. The model shown in Fig. 7.13 is originally due to Lord Kelvin. Consider each silicon atom to be positively charged and each oxygen double atom to be represented by the negative charge as shown in Fig. 7.13a. Under a compressional stress, the angle θ increases to θ' and the separation between 0 and $0'$ increases but the distance between A and B remains unchanged.

[12] A. Von Hippel, "Dielectric Materials and Applications," The Technology Press of the Massachusetts Institute of Technology, Cambridge, Massachusetts, 1954.

Now if we compare the new atom arrangement with the old, the positive charges have a net displacement toward the left and the negative charges toward the right, resulting in a net dipole moment pointing in the direction BA for the cluster of atoms $ABC'D'E'F'$ shown in Fig. 7.13b. The situation under a shear stress is illustrated in Fig. 7.13c. The movement of atom D toward the right is compensated by the movement of F toward the left; however, the downward movements of atoms A, D, and F are additive. Therefore, the cluster of atoms $A'B'C'D'E'F'$ as a whole has a net dipole moment pointing upward. It should be emphasized that the model just discussed is descriptive in nature and should not be scrutinized too rigorously. For quantitative analysis of the piezoelectric effect, we must know the mechanical, piezoelectric, and dielectric properties of a material.

Consider the stress (force per unit area) exerted on an elementary cube of material as shown in Fig. 7.14a. The stress exerted on the area $ABCD$ can be resolved into three components T_{xx}, T_{yx}, and T_{zx}, the first subscript denoting the direction of the stress and the second subscript indicating the direction of the normal to the plane $ABCD$. Obviously, T_{xx} represents the tensile stress and T_{yx} and T_{zx} the shear

Figure 7.13

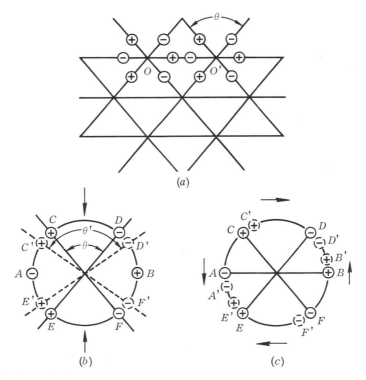

The model used by Lord Kelvin to explain the piezoelectric effect in quartz.

stresses. For torque balance around the z axis, $T_{xy} = T_{yx}$. Therefore, there are altogether three tensile stresses denoted by $T_1 (= T_{xx})$, $T_2 (= T_{yy})$, and $T_3 (= T_{zz})$, and three shear stresses denoted by $T_4 (= T_{yz} = T_{zy})$, $T_5 (= T_{xz} = T_{zx})$, and $T_6 (= T_{xy} = T_{yx})$. If the stresses vary with the coordinate, there will be a force acting on the elementary cube. The net force in the x direction, for example, is given by

$$
F_x = \left(\frac{\partial T_1}{\partial x}\,dx\right) dy\,dz + \left(\frac{\partial T_6}{\partial y}\,dy\right) dx\,dz + \left(\frac{\partial T_5}{\partial z}\,dz\right) dx\,dy
$$

$$
= \left(\frac{\partial T_1}{\partial x} + \frac{\partial T_6}{\partial y} + \frac{\partial T_5}{\partial z}\right) dx\,dy\,dz \tag{7.75}
$$

The expressions for F_y and F_z can easily be deduced from Eq. (7.75).

Next, we shall discuss the strain components. Consider two points originally located at $P(0,0,0)$ and $Q(x,y,z)$. Under stress, the two points are shifted to $P'(\xi_1, \eta_1, \zeta_1)$ and $Q'(\xi_2 + x,\ \eta_2 + y,\ \zeta_2 + z)$ as

Figure 7.14

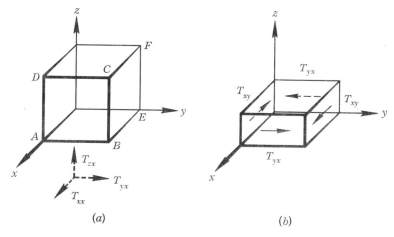

(a)　　　　　　　　　　(b)

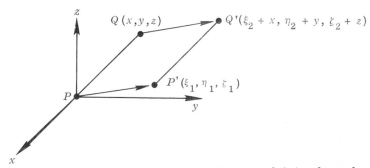

(c)

Diagrams defining the mechanical stresses and strains in a crystal.

shown in Fig. 7.14c. If $\zeta_1 - \zeta_2$, $\eta_1 - \eta_2$, and $\zeta_1 = \zeta_2$, there is a uniform translation but no strain. Therefore, for nonvanishing strain, the translation must depend upon the coordinates of the original point. In other words,

$$\xi_2 = \xi_1 + \frac{\partial \xi}{\partial x} x + \frac{\partial \xi}{\partial y} y + \frac{\partial \xi}{\partial z} z$$

$$\eta_2 = \eta_1 + \frac{\partial \eta}{\partial x} x + \frac{\partial \eta}{\partial y} y + \frac{\partial \eta}{\partial z} z \qquad (7.76)$$

$$\zeta_2 = \zeta_1 + \frac{\partial \zeta}{\partial x} x + \frac{\partial \zeta}{\partial y} y + \frac{\partial \zeta}{\partial z} z$$

The linear strains in the x, y, and z directions are given by

$$S_1 = \frac{\partial \xi}{\partial x} \qquad S_2 = \frac{\partial \eta}{\partial y} \qquad S_3 = \frac{\partial \zeta}{\partial z} \qquad (7.77)$$

and the shear strains around the x, y, and z axes are defined as

$$S_4 = \frac{\partial \zeta}{\partial y} + \frac{\partial \eta}{\partial z} \qquad S_5 = \frac{\partial \xi}{\partial z} + \frac{\partial \zeta}{\partial x} \qquad S_6 = \frac{\partial \xi}{\partial y} + \frac{\partial \eta}{\partial x} \qquad (7.78)$$

For small displacements, the stresses are proportional to the strains in view of Hooke's law, that is,

$$T_i = \sum_j c_{ij} S_j \qquad i, j = 1 \text{ to } 6 \qquad (7.79)$$

or the inverse relation

$$S_i = \sum_j s_{ij} T_j \qquad i, j = 1 \text{ to } 6 \qquad (7.80)$$

The dielectric properties of a material are best described by the relationship between D and E. However, most piezoelectric crystals are anisotropic; therefore, the simple relationship between D and E as given by Eq. (7.48) now should be replaced by the more general expressions

$$D_i = \sum_j \epsilon_{ij} E_j \qquad i, j = 1, 2, 3 \qquad (7.81)$$

The subscripts 1, 2, 3 refer to the three orthogonal directions x, y, z, respectively. For a crystal under stress, the electric displacement is also a function of stress (Fig. 7.13), and hence Eq. (7.81) should be replaced by

$$D_i = \epsilon_{i1} E_1 + \epsilon_{i2} E_2 + \epsilon_{i3} E_3 + d_{i1} T_1 + d_{i2} T_2$$
$$+ d_{i3} T_3 + d_{i4} T_4 + d_{i5} T_5 + d_{i6} T_6 \qquad (7.82)[13]$$

---

[13] Some of the piezoelectric constants d_{ij} can be zero on account of symmetry considerations in a given crystal structure. For such information, the reader is referred to W. G. Cady, "Piezoelectricity," pp. 190–192, Dover Publications, Inc., New York, 1964, and to W. P. Mason, "Piezoelectric Crystals and Their Applications to Ultrasonics," pp. 41–44, D. Van Nostrand Co., Inc., New York, 1950.

where $i = 1, 2, 3$. The proportionality constants d_{ij} are the *piezo-electric constants*.

It is to be noted that the internal energy U of a piezoelectric crystal is given by

$$dU = \sum_1^6 T_j \, dS_j + \sum_1^3 E_i \, dD_i + T \, d\sigma \tag{7.83}$$

where T is the temperature and σ the entropy. The enthalpy H is related to U by the following equation

$$dH = dU - \sum d(T_j S_j) - \sum d(E_i D_i)$$
$$= - \sum_1^6 S_j \, dT_j - \sum_1^3 D_i \, dE_i + T \, d\sigma \tag{7.84}$$

For dH to be a total differential, we have

$$S_j = \frac{\partial H}{\partial T_j} \qquad \text{and} \qquad D_i = \frac{\partial H}{\partial E_i}$$

or the following relationship

$$\frac{\partial S_j}{\partial E_i} = \frac{\partial D_i}{\partial T_j} \tag{7.85}$$

According to Eq. (7.85), the inverse relationship between S and E can be written as

$$S_j = d_{j1} E_1 + d_{j2} E_2 + d_{j3} E_3 + s_{j1} T_1 + s_{j2} T_2$$
$$+ s_{j3} T_3 + s_{j4} T_4 + s_{j5} T_5 + s_{j6} T_6 \tag{7.86}$$

After having established the general relations expressed in Eqs. (7.82) and (7.86), we can illustrate their application to the piezo-electric effect with a simple example. Consider a thin dielectric slab whose thickness and width are much smaller than its length as shown in Fig. 7.15. Electrodes are plated to the surfaces normal to the z axis and an alternating voltage is applied across the electrodes. Our task is to find an equivalent circuit for the dielectric slab, taking the piezoelectric effect into consideration. Our present discussion will be confined to the longitudinal vibrating mode (along the x direction) of the slab. Under such circumstances, the surfaces normal to the y and z axes can be considered free of stresses. Thus, $T_2 = T_3 = 0$, and $T_{yz} = T_{xy} = T_{xz} = 0$ or $T_4 = T_5 = T_6 = 0$. Since $E_1 = E_2 = 0$, Eq. (7.86) reduces to

$$S_1 = s_{11} T_1 + d_{13} E_3 \tag{7.87}$$

Now we consider the elementary cube (Fig. 7.14) again. The displacements of the cube in the x, y, and z directions are given by ξ, η, and ζ, respectively. Hence, according to Newton's law of motion, the net force acting on the cube in the x direction is given by

$$F_x = \rho \, dx \, dy \, dz \frac{d^2\xi}{dt^2} \qquad (7.88)$$

where ρ is the mass density of the material under consideration. Equating Eq. (7.88) with (7.75) and remembering $T_5 = T_6 = 0$, we have

$$\rho \frac{d^2\xi}{dt^2} = \frac{\partial T_1}{\partial x} \qquad (7.89)$$

By substituting Eq. (7.87) into Eq. (7.89) and letting $S_1 = (\partial\xi/\partial x)$, Eq. (7.89) is further transformed into

$$\rho \frac{d^2\xi}{dt^2} = \frac{1}{s_{11}} \frac{\partial^2\xi}{\partial x^2} - \frac{d_{13}}{s_{11}} \frac{\partial E_3}{\partial x} \qquad (7.90)^{14}$$

Since the applied field is independent of x, the last term in Eq. (7.90) vanishes. Thus, Eq. (7.90) is a simple wave equation, the solution of which is given by

$$\xi = A \sin kx + B \cos kx \qquad (7.91)^{15}$$

where $k = \omega/v$, ω is the angular frequency of oscillation and $v (= 1/\sqrt{\rho s_{11}})$ the velocity of sound propagation in a plated crystal.

The boundary conditions are those for a free crystal: at $x = 0$, and $x = l$, $T_1 = 0$ (stress free). Thus, from Eqs. (7.87) and (7.91),

[14] Note the similarity between Eqs. (7.90) and (4.159). The slight difference in appearance of these two equations is due to the fact that we define the piezo-electric constant differently in Eqs. (7.82) and (4.158).

[15] Here we are interested in the standing-wave type of solution in contrast to the traveling-wave solution discussed in Sec. 4.9.

Figure 7.15

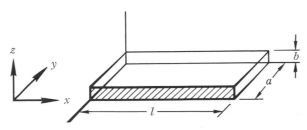

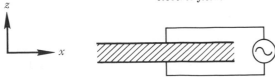

A thin piezoelectric crystal under an alternating electric field.

we have, at $x = 0$,

$$S_1 = \frac{\partial \xi}{\partial x} = Ak = d_{13}E_3 \qquad (7.92)$$

and at $x = l$,

$$Ak \cos kl - Bk \sin kl = d_{13}E_3 \qquad (7.93)$$

Solving for the constants A and B and substituting the values obtained in Eq. (7.91), we obtain

$$S_1 = \frac{d_{13}E_3 \cos\left[k(l - 2x)/2\right]}{\cos\left(kl/2\right)} \qquad (7.94)$$

The expression for D_3 can be obtained from Eqs. (7.82) and (7.87) as follows:

$$D_3 = \epsilon_{33}E_3 + d_{13}T_1 = \left(\epsilon_{33} - \frac{d_{13}^2}{s_{11}}\right)E_3 + \frac{d_{13}}{s_{11}}S_1 \qquad (7.95)$$

Using Eqs. (7.94) and (7.95) and integrating Eq. (7.95) over the surface normal to the z axis, we find, for the total current,

$$
\begin{aligned}
I_3 &= i\omega \int_0^l D_3 a \, dx \\
&= i\omega \left[\left(\epsilon_{33} - \frac{d_{13}^2}{s_{11}}\right)al + \frac{d_{13}^2}{s_{11}}\frac{2a}{k}\tan\left(\frac{kl}{2}\right)\right]E_3 \qquad (7.96)
\end{aligned}
$$

From Eq. (7.96), the equivalent admittance is given by

$$Y = \frac{I_3}{E_3 b} = i\omega \frac{al}{b}\left[\left(\epsilon_{33} - \frac{d_{13}^2}{s_{11}}\right) + \frac{d_{13}^2}{s_{11}}\frac{2}{kl}\tan\left(\frac{kl}{2}\right)\right] \qquad (7.97)$$

A quick look at Eq. (7.97) shows that Y possesses resonant and antiresonant frequencies. A simple equivalent circuit which has this property is shown in Fig. 7.16. Let us define

$$K = \frac{d_{13}}{\sqrt{\epsilon_{33}s_{11}}} \qquad (7.98)[16]$$

as the *electromechanical coupling coefficient*. At very low frequencies,

$$Y = i\omega\epsilon_{33}\frac{al}{b} = i\omega(C_0 + C_1) \qquad (7.99)$$

and at resonance frequency f_r, $Y = \infty$, $kl/2 = \pi/2$, or

$$f_r = \frac{1}{2l\sqrt{\rho s_{11}}} = \frac{1}{2\pi\sqrt{L_1 C_1}} \qquad (7.100)$$

[16] The electromechanical coupling constants K in Eqs. (7.98) and (4.166) are the same, realizing that $d_{13}/s_{11} = d$ and $1/s_{11} = c$. Strictly speaking, ϵ in Eq. (4.166) is related to ϵ_{33} through $\epsilon = \epsilon_{33} - d_{13}^2/s_{11}$. Since $K^2 \ll 1$, the difference in ϵ and ϵ_{33} can be neglected.

At antiresonance frequency f_a, $Y - 0$, or

$$\frac{1 - K^2}{K^2} = \frac{2v}{\omega_a l} \tan\left(\frac{\omega_a l}{2v}\right) \qquad (7.101)$$

If $f_a = f_r + \Delta f$, and $\Delta f/f_r \ll 1$, then $\tan(\omega_a l/2v)$ can be approximated by $2v/(2\pi\Delta fl)$, noting that $\omega_r l/2v = \pi/2$. Thus, Eq. (7.101) becomes

$$\frac{\Delta f}{f_r} = \frac{4}{\pi^2} \frac{K^2}{1 - K^2} \qquad (7.102)$$

From the equivalent circuit, we find

$$\omega_a{}^2 L_1 \left(\frac{C_0 C_1}{C_0 + C_1}\right) = 1 \qquad \text{and} \qquad \omega_r{}^2 L_1 C_1 = 1 \qquad (7.103)$$

and the ratio $\Delta f/f_r \approx (\omega_a{}^2 - \omega_r{}^2)/2\omega_r{}^2$ becomes

$$\frac{\Delta f}{f_r} = \frac{C_1}{2C_0} = \frac{4}{\pi^2} \frac{K^2}{1 - K^2} \qquad (7.104)$$

In principle, the parameters in the equivalent circuit (L_1, C_1, and C_0) can be completely determined from measurements of f_r, f_a, and Y at very low frequencies. The small resistance in Fig. 7.16c represents the internal losses in the crystal, which include the loss of energy at the clamping points and the loss of energy due to setting up of air waves by the crystal motion. From this equivalent circuit a quality factor is defined as $\omega L_1/R_1$. For quartz, values as high as 3×10^5 have been observed, losses being mainly internal losses.

It is also obvious from Eqs. (7.99), (7.100), and (7.102) that the parameters ϵ_{33}, s_{11}, and d_{13} can be determined from the same set of measurements. We should add that there are other more complicated modes of motion besides the simple longitudinal mode. Since different vibrational modes involve different elastic and electromechanical constants, a complete set of data concerning s_{ij} and d_{ij} can be obtained from measurements of f_r, f_a, and Y using different vibrational modes. Furthermore, it should be pointed out that for crystals

Figure 7.16

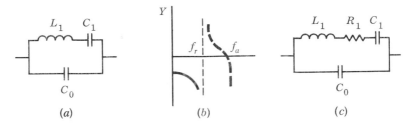

The equivalent circuit of a piezoelectric crystal under the applied voltage condition of Fig. 7.15 and also in the longitudinal mode of vibration.

possessing a certain kind of symmetry, some of the elastic and electromechanical constants may reduce to zero. Extensive discussions of various vibrational modes and methods of exciting them as well as information concerning crystal symmetry and electromechanical properties can be found in the book by W. P. Mason.[17]

Finally, it seems appropriate to mention some of the commonly used piezoelectric crystals. Besides quartz, they include Rochelle salt ($NaKC_4H_4O_6 \cdot 4H_2O$), ammonium dihydrogen phosphate ($NH_4H_2PO_4$), potassium dihydrogen phosphate (KH_2PO_4), ethylene diamine tartrate ($C_6H_{14}N_2O_6$), and dipotassium tartrate ($K_2C_4H_4O_6 \cdot \frac{1}{2}H_2O$), to name a few. Because of their narrow bandwidth compared with ordinary electric resonant circuits and the small temperature dependence of their properties if cut in certain crystal planes, these crystals are used in filter and oscillator circuits for greater frequency selectivity and stability. Piezoelectric crystals can also be used as electromechanical transducers and the medium for ultrasonic amplification. In the optical region, the amount of light transmitted through a piezoelectric crystal varies with the strain in the crystal, and hence piezoelectric crystals can be used for modulating light beams. We have already discussed applications with piezoelectric crystals at ultrasonic frequencies in Chapter 4 and shall discuss their applications at optical frequencies later in Chapter 12.

7.9 FERROELECTRIC AND ANTIFERROELECTRIC MATERIALS

A crystal is said to be ferroelectric if it exhibits a spontaneous polarization in the absence of an applied field. It is a necessary but not sufficient condition for ferroelectricity that the crystal lack an inversion center. This is illustrated in Fig. 7.17. If the point O is the center of inversion, atoms A and B possessing negative charges must be symmetrically located with respect to O, resulting in a cancellation of dipole moments. The same applies to atoms C and D possessing positive charges. Further, under isotropic stress atoms

[17] W. P. Mason, "Piezoelectric Crystals and Their Application to Ultrasonics," D. Van Nostrand Co., Inc., New York, 1950.

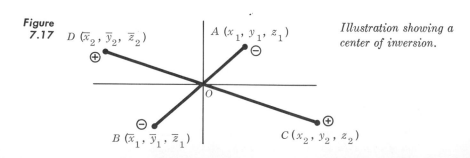

Figure 7.17

$D\ (\bar{x}_2,\ \bar{y}_2,\ \bar{z}_2)$ $A\ (x_1,\ y_1,\ z_1)$ *Illustration showing a center of inversion.*

$B\ (\bar{x}_1,\ \bar{y}_1,\ \bar{z}_1)$ $C\ (x_2,\ y_2,\ z_2)$

will be symmetrically displaced with respect to the center of sym
metry; hence, such a crystal cannot be piezoelectric. Among the 32
crystal classes, 12 crystal classes (Table 2.4) have this property and
are nonpiezoelectric. As an example, take the diamond structure in
the cubic system. For ease of discussion, let us cut the structure
halfway, as shown in Fig. 7.18a. The projections of the positions of
atoms upon the xy plane are shown in Fig. 7.18b and c, the z ordinates
being designated by numbers in circles. The center of inversion is
any point midway between two neighboring atoms. Suppose that
we choose the point $G(\frac{3}{8},\frac{3}{8},-\frac{3}{8})$ as indicated by a square in Fig.
7.18b. Atoms A and B are symmetrically located with respect to G.
Other pairs illustrated in the diagram are atoms CD and atoms EF.
For every atom in the diamond structure, there is always another
atom oppositely situated with respect to G. By reason of symmetry
as mentioned earlier, a crystal having a diamond structure cannot

**Figure
7.18**

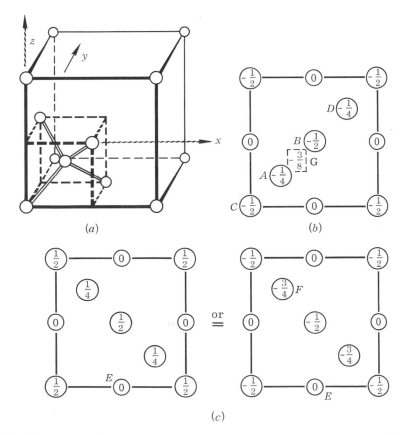

(a)

(b)

(c)

*The diamond lattice and illustrations showing that the point G is one of the
inversion centers in the structure.*

be ferroelectric. According to Table 2.4, other crystals possessing a center of inversion include, with one example for each of the seven crystal classes, $CuSO_4$ (triclinic), $CaSO_4 \cdot 2H_2O$ (monoclinic), $BaSO_4$ (orthorhombic), $ZrSiO_4$ (tetragonal), $CaCO_3$ (trigonal), $Be_3Al_2(SiO_3)_6$ (hexagonal), and $NaCl$ (cubic).

Besides the center of inversion, there are other elements of symmetry such as planes of symmetry and axes of symmetry. A combination of such symmetry elements can also eliminate the possible existence of a polar axis and hence the possibility of ferroelectricity. Suppose that a crystal has a twofold symmetry about the x and y axes. By n-fold symmetry, we mean that the crystal duplicates itself if it is rotated about an axis by $2\pi/n$ radians. Thus, for the example just mentioned, for a dipole moment with vector components (u,v,w), there must be another dipole moment $(u,-v,-w)$ by virtue of twofold symmetry about the x axis. The symmetry requirements about the y axis add further two vectors with components $(-u,v,-w)$ and $(-u,-v,w)$. The resultant dipole moment is obviously zero; therefore, a crystal possessing the said symmetry elements also cannot be ferroelectric. Later on, we shall see how structural changes which destroy the symmetry of a crystal can bring about ferroelectricity when we discuss the property of ferroelectric crystals as a function of temperature. Among the 32 crystal classes, 12 classes possess a center of inversion and an additional 10 classes have symmetry elements which exclude the possible existence of a polar axis.

The known ferroelectric materials mostly fall into three main groups, characterized by: (1) Rochelle salt, (2) potassium dihydrogen phosphate, (3) barium titanate. A fourth group is presently classified as "miscellaneous." The crystal first discovered to be ferroelectric is Rochelle salt of chemical composition $NaKC_4H_4O_6 \cdot 4H_2O$, a double tartrate of sodium and potassium with four molecules of water. Above 24°C, the crystal is orthorhombic and the three crystal axes are axes of twofold symmetry (Class 6 in Table 2.4); therefore, the crystal is not ferroelectric. Below the transition temperature, the crystal is slightly distorted and transformed into monoclinic structure. One axis, chosen to be the x axis, retains the twofold symmetry and becomes the polar axis while the y and z axes are now associated with no symmetry repetition. In Fig. 7.19, the relative dielectric constants of Rochelle salt along different crystal axes are shown as functions of temperature. In the temperature range from 24°C to −17°C, a hysteresis loop as shown in Fig. 7.20a appears, where the length of OC represents the *spontaneous polarization* P_s and OA the *coercive field* E_C. The magnitudes of P_s and E_C are plotted in Fig. 7.20b and c as functions of temperature, the measurements of course being taken along the polar axis. The critical temperature at which a crystal becomes ferroelectric is called the *Curie tempera-*

ture. For Rochelle salt, there exist an upper and a lower Curie temperature. The substitution of hydrogen by deuterium changes the properties of the crystal from solid to dotted curves in Figs. 7.19 and 7.20. Not only the Curie temperatures are different but also the spontaneous polarization and coercive force. There are other tartrates isomorphous with Rochelle salt with the replacement of potassium by Rb, Tl, or NH_4. However, none of these has been shown conclusively to be ferroelectric.

The most important member of the tetragonal phosphate group (group 2) of ferroelectric substances is KH_2PO_4. It has a Curie point at $-150°C$. Above the Curie temperature, the crystal has a tetragonal structure with the cell dimensions $a = b = 7.45$ Å and $c = 6.96$ Å. Below the Curie temperature, however, the original square base becomes slightly distorted into a rectangle and the crystal takes the form of an orthorhombic structure as shown in Fig. 7.21. It turns out that it is more convenient to consider the small crystal cell $ABCD$ than the large crystal cell $EFGH$. The small cell which is rhombic may be considered as being the result of shear stress applied to the cell $ABCD$. Because of the piezoelectric effect, the shear strain S_6 induces an electric field in the z direction. Based

Figure 7.19

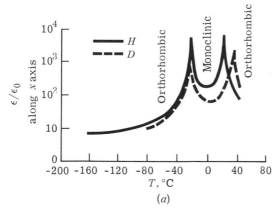

(a)

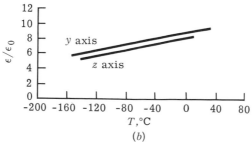

(b)

The relative dielectric constant of Rochelle salt as a function of temperature measured (a) along the x axis (the polar axis) and (b) along the y and z axes. The dashed curve in (a) is for Rochelle salt with H being substituted for by D. (J. Hablützel, Helv. Phys. Acta, vol. 12, p. 489, 1939.)

on this model, de Quervain[18] proposes that the spontaneous polarization can be estimated as follows:

$$P_s = \epsilon_c E_z = \epsilon_c \frac{S_6}{d_{36}}$$

(7.105)

In the above equation, ϵ_c is the dielectric constant along the c or z axis and d_{36} is the piezoelectric constant. These two quantities can be measured just above the Curie temperature, and the ratio ϵ_c/d_{36}

[18] M. de Quervain, *Helv. Phys. Acta*, vol. 17, p. 509, 1944.

Figure 7.20

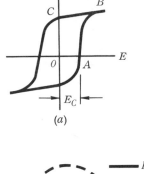

(a)

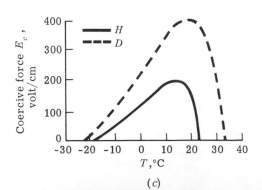

(b)

(c)

(a) The hysteresis loop, (b) the spontaneous polarization, and (c) the coercive field observed in Rochelle salt between 30°C and −20°C. The dashed curve is for Rochelle salt with H being substituted for by D. (J. Hablützel, Helv. Phys. Acta, vol. 12, p. 489, 1939.)

Figure
7.21

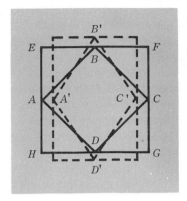

Structural change in KH_2PO_4 *during the transition from the nonferroelectric (solid line) to ferroelectric phase (dashed line).*

can be compared with the value of P_s/S_6 measured just below the Curie temperature. A good agreement between these two ratios, found experimentally, indicates that piezoelectricity and spontaneous polarization are caused by the same deformation of the crystal cell.

The spontaneous polarization and the relative dielectric constant of KH_2PO_4 are shown in Fig. 7.22a and b respectively. Above the Curie temperature, the relative dielectric constant along the polar

Figure
7.22

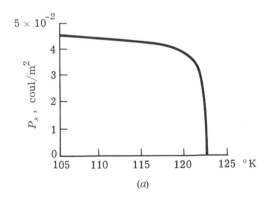

(a)

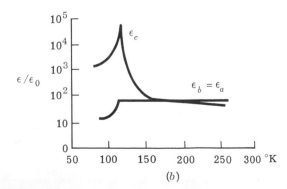

(b)

The spontaneous polarization and relative dielectric constants in KH_2PO_4 *measured as functions of temperature (M. de Quervain, Helv. Phys. Acta, vol. 17, p. 509, 1944, and G. Busch, Helv. Phys. Acta, vol. 11, p. 269, 1938.)*

axis obeys a Curie-Weiss law as follows:

$$\frac{\epsilon_c}{\epsilon_0} = 4.5 + \frac{C}{T - T_c} \qquad (7.106)$$

with $C = 3122$ and $T_c = 122°$K. A general derivation of the Curie-Weiss law for ferroelectric crystals will be found in the next section. Other ferroelectric crystals belonging to the tetragonal phosphate family include: KD_2PO_4 (213°K), KH_2AsO_4 (92°K), RbH_2PO_4

Figure 7.23

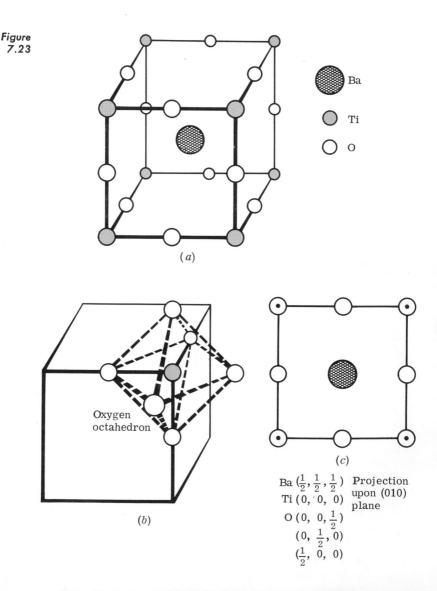

(a)

Ba

Ti

O

Oxygen octahedron

(b)

(c)

Ba $(\frac{1}{2}, \frac{1}{2}, \frac{1}{2})$ Projection
Ti $(0, 0, 0)$ upon (010) plane
O $(0, 0, \frac{1}{2})$
$(0, \frac{1}{2}, 0)$
$(\frac{1}{2}, 0, 0)$

The cubic structure of $BaTiO_3$ above the ferroelectric Curie temperature.

(146°K), RbD_2AsO_4 (178°K), and CsH_2AsO_4 (143°K), the temperature indicated in parentheses being the Curie temperature.

Among ferroelectric materials, barium titanate $BaTiO_3$ has been most thoroughly investigated and best understood. Above 120°C which is the Curie point, $BaTiO_3$ has the ideal perovskite structure, with a Ba atom sitting at the center and Ti atoms at the corners of a cube, and O atoms at the midpoints of the cube edges, as shown in Fig. 7.23. At the Curie point, the structure becomes slightly stretched parallel to one cubic edge. The new structure is tetragonal, and the stretched edge becomes the polar axis. In contrast to Rochelle salt and KDP in which only one ferroelectric phase exists, $BaTiO_3$ exhibits three different ferroelectric phases. The temperature dependence of the relative dielectric constant and that of the spontaneous polarization of $BaTiO_3$ are shown in Figs. 7.24 and 7.25 respectively. Between 5°C and −90°C, the structure becomes orthorhombic by slightly stretching the original cubic structure along one face diagonal and compressing it along the other. These two face diagonals are now the new orthorhombic axes and one of the two face diagonals becomes the new polar axis. Below −90°C, the structure changes again, this time into a rhombohedron formed by slightly stretching the original cube along its body diagonal which now becomes the new polar axis. It is to be noted that in Fig. 7.25, the quantity P measured experimentally is the component of P_s normal to the surface of the specimen, which coincides with the face of the cube. As the polar axis changes from the cubic edge to the face diagonal and then to the body diagonal, it is found that the value of P changes from P_s to $P_s/\sqrt{2}$ and then to $P_s/\sqrt{3}$, indicating that P_s is essentially constant in the three ferroelectric phases.

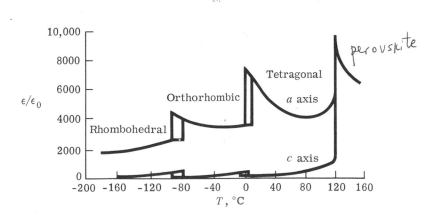

Figure 7.24

The dielectric constant of $BaTiO_3$ as a function of temperature. The three phases (tetragonal, orthorhombic, and rhombohedral) are all ferroelectric. (W. J. Merz, Phys. Rev., vol. 76, p. 1221, 1949.)

It may be instructive to see qualitatively how structural changes[19] can bring about ferroelectricity. The atomic positions of cubic barium titanate are shown in Fig. 7.23. In the tetragonal form, the titanium and oxygen atoms move relative to the barium atom as illustrated in Fig. 7.26 where the atomic movements are expressed again in units of the lattice constants a and c as in Fig. 7.23. Note that for easy reference, we have assigned subscripts I and II to indicate two different kinds of oxygen atoms, the former being oriented along the z axis while the latter are along the x and y axes with respect to the titanium atom. Several important points may be observed about this tetragonal structure. First, the actual bond lengths between the barium and the surrounding oxygen atoms are practically the same while those between the titanium and oxygen atoms are quite different, depending on the orientation of the oxygen atom. The various bond lengths and angles in tetragonal $BaTiO_3$ are given as follows:

Length	$Ba-O_I$ $Ba-O_{II}$	2.824 Å $\begin{cases} 2.800 \text{ Å} \\ 2.886 \text{ Å} \end{cases}$	$Ti-O_I$ $Ti-O_{II}$	$\begin{cases} 1.869 \text{ Å} \\ 2.167 \text{ A} \\ 1.999 \text{ Å} \end{cases}$
Angle			$Ti-O_{II}-Ti$	171°28′

[19] The structure factor discussed in Sec. 2.6 is extremely useful in analyzing the structural changes during ferroelectric transition. For a detailed discussion, refer to F. Jona and G. Shirane, "Ferroelectric Crystals," pp. 376–382, Pergamon Press, London, 1962.

Figure 7.25

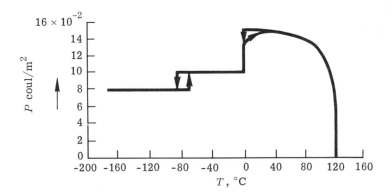

The component of the spontaneous polarization in $BaTiO_3$ along its cube edge plotted as a function of temperature. The change in the magnitude of P is indicative of the change of the polar axis (from cube edge to face diagonal to body diagonal). (W. J. Merz, Phys. Rev., vol. 76, p. 1221, 1949.)

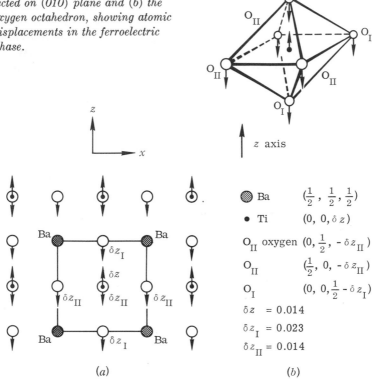

Figure 7.26 (a) The structure of $BaTiO_3$ projected on (010) plane and (b) the oxygen octahedron, showing atomic displacements in the ferroelectric phase.

⬤ Ba	$(\frac{1}{2}, \frac{1}{2}, \frac{1}{2})$
• Ti	$(0, 0, \delta z)$
O_{II} oxygen	$(0, \frac{1}{2}, -\delta z_{II})$
O_{II}	$(\frac{1}{2}, 0, -\delta z_{II})$
O_{I}	$(0, 0, \frac{1}{2} - \delta z_{I})$

$\delta z = 0.014$

$\delta z_{I} = 0.023$

$\delta z_{II} = 0.014$

(a) (b)

In other words, the oxygen octahedra shown in Fig. 7.23b are very little distorted even in the tetragonal structure but now the titanium atoms are displaced relative to the centers of their respective tetrahedra by about 0.15 Å. The situation is illustrated schematically in Fig. 7.27a, the positions of barium atoms being omitted to avoid

Figure 7.27

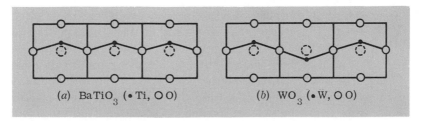

(a) $BaTiO_3$ ($\bullet$ Ti, $\circ$ O) (b) WO_3 ($\bullet$ W, $\circ$ O)

(a) Schematic representation showing the displacement of the titanium atom in the oxygen octahedron projected on the (010) plane. Since the displacements of titanium atoms in neighboring cells are parallel, the dipole moments are additive, meaning ferroelectricity. In this ferroelectric phase, the bond angle Ti–O–Ti changes. (b) The antiferroelectric arrangement in WO_3. In the antiferroelectric phase, the bond angle W–O–W remains $180°$.

confusion. Since the displacements of titanium atoms in neighboring cells are all in parallel, the dipole moments arising from the displacements are additive. This corresponds to the state of *ferroelectricity*.

The situation just discussed, however, is not unique. For example, in tungsten trioxide WO_3 the bond angle W—O—W remains at 180° as in the cubic phase, resulting in antiparallel displacements of tungsten atoms as shown in Fig. 7.27b. It is quite obvious that under such circumstances, the dipole moments from neighboring cells are antiparallel to each other, and consequently, a complete cancellation of polarization results. This corresponds to the state of *antiferroelectricity*. In the perovskite family,[20] besides $BaTiO_3$, known ferroelectric crystals, with their respective transition temperatures indicated in parentheses, include: $NaTaO_3$ (750°K), $RbTaO_3$ (520°K), $LiTaO_3$ (473°K), $PbTiO_3$ (763°K), $KNbO_3$ (698°K), and known antiferroelectric crystals include WO_3 (1010°K), $PbZrO_3$ (503°K), $PbHfO_3$ (488°K).

Other antiferroelectric crystals include: $NH_4H_2PO_4$ (148°K) and $NH_4H_2AsO_4$ (216°K) in the tetragonal phosphate group and possibly solid solutions of (K,NH_4) $NaC_4H_4O_6 \cdot 4H_2O$ in the Rochelle salt group. In the miscellaneous classification, $Cd_2Nb_2O_7$, $C(NH_2)_3Al(SO_4)_2 \cdot 6H_2O$, and $LiTlC_4H_4O_6 \cdot H_2O$ are known to be ferroelectric while $Pb_2Nb_2O_7$, $(NH_4)_2H_3IO_6$, and $Ag_2H_3IO_3$ are probably antiferroelectric. The essential difference between the electrical behavior of an antiferroelectric and a ferroelectric crystal is that the former does not show a hysteresis loop below its transition temperature. Above the transition temperature, however, the dielectric constant of an antiferroelectric crystal increases with decreasing temperature as in a ferroelectric crystal.

Another mark of ferroelectricity is the appearance of domain structure. Again take barium titanate as an example because it has been more extensively investigated than other ferroelectric crystals. When a single crystal of $BaTiO_3$ undergoes transition from cubic to tetragonal symmetry on cooling through the Curie temperature, the tetrad or the polar axis can be any one of the three original cube edges. If the crystal is small and perfect, it may transform to a single crystal having a single tetragonal axis. If the crystal is subject to local stresses such as due to inhomogeneous concentrations of impurities, the local mechanical stresses may be minimized by requiring one specific cube axis to be the new tetragonal axis. Therefore, depending on the local stress conditions, different parts of the

[20] For a compilation of ferroelectric and antiferroelectric materials, refer to C. Kittel, "Introduction to Solid State Physics," p. 184 and p. 203, John Wiley & Sons, Inc., New York, 1956. For a detailed discussion of structural changes and electric properties of these crystals, refer to H. D. Megaw, "Ferroelectricity in Crystals," Methuen & Co. Ltd., London, 1957, and to F. Jona and G. Shirane, "Ferroelectric Crystals," Pergamon Press, London, 1962.

crystal may have different cube axes as their new tetragonal axes. This means that different regions will have different directions of spontaneous polarization. A *domain* is defined as a region within which the direction of spontaneous polarization is constant, and the region which separates two domains is called the *domain wall.*

Because barium titanate has two different indices of refraction, parallel and perpendicular to the tetragonal axis, the domain structure in $BaTiO_3$ can be observed by optical methods. It is also found that chemical etching rates are vastly different for different domains, the surface having positive polarization being more rapidly attacked by the etchant HCl used in the experiment. Two kinds of domain walls are known: 90° and 180° walls, that means, the directions of polarization in the two domains separated by the wall are respectively at 90° and 180° to each other. Figure 7.28 shows the domain structure in tetragonal $BaTiO_3$ from the experimental work of Hooton and Merz. In this particular case, both the 90° and 180° walls are shown. Under certain circumstances, however, a ferroelectric crystal may contain only one kind of domain walls, either the 90° or the 180° wall.

It should be mentioned that domain structure also exists in ferromagnetic materials, but ferroelectric and ferromagnetic domain structures differ in many respects. Since ferromagnetic domains will be discussed in detail in Chapter 10, we shall concentrate our present discussion on ferroelectric domains. First, the ferroelectric domain wall is extremely thin, perhaps several lattice constants (as compared to several hundred lattice constants for a ferromagnetic domain wall). As we mentioned earlier, the domain structure exists on account of local stresses. More precisely, the domain wall is created to decrease the elastic energy associated with local stresses at the expense of increasing the electrostatic energy. The thickness of a domain wall is determined by minimizing the total energy. It turns out that the electrostatic energy arising from interaction between dipoles in neighboring domains is very small. As a matter of fact, the energies associated with the parallel and antiparallel phase are quite close so that a ferroelectric state can be induced in some antiferroelectric crystals by the application of a strong electric field.

Figure 7.28

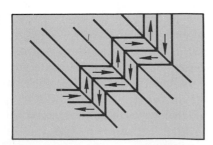

Domain formation in $BaTiO_3$ *showing 90° walls and antiparallel domains. (J. A. Hooton and W. J. Merz, Phys. Rev., vol. 98, p. 409, 1955.)*

Since the electrostatic energy increases as the separation between dipoles decreases, a thin domain wall enables the electrostatic energy to counterbalance the relatively large elastic energy. It should be further pointed out that because of small wall thickness, a high concentration of energy over a distance of a few lattice constants results. This is of practical importance when $BaTiO_3$ crystals are to be considered for use as memory devices.

If a negative electric field is applied to a single-domain $BaTiO_3$ crystal in a switching experiment to reverse the polarization vector, antiparallel domains appear in the form of sharp wedges with the blunt ends based on the edge of the crystal and extending several microns wide. As the field strength increases, new antiparallel domains are nucleated. Existing domains, however, advance forward in the direction of the pointed end instead of growing sideways. As is explained by Merz, the sidewise motion of domain walls is hampered because to move the wall by one unit cell requires an activation energy of the order of magnitude of the total wall energy itself (or 10^{-2} joule/m²). The energy gained by moving the wall one lattice constant is equal to $2EP_sa$ where E is the applied field, P_s is the spontaneous polarization and a is the lattice constant. For a field of 10^6 volt/m, the energy gained is only about 10^{-4} joule/m² which is much smaller than the wall energy. In the same switching experiment, the following empirical relationship is also found by Merz,[21]

$$t_s = \frac{\beta d}{E - E_c} \tag{7.107}$$

where t_s is the switching time, d the thickness of the crystal, E_c a constant of the order of the coercive force, and β is a proportionality constant. For $d = 5 \times 10^{-3}$ cm and $E = 10^4$ volt/cm, a switching time in the neighborhood of 2×10^{-6} sec is measured. The slowness of the switching speed coupled with the requirement of a high electric field certainly makes ferroelectric materials unattractive for use as memory devices as compared with ferromagnetic materials.

7.10 THEORY OF FERROELECTRICITY

From our discussion in the previous section, it seems clear that a shift of ionic positions, as for the Ti ion in $BaTiO_3$, is the key to ferroelectricity. The mechanism by which ferroelectricity may arise can be qualitatively described as follows. Imagine the situation in a ferroelectric crystal held above the Curie temperature just before the crystal goes into the ferroelectric phase. Suppose that an infinitesimal displacement of an ion from its equilibrium position occurs, then a local electric field is created on account of the dipole

[21] W. J. Merz, *Phys. Rev.*, vol. 95, p. 690, 1954.

moment associated with the displacement. If the electric force overcomes the elastic restoring force, a further displacement of the ion is expected. As displacement becomes appreciable, anharmonic restoring forces set in and the ion finds itself in a new equilibrium position corresponding to the position which the ion occupies in the ferroelectric phase. To explain the electric behavior of a specific ferroelectric crystal, a detailed knowledge of the particular arrangement of electric charges and forces in a given crystal structure is needed and then a specific model can be set up for the crystal under consideration. Such model theories are available for barium titanate, Rochelle salt, and potassium dihydrogen phosphate.

For the purpose of our present discussion, however, it seems instructive that we discuss the theory of ferroelectricity on a phenomenological basis originally due to Devonshire.[22] From the force balance just mentioned, we have the following equation

$$AE_{loc} = Bx + Cx^3 + Dx^5 \qquad (7.108)$$

where E_{loc} is the local field acting on, and x the displacement of, the ion under consideration and the constants A, B, C, and D are proportionality constants. The terms involving x^3 and x^5 represent the anharmonic components of the restoring force mentioned earlier. Since the dipole moment and in turn the polarization P created by the displacement are directly proportional to x, Eq. (7.108) can be expressed in terms of P as

$$E_{loc} = aP + bP^3 + cP^5 \qquad (7.109)$$

From our previous discussion on internal fields in Sec. 7.5, we also know that the local field can, by referring to Eq. (7.51), be written as

$$E_{loc} = E + fP \qquad (7.110)$$

where E is the applied field. Note that in Eq. (7.110) a general polarization field fP is used instead of the Lorentz field $P/3\epsilon_0$ for sake of clarity in subsequent discussions. Equating Eq. (7.110) with Eq. (7.109) gives

$$E = (a - f)P + bP^3 + cP^5 \qquad (7.111)$$

Therefore, the internal energy of the system is equal to

$$U = \int E\,dP = \frac{(a - f)P^2}{2} + \frac{bP^4}{4} + \frac{cP^6}{6} \qquad (7.112)$$

[22] A. F. Devonshire, Theory of Ferroelectrics, in *Advan. Phys.*, vol. 3, p. 85, 1954; also *Phil. Mag.*, vol. 40, p. 1040, 1949, and vol. 42, p. 1065, 1951. For ease of discussion, we follow the modified treatment by C. Kittel, "Introduction to Solid State Physics," pp. 194–197, John Wiley & Sons, Inc., New York, 1956.

First, let us study the situation in the ferroelectric phase. Two cases (Fig. 7.29) are of particular interest to us. In case I, $(a - f)$ and c are positive while b is negative. In this case, the spontaneous polarization jumps discontinuously from zero to a finite value at the Curie temperature T_c. The zero value corresponds to the dielectric phase and the nonzero value to the ferroelectric phase. Since these two states exist simultaneously at T_c,

$$\frac{(a - f)P_s{}^2}{2} + \frac{bP_s{}^4}{4} + \frac{cP_s{}^6}{6} = U_p = U_0 = 0 \qquad (7.113)$$

where P_s is the spontaneous polarization and the subscripts p and 0 refer to the polarized and nonpolarized state respectively. Furthermore, P_s exists in the absence of E, thus from Eq. (7.111), we have

$$(a - f) + bP_s{}^2 + cP_s{}^4 = 0 \qquad (7.114)$$

Equations (7.113) and (7.114) give the following results:

$$P_s{}^2(T_c) = -\frac{3b}{4c} = -\frac{4(a - f)}{b}$$

$$P_s{}^4(T_c) = \frac{3(a - f)}{c} \qquad (7.115)$$

In case II, $(f - a)$ and b are positive while c is negligible. It is obvious that with $c = 0$, the only solution of Eqs. (7.113) and (7.114) is that

$$P_s{}^2(T_c) = 0 \qquad \text{and} \qquad f - a = 0 \qquad \text{at } T = T_c \qquad (7.116)$$

Next, let us turn our attention to the dielectric phase. Differentiation of Eq. (7.111) gives

$$\epsilon - \epsilon_0 = \frac{\partial P}{\partial E} = \frac{1}{a - f} \qquad (7.117)$$

noting that the terms involving P^3 and P^5 are negligibly small compared to the linear term in P on account of the smallness of P in the

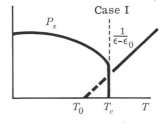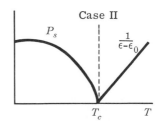

Figure 7.29

Schematic diagrams showing the variations of the spontaneous polarization (in the ferroelectric phase) and the reciprocal of dielectric constant (in the dielectric phase) as functions of temperature for cases I and II (defined in the text). Case I obeys Eqs. (7.115) and (7.119) whereas case II obeys Eqs. (7.116) and (7.118).

dielectric phase. For case II, we can expand $a - f$ into a power series in terms of $T - T_c$. If only the first-order term is kept, $\epsilon - \epsilon_0$ must be of the following form:

$$\epsilon - \epsilon_0 = \frac{\beta}{T - T_c} \tag{7.118}$$

since $a - f = 0$ at $T = T_c$. For case I, however, $a - f$ is not zero at $T = T_c$; therefore expansion at another temperature T_0 where $a = f$ is necessary. Thus for case I we have

$$\epsilon - \epsilon_0 = \frac{\beta}{T - T_0} \tag{7.119}$$

A summary of the behavior of P_s and ϵ as a function of temperature is illustrated in Fig. 7.29. Case I and case II are known thermodynamically as first- and second-order transitions, respectively, since case I involves a latent heat of transition while case II does not. Among the ferroelectric materials previously discussed, barium titanate shows the characteristics of a first-order transition while potassium dihydrogen phosphate and Rochelle salt probably belong to the second-order transition.

It may be interesting to discuss the relationship between the spontaneous polarization and the applied field. Equation (7.111) can be transformed into a universal equation containing normalized quantities as follows:

$$e = 2pt - 4p^3 + 2p^5 \tag{7.120}$$

where the new quantities are related to the old quantities by the following relations

$$p = \left(\frac{2c}{-b} \right)^{1/2} P$$

$$e = -\frac{4}{b} \left(\frac{2c}{-b} \right)^{3/2} E \tag{7.121}$$

$$t = \frac{-2}{\beta b} \frac{2c}{-b} (T - T_0)$$

A plot of p as a function of e for different values of t is shown in Fig. 7.30. Since regions in which $\partial e / \partial p$ is negative represent unstable states, two types of hysteresis loop may be observed, the type following the trace $ABCDEDFBA$ or the other type following the trace $A'B'C'OD'E'F'E'G'H'OI'J'A'$ as illustrated in Fig. 7.30, depending upon the value of t or $T - T_0$. Such butterfly loops have indeed been observed experimentally.[23]

Another point of considerable interest concerns the values of f and β. If ferroelectricity were due to an alignment of permanent dipoles, then Onsager's rather than Lorentz's approach should be used to

[23] W. J. Merz, *Phys. Rev.*, vol. 91, p. 513, 1953.

calculate the polarization field. In other words, the value of f should be different from $1/3\epsilon_0$ even for crystals with cubic symmetry. On the other hand, if polarization is induced by a strain-inspired field as our derivation of Eq. (7.109) from Eq. (7.108) implies, Lorentz's approach should be valid as for all induced polarization. Another confirmation that induced polarization is the cause of ferroelectricity comes from the magnitude of β. If ferroelectricity were the result of dipole alignment, Eqs. (7.31) and (7.54) predict a value of $\beta = 3T_c\epsilon_0$ which for $BaTiO_3$ is of the order of $10^3\epsilon_0$. However, the experimentally measured value from Eq. (7.119) is of the order of $10^5\epsilon_0$. A reasonable explanation is as follows. It is to be noted that Eqs. (7.54) and (7.117) are equivalent, if we realize that a in Eq. (7.109) is actually the same as $1/N\alpha$ in Eq. (7.54) and $f = 1/3\epsilon_0$. Expanding the temperature variation of $N\alpha$ as a power series, we have

$$1 - \frac{N\alpha}{3\epsilon_0} = 1 - \frac{N_0\alpha}{3\epsilon_0} - \frac{\alpha}{3\epsilon_0}\frac{\partial N}{\partial T}(T - T_0) \qquad (7.122)$$

or the value of β

$$\beta = \frac{T - T_0}{a - f} = \frac{(T - T_0)3\epsilon_0 N\alpha}{3\epsilon_0 - N\alpha} = \frac{3\epsilon_0}{-\dfrac{1}{N}\dfrac{\partial N}{\partial T}} = \frac{3\epsilon_0}{\dfrac{1}{V}\dfrac{\partial V}{\partial T}} \qquad (7.123)$$

Figure 7.30

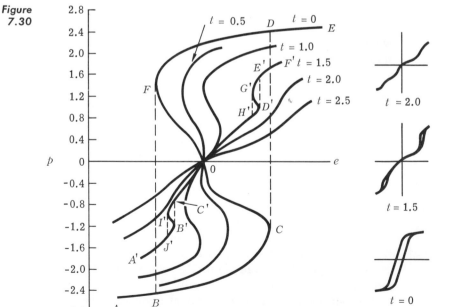

Normalized curves of polarization versus field at different temperatures. (W. J. Merz, Phys. Rev., vol. 91, p. 513, 1953.)

The volume expansion coefficient $V^{-1}(\partial V/\partial T)$ in $BaTiO_3$ is of the order of $10^{-5}/°K$ and this gives a correct order of magnitude for the value of β. In other words, the calculation shows that the temperature dependence of $N\alpha$ mainly comes from N. In contrast, the earlier result ($\beta = 3T_c\epsilon_0$) is obtained by expanding the temperature dependence of $\alpha(= p_0{}^2/3kT)$.

The above discussion shows that the phenomenological theory is able to explain the main features of ferroelectricity. Furthermore, we should emphasize that although the Curie-Weiss law is obeyed in ferroelectric and ferromagnetic materials above their respective Curie temperatures, the physical mechanisms responsible for ferroelectricity and ferromagnetism are very different. As we shall see in the next chapter, ferromagnetism arises due to a near perfect alignment of permanent magnetic dipoles. On the other hand, it is clear from Slater's detailed analysis that ferroelectricity in $BaTiO_3$ is a result of enhanced ionic polarization of Ti ions because of the perovskite structure. It may be worthwhile to give a brief account of Slater's theory.[24] Equation (7.117) or (7.54) predicts a polarization catastrophe when $N\alpha/3\epsilon_0 = 1$. The electronic contribution to the polarizability α can be estimated from the optical refractive index n of $BaTiO_3$ as follows:

$$\frac{n^2 - 1}{n^2 + 2} = \frac{N\alpha_e}{3\epsilon_0} \tag{7.124}$$

From a value of $n = 2.4$, the value of $N\alpha_e/3\epsilon_0$ is found to be 0.61. If $f = 1/3\epsilon_0$ in Eq. (7.117), this leaves a value of 0.39 for $fN\alpha_i(\text{Ti})$ due to ionic polarization of Ti ions. At this point, we should remind ourselves that the field E_1 in Eq. (7.43) is zero or $E_1 + E_3 = P/3\epsilon_0$ in Eq. (7.51) only for dipole arrangements having cubic symmetry. In barium titanate, the Ti ions, by their polarization, polarize the oxygen (type O_I) ions and these in turn act back on the Ti ions, resulting in linear chains of dipoles as illustrated in Fig. 7.26. Since the oxygen (type O_I) ions in this linear chain do not possess cubic symmetry, the field factor f is very much larger than $1/3\epsilon_0$. It is this great enhancement of the field at the Ti ion which causes the polarization of the Ti ion and makes the value of $fN\alpha_i(\text{Ti})$ large enough for the polarization catastrophe. A more detailed outline of Slater's calculation may be found in Prob. 7.21.

7.11 FURTHER DEVELOPMENTS IN THE THEORY OF DIELECTRIC CONSTANTS AND FERROELECTRICITY

The classical theory of dielectric polarizability developed in earlier sections explains well the qualitative behavior of dielectric materials. To test how successful is the classical theory, however, we must

---

[24] J. C. Slater, *Phys. Rev.*, vol. 78, p. 748, 1950.

express the dielectric constant of a given substance in terms of known properties of the substance so that a direct comparison with experiments can be made. The ionic crystals of alkali halides are especially suited for such studies because the classical model of an ionic crystal is relatively simple. According to Eq. (7.55) and our discussion in Sec. 7.7, the dielectric constant of an ionic crystal at optical frequencies can be written as

$$\frac{\epsilon_\infty - \epsilon_0}{\epsilon_\infty + 2\epsilon_0} = \frac{N}{3\epsilon_0}(\alpha_+ + \alpha_-)$$ (7.125)

where α_+ and α_- are the electronic polarizabilities of the positive and negative ions, respectively. In the low-frequency region, both electronic and ionic polarizabilities play a part; hence Eq. (7.125) should be replaced by

$$\frac{\epsilon_s - \epsilon_0}{\epsilon_s + 2\epsilon_0} = \frac{N}{3\epsilon_0}[(\alpha_+ + \alpha_-) + \alpha_i]$$ (7.126)

where α_i is the ionic polarizability. In Eqs. (7.125) and (7.126), the subscripts ∞ and s refer to the high-frequency and low-frequency (static) regions, respectively.

Under an applied field $\mathbf{E}_0$, the polarization $\mathbf{P}$ consists of two parts: $\mathbf{P}_e$ due to elastic displacement of electrons in each ion and $\mathbf{P}_i$ due to relative displacement of ions. Therefore, we can write

$$\mathbf{P} = \mathbf{P}_e + \mathbf{P}_i \quad \text{with } \mathbf{P}_i = N\alpha_i\mathbf{E}_0$$ (7.127)

The self-energy associated with the ionic polarization is equal to $W_1 = E_0P_i/2$ and the self-energy associated with elastic displacement is given by $W_2 = N\gamma(u - v)^2$ where u and v are the ionic displacements and they have the same meaning as in Eq. (7.24). Equating $W_1 = W_2$ and realizing that $\omega_0 = \sqrt{2\gamma/M}$, and $P_i = Ne(u - v)$, we have

$$N\alpha_i = \frac{Ne^2}{M\omega_0^2}$$ (7.128)

where M is the reduced mass of ions and it has the same meaning as in Eq. (7.24). Eliminating $N\alpha_i$ and $N(\alpha_+ + \alpha_-)$ from Eqs. (7.125), (7.126) and (7.128), we find

$$\epsilon_s - \epsilon_\infty = \frac{(\epsilon_s + 2\epsilon_0)(\epsilon_\infty + 2\epsilon_0)}{(3\epsilon_0)^2}\frac{Ne^2}{M\omega_0^2}$$ (7.129)

Next, we shall express ω_0 in terms of known vibrational frequency of the crystal. In doing so, we write Eq. (7.23) as

$$\frac{1}{\omega_0^2}\frac{d^2\mathbf{P}_i}{dt^2} + \mathbf{P}_i = N\alpha_i\mathbf{E}_0$$ (7.130)

by letting $\mathbf{P}_i - Ne(\mathbf{u} - \mathbf{v})$ and using Eq. (7.128). Now a distinction must be made between the polarization associated with the longitudinal mode and that associated with the transverse mode of vibration. It turns out[25] that for longitudinal waves,

$$\mathbf{P}_i = \frac{-3\epsilon_\infty}{\epsilon_\infty + 2\epsilon_0} \frac{3\epsilon_0}{2} \mathbf{E}_0 \tag{7.131a}$$

and for transverse waves,

$$\mathbf{P}_i = \frac{3\epsilon_0}{\epsilon_\infty + 2\epsilon_0} 3\epsilon_0 \mathbf{E}_0 \tag{7.131b}$$

Substituting $\mathbf{E}_0$ from Eq. (7.131) into Eq. (7.130) gives

$$\frac{1}{\omega^2} \frac{d^2\mathbf{P}_i}{dt^2} + \mathbf{P}_i = 0 \tag{7.132}$$

In Eq. (7.132),

$$\omega^2 = \omega_l{}^2 = \omega_0{}^2 \frac{\epsilon_s(\epsilon_\infty + 2\epsilon_0)}{\epsilon_\infty(\epsilon_s + 2\epsilon_0)} \qquad \text{for longitudinal waves} \tag{7.133a}$$

$$\omega^2 = \omega_t{}^2 = \omega_0{}^2 \frac{\epsilon_\infty + 2\epsilon_0}{\epsilon_s + 2\epsilon_0} \qquad \text{for transverse waves} \tag{7.133b}$$

In terms of ω_t, Eq. (7.129) becomes

$$\epsilon_s - \epsilon_\infty = \left(\frac{\epsilon_\infty + 2\epsilon_0}{3\epsilon_0}\right)^2 \frac{Ne^2}{M\omega_t{}^2} \qquad \text{farad/m} \tag{7.134}$$

The above equation is known as Szigeti relation and it is the same as Eq. (3.114). The other two equations of interest are

$$\gamma = \frac{e^2}{\alpha_i} = M\omega_0{}^2 = M\omega_t{}^2 \frac{\epsilon_s + 2\epsilon_0}{\epsilon_\infty + 2\epsilon_0} \tag{7.135}$$

$$M\omega_t{}^2 = \gamma - \frac{\epsilon_\infty + 2\epsilon_0}{(3\epsilon_0)^2} Ne^2 \tag{7.136}$$

The force constant γ is related to the compressibility K by $\gamma = 6R_0/K$ where R_0 is the nearest-neighbor distance. The calculated values e^* and K^* [from Eqs. (7.134) and (7.135) respectively] are compared with the measured values in Table 7.2 for a number of the alkali halides. In obtaining Eqs. (7.135) and (7.136), we define a force constant γ which has twice the value of γ defined in Eq. (7.22).

If the simple classical model of ionic crystals were correct, the value of e^*/e would be unity. Yet other experimental evidences strongly suggest that the alkali halides are highly ionic. For example, the simple Born-Mayer model which we used in Sec. 2.3 to calculate the cohesive energy of an ionic crystal is indeed very successful.

[25] The reader is referred to H. Fröhlich, "Theory of Dielectrics," pp. 150–154, Clarendon Press, Oxford, 1949.

Moreover, the fact that the Cauchy relation $c_{12} = c_{44}$ in Eq. (7.79) is approximately satisfied indicates that all forces act along the lines joining the centers of the atoms in these crystals. This is another characteristic of being an ionic crystal. In view of these considerations, the departure of e^*/e from unity must be due to inadequacy of the theory of dielectric constant itself.

A shell model for the ions in an alkali halide crystal, proposed by Dick and Overhauser,[26] has been found to give a satisfactory explanation of the dielectric properties of alkali halides. In the following, we shall present the highlights of their model. The classical model of an ionic crystal is still correct in that the positive and negative ions can be considered as having the closed-shell configurations of rare gases. However, those electrons far from the nucleus, being less tightly bound, are more profoundly affected by the application of an electric field than electrons in the inner shells. Sternheimer[27] has found that the polarizabilities of rare-gas ions are almost entirely due to electrons in the outermost shells.

In the model of Dick and Overhauser, the alkali-halide ions are considered as being constituted of an outer spherical shell and a core of the nucleus and the remaining electrons. Under the action of an applied field, the shell does not necessarily move together with the core, as illustrated in Fig. 7.31. Furthermore, the shells of the ions repel one another, and this repulsion causes additional displacements $d'_{\pm}$ in Fig. 7.31 of the shells. The polarization thus created is referred to as the *short-range polarization* by Dick and Overhauser. There is another kind of polarization called the *exchange-charge polarization*.

[26]   B. G. Dick and A. W. Overhauser, *Phys. Rev.*, vol. 112, p. 90, 1958.
[27]   R. M. Sternheimer, *Phys. Rev.*, vol. 96, p. 951, 1954; vol. 107, p. 1565, 1957.

TABLE 7.2   Values of K^*/K and e^*/e derived from experimental data

Alkali halide	K^*/K	e^*/e	ϵ_s/ϵ_0	$n^2 = \epsilon_\infty/\epsilon_0$
LiF	1.0	0.87	9.27	1.92
NaF	0.83	0.93	6.0	1.74
NaCl	0.99	0.74	5.62	2.25
NaBr	1.13	0.69	5.99	2.62
NaI	1.05	0.71	6.60	2.91
KCl	0.96	0.80	4.68	2.13
KBr	0.95	0.76	4.78	2.33
KI	0.99	0.69	4.94	2.69

[Reference: B. Szigeti, *Trans. Faraday Soc.*, vol. 45, p. 155, 1945, and *Proc. Roy. Soc.* (London), vol. A204, p. 51, 1950.]

As we recall our discussion of chemical bonds in Sec. 2.1, the exchange integral of the kind shown in Eq. (2.6) plays an important part. It is the change in the exchange energy, which is responsible for the repulsive energy [represented phenomenologically in Eq. (2.9)] of two ions. As the ions move under an applied field, not only does the exchange integral change, but associated with this change, there is a redistribution of charges. The net dipole moment per unit volume as a result of the charge redistribution is called the exchange-charge polarization. By incorporating the short-range-interaction and exchange-charge polarizations into the formulation, Dick and Overhauser not only are able to explain the deviations of e^*/e and K^*/K from unity but also succeed in generalizing the Szigeti and Clausius-Mossotti relations. For these generalized relations and the detailed mathematics used in their derivations, the reader is referred to the original article by Dick and Overhauser.

Let us now turn our attention to the theory of ferroelectricity. In the previous section, we have shown that the phenomenological theory of Devonshire can account for the qualitative behavior of the physical properties (Figs. 7.29 and 7.30) of a number of ferroelectric crystals. Among these, barium titanate is most extensively investigated and the best understood. However, there are other ferroelectric materials which can hardly be treated even with such a theory in its simplest form. The main difficulty in any attempt to formulate a unified theory of ferroelectricity arises from the necessity of identifying the fundamental mechanism which is responsible for the onset of

Figure
7.31

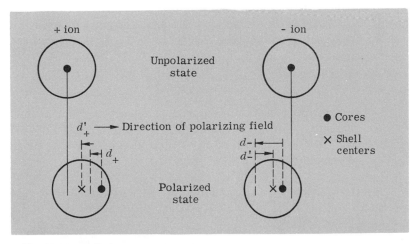

The polarized state of an alkali-halide crystal according to the model of Dick and Overhauser. The $d_\pm$ are the displacements of the shells with respect to the cores in the absence of repulsion between the ions. The $d'_\pm$ are the additional shell displacements resulting from repulsion.

spontaneous polarization. The recent work[28] of Anderson, Cochran, and Landauer in which they treat ferroelectricity as a problem of lattice dynamics is a major step toward establishing a unified theory.

The basic idea behind the approach is relatively simple. In the ideal case of zero damping, the dielectric constant of a crystal can be written in a general form[29] as

$$\epsilon(\omega) = \epsilon_\infty + \sum_i \frac{b_i{}^2}{\omega_i{}^2 - \omega^2} \qquad (7.137)$$

where ϵ_∞ is the high-frequency (optical) dielectric constant and the summation index i refers to the ith optical mode of lattice vibration. By comparing Eq. (7.137) with Eq. (7.25), we see that $\omega_i{}^2$ measures the strength of the restoring force and $b_i{}^2$ the strength of the coupling between the ith vibrational mode to the electric field. According to Eq. (7.137), the static dielectric constant can be written as

$$\epsilon_s = \epsilon_\infty + \sum_i \frac{b_i{}^2}{\omega_i{}^2} \qquad (7.138)$$

When the restoring force of one of the vibrational modes of the lattice becomes very small at a given temperature, the static dielectric constant can become very large.

Let us refer to Eq. (7.136). Based on the shell model of Dick and Overhauser, the value of γ in NaI has been calculated by Wood, Cochran, and Brockhouse[30] and is found to be about twice the value of the second term involving Ne^2. Furthermore, calculations by Wood *et al.* also show that the ω_t may approach zero without the crystal necessarily becoming unstable against other vibrational modes. Cochran[31] further extended the analyses and applied the results to ferroelectrics. When the two terms in Eq. (7.136) are nearly equal, we have

$$\frac{M\omega_t{}^2}{\gamma} = 1 - \frac{(\epsilon_\infty + 2\epsilon_0)N(ze)^2}{(3\epsilon_0)^2\gamma} = \eta(T - T_c) \qquad (7.139)$$

by expanding the temperature dependence of certain atomic parameters into a Taylor's series at $T = T_c$. Equation (7.139) differs from Eq. (7.136) in that an effective ionic charge of ze is assumed for sake of generality.

[28] P. W. Anderson, a paper presented at the Moscow Conference on Dielectrics, December, 1958; W. Cochran, *Phys. Rev. Letters*, vol. 3, p. 412, 1959; and R. Landauer and L. H. Thomas, *Bulletin Am. Phys. Soc.*, series II, vol. 4, p. 424, 1959.

[29] M. Born and K. Huang, "Dynamic Theory of Crystal Lattices," sec. 9, Clarendon Press, Oxford, 1958.

[30] A. D. B. Wood, W. Cochran, and B. N. Brockhouse, *Phys. Rev.*, vol. 119, p. 980, 1960.

[31] W. Cochran, *op. cit.*

Substituting Eq. (7.139) into Eq. (7.129), we find

$$\epsilon_s - \epsilon_\infty = \frac{(\epsilon_s + 2\epsilon_0)(\epsilon_\infty + 2\epsilon_0)}{(3\epsilon_0)^2} \frac{N(ze)^2}{\gamma\eta(T - T_c)} \tag{7.140}$$

Equation (7.140) is known as the Curie-Weiss law. The condition for ferroelectricity ($\epsilon_s \to \infty$) requires $T = T_c$ in Eq. (7.140), which in turn requires $\omega_t \to 0$ in Eq. (7.139). Therefore, the theory of ferroelectricity is intimately connected with the theory of lattice dynamics. Note that the condition for $\omega_t = 0$ from Eq. (7.139) can be written as

$$\frac{N(ze)^2}{3\epsilon_0\gamma} = 1 - \frac{\epsilon_\infty - \epsilon_0}{\epsilon_\infty + 2\epsilon_0} = 1 - \frac{N(\alpha_+ + \alpha_-)}{3\epsilon_0} \tag{7.141}$$

Realizing that $(ze)^2/\gamma$ is the ionic polarizability α_i [Eq. (7.25)] at zero frequency, Eq. (7.141) simply says

$$N(\alpha_i + \alpha_e)/3\epsilon_0 = 1 \tag{7.142}$$

Therefore, the condition for polarizability catastrophe [Eq. (7.54)] and the condition for $\omega_t = 0$ are the same. For a comprehensive review of the lattice-dynamics theory of ferroelectricity, the reader is referred to the article by Cochran.[32]

PROBLEMS

7.1 Find the magnitude and direction of the dipole moments associated with the following charge distributions: (a) $+ e$ at $(0, -a, 0)$ and $(0, 0, a)$, and $-e$ at $(0, a, 0)$ and $(0, 0, -a)$; (b) $+ 2e$ at $(0, 0, 0)$, and $-e$ at $(0, 0, -a)$ and $(0, 0, a)$.

7.2 Whether a molecule has a permanent dipole or not can be argued from symmetry considerations. For example, all diatomic molecules of the type AA (such as H_2, O_2, etc.), being electrically symmetrical, are nonpolar while those of the type AB (such as HCl, CO, etc.) are polar. By similar reasoning, determine whether the following molecules are polar or nonpolar: Triatomic molecules: CO_2, CS_2 (rectilinear arrangement ABA); H_2O, SO_2 (triangular arrangement ABA); HCN, KCN (linear arrangement ABC). Tetratomic molecules: BCl_3 (plane triangular AB_3 with A at center); NH_3 (triangular pyramid with A at apex). Pentatomic molecules: CH_4 (tetrahedral group AB_4 around A).

7.3 Consider a molecule having the form of a regular hexagon as shown. Find the dipole moment of the molecule. Now the molecule is compressed along the x direction so that the bond length a remains the same but the bond angle α changes from 120° to 120° − θ. Calculate the dipole moment of the distorted molecule for small θ.

[32] W. Cochran, Crystal Stability and the Theory of Ferroelectricity, in *Advan Phys*, vol. 9, p. 387, 1960.

P7.3

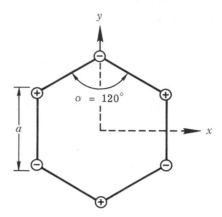

7.4 Using Eq. (7.2) and Poisson's equation div $\mathbf{D} = -\varrho$, show that the charge density ϱ can be separated into parts ϱ_1 and ϱ_2 of which ϱ_1 is the ordinary charge density due to free charges. The part $\varrho_2 = -\text{div } \mathbf{P}$ is due to polarization of charges and hence called the polarization charge density. By similar reasoning, the boundary condition at a vacuum-dielectric interface can be taken as consisting of two parts, the ordinary part and the part due to polarization of charges. Show that the surface charge density due to the latter is given by $\varrho_s = \mathbf{n} \cdot \mathbf{P}$ where $\mathbf{n}$ is an outward normal to the dielectric surface.

7.5 The electronic polarizability α_e of many ions can be empirically determined from the refractive indices of the substances which they form. According to J. Tessman, A. H. Kahn, and W. Shockley (*Phys. Rev.*, vol. 92, p. 890, 1953), the values of α_e of a selected number of alkali and halogen ions are given below (units in cgs $\times 10^{-24}$ cm³):

$$
\begin{array}{llllll}
\text{F}^- & 0.65 & \text{Cl}^- & 2.97 & \text{Br}^- & 4.17 \\
\text{Na}^+ & 0.41 & \text{K}^+ & 1.33 & \text{Rb}^+ & 1.98
\end{array}
$$

The value of α_e in MKS units is equal to the value quoted above multiplied by $4\pi\epsilon_0$. Find the refractive indices ($n = \sqrt{\epsilon/\epsilon_0}$) of NaCl and RbCl, using the Clausius-Mossotti equation. The distances between nearest Na atoms and Rb atoms are respectively 3.98 Å and 3.74 Å in NaCl and RbCl (CsCl structure). The static relative dielectric constants ϵ/ϵ_0 of NaCl and RbCl are 5.90 and 5.0. Explain the difference between n^2 and static ϵ/ϵ_0.

7.6 Using the Clausius-Mossotti equation, Eq. (7.55), show that the relative dielectric constant of a substance changes with frequency as

$$
\frac{\epsilon}{\epsilon_0} = 1 + \frac{Ne^2}{m\epsilon_0} \frac{1}{\omega_0^2 - \omega^2 + i\omega\beta/m - Ne^2/3m\epsilon_0}
$$

if it is assumed that the substance has electronic polarization only. From the value of α_e given in the previous problem, calculate ω_0 for Na$^+$ ion, taking e and m as the charge and mass of a free electron. Note that ϵ/ϵ_0 has a resonant frequency ω_1 which is experimentally measurable and different from ω_0. Estimate the error involved in taking ω_1 as ω_0 for the Na$^+$ ion in NaCl.

7.7 Show that the relative dielectric constant of an ionic crystal has a resonant frequency ω_1 related to ω_0 by

$$\left(1 - \frac{N\alpha_e}{3\epsilon_0}\right)(\omega_0{}^2 - \omega_1{}^2) = \frac{Ne^2}{3M\epsilon_0}$$

due to ionic displacements. Verify also that the static relative dielectric constant of an ionic crystal can be written as

$$\frac{\epsilon}{\epsilon_0} = 1 + \frac{3N(e^2/M\omega_0{}^2 + \alpha_e)}{3\epsilon_0 - N(e^2/M\omega_0{}^2 + \alpha_e)}$$

where M and ω_0 are the same as the symbols used in Eq. (7.25).

7.8 It is found experimentally from the absorption curve that $\omega_1 = 3.2 \times 10^{13}$ rad/sec in NaCl. Using the value of $N\alpha_e/3\epsilon_0 = 0.31$ and the results obtained in the previous problem, find the value of the static relative dielectric constant in NaCl. The masses of Na and Cl atoms are 23 and 35.5 times the mass of a proton. If the electronic charge e in the expressions derived in the previous problem is replaced by an effective electronic charge e^*, find the value of e^* such that the calculated static relative dielectric constant is equal to the experimental value of 5.90.

7.9 Show that the Clausius-Mossotti equation can be expressed in terms of the molecular weight M and the density D of a given substance as

$$\frac{M}{D} \frac{\epsilon - \epsilon_0}{\epsilon + 2\epsilon_0} = \frac{L\alpha}{3\epsilon_0}$$

where L is Avogadro's number. The quantity $L\alpha/3\epsilon_0$ is often referred to as the molar polarizability and is expressed in m³. It is found that the value of the molar polarizability changes from 6×10^{-5} at 400°K to 8×10^{-5} at 286°K in CH_3Cl and from 4×10^{-5} at 500°K to 6.8×10^{-5} at 300°K in H_2O. Calculate the dipole moment of CH_3Cl and H_2O.

7.10 Show that the electrostatic potential Φ at a point (r,θ,ϕ) due to a dipole p (with charge $+e$ at $z = +d$ and charge $-e$ at $z = -d$) along the z axis is

$$\Phi = \frac{p \cos \theta}{4\pi\epsilon_0 r^2}$$

P7.10

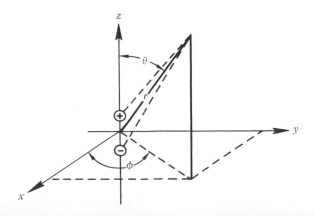

From Φ, evaluate the electric field components along $\mathbf{r}$, $\mathbf{\theta}$, and $\mathbf{\phi}$ directions. Check your result against Eq. (7.42).

7.11 A dielectric sphere of dielectric constant ϵ is placed in a uniform field E_0 as shown. First let us consider the case purely as a problem of electrostatics. Show that the potential functions Φ_0 outside the sphere and Φ inside the sphere as given by

$$\Phi_0 = E_0 r \cos \theta + \frac{A \cos \theta}{r^2} \qquad \Phi = Br \cos \theta$$

satisfy Laplace's equation in spherical coordinates

$$\frac{1}{r^2} \frac{\partial}{\partial r} \left(r^2 \frac{\partial \Phi}{\partial r} \right) + \frac{1}{r^2 \sin \theta} \frac{\partial}{\partial \theta} \left(\sin \theta \frac{\partial \Phi}{\partial \theta} \right) + \frac{1}{r^2 \sin^2 \theta} \frac{\partial^2 \Phi}{\partial \phi^2} = 0$$

By applying boundary conditions to Φ_0 and Φ, verify that the constants A and B are given by

$$A = - \frac{E_0(\epsilon - \epsilon_0)a^3}{\epsilon + 2\epsilon_0} \qquad \text{and} \qquad B = \frac{3E_0\epsilon_0}{\epsilon + 2\epsilon_0}$$

Using the definition of ϵ from Eq. (7.48) and that of $\mathbf{D}$ from Eq. (7.2), find the value of $\mathbf{P}$. Show that the value of $\mathbf{P}$ just found is consistent with $\mathbf{E} = \mathbf{E}_0 - N\mathbf{P}$ where the depolarization factor N is equal to $1/3\epsilon_0$ for a sphere.

P7.11

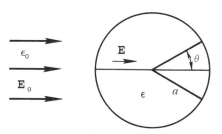

7.12 Consider an infinitely long dielectric cylinder placed in a uniform electric field E_0 with the direction of the field normal to the axis of the cylinder. Show that the potential functions Φ_0 outside the cylinder and Φ inside the cylinder given as follows

$$\Phi_0 = E_0 r \cos \theta + \frac{A \cos \theta}{r} \qquad \Phi = Br \cos \theta$$

satisfy Laplace's equation in cylindrical coordinates

$$\frac{\partial^2 \Phi}{\partial r^2} + \frac{1}{r} \frac{\partial \Phi}{\partial r} + \frac{1}{r^2} \frac{\partial^2 \Phi}{\partial \theta^2} + \frac{\partial^2 \Phi}{\partial z^2} = 0$$

By applying the boundary conditions to Φ and Φ_0, verify that $B = 2\epsilon_0 E_0/(\epsilon + \epsilon_0)$. Find the value of $\mathbf{P}$ inside the cylinder. Show that the electric field $\mathbf{E}$ inside the cylinder is given by $\mathbf{E} = \mathbf{E}_0 - N\mathbf{P}$ where $N = 1/(2\epsilon_0)$.

7.13 Consider an electric dipole **p** placed at the center of a fictitious cavity which is immersed in a dielectric medium with a dielectric constant ϵ. The situation is illustrated in Fig. 7.7*b*. Show that the electric potentials Φ_0 outside the cavity and Φ inside the cavity are given by

$$\Phi_0 = \frac{A \cos \theta}{r^2} \qquad \Phi = Br \cos \theta + \frac{p}{4\pi\epsilon_0} \frac{\cos \theta}{r^2}$$

where the polar angle θ refers to the direction of **p**.

Applying the boundary conditions to Φ_0 and Φ, show that the reaction field (represented by B in Φ) is given by Eq. (7.58).

7.14 Consider a hydrogen atom in the presence of an electric field E. The Schrödinger equation then becomes

$$\frac{\hbar^2}{2m} \nabla^2\psi + \left(\mathcal{E} + \frac{e^2}{4\pi\epsilon_0 r}\right)\psi + eE(r \cos \theta)\psi = 0$$

Explain the difference between the above equation and Eq. (1.68).

Show that the following solution

$$\psi = \left[1 - \frac{E4\pi\epsilon_0 a_1{}^2}{e} \cos \theta \left(\frac{r}{a_1} + \frac{r^2}{2a_1{}^2}\right)\right] \exp\left(-\frac{r}{a_1}\right)$$

satisfies the Schrödinger equation to the first order in E. The term $\exp(-r/a_1)$ is the wave function of the ground state and a_1 is the Bohr radius.

Evaluate the energy of the ground state in the presence of the electric field. Show that

$$\mathcal{E} = \mathcal{E}_0 - \frac{\alpha}{2} E^2$$

with $\alpha = (\tfrac{9}{2})\, 4\pi\epsilon_0 a_1{}^3$.

7.15 Consider the triangular arrangement of a water molecule as shown, in which the hydrogen ions are considered as being located at the shell of radius a. Find the expression for the electrostatic energy of the molecule. Show that for minimum energy

$$\sin^3 \theta = \frac{4\pi\epsilon_0 a^3}{8\alpha}$$

where α is the polarizability of the oxygen ion. Using the values of $\theta = 52.5°$ and $a = 1.02 \times 10^{-8}$ cm, find the value of α and the electric dipole moment associated with the water molecule.

P7.15

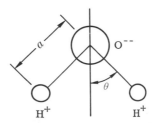

7.16 Read the book by W. P. Mason ("Piezoelectric Crystals and Their Applications to Ultrasonics," pp. 451–455, D. Van Nostrand Co., Inc., New York, 1950). Discuss the effect of crystal symmetry on the piezoelectric constant of a crystal. Illustrate with examples.

7.17 Consult the loose-leaf compilation by R. W. G. Wyckoff ("Crystal Structure," Interscience Publishers, New York, 1953). Find the crystal classes for the following crystals: CdS, GaAs, KH_2PO_4, quartz, and Rochelle salt. Refer either to the book by W. P. Mason (*ibid.*, pp. 41–46) or to the article on piezoelectric crystals (*Proc. IRE*, vol. 46, pp. 767–768, 1958). Write down the piezoelectric tensor for these crystals.

7.18 Refer to the book by F. Jona and G. Shirane ("Ferroelectric Crystals," pp. 376–378, Pergamon Press, London, 1962). Discuss how the structural changes in a ferroelectric crystal can be inferred from an analysis of the X-ray structure factor.

7.19 Find from Eqs. (7.110) and (7.111) the values of $\chi = dP/dE$ above and below the Curie temperature for case I.

7.20 Repeat Prob. 7.19 for case II.

7.21 Read the book by C. Kittel ("Introduction to Solid State Physics," pp. 192–194, John Wiley & Sons, Inc., New York, 1956). Explain the polarization catastrophe in $BaTiO_3$, based on Slater's calculation.

REFERENCES

Auty, R. P., and R. H. Cole: Dielectric Properties of Ice and D_2O, *Jour. Chem. Phys.*, vol. 20, p. 1309, 1952.

Cochran, W.: Crystal Stability and the Theory of Ferroelectricity, *Advan. Phys*, vol. 9, p. 387, 1960.

Devonshire, F.: Theory of Ferroelectrics, in *Advan. Phys*, vol. 3, p. 85, 1954.

Fröhlich, H.: "Theory of Dielectrics," Clarendon Press, Oxford, 1958.

Jona, F., and G. Shirane: "Ferroelectric Crystals," Pergamon Press, London, 1962.

Känzig, W.: Ferroelectrics and Antiferroelectrics, *Solid State Phys*, vol. 4, p. 1, 1954.

Kittel, C.: "Introduction to Solid State Physics," John Wiley & Sons, Inc., New York, 1956.

Mason, W. P.: "Piezoelectric Crystals and Their Applications to Ultrasonics," D. Van Nostrand Co., Inc., New York, 1950.

Megaw, H. D.: "Ferroelectricity in Crystals," Methuen and Co., Ltd., London, 1957.

Shirane, G., F. Jona, and R. Pepinsky: Some Aspects of Ferroelectricity, *Proc. IRE*, vol. 43, p. 1738, 1955.

Smyth, C. P.: "Dielectric Behavior and Structure," McGraw-Hill Book Company, New York, 1955.

Von Hippel, A. K.: "Dielectrics and Waves," John Wiley & Sons, Inc., New York, 1954.

8 THEORY OF MAGNETISM

8.1 MACROSCOPIC DESCRIPTION

In the macroscopic description of the electric properties of a substance, we have discussed in Chapter 7 the relationship between three physical quantities: the electric field intensity $\mathbf{E}$, the flux density $\mathbf{D}$, and the polarization $\mathbf{P}$. The three physical quantities which describe the macroscopic magnetic behavior of a substance are: the magnetic field intensity $\mathbf{H}$, the magnetic flux density $\mathbf{B}$, and the *magnetization* $\mathbf{M}$. As we recall, the electric polarization has its origin in the separation of the centers of negative and positive charge distributions from each other by a finite distance. Such a charge separation constitutes an electric dipole and the polarization represents the average density (per unit volume) of the electric dipole moment (a vector quantity) in a given direction. While the electric polarization has to do with the static positioning of the charges, the magnetization is due to the motion of these charges. It is instructive to demonstrate that an exact correspondence may be drawn between a magnetic dipole conceived of as two magnetic monopoles separated in space (which is the exact analogue of the electric dipole), and the magnetic dipole due to a current loop (which is the preferred physical model based on atomic structure).

Let us compare the magnetic field produced by a circular current loop with that by a magnetic dipole shown in Fig. 8.1. For simplicity, take point Q on the z axis. The magnetic field at Q is in the z direction in both cases and its magnitude is equal to

$$H_z = \frac{Ia^2}{2(a^2 + z^2)^{3/2}} \tag{8.1a}$$

for the current loop and

$$H_z = \frac{\psi}{4\pi(z - d)^2} - \frac{\psi}{4\pi(z + d)^2} \tag{8.1b}$$

for the magnetic dipole. If we let the radius of the current loop a and the magnetic dipole length $2d$ diminish while holding $I\pi a^2$ and $2\psi d$ constant, we find

$$\lim_{a \to 0} H_z = \frac{I\pi a^2}{2\pi z^3} \tag{8.2a}$$

453

for the current loop and

$$\lim_{d \to 0} H_z = \frac{2\psi d}{2\pi z^3} \qquad (8.2b)$$

for the magnetic dipole. We can extend the analysis to any arbitrary point $Q(x,y,z)$ and evaluate the magnetic field components (H_x, H_y, H_z) due to the current loop and the magnetic dipole. We find that the magnetic fields are the same everywhere if we set $I\pi a^2$ equal to $2\psi d$. (The general proof will be left to the reader in Prob. 8.2.) Thus, macroscopically, there is no distinguishable difference between a microscopic current loop and a microscopic magnetic dipole.

Next we shall examine the torque experienced by the current loop and the magnetic dipole in a uniform magnetic field. In Fig. 8.2, $\mathbf{B}$ is in the xz plane and the torque on the magnetic dipole about the y axis is given by $(T_y)_{\text{dipole}} = 2\psi d\, B \sin \theta$. For the case of the loop the z component of $\mathbf{B}$ produces a radial force and exerts no torque. The x component of $\mathbf{B}$ produces a downward force

$$dF = |\mathbf{I} \times \mathbf{B}\, dl| = IB_x(dl)_y = (IB \sin \theta)\, dy \qquad (8.3)$$

for $x > 0$ and a similar upward force for $x < 0$. Thus the torque is again about the y axis and is given by

$$(T_y)_{\text{loop}} = IB \sin \theta \int_{\text{loop}} x\, dy = IBA \sin \theta \qquad (8.4)$$

where $A = \pi a^2 =$ the area of the current loop.

From the above discussion, we conclude that a planar current loop is equivalent to a magnetic dipole so far as the magnetic effects both produced and experienced by them are concerned. Quantitatively,

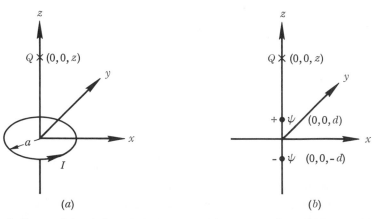

Figure 8.1

(a)   *(b)*

Equivalent models of elementary magnets: (a) a current loop in the xy plane; (b) a magnetic dipole pointing in the z direction. The figure is used to show the equivalence of the magnetic effects produced by (a) and (b).

Figure 8.2

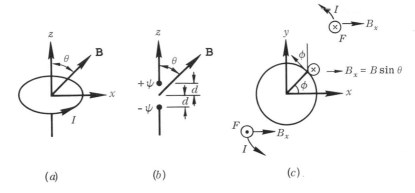

(a)   (b)   (c)

*A uniform magnetic flux density B is applied in the xz plane making an angle θ with the z axis: (a) situation applied to the current loop and (b) situation applied to the magnetic dipole. The situation with the current loop is further analyzed in diagram (c). Since the current loop is in the xy plane, the z component of **B** produces a force in the radial direction and hence does not contribute to the torque. Therefore, our attention is focused on the interaction between B_x and I_y.*

the *magnetic dipole moment* is given by

$$\mathbf{n}IA = \psi 2\mathbf{d}$$

where **n** is a unit vector normal to the plane of the loop as determined by the right-hand rule and 2**d** is the displacement vector of the dipole in the direction from the negative magnetic pole to the positive pole.

The macroscopic quantity **M**, called the magnetization, is defined as the average density per unit volume of the magnetic moment in a given direction. To find the relationship between **H**, **B**, and **M** let us illustrate with the simple case of a long solenoid of volume V which consists of N turns, each turn carrying a current I and having a width $2d$, as shown in Fig. 8.3. The magnetization as defined above is

Figure 8.3

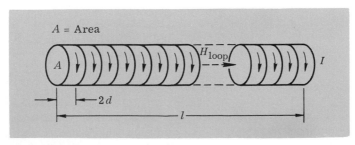

*A magnetic substance can be considered as consisting of many elementary current loops. The figure is used to find the relationship between **B**, **H**, and **M**.*

equal to

$$M = \frac{(IA)N}{V} = \frac{IAN}{Al} = \frac{I}{l}\frac{l}{2d} = \frac{I}{2d} \qquad \text{amp/m} \qquad (8.5)$$

The field inside the solenoid due to the circulating current is

$$H_{\text{loop}} = \frac{NI}{l} = \frac{I}{2d} = M \qquad \text{amp/m} \qquad (8.6)$$

The magnetic flux density B inside a magnetic substance is the sum of the flux produced by the external applied field and that by the internal magnetic dipole (i.e. the circulating current). Thus, we have

$$\mathbf{B} = \mu_0(\mathbf{H} + \mathbf{H}_{\text{loop}}) = \mu_0(\mathbf{H} + \mathbf{M}) \qquad \text{weber/m}^2 \qquad (8.7)$$

where μ_0 (henry/m) is the permeability of free space.

There are other ways of writing Eq. (8.7). Alternatively, we often see

$$\mathbf{B} = \mu_0\mu_r\mathbf{H} \qquad (8.8)$$

or

$$\mathbf{M} = (\mu_r - 1)\mathbf{H} = \chi\mathbf{H} \qquad (8.9)$$

where μ_r is the relative permeability and χ the *magnetic susceptibility* of the material. A word of caution is in order. In Eqs. (8.7) to (8.9), H is the external applied field, and all equations are vector equations. It should be pointed out that the field produced by the internal magnetic dipoles is not necessarily coincident in direction with and proportional to the external field. In that case, Eqs. (8.8) and (8.9) are no longer valid; and Eq. (8.7) is the only fundamental equation. However, for a large class of magnetic materials discussed later, we find that the net magnetization is either induced by the external field or due to a rearrangement caused by the external field of the orientation of the magnetic dipoles. Under such circumstances, the magnetization will be proportional to the external field but will not necessarily coincide with the applied field, and we can express the relation between $\mathbf{M}$ and $\mathbf{H}$ by a susceptibility tensor as

$$\begin{bmatrix} M_x \\ M_y \\ M_z \end{bmatrix} = \begin{bmatrix} \chi_{xx} & \chi_{xy} & \chi_{xz} \\ \chi_{yx} & \chi_{yy} & \chi_{yz} \\ \chi_{zx} & \chi_{zy} & \chi_{zz} \end{bmatrix} \begin{bmatrix} H_x \\ H_y \\ H_z \end{bmatrix} \qquad (8.10)$$

This reduces to Eqs. (8.8) and (8.9) when $\chi_{xx} = \chi_{yy} = \chi_{zz} = \chi$ and all the off-diagonal elements are zero.

8.2 PERMANENT AND INDUCED MAGNETIC DIPOLES

The fact that different materials have different magnetic properties can best be explained by examining what constitutes the magnetization in a given substance. We know that the electrons in an atom move in certain prescribed fashions about the nucleus of the atom, and in Chapter 1 we have described the electronic motion by assigning a certain angular momentum to each electron. The magnetic properties of a substance are determined by the state of the angular momentum which the electrons may possess.

To make an introductory connection between the magnetic moment and the angular momentum, we shall again use the classical model of circulating currents as shown in Fig. 8.4. Let us consider an electron with rest mass m_0 moving around the nucleus in a circular orbit with radius a and angular frequency ω. The angular momentum of the electron about the z axis is

$$\mathbf{J}_z = \mathbf{z} m_0 \omega a^2 \qquad (8.11)$$

where $\mathbf{z}$ is a unit vector in the z direction. Since the electron goes around every $2\pi/\omega$ second, the current generated by the electronic motion is given by

$$\mathbf{I}_\phi = -\boldsymbol{\phi} e\omega/2\pi \qquad (8.12)$$

where $\boldsymbol{\phi}$ is a unit vector in the ϕ direction. The magnetic moment associated with the electronic motion is thus given by

$$\mathbf{u}_z = -\mathbf{z}\pi a^2 I_\phi = -\mathbf{z}\frac{e\omega}{2}a^2 = -\frac{e}{2m_0}\mathbf{J}_z \qquad \text{amp-m}^2 \qquad (8.13)$$

A substance is said to possess permanent magnetic dipoles if the electrons of its constituent atoms have a net nonvanishing angular momentum.

Besides permanent dipoles, magnetic moments can also be induced by the application of an external magnetic field. We are familiar with the principle involved in the operation of a betatron in which the electrons are accelerated into a circular orbital motion by a time-varying magnetic field. Here we are interested in the induced magnetic moment due to the change induced in the orbital motion of the electron. The reader may wonder that since the force exerted by a steady magnetic field is always perpendicular to the path of an electron, no change in the electron motion probably would be expected by the application of a d-c magnetic field. However, we must remember that while the flux density B is building up from a value of zero to its steady-state value, a nonvanishing $\partial B/\partial t$ exists. Let

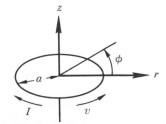

Figure 8.4

Classical model of a circulating electron. The model is used to establish the relation between the angular momentum and the magnetic dipole moment of the electron.

$B(t)$, the changing magnetic flux density, be in the z direction as shown in Fig. 8.5. Electrons describing a circular motion with radius a experience a tangential force

$$\mathbf{F} = \hat{\phi}(-e)E_\phi = \hat{\phi}(-e)\frac{-1}{2\pi a}\frac{d}{dt}(B\pi a^2) \tag{8.14}[1]$$

and the force in turn produces a tangential acceleration

$$\hat{\phi}\frac{dv_\phi}{dt} = \frac{\mathbf{F}}{m_0} = \hat{\phi}\frac{ea}{2m_0}\frac{dB}{dt} \tag{8.15}$$

By integrating Eq. (8.15), we have for the angular frequency

$$\omega_L = \omega - \omega_0 = \frac{v_\phi - v_{\phi 0}}{a} = \frac{eB}{2m_0} \tag{8.16}$$

where ω and ω_0 represent, respectively, the angular frequencies of the electron motion with and without B, and m_0 is the electron mass.

The induced orbital motion of the electron is commonly referred to as the *Larmor precession* and the angular frequency given by Eq. (8.16) as the *Larmor precession frequency*. The Larmor precession represents a slight modification of the electron motion by the action of a magnetic field. We further note that the Larmor frequency depends only on the final value of the magnetic flux density B irrespective of the process by which the final value of B is arrived at. The induced magnetic moment due to the precessional motion is given by

$$\mathbf{u}_z = \hat{z}\left(-e\frac{\omega}{2\pi}\right)(\pi a^2) = -\hat{z}\frac{e^2}{4m_0}a^2B \tag{8.17}$$

The induced magnetic moment is opposite to the applied magnetic field as expected from Lenz's law: that the induced current will produce a magnetic field tending to counteract the change of magnetic flux due to the original field.

In concluding this section, we add the following remarks. Equations (8.13) and (8.17) are derived on the basis of classical mechanics. These equations will be properly modified in later sections when we

[1] The symbol $\wedge$ indicates a unit vector.

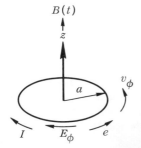

Figure 8.5

The effect induced by a magnetic flux density B upon the motion of an electron. The flux density B, in building up from a value of zero to its final value, induces a tangential electric field E_ϕ which acts on the electron to modify the electron motion. Note that the direction of the induced current produces a magnetic moment opposing **B**.

apply the modern theory of atomic structure to discuss the magnetic property of solids. However, based on Eqs. (8.13) and (8.17), we can make some estimate of the magnetic effects due to permanent and induced dipoles. The angular momentum of an electron according to modern atomic theory is measured in multiples of $h/2\pi$ where h is Planck's constant; thus the strength of a permanent magnetic dipole can be expressed in terms of an atomic unit which is called the *Bohr magneton* u_B:

$$1u_B = \frac{e}{2m_0} \frac{h}{2\pi} = 9.27 \times 10^{-24} \text{ amp-m}^2 \qquad (8.18)$$

According to Eq. (8.18), an elementary permanent dipole will have a strength of the order of 9×10^{-24} amp-m^2. The strength of an induced dipole is proportional to B. In making the estimate of its strength, let us take a as the Bohr radius for a hydrogen atom which according to Eq. (1.16) is equal to $a_1 = (\epsilon_0 h^2/\pi e^2 m_0) = 5.3 \times 10^{-11}$ m. For a magnetic flux density of 1 weber/m^2 which can be conveniently produced in the laboratory, we have

$$u = \frac{e^2}{4m_0} a_1{}^2 B = 2 \times 10^{-29} \text{ amp-m}^2 \qquad (8.19)$$

Thus, it is safe to say that for $B < 1$ weber/m^2, the strength of an induced dipole is many orders of magnitude smaller than that of a permanent dipole. It is worth carrying our preliminary discussion even further. We may ask how large the magnetization can get. There are approximately 10^{29} atoms/m^3 in a solid. If all the atoms possess a permanent dipole and all the dipoles are aligned in the same direction, then the total magnetization is equal to $M = 10^{29} \times 9 \times 10^{-24} = 9 \times 10^5$ amp/m which corresponds to a flux density of about 1 weber/m^2. These simple examples are only good for order-of-magnitude estimate; but they do confirm our earlier remark that different materials can have widely different magnetic properties. In the following sections, we shall explore more in detail the differences among different classes of magnetic materials.

8.3 CLASSIFICATION OF MAGNETIC MATERIALS

From the discussion at the end of the previous section, we see that the magnetic properties of a substance depend upon: (1) whether the substance has permanent dipoles and (2) if it does have permanent dipoles, how these dipoles are oriented with respect to each other. Based on these criteria, magnetic materials are generally classified into five categories: (1) *diamagnetic*, (2) *paramagnetic*, (3) *ferromagnetic*, (4) *antiferromagnetic*, and (5) *ferrimagnetic*.

In a diamagnetic substance, there are no permanent dipoles; hence the magnetic effects are very small. As we mentioned earlier, the induced magnetic moments always oppose the applied field; thus the diamagnetic susceptibility is negative. Moreover, it is almost temperature independent. In Eq. (8.17), all quantities are constants except that the orbital radius a may change slightly with temperature.

The atoms in a paramagnetic substance have a nonvanishing angular momentum, thus they possess permanent dipoles. However, in the absence of an applied field, the dipoles are randomly oriented and the net magnetization in any given direction is zero. We must remember that at a finite temperature, each atom has a thermal energy on the average equal to kT, and the resulting thermal agitation tends to randomize the dipole orientation. To align the dipoles in a given direction, we need a field strong enough to compete with the force due to thermal agitation. To state the problem differently, we may say that the energy involved in aligning an elementary dipole in the direction of $\mathbf{B}$ is equal to Bu and this energy must be greater than the thermal energy. At room temperature, $kT \approx 4.3 \times 10^{-21}$ joule, thus a field of the order of $kT/u \approx 5 \times 10^2$ weber/m^2 is required. The field commonly available in the laboratory is below 1 weber/m^2, far too small to turn the dipoles into the direction of the field. What the field does is to turn the dipoles slightly in the direction of the field so that on the average (statistically) there are slightly more dipoles pointing in the direction of $\mathbf{B}$ than those in the opposite direction. Thus, the net magnetization in a paramagnetic substance is due to a reorientation of the dipoles caused by the applied field. Since the applied field has to compete with thermal motion, paramagnetic susceptibility increases with decreasing temperature.

In a ferromagnetic substance, the permanent dipoles are strongly aligned in the same direction and consequently a large *spontaneous magnetization* results even in the absence of an applied field. Weiss postulated that the alignment is due to a very large internal field which is now referred to as the *Weiss molecular field*. The molecular field represents the interaction between neighboring magnetic ions. Since the interaction is of quantum-mechanical nature and has no classical counterpart, we shall explain the origin of this field in a later section. As long as the molecular field is strong enough to compete with thermal motion, nearly all the dipoles have parallel alignment and the substance is ferromagnetic. As the temperature is raised, thermal motion becomes stronger and fewer dipoles are aligned. Consequently, spontaneous magnetization decreases with increasing temperature. At a critical temperature, referred to as the Curie temperature, the internal field is no longer sufficient to overcome thermal agitation. Above the Curie temperature, the substance loses its spontaneous magnetization and becomes paramagnetic.

Many magnetic materials contain more than one kind of magnetic species because magnetic ions located at different crystal sites may behave differently. It is often convenient to divide magnetic ions according to their positions in the lattice into subgroups usually referred to as sublattices. We shall call the sublattices A and B. The important interaction to be considered here is the interaction between neighboring magnetic ions on sublattice A with those on sublattice B. The magnetic ions on lattice site A will be aligned parallel or antiparallel to those on lattice site B, depending on whether the molecular field is positive or negative. In antiferromagnetic and ferrimagnetic substances, the molecular field is negative, favoring an antiparallel arrangement. A one-dimensional model is illustrated in Fig. 8.6. If the magnetization on lattice site A, M_A, is exactly equal to that on lattice site B, M_B, then the net magnetization $M_A - M_B$ is zero and the substance is antiferromagnetic. If $M_A > M_B$, the resultant magnetization is equal to $M_A - M_B$ and the substance is ferrimagnetic. Both antiferromagnetic and ferrimagnetic substances have a critical temperature sometimes referred to as the *Néel temperature*. Below the critical temperature, the magnetic moments on both lattice sites have an orderly arrangement as shown in Fig. 8.6 while above the critical temperature, the magnetic moments become randomly oriented as in a paramagnetic substance.

In summary, the magnetic properties of a material depend on whether the substance possesses a permanent dipole or not. Although diamagnetic material usually refers to substances with no permanent dipoles, magnetic moments induced by an applied magnetic field do exist in all substances and so does diamagnetism. However, in substances having permanent dipoles, magnetic effects due to permanent dipoles usually mask those due to induced dipoles. Therefore, we may neglect diamagnetic effects in paramagnetic substances so far as the permeability is concerned. In substances possessing permanent dipoles, two physical states, ordered and disordered, exist regarding the orientational arrangements of the dipoles. Paramagnetism represents a state of disorderly arrangement. In ferromagnetic, antiferromagnetic, and ferrimagnetic substances, there is a strong internal field to align the magnetic moments in an orderly fashion. The state

Figure 8.6

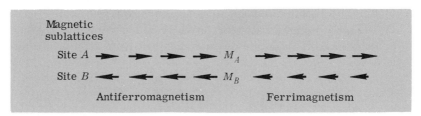

A one-dimensional schematic representation of the direction of magnetic dipoles in antiferromagnetic and ferrimagnetic materials.

of order prevails below a certain critical temperature. The critical temperature, thus, is a measure of the strength of the internal field; the higher the critical temperature, the stronger the field.

8.4 QUANTUM-MECHANICAL DESCRIPTION

In Sec. 1.7, we discussed the motion of an electron in a hydrogen atom in the light of quantum mechanics, and in this section we shall further review the fundamentals of quantum mechanics pertinent to subsequent discussions of the magnetic properties of materials. As we recall, the electron motion in quantum mechanics is described by the proper wave function which can be written as

$$\psi(r,\theta,\phi) = R(r)\Theta(\theta)\Phi(\phi) \tag{8.20}$$

and the product $\psi^*\psi$ is interpreted as the probability of finding the electron at a given position (r,θ,ϕ). To refresh our memory [Eqs. (1.74), (1.77), and (1.79)], we further recall that the functions $\Theta(\theta)$ and $\Phi(\phi)$ respectively have the following form:

$$\Theta(\theta) = P_l^m (\cos\theta) \tag{8.21}$$

and

$$\Phi(\phi) = A\cos m\phi + B\sin m\phi \tag{8.22}$$

where $P_l^m (\cos\theta)$ is the associated Legendre polynomial. Since the product $\psi^*\psi$ physically represents a distribution function for electrons, the function ψ must be single-valued and finite. Under such conditions, that is for ψ to behave properly, both l and m must be integers and $|m| \leq l$. Mathematically, we have [Eqs. (1.97) and (1.98)]:

$$l = 0, 1, \ldots (n-1) \tag{8.23}$$

$$m = -l, -(l-1), \ldots -1, 0, 1, \ldots l \tag{8.24}$$

where n is the principal quantum number, l the azimuthal quantum number, and m the magnetic quantum number.

Let us now get acquainted with the physical properties of the wave function. The radial function R determines the average distance of the electron from the nucleus. For our present discussion of the magnetic property of solids, we need not know the exact form of $R(r)$. On the other hand the other two functions $\Theta(\theta)$ and $\Phi(\phi)$ tell us the precise magnetic state of the electron. Hence, we list here a few of the associated Legendre polynomials for future reference:

$$P_0^0 (\cos\theta) = 1$$
$$P_1^0 = \cos\theta \qquad P_1^1 = \sin\theta \tag{8.25}$$
$$P_2^0 = \tfrac{1}{2}(3\cos^2\theta - 1) \qquad P_2^1 = 3\sin\theta\cos\theta \qquad P_2^2 = 3\sin^2\theta$$

The reader is asked to check the above functions against Eqs. (1.79) and (1.81). For $l = 0$, m must also be zero, thus ψ depends upon r

only, so the wave function is spherically symmetrical. The situation for $l = 1$ is more complicated and it needs some clarification.

From the theory of linear differential equations, we know that the solution of a linear differential equation can be expressed in any linear combination of its independent solutions. Thus, a particular form of the solution is not unique. However, two things should be kept in mind. First, the number of independent linear combinations must be equal to the number of independent solutions. Second, there is a definite physical significance attached to a given form of solution; in other words, the solution represents a physical state. Hence, once the final form of the solution is chosen, we cannot arbitrarily change its form without changing it from one physical state to another.

Let us now take the $l = 1$ state as an example. There are three possible values of m. For $m = 0$, we have $\Phi(\phi) = \text{const}$ and for $m = \pm 1$, we have $\sin \phi$ and $\cos \phi$ as two independent solutions for $\Phi(\phi)$. Combined with $\Theta(\theta)$, the product of $\Theta(\theta)\Phi(\phi)$ has three independent solutions: (1) $\cos \theta$, (2) $\sin \theta \cos \phi$, and (3) $\sin \theta \sin \phi$. Remembering that in rectangular coordinate,

$$r \sin \theta \cos \phi = x \qquad r \sin \theta \sin \phi = y \qquad r \cos \theta = z$$

we can rewrite the solutions in terms of x, y, and z simply as

$$\psi_A = xf(r) \qquad \psi_B = yf(r) \qquad \text{and} \qquad \psi_C = zf(r) \qquad (8.26)$$

where $r^2 = x^2 + y^2 + z^2$ and $f(r) = R(r)/r$. Note that the distributions represented by ψ_A, ψ_B, and ψ_C are no longer spherically symmetrical but have loops directed along the x, y, and z axes respectively. To help visualize qualitatively the charge distribution for $l = 1$ and $l = 0$, the angular dependence of functions ψ_A, ψ_B, ψ_C evaluated for a hydrogen atom is illustrated in Fig. 8.7. We can immediately see that if there are electrostatic charges distributed unevenly along the x, y, and z axes, the three states ψ_A, ψ_B, and ψ_C will have different electrostatic energies. In a later section, we shall discuss the effect of this electrostatic energy upon the magnetic properties of a substance.

Figure 8.7

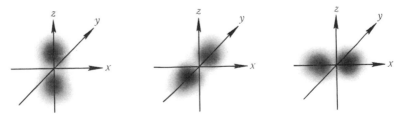

Schematic representation of the three electron distributions in the p state ($l = 1$). These distributions are given by ψ_C, ψ_B, and ψ_A of Eq. (8.26) in that order.

In the previous paragraph, we have tacitly stated that the functions sin ϕ and cos ϕ correspond to $m = \pm 1$ and yet tactfully avoided correlating the function with a specific value of m. The reason for doing so will become obvious shortly. In Sec. 1.7, as we recall, the differential equation in ϕ has the following form:

$$\frac{d^2\Phi}{d\phi^2} + m^2\Phi = 0 \tag{8.27}$$

The proper solution or the normal mode of Eq. (8.27) is

$$\Phi = \exp(im\phi) \tag{8.28}$$

Thus, for $m = 1$ $\qquad \Phi = \exp(+i\phi) \tag{8.29}$

and for $m = -1$ $\qquad \Phi = \exp(-i\phi) \tag{8.30}$

Note that sin ϕ and cos ϕ are linear combinations of Eqs. (8.29) and (8.30), that is to say, they represent a mixture of the $m = +1$ and $m = -1$ states. The proper wave functions which represent the pure $(l = 1, m = -1)$, $(l = 1, m = 0)$ and $(l = 1, m = +1)$ states are, respectively,

$$\psi_{-1} = \frac{1}{\sqrt{2}}(\psi_A - i\psi_B) \qquad \psi_0 = \psi_C$$

and $\qquad \psi_{+1} = \frac{-1}{\sqrt{2}}(\psi_A + i\psi_B) \tag{8.31}$

The factor $1/\sqrt{2}$ in Eq. (8.31) comes from the normalization process which makes

$$\int \psi_{-1}^* \psi_{-1}\, dx\, dy\, dz = \int \psi_A^* \psi_A\, dx\, dy\, dz = 1 \tag{8.32}$$

and so on, as required by the condition that the total probability of finding an electron should be unity. In Eq. (8.32), the integration limits are from $-\infty$ to ∞ over dx, dy, and dz.

The next question naturally arises as to how we can derive some information from the wave function about the physical state of the electron, for example the position and the momentum. We also wish to know whether the quantity m has any physical meaning. As stated in Sec. 1.5, since the product $\psi^*\psi$ is interpreted as the probability of finding an electron at a given position, the quantity $\langle \mathbf{r} \rangle$, defined as follows,

$$\langle \mathbf{r} \rangle = \int \psi^* \mathbf{r} \psi\, dx\, dy\, dz \tag{8.33}$$

represents the expectation value of the position of the electron. Similarly, if $\mathbf{p}$ and $\mathbf{L}$ represent the linear and angular momentum respectively, then their expectation values are

$$\langle \mathbf{p} \rangle = \int \psi^* \mathbf{p} \psi\, dx\, dy\, dz \tag{8.34}$$

and $\qquad \langle \mathbf{L} \rangle = \int \psi^* \mathbf{L} \psi\, dx\, dy\, dz \tag{8.35}$

In the theory of quantum mechanics, the expectation values have the same physical meaning as their counterparts in classical mechanics. In quantum mechanics, however, the quantities **p** and **L** are linear differential operators; therefore, in Eqs. (8.34) and (8.35), the order of multiplication must be maintained. It is also worth noting that since any physical quantity must be a real number, it is necessary to have the complex conjugate of the wave function ψ^* together with the wave function ψ in Eqs. (8.33) to (8.35). The above discussion is actually a recapitulation of the highlights of our discussion in Secs. 1.5 and 1.7, which are relevant to our present discussion of the magnetic properties of solids.

As far as the magnetic properties are concerned, the important physical quantity is the angular momentum. We shall show presently that the magnetic quantum number m is associated with the z component of angular momentum, the azimuthal quantum number l with the orbital angular momentum of an electron. The classical angular momenta are given by

$$L_x = yp_z - zp_y \qquad L_y = zp_x - xp_z \qquad L_z = xp_y - yp_x \qquad (8.36)$$

and the quantum mechanical counterparts are to replace p_x by $(\hbar/i)(\partial/\partial x)$, etc., in Eq. (8.36). Since we choose the z direction as the direction of the applied magnetic field, we are interested to know what is the expectation value of L_z in that direction. We shall proceed as follows:

$$L_z\psi_0 = \frac{\hbar}{i}\left(x\frac{\partial}{\partial y} - y\frac{\partial}{\partial x}\right)zf(r) = 0\psi_0$$

$$L_z\psi_{-1} = \frac{\hbar}{i}\left(x\frac{\partial}{\partial y} - y\frac{\partial}{\partial x}\right)\frac{1}{\sqrt{2}}[xf(r) - iyf(r)] = -\hbar\psi_{-1} \qquad (8.37)$$

$$L_z\psi_{+1} = \frac{\hbar}{i}\left(x\frac{\partial}{\partial y} - y\frac{\partial}{\partial x}\right)\frac{-1}{\sqrt{2}}[xf(r) + iyf(r)] = \hbar\psi_{+1}$$

noting that $(\partial/\partial x)f(r) = f'(r)(\partial r/\partial x) = f'(r)(x/r)$, etc. It is quite obvious from Eq. (8.37) together with Eqs. (8.32) and (8.35) that the expectation values of L_z associated with ψ_{-1}, ψ_0, and ψ_{+1} are respectively $-\hbar$, 0, and $\hbar$. Thus, for $l = 1$, we have three possible values for m: $-1, 0$, and 1; and the value of L_z is quantized in units of $\hbar$ as indicated. Next, we want to know the value of total angular momentum. It can easily be shown that the expectation value of $L^2(= L_x{}^2 + L_y{}^2 + L_z{}^2)$ for $l = 1$ is equal to $2\hbar^2$ associated with ψ_{-1}, ψ_0, and ψ_{+1}. The proof is left to the reader as a problem (8.5).

Our conclusion can be extended to other values of l. In general, we will find that there are $2l + 1$ proper wave functions, each being associated with a specific value of m ranging from $m = -l$ to $m = l$. The expectation value of L^2, however, is the same for all the $2l + 1$ wave functions and it is $l(l + 1)\hbar^2$. Thus, we may view the different values of m merely as the projection of the l vector upon the z axis (Fig. 8.8a). Because of quantization, the projection is changed in steps of unity. It is also to be noted that the magnitude of L is $\sqrt{l(l+1)}\ \hbar$ but the maximum value of L_z is $l\hbar$. The difference is a direct consequence of the quantum theory.

The Schrödinger equation which we have just considered accounts completely for the effects of the orbital motion of an electron. In addition to its orbital motion, however, the electron is considered as having a spinning motion. We may visualize[2] the electron as a minute solid rather than a dimensionless point and the spinning motion about its own axis as similar to that of a gyroscope. Thus, over and above the orbital angular momentum around the nucleus, an electron possesses an additional angular momentum about its own axis of spinning. It turns out that the spin angular momentum is quantized in one half-unit of $\hbar$. If we denote s as the spin quantum

[2] As mentioned earlier, in Sec. 1.8, it has been shown by Dirac that the electron spin is a purely relativistic property of an electron. For atomic theory and discussion of magnetic properties, however, the classical concept of an electron having a spinning charge distribution can still be retained without special inquiry into the origin of spin.

Figure 8.8

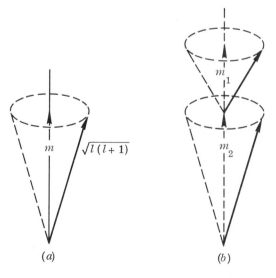

(a)   (b)

*Vector model showing the relation between the magnetic and azimuthal quantum numbers. (a) The magnetic quantum number m can be considered as the projection of the azimuthal quantum number l upon the z axis. (b) The diagram illustrates the use of Eq. (8.41b). If m_1 and m_2 are the maximum values of the magnetic quantum numbers allowed for the two electrons in Eq. (8.41b), the maximum projection of **L** upon the z axis is $|m_1 + m_2|$ and hence the value of L is given by $|m_1 + m_2|$.*

number, then the angular momentum associated with the spin is $\pm\frac{1}{2}\hbar$, that is to say $s_z = \frac{1}{2}$ or $-\frac{1}{2}$. In the case of the orbital quantum number, the degree of multiplicity is $(2l + 1)$-fold; that means, for each value of l there are $(2l + 1)$ possible values for m. In the case of spin, the multiplicity is $(2s + 1)$-fold; thus for $s = \frac{1}{2}$, we have two states $\pm\frac{1}{2}$.

In summary, the atomic state in the light of the quantum theory is based on four quantum numbers. They are (1) the principal quantum number n, (2) the azimuthal quantum number l, which determines the total orbital angular momentum, (3) the magnetic quantum number m, which determines the component of orbital angular momentum along the direction of the applied magnetic field, and (4) the spin quantum number s, which determines the angular momentum due to the spinning motion. The magnetic properties of a solid depend on what atomic states its constituent atoms may occupy. This constitutes the subject of discussion in the following section.

8.5 ATOMIC STATE AND MAGNETIC MOMENT[3]

The purpose of this section is to make connections between the various quantum numbers which an electron has and the magnetic moment of that electron. In the case of orbital motion, the angular momentum is quantized in units of $h/2\pi$ so the magnetic moment due to orbital motion is measured in units of Bohr magnetons. Let us assume that the external field is in the z direction; then according to Eq. (8.13) we have

$$u_z = -\frac{e}{2m_0}\frac{h}{2\pi}m = -mu_B \qquad (8.38a)$$

where u_B is the Bohr magneton as given by Eq. (8.18). For $l = 1$, there are three possible values for m: 1, 0, and -1. In the case of spinning motion, however, the relation between angular momentum and magnetic moment can no longer be derived on the basis of a classical model as outlined in Sec. 8.2. The spin quantum number s_z is quantized in half integers, that is $\frac{1}{2}$ or $-\frac{1}{2}$, but the elementary magnetic moment associated with electron spin is found to be an integral multiple of the Bohr magneton. For example: the energy split between the $s_z = \frac{1}{2}$ and $s_z = -\frac{1}{2}$ states in a magnetic field H is found to be $2\mu_0 u_B H$ in contrast to $\mu_0 u_B H$ as predicted by the classical model. Thus, the electron spin, for the same amount of angular momentum, contributes twice the amount of magnetic moment as the orbital motion. Quantitatively, we have

$$u_z = -2\frac{e\hbar}{2m_0}s_z = -2u_B s_z \qquad (8.38b)$$

[3] The reader may find it helpful to refer to our discussion in Sec. 1.9 in connection with Table 1.2, p. 38.

The magnetic properties of a substance depend on the atomic states of its constituent atoms. The situation is rather complicated because there exist various interactions which we have not taken into consideration. First, even in the atomic state, if an atom has two or more electrons, we have to know precisely how the magnetic moments due to the spin and orbital momentum of each individual electron combine. Second, electrons in a solid may behave quite differently as a result of interactions between electrons among themselves and between electrons and neighboring ions. An extensive treatment of the subject is clearly beyond the scope of this book. However, we shall attempt to illustrate with some simple examples the principles involved in treating complex cases.

The simplest example is the atom of an inert gas, say He. The ground state of an He atom is $1s^2$. This means that $n = 1$ and $l = 0$ so the two electrons do not possess any orbital momentum. As for spin, one electron has $s_z = \frac{1}{2}$ and the other $s_z = -\frac{1}{2}$, so the resultant is again zero. The fact that one electron has $s_z = \frac{1}{2}$ and the other $s_z = -\frac{1}{2}$ is required by the Pauli exclusion principle which states that no two electrons can have the same set of quantum numbers in an atom. If we assign $s_z = \frac{1}{2}$ or $s_z = -\frac{1}{2}$ simultaneously to both electrons, we obviously are violating the exclusion principle. Next we consider Ne. Its ground state is given by $1s^2\, 2s^2\, 2p^6$. For the $2s$ electrons, $n = 2, l = 0$, and by the same argument we conclude that the resultant m and s are zero. For the $2p$ electrons, $n = 2, l = 1$. If we again attempt to assign values of magnetic and spin quantum numbers to the electrons, we find that the only combination consistent with the Pauli exclusion principle is the following:

Electrons	i	ii	iii	iv	v	vi
m	-1	-1	0	0	1	1
s_z	$-\frac{1}{2}$	$\frac{1}{2}$	$-\frac{1}{2}$	$\frac{1}{2}$	$-\frac{1}{2}$	$\frac{1}{2}$

Thus, for the ground-state neon atom, there are no net magnetic moments. We can extend our arguments to all atoms and ions which have complete-shell configurations and conclude that they are diamagnetic. Since a complete shell represents the most stable structure in a chemical bond, diamagnetism includes a large class of materials. This applies to A, Kr, Cl^-, Na^+, Ca^{2+}, CH_4 for example.

For diamagnetic substances, the induced magnetic moment derived on the basis of a classical model can easily be modified in the light of quantum mechanics. In Eq. (8.17), a^2 represents the expectation value of $x^2 + y^2$. If $\langle r^2 \rangle$ is the mean square distance of the electron from the nucleus, then $\langle r^2 \rangle = \langle x^2 + y^2 + z^2 \rangle$. For a spherically-symmetric charge distribution, $a^2 = \frac{2}{3}\langle r^2 \rangle$ and for z electrons, the diamagnetic susceptibility is

$$\chi = \frac{M}{H} = -\frac{e^2 z}{6m_0} \mu_0 \langle r^2 \rangle N \tag{8.39}$$

where N is the number of atoms per unit volume.

In contrast to diamagnetic materials, permanent magnetic dipoles exist in substances whose constituent ions or atoms have incomplete shells so that they possess some uncancelled orbital momentum and unpaired spin. Some arbitrary examples are: free Na atoms, Fe, Co, Fe^{2+}, Mn^{2+}, Gd^{3+}, Sm^{3+}, to name a few. The situation with substances having permanent dipoles, however, is much more complicated. To be specific, take the Mn^{2+} ion as an example. It has an electronic configuration $1s^2 2s^2 2p^6 3s^2 3p^6 3d^5$, that is, the argon-core configuration plus five $3d$ electrons. The argon-core electrons, behaving in the same way as in argon, do not contribute to any permanent dipole. With the five $3d$ electrons, two important questions arise: (1) Of the ten quantum states available to the $3d$ electrons, which five states should be assigned to the five $3d$ electrons of Mn^{2+} ion? (2) Given the five quantum states to the Mn^{2+} ion, how should the contributions to the magnetic moment be added? Before we can answer these questions, we must know the relative importance of various interactions which may exist between electrons belonging to a given quantum state.

Such basic understanding of material properties is not only of theoretical interest but of practical importance. The recent rapid advance in technology has received its impetus from discoveries in the basic sciences. A striking example of such an advance in the field of magnetics is the development of the maser, a solid-state device capable of amplifying extremely small signals because of its low inherent noise level. In order to be able to select materials for maser operation, we must understand how the constituent atoms in a solid behave. We shall actually discuss the principle of a maser in the next chapter. Our present task is to point the way, step by step, in the order of complexity toward understanding the complicated magnetic behavior of a solid. In this section, we have discussed how the magnetic state of a single electron in a free atom is related to the various quantum numbers of the electron. In the next section, we shall introduce the theory of atomic spectroscopy and explain how we can treat the combined magnetic effects of several electrons in a free atom. Following that, we shall discuss the magnetic behavior of the ions of the rare earth and iron group in a solid. The reason that we choose these ions is because information regarding these ions is relatively abundant as a result of many theoretical studies and experimental investigations.

8.6 ATOMIC SPECTROSCOPY

When an atom has more than one electron, the question arises as to how the electrons will occupy the various quantum states available to them. Since the situation with a closed shell has been discussed in the previous section, we shall concentrate our attention on those electrons outside closed shells. The simplest atomic model is that of hydrogen and hydrogen-like ions which have only a single electron outside the closed-shell configuration. As pointed out earlier in Sec. 1.8, the energy of an atomic level depends not only on the quantum numbers n and l (Eq. 1.105) but also on the alignment of the spin with respect to the orbital angular momentum. As a result, a doublet appears with $j_1 = l + \frac{1}{2}$ and $j_2 = l - \frac{1}{2}$. The energy split between the doublet is a direct consequence of the *spin-orbit interaction*, that is, the interaction of the magnetic moment of the electron spin with the effective magnetic field set up by its orbital motion around the nucleus. The energy involved in the spin-orbit interaction can be represented by a term $\xi \mathbf{l} \cdot \mathbf{s}$ where $\xi(r)$ is given by Eq. (12.44). For our present qualitative discussion, however, the parameter ξ can simply be considered as a proportionality constant.

In addition to the interaction of the orbital motion of an electron with its own spin, electrons can also interact with each other through the coulomb electrostatic forces. Therefore, for an atom having N electrons and an effective nuclear charge ze, the energy can be found from a generalization of Eq. (1.67), leading to

$$\mathcal{H} = \sum_{i=1}^{N} \left[\frac{1}{2m_0} p_i{}^2 - \frac{ze^2}{4\pi\epsilon_0 r_i} + \xi(r_i)\mathbf{l}_i \cdot \mathbf{s}_i \right] + \sum_{i \neq j = 1}^{N} \frac{e^2}{4\pi\epsilon_0 r_{ij}} \quad (8.40)$$

where the term $e^2/4\pi\epsilon_0 r_{ij}$ represents the coulomb interaction energy between the ith and the jth electron, r_{ij} being the distance between the two electron distributions. The way in which electrons select their quantum states is to minimize the energy. It turns out that for orbits with low principal quantum number n, the coulomb interaction energy is larger than the spin-orbit interaction energy. Therefore, the matter of first-order importance is to avoid putting electrons into states with identical orbits. In other words, even though two electrons are allowed for a given azimuthal quantum number l and magnetic quantum number m on account of two possible spin orientations, the two electrons energetically will prefer to have two different values of m.

As discussed in Sec. 8.4, the allowable values of m can be considered as the projection of the l vector upon a certain direction of quantization. This is illustrated in Fig. 8.8a. On top of the orbital angular momentum, we must add the spin angular momentum. The question naturally arises as to the order in which these vectors should be added. This is again decided by energy considerations. In cases where the coulombic interaction is much stronger than the spin-orbit interaction, the spins on different electrons should be added together and so should the orbital angular momenta. Thus, for two electrons, we have

$$\mathbf{S} = \mathbf{s}_1 + \mathbf{s}_2 \tag{8.41a}$$

$$\mathbf{L} = \mathbf{l}_1 + \mathbf{l}_2 \tag{8.41b}$$

where S and L represent, respectively, the total spin and azimuthal quantum numbers. Since the orbital angular momentum and the azimuthal quantum number differ only by a factor $\hbar$, we have used L to represent both quantities in order to avoid the use of too many symbols. The same practice applies to S.

The equation for $\mathbf{L}$, Eq. (8.41b), needs further clarification. According to the vector model of Fig. 8.8b, the maximum quantized value of $\mathbf{L}$, represented by M, is $m_1 + m_2$. Hence, Eq. (8.41b) can be replaced by a scalar equation $L = |m_1 + m_2|$ where m_1 and m_2 are the maximum values of m allowed for the two electrons consistent with the exclusion principle. Take the combination of two $2p$ electrons as an example. The various combinations can be classified into three groups with $L = 0$, 1, and 2 in spectroscopic terms. The value of $L = 1$ is obtained by taking $m_1 = 1$ and $m_2 = 0$ in Eq. (8.41b). For $L = 2$, $m_1 = 1$ and $m_2 = 1$; for $L = 0$, $m_1 = 1$ and $m_2 = -1$.

If no external field is applied or if the external field is not strong enough to break the spin-orbit coupling, then the spin vector $\mathbf{S}$ and the orbital momentum vector $\mathbf{L}$ can be compounded to form a resultant $\mathbf{J}$, the value of which is often called the inner quantum number

$$\mathbf{J} = \mathbf{S} + \mathbf{L} \tag{8.42}[4]$$

The quantity $\mathbf{J}$ represents the total angular momentum of the atom. The coupling scheme represented by Eq. (8.42) is called the *Russell-Saunders coupling*[5] and it is the only case which we shall consider here. Under the Russell-Saunders coupling, it is assumed that the various interactions have their relative importance listed in the following order: (1) the electrostatic interaction between electrons, (2) the magnetic interaction between the spin and orbital magnetic moments of an electron, and (3) the magnetic interaction of the magnetic moment of an electron with the applied magnetic field.

---

[4] Here we use J to represent both the inner quantum number and the total angular momentum. These two quantities differ by a factor $\hbar$. The reader can easily determine which quantity J represents by referring to the context.

[5] For a more thorough and advanced treatment, the reader is referred to E. U. Condon and G. H. Shortley, "The Theory of Atomic Spectra," Cambridge University Press, Cambridge, 1959. Also J. H. Van Vleck, "The Theory of Electric and Magnetic Susceptibilities," Oxford University Press, London, 1932.

The said coupling scheme applies well to the ground states of free atoms with low and moderate atomic numbers.

Note that Eqs. (8.41) and (8.42) are in vector form. Therefore, it is essential to know the rules governing the addition of the vectors in these equations. According to the Hund theory of spectral lines, the lowest energy (ground) state of an atom has the maximum multiplicity, which means the largest value of S. If there are several choices of L for the same multiplicity, the state with the greatest L value is the lowest energy state. The above statement is of course subject to restrictions imposed by the Pauli exclusion principle. The *Hund rule* further says that the resultant J value is equal to $L - S$ if the shell is less than half full and $L + S$ if the shell is more than half full.

For the following discussion, the spectroscopic notation adopted in Secs. 1.7 and 1.9 is used to describe the electron configuration in an atom. For example: a vanadium atom in the iron group has $1s^2 2s^2 2p^6 3s^2 3p^6 3d^3 4s^2$. To refresh our memory, the letters s, p, d, indicate the orbital quantum number; that is to say, $l = 0, 1, 2$ for s, p, d, respectively. The numerals preceding the letters give the principal quantum number and the numerals in the superscript the number of electrons occupying the state. Thus, there are two electrons with $n = 1$ and $l = 0$ in vanadium. The ground state of a free vanadium atom is represented by $^4F_{3/2}$. Since a complete shell does not contribute any net S or L, we only need to consider the $3d$ electrons. The maximum value of S is thus $3/2$ so the multiplicity is $2S + 1 = 4$ and it is indicated by the superscript. The maximum value of m allowed for the first electron is 2 for a $3d$ electron. Since we have assigned $s_z = 1/2$ to all three electrons, the maximum values consistent with the Pauli exclusion principle are 1 and 0 for the second and third electrons respectively. The combination of $m_1 = 2, m_2 = 1$, and $m_3 = 0$ gives the largest value of L of 3 and it is symbolized by the capital letter F with the understanding that S, P, D, F, G, H mean, respectively, $L = 0, 1, 2, 3, 4, 5$. The J value is denoted by the subscript, in this case $3/2$. Since the shell is less than half full, we have $J = L - S = 3/2$ in agreement with the spectroscopic notation.

In the presence of a weak magnetic field, the energy of an atom is quantized into $2J + 1$ levels. Since the spin and orbital motion contribute differently to the magnetic moment for the same amount of angular momentum, the energy split between successive magnetic levels will not be, in general, an integral multiple of $\mu_0 u_B H$. However, we can define an effective magnetic moment u_{eff} in the following manner:

$$u_{\text{eff}} = -\frac{e\hbar}{2m_0} g[J(J + 1)]^{1/2} = -g[J(J + 1)]^{1/2} u_B \qquad (8.43a)$$

whose projection upon the direction of the magnetic field is given by

$$u_z = -gJu_B \qquad (8.43b)$$

where g is called the *Landé splitting factor*. Since the change in J is quantized or $\Delta J = 1$, the splitting in energy between successive levels is

$$\Delta \varepsilon = -\mu_0 g u_B H = -\frac{\mu_0 H u_{\text{eff}}}{J} \tag{8.44}$$

For simplicity, we have taken the classical approach in making $[J(J + 1)]^{\frac{1}{2}}$ equal to J in the above equation and we shall make the necessary modification of the classical result later on. For given values of S, L, and J, the value of g is known. In the following, we shall again use the classical model of vector addition to find g.

Referring to Fig. 8.9 in connection with Eq. (8.38), the projection of the magnetic moment vector upon the J axis is equal to

$$u_{\text{eff}} = \frac{e\hbar}{2m_0} (L \cos \theta_{LJ} + 2S \cos \theta_{SJ}) \tag{8.45}$$

where θ_{LJ} and θ_{SJ} are the angles between the **J** vector and the **L** and **S** vectors, respectively. By the law of cosines, we have

$$\cos \theta_{LJ} = \frac{J^2 + L^2 - S^2}{2JL} \quad \text{and} \quad \cos \theta_{SJ} = \frac{J^2 + S^2 - L^2}{2JS} \tag{8.46}$$

Substituting Eq. (8.46) into Eq. (8.45) and comparing it with Eq. (8.43), we find

$$g = 1 + \frac{J^2 + S^2 - L^2}{2J^2} \tag{8.47}$$

From the correspondence principle, the quantum-mechanical g factor is given by

$$g = 1 + \frac{J(J + 1) + S(S + 1) - L(L + 1)}{2J(J + 1)} \tag{8.48}$$

Since the value of g determines the fine energy splitting in the spectral lines, it is also referred to as the *spectroscopic splitting factor*. Note that for $L = 0$, $g = 2$ while for $S = 0$, $g = 1$. Therefore, the value of g is a measure of the relative contribution of the spin and orbital motion to the magnetic moment.

Figure 8.9

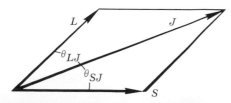

Vector model used to calculate the Landé splitting factor.

8.7 IONS OF THE RARE EARTH AND IRON GROUP[6]

In the previous section, we discussed the general principles related to determining the electronic configuration for the lowest energy state of a free atom. That is the atomic state. In a solid, as we mentioned earlier, a magnetic ion may be subject to electrostatic forces due to neighboring ions and hence its magnetic properties may be affected. There are places, however, in the periodic table where we may find elements of which the magnetic electrons are in an inner shell and consequently are well shielded by electrons in outer shells. One of such series is the rare earth group.

The elements in the rare earth group (Table 1.2) all have an incomplete $4f$ shell, a complete $5s^2 5p^6$ shell, and some outer valence electrons. In a compound, most of the rare earth group are trivalent, losing the valence electrons and if necessary one of the $4f$ electrons, but leaving the $5s^2 5p^6$ electrons intact. Take Pr^{3+}. Its ground-state configuration is $4f^2 5s^2 5p^6$, and only the $4f^2$ electrons contribute to permanent dipoles. Since the $4f^2$ electrons are well shielded from any electrostatic field by the $5s^2 5p^6$ electrons, they behave quite freely as in the atomic state of a gas. Thus, according to the Hund rule and the Russell-Saunders coupling, the basic level of Pr^{3+} in a salt is $^3H_4$, that is $S = 1$, $L = 5$, and $J = 4$; and the effective number of Bohr magnetons for Pr^{3+} is $g[J(J + 1)]^{1/2} = 3.58$. Similar calculations hold for other members of the rare earth group from La^{3+} to Lu^{3+}. Figure 8.10 shows a comparison of effective magnetic moments (in Bohr magnetons) calculated from Eq. (8.43a) with experimental values. A general agreement is obtained. The only exceptions are Sm^{3+} and Eu^{3+}. In these two cases, the excited states are relatively close to the ground state and hence they also contribute to magnetism. So far as the values of S, L, and J of the ground state of these ions are concerned, however, our discussion still holds.

The iron group constitutes an entirely different story. All the elements in the group have an incomplete $3d$ shell and two $4s$ electrons. In a compound, they lose their outer valence ($4s$) electrons and possibly one of the $3d$ electrons, depending on whether they are divalent or trivalent. Thus, the $3d$ electrons which are responsible for magnetism now become the outermost electrons and they are subject to a strong electric field due to neighboring ions. We shall discuss the role of the crystalline field presently.

It is to be remembered that a specific wave function represents a definite electronic charge distribution, and hence involves a definite amount of interaction energy with the crystalline field. For simplicity, we assume the crystalline electrostatic potential to be of the following form:

$$V = Ax^2 + By^2 + Cz^2 \tag{8.49}$$

---

[6]   For discussions in this section, the reader will find frequent references to Table 1.2, p. 38.

If the electron wave function is ψ, then we have

$$\Delta\mathcal{E} = \int\!\!\int\!\!\int_{-\infty}^{\infty} \psi^* e V \psi \, dx \, dy \, dz \tag{8.50}$$

as the interaction energy. For illustrative purpose, let us take the p state; and the three wave functions which we choose are ψ_A, ψ_B, and ψ_C of Eq. (8.26). Substituting these wave functions into Eq. (8.50), we find

$$
\begin{aligned}
(\Delta\mathcal{E})_A &= AI_1 + (B + C)I_2 \\
(\Delta\mathcal{E})_B &= BI_1 + (C + A)I_2 \\
(\Delta\mathcal{E})_C &= CI_1 + (A + B)I_2
\end{aligned}
\tag{8.51}
$$

where $I_1 = e\int\!\int\!\int x^4 |f(r)|^2 \, dx \, dy \, dz$ and $I_2 = e\int\!\int\!\int x^2 y^2 |f(r)|^2 \, dx \, dy \, dz$. It is obvious that the value of I_1 is the same if x is replaced by y or z and the value of I_2 is the same if xy is replaced by yz or zx.

From Eq. (8.51) we see that the energies represented by $(\Delta\mathcal{E})_{A,B,C}$ will be different if A, B, and C are different. In other words, the $2L + 1$ levels which have the same energy (*degenerate levels*) in a free atom are split up in a crystalline field. When the energy split, that is, $(\Delta\mathcal{E})_A - (\Delta\mathcal{E})_C$ for example, is large, it will produce two important effects. First, the spin-orbit coupling is broken up so that $\mathbf{S}$ and $\mathbf{L}$ will be separately quantized in the direction of the field. Second, the three components of orbital angular momentum L_x, L_y, and L_z are essentially quenched by the crystalline field. Here we need a brief explanation of what is meant by quenching.

Figure 8.10

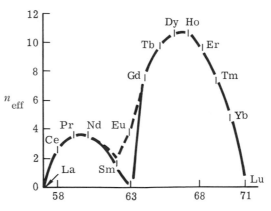

Effective magnetic moments of trivalent rare earth ions expressed in terms of number of Bohr magnetons. The curve is calculated from Eq. (8.43a), while the values measured experimentally are represented by vertical lines. The dashed curve is the theoretical value calculated by J. H. Van Vleck and A. Frank (Phys. Rev., vol. 34, pp. 1494 and 1625, 1929), taking into account the excited states.

In a free atom, the $2J + 1$ levels which have the same energy in the absence of a magnetic field will have different energies in the presence of an applied field. The difference in energy is given by Eq. (8.44). Whenever there is an energy difference, there is a difference in population as demanded by the Boltzmann statistics. Without the field, the $2J + 1$ levels ranging from $-J$ to $+J$ are equally populated. The effect of the magnetic field is to bring about the difference in energy and in turn the difference in population of the states having different J values. In the presence of a strong crystalline field, as discussed earlier, the **S** vector and the **L** vector are decoupled so we treat them separately. Since the spin is not coupled directly to the crystalline field, the **S** vector is not affected. However, so far as the orbital momentum is concerned, the effect of the crystalline field is to split the levels, the lowest of which then becomes a single ground state. For example: for $L = 1$, one state, say ψ_C, becomes the ground state and the other two states, say ψ_A and ψ_B, become excited states. If the energy separation between the ground and excited states is several orders of magnitude larger than the magnetic energy $\mu_0 u_B H$, the application of a magnetic field will not change the nature of the ground state. In other words, ψ_C still represents the orbital wave function of the ground state in the presence of a magnetic field. Using Eqs. (8.35) and (8.36), we find that the expectation values of L_z, L_x, and L_y are identically zero for ψ_C. That means, even though the state ψ_C has a nonvanishing value of $L = 1$, the time average of the projection of the orbital momentum vector upon a specified axis is zero. Under such circumstances, the application of a magnetic field affects only the spin part of the magnetic energy, and in Eq. (8.44), we can replace J by S. This is equivalent to saying that we have a nonmagnetic orbital state or the orbital momentum is quenched. We find the situation prevalent in the salts of the iron group.

For ions of the iron group, magnetic dipoles are almost entirely due to the spin of the $3d$ electrons. Take Cr^{3+} which has a $3d^3$ configuration so $S = \frac{3}{2}$. The number of Bohr magnetons due to spin only is equal to $2[S(S + 1)]^{\frac{1}{2}} = 3.87$ and the experimental value is 3.8. The Landé g factor for a pure spin state is 2 and the observed value for Cr^{3+} is 1.97. The slight discrepancy between the pure spin value and the experimental value is a manifestation that the spin may carry some orbital momentum with it. In the spin-only analysis, we neglected the spin-orbit coupling entirely. Due to spin-orbit coupling, the ground state will contain a small percentage of the excited states and this admixture of the excited states with the ground state causes the value of g to differ from 2. The mixing of states is very important in maser action and we shall discuss it further in the next chapter. What we have done in this section is to present a general picture which is a first-order approximation and this general picture provides a foundation for future refinement.

8.8 PARAMAGNETIC SUSCEPTIBILITY

In a paramagnetic substance, permanent dipoles are oriented randomly in all directions so the net magnetization is zero. In a magnetic field, however, the dipoles tend to align themselves in the direction of the applied field to minimize the energy of the system. Let us refer to the dipole shown in Fig. 8.2b and calculate its energy. The torque exerted on the dipole is

$$T = u\mu_0 H \sin \theta \tag{8.52}$$

If we choose the plane $\theta = \pi/2$ as the plane of reference energy, then the energy of the dipole at any angle θ is equal to the work done in rotating the dipole position from $\pi/2$ to θ. Thus,

$$W = \int_{\pi/2}^{\theta} T \, d\theta = -u\mu_0 H \cos \theta \tag{8.53}$$

Hence the position with **u** in the direction of **H** is energetically more favorable than that with **u** in the opposite direction, resulting in a net magnetization. Following exactly the steps in deriving the dielectric polarizability in Sec. 7.4 and substituting u for p and H for E, we have

$$M = Nu \left[\coth \left(\frac{u\mu_0 H}{kT} \right) - \frac{kT}{u\mu_0 H} \right] \tag{8.54}[7]$$

The function in the brackets is the Langevin function. For $u\mu_0 H/kT \ll 1$, an expansion of the Langevin function gives the magnetization per unit volume,

$$M = \frac{Nu^2\mu_0 H}{3kT} \tag{8.55}$$

and the paramagnetic susceptibility per unit volume,

$$\chi = \frac{M}{H} = \frac{Nu^2\mu_0}{3kT} = \frac{C}{T} \tag{8.56}$$

where N is the number of magnetic dipoles per unit volume.

From Eq. (8.56), the strength of a magnetic dipole can be obtained by measuring the paramagnetic susceptibility as a function of temperature. The theoretical model which we discussed in the previous section is based on experimental information obtained from the susceptibility measurement and other measurements such as the paramagnetic resonance experiment which we shall discuss later. For ions of the rare earth group, it appears that the inner quantum number is a good quantum number and $u = g[J(J+1)]^{1/2} u_B$. For

---

[7] At very low temperatures and high fields such that $\mu_0 uH/kT \gg 1$, all dipoles are expected to be aligned in the direction of the applied field. For experimental observation of the saturation effect, the reader is referred to W. E. Henry, *Phys. Rev.*, vol. 88, p. 559, 1952.

ions of the iron group, the spin-only value agrees well with experimental results, that is $u = 2[S(S + 1)]^{1/2}u_B$.

It may be instructive to derive the paramagnetic susceptibility from the quantum-mechanical point of view. For simplicity, let us take the free spin system with $s_z = \frac{1}{2}$ or $-\frac{1}{2}$ so that the energy of the spin system is split into two levels as shown in Fig. 8.11. Note that the quantum-energy levels are discrete in contrast to the classical (continuous) energy expression of Eq. (8.53). The number of electrons having $s_z = -\frac{1}{2}$ is only slightly larger than that having $s_z = +\frac{1}{2}$, the ratio being determined by the Boltzmann factor. If N_1 and N_2 are the number of electrons in the upper and lower levels, then we have

$$\frac{N_1}{N_2} = \exp\left(-\frac{2\mu_0 u_B H}{kT}\right) \tag{8.57}$$

The net magnetization is

$$M = u_B(N_2 - N_1) = N u_B \tanh\left(\frac{\mu_0 u_B H}{kT}\right) \tag{8.58}$$

realizing that $N_1 + N_2 = N$ (the total number of electrons). Equation (8.58) is a special case of a general expression known as the *Brillouin function* which holds for all values of S. The reader is referred to Prob. 8.11 for the derivation of the Brillouin function. If $\frac{\mu_0 u_B H}{kT} \ll 1$,

$$M = \frac{N \mu_0 u_B^2 H}{kT} \tag{8.59}$$

Note that for $s = \frac{1}{2}$, $u^2 = g^2 s(s + 1)u_B^2 = 3u_B^2$ in Eq. (8.56), so Eqs. (8.59) and (8.56) yield the same result.

The functional dependence of χ on $1/T$ as given by Eq. (8.56) is known as the *Curie law* and the proportionality constant $C = N\mu_0 u^2/3k$ as the *Curie constant*. It should be pointed out that Eqs. (8.56) and (8.59) hold only when there is no internal field. As we mentioned earlier in Sec. 8.3, there is a strong internal field in ferro-

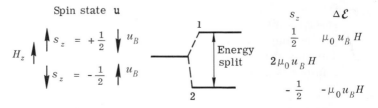

Figure 8.11

Energy level diagram for a spin system with $s = \frac{1}{2}$ in the presence of a magnetic field H.

magnetic, ferrimagnetic, and antiferromagnetic substances. Above an appropriate critical temperature, these substances all become paramagnetic. The functional dependence of paramagnetic susceptibility upon $1/T$ in these cases is different, as we shall see later. Therefore, strictly speaking, Eqs. (8.56) and (8.59) hold only for ideal paramagnetic substances in which no internal field is assumed.

Let us estimate the order of magnitude of the paramagnetic susceptibility from Eq. (8.56). If we take $N = 5 \times 10^{28}$ m^{-3}, $u^2 = 3u_B{}^2 \approx 2 \times 10^{-46}$ amp^2-m^4, $T = 300°$K, we have $\chi \approx 10^{-3}$. Thus, for ordinary magnetic substances, we can neglect the minute contribution of paramagnetic ions to permeability and take the permeability of a paramagnetic substance as equal to the permeability of free space, μ_0. However, there are circumstances where it is exactly this minute contribution from the paramagnetic ions that performs marvels. Such is the case in the use of paramagnetic ions in solid-state masers, which are very low-noise amplifiers, and in the use of paramagnetic salts in achieving temperatures near absolute zero ($<1°$K) by adiabatic demagnetization.[8]

8.9 FERROMAGNETISM AND SPONTANEOUS MAGNETIZATION

In the case of a ferromagnetic substance, we can no longer treat each magnetic ion individually as a free bar magnet as in the case of a paramagnetic substance. For ferromagnetic materials, as we mentioned in Sec. 8.3, there is a strong coupling which we call the molecular field between the magnetic ions so that all the elementary dipoles tend to line up in the same direction. The concept of a molecular field was first introduced by Weiss. He proposed in a phenomenological way that the field acting on the bar magnet actually consisted of two parts: (1) externally applied and (2) internally existing as quantitatively represented by the following equation,

$$H_{\text{eff}} = H_{\text{ext}} + \lambda M \tag{8.60}$$

where λM is the phenomenological internal field. Hence, for ferromagnetic materials, the Langevin equation now reads

$$M = Nu\mathcal{L}\left[\frac{(H_{\text{ext}} + \lambda M)u\mu_0}{kT}\right] \tag{8.61}$$

where $\mathcal{L}[x]$ represents the Langevin function.

The important feature of Eq. (8.61) is that it may possess a solution for M even in the absence of any applied field. Let $x = \lambda M u \mu_0 / kT$ and $y = M$. We plot the Langevin function $y = Nu\mathcal{L}(x)$

---

[8] The reader is referred to E. Ambler and R. P. Hudson, "Magnetic Cooling," *Report Progr. Phys.*, vol. 18, p. 251, 1955, and to C. G. Garrett, "Magnetic Cooling," Harvard University Press, Cambridge, Mass., 1954.

and the straight line $y = (kT/\lambda u \mu_0)x$ in Fig. 8.12. The intercept of
the line with the curve gives the solution of Eq. (8.61) for M. We
call the value of M the *spontaneous magnetization* because it exists
by virtue of its own field λM.

Two essential features of the graph stand out. First, a solution of
M exists only when the temperature T is below a certain critical
temperature T_c which is known as the Curie temperature. At T_c, the
initial slope of the curve is equal to the slope of the line. Thus, we
have

$$Nu \left. \frac{d\mathcal{L}(x)}{dx} \right|_{x=0} = \frac{kT_c}{\lambda u \mu_0} \quad \text{or} \quad T_c = \frac{N \mu_0 u^2 \lambda}{3k} \quad (8.62)$$

Second, the value of M depends on temperature. At $T = 0°\text{K}$, the
spontaneous magnetization is equal to Nu, a value which corresponds
to a complete alignment of the dipoles and hence is called the *saturation magnetization*. In Fig. 8.13 the value of $M(T)$ determined from
a numerical solution of the Brillouin function[9] (Prob. 8.11) is plotted
as a function of temperature, the ordinate and abscissa being
normalized by Nu and T_c respectively to make the curve universal;
it can be seen that the experimental points for Fe, Co, and Ni fall on
the theoretical curve with $J = S = \frac{1}{2}$. However, the good agreement between theory and experiment should not be taken literally
to mean that Fe, Co, and Ni all have an effective number of Bohr
magnetons, $n_{\text{eff}} = 2\sqrt{S(S+1)} = 1.73$. The values of n_{eff} for these
elements obtained from the saturation magnetization (M_s) are found
to be 2.22, 1.72, and 0.606 (Table 8.1, p. 483). Therefore, the general
agreement can only be interpreted as meaning the general validity
of the phenomenological Weiss molecular field.

Above the Curie temperature, the permanent dipoles become
randomly oriented; thus, the value of λM becomes very small but it

[9] L. Brillouin, *Jour. Phys. Radium*, vol. 8, p. 74, 1927.

**Figure
8.12**

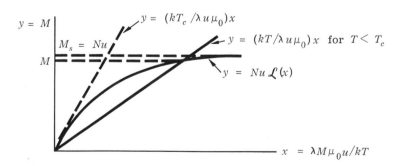

*Diagram illustrating how Eq. (8.61) can be solved graphically for $H_{\text{ext}} = 0$.
A solution for the value of spontaneous magnetization exists only when the
straight line whose slope is a function of temperature intersects the curve.*

is not zero. After expansion, Eq. (8.61) reads

$$\frac{M}{H_{\text{eff}}} = \frac{M}{H + \lambda M} = \frac{Nu^2\mu_0}{3kT} \quad \text{or} \quad \chi = \frac{M}{H} = \frac{C}{T - T_c} \quad (8.63)$$

where $C = Nu^2\mu_0/3k$ and $T_c = \lambda C$, the quantities C and T_c being the same as those given respectively in Eqs. (8.56) and (8.62). The expression, Eq. (8.63), which is known as the Curie-Weiss law, is different from Eq. (8.56) on account of the internal field.

The phenomenological Weiss theory is quite successful in explaining the essential features of ferromagnetism. However, there are two fundamental questions remaining unanswered in the theory: (1) the origin of the molecular field and (2) the number of Bohr magnetons associated with the permanent dipole. To answer the first question, let us estimate the order of magnitude of the Weiss field. From Eq. (8.62), we have $\lambda M = \lambda(Nu) = 3kT_c/\mu_0u$. If we take $T_c = 1000°\text{K}$ and $u = u_B \approx 10^{-23}$ amp-m^2, we have $\lambda M \approx 3 \times 10^9$ amp/m! In Sec. 8.1, we calculated the internal field due to magnetic dipoles and found $H_{\text{int}} = M$. For most ferromagnetic materials like Fe, Ni, and Co, $M \approx 10^6$ amp/m; so the Weiss field, being larger by a factor of a thousand, cannot be due to the magnetic dipoles themselves. It turns out that the Weiss field (λM) is of electrostatic origin and can only be explained by quantum mechanics. Here we shall give a brief descriptive account of its origin.

As discussed in Secs. 1.8 and 2.1, the complete wave function to describe an atomic state actually consists of the product of two parts: (1) the part due to the orbital motion as described in Sec. 8.4 and

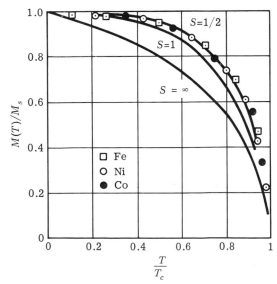

Figure 8.13

Spontaneous magnetization as a function of temperature.

(2) the part due to the spinning motion. Due to the Pauli exclusion principle, the total wave function describing a system of atoms has to be antisymmetrical, that is, a combination of a symmetrical orbital wave function and an antisymmetrical spin wave function or vice versa. The two spin wave functions, symmetrical and antisymmetrical, represent two different states of affairs, one representing a parallel and the other an antiparallel alignment of the spin on the ith and jth atom, for example. Referring to Eqs. (2.4), (2.5), and (2.6), the two orbital wave functions have different coulombic interaction energy (electrostatic in nature) because the two integrals $\iiint \psi^* V \psi \, dx \, dy \, dz$ with ψ being symmetrical and antisymmetrical are different. Thus, the restriction on the nature of the total wave function makes the spin have an indirect effect on the energy of the system through its association with the orbital wave function. The energy difference between the two different combinations of the orbital and spin wave function is often called the *exchange energy* and it can be stated quantitatively as follows:

$$\mathcal{E}_{\text{exch}} = -2J\mathbf{s}_i \cdot \mathbf{s}_j \tag{8.64}$$

where J is a proportionality constant generally referred to as the *exchange integral*. In Eq. (8.64), the interaction is the strongest when the ith and jth atoms are nearest neighbors. The magnitude of J can be estimated as follows. If there are Z nearest neighbors, the total energy required in turning one elementary dipole from parallel to antiparallel position is $2\mu_0 \lambda M u$ in view of Eq. (8.60). Equating

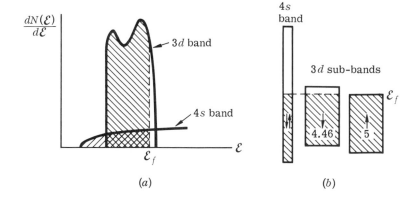

Figure 8.14

(a)   (b)

(a) *Schematic representation showing that the density-of-states functions $N(\mathcal{E})$ overlap for the 3d and 4s bands in transition metals of the iron group. The shaded area represents the occupied states. (b) The 3d band can further be considered as consisting of two sub-bands with spins pointing in opposite directions in the two sub-bands. The diagram illustrates the situation believed to exist in nickel.*

$\mathcal{E}_{\text{exch}}$ to $\mu_0\lambda Mu$, we have

$$2ZJs(s+1) = \mu_0\lambda Mu = \lambda N\mu_0 u^2 = 3kT_c \qquad (8.65)$$

noting that $|s^2| = s(s+1)$. The value of J is expressed in units of energy and from Eq. (8.65), its magnitude is of the order of three-tenths of kT_c for $Z = 6$ (a simple cubic crystal) and $s = \frac{1}{2}$.

Most of the ferromagnetic materials, like Fe, Co, and Ni, are conductors so their magnetic property is intimately connected with the band structure of the substance. Let us take the elements in the iron group whose energy-band structure is schematically shown in Fig. 8.14a. The energy bands of interest are the 3d band and the 4s band. However, due to the exchange energy, the 3d band is split into two sub-bands (Fig. 8.14b), one band with spin pointing in one direction and the other in the opposite direction. How the electrons distribute themselves among the different bands depends on the relative energy and the overlap of these bands. In Ni, there are, on the average, 5 electrons in one 3d sub-band, 4.46 electrons in the other 3d sub-band, and 0.54 electron in the 4s band. Since the electrons in the 4s band are equally divided between two spin orientations, there is a net $5 - 4.46 = 0.54$ Bohr magneton per nickel atom. The fact that many ferromagnetic substances have a nonintegral number of Bohr magnetons can be explained in this manner.[10]

[10] The reader is referred to C. Kittel, "Introduction to Solid State Physics," pp. 329–337, John Wiley & Sons, Inc., New York, 1956, for a review of the band-model theory of ferromagnetism, and to a series of articles by Slater, Zener and Heikes, and Van Vleck in *Rev. Modern Phys.*, vol. 25, January, 1953, for a detailed discussion of the band model.

TABLE 8.1 *The spontaneous magnetization, Curie temperature, and effective number of Bohr magnetons in a selected number of magnetic materials*

Substance	Spontaneous magnetization (10^3 amp/m)		Curie temperature, $T_c\,°K$	Effective No. of Bohr magnetons, n_{eff} (0°K)
	300°K	0°K		
Fe	1707	1752	1043	2.22
Co	1400	1446	1400	1.72
Ni	485	510	631	0.606
Gd	1090	1980	289	7.10
Dy	....	1830 (80°K)	105	
$NiOFe_2O_3$	240	....	863	2.3
Cu_2MnAl	430	580	603	(4.0)
MnAs	670	870	318	3.4

[Reference: C. Kittel, "Introduction to Solid State Physics," John Wiley & Sons, Inc., New York, 1956.]

For completeness, the values of the saturation magnetization, the Curie temperature, and the effective number of Bohr magnetons per atom for some common ferromagnetic elements and alloys are listed in Table 8.1. It is to be noted that the effective magneton number is quite different from what is expected on the basis of free ions. For a more detailed discussion of the application of band theory to ferromagnetism and an extensive discussion of other related subjects, such as the effect of added impurities and alloying on the value of saturation magnetization, the reader is referred to other books.[11]

8.10 ANTIFERROMAGNETISM

In the previous section, we discussed the origin of the Weiss molecular field from the quantum-mechanical point of view. The exchange integral J, actually being the same in nature as $\mathcal{E}_b$ of Eq. (2.6), involves the product of the electron wave function and the electrostatic potential energy V of Eq. (2.3). For J to be positive, the term e^2/r_{12} must be relatively large. This happens when the ratio of the interatomic distance to the effective mean radius of the electron-density distribution involved in the integral is large. As pointed out by Slater,[12] this condition is met only in a limited number of materials, notably Fe, Co, and Ni. Therefore, ferromagnetism is the exception rather than a general rule. There is a large class of materials of which the value of J is negative, favoring antiparallel arrangement of the spins. Common examples are certain oxides and compounds of the iron group, such as MnO, CoO, Cr_2O_3, MnF_2, and $NiCl_2$, to name a few.

To have a concrete physical picture, we present the model[13] of MnO in Fig. 8.15. The MnO crystal has the simple NaCl structure. It turns out that the strongest exchange interaction is between the next-nearest neighbors, for example Mn atoms 1 and 2 in Fig. 8.15, through the intervening oxygen anion (negative ion). Furthermore,

[11] See, for example, R. M. Bozorth, "Ferromagnetism," D. Van Nostrand Co., Inc., New York, 1951; F. Seitz, "Modern Theory of Solids," McGraw-Hill Book Company, New York, 1940; N. F. Mott and H. Jones, "The Theory of the Properties of Metals and Alloys," Oxford University Press, London, 1936; W. Hume-Rothery and G. V. Raynor, "The Structure of Metals and Alloys," The Institute of Metals, London, 1962.

[12] J. C. Slater, *Phys. Rev.*, vol. 35, p. 509, 1930, and vol. 36, p. 57, 1930.

[13] Neutron-diffraction studies have been extremely useful in ascertaining the magnetic structures of antiferromagnetic and ferrimagnetic materials. For a general discussion of neutron diffraction theories, the reader is referred to G. E. Bacon, "Neutron Diffraction," Clarendon Press, Oxford, 1962. The list of literature concerning neutron-diffraction studies is extensive. For recent developments in the neutron-diffraction study of magnetic ordering, the reader is referred to the *Proceedings* of the International Conference on Magnetism and Crystallography (held at Kyoto, September 1961) published by the Physical Society of Japan, 1962.

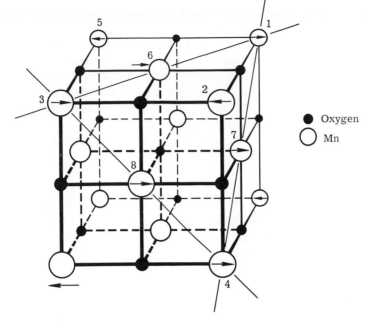

Figure 8.15

● Oxygen
○ Mn

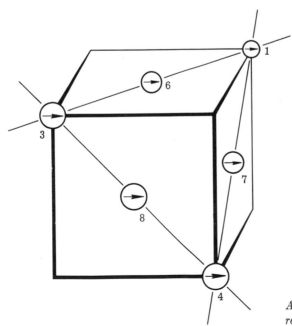

Antiferromagnetic arrangement of spins in MnO. (C. G. Shull, W. A. Strauser, and E. O. Wollan, Phys. Rev., vol. 83, p. 333, 1951.)

neutron-diffraction experiments have shown that the spins on the same (111) planes (such as atoms 6 and 1) point in the same direction and the spins in the alternate (111) planes (such as atoms 6 and 2) point in the opposite direction. Thus, we may conveniently think of an antiferromagnet as made of two sublattices, A and B, with $\mathbf{M}_A$ opposing $\mathbf{M}_B$. Since $\mathbf{M}_A$ is equal in magnitude but opposite in direction to $\mathbf{M}_B$, the net magnetization of an antiferromagnet is zero in the absence of an applied field (Fig. 8.6).

The effect of an external field on the net magnetization depends upon the direction of the field with respect to that of $\mathbf{M}_A$ and $\mathbf{M}_B$. Let us first consider the case where the applied field is perpendicular to $\mathbf{M}_A$ and $\mathbf{M}_B$ as shown in Fig. 8.16. It is to be remembered that magnetic energy of the system is a minimum when the magnetization vector lies in the direction of the effective field. Here, the effective field, which is the significant quantity, consists of both the applied field and the molecular field. Thus, the effective fields acting on $\mathbf{M}_A$ and $\mathbf{M}_B$ are, respectively,

$$(\mathbf{H}_{\text{eff}})_A = \mathbf{H} - \alpha\mathbf{M}_A - \lambda\mathbf{M}_B \tag{8.66a}$$

and $$(\mathbf{H}_{\text{eff}})_B = \mathbf{H} - \lambda\mathbf{M}_A - \alpha\mathbf{M}_B \tag{8.66b}$$

where α represents the exchange interaction between the magnetization on the same sublattice (AA or BB) while λ represents the A-B type exchange interaction. A comparison of Eq. (8.66) with Eq. (8.60) shows that the negative sign in Eq. (8.66) is indicative of the antiferromagnetic nature of the exchange interaction.

What the applied field does is to turn the magnetization $\mathbf{M}_A$ and $\mathbf{M}_B$ toward the direction of $\mathbf{H}$ so that the resultant $(\mathbf{H}_{\text{eff}})_A$ and $(\mathbf{H}_{\text{eff}})_B$ will be in the new direction of $\mathbf{M}_A$ and $\mathbf{M}_B$, respectively. The arrangement is illustrated in Fig. 8.16b. Since the magnitude of H is much smaller than that of λM_B and αM_A, the angle ϕ is very small. Thus we have

$$\phi \approx \tan\phi \approx \sin\phi = \frac{H}{2\lambda M_B} \tag{8.67}$$

Figure 8.16

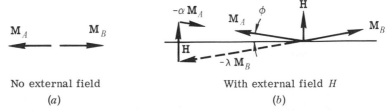

$\mathbf{M}_A$ $\mathbf{M}_B$

No external field

(a)

With external field H

(b)

Arrangements of magnetization vectors on two sublattices of an antiferromagnet: (a) under no applied magnetic field, and (b) under an applied magnetic field. Because of strong internal molecular fields, all that an external field can do is to turn the magnetization vectors slightly in the direction of the applied field.

and the net magnetization

$$M_\perp = M_A\phi + M_B\phi = \frac{H}{\lambda} \qquad (8.68)$$

or

$$\chi_\perp = \frac{M_\perp}{H} = \frac{1}{\lambda} \qquad (8.69)$$

From Eqs. (8.68) and (8.69), we note that although the values of M_A and M_B are large, the values of M and χ are small because of the strong influence of the molecular field.

The analysis of the effect of an applied field parallel to $\mathbf{M}_A$ and $\mathbf{M}_B$ is much more involved. Here, we shall only attempt to give a qualitative description of how $\chi_\parallel$ will behave as a function of temperature. At absolute zero, all the dipoles are in perfect alignment; thus the applied field can neither subtract anything from nor add anything to the saturation values of M_A and M_B. Consequently

$$\chi_\parallel = 0 \qquad \text{at } T = 0°\text{K} \qquad (8.70)$$

At a finite temperature, the values of M_A and M_B are given by the solutions to the Langevin equation

$$M_A = Nu\mathcal{L}\left[(\lambda M_B - \alpha M_A + H)\frac{u\mu_0}{kT}\right] \qquad (8.71a)$$

$$M_B = Nu\mathcal{L}\left[(\lambda M_A - \alpha M_B - H)\frac{u\mu_0}{kT}\right] \qquad (8.71b)$$

assuming that $\mathbf{H}$ is in the direction of $\mathbf{M}_A$. Referring to Fig. 8.12 we see that M_A will have a larger value than M_B on account of H. Moreover, for the same H, the effect will be larger as temperature increases. At the critical temperature, the antiferromagnet becomes paramagnetic, and the magnetization no longer has any preferred direction. Hence

$$\chi_\parallel = \chi_\perp = \frac{1}{\lambda} \qquad \text{at } T = T_c \qquad (8.72)$$

A calculation by Van Vleck[14] shows that the value of $\chi_\parallel$ increases smoothly from zero to $1/\lambda$ as the temperature varies from absolute zero to the critical temperature T_c.

For an applied field with an arbitrary direction, we can always resolve the field into two components: (1) $H \cos \theta$ parallel to the magnetization and (2) $H \sin \theta$ perpendicular to the magnetization, θ being the angle between the direction of the field and the magnetization. Thus, we have

$$M_\parallel = \chi_\parallel H \cos \theta \qquad \text{and} \qquad M_\perp = \chi_\perp H \sin \theta \qquad (8.73)$$

[14] J. H. Van Vleck, *Jour. Chem. Phys.*, vol. 9, p. 85, 1941.

and the resultant magnetization in the direction of $\mathbf{H}$

$$M = M_\parallel \cos\theta + M_\perp \sin\theta = (\chi_\parallel \cos^2\theta + \chi_\perp \sin^2\theta)H \qquad (8.74)$$

It should be pointed out that Eq. (8.74) is meaningful only for single crystals. In a randomly-oriented polycrystalline material, each tiny single crystal has its crystal axis pointing in a different direction. Thus, for each tiny single crystal, there is a different value of θ and consequently to obtain the net magnetization for a polycrystalline substance, we have to perform a statistical average of Eq. (8.74) over a given distribution of θ.

Let us now discuss the magnetic property of an antiferromagnet above the critical temperature T_c. Since the dipoles are randomly oriented, both M_A and M_B exist on account of H. Thus, not only are M_A and M_B small but they are in the direction of H as well. The effective fields acting on M_A and M_B are, respectively,

$$
\begin{aligned}
(H_{\text{eff}})_A &= H - \alpha M_A - \lambda M_B \\
(H_{\text{eff}})_B &= H - \lambda M_A - \alpha M_B
\end{aligned}
\qquad (8.75)
$$

It is noteworthy to point out the difference between Eqs. (8.75) and (8.66). Below T_c, M_A and M_B are large and they point in different directions; hence Eq. (8.66) is in vector form. Above T_c, the net magnetization M_A or M_B, being due to a reorientation of dipoles on sublattices A or B, lies in the direction of the applied field; and consequently Eq. (8.75) can be written in scalar form. To find the net magnetization, we again expand the Langevin function and obtain

$$M_A = \frac{C}{T}(H - \alpha M_A - \lambda M_B) \qquad (8.76a)$$

$$M_B = \frac{C}{T}(H - \lambda M_A - \alpha M_B) \qquad (8.76b)$$

Adding Eq. (8.76a) to Eq. (8.76b) gives

$$M = M_A + M_B = \frac{2CH}{T + \Theta} \qquad (8.77)$$

where
$$\Theta = C(\lambda + \alpha) \qquad (8.78)$$

and C is the same Curie constant as was introduced in conjunction with Eq. (8.63). To find the critical temperature T_c, we set $H = 0$ in Eq. (8.76). For nontrivial solutions of M_A and M_B, the determinant must vanish. Thus, we have

$$T_N = T_c = C(\lambda - \alpha) \qquad (8.79)$$

We call the critical temperature of an antiferromagnet the Néel temperature denoted by T_N to avoid confusion with the critical temperature of a ferromagnet. Note that at $T = T_N$, Eq. (8.77) gives the value $M = H/\lambda$, which checks with the result of Eq. (8.72).

To summarize the discussion, we shall compare the results of our analysis for paramagnetic, ferromagnetic, and antiferromagnetic substances. In Fig. 8.17, the inverse of the d-c susceptibility is plotted as a function of temperature. The variation in behavior is a manifestation of the existence of a strong internal field. In a ferromagnet, the magnetization suddenly jumps to a very high value at $T \leq T_c$ and this physical phenomenon is accounted for analytically by the singularity which χ has at $T = T_c$. The behavior of $1/\chi$ for an antiferromagnet is different. First, the sign of the internal field is negative [Eq. (8.66)] rather than positive [Eq. (8.60)]. This explains why we have $T + \Theta$ in $1/\chi$ instead of $T - T_c$. Second, although there is a sudden jump in M_A and M_B at $T = T_N$, the change in net

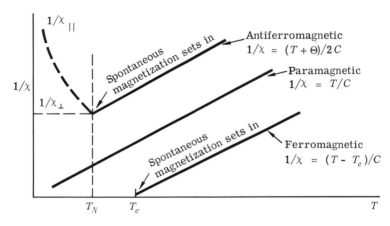

Figure 8.17

Comparison of the behaviors of paramagnetic susceptibility as a function of temperature in antiferromagnetic, ideal paramagnetic, and ferromagnetic materials.

magnetization at $T = T_N$ is smooth. Finally, we should point out the difference in Θ and T_N and its physical interpretation. In the paramagnetic state, M_A and M_B both point in the direction of H and hence the resultant internal fields are additive as shown by Eq. (8.78). The critical temperature, on the other hand, is associated with the existence of an antiferromagnetic state and consequently the two internal fields are subtractive as implied in Eq. (8.79).

8.11 FERRIMAGNETISM

The best known ferrimagnetic substance is the ferrite. The chemical composition of *ferrites* is generally represented by $MOFe_2O_3$ where M is a divalent metal such as Mn, Fe, Co, Ni, Cu, Zn, Mg, or Cd. The structure of ferrites is of the spinel type $(MgAl_2O_4)$ shown in

Fig. 8.18. In a unit cell, there are two types of lattice sites: (1) the tetrahedral sites surrounded by four O^{2-} ions and (2) the octahedral sites surrounded by six O^{2-} ions. The magnetic properties of a given ferrite depend upon how the various cations distribute themselves among the various lattice sites. According to Fig. 8.18, the ratio of the number of occupied tetrahedral sites to that of occupied octahedral sites is 1 to 2. In a normal spinel, the divalent metal ions occupy the tetrahedral sites while the trivalent metal ions (Fe^{3+}) occupy the octahedral sites. In an inverted spinel, the octahedral sites are shared by the divalent and the trivalent metal ions while the tetrahedral sites are occupied by the trivalent metal ions. Both zinc and cadmium ferrites have the normal spinel structure while the other simple ferrites, such as $MnFe_2O_4$, $NiFe_2O_4$, etc., have the inverted spinel structure.

According to Néel,[15] all the exchange interactions (A-A, B-B, and

[15] L. Néel, *Ann. Phys.*, vol. 3, p. 137, 1948.

Figure 8.18

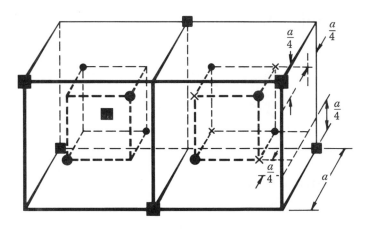

● Oxygen
■ Occupied tetrahedral site
✕ Occupied octahedral site

Tetrahedral site

Octahedral site

The spinel structure of ferrite. (*E. W. Gorter, Philips Res. Rept., vol. 9, p. 295, 1954.*)

A-B) in ferrites are antiferromagnetic with the A-B type interaction being the strongest. Let us take $MnOFe_2O_3$ as an example. Since $MnFe_2O_4$ has an inverted spinel structure, the various cations (positive ions) are distributed among the various sites in the manner shown in Fig. 8.19. Furthermore, since the A-B type interaction overrules the other types of interaction, the magnetic dipoles on sites B must be all aligned in opposition to the magnetic dipoles on sites A, which are also aligned. The net magnetization per molecule is thus $10u_B - 5u_B = 5$ Bohr magnetons. In zinc ferrite, the zinc ion is nonmagnetic. The only existent interaction is of the B-B type which is also antiferromagnetic. This explains why zinc ferrites have zero saturation magnetization.

To give further support of the Néel theory, the case of mixed ferrites may be considered. Let us take $Mn_{(1-x)}Zn_xFe_2O_4$, x being the percentage of zinc ferrite in manganese ferrite. Ordinarily when we mix a nonmagnetic substance with a magnetic substance, we would expect the saturation magnetization to decrease because of dilution. In this case, however, the experimental result is just the opposite. Since the zinc ion is found only in tetrahedral sites, the distribution of the cations is as follows: (1) tetrahedral site: xZn^{2+} and $(1-x)Fe^{3+}$ and (2) octahedral site: $(1+x)Fe^{3+}$ and $(1-x)Mn^{2+}$. Thus the saturation magnetization of the mixed ferrite is $M_B - M_A = 5(1+x)u_B$. The replacement of one manganese ion by a zinc ion apparently decreases the local magnetic contribution by $5u_B$, but in actual fact, a net increase occurs by a series of forced displacements. Zinc entering the tetrahedral sites displaces trivalent Fe ions to the octahedral sites so that while the latter lose a Mn^{2+} ion they gain a Fe^{3+} ion. The magnetic moment of the octahedral sites thus remains constant while the opposing magnetic moment of the tetrahedral sites diminishes with increasing zinc content. The net result is that for each initial replacement of a manganese ion by a zinc ion, the saturation magnetization is increased by 5 Bohr magnetons. In Fig. 8.20, the experimental value of the saturation magnetization of various mixed ferrites is plotted as a function of the zinc ferrite content. The initial slope of the curve can be quantitatively accounted for in the manner we have just discussed.

Figure 8.19

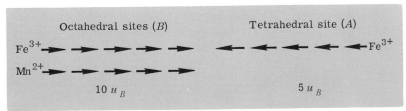

Schematic representation of the distribution of cations among the octahedral and tetrahedral sites in $MnOFe_2O_3$.

So far as the general magnetic property of a ferrite is concerned, we do not have anything new to add. It possesses a critical temperature, T_N, above which it becomes paramagnetic and below which it shows a saturation magnetization.[16] We should mention, however, one feature of ferrites that is of practical importance. All ferrites except magnetite (Fe_3O_4) have very high resistivity. Although ferrites were originally used at low frequencies as a replacement for conducting ferromagnetic substances to reduce the eddy current loss, they are found nowadays in many high frequency applications, ranging from the magnetic core in a computer to the isolator in a microwave system. The application of magnetic materials will be discussed in Chapters 9 and 10.

Finally, we should like to discuss another class of ferrimagnetic insulators known as *ferrimagnetic garnets*. The term garnet originally refers to a group of minerals having the general chemical formula $M_3Q_2R_3O_{12}$. In natural garnets, the R ions are always tetravalent silicon Si^{4+} while the M and Q ions are respectively divalent and trivalent metal ions. An example of the natural garnet is Mg_3Al_2-$(SiO_4)_3$ in which Mg^{2+} ions may be substituted for by Ca^{2+} or Mn^{2+} ions, and Al^{3+} ions by Fe^{3+} or Cr^{3+} ions. The ferrimagnetic garnets are, however, synthetic with the general chemical formula M_3Fe_2-Fe_3O_{12}[17] where M is a rare earth element such as Sm, Eu, Gd, Tb, Dy, Ho, Er, Tm, Yb, Lu, or the element Y (yttrium). Since the tetravalent Si is replaced by trivalent Fe in ferrimagnetic garnets,

[16] The spontaneous magnetization of a mixed ferrite varies in a rather complicated fashion with temperature. For a theoretical discussion, the reader is referred to E. W. Gorter, *Proc. IRE*, vol. 43, p. 1945, 1955.

[17] F. Bertaut and F. Forrat, *Compt. Rend.* (Paris), vol. 242, p. 382, 1956; S. Geller and M. A. Gilleo, *Acta Cryst.*, vol. 10, p. 239, 1957.

Figure 8.20

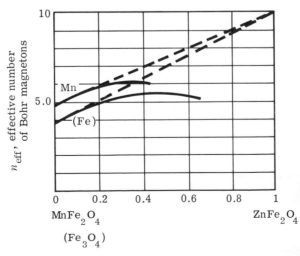

Variation of the saturation magnetization with composition in mixed ferrite. (E. W. Gorter, Philips Res. Rept., vol. 9, p. 321, 1954.)

the metal ion M is now trivalent instead of being divalent as in natural garnets.

The basic crystal structure of a garnet is cubic with 96 oxygen ions per unit cell.[18] Although the complete unit cell is rather complicated, the lattice positions can be divided (in a way similar to that done for the spinel structure of ferrites) into three sublattices: (1) The a sites (octahedral site) are occupied by Fe^{3+} (corresponding to

[18] The basic structure is obviously too complicated to be drawn in full. For a simplified representation of the structure, the reader is referred to M. A. Gilleo and S. Geller, *Phys. Rev.*, vol. 110, p. 73, 1958.

Figure 8.21

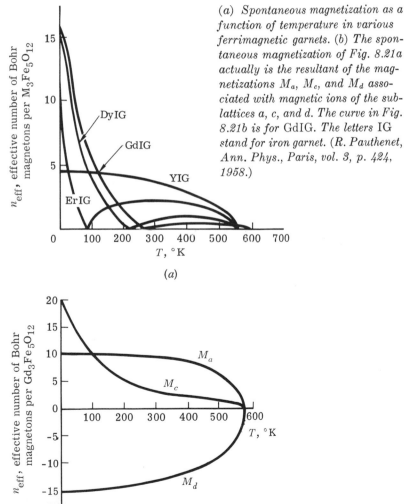

(a) *Spontaneous magnetization as a function of temperature in various ferrimagnetic garnets.* (b) *The spontaneous magnetization of Fig. 8.21a actually is the resultant of the magnetizations M_a, M_c, and M_d associated with magnetic ions of the sublattices a, c, and d. The curve in Fig. 8.21b is for GdIG. The letters IG stand for iron garnet.* (R. Pauthenet, *Ann. Phys., Paris*, vol. 3, p. 424, 1958.)

Q) ions, (2) the d sites (tetrahedral site) are occupied by the other Fe^{3+} (corresponding to R) ions, and (3) the c sites (the center of a dodecahedron) are occupied by the metal ions M^{3+}. Figure 8.21a shows the spontaneous magnetization in various iron garnets as a function of temperature. The magnetization curve can be understood in terms of magnetizations M_a, M_c, and M_d associated with the sublattices a, c, and d respectively.

The simplest example among the ferrimagnetic garnets is YIG. Since the Y^{3+} ion has a closed-shell configuration, the magnetization is due to the Fe^{3+} ions, with $M_a = 10u_B$ and $M_d = 15u_B$ expressed in terms of the number of Bohr magnetons (u_B) per $Y_3Fe_5O_{12}$ molecule. If we assume that M_a and M_d are antiparallel, then a net magnetization $M_d - M_a = 5u_B$ is expected. The experimental value is $4.96u_B$. To explain the other magnetization curves in Fig. 8.21a, we shall refer to Fig. 8.21b.

If the Y^{3+} ion is substituted for by a rare-earth trivalent ion, M_c will no longer be zero. The negative exchange interaction between sublattices d and a and between d and c makes M_d antiparallel to M_a and M_c which are by necessity parallel to each other. Thus, the net magnetization is given by $M_c + M_a - M_d$. At low temperatures, M_c is large and the sum of M_c and M_a determines the direction of the net magnetization. As the temperature increases, however, the value of M_c drops quickly because of the relatively weak interaction between sublattices d and c. Since $M_d > M_a$, there will be a temperature, called the compensation temperature, at which

$$M = M_c + M_a - M_d = 0$$

The curve shown in Fig. 8.21b is for GdIG calculated by Pauthenet[19] in his analysis of the magnetization curve.

PROBLEMS

8.1 Show that the magnetic potential at any point $Q(x,y,z)$ due to a magnetic dipole of Fig. 8.1b is given by

$$\Phi_m = \frac{(2\psi d)z}{4\pi r^3}$$

where $r^2 = x^2 + y^2 + z^2$. From Φ_m, obtain the following expressions for the magnetic fields:

$$H_x = \frac{(2\psi d)3zx}{4\pi r^5} \qquad H_y = \frac{(2\psi d)3zy}{4\pi r^5} \qquad H_z = \frac{(2\psi d)[(3z^2/r^2) - 1]}{4\pi r^5}$$

Check the above results against Eq. (7.42) if $2\psi d$ is substituted for p and H for E.

[19]   R. Pauthenet, *Ann. Phys.* (Paris), vol. 3, p. 424, 1958.

8.2 The vector potential $\mathbf{A}$ established by a current $\mathbf{I}$ is given by

$$\mathbf{A} = \oint \mathbf{I}\,\frac{dl}{4\pi r}$$

where $dl = a\,d\phi$ represents the elementary length of the current loop (Fig. 8.2) and r is the distance between the point $Q(x,y,z)$, at which $\mathbf{A}$ is measured, and the element dl. Express A_x and A_y in integral form in terms of a, ϕ, x, y, and z. The magnetic field $\mathbf{H}$ is related to $\mathbf{A}$ by $\mathbf{H} = \nabla \times \mathbf{A}$. Evaluate H_x (or H_y or H_z) by expanding r in terms of a Taylor's series in a and keeping only first-order terms in a. Verify that the result is identical to the one obtained in the previous problem if $2\psi d = I\pi a^2$. In evaluating the integral, we note that integrals $\oint \cos\phi\,d\phi = \oint \sin\phi\,d\phi = 0$ around the current loop.

8.3 Consider an electron moving in a circular orbit around a nucleus. Obtain equations [similar to Eq. (1.14)] of balance of force with and without a magnetic flux density $\mathbf{B}$ perpendicular to the orbit. Show that the new angular frequency ω with B is related to the old frequency ω_0 without B by

$$\omega = \omega_0 + \omega_L$$

where $\omega_L = eB/2m_0 \ll \omega_0$, and the sense of electron motion is indicated in Fig. 8.5.

8.4 Check the dimension of Eqs. (8.16), (8.17), (8.18), and (8.19). Apply Eqs. (8.18) and (8.19) to two contrasting cases: (a) iron, which is ferromagnetic, and (b) sodium chloride, whose ions are diamagnetic. Using the lattice constant of Fe in Table 2.3 and assuming that each iron atom has a permanent dipole of 2 Bohr magnetons, estimate the value of M and compare the calculated value with the observed value ($\mu_0 M = 2.2$ weber/m^2 from Table 8.1). Find the magnetic susceptibility χ [defined in Eq. (8.9)] of NaCl from Eq. (8.19), assuming that each electron contributes equally to the susceptibility and each has an orbital radius equal to the Bohr radius. Compare the calculated value with the observed value of -1.2×10^{-5}. Using the observed value of χ, calculate the induced magnetization at $B = 1$ weber/m^2.

8.5 Verify the following expressions:

$$L_x\psi_0 = \frac{\hbar}{i}\,yf(r) \qquad L_y\psi_0 = -\frac{\hbar}{i}\,xf(r)$$

$$L_x^2\psi_0 = \hbar^2 zf(r) \qquad \text{and} \qquad L_y^2\psi_0 = \hbar^2 zf(r)$$

From the above relations, calculate the expectation values of L_x, L_y, and L^2 for the state ψ_0 [defined in Eq. (8.31)].

8.6 Show that

$$(L_x^2 + L_y^2 + L_z^2)\psi_A = 2\hbar^2\psi_A = \hbar^2 l(l+1)\psi_A$$

where ψ_A is given by Eq. (8.26) and L_x, L_y, and L_z by Eq. (8.36).

8.7 Express the ground states of free Cr^{3+}, Mn^{2+}, and Co^{2+} ions in spectroscopic notations.

8.8 Find the g factor and the effective number of Bohr magnetons associated with a Nd^{3+} ion.

8.9 Calculate the effective number of Bohr magnetons associated with Cr^{3+}, Mn^{2+}, and Co^{2+} ions in a crystalline solid. Show that the susceptibility of $MnSO_4$ (gram molecular weight $= 151$ and specific density $= 3.25$) at $300°K$ is equal to 3.7×10^{-3}. Find the permeability of $MnSO_4$ at $300°K$.

8.10 Consider the magnetic moments of V^{4+} ion (in VO_2 for example) with contributions from spin ($S = 1$) only. By extending the quantum mechanical treatment developed in the text [Eqs. (8.57) to (8.59)] to the present case, express the net magnetization as a function of the applied magnetic field. Find the distribution of electrons [that is, the ratio $N_1(S_z = -1)/N$ (total)] at the following temperatures: $T = 2°K$, $20°K$, and $300°K$.

8.11 Show that for a substance having an inner quantum number J, a generalization of Eq. (8.58) leads to

$$M = \frac{N\Sigma y \exp(-y/kT)}{\Sigma \exp(-y/kT)}$$

where $y = M_J g u_B \mu_0 H$ and the summation is over M_J from $M_J = J$, $(J - 1) \cdots$ to $-J$. The above summation can be evaluated exactly and is equal to

$$M = Ng u_B J \left[\frac{2J + 1}{2J} \coth \frac{(2J + 1)x}{2J} - \frac{1}{2J} \coth \frac{x}{2J} \right]$$

where $x = g u_B \mu_0 H/kT$ and the function in the brackets is known as Brillouin function. Verify that at high temperatures ($x \ll 1$), the quantum-mechanical result reduces to the classical result if $u^2 = g^2 J(J + 1)u_B^2$ in Eq. (8.55).

8.12 Using the result of the preceding problem, plot the value of $M/(Nu_B)$ as a function of x for Fe^{3+} and Gd^{3+} ions in ferric ammonium alum and gadolinium sulfate octahydrate, respectively. Compare the calculated values of Bohr magnetons/ion with the experimental values reported by W. E. Henry, *Phys. Rev.*, vol. 88, p. 559, 1952.

8.13 Read the book by J. H. Van Vleck ("Theory of Electric and Magnetic Susceptibilities," secs. 41–43, Oxford University Press, London, 1932) or the book by A. H. Morrish ("The Physical Principles of Magnetism," pp. 51–54, John Wiley & Sons, Inc., New York, 1965). Discuss the quantum mechanical treatment of paramagnetic and diamagnetic susceptibilities.

8.14 Read the articles by C. G. Shull *et al.* (*Phys. Rev.*, vol. 83, p. 333, 1951) and by R. A. Erickson (*Phys. Rev.*, vol. 90, p. 779, 1953). Discuss the application of neutron-diffraction experiments in the determination of magnetic arrangements in antiferromagnetic crystals. References to structure factors should be made in the discussion.

8.15 Show that the susceptibility $\chi_\|$ of an antiferromagnet with the applied field in the direction of M_A is given by

$$\frac{N\mu_0 u^2 \mathcal{L}'(x)}{kT + \frac{1}{2}(\alpha + \lambda)N\mu_0 u^2 \mathcal{L}'(x)}$$

where $\mathcal{L}'(x)$ is the derivative of the Langevin function with respect to its argument x. The above equation is obtained from Eq. (8.71) by expanding the Langevin function as a Taylor series in H.

8.16 Read the articles by P. W. Anderson (*Phys. Rev.*, vol. 79, p. 350, 1950) and by J. H. Van Vleck (*J. Phys. Rad.*, vol. 12, p. 262, 1951). Outline the model of superexchange interaction in explaining antiferromagnetism in MnO. Explain why the next-nearest-neighbor interactions are stronger than the nearest-neighbor interactions.

8.17 Read the article by E. W. Gorter (*Proc. IRE*, vol. 43, p. 1945, 1955). Describe the various M versus T curves possible as the chemical composition of a ferrite is varied.

REFERENCES

Bates, L. F.: "Modern Magnetism," 4th ed., Cambridge University Press, New York, 1961.

Born, M.: "Atomic Physics," Hafner Publishing Co., New York, 1957.

Bozorth, R. M.: "Ferromagnetism," D. Van Nostrand Co., Inc., New York, 1951.

Foex, G.: "Diamagnetism and Paramagnetism," Pergamon Press, London, 1957.

Kittel, C.: "Introduction to Solid State Physics," 2d ed., John Wiley & Sons, Inc., New York, 1956.

Morrish, A. H.: "The Physical Principles of Magnetism," John Wiley & Sons, Inc., New York, 1965.

Smart, J. S.: "The Néel Theory of Ferrimagnetism," *Am. Jour. Phys.*, vol. 23, p. 356, 1955.

Snoek, J. L.: "New Developments in Ferromagnetic Materials," 2d ed., Elsevier, Amsterdam, 1949.

Van Vleck, J. H.: "The Theory of Electric and Magnetic Susceptibilities," Oxford University Press, London, 1932.

Wolf, W. P.: "Ferrimagnetism," in *Reports on Progress in Physics*, vol. 24, p. 212, 1961.

9 PARAMAGNETIC RESONANCE PHENOMENA AND MASERS

9.1 PARAMAGNETIC MATERIALS

Most electrical engineers are familiar with the properties of ferromagnetic substances such as iron, nickel, and cobalt, because these substances have been used in electrical engineering applications for decades. Paramagnetic materials, on the other hand, have been completely neglected by electrical engineers until very recently. The reason is obvious. In the old days, electrical engineers were very much occupied with things big, powerful, and of low frequency. As we recall, the magnetization of a paramagnet is many orders of magnitude smaller than that of a ferromagnet, and hence the magnetic effect produced by a paramagnetic substance would be of very little interest to an electrical engineer. However, the trend has been changing. In order to increase the number of communication channels and to lower the threshold of detectability for signals, our interest has shifted toward high-frequency and low-level signal detection schemes.

During the course of this development, however, we should not forget the role played by physicists and the impact of their discoveries on engineering. In the meantime, we should also emphasize the role played by engineers in bringing advances in modern technology and making them available to physicists. The development of the maser[1] serves as an instructive example of the cooperative roles played by both disciplines. Microwave techniques have made possible many basic investigations on molecular spectroscopy and on paramagnetic resonance phenomena, which have been instrumental in bringing about a practical, working maser. The interplay between basic and applied sciences has become increasingly fruitful and important.

It goes without saying that masers are finding many important applications as microwave, low-level amplifiers or oscillators. There are two types of masers, the gaseous[2] and the solid-state, classified

---

[1]   The idea of using stimulated emission for amplification was first conceived by C. H. Townes and the word *maser* was coined in a paper by J. P. Gordon, H. J. Zeiger, and C. H. Townes, *Phys. Rev.*, vol. 99, p. 1264, 1955.

[2]   For the development of gaseous masers, the reader is referred to a series of articles by C. H. Townes and coworkers, *Phys. Rev.*, vol. 95, p. 282, 1954; *Phys. Rev.*, vol. 99, p. 1264, 1955; *Phys. Rev.*, vol. 102, p. 1308, 1956.

according to whether the maser action takes place in a gaseous or solid medium. We shall concentrate our discussion on the solid-state type which is made of a single crystal of a paramagnetic solid. In the remaining part of this section, we shall first acquaint the reader with some of the commonly known paramagnetic substances and then discuss the pertinent properties of a paramagnetic material.

A large class of paramagnetic substances exists in the form of (1) hydrated sulphates such as $Gd_2(SO_4)_3 \cdot 8H_2O$, $MnSO_4 \cdot 4H_2O$, (2) oxides such as Pr_2O_3, $Cr_2O_3 \cdot 7H_2O$, or (3) halide salts such as CeF_3, $NiCl_2$. Since the negative ions such as O^{2-}, F^- and the radical SO_4^{2-} have only complete shells, paramagnetism is due to the positive ions of the rare-earth or the iron group elements such as Gd^{3+} and Mn^{2+}. In a crystal, as discussed in the previous chapter, magnetic ions are subject to various internal fields. What we are particularly concerned with are the fields resulting from mutual interactions among neighboring magnetic ions, that is to say (1) the Weiss molecular or the exchange field of electrostatic origin and (2) the dipole field of purely magnetic type. If a paramagnet is visualized as an assemblage of individual magnetic dipoles, then each dipole is coupled to neighboring dipoles through the above-mentioned internal fields. From electric circuit theory, one is familiar with the fact that couplings between identical resonant circuits will split the single resonant frequency into closely-spaced resonances. Likewise couplings between magnetic ions tend to spread the magnetic energy from a sharply defined value into a band. In a maser, the energy difference between two magnetic energy levels $\mathcal{E}_1$ (upper) and $\mathcal{E}_2$ (lower) determines the operating frequency $\nu = (\mathcal{E}_1 - \mathcal{E}_2)/h$. Hence, a broadening of the magnetic energies means a spreading in ν. There are also other undesirable aspects associated with this broadening as we shall discuss later in this chapter. An obvious way to reduce the mutual interaction is to disperse the magnetic ions so that they cannot interact strongly. This may be accomplished in a way similar to what is commonly done in semiconductors. We can use a nonmagnetic substance as the host crystal and dope it with magnetic impurities. The common maser material ruby is a good example. It is composed of mainly nonmagnetic aluminum oxide Al_2O_3 and a very small percentage of magnetic chromium oxide Cr_2O_3.

Any compound whose constituent ions or radicals have only complete shells is nonmagnetic. Substances having such properties are numerous, such as Al_2O_3, TiO_2, $CaWO_4$, $BaSO_4$, $CaCO_3$, La-$(C_2H_5SO_4)_3 \cdot 9H_2O$, $ZnSiF_6 \cdot 6H_2O$, and so on, the last two being suggested as host crystals in Bloembergen's proposal of a three-level maser. Among paramagnetic ions, transition elements of the iron and rare-earth group have been most extensively discussed and investigated. Host crystals have an important effect upon the magnetic behavior of paramagnetic ions because of the crystalline electrostatic field. In the following sections, we shall first introduce the simple theory of the paramagnetic resonance experiment assuming that paramagnetic ions are free of the crystalline field. Then the role of the crystalline field will be discussed. Finally, the principles of masers will be explained and our knowledge from the resonance experiment will be applied to masers.

9.2 THE CLASSICAL EQUATION OF MOTION FOR A SINGLE DIPOLE

Let us consider an elementary magnetic dipole having dipole moment u placed in a d-c magnetic field B_0. Under equilibrium conditions, the dipole moment vector $\mathbf{u}$ lies in the direction of $\mathbf{B_0}$ which is assumed to be in the z direction. Now let us suppose that the magnetic dipole is tilted by a small external force so that it makes a small angle θ with B_0 as shown in Fig. 9.1. Since an ideal situation of an isolated single dipole is assumed, the only field acting on $\mathbf{u}$ is B_0. Thus, the torque [Eq. (8.52)] exerted on $\mathbf{u}$ is equal to

$$\mathbf{T} = \mathbf{u} \times \mathbf{B} \tag{9.1}$$

Associated with the magnetic dipole $\mathbf{u}$, as shown in Sec. 8.6, there is an angular momentum $\mathbf{J}$. Furthermore, a general expression relating $\mathbf{u}$ and $\mathbf{J}$ exists as given by Eq. (8.43). Remembering that the angular

Figure 9.1

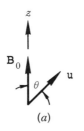

(a)

The motion of a magnetic dipole $\mathbf{u}$ in a d-c magnetic field B_0. Diagram (a) defines the spatial relation between $\mathbf{u}$ and $\mathbf{B_0}$. Diagram (b) shows the angular momentum $\mathbf{J}$ at two instants ($t_2 > t_1$) as a result of the torque $\mathbf{T}$ acting on the dipole. Diagram (c) shows the precessional motion of the dipole.

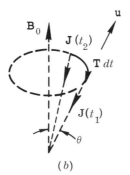

(b)

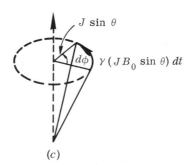

(c)

momentum is equal to $\hbar\mathbf{J}$, we can rewrite Eq. (8.43) as

$$\mathbf{u} = -\frac{e}{2m_0}\, g\mathbf{J} = -\gamma\mathbf{J} \qquad (9.2)^3$$

where $\gamma = ge/2m_0$ is the numerical ratio between $\mathbf{u}$ and $\mathbf{J}$, usually referred to as the *gyromagnetic ratio*. From Eqs. (9.1) and (9.2), we have the classical equation of motion

$$\frac{d\mathbf{J}}{dt} = \mathbf{T} = \mathbf{u} \times \mathbf{B} \qquad \text{or} \qquad \frac{d\mathbf{u}}{dt} = -\gamma\mathbf{u} \times \mathbf{B} \qquad (9.3)$$

It is instructive to examine physically the motion of the elementary dipole. The incremental angular momentum represented by $\mathbf{T}\, dt$ in Fig. 9.1 is always perpendicular to $\mathbf{J}$, thus the magnitude of $\mathbf{J}$ remains the same but its direction changes continuously. This results in a uniform precessional motion of the dipole around the d-c magnetic field as shown in Fig. 9.1b. The problem can also be analyzed from the energy point of view. Since the elementary dipole is alone by itself in a magnetic field, it can neither acquire nor lose energy. Therefore, once it makes an angle θ with $\mathbf{B}$, the conical angle θ of the precessional motion must not vary. The angular frequency of precession can be determined in the following manner. Referring to Fig. 9.1c, the magnitude of the projection of $\mathbf{J}$ in the xy plane is $J \sin \theta$ and the incremental angular momentum is equal to $uB_0 \sin \theta\, dt = \gamma JB_0 \sin \theta\, dt$; thus we have

$$(J \sin \theta)\, d\phi = \gamma JB_0 \sin \theta\, dt \qquad \text{or} \qquad \omega_0 = \frac{d\phi}{dt} = \gamma B_0 \qquad (9.4)$$

where ω_0 is commonly referred to as the Larmor precessional frequency. For magnetic dipoles resulting from electron spin, $g = 2$ and $\gamma = 1.76 \times 10^{11}$ coul/kg. Thus, for a magnetic flux density $B_0 = 0.3$ weber/m², $\omega_0 = 5.28 \times 10^{10}$ rad/sec corresponding to a frequency in the microwave region.

From the above analysis, we see that the magnetization vector has a time-varying component in the xy plane. Hence an a-c magnetic field of angular frequency $\omega = \gamma B_0$ in the xy plane will interact synchronously with the elementary dipole. The phenomenon is analogous to that of forced oscillation in electric circuits. To make our statement more definitive the problem must be formulated mathematically. For generality, we assume that the a-c fields H_x and H_y vary sinusoidally with an angular frequency ω not necessarily equal to ω_0 and that the phase constants in H_x and H_y are unspecified.

---

[3] We use J to indicate both the quantum number and the total angular momentum. These two quantities differ by a factor $\hbar$. In Eq. (9.2), $\mathbf{J}$ represents the total angular momentum. The reader can easily determine which quantity J represents by referring to the context.

The three component equations of Eq. (9.3) read:

$$\frac{du_x}{dt} = i\omega u_x = -\gamma(u_y B_0 - u_z B_y)$$

$$\frac{du_y}{dt} = i\omega u_y = -\gamma(u_z B_x - u_x B_0) \qquad (9.5)$$

$$\frac{du_z}{dt} = 0 = -\gamma(u_x B_y - u_y B_x)$$

Note that the projection of the magnetization vector upon the z axis is constant and hence $du_z/dt = 0$ in Eq. (9.5). From the first two equations, we obtain

$$u_x = \frac{\gamma u_z}{\omega_0{}^2 - \omega^2}(\omega_0 B_x + i\omega B_y)$$

$$u_y = \frac{\gamma u_z}{\omega_0{}^2 - \omega^2}(-i\omega B_x + \omega_0 B_y) \qquad (9.6)$$

If there are N free dipoles per unit volume, then the a-c magnetization has components $M_x = Nu_x$ and $M_y = Nu_y$. According to Eq. (8.10), the a-c susceptibility for such a system is given by

$$\chi_{xx} = \chi_{yy} = \frac{\omega_0 \gamma \mu_0 N u_z}{\omega_0{}^2 - \omega^2} \qquad (9.7a)$$

$$\chi_{xy} = -\chi_{yx} = \frac{i\omega \gamma \mu_0 N u_z}{\omega_0{}^2 - \omega^2} \qquad (9.7b)$$

From Eq. (9.7), it is clear that the a-c susceptibility is not a scalar but a tensor. Two important results of our analysis are: (1) The susceptibility shows a resonance phenomenon at $\omega = \gamma B_0$ referred to as the paramagnetic resonance. (2) The susceptibility tensor is antisymmetrical, and its consequence will be examined later on. As stated earlier, the dipole moment vector precesses around the d-c magnetic field with an angular frequency ω_0. With B_0 pointing in the z direction, the direction of the precession in the xy plane is counterclockwise as shown in Fig. 9.2a. Qualitatively, resonance will occur only when the resultant magnetic field is synchronous with the magnetization vector. In other words, the a-c magnetic field must be right-hand circularly polarized. The sense right-hand refers to the direction of $\mathbf{B}_0$ and not to the z axis, although in the present case, $\mathbf{B}_0$ happens to be in the z direction. Thus according to Fig. 9.2b we have $B_x = B_1 \cos \omega t$ and $B_y = B_1 \sin \omega t$ which may be expressed in complex notation as

$$B_y = -iB_x \qquad (9.8a)$$

For a left-hand circularly polarized wave,

$$B_y = iB_x \qquad (9.8b)$$

Substituting Eq. (9.8) into Eq. (9.6), we find

$$u_x{}^L = \frac{\gamma u_z}{\omega_0 + \omega} B_x \qquad u_y{}^L = -i\frac{\gamma u_z}{\omega_0 + \omega} B_x \qquad (9.9)$$

$$u_x{}^R = \frac{\gamma u_z}{\omega_0 - \omega} B_x \qquad u_y{}^R = -i\frac{\gamma u_z}{\omega_0 - \omega} B_x \qquad (9.10)$$

The superscripts L and R refer to the left-hand and right-hand polarized waves, respectively. From Eqs. (9.9) and (9.10), we see that only the right-hand polarized field will show resonance phenomena. The result confirms our earlier conclusion reached qualitatively (by referring to Fig. 9.2) that the a-c field must be synchronous with the precessional motion of the dipole in order to have magnetic resonance.

Although the classical equation of motion based on the simple model of a single dipole correctly predicts the resonance condition, it is fundamentally inadequate to describe a physical system because it does not embody any loss mechanism to maintain a proper energy balance for the system. To be specific, in Eq. (9.3), no term is provided for the magnetic dipole to transfer part of its energy gained from the a-c field either to another magnetic dipole or to the host

Figure 9.2

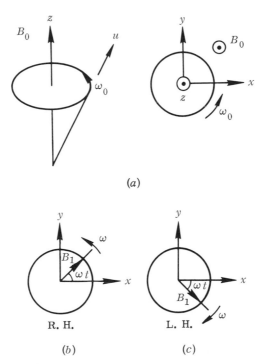

(a) *The precessional motion of a dipole;* (b) *a right-hand circularly polarized field; and* (c) *a left-hand circularly polarized field. The right- and left-hand senses use the direction of* **B₀** *as the reference axis.*

lattice. As a consequence, the solution attained thus far has certain peculiarities. First, the opening of the conical angle θ, being equal to $\tan^{-1} \gamma B_1/(\omega_0 - \omega)$ from Eq. (9.10), is caused by the a-c magnetic field, and yet at resonance $\theta = \pi/2$ irrespective of the magnitude of B_1. Second, the third equation involving du_z/dt of Eq. (9.5) simply cannot be valid in its present form because its right-hand side is not zero. Most important of all, we find that once the system is disturbed from its equilibrium position, there is no restoring torque to return the dipole to its original position. The inadequacies of the simple classical equation will be amended in the following section.

9.3 THE BLOCH EQUATION OF MOTION

We shall now extend our analysis to an assemblage of magnetic dipoles as they exist in solids. If there are N dipoles in a volume V, by substituting $\mathbf{M}(= N\mathbf{u}/V)$ for $\mathbf{u}$ in the equation of motion [Eq. (9.3)], we have

$$\frac{d\mathbf{M}}{dt} = -\gamma \mathbf{M} \times \mathbf{B} \tag{9.11}$$

However, Eq. (9.11) is incomplete for the same reason that Eq. (9.3) was incomplete. First, there is no mechanism in the equation to take care of the mutual interaction between neighboring magnetic ions, and second, the question of the rate balance of energy transfer is not considered. The Bloch equation of motion[4] takes care of these two considerations by incorporating two relaxation terms into the original equation as follows:

$$\frac{dM_{x,y}}{dt} = -\gamma(\mathbf{M} \times \mathbf{B})_{x,y} - \frac{M_{x,y}}{T_2} \tag{9.12}$$

$$\frac{dM_z}{dt} = -\gamma(\mathbf{M} \times \mathbf{B})_z - \frac{M_z - M_0}{T_1} \tag{9.13}$$

where T_1 and T_2 are generally referred to as the *spin-lattice* and *spin-spin relaxation times*. The terms involving T_1 and T_2 are introduced purely on a phenomenological basis; therefore, the best way to justify their inclusion in Eqs. (9.12) and (9.13) and to interpret their physical meaning is by examining the solution of these equations.

Equation (9.12) can easily be solved by noting that it differs from Eq. (9.5) in that $i\omega$ in the latter is now replaced by $i\omega + T_2^{-1}$ in the former. Again assuming a sinusoidal time variation for M_x and M_y,

---

[4] The set of phenomenological equations was originally proposed by F. Bloch (*Phys. Rev.*, vol. 70, p. 460, 1946) to describe dynamic magnetic behavior of interacting nuclear paramagnets. The equations apply equally well to interacting electronic paramagnets.

we have

$$M_x = \frac{\gamma M_z}{\omega_0{}^2 - (\omega - iT_2{}^{-1})^2} [\omega_0 B_x + (i\omega + T_2{}^{-1})B_y] \qquad (9.14a)$$

$$M_y = \frac{\gamma M_z}{\omega_0{}^2 - (\omega - iT_2{}^{-1})^2} [(i\omega + T_2{}^{-1})(-B_x) + \omega_0 B_y] \qquad (9.14b)$$

For simplicity we shall assume a right-hand polarized wave, $B_y = -iB_x$, thus,

$$M_x = (\chi' - i\chi'')H_x = \frac{(\omega_0 - \omega) - iT_2{}^{-1}}{(\omega_0 - \omega)^2 + T_2{}^{-2}} (\gamma\mu_0 M_z)H_x \qquad (9.15)$$

The equation relating M_y to H_y is exactly the same. It is to be noted that in getting Eq. (9.15) from Eq. (9.14), B is approximated by $\mu_0 H$ since $M \ll H$ for paramagnetic substances. Now a brief explanation of the physical meaning of real and imaginary susceptibility is in order.

Consider the voltage across and the current through a coil containing a paramagnetic substance. In Eq. (9.15), H is proportional to the current while dB/dt is proportional to the voltage. Consequently, the power delivered to a unit volume of the sample is given by $\mu_0 H\, dM/dt$. For a two-dimensional case as we have here, the instantaneous power delivered to the sample is given by

$$P = \mu_0 \left(H_x \frac{dM_x}{dt} + H_y \frac{dM_y}{dt}\right) = P_x + P_y \qquad (9.16)$$

In Eq. (9.16) all quantities are real. Thus, in substituting Eq. (9.15) into Eq. (9.16), Eq. (9.15) must be expressed in real notation. That is, for $H_x = \text{Re } [H_1 \exp (i\omega t)]$ and $H_y = \text{Re } [-iH_1 \exp (i\omega t)]$ where Re stands for taking the real part, then

$$M_x = \chi' H_1 \cos \omega t + \chi'' H_1 \sin \omega t \qquad (9.17a)$$

and
$$M_y = \chi' H_1 \sin \omega t - \chi'' H_1 \cos \omega t \qquad (9.17b)$$

The instantaneous power is given by

$$P_y = \mu_0 \left(\frac{\omega}{2} \chi' H_1{}^2 \sin 2\omega t + \omega\chi'' H_1{}^2 \sin^2 \omega t\right) \qquad (9.18)$$

and
$$P = \mu_0 \omega \chi'' H_1{}^2$$

From Eq. (9.18), we see that the real susceptibility χ' accounts for the reactive power while the imaginary susceptibility leads to an average power absorbed in the sample. Hence χ' and χ'' are often referred to as the *dispersive* and *absorptive* parts of the susceptibility.

In Fig. 9.3, the values of χ' and χ'' normalized by the value of χ'' at resonance are plotted as functions of B_0 for a given ω. With losses accounted for, the value of susceptibility remains finite at resonance in contrast to the result predicted by Eq. (9.6). Furthermore, the absorption curve has a finite width, usually referred to as the *line width*, as indicated by ΔB_0 in Fig. 9.3. The value of χ''/χ''_{res} drops to one-half when $B_0 - (B_0)_{res} = \pm\Delta B_0/2$. Thus, the center of the absorption curve is determined by

$$(\omega_0)_{res} = \omega \qquad \text{or} \qquad \gamma(B_0)_{res} = \omega \tag{9.19}$$

while the spread or the line-width of the absorption curve is related to the relaxation time T_2 by

$$[\omega_0 - (\omega_0)_{res}]^2 = T_2^{-2} \qquad \text{or} \qquad \gamma\,\Delta B_0 = \frac{2}{T_2} \tag{9.20}$$

The physical meaning of ΔB_0 and T_2 will be explained in detail in Sec. 9.7. Qualitatively, however, the relaxation time T_2 does seem to bring out phenomenologically the same effect as expected from the mutual interaction between neighboring magnetic ions, as the absorption curve now spreads over a band of B_0. As to the dispersion curve χ', its principal features are illustrated and self-evident from Fig. 9.3. The derivation of magnetic susceptibilities for a linearly polarized wave is more involved algebraically, so it is left to the reader as a problem.

Let us now turn our attention to Eq. (9.13) and examine the physical significance of each term. For simplicity, a right-hand circularly polarized wave is assumed. Using the results of Eq. (9.17), we find

$$\gamma(M_z B_y - M_y B_z) = \gamma\mu_0\chi'' H_1^2 = \bar{P}\frac{\gamma}{\omega} \tag{9.21}$$

where $\bar{P}$ is the average power delivered to the sample. Substituting Eq. (9.21) into Eq. (9.13) and noting that at resonance $\omega = \omega_0 = \gamma B_0$, we have

$$\frac{d}{dt}(B_0 M_z) = -\bar{P} - \frac{B_0 M_z - B_0 M_0}{T_1} \tag{9.22}$$

Note that $\mathcal{E} = -B_0 M_z$ represents the magnetic energy of a paramag-

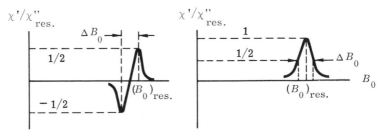

Figure
9.3

The real and imaginary parts of susceptibility χ' and χ'' as functions of the d-c induction field B_0 for a fixed angular frequency ω of the a-c field.

netic substance in the presence of B_0 while $\mathcal{E}_0 = -B_0 M_0$ is the equilibrium value of the magnetic energy. Thus, both Eq. (9.13) and Eq. (9.22) are really a statement of energy balance,

$$\frac{d}{dt}\mathcal{E} = \bar{P} - \frac{\mathcal{E} - \mathcal{E}_0}{T_1} \tag{9.23}$$

that is to say, the rate of increase in the magnetic energy is equal to the power supplied by the a-c field minus the power dissipated by the paramagnetic substance. One obvious way for dissipation to occur is by conversion of the magnetic energy into the heat energy of the lattice. For this reason, the relaxation time T_1 is usually referred to as the spin-lattice relaxation time.

To solve Eq. (9.13) together with Eq. (9.14), it is convenient to use complex notations. Thus, under equilibrium conditions, Eq. (9.13) can be written as

$$\mathrm{Re}\frac{\gamma}{2}(M_x B_y^* - M_y B_x^*) = \frac{M_0 - M_z}{T_1} \tag{9.24}$$

Substituting Eq. (9.14) into Eq. (9.24) and solving for M_z, we have

$$M_z = \frac{M_0[1 + (\omega_0 - \omega)^2 T_2^2]}{1 + (\omega_0 - \omega)^2 T_2^2 + T_1 T_2 (\gamma\mu_0 H_1)^2} \tag{9.25}$$

Note that Eq. (9.25) can also be obtained by directly substituting Eq. (9.21) into Eq. (9.22) and using the value of χ'' from Eq. (9.15). The factor $\frac{1}{2}$ in Eq. (9.24) is used for average power. From Eq. (9.21), the average power absorbed by the sample is

$$\bar{P} = \mu_0\omega H_1^2 \frac{\gamma\mu_0 M_0 T_2}{1 + (\omega_0 - \omega)^2 T_2^2 + T_1 T_2 (\gamma\mu_0 H_1)^2} \tag{9.26}$$

For clarity, we shall only discuss the case at resonance, where Eq. (9.26) reduces to

$$\bar{P} = \mu_0\omega H_1^2 \frac{\gamma\mu_0 M_0 T_2}{1 + T_1 T_2 (\gamma\mu_0 H_1)^2} \tag{9.27a}$$

From Eq. (9.21), the imaginary susceptibility is

$$\chi'' = \frac{\gamma\mu_0 M_0 T_2}{1 + T_1 T_2 (\gamma\mu_0 H_1)^2} \tag{9.27b}$$

For small values of the a-c field H_1, χ'' is a constant and increases with increasing T_2. That means, the narrower the resonance absorption curve, the larger the value of χ''. For large values of H_1, the value of $\bar{P}$ tends to saturate. In Fig. 9.4, the values of $\bar{P}$ and χ'' are plotted as functions of H_1^2 and the measurement leading to these curves is called the *saturation measurement*. It is seen that the value of χ'' drops to half its value at low values of H_1 when $T_1 T_2 (\gamma\mu_0 H_1)^2 = 1$. Thus, from the saturation measurement, the value of T_1 may be determined.

In summary, the Bloch equation of motion has incorporated two basic physical considerations: the mutual interaction and the energy transfer in a magnetic system. The resonance absorption curve has a finite line-width on account of dipole interactions which result in a relaxation parameter T_2. The average power absorbed tends to saturate because of lattice interactions which result in a relaxation time T_1. In order to understand the basic physical processes which account for T_1 and T_2, however, we need to examine the resonance phenomenon from a quantum-mechanical viewpoint. In the following sections, we shall try to interpret the meaning of such classical

Figure 9.4

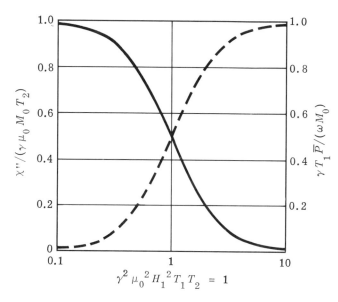

The saturation curves of the susceptibility χ'' (*solid curve*) and the power absorption (*dashed curve*) as functions of the a-c magnetic field. To make the curves universal, both the abscissa and ordinate are normalized and made dimensionless.

quantities as M_x, M_y, T_1, and T_2 in terms of the quantum-mechanical language. Such an understanding will provide the necessary background for explaining the principles involved in a maser.

9.4 ORBITAL AND SPIN ANGULAR MOMENTUM OPERATORS[5]

In Sec. 8.4, we discussed the solution of the Schrödinger wave equation, and showed that the property of the wave function $\psi(r,\theta,\phi)$ is

[5] For quantum-mechanical foundation, see for example J. L. Powell and B. Crasemann, "Quantum Mechanics," Addison-Wesley Publishing Co., Inc., Reading, Mass., 1961; N. F. Mott and I. N. Sneddon, "Wave Mechanics and Its Applications," Clarendon Press, Oxford, 1948; L. I. Schiff, "Quantum Mechanics," McGraw-Hill Book Company, New York, 1949.

determined by a set of quantum numbers, n, l, and m; or mathematically

$$\psi(r,\theta,\phi) = R(r)P_l{}^m(\cos\theta)\exp(im\phi) \qquad (9.28)$$

To be more indicative of this dependence, the wave function ψ is usually denoted as $\psi(n,l,m)$. In terms of the new notation, Eq. (8.37) can be written as

$$L_z\psi(n,l,0) = 0\psi(n,l,0)$$
$$L_z\psi(n,l,-1) = -\hbar\psi(n,l,-1) \qquad (9.29)$$
$$L_z\psi(n,l,+1) = \hbar\psi(n,l,+1)$$

Since the momentum operators, L_x, L_y, and L_z, are known differential operators as given by Eq. (8.36), and the wave function is a product of known functions as given by Eq. (9.28), we can work out expressions for L_x and L_y similar to Eq. (9.29). However, the notation just adopted is still too cumbersome, especially in a mathematical operation.

In quantum mechanics, as discussed in Sec. 8.4, the value of a physical quantity is given by its expectation value which involves a product such as $\psi_1^*L_x\psi_2$. Since the wave functions are orthogonal and normalized, the integral $\int\int\int\psi_i^*\psi_j\,dx\,dy\,dz$ is zero if $i \neq j$ and is unity if $i = j$. Thus, Eq. (9.29) can be transformed into a new set of equations by adopting the following notation:

$$(n,l,-1|L_z|n,l,-1)$$
$$= \int\int\int\psi^*(n,l,-1)L_z\psi(n,l,-1)\,dx\,dy\,dz = -\hbar$$
$$(n,l,0|L_z|n,l,-1) \qquad (9.30)$$
$$= \int\int\int\psi^*(n,l,0)L_z\psi(n,l,-1)\,dx\,dy\,dz = 0, \ldots$$

In general, we have

$$(n,l,m|L_z|n',l',m') = \int\int\int\psi^*(n,l,m)L_z\psi(n',l',m')\,dx\,dy\,dz \qquad (9.31)$$

and similar expressions for L_x and L_y. The various quantities $(n,l,m|L_z|n',l',m')$ are called *matrix elements* of operator L_z connecting the states $\psi(n,l,m)$ and $\psi(n',l',m')$.

To evaluate the matrix elements of the angular momentum operators, L_x, L_y, and L_z, it is convenient to use polar coordinates, for which Eq. (8.36) becomes

$$L_x = i\hbar\left(\sin\phi\frac{\partial}{\partial\theta} + \cot\theta\cos\phi\frac{\partial}{\partial\phi}\right)$$

$$L_y = i\hbar\left(-\cos\phi\frac{\partial}{\partial\theta} + \cot\theta\sin\phi\frac{\partial}{\partial\phi}\right) \qquad (9.32)$$

$$L_z = -i\hbar\frac{\partial}{\partial\phi}$$

The matrix elements with which we are concerned in paramagnetic resonance involve states having the lowest principal quantum number, and hence we can omit n by implication in Eq. (9.31). In other words, we can integrate the radial part of $\psi^* L_z \psi$ in Eq. (9.31) and rewrite Eq. (9.31) in polar coordinates as

$$(l,m|L_z|l',m') = \int_0^\pi \int_0^{2\pi} P_l^m(\cos\theta)\exp(-im\phi)\left(-i\hbar\frac{\partial}{\partial\phi}\right)$$
$$P_{l'}^{m'}(\cos\theta)\exp(im'\phi)\sin\theta\,d\theta\,d\phi \quad (9.33)$$

where $P_l^m(\cos\theta)$ is the normalized associated Legendre polynomial such that

$$\int_0^\pi [P_l^m(\cos\theta)]^2 \sin\theta\,d\theta = \frac{1}{2\pi} \quad (9.34)$$

The factor $1/2\pi$ takes care of the normalization of the function $\exp(im\phi)$. Thus, except for a normalization factor A, the functions $P_l^m(\cos\theta)$ correspond to those given by Eq. (8.25). After evaluating the factor A from Eq. (9.34), we find

$$P_1^0 = \sqrt{\frac{3}{4\pi}}\cos\theta \quad \text{and} \quad P_1^1 = -\sqrt{\frac{3}{8\pi}}\sin\theta \quad (9.35)$$

As an example, we shall evaluate $(1,1|L_x|1,0)$ as follows:

$$(1,1|L_x|1,0) = \int_0^\pi \int_0^{2\pi} \sqrt{\frac{3}{8\pi}}\sin\theta\exp(-i\phi)$$
$$\left[-i\hbar\left(\sin\phi\frac{\partial}{\partial\theta} + \cot\theta\cos\phi\frac{\partial}{\partial\phi}\right)\sqrt{\frac{3}{4\pi}}\cos\theta\right]\sin\theta\,d\theta\,d\phi$$
$$= \int_0^\pi \int_0^{2\pi} \frac{-3}{4\pi\sqrt{2}}\sin\theta\exp(-i\phi)(-i\hbar\sin\theta\sin\phi)$$
$$\sin\theta\,d\theta\,d\phi$$
$$= \frac{\hbar}{\sqrt{2}} \quad (9.36)$$

In general, we have

$$(l, m+1|L_x|l,m) = \frac{\hbar}{2}\sqrt{l(l+1) - m(m+1)} = (l, m|L_x|l,m+1)$$

$$(l, m+1|L_y|l,m) = -\frac{\hbar i}{2}\sqrt{l(l+1) - m(m+1)} \quad (9.37)[6]$$
$$= -(l,m|L_y|l, m+1)$$
$$(l,m|L_z|l,m) = \hbar m$$

with all other matrix elements vanishing.

Now a brief comment on the meaning of the matrix elements follows. Mathematically, we may view a system of orthogonal

[6] See for example J. L. Powell and B. Crasemann, "Quantum Mechanics," p. 343, Addison-Wesley Publishing Co., Inc., Reading, Mass., 1961; N. F. Mott and I. N. Sneddon, "Wave Mechanics and Its Applications," p. 97, Clarendon Press, Oxford, 1948.

wave functions as a multidimensional linear vector space, each state (n,l,m) representing one of the orthogonal coordinates. Referring to the example given in Sec. 8.4, that is $l = 1$, the three orthogonal coordinates are ψ_{-1}, ψ_0 and ψ_{+1}. Thus, the state ψ_A can be interpreted as a unit vector having projections upon the ψ_{-1} and ψ_{+1} axes according to Eq. (8.31); in other words, it is a mixture of the ψ_{-1} and ψ_{+1} states. What the differential operator L_x, L_y, or L_z does to ψ is to bring the ψ vector into a new orientation in the vector space. Physically, this means a change in the composition of the state. For example, from Eq. (9.36), we see that the operator L_x transforms the old state ψ (1,0) into a new state which has a component along the ψ (1,1) axis. A nonvanishing matrix element for an operator connecting two states means that the operator can cause *transitions* between these two states. From Eq. (9.37), we see that the operator L_x and L_y can cause transitions $\Delta m = \pm 1$ and the operator L_z leads to $\Delta m = 0$. The condition for a nonvanishing matrix element is known as the quantum-mechanical *selection rule*. It is worth noting that the matrix elements of $L_x + iL_y$ and $L_x - iL_y$ have the following property:

$$(l, m + 1|L_x + iL_y|l,m) = \hbar \sqrt{l(l + 1) - m(m + 1)}$$
$$= (l,m|L_x - iL_y|l, m + 1)$$
$$(l,m|L_x + iL_y|l, m + 1) = 0 = (l, m + 1|L_x - iL_y|l,m)$$

(9.38)

Consequently, the operator $L_x + iL_y$ raises while $L_x - iL_y$ lowers the magnetic quantum number m by one. We shall use this property quite often later on.

The properties of the spin angular momentum operators s_x, s_y, and s_z are quite similar to those of the corresponding orbital angular momentum operators except that the former are quantized in $\hbar/2$. Denote α and β as the state having $s_z = +\hbar/2$ and $-\hbar/2$ respectively. Since there are only two possible states, the quantities s_x, s_y, and s_z are usually expressed in matrix form for convenience as follows:

$$s_x = \frac{\hbar}{2}\begin{pmatrix} 0 & 1 \\ 1 & 0 \end{pmatrix} \qquad s_y = \frac{\hbar}{2}\begin{pmatrix} 0 & -i \\ i & 0 \end{pmatrix} \qquad s_z = \frac{\hbar}{2}\begin{pmatrix} 1 & 0 \\ 0 & -1 \end{pmatrix}$$

(9.39)

The convention is that the wave function is represented by a column matrix $\begin{pmatrix} \alpha \\ \beta \end{pmatrix}$. Thus, we have as an example

$$s_x\beta = \frac{\hbar}{2}\alpha \qquad s_x\alpha = \frac{\hbar}{2}\beta$$

(9.40a)

Expressed in the notation used in Eq. (9.37), Eq. (9.40a) becomes

$$(+|s_x|-) = \frac{\hbar}{2} \qquad \text{and} \qquad (-|s_x|+) = \frac{\hbar}{2}$$

(9.40b)

If s_z is substituted for L_z and the values of $m = -\frac{1}{2}$, $l = \frac{1}{2}$ are used in Eq. (9.37), we find

$$(+\tfrac{1}{2}|s_z|-\tfrac{1}{2}) = \frac{\hbar}{2} \quad \text{and} \quad (-\tfrac{1}{2}|s_z|+\tfrac{1}{2}) = \frac{\hbar}{2} \qquad (9.41)$$

in agreement with the result of Eq. (9.40). In other words, Eqs. (9.37) and (9.38) can also be applied to spin angular momentum operators s_x, s_y, and s_z if the spin quantum number s and its z component s_z are used in place of l and m in these equations.

If there are several electrons in an atom and the spins on each individual electron are combined to form a resultant spin, then in Eq. (9.37) the value of l should be substituted for by the total spin quantum number and m should be replaced by the z component of the spin quantum number. Take Ni^{2+} as an example. According to the discussion in Sec. 8.7, the ground state of Ni^{2+} has $S = 1$ and thus three possible values of $S_z = +1, 0,$ and -1. From Eq. (9.37), we find

$$(1|S_z|0) = \frac{\hbar}{2}\sqrt{2} = (0|S_z|1) \qquad (9.42)$$

The proof of Eq. (9.42) and other similar expressions is left for the reader as a problem. We should emphasize again that the general problem of the addition of angular momentum operators is complicated, as already discussed in Secs. 8.6 and 8.7. Therefore, the use of Eqs. (9.37) and (9.38) can only be extended to special cases where such an extension is possible as in the case of the paramagnetic ions of the rare-earth and the iron group. We shall further refer to Eqs. (9.37) and (9.38) when we discuss maser materials later in this chapter. The following notations are adopted for paramagnetic ions having more than one electron. The capital L and S respectively stand for the resultant orbital and spin angular-momenta quantum numbers, and the capital M and S_z are the respective magnetic quantum numbers (i.e., z component). In that case, Eqs. (9.37) and (9.38) should be modified accordingly by replacing l, m, s, s_z respectively, with L, M, S, and S_z.

9.5 THE EFFECTS OF A RADIATION FIELD ON A PARAMAGNETIC SUBSTANCE

In Sec. 1.7, we have presented the time-dependent Schrödinger equation for a single hydrogen atom and have shown that its solution can be expressed as the product of a spatial-dependent and a time-dependent function. Similarly, the electron state in a solid can be described by a proper time-dependent Schrödinger equation as

$$i\hbar\frac{\partial\psi}{\partial t} = \mathcal{H}\psi \qquad (9.43)$$

In Eq. (9.43), $\mathcal{3C}$ is the Hamiltonian operator which for electrons in a solid consists not only of the usual kinetic and potential energy of the electron as in Eq. (1.68), but also of the various interaction energies as discussed in Chapter 8. At present, however, we are concerned with the mathematical formulation of the problem, so we shall leave the discussion of how $\mathcal{3C}$ is determined to a later section. Therefore, in Eq. (9.43), we assume that $\mathcal{3C}$ is known and the solution of Eq. (9.43) has the following general form

$$\psi(q,t) = \psi(q) \exp\left(-\frac{i\mathcal{E}t}{\hbar}\right) \tag{9.44}$$

where for brevity q stands for the spatial coordinates x, y, and z and $\mathcal{E}$ is the energy of the state ψ. For clarity, we shall consider only two states ψ_A and ψ_B with energies $\mathcal{E}_A$ and $\mathcal{E}_B$, respectively. Thus, Eq. (9.43) becomes

$$i\hbar \frac{\partial \psi_A}{\partial t} = \mathcal{3C}\psi_A = \mathcal{E}_A\psi_A \tag{9.43a}$$

$$i\hbar \frac{\partial \psi_B}{\partial t} = \mathcal{3C}\psi_B = \mathcal{E}_B\psi_B \tag{9.43b}$$

The simplest example is an electron spin. In a magnetic field B_0, as shown in Sec. 8.8, the energy of the state $\psi(+) = \alpha$ with $s_z = +\frac{1}{2}$ is B_0u_B while that of $\psi(-) = \beta$ with $s_z = -\frac{1}{2}$ is $-B_0u_B$ where u_B is the Bohr magneton.

Suppose that in the presence of a radiation field, the Schrödinger wave equation becomes

$$i\hbar \frac{\partial \psi}{\partial t} = (\mathcal{3C} + U)\psi \tag{9.45}$$

where the quantity U is a small addition to the Hamiltonian due to the perturbing field. We further propose that the solution of Eq. (9.45) takes the following form:

$$\psi(q,t) = a(t)\psi_A(q,t) + b(t)\psi_B(q,t) \tag{9.46}$$

Note that $a^2 + b^2 = 1$ on account of the normalization of $\psi(q,t)$. Substituting Eq. (9.46) into Eq. (9.45) and utilizing Eq. (9.43), we find

$$i\hbar \left[\psi_A(q,t)\frac{da}{dt} + \psi_B(q,t)\frac{db}{dt}\right] = U\psi(q,t) \tag{9.47}$$

Using the orthogonality of ψ_A and ψ_B, we further obtain

$$i\hbar \frac{da}{dt} = \int \psi_A^* U\psi \, dq \tag{9.48}$$

and a similar expression for db/dt. In Eq. (9.48), $dq = dx \, dy \, dz$.

To see clearly the physical meaning of Eq. (9.48), we shall make the following simplifying assumptions. The state ψ_B is taken to be the lower energy state and at $t = 0$, $a = 0$ and $b = 1$. Furthermore, the perturbation term U is very small so that ψ will not differ appreciably from $\psi(t = 0) = \psi_B$. Thus, Eq. (9.48) becomes

$$a(t) = \frac{1}{i\hbar} \iint \psi_A^*(q) \exp\left(\frac{i\mathcal{E}_A t}{\hbar}\right) U \psi_B(q) \exp\left(\frac{-i\mathcal{E}_B t}{\hbar}\right) dq\, dt \quad (9.49)$$

For illustration, let us now consider the elementary dipole due to electron spin alone. The energy of the dipole in a magnetic field is equal to $B_0 u_z + B_x u_x + B_y u_y$ where B_0 is the d-c field in the z direction and B_x and B_y are the a-c radiation fields. Thus, the perturbing term U is given by

$$U = \gamma(B_x S_x + B_y S_y) \quad (9.50)$$

where $\gamma = ge/2m_0$ is the gyromagnetic ratio. Substituting Eq. (9.50) into Eq. (9.49), we find that the function $a(t)$ depends upon the product of two integrals I_1 and I_2 of the form

$$I_1 = \int \psi_A^*(q) S_{x,y} \psi_B(q)\, dq \quad (9.51)$$

$$I_2 = \int \exp\left(\frac{i\mathcal{E}_A t}{\hbar}\right) B_{x,y} \exp\left(\frac{-i\mathcal{E}_B t}{\hbar}\right) dt \quad (9.52)$$

It is to be noted that if the elementary dipole is due to the orbital motion of an electron, then in Eq. (9.51) S_x or S_y should be replaced by L_x or L_y respectively. If the value of $|a(t)|^2$ is nonzero, the dipole is capable of making a transition from the state B to state A.

The transition under consideration is caused by the interaction of the magnetic dipole with the a-c magnetic field, that is of the type $\mathbf{B} \cdot \mathbf{u}$, and hence it is referred to as the *magnetic-dipole transition*. The value of the integral I_2 is stationary only when $\mathcal{E}_A - \mathcal{E}_B = \hbar\omega$ where ω is the angular frequency of $B_{x,y}$. This establishes the resonance condition and ω is generally in the microwave region. For a free magnetic dipole, $\omega = \gamma B_0$ as shown in Sec. 9.2. The value of the integral I_1 is nonvanishing only if the matrix elements of $S_{x,y}$ or $L_{x,y}$ connecting the states ψ_A and ψ_B are nonzero. This establishes the selection rule for possible transitions between states. For a free magnetic dipole, $\Delta M = \pm 1$ for L_x or L_y, $\Delta M = 0$ for L_z, $\Delta S_z = \pm 1$ for S_x or S_y, and $\Delta S_z = 0$ for S_z, as discussed in the previous section. In a solid, however, the magnetic dipole is subject to the influence of the crystalline field. We shall discuss such cases extensively in later sections and work out the resonance condition and the selection rule for some typical cases.

9.6 THE QUANTUM-MECHANICAL TREATMENT OF PARAMAGNETIC RESONANCE

In this section, we shall derive the classical equation of motion from the quantum-mechanical viewpoint,[7] and again start with Eqs. (9.45) and (9.46). However, it is convenient to alter Eq. (9.46) so that the time-dependent parts of $\psi_A(q,t)$ and $\psi_B(q,t)$ are grouped with $a(t)$ and $b(t)$, respectively, to form $A(t)$ and $B(t)$. Thus, we have

$$\psi(q,t) = A(t)\psi_A(q) + B(t)\psi_B(q) \tag{9.53}$$

For generality, the simplifying assumptions imposed on $a(t)$ and $b(t)$ in the previous section will be removed. Substituting Eq. (9.53) into Eq. (9.45) gives

$$i\hbar \left[\frac{\partial A}{\partial t} \psi_A(q) + \frac{\partial B}{\partial t} \psi_B(q) \right] = A\mathcal{E}_A\psi_A(q) + AU\psi_A(q)$$
$$+ B\mathcal{E}_B\psi_B(q) + BU\psi_B(q) \tag{9.54}$$

Multiplying both sides by $\psi_A^*(q)$ or $\psi_B^*(q)$ and utilizing the orthogonality of ψ_A and ψ_B, we obtain

$$i\hbar \frac{\partial A}{\partial t} = A(\mathcal{E}_A + \mathcal{E}_{AA}) + B\mathcal{E}_{AB}$$
$$\tag{9.55}$$
$$i\hbar \frac{\partial B}{\partial t} = A\mathcal{E}_{BA} + B(\mathcal{E}_B + \mathcal{E}_{BB})$$

In the above equation,

$$\begin{aligned}
\mathcal{E}_{AA} &= \int\psi_A^*(q)U\psi_A(q)\, dq \\
\mathcal{E}_{BB} &= \int\psi_B^*(q)U\psi_B(q)\, dq \\
\mathcal{E}_{AB} &= \int\psi_A^*(q)U\psi_B(q)\, dq = (A|U|B) = \mathcal{E}_{BA}^* \\
\mathcal{E}_{BA} &= \int\psi_B^*(q)U\psi_A(q)\, dq = (B|U|A) = \mathcal{E}_{AB}^*
\end{aligned} \tag{9.56}$$

For most cases of interest, $\mathcal{E}_{AA}$ or $\mathcal{E}_{BB}$ is either zero or negligibly small compared with $\mathcal{E}_A$ or $\mathcal{E}_B$ respectively, so these terms will be dropped in subsequent discussions. The relation $(A|U|B) = (B|U|A)^*$ is observed only for certain quantum-mechanical operators known as *Hermitian operators*. A proof of this relationship for the specific example of an elementary dipole to be treated below can be found in Prob. 9.9.

In Eq. (9.53), the coefficients $A(t)$ and $B(t)$ are complex quantities. In order to interpret these coefficients physically, we shall form the real combinations of $A(t)$ and $B(t)$. These combinations are

---

[7] The discussion here follows the outline of a treatment given by R. P. Feynman, F. L. Vernon, Jr., and R. W. Hellwarth, *Jour. Appl. Phys.*, vol. 28, p. 49, 1957.

$$u_1 = -(AB^* + BA^*)$$
$$u_2 = i(BA^* - AB^*)$$
$$u_3 = BB^* - AA^*$$

(9.57)

Differentiating Eq. (9.57) with respect to time and using Eq. (9.55), we find

$$i\hbar \frac{du_1}{dt} = u_3(\mathcal{E}_{BA} - \mathcal{E}_{AB}) - iu_2(\mathcal{E}_A - \mathcal{E}_B)$$

$$i\hbar \frac{du_2}{dt} = -iu_3(\mathcal{E}_{BA} + \mathcal{E}_{AB}) + iu_1(\mathcal{E}_A - \mathcal{E}_B)$$

(9.58)

$$i\hbar \frac{du_3}{dt} = u_1(\mathcal{E}_{AB} - \mathcal{E}_{BA}) + iu_2(\mathcal{E}_{AB} + \mathcal{E}_{BA})$$

For simplicity, the following substitutions are made:

$$\omega_3 = \frac{\mathcal{E}_A - \mathcal{E}_B}{\hbar} \qquad \omega_1 = \frac{\mathcal{E}_{AB} + \mathcal{E}_{BA}}{\hbar} \qquad \omega_2 = \frac{i(\mathcal{E}_{AB} - \mathcal{E}_{BA})}{\hbar}$$

(9.59)

then Eq. (9.58) becomes

$$\frac{du_1}{dt} = -u_2(\omega_3) + u_3(\omega_2)$$

$$\frac{du_2}{dt} = u_1(\omega_3) - u_3(\omega_1)$$

(9.60)

$$\frac{du_3}{dt} = -u_1(\omega_2) + u_2(\omega_1)$$

The above equation is easily recognizable if it is expressed in vector notation as

$$\frac{d}{dt}(-\mathbf{u}) = \mathbf{u} \times \boldsymbol{\omega}$$

(9.61)

To illustrate, again let us use the elementary dipole as a specific example. The perturbation term U is given by

$$U = B_x u_x + B_y u_y = \frac{u^+ B^- + u^- B^+}{2}$$

(9.62a)

where
$$u^\pm = u_x \pm iu_y \qquad \text{and} \qquad B^\pm = B_x \pm iB_y$$

(9.62b)

Furthermore, the state ψ_B has $s_z = -\frac{1}{2}$ and ψ_A has $s_z = +\frac{1}{2}$. From Eqs. (9.56) and (9.38), we find

$$\mathcal{E}_A - \mathcal{E}_B = \gamma\hbar B_0 \qquad \mathcal{E}_{AB} = \frac{\gamma\hbar B^-}{2} \qquad \text{and} \qquad \mathcal{E}_{BA} = \frac{\gamma\hbar B^+}{2}$$

(9.63)

Substituting Eq. (9.63) into Eq. (9.59), we have

$$\omega_3 = \gamma B_0 = \omega_0 \qquad \omega_1 = \gamma B_x \qquad \text{and} \qquad \omega_2 = \gamma B_y$$

(9.64)

The identification of Eq. (9.61) with Eq. (9.3) is complete if we set

$$u_x = \gamma\hbar u_1 \qquad u_y = \gamma\hbar u_2 \qquad u_z = \gamma\hbar u_3$$

From the above analysis, we conclude that although the classical equation is based on the model of an elementary dipole, its validity can be extended to all magnetic dipole transitions, if the classical equation is correctly interpreted and properly modified. First, the resonance condition is now given by $\mathcal{E}_A - \mathcal{E}_B = \hbar\omega$; that is to say, the magnetic dipole can be viewed as precessing around an effective d-c magnetic field $B_{eff} = (\mathcal{E}_A - \mathcal{E}_B)/\gamma\hbar$. Second, the effectiveness of the a-c magnetic field depends upon the nature of the two states connected by the dipole moment operator U, that is, the magnitude of the matrix element $(A|U|B)$ or $(B|U|A)$. These ideas will become clear when we discuss the operational principle of a maser in later sections.

It is worthwhile to note that by the orthogonality and normalization condition of $\psi(q,t)$, $A^2 + B^2 = 1$. If N electrons are distributed between the two states ψ_A and ψ_B, say ψ_A with N_A and ψ_B with N_B electrons, $(N_A + N_B) = N$. We see that $|A|^2$ and $|B|^2$ can be interpreted as the probability of finding a given electron in the state ψ_A and ψ_B respectively. In other words, the quantity u_3 which reads

$$u_3 = |B|^2 - |A|^2 = \frac{N_B - N_A}{N} \tag{9.65}$$

is a measure of the population difference $N_B - N_A$ of the two states. Under equilibrium conditions, the ratio N_B/N_A is equal to the Boltzmann factor $\exp[(\mathcal{E}_A - \mathcal{E}_B)/kT]$, and the value of M_z or $N\gamma\hbar u_3$ is equal to M_0 in Eq. (9.13). This is exactly how we have defined the net magnetization of a paramagnetic substance in Sec. 8.8.

It is also to be noted from Eq. (9.55) that if $U = 0$, A and B will vary as $\exp(-i\mathcal{E}_A t/\hbar)$ and $\exp(-i\mathcal{E}_B t/\hbar)$ respectively and hence the value of $N_B - N_A = N(|B|^2 - |A|^2)$ is stationary. The effect of an a-c magnetic field is to cause transitions between the states ψ_A and ψ_B as evidenced from Eq. (9.49). Since $\mathcal{E}_A > \mathcal{E}_B$, $N_B > N_A$ according to the Boltzmann distribution. Hence, there will be more transitions from the state ψ_B to ψ_A than from ψ_A to ψ_B and consequently a net absorption of energy from the radiation field results. This is what happens in a paramagnetic resonance experiment. When N_A approaches N_B, M_z and in turn χ'' decrease; and this is known as the saturation effect. On the other hand, if $N_B < N_A$, we expect that the radiation field will receive energy from the paramagnetic substance. This is exactly what is required for a maser. Since the condition $N_B < N_A$ is opposite to the equilibrium distribution, we call the process a *population inversion*. When a microwave signal with a proper frequency passes through a paramagnetic substance whose population is inverted, the signal will be amplified after it receives energy from the paramagnetic substance. The word *maser* stands for microwave amplification by stimulated emission of radiation.

9.7 THE QUANTUM-MECHANICAL INTERPRETATION OF THE BLOCH EQUATION

The quantum-mechanical equation of motion, Eq. (9.60) or Eq. (9.61), can be solved in the same manner as the classical equation of motion, Eq. (9.11). We shall again introduce the relaxation terms involving T_1 and T_2 in Eq. (9.60) as in the Bloch equation. Furthermore, for simplicity, the a-c field is assumed to be right-hand circularly polarized so

$$U = B_x u_x + B_y u_y = \frac{B_1[u^+ \exp(-i\omega t) + u^- \exp(i\omega t)]}{2} \tag{9.66}$$

where $u^+ = u_x + iu_y$, $u^- = u_x - iu_y$, $B_x = B_1 \cos \omega t$, and $B_y = B_1 \sin \omega t$. Following exactly the same procedure as given in Sec. 9.3, we find (1) for the $u^- \exp(i\omega t)$ term

$$u_2 \omega_1^* - u_1 \omega_2^* = u_3 \frac{-i\omega_4(|\omega_2|^2 + |\omega_1|^2) + \omega_3(\omega_2 \omega_1^* - \omega_2^* \omega_1)}{\omega_3^2 - \omega_4^2} \tag{9.67}$$

where

$$\omega_1 = (B|u^-|A)B_1 \frac{\exp(i\omega t)}{2\hbar} \qquad \omega_2 = -i\omega_1 \qquad \omega_4 = \omega - \frac{i}{T_2} \tag{9.68}$$

and (2) for the $u^+ \exp(-i\omega t)$ term

$$u_2 \omega_1^* - u_1 \omega_2^* = u_3 \frac{i\omega_4(|\omega_2|^2 + |\omega_1|^2) + \omega_3(\omega_2 \omega_1^* - \omega_2^* \omega_1)}{\omega_3^2 - \omega_4^2} \tag{9.69}$$

where

$$\omega_1 = (A|u^+|B)B_1 \frac{\exp(-i\omega t)}{2\hbar} \qquad \omega_2 = i\omega_1 \qquad \omega_4 = \omega + \frac{i}{T_2} \tag{9.70}$$

Substituting Eqs. (9.68) and (9.70), respectively, into Eqs. (9.67) and (9.69) and taking the real part of Eqs. (9.67) and (9.69), we find that the third equation of Eq. (9.60) becomes

$$\frac{du_3}{dt} = -2u_3 \frac{T_2}{1 + (\omega - \omega_3)^2 T_2^2} \frac{B_1^2}{4\hbar^2} [|(A|u^+|B)|^2$$

$$+ |(B|u^-|A)|^2] - \frac{u_3 - u_0}{T_1} \tag{9.71}$$

In Eq. (9.71), the quantity u_0 is the equilibrium value of u_3 and the term $(u_3 - u_0)/T_1$ is the relaxation term mentioned earlier.

Now let us consider a two-state system with a population N_A in the state ψ_A and N_B in the state ψ_B, ψ_A being the higher energy state. If W_{AB} and W_{BA} are the probability of a single transition in unit time from the state ψ_B to ψ_A and that from the state ψ_A to ψ_B

respectively, then the rate equation says

$$\frac{dN_A}{dt} = -W_{BA}N_A + W_{AB}N_B$$

$$\frac{dN_B}{dt} = -W_{AB}N_B + W_{BA}N_A$$

(9.72)

For the time being, let us take for granted that $W_{AB} = W_{BA}$. Subtracting the first equation from the second, we obtain

$$\frac{du_3}{dt} = \frac{1}{N}\frac{d}{dt}(N_B - N_A) = -u_3(W_{AB} + W_{BA})$$

(9.73)

A comparison of Eq. (9.73) with Eq. (9.71) suggests that

$$W_{AB} = \frac{T_2}{1 + (\omega - \omega_3)^2 T_2^2}\frac{B_1^2}{2\hbar^2}|(A|u^+|B)|^2$$

$$W_{BA} = \frac{T_2}{1 + (\omega - \omega_3)^2 T_2^2}\frac{B_1^2}{2\hbar^2}|(B|u^-|A)|^2$$

(9.74)

It is to be noted that from Eq. (9.38), $(A|u^+|B) = (B|u^-|A)$; therefore, the earlier assumption that $W_{AB} = W_{BA}$ is justified.

The *probability of transition* from a given state ψ_A to another ψ_B can be calculated from Eq. (9.49) and the derivation of Eq. (9.74) can be found in textbooks on quantum mechanics.[8] The presentation here is intended to show that the classical and the quantum-mechanical approaches to the problem are closely related. We have a better physical feeling of the problem if we examine it from both viewpoints. In general, if the perturbing term U can be expressed as

$$U = U_1 \exp(i\omega t) + U_1^* \exp(-i\omega t)$$

(9.75)

then the transition probabilities are

$$W_{AB} = \frac{2\pi}{\hbar}|(A|U_1|B)|^2 \rho(\mathcal{E})$$

$$W_{BA} = \frac{2\pi}{\hbar}|(B|U_1^*|A)|^2 \rho(\mathcal{E})$$

(9.76)[9]

where $\rho(\mathcal{E})$ is so defined that $\rho(\mathcal{E})\,d\mathcal{E}$ is the density-of-states in transition having energy difference $\mathcal{E}_A - \mathcal{E}_B$ between $\mathcal{E}$ and $\mathcal{E} + d\mathcal{E}$. For the case just considered, $U_0 = B_1 u^+/2$ and $U_0^* = B_1 u^-/2$. Hence, by identification, we find

$$2\pi\hbar\rho(\mathcal{E}) = \frac{2T_2}{1 + (\omega - \omega_3)^2 T_2^2} = \rho(\nu)$$

(9.77)

[8] The transition probability is discussed either under perturbation theory or under interaction of radiation with matter.

[9] See, for example, J. L. Powell and B. Crasemann, *op. cit.*, p. 407; N. F. Mott and I. N. Sneddon, *op. cit.*, p. 273; L. I. Schiff, *op. cit.*, p. 193.

By definition, both the functions $\rho(\nu)$ and $\rho(\mathcal{E})$ are normalized so that $\int_{-\infty}^{\infty} \rho(\nu) \, d\nu = 1 = \int_{0}^{\infty} \rho(\mathcal{E}) \, d\mathcal{E}$. The factor $2\pi\hbar$ arises because of the change of variable from ν to $\mathcal{E}$ where $\nu = \omega/2\pi$.

Some supplementary remarks seem appropriate. A quantum-mechanical operator L is said to be Hermitian if for any two wave functions Φ and Ψ which vanish at infinity, the operator has the following property:

$$\int \Psi^* L \Phi \, dq = \int \Phi L^* \Psi^* \, dq \qquad \text{or} \qquad (A|L|B) = (B|L|A)^* \qquad (9.78)[10]$$

The operator U possesses such property; therefore, the earlier assumption that $W_{AB} = W_{BA}$ is true as is now obvious from Eqs. (9.76), (9.74), and (9.38). The function $\rho(\nu)$ or $\rho(\mathcal{E})$ takes care of the spread in resonance frequency. In a solid, the magnitude of the local magnetic field due to the dipole field of neighboring magnetic ions varies from place to place, depending on the specific magnetic state which the neighboring ions may possess. Such a dipole field may come from the electron or from the nuclear magnetic moment of the constituent atoms. As a result, the difference $\mathcal{E}_A - \mathcal{E}_B$ spreads over a finite band. The function $\rho(\mathcal{E})$ describes the distribution of states as a function of $\mathcal{E}_A - \mathcal{E}_B$ and consequently it is a measure of the relative strength of the absorption line at different values of ω according to Eqs. (9.15) and (9.18).

Quantum-mechanically, the spin-spin relaxation time T_2 can be interpreted as follows. The dipole-dipole interaction between neighboring magnetic ions causes a simultaneous change of the magnetic states of these ions, that is, from ψ_B to ψ_A for magnetic ion i and from ψ_A to ψ_B for magnetic ion j. The rate of such transition per unit time is represented phenomenologically in the Bloch equation by the reciprocal of the spin-spin relaxation time, that is, $1/T_2$. Because of the dipole-dipole interaction, therefore, a magnetic ion can stay in a given magnetic state only for a finite amount of time T_2. From the well-known uncertainty relation between $\Delta \mathcal{E}$ and Δt, if the lifetime of a given state is finite, a finite amount of uncertainty in the energy of that state results. The faster T_2 is, the broader the band is, as is evident from Eq. (9.20).

The term $(u_3 - u_0)/T_1$ in Eq. (9.71) may be viewed as a change in u_3 due to transitions induced by interactions with the host lattice. The underlying physical mechanism is quite involved but may be summarized as follows: The crystalline potential given in Eq. (8.49) depends upon the position of the surrounding charge distribution with respect to the magnetic ion under consideration. At a given temperature, atoms inside a solid are not stationary but vibrate about their equilibrium positions. The vibrational modes of the host lattice modulate the crystalline field which in turn affects the energy

---

[10]   See, for example, N. F. Mott and I. N. Sneddon, *op. cit.*, pp. 37 and 364.

of a given magnetic state. In other words, in the Hamiltonian, a small perturbing term proportional to the amplitude of the vibrational modes of the host lattice must be included. This perturbing term brings about an energy exchange between the magnetic ions and the host lattice, that is, a simultaneous change of the state of the magnetic ion and that of the lattice vibrational mode. More precisely, a change of magnetic state from ψ_B to ψ_A (raising the magnetic energy) is accompanied by an absorption of lattice vibrational energy through the interaction of the orbital wave function with the crystalline field. The reverse process lowers the magnetic energy (downward transition from ψ_A to ψ_B) but simultaneously raises the lattice energy. Such a process is treated quantum-mechanically in terms of phonon creation and annihilation operators, phonon being the unit of lattice vibrational energy. A discussion of the quantization of lattice vibrational mode may be found in Sec. 1.6. For a more thorough discussion of the spin-spin and spin-lattice relaxation mechanisms, the reader is referred to the book by Pake.[11]

At present, it seems desirable to examine the term $(u_3 - u_0)/T_1$ from a purely phenomenological viewpoint. Let w_{AB} and w_{BA} be the probability of transition per unit time from state ψ_B to ψ_A and that from ψ_A to ψ_B respectively, the transition being induced by lattice vibration. Under thermal equilibrium without external excitation ($B_1 = 0$),

$$w_{BA}N_{A0} = w_{AB}N_{B0} \qquad (9.79)$$

where the subscript 0 refers to the state under thermal equilibrium, and $N_{A0}/N_{B0} = \exp\left[(\mathcal{E}_B - \mathcal{E}_A)/kT\right] < 1$ according to the Boltzmann distribution law. Note that $w_{BA} > w_{AB}$ in contrast to the equality $W_{AB} = W_{BA}$ applicable to transitions induced by the radiation field B_1. Under external excitation ($B_1 > 0$), because the microwave field induces more upward than downward transitions,

$$N_A = N_{A0} + \Delta N \qquad \text{and} \qquad N_B = N_{B0} - \Delta N \qquad (9.80)$$

The net rate of downward transitions caused by lattice vibration now becomes

$$-\frac{d\,\Delta N}{dt} = w_{BA}N_A - w_{AB}N_B = \Delta N(w_{AB} + w_{BA}) \qquad (9.81)$$

In view of the fact that $N(u_3 - u_0) = -2\Delta N$, Eq. (9.81) is readily transformed into the third term of Eq. (9.71) if the following substitution is made:

$$\frac{1}{T_1} = w_{AB} + w_{BA} \qquad (9.82)$$

The effect of lattice vibration upon the magnetic system can be

[11] G. E. Pake, "Paramagnetic Resonance," W. A. Benjamin, Inc., New York, 1962.

summarized as follows: Under equilibrium conditions, $u_3 = u_0$, the lattice causes as many upward transitions as downward transitions so the term $(u_3 - u_0)/T_1 = 0$. From the thermodynamical viewpoint, the lattice can be considered as a heat reservoir. It tends to bring the magnetic system with which it is in contact back to the equilibrium condition. If $u_3 < u_0$, the magnetic system is excited, the sign of the term $(u_3 - u_0)/T_1$ in Eq. (9.71) says that the lattice will induce more downward transitions than upward transitions, thus taking energy away from the magnetic system. Therefore, Eq. (9.71) can be viewed either as a population equation in $(N_B - N_A)$ or an energy balance equation as discussed in Sec. 9.3.

Now let us consider the power transfer between the microwave field and the magnetic system. The transition from the state ψ_B to ψ_A means an absorption of energy $\hbar\omega_3$ while that from the state ψ_A to ψ_B an emission of energy $\hbar\omega_3$ where ω_3 is given by Eq. (9.59). Therefore, the total net power absorption by the paramagnetic substance is given by

$$P = \hbar\omega_3 W_{AB} N_B - \hbar\omega_3 W_{BA} N_A = \hbar\omega_3 N u_3 W_{AB} \tag{9.83}$$

The total energy of the magnetic system is equal to

$$\mathcal{E} = N_A \mathcal{E}_A + N_B \mathcal{E}_B = \frac{(N_A + N_B)(\mathcal{E}_A + \mathcal{E}_B)}{2} - \frac{(N_B - N_A)(\mathcal{E}_A - \mathcal{E}_B)}{2}$$

Thus, by differentiating the above equation with respect to time, we have

$$\frac{d\mathcal{E}}{dt} = -\tfrac{1}{2} N \hbar\omega_3 \frac{du_3}{dt} = P = \hbar\omega_3 N u_3 W_{AB} \tag{9.84}$$

Note that Eq. (9.84) is identical to Eq. (9.73), justifying our earlier statement that Eq. (9.73) [or Eq. (9.71)] is actually an energy balance equation.

The power transfer P is absorptive or emissive depending upon whether $N u_3 = N_B - N_A$ is positive or negative. In a maser, energy is extracted from the paramagnetic substance by a radiation field; thus we must have $N_A > N_B$. This means a population inversion with the higher energy state being more populated than the lower energy state. Under thermal equilibrium $(N_A)_0 < (N_B)_0$ and absorption of energy prevails. Therefore, the problem of achieving a population inversion in a magnetic system is of first-order importance in maser operations. The two remaining and equally important problems are (1) to select a pair of magnetic levels such that the energy difference gives the proper frequency, that is $\mathcal{E}_A - \mathcal{E}_B = \hbar\omega_3$, and (2) to ensure a nonvanishing transition probability W_{AB} between these two levels. These topics will be the subjects of our discussion in the following sections.

9.8 THE ENERGY LEVELS OF PARAMAGNETIC IONS IN A SOLID[12]

The crystalline electric field plays a very important role in determining the energy levels of a paramagnetic ion in a solid. As shown in Sec. 8.7, the three atomic states $l = 1$ have different electrostatic energies in the presence of an assumed lattice electrostatic potential of the form given by Eq. (8.49). Here, we shall discuss some concrete cases involving the magnetic ions of the iron group. From paramagnetic resonance studies, both the magnitude and the form of the crystalline field can be deduced. According to the relative influence of the field on the magnetic behavior of a paramagnet, paramagnetic substances are generally classified into three groups: (1) The weak-field group includes primarily ions of the rare-earth elements, (2) the medium-field group covers most salts and oxides of the iron group, and (3) the strong-field group consists of covalent complexes such as $[Co(CN)_6]^{3-}$.

First, let us take ions of the iron group and examine the charge distribution for the $3d$ electrons. The proper combinations of $\Theta(\theta)\Phi(\phi)$ of the wave function associated with $l = 2$ are listed below:

$$P_2{}^0(\cos\theta) = 3\cos^2\theta - 1, \quad P_2{}^1(\cos\theta) \times [\cos\phi \text{ or } \sin\phi],$$
$$P_2{}^2(\cos\theta) \times [\cos 2\phi \text{ or } \sin 2\phi] \quad (9.85a)$$

In terms of rectangular coordinates, the five proper orbital wave functions can be expressed as a product of a radial part $R(r)$ and an angular part of the following form:

$$\Theta(\theta)\Phi(\phi) \sim \frac{2z^2 - (x^2 + y^2)}{r^2}, \frac{x^2 - y^2}{r^2}, \frac{xy}{r^2}, \frac{zx}{r^2}, \text{ and } \frac{yz}{r^2} \quad (9.85b)$$

The charge-cloud distribution of these five orbitals is illustrated schematically in Fig. 9.5, the first two and the remaining three orbitals being designated as the $d\gamma$ and $d\epsilon$ orbitals, respectively.

Now consider a crystalline field of cubic symmetry in which the paramagnetic ion is surrounded by six negative ions situated at the corners of a regular octahedron, that is, $(\pm a,0,0)$ $(0,\pm a,0)$ and $(0,0,\pm a)$, as shown in Fig. 9.6a. From Figs. 9.5 and 9.6a we see qualitatively that the $d\gamma$ orbitals will have a much higher electrostatic energy than the $d\epsilon$ orbitals because the charge distribution of the former has maxima along the coordinate axes where the electric potential is a maximum. Quantitatively, the electrostatic potential V in Fig. 9.6a can be expressed in cartesian coordinates as $V_{\text{crystal}} = C_1[(a - x)^2 + y^2 + z^2]^{-1/2} +$ similar terms. Expanding this expres-

[12] For a more thorough and detailed discussion, the reader is referred to the following articles: B. Bleaney and K. W. H. Stevens, Paramagnetic Resonance, *Rept. Progr. Phys.*, vol. 16, p. 108, 1953; K. D. Bowers and J. Owen, Paramagnetic Resonance II, *Rept. Progr. Phys.*, vol. 18, p. 304, 1955; J. W. Orton, Paramagnetic Resonance Data, *Rept. Progr. Phys.*, vol. 22, p. 204, 1959.

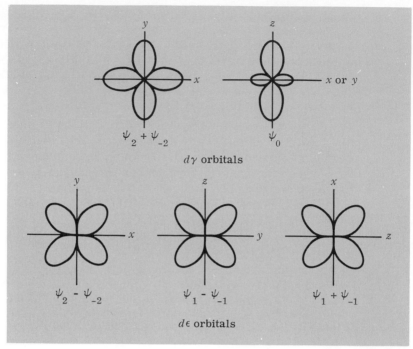

Schematic representation of the five d-orbital wave functions.

sion in a power series of $1/a$, and retaining only the first-order non-vanishing terms, we find

$$V_{\text{crystal}} = C_2(x^4 + y^4 + z^4 - \tfrac{3}{5}r^4) \qquad (9.86)$$

where C_1 and C_2 are proportionality constants and $r^2 = x^2 + y^2 + z^2$. Take the single $3d^1$ electron of the Ti^{3+} ion as an example. As a free ion, the five orbital functions given by Eq. (9.85*b*) have the same

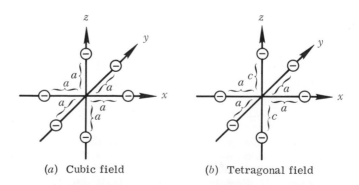

Illustrations showing (a) cubic and (b) tetragonal crystalline fields.

energy as shown in Fig. 9.7a. In a cubic crystalline field, the electrostatic energy associated with different orbitals can be calculated
quantum-mechanically by using the quantity V given by Eq. (9.86)
as the perturbation potential and the orbitals of the form represented
by Eq. (9.85b) as the wave functions. It is found as expected that
the three $d\epsilon$ orbitals have the same energy and so do the two $d\gamma$
orbitals, the $d\epsilon$ orbitals being the lower energy states. The energy
difference between the $d\epsilon$ and $d\gamma$ orbitals is usually referred to as the
splitting in a cubic field and it is designated by Δ in Fig. 9.7a.

The above discussion can be extended to ions having more than
one 3d electron. Figure 9.7b and c shows schematically how to get
the ground state of such ions by assigning the 3d electrons to the
available boxes in the diagram. Here, we have to distinguish between
the medium-field and the strong-field case. The Hund rule mentioned
in Sec. 8.6 is based on energy consideration of a free ion and is no
longer applicable in these cases. In the strong-field case, the separation between the $d\epsilon$ and $d\gamma$ orbitals is so high that the reduction in

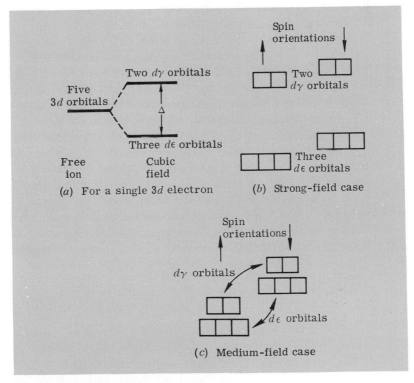

*Splitting of the energies of the five d-orbitals by a crystalline field. Diagram (a)
is for a single 3d electron. For two or more electrons, we must also take the spin
orientation into consideration as illustrated in (b) and (c) with (b) representing
the strong-field case and (c) representing the medium-field case.*

energy by maximizing the resultant spin vector S is not enough to compensate the gain in energy by putting one electron into the $d\gamma$ orbital. This corresponds to the situation shown in Fig. 9.7b and hence the Hund rule is clearly violated. Take $K_3Co(CN)_6$ as an example. The Co^{3+} ion has six $3d$ electrons. According to the Hund rule consistent with the Pauli exclusion principle, the six electrons would have five electron spins in one direction and one electron spin in the opposite direction, resulting in $S = \frac{4}{2}$. In a strong field, however, the six electrons are divided into two groups of three electrons each, occupying the three $d\epsilon$ orbitals with two opposite spin orientations to satisfy the exclusion principle. Consequently, $K_3Co(CN)_6$ is diamagnetic. Therefore, in the strong-field case, the six lowest energy states are constituted by three $d\epsilon$ orbitals with two possible spin directions (3×2) and these six states may be considered to form a subshell. The $d\epsilon$ subshell will be filled first.

The medium-field group, corresponding to the situation shown in Fig. 9.7c, has been most extensively studied. Take $Cr(NH_4)_2(SO_4)_2 \cdot 6H_2O$ (Tutton salt) as an example. The Cr^{2+} ion has a $3d^4$ configuration. According to Fig. 9.7c, in one spin orientation group, all of the three $d\epsilon$ orbitals and one of the two $d\gamma$ orbitals are occupied, resulting in $S = \frac{4}{2}$. Thus, the ground state has a spin degeneracy of 5 ($= 2S + 1$) and an orbital degeneracy of 2 (two possible choices of the $d\gamma$ orbitals), giving a total degeneracy of 10 (2×5). Therefore, for the Cr^{2+} ion in a cubic field alone, the ground state (state of lowest energy) has ten arrangements which have the same energy. There are other mechanisms, however, to split further the energy of these ten degenerate (having the same energy) states. Obviously, a crystalline field of lower symmetry should be able to differentiate the three $d\epsilon$ orbitals or the two $d\gamma$ orbitals. Figure 9.6b shows a crystalline field of tetragonal symmetry in which the dimension along the z axis is different from those along the other two axes, but the field about the z axis still preserves a fourfold symmetry, that is, being the same for every 90 degrees. Such a crystalline field gives rise to an electrostatic potential which can be represented mathematically in the following form:

$$V_{crystal} = C_3(3z^2 - r^2) + C_4(35z^4 - 30r^2z^2 + 3r^4)$$
$$+ C_5(x^4 - 6x^2y^2 + y^4) \quad (9.87)$$

Now V is symmetrical only in x and y so from an energy viewpoint it matters which of the two $d\gamma$ orbitals the fourth electron may have. Thus, in a tetragonal field, the 10 degenerate states of Cr^{2+} ions are split into two groups of 5 each as shown in Fig. 9.8b. The second mechanism for further energy split is the magnetic coupling between the spin and the orbital magnetic moment. In a free atom, it is due to this coupling that the S and L vectors are compounded to form a resultant J. In solids belonging to either the medium- or the strong-field group, this type of interaction is too weak in comparison with the crystalline field to couple **S** and **L** together. However, the interaction still exists and causes a further split in the energy of the states which are originally degenerate.

Figure 9.8[13] shows the lowest energy levels of several paramagnetic ions in the iron group in a crystal field of medium strength. The letters C, $S\text{-}O$, T, and H stand for the effects produced by a cubic field, the spin-orbit interaction, a tetragonal field, and an applied magnetic field, respectively. The number in parentheses represents the total degeneracy; for example, in the case of Cr^{2+} it is 10 (2×5) as stated in the previous paragraph. The energy splitting caused by the cubic part of a crystalline field, that is, Δ in Fig. 9.7a, is of the magnitude of 10,000 cm^{-1} (in reciprocal wavelength), and this energy split is much too large to be drawn in Fig. 9.8. The spin-

[13] For other $3d$ configurations, the reader is referred to **K. D. Bowers** and **J. Owen**, *Rept. Progr. Phys.*, vol. 18, p. 311, 1955. The book by **C. J. Gorter** ("Paramagnetic Relaxation," Elsevier Publishing Co., Inc., Amsterdam, 1947) contains valuable information on the energy levels and relaxation mechanisms in paramagnetic salts from dispersion and absorption measurements which predated the paramagnetic resonance experiment.

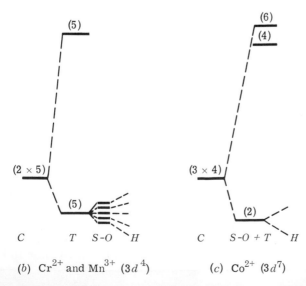

Figure 9.8

(2)
(1 × 4)
(2)

C $T + S\text{-}O$ H

(a) V^{2+} and Cr^{3+} ($3d^3$)

Further splitting of the magnetic energy levels by spin-orbit coupling (denoted by $S - O$) and tetragonal field (denoted by T) for ions of the iron group having (a) $3d^3$, (b) $3d^4$, and (c) $3d^7$ configurations. The diagrams are for the medium-field case.

(5)

(2 × 5)

(5)

C T $S\text{-}O$ H

(b) Cr^{2+} and Mn^{3+} ($3d^4$)

(6)
(4)

(3 × 4)

(2)

C $S\text{-}O + T$ H

(c) Co^{2+} ($3d^7$)

orbit coupling and the noncubic part of the crystalline field produce a further but much smaller splitting, usually in the range 100 cm⁻¹ to 1000 cm⁻¹ when there are both orbital and spin degeneracy and about 1 cm⁻¹ when there is only spin degeneracy. The ions having $3d^3$, $3d^5$, and $3d^8$ configurations belong to the latter category. Thus, the splitting in Fig. 9.8a is much smaller than those in Fig. 9.8b and c although the separations are not drawn to the scale.

For the rare-earth group, the crystal-field splitting is much smaller than that for the iron group. It is usually of the order of 100 cm⁻¹, an order of magnitude smaller than the spin-orbit coupling. Therefore, even in a crystal field, the orbital and spin angular momenta $\mathbf{L}$ and $\mathbf{S}$ are still additive according to the Hund rule to form a resultant $\mathbf{J}$, the total angular momentum. That is to say, the lowest energy level has $J = L - S$ if the $4f$ shell is less than half-filled and $J = L + S$ if it is more than half-filled. In a free ion, the lowest energy level is $(2J + 1)$-fold degenerate. The effect of the crystal field is to further split the energy of the degenerate levels into $2J + 1$ separate levels, the separation being of the order of 100 cm⁻¹. The only exceptions are ions which have a $4f^7$ configuration. Since the ground state in this case is in the S state ($L = 0$), the crystal-field splitting between the eight levels of the ^8S ground state are very small (of the order of 0.1 cm⁻¹).

At this point, it seems appropriate to bring into our discussion the frequency and temperature range in which a paramagnetic substance is used in maser operations. For comparison, we express energy in terms of reciprocal wavelength, temperature, and frequency in Table 9.1. It is to be noted that for masers operated in the microwave region, only paramagnetic ions having multienergy levels with an energy separation about 1 cm⁻¹ are important; that means ions having only spin degeneracy. To have a significant population difference between these states, $h\nu/kT$ must be large; that is, a low operating temperature is required. We further note that a small crystal-field splitting implies a weak coupling between the magnetic system and the host lattice and hence a long spin-lattice relaxation time T_1. Therefore, in subsequent discussions, we shall be concerned with those low-lying states only, which have a relatively small crystal-field splitting, or in other words, ions having an orbital singlet.

TABLE 9.1 *Comparison of energies expressed in different units*

Energy	4.14×10^{-6}	4.14×10^{-5}	4.14×10^{-4}	2.59×10^{-2} ev
in λ^{-1} (cm⁻¹)	0.033	0.33	3.3	206
in T (°K)	0.048	0.48	4.8	300
in ν (cps)	10^9	10^{10}	10^{11}	6.25×10^{12}

These include the ions V^{2+}, Cr^{3+}, Mn^{2+}, Fe^{3+}, Ni^{2+}, and possibly Cr^{2+} in the iron group and the ions Gd^{3+} and Eu^{2+} in the rare earth group.

9.9 THE SPIN HAMILTONIAN

In the previous section, we have shown that the energy of a paramagnetic ion in a solid is determined in a complicated way by the atomic configuration of the ion, the form and the strength of the crystalline field, and the spin-orbit coupling. With a precise knowledge of the above quantities, the energy levels can be calculated quantum-mechanically by perturbation techniques. In a microwave transition, however, we are only concerned with the group of electronic states having the lowest energy. For these states, the energy levels can be expressed in terms of a *spin Hamiltonian* which consists of two parts: (1) due to an external magnetic field and (2) due to the internal crystal field.

For a free magnetic dipole, the splitting in energy between successive magnetic levels due to an applied magnetic field is given by Eq. (8.44) and equal to $g u_B B$ where u_B is a Bohr magneton and g is the Landé splitting factor. In a solid, the corresponding term is $u_B(g_x B_x + g_y B_y + g_z B_z)$. The g factor is now anisotropic because the crystalline field is orientation-dependent and so are the orbital and spin contributions to the magnetic moment owing to spin-orbit and orbital-crystal-field interactions. For the ions mentioned at the end of the previous section, the lowest energy states have either an orbital singlet or an S state. Hence, the crystal-field-orbital interaction is relatively weak and the orbital contribution to the magnetic moment is small. For these ions, the g factor is almost isotropic (that is, $g_x = g_y = g_z = g$) and its value is close to the free spin value. Thus, the part of the spin Hamiltonian due to an applied magnetic field is equal to $g u_B \mathbf{B} \cdot \mathbf{S}$ where S is the value of the total spin quantum number and is equal to the number which gives the spin degeneracy of the state. Take Fe^{3+} as an example. In a medium crystal field, $S = \frac{5}{2}$, and hence in a magnetic field, the term $g u_B \mathbf{B} \cdot \mathbf{S}$ has six possible values according to $S_z = -\frac{5}{2}, -\frac{3}{2}, -\frac{1}{2}, \frac{1}{2}, \frac{3}{2},$ or $\frac{5}{2}$.

The spin of a magnetic ion in a solid interacts indirectly with the crystalline field through the combined mechanism of spin-orbit coupling and crystal-field-orbital interaction. Thus, even in the absence of an applied magnetic field, the six states of Fe^{3+} corresponding to different values of S_z have different energies. This splitting of energy values is called the *zero-field splitting* and is represented in the spin Hamiltonian by terms which are functions of S_x, S_y, and S_z but are independent of B. It should be pointed out that the axes x, y, and z in this case are no longer arbitrary and must coincide with the axes of the crystalline field with reference to which the crystal symmetry is defined. For a crystal field having an axial symmetry about the z axis as in the case for the tetragonal field shown in Fig. 9.6b, a term $DS_z{}^2$ in the spin Hamiltonian represents the zero-field splitting. For a crystal field of lower symmetry, an additional term $E(S_x{}^2 - S_y{}^2)$ is further needed.

From the above discussion, the spin Hamiltonian, denoted by $\mathfrak{IC}$, takes the following general form:

$$\mathfrak{IC} = gu_B(B_xS_x + B_yS_y + B_zS_z) + D[S_z{}^2 - \tfrac{1}{3}S(S + 1)] \\ + E(S_x{}^2 - S_y{}^2) \quad (9.88)$$

The term $S(S + 1)/3$ in the above equation simply shifts the reference energy level and does not affect the relative energy difference between two levels. Also in Eq. (9.88), the applied magnetic field can be in any arbitrary direction with reference to the crystal axes and the quantities B_x, B_y, and B_z are the respective projections of the applied field upon the chosen crystal axes. The above form of spin Hamiltonian applies to the $3d^3$ configurations of V^{2+} and Cr^{3+}, the $3d^4$ configuration of Cr^{2+}, and the $3d^8$ configuration of Ni^{2+}. For the S-state ions such as Mn^{2+} and Fe^{3+} of the iron group in a medium crystal field and Gd^{3+} and Eu^{2+} of the rare-earth group, the coefficients D and E are extremely small, so higher order terms in S_x, S_y, and S_z must also be included. In the former two cases, terms with fourth power in S_x, S_y, and S_z are sufficient while in the latter two cases, terms up to the sixth power are necessary. For detailed information about the specific form of these spin Hamiltonians, the reader is referred to two reports by Bleaney and Stevens[14] and by Bowers and Owen.[15]

Our next step is to show how the energy levels of the ground state can be found from a given spin Hamiltonian. To be specific, take the Cr^{3+} ion in ruby as an example. For simplicity, we shall assume that the applied magnetic field is along the c axis of the crystal which is chosen as the z direction. The spin Hamiltonian then reads

$$\mathfrak{IC} = gu_BBS_z - D(S_z{}^2 - \tfrac{5}{4}) \quad (9.89)$$

We let M ($= S_z$) be the spin magnetic quantum number and Φ_M be the set of spin wave functions pertaining to the Cr^{3+} ion. Since $S = \tfrac{3}{2}$ for Cr^{3+}, possible values for M are $\tfrac{3}{2}$, $\tfrac{1}{2}$, $-\tfrac{1}{2}$, and $-\tfrac{3}{2}$. It should be pointed out that in Eqs. (9.88) and (9.89), the factor $\hbar$ has been removed from the operators S_x, S_y, and S_z and incorporated into the constants u_B, D, and E for neatness. Furthermore, in a maser transition, only the spin levels are involved. Therefore, the symbol $M (= S_z)$ represents the spin, but not the orbital, magnetic-

[14]   B. Bleaney and K. W. H. Stevens, *op. cit.*
[15]   K. D. Bowers and J. Owen, *op. cit.*

quantum-number. Following the discussion in Sec. 9.4, we have

$$S_z\Phi_M = M\Phi_M$$
$$S_+\Phi_M = [S(S + 1) - M(M + 1)]^{1/2}\Phi_{M+1} \qquad (9.90)$$
$$S_-\Phi_M = [S(S + 1) - M(M - 1)]^{1/2}\Phi_{M-1}$$

where $S_+ = S_x + iS_y$ and $S_- = S_x - iS_y$. Substituting Eq. (9.90) into Eq. (9.89), we find

$$\mathcal{H}\Phi_M = [gu_B BM - D(M^2 - 5/4)]\Phi_M \qquad (9.91)$$

and the energies, $\mathcal{E}(M)$, associated with Φ_M as given below

$$\mathcal{E}(3/2) = \tfrac{3}{2}gu_B B - D$$
$$\mathcal{E}(1/2) = \tfrac{1}{2}gu_B B + D$$
$$\mathcal{E}(-1/2) = -\tfrac{1}{2}gu_B B + D \qquad (9.92)$$
$$\mathcal{E}(-3/2) = -\tfrac{3}{2}gu_B B - D$$

The energies given by Eq. (9.92) are plotted as a function of the magnetic field B in Fig. 9.9.[16] Thus, we see that a four-level system with adjustable energy differences between different levels is predicted by theory.

If an a-c magnetic field is applied in the xy plane, additional terms $S_x B_x + S_y B_y$ must be included in the spin Hamiltonian, Eq. (9.89). We can treat these terms as perturbations and calculate the transition probability between two different states as outlined in Secs. 9.5 and 9.6. Owing to the selection rule for S_x and S_y, however, only transitions with $\Delta M = \pm 1$ are allowed. Consider the possibility of designing a maser system involving the states $+3/2$, $+1/2$, and $-1/2$. As discussed in Sec. 9.7, the population difference between any

---

[16] For other orientations of **B** with respect to the crystal axes, the reader is referred to J. E. Geusic and H. E. D. Scovil, "Microwave and Optical Masers," *Rept. Progr. Phys.*, vol. 27, p. 241, 1964, for the calculation of the eigen energies and for the plots of $\mathcal{E}$ versus B (p. 260).

Figure 9.9

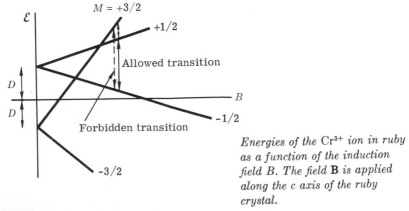

*Energies of the Cr^{3+} ion in ruby as a function of the induction field B. The field **B** is applied along the c axis of the ruby crystal.*

pair of states can be significantly changed by the application of a microwave field with proper frequency. In the next section, it will be shown that a population inversion between two levels can actually be achieved in a three-level system. However, in the present case, the transition between the states $+\frac{3}{2}$ and $-\frac{1}{2}$ requires $\Delta M = \pm 2$ and hence is forbidden as indicated by a broken arrow in Fig. 9.9. Such blocking of the circulation of traffic, so to speak, prevents the present system from being capable of continuous operation. Notice that the energies in Eq. (9.92) are associated with pure states Φ_M. In order to make transitions possible between any pair of energy states, we have to have an admixture of these pure spin states.

There are two possible ways to mix the spin states: (1) by using a host crystal of lower symmetry, that is rhombic, so that the term $E(S_x{}^2 - S_y{}^2)$ is included or (2) by applying the d-c magnetic field at an angle to the crystal axis of symmetry. Since the procedure to get the new admixture of states is quite similar in the two cases, only the former case will be treated here. Furthermore, for simplicity, we shall consider a magnetic ion, for example Ni^{2+}, with $S = 1$. The spin Hamiltonian in a rhombic field reads $\mathcal{H} = g u_B B S_z - D(S_z{}^2 - \frac{2}{3}) + E(S_x{}^2 - S_y{}^2)$, or in terms of S_+ and S_-

$$\mathcal{H} = g u_B B S_z - D(S_z{}^2 - \tfrac{2}{3}) + \frac{E}{2}(S_+{}^2 + S_-{}^2) \qquad (9.93)$$

The problem is to find the new proper wave function which can be expressed as an admixture of the old spin states as follows:

$$\Psi(N) = \sum_M a(N,M)\Phi_M \qquad (M = 1, 0, \text{ and } -1) \qquad (9.94)$$

and which satisfies the wave equation

$$\mathcal{H}\Psi(N) = \mathcal{E}_N\Psi(N) \qquad (9.95)$$

where $\mathcal{E}_N$ is the energy of the new proper state to be determined. Substituting Eq. (9.94) into Eq. (9.95) and utilizing the relations in Eq. (9.90), we have the following equations:

$$
\begin{aligned}
\mathcal{E}_N a(N,1) &= (b - d)a(N,1) + Ea(N,-1) \\
\mathcal{E}_N a(N,0) &= 2da(N,0) \\
\mathcal{E}_N a(N,-1) &= Ea(N,1) - (b + d)a(N,-1)
\end{aligned}
\qquad (9.96)
$$

by collecting the coefficients of Φ_M (in the order of $M = 1, 0,$ and -1). In Eq. (9.96), $b = g u_B B$ and $d = D/3$. For nontrivial solutions, the following determinant must be zero:

$$
\begin{vmatrix}
b - d - \mathcal{E}_N & 0 & E \\
0 & 2d - \mathcal{E}_N & 0 \\
E & 0 & -b - d - \mathcal{E}_N
\end{vmatrix} = 0
$$

which gives the three roots of $\mathcal{E}_N$ as follows:

$$\mathcal{E}_I = c - d \qquad \mathcal{E}_{II} = 2d \qquad \mathcal{E}_{III} = -c - d \qquad (9.97)$$

where $c = (b^2 + E^2)^{\frac{1}{2}}$. The three energy levels are plotted in Fig. 9.10 as functions of B.

Note that for $\mathcal{E}_{I,III}$, $a(N,0) = 0$ while for $\mathcal{E}_{II}$, $a(N,1) = a(N,-1) = 0$ because $\mathcal{E}_{I,III}$ is the solution of the first and third equations of Eq. (9.96) while $\mathcal{E}_{II}$ is the solution of the second equation of Eq. (9.96). Thus, the various coefficients can easily be found by taking their ratio from Eq. (9.96), for example,

$$\frac{a(I,1)}{a(I,-1)} = \frac{E}{\mathcal{E}_I - b + d}$$

and by noting the normalization condition $|\Psi(N)|^2 = 1$, that is, $a^2(I,1) + a^2(I,-1) = 1$. After some algebraic manipulations, we find

$$\begin{aligned}
\Psi_I &= k[E\Phi_{+1} + (c - b)\Phi_{-1}] \\
\Psi_{II} &= \Phi_0 \\
\Psi_{III} &= k[(b - c)\Phi_{+1} + E\Phi_{-1}]
\end{aligned} \qquad (9.98)$$

where $k^{-2} = 2c(c - b)$. From Eq. (9.98), it is obvious that a rhombic

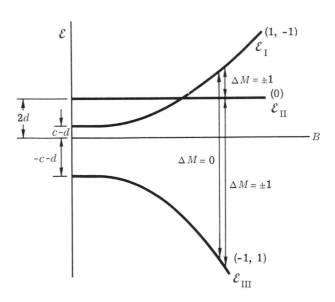

Figure 9.10

The three-level system of Ni^{2+} ion in a rhombic field with the d-c induction field **B** applied along the c axis. If quantum-mechanical selection rules are applied to these levels, we see that the transitions between $\mathcal{E}_{II}$ and $\mathcal{E}_{I,III}$ require $\Delta M = \pm 1$, and the transition between $\mathcal{E}_I$ and $\mathcal{E}_{III}$ requires $\Delta M = 0$. Therefore, to make all transitions possible, the a-c microwave field should have components B_z and B_x (or B_y).

field causes a mixing of states; in other words, both Ψ_I and Ψ_{III} are now a mixture of the pure spin states $M = +1$ and -1.

If a microwave field is applied in the xz plane, for example, the perturbation term consists of $B_{x1}S_x + B_{z1}S_z$ where B_{x1} and B_{z1} are the components of the microwave field in the x and z directions respectively. Now the S_z operator makes the transition between Ψ_I and Ψ_{III} possible ($\Delta M = 0$) while the S_x operator makes the transitions between Ψ_{II} and $\Psi_{I,III}$ possible ($\Delta M = \pm 1$). In other words, transitions are allowed between any two of the three levels. Hence we have an example of a practical multilevel system which is capable of continuous operation.

9.10 POPULATION INVERSION IN MASERS

As shown in Sec. 9.7, a paramagnetic system is capable of delivering energy to an external microwave field if its higher energy state is more populated than its lower energy state. To realize such an active device, we need first a multilevel system and then a population inversion in the system. In the previous section, we have discussed the various possibilities of achieving a multilevel system and naturally, in this section, we shall exploit the scheme to achieve a population inversion in the system. Consider a three-level paramagnetic system shown in Fig. 9.11 with energy levels at $\mathcal{E}_1$, $\mathcal{E}_2$, and $\mathcal{E}_3$, and their respective populations of N_1, N_2, and N_3 where $\mathcal{E}_3 > \mathcal{E}_2 > \mathcal{E}_1$. Under thermal equilibrium, their relative population distribution is governed by the Boltzmann relation[17]

$$\left(\frac{N_i}{N_j}\right)_0 = \exp\left(-\frac{\mathcal{E}_i - \mathcal{E}_j}{kT}\right) \tag{9.99}$$

where $i, j = 1, 2, 3$ and k is the Boltzmann constant. If N is the total number of electrons, the constraint is that $N_1 + N_2 + N_3 = N$.

[17] As pointed out in Sec. 1.12, the Boltzmann statistics is appropriate for a system of noninteracting or weakly interacting particles. Such is the case with the paramagnetic ions in a maser material.

Figure
9.11

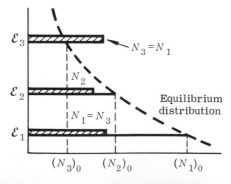

The population distribution in a three-level system. The dashed line represents the equilibrium distribution as governed by the Boltzmann statistics. A population inversion can be achieved by making $N_1 = N_3$ in a saturation experiment.

The problem is to create a nonequilibrium situation in which $N_3 > N_2$, for example, so that an active device may be realized at a frequency $\nu_{32} = (\mathcal{E}_3 - \mathcal{E}_2)/h$. For this purpose, it is necessary to establish a set of dynamic rate equations similar to Eq. (9.72).

We shall follow the original analysis by Bloembergen.[18] Let W_{ij} be the transition probability per unit time from the energy state $\mathcal{E}_j$ to $\mathcal{E}_i$ caused by the microwave field and w_{ij} be the corresponding transition probability caused by the thermal motion of the host lattice. The former is the same as the quantities W_{AB} and W_{BA} in Eq. (9.72), while the latter is related to $1/T_1$ in Eq. (9.71) and is actually the same as w_{AB} in Eq. (9.81). Thus, we have

$$\frac{dN_1}{dt} = -(W_{31} + W_{21} + w_{31} + w_{21})N_1 + (W_{12} + w_{12})N_2$$
$$+ (W_{13} + w_{13})N_3 \quad (9.100)$$

and similar equations for N_2 and N_3. Note that Eq. (9.100) is a combination of Eq. (9.72) and Eq. (9.81) modified to suit a three-level system. Under thermal equilibrium, that is $W_{ij} = 0$, the following Boltzmann relations must be observed:

$$\frac{w_{31}}{w_{13}} = \left(\frac{N_3}{N_1}\right)_0 = \exp\left(-\frac{h\nu_{13}}{kT}\right) \quad (9.101)$$

where $\nu_{ij} = |\mathcal{E}_i - \mathcal{E}_j|/h$. Further, from our previous discussion in Sec. 9.7, $W_{ij} = W_{ji}$ in Eq. (9.100).

Before actually solving Eq. (9.100), we shall examine the situation physically. Figure 9.11 shows schematically the populations of a three-level system in relation to their energies. If we saturate the $\mathcal{E}_1$ and $\mathcal{E}_3$ levels with a strong microwave field to bring $N_1 = N_3 = (N_1 + N_3)/2$, then we achieve a population inversion irrespective of the new value of N_2. If $N_2 < (N_1 + N_3)/2 = N_3$, a population inversion occurs between the levels $\mathcal{E}_2$ and $\mathcal{E}_3$. On the other hand, if $N_2 > (N_1 + N_3)/2 = N_1$, a population inversion occurs between the levels $\mathcal{E}_1$ and $\mathcal{E}_2$. The microwave field used to saturate the $\mathcal{E}_1$ and $\mathcal{E}_3$ levels is usually very strong. Since it supplies energy to the magnetic system, it is called the pumping field and the frequency ν_{31} the pump frequency. If maser action (or population inversion) takes place between $\mathcal{E}_2$ and $\mathcal{E}_3$, the frequency ν_{32} is called the signal frequency, and the remaining frequency ν_{21} the idling frequency. Since the microwave field at pump frequency is much stronger than those at signal and idling frequencies, in Eq. (9.100), $W_{13} \gg W_{23}$ and W_{12}.

Only the steady-state solution of Eq. (9.100) will be discussed so the time derivatives of N_1, N_2, and N_3 can be set to zero. Furthermore, we shall assume that $W_{13} \gg w_{ij}$ where $i, j = 1, 2,$ and 3; in

[18] The first proposal for a three-level maser capable of continuous operation was made by N. Bloembergen, *Phys. Rev.*, vol. 104, p. 324, 1956.

other words, the transition induced by the pumping field is strong enough to overcome the transitions induced by interactions with the host lattice. That is to say, a condition of saturation between levels 1 and 3 prevails. Thus, in Eq. (9.100), we neglect all terms except W_{13} and W_{31}, and find

$$N_3 - N_1 = 0 \quad \text{or} \quad N_1 = N_3 \tag{9.102}$$

This is just what is expected by physical reasoning. The equation for dN_2/dt reads

$$0 = \frac{dN_2}{dt} = (W_{21} + w_{21})N_1 - (W_{12} + W_{32} + w_{12} + w_{32})N_2$$
$$+ (W_{23} + w_{23})N_3 \tag{9.103}$$

Noting that $N_1 + N_2 + N_3 = N$, we can solve for N_1, N_2, and N_3 from Eqs. (9.102) and (9.103) and thus obtain

$$N_1 = N_3 = \frac{N(W_{12} + W_{32} + w_{12} + w_{32})}{G}$$

$$N_2 = \frac{N(W_{21} + W_{23} + w_{21} + w_{23})}{G} \tag{9.104}$$

$$G = 3W_{12} + 3W_{32} + 2w_{12} + 2w_{32} + w_{21} + w_{23}$$

The population difference under the nonequilibrium condition is

$$N_3 - N_2 = N_1 - N_2 = \frac{N(w_{12} - w_{21} + w_{32} - w_{23})}{G} \tag{9.105}$$

From the above equation, it is clear that a population inversion is established between either states $\mathcal{E}_3$ and $\mathcal{E}_2$ or states $\mathcal{E}_2$ and $\mathcal{E}_1$ depending on whether $w_{12} - w_{21} + w_{32} - w_{23} > $ or < 0. Noting that under ordinary operating temperature of a maser, $kT > h\nu_{ij}$ so we can expand the exponential factor in Eq. (9.101) and obtain the following relations:

$$w_{ij} = (w_{ij})_0 \left(1 + \frac{1}{2}\frac{h\nu_{ij}}{kT}\right)$$

$$w_{ji} = (w_{ij})_0 \left(1 - \frac{1}{2}\frac{h\nu_{ij}}{kT}\right) \tag{9.106}$$

where $i, j = 1, 2, 3$ with $j > i$ and $(w_{ij})_0 = (w_{ij} + w_{ji})/2 = 1/2(T_1)_{ij}$, the quantity $(T_1)_{ij}$ being the spin-lattice relaxation time T_1 between levels i and j. Thus, the condition for a population inversion between the levels $\mathcal{E}_2$ and $\mathcal{E}_3$ becomes

$$(w_{12})_0 \nu_{12} - (w_{23})_0 \nu_{23} > 0 \quad \text{or} \quad (T_1)_{23}\nu_{12} > (T_1)_{12}\nu_{23} \tag{9.107}$$

Obviously, the two spin-lattice relaxation times play a decisive role in selecting the levels between which maser action takes place.

To be specific, let us assume that Eq. (9.107) is true and hence microwave power can be extracted from the paramagnetic ions at the frequency ν_{23}. According to Eqs. (9.84) and (9.105) the available microwave power is given by

$$P = \frac{(w_{12})_0 \nu_{12} - (w_{23})_0 \nu_{23}}{W_{12} + W_{32} + (w_{12})_0 + (w_{23})_0} \frac{h^2 \nu_{23}}{3kT} W_{23} \qquad (9.108)$$

It should be pointed out that Eq. (9.108) is based on the assumption that a complete saturation is obtained between levels $\mathcal{E}_1$ and $\mathcal{E}_3$. If the saturation is only partial, we have to keep the other terms besides W_{13} and W_{31} in Eq. (9.100) and as a consequence, the population difference $N_3 - N_2$ will be a function of the pumping power. It is also evident from Eq. (9.108) that a maser is a linear device for so long as the signal power is low so that $W_{23} < (w_{12})_0$ and $(w_{23})_0$. Therefore, a maser is inherently a low-level active device. Physically, a strong signal will bring equalization of population between the levels $\mathcal{E}_2$ and $\mathcal{E}_3$ and this counters the maser action.

We should emphasize that in Eq. (9.108), the quantity W_{23} is the transition probability between the levels $\mathcal{E}_2$ and $\mathcal{E}_3$ caused by the signal microwave field and it can be calculated from Eq. (9.76). The emission of the microwave power P is, therefore, stimulated by a radiation field and is referred to as the *stimulated emission*. It is distinguished from the spontaneous emission in that the emitted field maintains a definite phase relationship with the stimulating field while in the spontaneous emission, the phase of the emitted radiation field is random. Hence, a maser is a coherent active device utilizing the quantum energy levels of paramagnetic ions.

The three-level system discussed above can be operated continuously. For completeness, we shall discuss briefly a two-level system which can be used for pulse operations. There are many ways to achieve a population inversion for a short period of time. Consider the spin states with $S = \frac{1}{2}$. In a d-c magnetic field B the two states $M = S_z = -\frac{1}{2}$ and $+\frac{1}{2}$ will have an equilibrium population ratio of $\exp(2u_B B/kT)$. Suppose the direction of the magnetic field is suddenly reversed, then the role of these two states becomes reversed, that is to say, the former lower energy state now becomes the higher energy state and vice versa. It takes a finite time for the spin system to readjust itself to the new situation and establish a new equilibrium population relation. If we can reverse the field in a time faster than the spin system can respond, we have a transient population inversion. Other possible ways to achieve the same include the adiabatic fast passage and the 180° pulse inversion which were employed originally by Bloch[19] and Hahn[20] in nuclear magnetic-resonance experiments at radio frequencies. At microwave frequencies, however, some of the inversion schemes present practical difficulties;

[19]   F. Bloch, *Phys. Rev.*, vol. 70, p. 464, 1946 and vol. 102, p. 104, 1956.
[20]   E. L. Hahn, *Phys. Rev.*, vol. 80, p. 580, 1950; H. Y. Carr and E. M. Purcell, *Phys. Rev.*, vol. 94, p. 630, 1954.

for example, a microwave field cannot change in a time less than the ringing time of the cavity. It is evident that a maser capable of continuous operation is much superior to one based on pulse operation.

We should further add that it is also possible and sometimes may even be advantageous to employ more than three quantum levels in a maser system. A typical example is the push-pull scheme shown in Fig. 9.12a. The pumping field with frequency ν_p completely saturates the transitions between levels 1 and 3 and levels 2 and 4. From Fig. 9.12a, it is clear that a population inversion is established

Figure 9.12

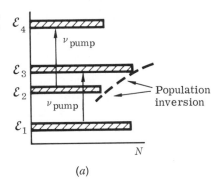

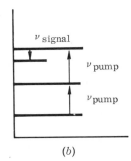

(a)

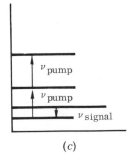

(b)   (c)

Four-level maser systems with (a) $h\nu_{\text{pump}} = \mathcal{E}_4 - \mathcal{E}_2 = \mathcal{E}_3 - \mathcal{E}_1$; (b) $h\nu_{\text{pump}} = \mathcal{E}_4 - \mathcal{E}_2 = \mathcal{E}_2 - \mathcal{E}_1$; and (c) $h\nu_{\text{pump}} = \mathcal{E}_4 - \mathcal{E}_3 = \mathcal{E}_3 - \mathcal{E}_1$.

between levels 2 and 3. The analysis of such a scheme is left for the reader as a problem. Other maser schemes using four quantum levels are shown in Fig. 9.12b and c. Usually a better population inversion ratio may be achieved with the four-level scheme than the three-level scheme.

9.11   NOISE AND MATERIAL CONSIDERATIONS

A general discussion of quantum electronic devices, to which microwave masers belong, is incomplete without a discussion of noise associated with quantum processes. It is instructive to present the original interpretation due to Einstein of Planck's law of black-body radiation. Consider the atomic transitions between any two states, say i and j, of an atomic system, the state i being the higher energy state. In an emission process, transitions from a higher to a lower energy state may take place spontaneously (spontaneous emission) or may be induced by a radiation field (stimulated emission). Let A_{ji} be the probability per unit time for a single state i to make a spontaneous transition and $B_{ji}I(\nu_{ij})$ be the corresponding probability for stimulated emission where $I(\nu_{ij})$ is the energy density of radiation per unit frequency interval at frequency $\nu_{ij} = (\mathcal{E}_i - \mathcal{E}_j)/h$. In the reverse process, that is the absorption process, transitions from a lower to a higher energy state are possible only through the action of a radiation field, and the probability per unit time for a single state j to make such a transition is, by analogy, equal to $B_{ij}I(\nu_{ij})$. If the atomic system is in thermal equilibrium with a black body at temperature T, then the total number of upward and downward transitions per unit time must be equal, that is

$$[A_{ji} + B_{ji}I(\nu_{ij})]N_i = B_{ij}I(\nu_{ij})N_j \qquad (9.109)$$

where N_i and N_j are the populations of the states i and j respectively. In view of the Boltzmann relation given by Eq. (9.99), Eq. (9.109) can be restated as follows:

$$I(\nu_{ij}) = \frac{A_{ji}N_i}{B_{ij}N_j - B_{ji}N_i} = \frac{A_{ji}}{B_{ij} \exp(h\nu_{ij}/kT) - B_{ji}} \qquad (9.110)$$

A comparison of Eq. (9.110) with Planck's law [Eq. (1.135)] shows that

$$B_{ji} = B_{ij} \qquad \text{and} \qquad \frac{A_{ji}}{B_{ji}} = \frac{8\pi h}{\lambda^3} \qquad (9.111)$$

where λ is the wavelength of the radiation field. The relationship $B_{ji} = B_{ij}$ is consistent with our earlier conclusion about transition probabilities ($W_{AB} = W_{BA}$) made in Sec. 9.7. The other identity relating A_{ji} to B_{ji} can be derived quantum-mechanically and will be further discussed.

Next, the quantum-mechanical derivation of Eq. (9.109) will be discussed. In Sec. 9.7, the problem of emission and absorption is treated in a semiquantum-mechanical and semiclassical manner. It is clear from Eq. (9.74) that the transition probability W_{AB} describes the rate of a quantum-mechanical transition from the atomic state B to A. Since the state A is the higher energy state, the probability W_{AB} is associated with the absorption process. However, in the formulation given in Sec. 9.7, it is not clear how the radiation field changes its state in the process of transferring its energy to the atomic system. The reason that we have treated the radiation field in a classical manner is because we do not want to confuse the reader by introducing too many concepts at one time. The quantization of radiation field is similar to the quantization of lattice vibration discussed in Sec. 1.6 as a simple harmonic oscillator. Here, we shall attempt to summarize the relevant results derived from the quantum mechanical treatment of radiation field.[21]

First, the energy of a radiation field is quantized and given by $(n + \frac{1}{2})h\nu$ where n is an integer and ν is the frequency of the radiation field. The state of the radiation field can be described by a proper wave function; for example, the eigenfunction ψ_n corresponds to the state having an energy $(n + \frac{1}{2})h\nu$. In an absorption process, a quantized unit of energy, called photon, of $h\nu$ is taken away from the radiation field by the atomic system, and the radiation field changes its state from ψ_n to ψ_{n-1}. Likewise, in an emission process, a photon of energy $h\nu$ is given to the radiation field, changing its state from ψ_n to ψ_{n+1}. Mathematically, these processes can be described by two operators: the photon-annihilation operator q^* and the photon-creation operator q such that their matrix elements are given as follows:

$$(n + 1|q|n) = \sqrt{2(n + 1)}$$
$$(n - 1|q^*|n) = \sqrt{2n}$$

(9.112)[22]

Since the transition probabilities are proportional to the square of the magnitudes of the matrix elements, the total rate for the emission process is equal to $CN_i(n + 1)$ where C is a proportionality constant and can be calculated quantum-mechanically. Similarly, the total rate for absorption process is equal to $CN_j n$. The proportionality constants in the two processes must be the same because of Eq. (9.112) and $W_{ij} = W_{ji}$, and the proof of this will be saved for a later chapter on lasers (optical masers).

The dynamic equation for the energy $\mathcal{E}$ of the radiation field can be expressed in terms of n as follows:

$$\frac{1}{h\nu} \frac{d\mathcal{E}}{dt} = \frac{dn}{dt} = CN_i(n + 1) - CN_j n$$

(9.113)

Since the terms $CN_i n$ and $CN_j n$ are proportional to the energy of the radiation field, a comparison of Eq. (9.113) with Eq. (9.109) shows that these terms represent the stimulated emission and absorption respectively. The term CN_i obviously is responsible for the spontaneous emission process. The ratio A/B given by Eq. (9.111) can be derived as follows. Referring to Eq. (1.134), the number of normal modes of electromagnetic radiation in a volume V with frequencies between ν and $\nu + d\nu$ is equal to

$$\frac{V2(4\pi k^2 \, dk)}{(2\pi)^3} = \frac{V8\pi\nu^2 \, d\nu}{c^3}$$

(9.114)

[21] For a discussion of the quantization of the radiation field, the reader is referred to L. I. Schiff, *op. cit.*, chap. 13, and to N. F. Mott and I. N. Sneddon, *op. cit.*, sec. 47.
[22] Note the similarity between Eqs. (9.112) and (1.66).

where k is the wave number, c the velocity of light, and the factor 2 accounts for the two directions of polarization. Note that the left side of Eq. (9.114) is the same as the density-of-states calculated for semiconductors except that in the latter case, the factor 2 is for two spin orientations. The energy associated for each normal mode is $nh\nu$; hence the total energy is given by

$$\text{Total energy} = \int \frac{nh\nu V 8\pi \nu^2 \, d\nu}{c^3} = \int VI(\nu) \, d\nu \qquad (9.115)$$

A differentiation of Eq. (9.115) with respect to ν yields

$$\frac{I(\nu)}{n} = \frac{8\pi h}{\lambda^3} \qquad (9.116)$$

which is identical to the result of Eq. (9.111) by noting that $BI/A = n$ from Eqs. (9.109) and (9.113).

In the black-body radiation, no advantage is taken of the differences between the spontaneous and stimulated emission processes because the radiation field itself is incoherent and hence not different from the field resulting from spontaneous emission. In a maser system, however, the radiation field is a coherent source. Suppose a microwave signal enters the input side of a maser at time t_0 and leaves the output end at time t_1. The solution of Eq. (9.113) shows

$$n(t) = \left[n(t_0) + \frac{CN_i}{\alpha} \right] \exp[\alpha(t - t_0)] - \frac{CN_i}{\alpha} \qquad (9.117)$$

where $\alpha = C(N_i - N_j)$. Note that the signal has a gain only when $N_i > N_j$, or in other words, a population inversion is established. The factor $\exp[\alpha(t_1 - t_0)]$ can be considered as the gain of the device and is denoted by G. Thus, Eq. (9.117) becomes

$$n_1 = n_0 G + (G - 1) \left(\frac{CN_i}{\alpha} \right) \qquad (9.118)$$

The first term on the right-hand side of Eq. (9.118) is the amplified output of the incoming signal and hence it is useful. The second term is a result of spontaneous emission and consequently represents noise generated in the device. Multiplying Eq. (9.118) by $h\nu$, we have for the total noise output NO,

$$\text{NO} = (G - 1) \frac{h\nu}{1 - N_j/N_i} \qquad (9.119)$$

For complete population inversion $N_j \ll N_i$ the noise output has an equivalent noise temperature T_n of $h\nu/k$. Take $\nu = 10^{10}$ cps; this gives a theoretical minimum value of $T_n = 0.5°$K. If $N_j/N_i = 0.9$, a value easily obtainable in practical maser operations, $T_n = 5°$K.

The inherent noise in a maser, therefore, is extremely low as compared with that in conventional amplifiers.

Because of its low noise figure, masers find many important applications[23] in amplifying weak signals. In radio astronomy, the hydrogen emission line of distant stars and nebulae at 21 cm wavelength and the thermal emission from planets of the solar system are studied. In radar astronomy, radar techniques are applied to obtain radio echoes from neighboring planets. In these applications, to sort out the desired signal which is almost buried in noises, not only an extremely low-noise amplifier but an extremely sensitive detecting system must be used. Masers are also used to improve the performance of present communication systems, especially for space exploration and radio relay by communications satellites.

It may be worthwhile to restate some of the guidelines in the choice of maser material. The magnitude of the zero-field splitting, or D and E in Eq. (9.88), should be such that allowable transitions are available at the chosen pump and signal frequencies. A long spin-lattice relaxation time T_1 means that saturation of the pump transition levels can be achieved with a reasonable amount of pump power, say of the order of 10 milliwatts. The line-width of the spin-resonance line should be narrow; that means a long T_2. The magnetic ions having resonant frequencies outside the designed bandwidth of a maser do not participate in the maser action; hence a broad line reduces the effective density of useful paramagnetic ions. In view of these considerations and our earlier discussion in Sec. 9.8, the magnetic ions of Cr^{3+}, Fe^{3+}, and Gd^{3+} are considered to be most suitable for maser use and the ions of Ni^{2+}, Fe^{2+}, and Cr^{2+} have also been suggested, but these appear to have less promise.

Some of the above considerations also apply to host crystals. As stated in Sec. 9.1, a host crystal must be diamagnetic. Furthermore, nuclear magnetic moments belonging to the nuclei of the host crystal also contribute to the broadening of the resonance line. The host crystal should have a certain amount of asymmetry to provide a large enough zero-field splitting. The unit cell of the host lattice should be small so that there is only one available lattice site per unit cell for substitution. This is to ensure that paramagnetic ions will occupy magnetically equivalent sites when they substitute the original cations in the host lattice. It is also desirable that the cations have the same valence and approximately the same ionic radii as the intended paramagnetic substitute. The above considerations rule out a very large number of the most commonly available crystals.

[23] The following references may serve to illustrate the practical applications of masers: J. J. Cook *et al.*, *Proc. IRE*, vol. 49, p. 768, 1961; R. W. DeGrasse *et al.*, *Bell System Tech. J.*, vol. 40, p. 1117, 1961; E. A. Ohm, *Bell System Tech. J.*, vol. 40, p. 1065, 1961; T. Sato and C. T. Stelreid, *IRE Trans.*, SET-8, p. 164, 1962.

The host crystals which have been successfully used in maser opera-tions include $La(C_2H_5SO_4)_3 \cdot 9H_2O$,[24] $K_3Co(CN)_6$,[25] Al_2O_3[26] and TiO_2.[27] Various tungstates of the form XWO_4 where X stands for diamagnetic ions such as Mg, Cd, Ca, Pb, various oxides such as MgO, Y_2O_3, ThO, and various yttrium garnets such as $3Y_2O_3 \cdot 5Al_2O_3$, $3Y_2O_3 \cdot 5Ga_2O_3$, doped with paramagnetic ions, have also been in-vestigated for possible use in maser applications.

In concluding this chapter, it seems appropriate to describe the essential components of a maser system. A typical apparatus is shown in Fig. 9.13. The double-wall, silver-plated and evacuated Dewar flask is intended to cut down the loss of liquid helium through heat conduction, convection, and radiation. The best insulation against heat conduction and convection is a good vacuum. The heat radiation is reduced by reflections at silver-plated walls and by radiation shield provided by the outer liquid nitrogen. The boiling point of liquid helium goes down as the vapor pressure is reduced. The purpose of the vacuum pump in Fig. 9.13 is to control the helium vapor pressure so that the temperature of the liquid helium bath can

[24]   H. E. D. Scovil *et al.*, *Phys. Rev.*, vol. 105, p. 762, 1957.
[25]   A. L. McWhorter and J. W. Meyer, *Phys. Rev.*, vol. 109, p. 312, 1958.
[26]   J. E. King and R. W. Terhune, *Jour. Appl. Phys.*, vol. 30, p. 1844, 1959.
[27]   H. J. Gerritsen and H. R. Lewis, *Jour. Appl. Phys.*, vol. 31, p. 608, 1960.

Figure 9.13

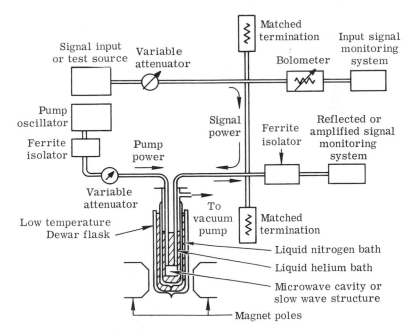

Experimental setup for maser operation. (*A. L. McWhorter and J. W. Meyer, Phys. Rev., vol. 109, p. 312, 1958.*)

be adjusted from 4.2°K down to about 1°K. The ferrite isolator used in the microwave circuit is a nonreciprocal device. It only allows the flow of microwave power in one direction and prevents power flow in the opposite direction. The principles involved in nonreciprocal microwave magnetic devices will be discussed in Chapter 10.

The maser crystal is put inside a microwave cavity or a slow-wave structure which provides microwave fields for interaction with the paramagnetic ions. A slow-wave structure is a broad-band circuit element and consequently traveling-wave masers in general have a larger bandwidth than cavity masers. The analysis of gain-bandwidth product of a maser system is similar to that involving negative resistance circuit elements. Such an analysis is outside the scope of the present book. The same can be said about microwave circuit design for maser operations, such as cavity design, scheme for broadbanding, design of traveling wave structure, etc. For information on these subjects, the reader should consult a book[28] on microwave-circuit design and analysis.

To test whether a maser system is in operating condition or not, a test signal from a local klystron is sent into the input end of Fig. 9.13. The reflected power is detected and monitored on the vertical axis of an oscilloscope, and the frequency of the klystron is swept by the time base of the oscilloscope. In Fig. 9.14a, the dotted curve shows the envelope of the klystron mode representing the incident power on the cavity while the solid curve shows the power absorbed by the cavity. Therefore, the reflected power from the cavity is the difference of these two curves as shown in Fig. 9.14b. The maser cavity shows absorption as long as the solid curve is below the dotted curve in Fig. 9.14b. As we gradually increase the pump power, the population difference $N_j - N_i$ begins to decrease, the state i being the higher energy state. When the pump power is large enough to cause a population inversion, the maser material becomes emissive and in terms of language familiar to electrical engineers, the maser crystal changes from a positive to a negative resistance element. In Fig. 9.14c, it is shown that the power coming out from the cavity exceeds the power going into the cavity; that means, the maser is in an amplifying mode of operation. Note that the bandwidth of the amplified signal is much narrower than the original bandwidth of the cavity due to the regenerative action of the field; that means, the stronger the microwave signal field, the larger the transition probability which in turn causes the field to grow. As the pump power is further increased, the maser may become self-oscillating if the rate

---

[28] See, for example, D. A. Watkins, "Topics in Electromagnetic Theory," John Wiley & Sons, Inc., New York, 1958; The MIT Radiation Lab. Series, vols. 8, 9, and 11, McGraw-Hill Book Company, New York, 1947. For traveling-wave masers, the article by R. W. DeGrasse et al. (*Bell System Tech. J.*, vol. 38, p. 305, 1959) contains design information.

of energy emission from quantum transitions exceeds the power loss associated with microwave circuits. The condition for oscillation will be discussed in a later chapter on lasers.

Many technical subjects are common to both masers and lasers. In selecting topics for these two chapters, we consider it desirable that each chapter be reasonably complete by itself and yet without undue repetition.

Figure 9.14

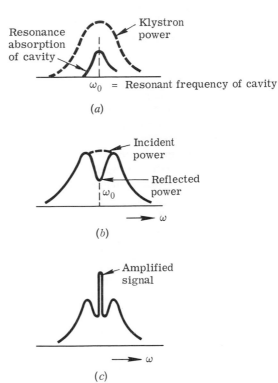

(a)

(b)

(c)

Oscilloscope traces showing the operating condition of a maser system. Diagram (a) shows resonance absorption in the cavity, which is the difference of the incident and reflected power as illustrated in (b). Diagram (c) shows that the reflected power exceeds the incident power, and hence the maser is in an amplifying mode of operation.

PROBLEMS

9.1   Derive Eq. (9.15) from Eq. (9.14) for a right-hand circularly polarized wave. Show also that the real and imaginary parts of the susceptibility respectively are given by

$$\chi' = \frac{\gamma\mu_0 M_0 T_2{}^2(\omega_0 - \omega)}{1 + T_2{}^2(\omega_0 - \omega)^2 + T_1 T_2(\gamma\mu_0 H)^2}$$

$$\chi'' = \frac{\gamma\mu_0 M_0 T_2}{1 + T_2{}^2(\omega_0 - \omega)^2 + T_1 T_2(\gamma\mu_0 H)^2}$$

9.2 Note that in the preceding problem, the real and imaginary parts of the susceptibility are determined by the same set of parameters M_0, T_2, and T_1. When $T_1 T_2 (\gamma \mu_0 H)^2 \ll 1$, the two susceptibilities χ' and χ'' are related to each other by the Kramers-Kronig relations. Read the book by C. Kittel ("Elementary Statistical Physics," sec. 44, John Wiley & Sons, Inc., New York, 1958) and outline the derivation of the relations. Explain why the condition $T_1 T_2 (\gamma \mu_0 H)^2 \ll 1$ is necessary in the derivation.

9.3 Read the articles by G. Feher (*Bell System Tech. J.*, vol. 36, p. 449, 1957) and by J. P. Goldsborough and M. Mandel (*Rev. Sci. Instr.*, vol. 31, p. 1044, 1960). Describe the essential components in a microwave spectrometer and discuss the sensitivity of such a spectrometer. By spectrometer, we mean apparatus which measures χ' and χ''.

9.4 Read the article by N. Bloembergen, E. M. Purcell, and R. V. Pound (Relaxation Effects in Nuclear Magnetic Resonance Absorption, *Phys. Rev.*, vol. 73, p. 679, 1948). Discuss how the spin-spin and spin-lattice relaxation times are obtained from the resonance absorption and dispersion curves.

9.5 Work out an integration similar to Eq. (9.36) for $(2,-2|L_y|2,-1)$ and check your result against Eq. (9.37). Refer to Eq. (1.81) for the spherical harmonics needed in the computation.

9.6 Consider the following combinations for the spin states of two electrons

$$\psi_a(+)\psi_b(+) \qquad \frac{1}{\sqrt{2}}[\psi_a(+)\psi_b(-) + \psi_a(-)\psi_b(+)] \qquad \psi_a(-)\psi_b(-)$$

the $+$ and $-$ signs meaning $s_z = \frac{1}{2}$ and $-\frac{1}{2}$, respectively. Show that the expectation values of S_z for these states are respectively 1, 0, and -1 and that

$$(0|S_z|1) = (1|S_z|0) = (-1|S_z|0) = \frac{\hbar}{\sqrt{2}}$$

In verifying the above statements, make use of the fact that $S_z(\psi_a\psi_b) = \psi_a(S_z\psi_b) + \psi_b(S_z\psi_a)$, etc.

9.7 Illustrate with the wave functions given in the preceding problem the following quantum-mechanical selection rule. In order to have a nonvanishing matrix element connecting two states, the values of M for these two states must be such that $\Delta M = 0$ for S_z and $\Delta M = \pm 1$ for S_x and S_y where M represents the z component of the spin quantum number.

9.8 Using the definition of u_1, u_2, and u_3, derive Eq. (9.58) from Eq. (9.55).

9.9 Verify Eqs. (9.63) and (9.64) used in the text for the spin state with $S = \frac{1}{2}$. Also show that the relation $(A|U|B) = (B|U|A)^*$ is obeyed. Note that expressions such as $(S_x + iS_y)\beta = \hbar\alpha$ and $(S_x - iS_y)\beta^* = \hbar\alpha^*$ can be deduced from Eqs. (9.37) and (9.38).

9.10 Consider a charge distribution having two positive charges situated along the x axis and two negative charges situated along the y axis. The four charges are all equidistant from the origin. Show that the lowest order term in the expansion of the crystal potential has the form

$$V_{crystal} = A(x^2 - y^2)$$

9.11 Derive Eq. (9.86).

9.12 Derive Eq. (9.87).

9.13 Show that the five d-orbital wave functions of Eq. (9.85b) can be expressed in terms of the spherical harmonics as $Y_2{}^2 \pm Y_2{}^{-2}$, $Y_2{}^1 \pm Y_2{}^{-1}$, and $Y_2{}^0$. Identify them.

9.14 Refer to the article by K. D. Bowers and J. Owen (*Rept. Progr. Phys.*, vol. 18, p. 311, 1955). Draw the energy-level diagrams, similar to the ones shown in Fig. 9.8, for other $3d$ configurations. Explain the causes for energy splitting.

9.15 Consider the case in which the d-c magnetic field is applied in the xz plane of a ruby crystal with the crystal c axis being chosen as the z axis. Show that Eq. (9.91) now must be replaced by the following equation

$$\mathfrak{H}\Phi_M = [gu_BBM \cos\theta - D(M^2 - \tfrac{5}{4})]\Phi_M$$
$$+ \tfrac{1}{2}gu_BB \sin\theta\, [\sqrt{\tfrac{15}{4} - M(M+1)}\ \Phi_{M+1} + \sqrt{\tfrac{15}{4} - M(M-1)}\ \Phi_{M-1}]$$

where θ is the angle between $\mathbf{B}$ and the c axis. Comment on the composition of the new proper wave functions.

9.16 Show that the spin operator $S_x{}^2$ can be expressed as

$$S_x{}^2 = \tfrac{1}{4}(S_+{}^2 + S_-{}^2) - \tfrac{1}{2}(S_z{}^2 - \tfrac{15}{4})$$

for $S = \tfrac{3}{2}$. Explain what happens to the cross-terms in $S_+{}^2$ and $S_-{}^2$. Consider the case in which the applied field (chosen to be the z direction) is perpendicular to the crystal c axis (chosen to be the x direction) of a ruby crystal. Show that the spin Hamiltonian now becomes

$$\mathfrak{H} = gu_BBS_z + \frac{D}{2}\,(S_z{}^2 - \tfrac{5}{4}) - \frac{D}{4}\,(S_+{}^2 + S_-{}^2)$$

The above set of axes is chosen because it results in a simple form for the secular equation.

9.17 Using the result of the preceding problem, show that the four energy levels have energies given below

$$\mathcal{E}_{A,B} = \left(\pm\frac{x}{2} + \sqrt{1 + x + x^2}\right) D$$

$$\mathcal{E}_{C,D} = \left(\pm\frac{x}{2} - \sqrt{1 - x + x^2}\right) D$$

where $x = gu_BB/D$. With $g = 1.985$ and $D = 5.75$ Gc, plot the energies as functions of the applied magnetic field B. In setting up the determinant, it is advisable to use the following order for the wave functions: $\Phi_{3/2}$, $\Phi_{-1/2}$, $\Phi_{1/2}$, $\Phi_{-3/2}$.

9.18 Following the same procedure as used in deriving Eq. (9.98), find the expressions for the wave functions of the various energy states discussed in Prob. 9.17. As a check for the reader, the Ψ_A for the state $\mathcal{E}_A$ is given below:

$$\Psi_A = A\frac{\sqrt{3}}{2}\Phi_{3/2} + (x + \tfrac{1}{2} - \sqrt{1 + x + x^2})\,\Phi_{-1/2}$$

where A is a normalization constant. Discuss the scheme to make transitions possible between any two of the four levels.

9.19 Read the article by N. Bloembergen, E. M. Purcell, and R. V. Pound (*Phys. Rev.*, vol. 73, p. 679, 1948) on relaxation effects in nuclear magnetic resonance. Applying their model to electron spin resonance, discuss qualitatively how the dipolar and exchange interactions of the type $\mathbf{S}_i \cdot \mathbf{S}_j$ can cause the interchange of spin states between neighboring electrons and hence the spin-spin relaxation time T_2.

9.20 Read the book by G. E. Pake ("Paramagnetic Resonance," sec. 6-2, W. A. Benjamin, Inc., New York, 1962). Describe qualitatively the process of spin-lattice relaxation.

9.21 Read the articles by R. W. DeGrasse, J. J. Kostelnick and H. E. D. Scovil (*Bell System Tech. J.*, vol. 40, p. 1117, 1961) and by J. V. Jelley and B. F. C. Cooper (*Rev. Sci. Instr.*, vol. 32, p. 166, 1961). Describe the maser systems used for space communication and radio astronomy.

9.22 Read the article by H. E. D. Scovil, G. Feher and H. Seidel (*Phys. Rev.*, vol. 105, p. 762, 1957). Describe the energy levels of Gd^{3+} in $La(C_2H_5SO_4)_3 \cdot 9H_2O$ which is the material reported successful early in the maser development.

9.23 Consider the four-level system shown in Fig. 9.12a. It is assumed that saturating transitions are applied simultaneously between levels $\mathcal{E}_1$ and $\mathcal{E}_3$ and between levels $\mathcal{E}_2$ and $\mathcal{E}_4$. Show that the difference in population between levels $\mathcal{E}_3$ and $\mathcal{E}_2$ is given by

$$N_3 - N_2 = \frac{Nh}{4kT}\frac{(w_{12})_0\nu_{12} - (w_{23})_0\nu_{23} + (w_{14})_0\nu_{14} + (w_{34})_0\nu_{34}}{(w_{12})_0 + (w_{23})_0 + (w_{14})_0 + (w_{34})_0 + W_{23}}$$

The above equation is obtained by first considering the set of equations $dN_1/dt = 0$ and $dN_3/dt = 0$ (or the alternative set $dN_2/dt = 0$ and $dN_4/dt = 0$) and then setting $N_1 = N_3$ and $N_2 = N_4$. Show that the error introduced in considering only two equations instead of three in the exact solution will involve terms of the order of $h\nu_{ij}/kT$ which is assumed to be small.

9.24 Consider the push-pull mode of maser operation shown in Fig. 9.12b for which saturating transitions are employed between three levels. Show that the population difference between the two top levels is given by

$$N_4 - N_3 = \frac{Nh}{4kT}\frac{(w_{13})_0\nu_{13} + (w_{23})_0\nu_{23} - (w_{34})_0\nu_{34}}{(w_{13})_0 + (w_{23})_0 + (w_{34})_0 + W_{34}}$$

Discuss physically why a better population inversion ratio is achieved in the present four-level system than in the three-level system.

REFERENCES

Bleaney, B., and K. W. H. Stevens: Paramagnetic Resonance, *Rept. Progr. Phys.*, vol. 16, p. 108, 1953.

Bowers, K. D., and J. Owen: Paramagnetic Resonance II, *Rept. Progr. Phys.*, vol. 18, p. 304, 1955.

Geusic, J. E., and H. E. D. Scovil: Microwave and Optical Masers, *Rept. Progr. Phys.*, vol. 27, p. 241, 1964.

Gorter, C. J.: "Paramagnetic Relaxation," Elsevier Publishing Co., Inc., Amsterdam, 1947.

Louisell, W.: "Radiation and Noise in Quantum Electronics," McGraw-Hill Book Company, New York, 1964.

Low, W.: "Paramagnetic Resonance in Solids," Academic Press, Inc., New York, 1960.

Orton, J. W.: Paramagnetic Resonance Data, *Rept. Progr. Phys.*, vol. 22, p. 204, 1959.

Pake, G. E.: "Paramagnetic Resonance," W. A. Benjamin, Inc., New York, 1962.

"Quantum Electronics," Columbia University, New York, 1960, 1961, and 1964. Proceedings of the first, second, and third International Congress on quantum electronics.

Siegman, A. E.: "Microwave Solid-State Masers," McGraw-Hill Book Company, New York, 1964.

Singer, J. R.: "Masers," John Wiley & Sons, Inc., New York, 1959.

Townes, C. H.: "Masers," in A. Gozzini (ed.), "Topics of Radiofrequency Spectroscopy," Academic Press, Inc., New York, 1962.

Vuylsteke, A. A.: "Elements of Maser Theory," D. Van Nostrand Co., New York, 1960.

10 FERROMAGNETISM AND FERROMAGNETIC DEVICES

10.1 THE *B-H* LOOP

One of the most informative experiments about the properties of a ferromagnetic substance is to measure the characteristic *B-H* curve of the material. Figure 10.1 shows a simple circuit arrangement in which a magnetic core is driven by an a-c current I flowing through a primary coil of N turns. If the core has a magnetic path length l, the magnetic field H inside the core is equal to NI/l. The voltage induced in the secondary coil, on the other hand, is equal to the time rate of change in the total flux linkage or proportional to dB/dt. To obtain B as a function of time, the induced voltage must be integrated with respect to time and this is the function of the capacitor. From Fig. 10.1, it is clear that the oscilloscope trace will show a *B-H* curve with the horizontal axis representing H and the vertical axis B.

Before discussing the details of a *B-H* curve, however, we must know something about the internal structure of a ferromagnetic substance. From our discussion in Sec. 8.9, it appears that every ferromagnetic substance should possess a nonvanishing and constant magnetization which is determined by temperature alone. Yet in reality, a ferromagnetic specimen can be demagnetized upon the proper application of a magnetic field. We also have pointed out that a powerful internal field of the order of 3×10^9 amp/m exists which is responsible for spontaneous magnetization. Yet experiments show that a magnetic field anywhere from 1 amp/m to 10^4 amp/m, depending on the specimen, can change the apparent magnetization from a

Figure 10.1

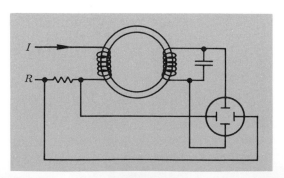

A simple circuit measuring the B-H curve of a ferromagnetic material.

value of zero to its full value given by the spontaneous magnetization. The explanation of this behavior rests in the fact that a magnetic specimen is actually composed of many small regions called *domains*. Within each domain, the magnetization actually attains its full value, but the magnetization vectors in different domains may be oriented in different directions. *A demagnetized state* is a state in which the magnetization vectors of each domain are so oriented that their resultant is zero. This is illustrated in Fig. 10.2a. The existence of the ferromagnetic domain structure was first predicted by Weiss and direct observation of such a structure was first made by Bitter.

In the presence of a magnetic field H, a specimen can become magnetized by two distinct processes. As illustrated in Fig. 10.2b, the domain of which the magnetization is oriented favorably with respect to the field grows in size at the expense of other domains with unfavorable orientations. This process therefore involves the movement of domain boundaries. A net magnetization will also result if the magnetization vector in each domain turns in the direction of the applied field. The process known as *magnetization rotation* is illustrated in Fig. 10.2c. In weak fields the magnetization changes proceed through the process of *domain boundary movements* and in high fields through the process of magnetization rotation as indicated in Fig. 10.3a. Had all the processes been reversible, the magnetization curve would follow the curve $CBAO$ upon removal of the field. This however is not the case.

What we see on the oscilloscope is not the backward trace of curve $OABC$ and its counterpart for negative H but a loop $CDEFGC$ as shown in Fig. 10.4a. The closed B-H loop is generally known as the *hysteresis loop*. To explain why part of the magnetization process may be irreversible, we refer to Fig. 10.3b. The region which separates two domains actually is not abrupt but finite. The boundary region

Figure 10.2

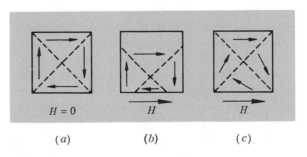

(a) (b) (c)

Schematic illustration showing domain structure and state of magnetization in a ferromagnet. (a) The demagnetized state. Under the influence of an applied magnetic field, a specimen can become magnetized either by (b) expanding the size of the domain with favorable orientation or by (c) turning the magnetization vector in each domain in the direction of the applied field.

Figure 10.3

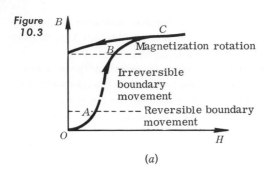

(a)

A qualitative explanation of the B-H curve in terms of the domain structure. Diagram (a) shows that different mechanisms are responsible for the change in the net magnetization M (and hence B) in different regions of H. Diagram (b) shows the energy associated with a domain wall as a function of the position of the wall.

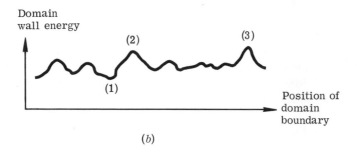

(b)

between two domains is generally referred to as a *domain wall*. Associated with the wall, as we shall see later, there is a finite amount of magnetic energy. As the position of the wall moves, the wall energy changes. This is the meaning of Fig. 10.3*b*. Suppose that the initial wall position is at point 1. It takes a magnetic field of sufficient strength to push the wall over the energy barrier at point 2. Once the wall is over to the right side of point 2, it can not return to its original position unless it is pushed by a magnetic field of sufficient strength in the opposite direction. Therefore, the reversible

Figure 10.4

$$B = \mu_0 (H + M)$$

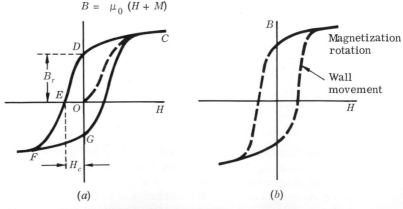

(a)   (b)

The hysteresis loop of a ferromagnet.

movement of a domain wall is confined to a small displacement about a minimum energy position and it involves a relatively small change of magnetization as indicated by portion OA of curve $OABC$ in Fig. 10.3a. A close examination of the B-H curve shows that the process of magnetization rotation also has a reversible and an irreversible part. A meaningful discussion of these pro\`esses, however, must wait until a full discussion of the various energies associated with the domain wall is made.

Now let us discuss in general terms the technical information which can be obtained from the B-H curve. At the saturated state C, the flux density B is given by $\mu_0(H + M_0)$ where M_0 is the value of the spontaneous magnetization plotted as a function of temperature in Fig. 8.13. When the field is decreased, the apparent magnetization decreases also. The part of magnetization which a magnetic substance apparently still retains at $H = 0$ is called the *residual magnetization* or *remanence* $B_r(= OD)$. It turns out that the spontaneous magnetization has a preferred orientation along certain crystal directions. As will be discussed later, this tendency for M to be directed along a preferred orientation can be expressed quantitatively in terms of an equivalent field called the *anisotropy field*. The value of B_r may vary anywhere from μ_0 times the full spontaneous magnetization to a value close to zero, depending on the nature of the anisotropy field and the direction of the magnetic field. If the magnetic field is increased now in the negative direction, the apparent magnetization can be brought to a value zero and the field $H_c(= OE)$ required for this purpose is called the *coercive force*. The value of H_c may range from a value of 10^6 amp/m in a permanent magnet to a value of 0.2 amp/m for supermalloy (a 5 per cent Mo, 79 per cent Ni and 16 per cent Fe alloy). A substance with a high coercive force is generally referred to as being *magnetically hard* and that with a low coercive force, *magnetically soft*.

We should point out that although the title of this chapter is ferromagnetism and ferromagnetic devices, we shall discuss ferrimagnetic and antiferromagnetic materials as well. All these materials have one property in common, that is, a large Weiss molecular field. In addition, ferromagnetic and ferrimagnetic substances both possess a net spontaneous magnetization. The following discussions regarding anisotropy, magnetic domains, and magnetization processes apply equally well to ferrimagnetic substances.

Finally, we should say a few words about the units which the reader at times may find confusing. In the cgs system, the equation corresponding to Eq. (8.7) which relates B to H and M is often written as

$$B = \mu_0(H + 4\pi M) = H + 4\pi M \qquad \text{(cgs units)} \qquad (10.1)$$

where μ_0, the permeability of free space, is equal to unity. The quantity B is expressed in gauss which is equal to 10^{-4} weber/m² and H in oersted which is equal to $10^3/4\pi$ amp/m. Thus, in the cgs system, a field of 1 oe produces 1 gauss in free space. In terms of units in the mks system,

$$10^{-4} \text{ weber/m}^2 = (4\pi \times 10^{-7} \text{ henry/m}) \times \frac{10^3}{4\pi} \text{ amp/m} \quad (10.2)$$

What may cause the confusion is the unit used for M. In the cgs system, M is usually expressed in gauss, not oersted. According to Eqs. (8.7) and (10.1), the conversion formula should read

$$\mu_0 M \text{ (weber/m}^2, \text{ mks)} = 4\pi \times 10^{-4} M \text{ (gauss, cgs)} \quad (10.3)$$

The factor 4π comes in because of the same factor in Eq. (10.1). Since most magnetic data are expressed in cgs units, the reader will find the conversion factors $10^3/4\pi$ for H and $4\pi \times 10^{-4}$ for M used quite often in the text.

10.2 THE HEISENBERG THEORY OF FERROMAGNETISM AND THE EXCHANGE ENERGY

Before discussing any theory of ferromagnetism, we must know the origin of magnetic moment in a ferromagnetic substance. As we recall our discussion in Sec. 8.5, the magnetic moment and angular momentum are related to each other by a constant $ge/(2m_0)$ called the *magnetomechanical ratio*. For electron spin, $g = 2$, and for orbital motion, $g = 1$. The Landé splitting factor g can be determined by gyromagnetic experiments using the Barnett or Einstein-de Haas method and by magnetic resonance experiments. Experimental values of g for a selected number of substances are given in Table 10.1. The difference in the two g values was first predicted by Kittel

TABLE 10.1 *Experimental values of g*

Substance	g' (gyromagnetic experiment)	g (resonance experiment)
Iron	1.93	2.12–2.17
Cobalt	1.85	2.22
Permalloy (78% Ni, 22% Fe)	1.90	2.07–2.14
Supermalloy	1.91	2.12–2.20
$MnFe_2O_4$	....	2

and the reader should refer to Kittel's book[1] for the explanation. In any event, from the wealth of experimental data accumulated so far,

[1] C. Kittel, "Introduction to Solid State Physics," pp. 408–410, John Wiley & Sons, Inc., New York, 1956.

it is safe to conclude that the orbital motion contributes very little, of the order of 10 per cent or less, to the magnetic moment. Therefore, the origin of ferromagnetism must lie in the fact that spins on neighboring atoms favor parallel alignment.

Let us now retrace our discussion in Sec. 2.1. For two atoms a and b which have one electron each, the total energy is given by Eq. (2.5) or

$$\mathcal{E}_{total} = 2\mathcal{E} + \mathcal{E}_a \pm \mathcal{E}_b \qquad (10.4)$$

where $\mathcal{E}$ is the atomic energy and $\mathcal{E}_a$ is given by Eq. (2.4). The quantity $\mathcal{E}_b$ defined in Eq. (2.6) as

$$\mathcal{E}_b = J = \int \psi_a^*(2)\psi_b^*(1) V \psi_a(1)\psi_b(2) \, dx_1 \, dy_1 \, dz_1 \, dx_2 \, dy_2 \, dz_2 \qquad (10.5)$$

is known in the theory of ferromagnetism as the *exchange energy* denoted by J, and the quantity V is the electrostatic interaction energy given by

$$V = \frac{1}{4\pi\epsilon_0}\left(\frac{e^2}{r_{ab}} + \frac{e^2}{r_{12}} - \frac{e^2}{r_{1b}} - \frac{e^2}{r_{2a}}\right) \qquad (10.6)$$

where r_{ab} and r_{12} are respectively the interatomic and interelectronic distances while r_{1b} and r_{2a} are the distances between one nucleus and the electron on the other atom. We also recall that the exchange charge density $\rho_{12}^2 = \psi_a^*(2)\psi_b^*(1)\psi_a(1)\psi_b(2)$ can be considered as being shared by two neighboring atoms. The positive and negative signs in Eq. (10.4) correspond respectively to symmetrical and antisymmetrical orbital wave functions and hence to spin antiparallel and parallel arrangements. In most cases, the value of $\mathcal{E}_b$ is negative, favoring spin antiparallel arrangement. As discussed in Sec. 2.1, a covalent bond is a result of minimizing the energy by sharing two electrons with antiparallel spins between two neighboring nuclei. However, in trying to see whether the exchange energy can be positive or not, we must remember that the magnitude of $\mathcal{E}_b$ changes quite drastically with the interatomic distance r_{ab}.

First of all, the energy involved in the bond of a hydrogen molecule which is shown in Fig. 2.2 is of the order of the ionization energy or 13.6 ev. The exchange energy responsible for ferromagnetism as shown in Sec. 8.9 is of the order kT_c, T_c being the Curie temperature. For $T_c = 1000°$K, J is of the order of 0.085 ev. Therefore, so far as ferromagnetism is concerned, we are interested in a different region of r_{ab} or specifically in the tail end of Fig. 2.2. In this far-out region, two possibilities exist. The value of J either changes its sign or remains negative as shown respectively in Fig. 10.5a and b. Dehlinger[2] studied the paramagnetic susceptibilities of many substances and deduced that the elements Cr, Mn, Fe, Co, Ni, Pd, and Pt possess a crossover point near $r_{ab} = d$, d being the distance between the near-

[2] U. Dehlinger, *Z. Metallkunde*, vol. 28, pp. 116 and 194, 1936; vol. 29, p. 388, 1937.

est neighbors, while the elements V, Mo, and W do not. With the former group of elements, ferromagnetism is possible if the distance d happens to be in the region where J is positive. The elements Co, Ni, Pd, and Pt have this property. Iron has two phases with the face-centered phase being nonferromagnetic and the body-centered phase which has a slightly larger d being ferromagnetic. Although Cr and Mn are not ferromagnetic, they potentially can become ferromagnetic if the distance d can be changed to the region of positive J by alloying. This indeed happens in the case of MnAs. Other examples of ferromagnetic alloys include Cu_2MnAl, MnSb, and CrTe, to name a few.

The idea which links the quantum-mechanical exchange integral with ferromagnetism was first expounded by Heisenberg.[3] The condition that J is positive is met, however, only under certain circumstances. According to a qualitative analysis by Bethe,[4] the integral J is most likely to be positive if (1) the interatomic distance r_{ab} is large compared with the radius r_0 of the electron orbit and (2) the electron wave function is small near the nucleus. These conditions are to ensure that the exchange charge density $\rho_{12}{}^2$ is small at the nuclei and comparatively large at midpoints so that the expectation values of $1/r_{1b}$ and $1/r_{2a}$ of Eq. (10.6) can be made small in comparison with the expectation value of $1/r_{12}$. The transition metals of the iron group and the rare-earth group are most favorably disposed in meeting these conditions. In these elements, the interatomic distances are determined by the outer s and p electrons, and the wave functions of the inner d or f electrons have relatively small amplitudes near the nucleus because of their large quantum numbers. The above

[3] W. Heisenberg, *Z. Physik*, vol. 49, p. 619, 1928.
[4] H. A. Bethe, *"Handbuch der Physik,"* vol. XXIV/2, p. 596, J. Springer, Berlin, 1933.

Figure 10.5

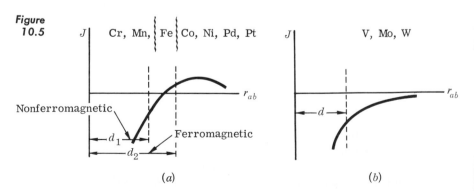

The exchange energy J as a function of interatomic distances in two distinct groups of materials: (a) where ferromagnetism is possible if the interatomic distance is in the right region, and (b) where ferromagnetism will not occur.

discussion serves to show that the value of J may be positive as shown in Fig. 10.5a under favorable conditions.

Finally, we should mention that a different type of exchange coupling called *superexchange* exists in certain antiferromagnets with face-centered cubic structure. As mentioned in Sec. 8.10, evidences from neutron diffraction experiments show that the coupling between nearest Mn atoms which form a collinear Mn—O—Mn bond is much weaker than that between next-to-nearest Mn atoms which have the two Mn—O bonds at right angles. Based on quantum-mechanical calculations, the coupling is through the excited state of MnO, with the oxygen ion acting as an intermediary. For a further discussion of the theory of superexchange and a general review of existing theories on ferromagnetism and antiferromagnetism, the reader is referred to two papers by Van Vleck.[5]

In summary, the exchange energy is an energy of electrostatic origin which is a result of quantum mechanics, having no classical counterpart. For our present purpose, the exchange energy represents the coupling energy between spins of neighboring atoms and this energy is usually written as

$$\mathcal{E}_{ex} = -2J\mathbf{S}_i \cdot \mathbf{S}_j \qquad (10.7)$$

where $\mathbf{S}_i$ and $\mathbf{S}_j$ are respectively the spin quantum numbers of atoms i and j. We must be careful in evaluating the scalar product of $\mathbf{S}_i$ and $\mathbf{S}_j$ because in quantum mechanics $\mathbf{S} \cdot \mathbf{S} = S(S + 1)$. By the law of cosines, we have

$$2\mathbf{S}_i \cdot \mathbf{S}_j = |\mathbf{S}_{i+j}^2| - |\mathbf{S}_i^2| - |\mathbf{S}_j^2| \qquad (10.8)$$

where $\mathbf{S}_{i+j}$ is the resultant of $\mathbf{S}_i$ and $\mathbf{S}_j$. For spin $\frac{1}{2}$, $S_{i+j} = 1$ in the parallel arrangement and $S_{i+j} = 0$ in the antiparallel arrangement, so $|\mathbf{S}_{i+j}^2| = 2$ and 0 respectively for these two arrangements while $|\mathbf{S}_i^2| = |\mathbf{S}_j^2| = \frac{3}{4}$. Substituting these values into Eqs. (10.8) and (10.7), we find $\mathcal{E}_{ex} = -J/2$ for the parallel arrangement and $\mathcal{E}_{ex} = 3J/2$ for the antiparallel arrangement. The difference between the two energies is $2J$, exactly equal to the difference given by Eq. (10.4).

For the following discussion, however, we shall not be concerned with the quantum-mechanical interpretation of Eq. (10.7) but instead shall treat the spin operators simply as classical vectors. This approximation is justifiable only when neighboring spins make small angles with each other. If θ is the angle between the directions of neighboring spins and $\theta \ll 1$, then Eq. (10.7) becomes

$$\Delta\mathcal{E}_{ex} = -2JS^2(1 - \cos\theta) = -JS^2\theta^2 \qquad (10.9)$$

The above expression will be used in our discussion of domain walls.

[5] J. H. Van Vleck, *J. Phys. Radium*, vol. 12, p. 262, 1951; *Rev. Mod. Phys.*, vol. 17, p. 27, 1945.

There is another form of Eq. (10.7) which is used in the treatment of spin waves. If $\mathbf{S}_j$ varies smoothly from lattice point to lattice point, it can be expanded into a Taylor series. For a simple cubic lattice with lattice constant a, summing $\mathbf{S}_j$ over the six nearest neighbors of lattice point i results in the following expression:

$$\sum_j \mathbf{S}_j = 6\mathbf{S}_i + a^2 \left(\frac{\partial^2 \mathbf{S}_i}{\partial x^2} + \frac{\partial^2 \mathbf{S}_i}{\partial y^2} + \frac{\partial^2 \mathbf{S}_i}{\partial z^2} \right) \tag{10.10}$$

The first derivatives of $\mathbf{S}_i$ cancel each other out for a pair of lattice points with coordinates $\mathbf{r}_j = \mathbf{r}_i \pm \mathbf{a}$ where $\mathbf{a}$ can be along any one of the three axes of the cube. Substituting Eq. (10.10) into (10.7), we find the change (per unit cell) in the exchange energy due to the spatial variation of $\mathbf{S}_i$ as

$$\Delta \mathcal{E}_{ex} = -(ZJa^2)\mathbf{S}_i \cdot (\nabla^2 \mathbf{S}_i) = (ZJa^2) \sum_{x,y,z} (\nabla S_x)^2 \tag{10.11}$$

The last step is worked out by using $\nabla^2(S_x{}^2 + S_y{}^2 + S_z{}^2) = 0$. Equation (10.11) is generalized to include other lattices of cubic symmetry with $Z = 1$, 2, and 4 respectively for simple, body-centered, and face-centered cubic structures.

Finally, we shall estimate the value of J. Calculations by Weiss[6] yield the following result:

$$\begin{aligned} J &= 0.54kT_c && \text{(simple cubic; spin } \tfrac{1}{2}\text{)} \\ J &= 0.34kT_c && \text{(body-centered cubic; spin } \tfrac{1}{2}\text{)} \\ J &= 0.15kT_c && \text{(body-centered cubic; spin 1)} \end{aligned} \tag{10.12}$$

The value of the exchange integral can also be estimated from the temperature variation of the spontaneous magnetization at low temperatures. According to the calculation [Eq. (10.140)] based on the spin-wave theory of ferromagnetism,

$$M_0 = M_s \left[1 - \frac{C}{2S} \left(\frac{kT}{2SJ} \right)^{3/2} \right] \tag{10.13}$$

where M_s is the saturation magnetization and C is a constant equal to 0.1174, 0.0587, or 0.0294, respectively, for a simple, a body-centered, or a face-centered cubic structure. Taking iron as an example, $T_c = 1043°K$ and the value S from Table 8.1 is close to 1. Thus, from the third equation of Eq. (10.12), we have the value of $J = 0.013$ ev $= 2.1 \times 10^{-21}$ joule. For an order-of-magnitude estimate of the value of J, Eq. (8.65) is quite adequate even though Eqs. (10.12) and (10.13) give more accurate results.

[6] P. R. Weiss, *Phys. Rev.*, vol. 74, p. 1493, 1948. For a different calculation, refer to H. A. Brown, *Phys. Rev.*, vol. 104, p. 624, 1956.

10.3 MAGNETIC ANISOTROPY AND MAGNETOSTRICTION

In the previous section, we see that as a result of the exchange energy, the spins on neighboring atoms have a tendency to be aligned along the same direction, producing a resultant magnetization. However, nothing has been said about the specific direction which the spin and hence the magnetization will take in an actual crystal. Note that Eq. (10.7) depends only on the relative direction of S_i and S_j but not on the absolute direction of S_i or S_j. Figure 10.6 shows the magnetization curves of single crystals of iron and cobalt taken with H along certain crystallographic directions. In Fe, the magnetization is more easily aligned along the [100] axis of the cubic structure than any other directions, while in Co, along the c axis of the hexagonal structure. These directions are called directions of *easy magnetization*. The axis along which it is most difficult to magnetize the crystal is called the *hard axis*. In Fe, the hard axis is along the body diagonal, while in Co it is along any direction in the basal plane.

The extra energy required to magnetize a crystal in its hard direction is known as the *anisotropy energy*. The simplest form of anisotropy energy occurs in uniaxial crystals of which the hexagonal structure of cobalt is an example. Actually the anisotropy energy does not start or stop with the hard axis. As the magnetization vector turns away from the easy axis, the internal energy of the crystal increases. The energy is a maximum when the magnetization vector is directed along the hard axis which in the case of uniaxial anisotropy is perpendicular to the easy axis. It is natural to suggest, therefore, that the anisotropy energy changes with even powers of

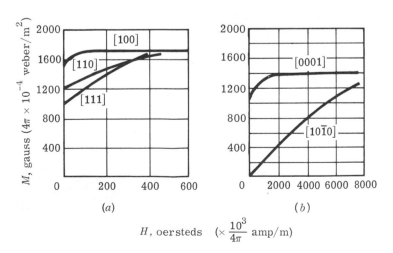

Figure 10.6

Magnetization curves with magnetic fields H applied along different crystal directions in (a) Fe and (b) Co.

$\sin \theta$ where θ is the angle between the easy axis and the magnetization vector. For cobalt, it is found that the anisotropy energy $\mathcal{E}_a$ can be well approximated by

$$\mathcal{E}_a = K_1 \sin^2 \theta + K_2 \sin^4 \theta \qquad \text{joule/m}^3 \qquad (10.14)$$

where K_1 and K_2 are known as *anisotropy constants*. Since $\mathcal{E}_a(\theta) = \mathcal{E}_a(180° - \theta)$, odd powers of $\sin \theta$ do not exist.

In a cubic crystal, the form of the anisotropy energy can be deduced from the following reasoning. If $\mathcal{E}_a$ is a function of the direction cosines α_1, α_2, and α_3 of the magnetization vector referred to the cube edges, any interchange of α_i with α_j must yield the same $\mathcal{E}_a$. The simplest dependence with even powers of α is of the form $\alpha_1^2 + \alpha_2^2 + \alpha_3^2$. Since $\alpha_1^2 + \alpha_2^2 + \alpha_3^2 = 1$, no angular dependence of $\mathcal{E}_a$ is expected. The next combination is of the $\Sigma\alpha_i^4$ and $\Sigma\alpha_i^2\alpha_j^2$ variety. Because of the relation $(\alpha_1^2 + \alpha_2^2 + \alpha_3^2)^2 = 1$, however, only one of the two kinds will suffice. The reader can easily show that out of the next group containing $\Sigma\alpha_i^6$, $\Sigma\alpha_i^4\alpha_j^2$ and $\alpha_1^2\alpha_2^2\alpha_3^2$, only the last term contributes to a new angular dependence. Usually the terms just considered are sufficient to fit experimental results; so for a cubic crystal

$$\mathcal{E}_a = K_1(\alpha_1^2\alpha_2^2 + \alpha_2^2\alpha_3^2 + \alpha_3^2\alpha_1^2) + K_2\alpha_1^2\alpha_2^2\alpha_3^2 \qquad \text{joule/m}^3$$
$$(10.15)$$

It is sometimes useful to think of anisotropy energy as resulting from an effective field called the *anisotropy field* H_a acting along the direction of the easy axis. If M makes an angle θ with H_a as shown in Fig. 10.7a, the energy involved according to Eq. (8.53) is given by $-\mu_0 M_0 H_a \cos \theta$ which except for an integration constant should be equal to $\mathcal{E}_a$ of Eq. (10.14) or (10.15). We can eliminate the constant by differentiating the energy expression with respect to θ and thus obtain

$$H_a = \frac{1}{\mu_0 M_0 \sin \theta} \frac{d\mathcal{E}_a}{d\theta} \qquad (10.16)$$

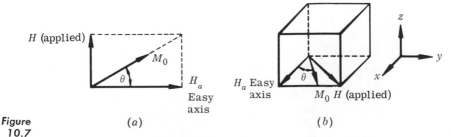

Figure 10.7

(a)   (b)

Diagram used in calculating the effective anisotropy field H_a in (a) a uniaxial crystal and (b) a cubic crystal.

For a uniaxial crystal, substitution of Eq. (10.14) into Eq. (10.16) gives

$$H_a = \frac{2K_1 \cos \theta + 4K_2 \sin^2 \theta \cos \theta}{\mu_0 M_0} \tag{10.17}$$

The situation in a cubic crystal is much more complicated; therefore, we consider only the following specific case as an example. If the magnetization was originally along the [100] direction and the applied field is along the [110] direction, the new magnetization must remain in the (001) plane as shown in Fig. 10.7b. In terms of the angle θ which M_0 makes with the x axis, $\alpha_1 = \cos \theta$, $\alpha_2 = \sin \theta$, and $\alpha_3 = 0$. Substituting these values into Eq. (10.15) and using Eq. (10.16), we find

$$H_a = \frac{1}{\mu_0 M_0 \sin \theta} \frac{d}{d\theta} K_1 \sin^2 \theta \cos^2 \theta = \frac{2K_1 \cos \theta \cos 2\theta}{\mu_0 M_0} \tag{10.18}$$

From Eqs. (10.17) and (10.18), we see that the anisotropy field H_a is not constant but changes with θ.

The value of the anisotropy constants can be determined most conveniently either from a magnetic resonance experiment or from the magnetization curve. Since the former will be discussed in a later section, only the latter case will be treated here. Let us take the magnetization curve of cobalt (Fig. 10.6) as an example. At a given applied field H, the magnetization must lie in the direction of the resultant field of H and H_a. Referring to Fig. 10.7a, we have

$$\frac{H}{H_a} = \tan \theta \tag{10.19}$$

It turns out, as will be explained later, that the magnetization shown in Fig. 10.6 is the component of spontaneous magnetization M_0 along the direction of the applied field or $M = M_0 \sin \theta$. For $M = 800$ gauss, $\theta = 35°$. Using the known values of $K_1 = 4.3 \times 10^6$ ergs/cm³, $K_2 = 1.2 \times 10^6$ ergs/cm³, and $M_0 = 1400$ gauss, we find from Eq. (10.17) the value of $H_a = 5830$ oe. From Eq. (10.19), $H = 4000$ oe in agreement with the value shown in Fig. 10.6. The reader is asked to check the magnetization curve of iron along the [110] direction by using Eqs. (10.18) and (10.19). Needless to say, if we work the other way by choosing two sets of values for M and H from the magnetization curve, the values of K_1 and K_2 can be solved from two simultaneous equations obtained from Eqs. (10.17) and (10.19). The values of the anisotropy constants are given in Table 10.2 for a selected number of materials. The negative sign in

TABLE 10.2 *Anisotropy constants for a selected number of materials*

Substance	Temperature, °C	K_1	K_2 (10^5 ergs/cm^3)	Types
Co	20	43.0	12	uniaxial
Co	−176	79.0	10	uniaxial
Fe	20	4.8	±0.5	cubic
Ni	20	−4.5	2.3	cubic
NiFe$_2$O$_4$	20	−0.67	.....	cubic
MnFe$_2$O$_4$	17	−0.4	.....	cubic
MnFe$_2$O$_4$	−183	−2.0	.....	cubic

front of K_1 simply means that the easy axis is along the $\langle 111 \rangle$ direction instead of the $\langle 100 \rangle$ direction, for the value of $\mathcal{E}_a$ is a minimum from Eq. (10.15) when $\alpha_1 = \alpha_2 = \alpha_3 = \pm 1/\sqrt{3}$. A compilation of experimental values of K_1 and K_2 can be found in the book by Chikazumi.[7]

The origin of the magnetic anisotropy is believed to be the result of spin-orbit coupling mentioned briefly in Sec. 8.6. In the idealized model discussed in Sec. 8.7, we say that the orbital contribution of the 3d electrons to the magnetic moment is completely quenched by the crystalline electric field. However, if the energy splitting, for example, $\Delta\mathcal{E}_A - \Delta\mathcal{E}_C$ from Eq. (8.51), is small, then the angular momentum is only partially quenched. Thus the spin can feel the influence of the crystalline field through the combined effects of spin-orbit coupling and orbit-crystal-field interaction. For further theoretical discussions on the subject, the reader is referred to the paper by Van Vleck.[8] Based on this model, certain qualitative observations can be made. The magnitude of the anisotropy constant should depend on the symmetry of the crystal. Generally speaking, the anisotropy constant is larger in crystals having lower symmetry. In complex crystals such as the spinel structure of ferrites, the anisotropy constant then depends on the symmetry of the lattice sites occupied by magnetic ions. For further discussions of anisotropy energy and its temperature dependence in ferrites, the reader is referred to the original article by Slonczewski.[9]

During the process of magnetization, a small change in the shape of a magnetic specimen occurs. This phenomenon is known as *magnetostriction*. It originates from the fact that the anisotropy energy depends upon the strain in a crystal and hence a state of minimum energy is achieved if the crystal is strained. In the derivation of Eq. (10.15), we have argued that the lowest order term $K_0(\alpha_1{}^2 + \alpha_2{}^2 +$

[7]   S. Chikazumi, "Physics of Magnetism," John Wiley & Sons, Inc., New York, 1964.

[8]   J. H. Van Vleck, *Phys. Rev.*, vol. 52, p. 1178, 1937.

[9]   J. C. Slonczewski, *Phys. Rev.*, vol. 110, p. 1341, 1958.

$\alpha_3{}^2$) is independent of the orientation of the magnetization and hence it need not be considered. In a strained crystal, however, because there are six strain components e_{ii} and e_{ij} where $i, j = x, y,$ or z, not only may the interatomic distance of which K_0 is a function change, but also the direction cosines may change. The strain components e_{ii} and e_{ij} are defined by Eqs. (7.77) and (7.78) in Sec. 7.8 and the present symbols are used instead of the old symbol S to be consistent with the convention used in the literature on magnetostriction and to avoid confusion with the spin quantum number. To explain the origin of magnetostriction, we shall follow the outlines of Néel's theory[10] which is based on the pair model first developed by Van Vleck. Consider a linear chain of magnetic moments shown in Fig. 10.8. In a ferromagnetic material, the magnetic moments are assumed parallel; therefore, only two parameters, the distance r between neighboring magnetic ions and the angle θ between the directions of the magnetic moment and the linear chain, will enter into the calculation of magnetic energy.

As mentioned earlier in Sec. 1.7, any given function of θ can be expanded into a series of Legendre polynomials. Thus, the magnetic interaction energy can be written as

$$\mathcal{E}_m = d_0(r) + d_2(r)(3 \cos^2 \theta - 1)$$
$$+ d_4(r)(35 \cos^4 \theta - 30 \cos^2 \theta + 3) + \cdots \quad (10.20)$$

Since a rotation of the magnetic moment by 180 degrees should not change the value of $\mathcal{E}_m$, only Legendre polynomials containing even powers of $\cos \theta$ are included. For a three-dimensional lattice, generalization of Eq. (10.20) leads to

$$\mathcal{E}_m = D_0(r) + \Sigma D_{2i}(r)(3\alpha_i{}^2 - 1)$$
$$+ \Sigma D_{4i}(r)(35\alpha_i{}^4 - 30\alpha_i{}^2 + 3) + \cdots \quad (10.21)$$

where α_i is the direction cosine of the magnetization vector with respect to the coordinate axes $x, y,$ or z, and the summation is over $i = x, y,$ and z. Since the first term is independent of α_i, phenomenologically it can be interpreted as representing the exchange energy.

[10] L. Néel, *J. Phys. Radium*, vol. 15, p. 227, 1954.

Figure 10.8

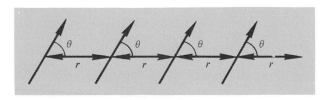

A linear chain of magnetic moments. The model was used by Van Vleck (Phys. Rev., vol. 52, p. 1178, 1937) to explain the origin of magnetic anisotropy.

A change in the volume of a magnetic specimen will change $D_0(r)$ and hence the magnetization. The reverse process is known as volume magnetostriction in which the volume of the specimen is changed by magnetization. Because of lack of space, however, the phenomenon will not be discussed.

Note that the angular dependence of Eq. (10.20) and that of Eq. (10.14) are the same if $\cos \theta$ is converted to $\sin \theta$ and the various terms are regrouped. Therefore, for a uniaxial crystal such as cobalt, Eq. (10.20) is used as a starting point for a discussion of the magnetostriction effect. For a cubic crystal, however, Eq. (10.21) must be used with $D_{2x} = D_{2y} = D_{2z} = D_2$ and $D_{4x} = D_{4y} = D_{4z} = D_4$. When a crystal is not strained, the term involving D_2 does not contribute to anisotropy because $\alpha_1{}^2 + \alpha_2{}^2 + \alpha_3{}^2 = 1$. When a crystal is under strain, we must be careful about this term. First, the direction cosines refer to the bond axes connecting neighboring magnetic ions, which under strain may not coincide with the original axes without strain. If β_1, β_2, and β_3 denote the direction cosines of the bond axes with respect to the original axes, the term $\mathcal{E}_2$ involving D_2 should be rewritten as

$$\mathcal{E}_2 = D_2(r)\Sigma[3(\alpha_1\beta_1 + \alpha_2\beta_2 + \alpha_3\beta_3)^2 - 1] \tag{10.22}$$

As a check, we consider the bond pairs along the x axis. For an unstrained crystal, $\beta_1 = 1$ and $\beta_2 = \beta_3 = 0$; thus, Eq. (10.22) yields the same result as Eq. (10.20). For a crystal under strain, β_1 will still be close to one, but β_2 and β_3 will no longer be zero. According to the definition of strain given in Eqs. (7.77) and (7.78), we have $\beta_2 = \partial\xi/\partial y = e_{xy}/2$, $\beta_3 = \partial\xi/\partial z = e_{xz}/2$ and $r = r_0(1 + e_{xx})$. For bond pairs aligned along the y axis, β_2 will be close to one, and β_1 and β_3 will be small but not zero.

Expanding Eq. (10.22) and considering bond pairs along the three cube axes, we find

$$\mathcal{E}_2 = D_2(r_0)\Sigma(3\alpha_i{}^2 - 1) + \Sigma B_1\alpha_1{}^2 e_{xx} + \Sigma B_2\alpha_1\alpha_2 e_{xy} \tag{10.23}$$

where $B_1 = 3r_0(\partial D_2/\partial r)$ and $B_2 = 3D_2$. The last two terms in Eq. (10.23), which represent the extra energy terms as a result of strain, are called the *magnetoelastic energy*. To the anisotropy energy considered in Eq. (10.15), we now must add the magnetoelastic energy which according to Eq. (10.23) is given by

$$\mathcal{E}_{me} = B_1(e_{xx}\alpha_1{}^2 + e_{yy}\alpha_2{}^2 + e_{zz}\alpha_3{}^2)$$
$$+ B_2(e_{xy}\alpha_1\alpha_2 + e_{yz}\alpha_2\alpha_3 + e_{zz}\alpha_3\alpha_1) \tag{10.24}$$

and the elastic energy which according to Eq. (7.83) is given by

$$\mathcal{E}_{el} = \frac{c_{11}(e_{xx}{}^2 + e_{yy}{}^2 + e_{zz}{}^2)}{2} + \frac{c_{44}(e_{xy}{}^2 + e_{yz}{}^2 + e_{zz}{}^2)}{2}$$
$$+ c_{12}(e_{xx}e_{yy} + e_{yy}e_{zz} + e_{zz}e_{xx}) \tag{10.25}$$

It is understood that Eqs. (10.24) and (10.25) are derived for crystals with cubic symmetry in which $c_{11} = c_{22} = c_{33}$, $c_{12} = c_{23} = c_{31}$, $c_{44} = c_{55} = c_{66}$, and other elastic constants are zero. Minimization of $\mathcal{E}_a + \mathcal{E}_{me} + \mathcal{E}_{el}$ with respect to e_{ii} and e_{ij} yields the following:

$$e_{ii} = \frac{B_1[c_{12} - \alpha_i^2(c_{11} + 2c_{12})]}{[(c_{11} - c_{12})(c_{11} + 2c_{12})]} \qquad e_{ij} = \frac{-B_2\alpha_i\alpha_j}{c_{44}} \qquad (10.26)$$

The elongation observed in the direction of a unit vector $\gamma = (\gamma_x, \gamma_y, \gamma_z)$ is given by

$$\frac{\delta l}{l} = e_{xx}\gamma_x^2 + e_{yy}\gamma_y^2 + e_{zz}\gamma_z^2 + e_{xy}\gamma_x\gamma_y + e_{yz}\gamma_y\gamma_z + e_{zx}\gamma_z\gamma_x \quad (10.27)$$

In terms of B_1 and B_2, Eq. (10.27) becomes

$$\frac{\delta l}{l} = -\frac{B_1}{c_{11} - c_{12}} \sum \alpha_x^2\gamma_x^2 - \frac{B_2}{c_{44}} \sum \alpha_x\alpha_y\gamma_x\gamma_y$$
$$+ \frac{3c_{12}B_1}{(c_{11} + 2c_{12})(c_{11} - c_{12})} \quad (10.28)$$

The summation is over terms of similar form but with different subscripts. The values of $B_1/(c_{11} - c_{12})$ and B_2/c_{44} are found experimentally to be -20.7×10^{-6} and 21.2×10^{-6} in iron and 45.9×10^{-6} and 24.3×10^{-6} in nickel respectively. These constants are generally known as the *magnetostriction constants*. For values of magnetostriction constants in various ferrites, the reader is referred again to the aforementioned book by Chikazumi. The magnetostrictive or magnetoelastic effect will be further discussed later in connection with domain structure and with the instabilities observed in a magnetic resonance experiment.

10.4 MAGNETIC DOMAINS

Domain walls. The transition layer which separates adjacent magnetic domains is called the domain wall. It is sometimes referred to as the *Bloch wall* named after F. Bloch[11] who was the first to investigate the property of the transition layer. In Sec. 2.8, we have seen that dislocation lines are created so that slips of atomic planes can be made in many small steps, greatly reducing the stress required to produce a slip. The same physical principle is behind the creation of domain walls. A large exchange energy is involved if the magnetization vector changes its direction suddenly. To lower the exchange energy, the change of direction must take place gradually over many atomic planes. The simplest kind of domain wall is a 180° wall in which the spins rotate by 180° through N atomic planes as illustrated in Fig. 10.9.

[11] F. Bloch, *Z. Physik.*, vol. 74, p. 295, 1932.

First let us make an order-of-magnitude estimate of the energy and thickness of the wall. For simplicity, we assume a simple cubic lattice of which the direction of easy magnetization is along either one of the cube edges. If a is the lattice constant, the number of atoms per unit area is $1/a^2$. Thus, according to Eq. (10.9), the exchange energy stored per unit area of the transition region is

$$\mathcal{E}_1 = \frac{NJS^2(\pi/N)^2}{a^2} = \frac{JS^2\pi^2}{Na^2} \tag{10.29}$$

for a total of N atomic planes. However, we must not forget that in turning away from the easy axis the magnetization must increase its anisotropy energy which is roughly equal to KNa per unit area of the transition region, K being the anisotropy constant. The thickness Na of the transition region is determined by minimizing the total energy $\mathcal{E}_w$ per unit area

$$\mathcal{E}_w = \frac{JS^2\pi^2}{Na^2} + KNa \tag{10.30}$$

with respect to N. From Eq. (10.30), we find

$$Na = \left(\frac{JS^2\pi^2}{Ka}\right)^{\frac{1}{2}} \quad \text{and} \quad \mathcal{E}_w = 2\pi\left(\frac{JS^2K}{a}\right)^{\frac{1}{2}} \tag{10.31}$$

For iron, $J = 2.1 \times 10^{-21}$ joule, $S = 1$, $K = 4.8 \times 10^4$ joule/m^3 and $a = 2.86 \times 10^{-10}$ m. Substituting these values into Eq. (10.31), we find $Na = 3.9 \times 10^{-8}$ m = 136 lattice constants and $\mathcal{E}_w = 3.7 \times 10^{-3}$ joule/m^2.

Now we shall proceed with a detailed calculation of the thickness and energy of the 180° wall which is taken to be parallel to the (010) plane. Let θ be the angle between the spin direction and the z axis as shown in Fig. 10.9a. Since $\alpha_1 = \sin\theta$, $\alpha_2 = 0$, and $\alpha_3 = \cos\theta$, the anisotropy energy per unit area per atomic spacing is equal to

Figure 10.9

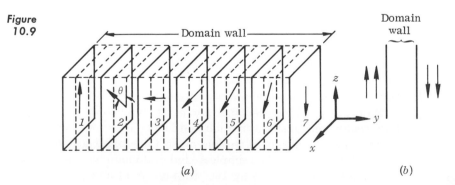

(a)   (b)

A 180° domain wall. Inside the wall, the direction of magnetization changes gradually from one atomic plane to another.

$K_1 \sin^2 \theta \cos^2 \theta$ according to Eq. (10.15). The corresponding exchange energy is given by $A(d\theta/dy)^2$ which is obtained from Eq. (10.11) by setting $S_x = S \sin \theta$, $S_z = S \cos \theta$, and $A = ZJS^2/a$. In view of the above discussion, the wall energy is

$$\mathcal{E}_w = \int_{-\infty}^{\infty} \left[K_1 \sin^2 \theta \cos^2 \theta + A \left(\frac{d\theta}{dy} \right)^2 \right] dy \qquad (10.32)$$

For a stable situation, the wall energy $\mathcal{E}_w$ should be a minimum with respect to changes in θ. If θ is changed to $\theta + \delta\theta$, the change in $\mathcal{E}_w$ or $\delta\mathcal{E}_w$ can be split into two parts of which the part due to the second term in Eq. (10.32) is

$$\delta\mathcal{E}_2 = \int_{-\infty}^{\infty} 2A \frac{d\theta}{dy} \frac{d\,\delta\theta}{dy} dy = 2A \frac{d\theta}{dy} \delta\theta \Big|_{y=-\infty}^{y=\infty} - \int_{-\infty}^{\infty} 2A \frac{d^2\theta}{dy^2} \delta\theta \, dy \qquad (10.33)$$

At $y = \pm \infty$, $d\theta/dy = 0$; thus we have

$$\delta\mathcal{E}_w = \int_{-\infty}^{\infty} \left(\frac{dg(\theta)}{d\theta} - 2A \frac{d^2\theta}{dy^2} \right) \delta\theta \, dy = 0 \qquad (10.34)$$

where $g(\theta) = K_1 \sin^2 \theta \cos^2 \theta$.

For a stable wall, Eq. (10.34) should be satisfied for any atomic plane in the transition layer. That means, the integrand must be identically zero, leading to

$$\frac{dg(\theta)}{d\theta} = 2A \frac{d^2\theta}{dy^2} \qquad \text{or} \qquad dg(\theta) = Ad \left(\frac{d\theta}{dy} \right)^2 \qquad (10.35)$$

At $y = -\infty$, $g(\theta) = 0$ and $d\theta/dy = 0$ so the integration constant is zero in Eq. (10.35). Thus, we obtain the wall thickness

$$y = \int dy = \int_0^{\pi} \frac{\sqrt{A}\, d\theta}{\sqrt{g(\theta)}} = \sqrt{\frac{A}{K_1}} \log \tan \theta \qquad (10.36)$$

and the wall energy

$$\mathcal{E}_w = 2 \int_{-\infty}^{\infty} |g(\theta)| \, dy = 2 \int_0^{\pi} |\sqrt{Ag(\theta)}| \, d\theta$$

$$= 2 \sqrt{\frac{ZK_1JS^2}{a}} = 2 \sqrt{AK_1} \qquad (10.37)$$

For iron, $Z = 2$. Substituting the values of K_1, J, S, and a used in the previous calculation into Eq. (10.37), we find $\mathcal{E}_w = 1.7 \times 10^{-3}$ joule/m^2. It is obvious that Eq. (10.36) does not lead to a finite thickness of the domain wall. The difficulty can be resolved by including the magnetostrictive energy into the wall energy. For a discussion of such a treatment, the reader is referred to a review paper by Kittel. For practical purposes, we may take the value of $dy/d\theta$ at $\theta = 45°$ as a typical value which is equal to $2 \sqrt{A/K_1}$. For a 180° wall, $\Delta y = 2\pi \sqrt{A/K_1}$.

Domain structures. Let us now turn to a discussion of the domain itself. The existence of domains of finite sizes can again be explained on the basis of energy considerations. This time the energy involved is the magnetostatic energy. Since the reader may be more familiar with the electrostatic case, we shall discuss the electrostatic energy first and then obtain the expression for the magnetostatic energy by analogy. Consider a bounded medium which is charged. If ϱ is the volume-charge density (coul/m³) and ϱ_s is the surface-charge density (coul/m²), the electrostatic energy U_e is

$$U_e = \tfrac{1}{2}\int \varrho\Phi\,dV + \tfrac{1}{2}\int \varrho_s\Phi\,dS \qquad (10.38)$$

where Φ is the electrostatic potential in volts set up by the electric charges in the medium and the two integrations are over the whole volume V and surface S of the medium.

The simplest magnetic domain structure is the laminated type shown in Fig. 10.10a with positive and negative signs indicating magnetic north and south poles respectively. Realizing that ϱ_s is equal to the normal component of polarization whose magnetic counterpart is $\mu_0 M_0$, we have for the magnetostatic energy U_m,

$$U_m = \frac{\mu_0}{2}\int M_0\Phi_m\,dS \qquad (10.39)$$

Figure 10.10

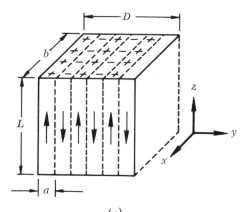

(a)

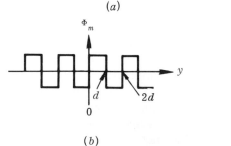

(b)

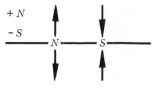

(c)

(a) A laminated structure of magnetic domains with alternate magnetic north and south poles, and (b) the magnetic potential Φ_m as a function of distance. Diagram (c) shows that the magnetic flux due to surface poles goes out in both directions.

where Φ_m is the magnetic potential. Except in the domain-wall region, the volume contributions to both Φ_m and U_m are independent of whether the magnetic medium consists of a single domain or many coplanar domains. Hence, in Eq. (10.39), only the contribution from surface magnetic poles to Φ_m will be considered. The magnetic potential Φ_m, like the electrostatic Φ, satisfies the Laplace equation,

$$\frac{\partial^2 \Phi_m}{\partial x^2} + \frac{\partial^2 \Phi_m}{\partial y^2} + \frac{\partial^2 \Phi_m}{\partial z^2} = 0 \qquad (10.40)$$

Since $\partial \Phi_m / \partial x = 0$, the solution of Eq. (10.40) appropriate to the distribution shown in Fig. 10.10b can be expressed in terms of a Fourier series in y as follows:

$$\Phi_m = \sum_{n=1}^{\infty} A_n \exp\left(\frac{n\pi z}{d}\right) \sin\left(\frac{n\pi y}{d}\right) \qquad (10.41)$$

where n is an odd integer.

Since Φ_m is due to surface poles, the boundary condition at $z = 0$ is such that

$$\left.\frac{\partial \Phi_m}{\partial z}\right|_{z=0} = \pm \frac{M_0}{2} \qquad (10.42)$$

The factor $\frac{1}{2}$ comes from the fact that the magnetic flux due to a surface pole comes in or goes out in both directions normal to the $z = 0$ plane (Fig. 10.10c). The positive and negative signs are for $2kd < y < (2k+1)d$ and $(2k+1)d < y < (2k+2)d$, respectively, k being an integer. Applying Eq. (10.42) to Eq. (10.41) and using the standard techniques for evaluating the Fourier coefficients A_n, we find

$$\Phi_m(z = 0) = \frac{2M_0 d}{\pi^2} \sum_{n=1}^{\infty} \frac{1}{n^2} \sin \frac{n\pi y}{d} \qquad (10.43)$$

Substituting Eq. (10.43) into Eq. (10.39) gives

$$U_m = \frac{\mu_0 M_0 d}{\pi^2} \int_0^b \int_0^d M_0 \Phi_m \, dx \, dy = \frac{2\mu_0 M_0^2 d}{\pi^3} (db) \sum \frac{1}{n^3} \qquad (10.44)$$

for a strip of area bd. Since the value of spontaneous magnetization is usually expressed in gauss, it is convenient to use $I_0 = \mu_0 M_0$ weber/m² in the mks system. The summation $1/n^3$ has a value 1.05; thus the magnetostatic energy per unit area of surface is equal to

$$\mathcal{E}_m = 5.40 \times 10^4 I_0^2 d \qquad \text{joule/m}^2 \qquad (10.45)$$

It is clear from Eq. (10.45) that magnetostatic energy favors small domains. However, as the value of d decreases, there will be more domain walls. Since there are D/d domain walls in a volume $V =$

bDL (Fig. 10.10a), the energy associated with domain walls is equal to $U_w = \mathcal{E}_w(D/d)(bL)$ where $\mathcal{E}_w$ is the wall energy per unit area given by Eq. (10.37). Counting the fact that there are two surfaces, the total energy per unit surface area (bD) is

$$\mathcal{E} = \mathcal{E}_w \left(\frac{L}{d}\right) + 1.08 \times 10^5 I_0^2 d \qquad \text{joule/m}^2 \qquad (10.46)$$

Minimizing $\mathcal{E}$ with respect to the domain width d gives

$$d = 3 \times 10^{-3} \frac{(\mathcal{E}_w L)^{1/2}}{I_0} \qquad (10.47)$$

For iron, $I_0 = 1707$ gauss $= 2.14$ weber/m^2 and $\mathcal{E}_w = 1.7 \times 10^{-3}$ joule/m^2; thus, the value of d is 5.8×10^{-6} m if $L = 1$ cm. Note that the width of the domain is larger than the thickness of wall by about a factor of hundred. Using the calculated value of d in Eq. (10.46) yields an energy $\mathcal{E}$ of the order of 6 joule/m^2. To calculate the magnetostatic energy density for a single domain, we must know the demagnetizing factor which is quite similar in nature as the depolarizing factor discussed in Sec. 7.5. Assuming that the demagnetizing factor is approximately unity, the energy density per unit surface area is of the order of $I_0^2 L/(2\mu_0)$ or 10^4 joule/m^2. Hence, the formation of finely divided domains results in a reduction of energy density by a factor of about a thousand. This is obvious from Fig. 10.10a as the contributions to Φ_m from different strips partly cancel each other.

Magnetic domain structures can be observed experimentally by powder techniques first introduced by Bitter[12] and independently by Hamos and Thiessen.[13] If a drop of colloidal suspension of a fine ferromagnetic powder is applied to the surface of a ferromagnetic crystal, a great concentration of the powder particles is expected at the edge of a domain wall because according to Eq. (10.41), $\partial \Phi_m / \partial y$ or H_y has the highest intensity at $y = 0$ or any of the domain boundaries. Figure 10.11 shows domain patterns in single crystals of silicon iron reported by H. J. Williams and his coworkers.[14] The more complicated tree-like structure was observed in crystals whose surface was inclined slightly with respect to a cube face. For an explanation of the treelike structure and a general discussion of other complicated structures, the reader is referred to the original article by Williams and to a review article by Kittel.[15] The direction of

[12] F. Bitter, *Phys. Rev.*, vol. 38, p. 1903, 1931, and vol. 41, p. 507, 1932.
[13] L. V. Hamos and P. A. Thiessen, *Z. Physik.*, vol. 71, p. 442, 1932.
[14] H. J. Williams, R. M. Bozorth, and W. Shockley, *Phys. Rev.*, vol. 75, p. 155, 1949.
[15] C. Kittel, *Rev. Modern Phys.*, vol. 21, p. 541, 1949; also C. Kittel and J. K. Galt, Ferromagnetic Domain Theory, in *Solid-State Physics*, vol. 3, p. 437, 1956.

magnetization in each domain can be determined by the scratch technique.[16] When the path of magnetic flux is interrupted by a scratch as shown in Fig. 10.12a, magnetic poles are formed at the surface of the groove, collecting colloidal particles. If the scratch is parallel to the direction of magnetization (Fig. 10.12b), no leakage path is needed by the magnetic flux and hence no visible effect on the powder is expected.

[16] H. J. Williams, *Phys. Rev.*, vol. 71, p. 646, 1947.

Figure 10.11

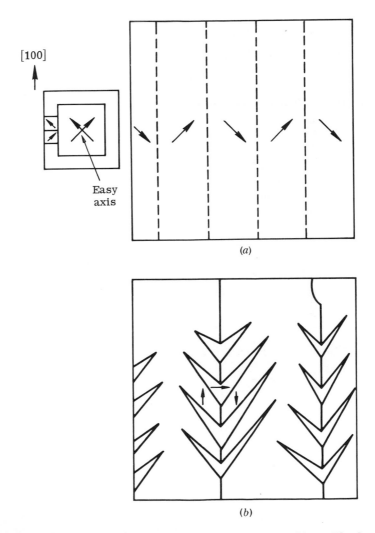

(a)

(b)

Schematic illustration of (a) simple domain pattern and (b) treelike domain pattern in single crystals of silicon iron. (H. J. Williams, R. M. Bozorth, and W. Shockley, Phys. Rev., vol. 75, p. 155, 1949.)

Figure 10.12

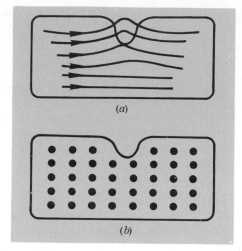

(a)

(b)

Scratch techniques used to determine the direction of magnetization: (a) a scratch perpendicular to the domain magnetization, and (b) a scratch parallel to the domain magnetization.

10.5 MAGNETIZATION PROCESSES

Magnetization rotation. In our discussion of the anisotropy energy and the effective anisotropy field, we have used the magnetization curve to evaluate the anisotropy constant. Referring to Fig. 10.7, we see that the magnetization vector will come to an equilibrium position which points in the direction of the resultant field. As the magnitude of the applied field increases, the magnetization vector turns more and more toward the direction of the applied field. This process is known as magnetization by rotation. In the following, we shall discuss the process in a uniaxial crystal. If ϕ_0 and ϕ represent respectively the angles between the direction of **H** and the easy axis and between **H** and **M** (Fig. 10.13a), the energy of the system is

$$\mathcal{E} = K_1 \sin^2 (\phi_0 - \phi) - \mu_0 M_0 H \cos \phi \qquad (10.48)$$

Under equilibrium, $\mathcal{E}$ is a minimum or

$$\frac{\partial \mathcal{E}}{\partial \phi} = -K_1 \sin 2(\phi_0 - \phi) + \mu_0 M_0 H \sin \phi = 0 \qquad (10.49)$$

where M_0 is the spontaneous magnetization.

For given values of ϕ_0 and $\mu_0 M_0 H / K_1$, the value of ϕ can be solved from Eq. (10.49). Two simple cases of interest are $\phi_0 = \pi/4$ and $\pi/2$. For $\phi_0 = \pi/4$,

$$\cos 2\phi = A \sin \phi \qquad (10.50)$$

and for $\phi_0 = \pi/2$, $$\cos \phi = \frac{A}{2} \qquad (10.51)$$

where $A = \mu_0 M_0 H / K_1$. The value of magnetization M along the direction of the applied field is $M = M_0 \cos \phi$. Thus, for $\phi_0 - \pi/2$,

$$M = \left(\frac{\mu_0 M_0^2}{2K_1}\right) H = \frac{AM_0}{2} \tag{10.52}$$

and for $\phi_0 = \pi/4$,

$$M = \left(\frac{M_0}{4}\right) [8 + 2A \sqrt{A^2 + 8} - 2A^2]^{\frac{1}{2}} \tag{10.53}$$

The M versus H curve is shown in Fig. 10.14 for three values of $\phi_0 = 0$, $\pi/4$, and $\pi/2$. The *initial susceptibility* which is defined as

$$\chi_i = \frac{\partial M}{\partial H}\bigg|_{H=0} = \frac{\partial M}{\partial A}\frac{\partial A}{\partial H}\bigg|_{A=0} \tag{10.54}$$

is found to be $\mu_0 M_0^2/(2K_1)$ for $\phi_0 = \pi/2$ and $\mu_0 M_0^2/(4K_1)$ for $\phi_0 = \pi/4$.

As the field is increased in the opposite direction, the magnetization vector will rotate beyond the easy axis as illustrated in Fig. 10.13b. The situation with $\phi_0 = \pi/2$ is made quite straightforward by merely extending the M versus H curves in the neagtive direction. It is clear from Fig. 10.14 that the two cases with H in the positive and negative directions are symmetrical with respect to the easy axis. The situation is not the same with other values of ϕ_0. The simplest is the case in which $\phi_0 = 0$. The magnetization vector, once it is along the positive direction of the easy axis, will remain so even though the applied field reverses its direction, provided that the applied field is smaller than the effective anisotropy field. As soon as the applied field becomes larger than the effective anisotropy field, however, the existing configuration is no longer energetically

Figure 10.13

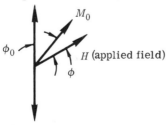

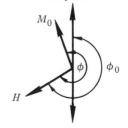

Easy axis (positive direction)

M_0

ϕ_0

H (applied field)

ϕ

Easy axis (negative direction)

(a)

Easy axis

M_0

ϕ

ϕ_0

H

(b)

*Diagrams showing the orientation of the applied field **H** with respect to the easy axis of a uniaxial crystal. (a) As **H** decreases, the vector **M** turns toward the easy axis. (b) When **H** reverses its direction, the magnetization goes to the other side of the easy axis but remains in the positive direction of the easy axis as long as the applied field is smaller than the effective anisotropy field.*

stable and the magnetization suddenly switches its direction, as illustrated by the line AB in Fig. 10.14. When the applied field again changes its direction, a square hysteresis loop $ABEBCDFA$ forms. Hysteresis loops also form for all other values of ϕ_0 except $\pi/2$.

The reason for the sudden switch of the direction of magnetization is that at the point of switching, the energy versus ϕ plot for the existing configuration is no longer at a minimum point but rather at a point of inflection. Mathematically, the equilibrium state is an unstable one if $d^2\mathcal{E}/d\phi^2 < 0$. Differentiating Eq. (10.49) again with respect to ϕ defines a value of ϕ for switching such that

$$\frac{d^2\mathcal{E}}{d\phi^2} = 2K_1 \cos 2(\phi_0 - \phi) + \mu_0 M_0 H \cos \phi = 0 \qquad (10.55)$$

For a given ϕ_0, the values of ϕ and A at switching can be found by solving Eqs. (10.49) and (10.55). Actually, two solution pairs exist but only the pair with a real value for A corresponds to a physical situation. Again we shall discuss the two simplest cases with $\phi_0 = 0$ and $\pi/4$. The results are: for $\phi_0 = 0$, $A = 2$ and $\phi = 0$ or $180°$; for $\phi_0 = \pi/4$, $A = 1$ and $\phi = 90°$ or $270°$.

References to Figs. 10.7a and 10.13a show that the angle θ in Eq. (10.17) is equal to $\phi_0 - \phi$. For $\phi_0 = 0$, $\theta = 0$ in Eq. (10.17); hence the effective anisotropy field H_a is equal to $H_a = 2K_1/\mu_0 M$. The direction of $\mathbf{M}$ switches from $\phi = 0°$ to $180°$ when $A = \mu_0 M_0 H/K_1 = 2$ or when the applied field barely exceeds the effective anisotropy field. The result is just what we expected from the qualitative argument presented earlier. For $\phi_0 = \pi/4$, the magnetization vector makes an angle $\theta = 270° - 225° = 45°$ with the easy axis. The effective

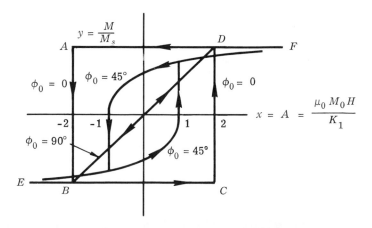

Figure 10.14

The magnetization versus magnetic field curve for several orientations of the field with respect to the easy axis. The magnetization curve is obtained by considering only the magnetization-rotation process; therefore, the curve is for materials which have a single domain.

anisotropy field from Eq. (10.17) is $K_1 \sqrt{2}/\mu_0 M_0$. The external field required to make the resultant field parallel to **M** is $H/\cos 45° = K_1 \sqrt{2}/(\mu_0 M_0)$ or $H = K_1/(\mu_0 M_0)$. When the applied field exceeds this value, switching occurs.

The reader is asked to show that for other values of ϕ_0, the value of A required for switching is determined by

$$(4 - A^2)^3 = 27 A^4 \sin^2 2\phi_0 \tag{10.56}$$

and that after the value of A is known, the direction of magnetization can be obtained from

$$\sin^2 \phi = \frac{4 - A^2}{3A^2} \tag{10.57}$$

We should also mention that the above equations remain the same if ϕ is changed to $\phi + \pi$. This is the reason why the M versus H curve (Fig. 10.14) is symmetrical. It is because of this fact that we did not say $\phi_0 = \pi + \pi/4$ for the $\phi_0 = \pi/4$ case when the field reverses its direction in Fig. 10.14. The relative orientation of the magnetization vector with respect to the easy axis and the direction of **H** during the switching process will be further discussed in the next section. The magnetization process by rotation is much more complex in a cubic crystal and the reader is referred again to the aforementioned review article by Kittel for such a treatment.

Domain wall movement. Now we shall consider a magnetization process of a different kind, the process of domain wall movement. As pointed out in Sec. 10.1, the magnetic energy of a domain wall in general depends on the position of the wall. This is illustrated in Fig. 10.3b. A variation of the magnetic energy such as that seen in Fig. 10.3b may be caused by inhomogeneous internal strains or else by aggregates or inclusions of impurities. Figure 10.15a and b shows the coercive force H_c of iron with heterogeneous inclusions of precipitated copper and the coercive force H_c of nickel under internal stress as functions of the copper concentration and the magnitude of the stress respectively. The fact that the coercive force is structure-sensitive indicates that it is controlled by the process of domain-wall movement. The magnitude of the coercive force is also consistent with this interpretation. The values of $2K_1/(\mu_0 M_0)$ in Fe and Ni are 500 and 135 oe respectively. Had the process by magnetization rotation been the dominant process, not only should the coercive force be independent of impurities and stresses but it should also have a much larger value [of the order of $K_1/(\mu_0 M_0)$]. Hence, except for substances with low anisotropy constants, the magnetization process at low fields is predominantly by domain-wall movements, as shown in Fig. 10.4b

According to the inclusion theory proposed by Kersten,[17] if non-

[17] M. Kersten, Z. Angew. Phys., vol. 8, pp. 382 and 496, 1956.

magnetic inclusions are introduced into a magnetic substance as illustrated in Fig. 10.16, the energy associated with a domain wall will vary with the position of the wall. First, because the inclusions are nonmagnetic, the effective area of a domain wall is reduced when the wall intersects a number of inclusions. Second, local strain may be created near the inclusions so that the local anisotropy constant is modified. In either case, we shall assume that the spatial variation

Figure 10.15

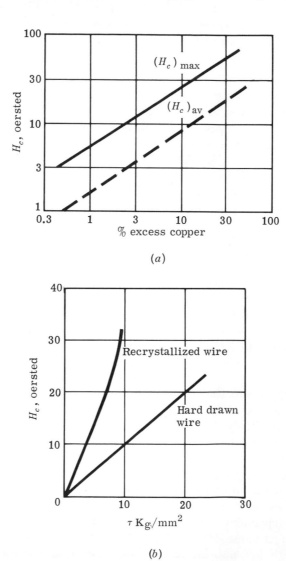

(a)

(b)

(a) Coercive force H_c of iron with heterogeneous inclusion of copper and (b) coercive force H_c of nickel under internal stress. (M. Kersten, "Grundlagen einer Theorie der ferromagnetischen Hysterese und der Koerzitivkraft," J. W. Edwards, Publisher, Incorporated, Ann Arbor, Mich., 1943.)

of the wall energy illustrated in Fig. 10.3b is a direct consequence of nonmagnetic inclusions and that a minimum energy results if the wall intersects a maximum number of inclusions. For simplicity, we shall take an idealized model in which a wall is fixed at inclusions represented by points $BDEF$ in Fig. 10.16b. The wall may be thought of as an elastic membrane. Under small applied fields, the wall surface is no longer flat but becomes cylindrical. If r is the radius of curvature of the cylinder and the angle 2ϕ is the angle subtended at the center C of the cylinder by points E and B, the area A added to the domain which has a favorable orientation is

$$A = r^2\phi - (r \cos \phi)(r \sin \phi) \simeq \frac{2r^2\phi^3}{3} \qquad (10.58)$$

for small ϕ, and the change L in the length of the wall is

$$L = 2r\phi - l \simeq \frac{r\phi^3}{3} \qquad (10.59)$$

For a given field H, the parameters r and ϕ are determined again by energy considerations. The net change in energy is equal to

$$\mathcal{E} = -2\mu_0 M_0 H A h \cos \theta + \mathcal{E}_w L h \qquad (10.60)$$

Figure 10.16

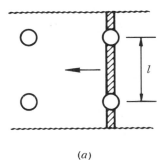

(a)

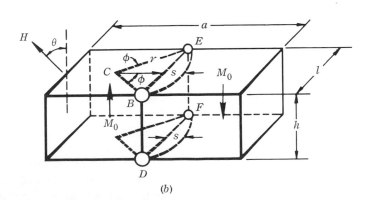

(b)

Nonmagnetic inclusions in a magnetic material: (a) a two-dimensional view and (b) a three-dimensional view.

where the first and second terms represent, respectively, the magnetic and domain wall energies, and $\mathcal{E}_w$ is the wall energy per unit area given by Eq. (10.31). To minimize $\mathcal{E}$, we eliminate ϕ by using $l = 2r\phi$ in Eq. (10.60) where l is the separation between two inclusions. Differentiating Eq. (10.60) with respect to r, we find

$$\frac{\partial \mathcal{E}}{\partial r} = 0 = \frac{\mu_0 M_0 H h l^3 \cos \theta}{6r^2} - \frac{\mathcal{E}_w l^3 h}{12r^3} \tag{10.61}$$

The magnetization change M owing to the movement of the domain wall is

$$M = \frac{2M_0 A}{al} \tag{10.62}$$

which has a component $M \cos \theta$ in the direction of H. Using the values of r and ϕ obtained from Eq. (10.61) and $\phi = l/(2r)$ in Eq. (10.58) and substituting the value of A from Eq. (10.58) into Eq. (10.62), we have

$$M = \frac{\mu_0 M_0^2 l^2 \cos \theta}{3a\mathcal{E}_w} H \tag{10.63}$$

and the initial susceptibility χ_i

$$\chi_i = \frac{M \cos \theta}{H} = \frac{\mu_0 M_0^2 l^2 \cos^2 \theta}{3a\mathcal{E}_w} \tag{10.64}$$

Let us apply Eq. (10.64) to a cubic crystal. If we assume that the substance is initially in the demagnetized state and that the direction of magnetization in each domain is randomly oriented along any one of the three cube axes which are taken to be the direction of easy magnetization, then the average value of $\cos^2 \theta$ is given by $(\alpha_1^2 + \alpha_2^2 + \alpha_3^2)/3 = \frac{1}{3}$ where α_i is the direction cosine of $\mathbf{H}$ with respect to the three cube axes. For a random distribution of nonmagnetic inclusions, the value of a can be taken as that of l and their average value is $\bar{a} = \bar{l} = (N_i)^{-\frac{1}{3}}$ where N_i is the concentration of nonmagnetic inclusions. In view of these considerations, Eq. (10.64) becomes

$$\chi_i = \frac{\mu_0 M_0^2 l}{9\mathcal{E}_w} = \frac{I_0^2 l}{9\mathcal{E}_w \mu_0} \tag{10.65}$$

For iron, $\mathcal{E}_w = 1.7 \times 10^{-3}$ joule/m^2 from Eq. (10.37), $I_0 = \mu_0 M_0 = 2.14$ weber /m^2, and a value of χ_i as high as 2.5×10^4 has been reported. Substituting these values into Eq. (10.65) gives $l = 8 \times 10^{-5}$ m which is larger than the lattice constant of iron by many orders of magnitude. For high initial magnetic susceptibility, it is essential to keep the concentration of impurities low.

The process of domain-wall movement discussed above is confined only to a small displacement of the wall near an equilibrium position such as position 1 of Fig. 10.3b. As the applied field is removed the wall returns to its initial position; hence the process is reversible. For large applied fields, the wall will no longer behave like an elastic membrane but will move as a whole. If $\mathcal{E}_w(x)$ is the wall energy as a function of the position x of the wall, the change in energy per unit area ($lh = 1$) is given by

$$d\mathcal{E} = -2\mu_0 M_0 H(\cos\theta)\, dx + d\mathcal{E}_w(x) \qquad (10.66)$$

for a displacement dx of the wall. The maximum field needed to sweep the wall across the energy barrier is given by

$$H_{max} = \frac{1}{2\mu_0 M_0 \cos\theta}\left[\frac{d\mathcal{E}_w(x)}{dx}\right]_{max} \qquad (10.67)$$

If H exceeds H_{max}, $d\mathcal{E}/dx$ will always be negative, meaning that the magnetic force has overcome the force resisting the wall movement. One conceivable origin for the variation of the wall energy $\mathcal{E}_w$ is the local stress.

As discussed in Sec. 10.3, magnetization may induce local strain through the magnetostrictive coupling terms. The opposite is also true. If a tensile stress τ(newton/m²) exists in a ferromagnetic material, it can be shown that the additional energy term due to the magnetostrictive effect is given by

$$\mathcal{E}_\tau = \tfrac{3}{2}\lambda\tau \cos^2\phi \qquad (10.68)$$

where ϕ denotes the angle between the direction of magnetization and the tensile stress τ. The quantity λ is equal to $2B_1/3(c_{11} - c_{12})$ if the domain magnetization is along the $\langle 100 \rangle$ direction and equal to $B_2/3c_{44}$ if it is along the $\langle 111 \rangle$ direction where B_1 and B_2 are the magnetoelastic coefficients defined in Eq. (10.24). For simplicity, we assume that the internal tensile stress is along the direction of easy magnetization. As the magnetization vector turns away from the easy axis, the total energy changes because of both the anisotropy and magnetoelastic energies. The two effects can be combined together to give an effective anisotropy constant K_e:

$$K_e = K_1 - \tfrac{3}{2}\lambda\tau \qquad (10.69)$$

To apply Eq. (10.69) to the analysis of the magnetization process, we shall discuss a much simplified model in which the local stress assumes a sinusoidal variation

$$\tau = \tau_0 \sin\left(\frac{2\pi x}{l}\right) \qquad (10.70)$$

where l is a characteristic length. Using Eq. (10.37) for $\mathcal{E}_w$, we find

$$\frac{d\mathcal{E}_w(x)}{dx} = \frac{d\mathcal{E}_w(x)}{dK}\frac{dK}{dx} = -\frac{\mathcal{E}_w}{2K_1}\frac{3\pi\lambda\tau_0}{l}\cos\left(\frac{2\pi x}{l}\right) \qquad (10.71)$$

From Eqs. (10.67) and (10.71), the maximum field required for irreversible domain wall movement is

$$H_{\max} = \frac{3\lambda\tau_0}{2\mu_0 M_0 \cos\theta}\left(\frac{\Delta y}{l}\right) \tag{10.72}$$

where $\Delta y = \pi \mathcal{E}_w/K_1$ is the wall thickness as deduced from Eqs. (10.36) and (10.37). Although the above equation predicts that $H_{\max}$ increases as l decreases, there must be a lower limit for l. If l becomes smaller than Δy, then the effect on wall energy of local variation of stress should be averaged over the wall thickness in the expression of $\mathcal{E}_w(x)$. Thus, it can be argued that the effect of the stress is a maximum when l is about equal to the wall thickness. Therefore, if irreversible domain-wall movement is mainly responsible for the coercive force H_c, Eq. (10.72) predicts a value of the order of $3\lambda\tau_0/(2\mu_0 M_0)$ for H_c. In nickel, the average value of λ is about 10^{-5} and $I_0 = \mu_0 M_0 = 0.69$ weber/m^2. For a stress $\tau = 10$ kg/mm^2 = 9.8×10^7 newton/m^2, $H_c = 2 \times 10^3$ amp/m = 25 oe, in good agreement with the experimental result of Fig. 10.15b.

In summary, we have discussed two basic magnetization processes: magnetization by rotation and magnetization by domain wall movement. As indicated in Fig. 10.3a, in a given magnetic material, both processes can take place but they are important in different regions of the applied magnetic field. According to the relative importance of these two processes, technically important magnetic materials can be divided into two general categories: magnetically hard and magnetically soft. The soft materials, which are characterized by their high permeability and low coercive force, are used as cores of transformers and inductors while the hard materials which have a high coercive force are used in permanent magnets. In a soft material, movements of domain walls can progress with relative ease. Such a

TABLE 10.3 *Magnetic properties of a selected number of soft and hard magnetic materials (Numbers with asterisk represent the value of B_r)*

Material	*Composition (%)*	H_c(oe)	I_0 or B_r* (*gauss*)	$\mu_{\max}/\mu_0$	T_c (°C)
Iron	0.2 impurity	1.0	1710	5,000	770
Iron	0.05 impurity	0.05	1710	200,000	770
78 Permalloy	78.5 Ni, 21.5 Fe	0.05	860	100,000	600
Supermalloy	5 Mo, 79 Ni, 16 Fe	0.002	629	1,000,000	400
Mn-Zn ferrite	50 Mn, 50 Zn	0.1	200	2,000 (av.)	110
Mg-Zn ferrite	50 Mg, 50 Zn	1.0	207	500 (av.)	120
Carbon steel	0.9 C, 1 Mn, 98.1 Fe	50	795*⎱	lattice transformation	
Vicalloy	52 Co, 14 V, 34 Fe	450	785*⎰	hardening	
Silmanal	9 Mn, 4 Al, 87 Ag	6000	44*	precipitation hardening	
Ba-ferrite	BaO·6Fe$_2$O$_3$	1500	159*⎱	fine powder	
Iron powder	100 Fe	770	453*⎰		

material must be pure and homogeneous. In a hard material, a high coercive force is achieved by suppressing the domain-wall movement. The suppression can be accomplished either by using very fine powders which are made of particles of a single domain or by introducing impurities and internal stresses to hinder wall movements. The pertinent data of a number of selected ferromagnetic materials are given in Table 10.3. In Table 10.3, B_r is the residual magnetization (Fig. 10.4a) and the various hardening processes are employed to create internal stress and magnetic heterogeneity for high coercive force.

10.6 FERROMAGNETIC THIN FILMS

Thin films of ferromagnetic alloys have been widely used for computer quick-access memories because with these films it is possible to attain fast switching times (of the order of 10^{-8} second). A commonly used material is the Ni-Fe alloy of permalloy. A thin film (about 1000 Å thick) can be made by evaporating the alloy in a vacuum and depositing it onto a heated glass substrate. By applying a magnetic field during deposition, it is possible to produce uniaxial anisotropy with the easy axis parallel to the applied field. Because of the combination of a low anisotropy constant and its uniaxial nature, these thin films are also potentially useful for detecting weak magnetic fields. Since the thickness of the film is very thin compared with the other dimensions, a large demagnetizing field keeps the magnetization vector in the plane of the film. For the film to possess desirable characteristics, the easy axis must lie in the plane of the film. The situation is illustrated in Fig. 10.17.

In the presence of an applied magnetic field **H** which is in the plane of the film and which makes an angle β with the easy axis, the magnetization vector assumes an equilibrium position making an angle θ with the easy axis. Note that the situation represented in Fig. 10.17 is exactly the same as that in Fig. 10.13. Hence, Eqs. (10.49)

Figure 10.17

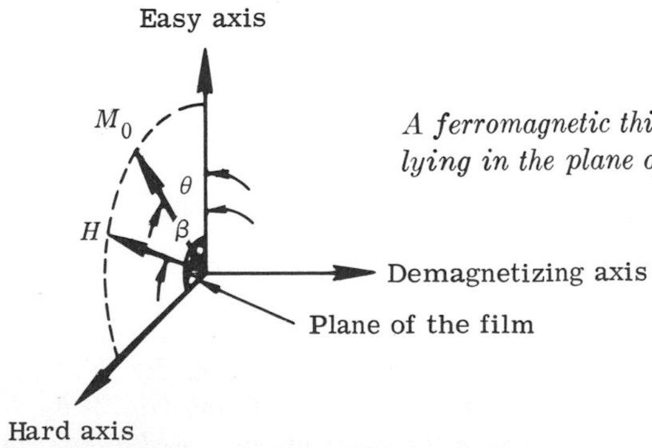

A ferromagnetic thin film with its easy axis lying in the plane of the film.

and (10.55) are directly applicable to the present situation:

$$h \sin (\beta - \theta) = \tfrac{1}{2} \sin 2\theta \qquad \text{(torque balance)} \qquad (10.73)$$

$$h \cos (\beta - \theta) = -\cos 2\theta \qquad \text{(point of inflection)} \qquad (10.74)$$

where $h = H/H_a$ is the normalized field with respect to the effective anisotropy field $H_a = 2K_1/(\mu_0 M_0)$. Since the easy axis is used as the reference axis in the following discussion, new angles θ and β are used instead of the old angles ϕ and ϕ_0. The θ versus h curve from Eq. (10.73) is plotted in Fig. 10.18 for various β. It is obvious that if $\mathbf{M}$ and $\mathbf{H}$ are in the same quadrant, the situation is energetically stable and the $\mathbf{M}$ vector simply lies between $\mathbf{H}$ and the easy axis. This corresponds to the lower half-plane with $\theta > 90°$. However, if $\mathbf{H}$ and $\mathbf{M}$ start out in different quadrants, a point of inflection predicted by Eq. (10.74) exists.

In Fig. 10.18, the dashed curve represents the locus of the h and θ values given by

$$4h^2 = \sin^2 2\theta + 4 \cos^2 2\theta \qquad (10.75)$$

which is obtained from Eqs. (10.73) and (10.74) by eliminating β. As the field increases from zero with an angle $\beta = 120°$ for example, the direction of the magnetization vector follows the curve OA or Eq. (10.73) until it reaches the point A. At point A which is the point of inflection, the magnetization vector suddenly switches from $\theta = 52°$ which represents an unstable equilibrium to $\theta = 159°$ (point B)

Figure 10.18

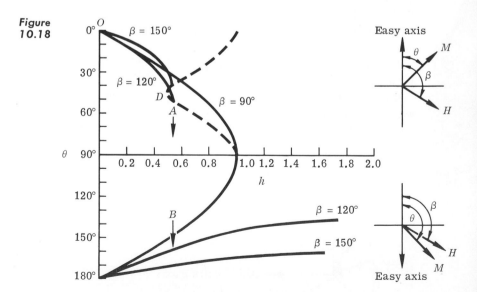

The θ versus h plot. The solid curves are obtained from Eq. (10.73) for various β. The dashed curve is obtained from Eqs. (10.73) and (10.74) by eliminating β. The field at point D represents the minimum field needed for switching.

which is the stable equilibrium point. Now let us discuss the field required for switching. Multiplying Eq. (10.73) by cos θ and Eq. (10.74) by sin θ and adding them together, we obtain

$$h \sin \beta = \sin^3 \theta \tag{10.76}$$

Multiplying Eq. (10.73) by sin θ and Eq. (10.74) by cos θ and subtracting them, we have

$$-h \cos \beta = \cos^3 \theta \tag{10.77}$$

The condition for switching is

$$h^{\frac{2}{3}}(\cos^{\frac{2}{3}} \beta + \sin^{\frac{2}{3}} \beta) \geq 1 \quad \text{or} \quad H_l^{\frac{2}{3}} + H_t^{\frac{2}{3}} \geq \left(\frac{2K_1}{\mu_0 M_0}\right)^{\frac{2}{3}} \tag{10.78}$$

where $H_l = H \cos \beta$ and $H_t = H \sin \beta$.

Equation (10.78) says that the critical longitudinal field for switching is reduced by the presence of a transverse field. In other words, the values of H_l and H_t can be properly chosen such that H_l causes switching[18] if H_t is present. Thus, by using two mutually perpendicular fields, we can select one element from an array and cause that element to switch. This fact makes thin films useful as memory elements in computers. The function $\rho(H,\beta)$

$$\rho(H,\beta) = H(\cos^{\frac{2}{3}} \beta + \sin^{\frac{2}{3}} \beta)^{\frac{3}{2}} \tag{10.79}$$

is plotted in Fig. 10.19a as a function of β in polar coordinates. As is evident, the plot resembles a four-leaf clover. Now we can mark the directions of H and the easy axis on this plot as shown in Fig. 10.19b. Switching occurs if the four-leaf clover includes the point H_a marked on the easy axis. The dotted part of the clover governs the condition for switching for a field direction opposite to that indicated in Fig. 10.19b. Note that the field required for switching is a minimum when $\beta = 45°$. From Fig. 10.18, this field is equal to one-half the effective anisotropy field or $K_1/(\mu_0 M_0)$ in agreement with the magnetization curve of Fig. 10.14.

Let us now consider the case in which there is a small perturbing field H_l acting along the easy axis. A new term $-\mu_0 M_0 H_l \cos(\phi_0 - \phi)$ should be added to the energy equation, Eq. (10.48). The new torque balance equation, Eq. (10.49) or Eq. (10.73), now becomes

$$h \sin(\beta - \theta) = \tfrac{1}{2} \sin 2\theta + h_l \sin \theta \tag{10.80}$$

by noting that $\phi_0 - \phi = \theta$ from Figs. 10.13 and 10.18 and letting $h_l = \mu_0 M_0 H_l / 2K_1$. If both the magnitude h and the direction β of the bias field are held constant, a small change in the perturbing field

[18] D. O. Smith, *J. Appl. Phys.*, vol. 29, p. 264, 1958; C. D. Olson and A. V. Pohm, *J. Appl. Phys.*, vol. 29, p. 274, 1958.

δh_l produces a change in θ

$$\delta\theta = \frac{-\sin\theta\,\delta h_l}{\cos 2\theta + h\cos(\beta - \theta) + h_l\cos\theta} \tag{10.81}$$

Equation (10.81) is obtained by differentiating Eq. (10.80). We can define a *longitudinal permeability*[19] as the ratio of change in M along H_l to change in H_l or

$$\begin{aligned}
\mu_l &= \frac{\delta M_l}{\delta H_l} = \frac{-M_0\sin\theta\,\delta\theta}{\delta H_l}\\
&= \frac{\mu_0 M_0{}^2}{2K_1}\frac{\sin^2\theta}{\cos 2\theta + h\cos(\beta - \theta) + h_l\cos\theta}
\end{aligned} \tag{10.82}$$

If $h_l \ll h$ and $\beta = 90°$, Eq. (10.80) yields $h = \sin\theta$ which obviously is true only for $h < 1$ and should be replaced by $\theta \doteq 90°$

[19]   E. J. Torok and R. A. White, *Jour. Appl. Phys.*, vol. 34, p. 1163, 1963.

Figure 10.19

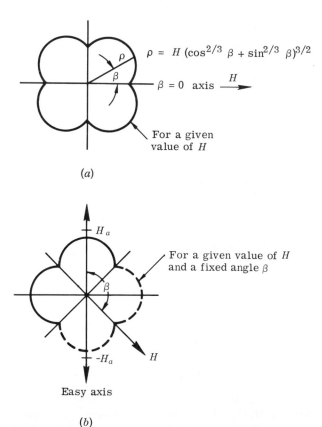

$$\rho = H(\cos^{2/3}\beta + \sin^{2/3}\beta)^{3/2}$$

$\beta = 0$ axis $H \longrightarrow$

For a given value of H

(a)

For a given value of H and a fixed angle β

Easy axis

(b)

(a) The ρ versus β plot in polar coordinates and (b) the same plot in reference to the easy axis of the film.

for $h > 1$. Substituting these values into Eq. (10.82), we find

$$
\mu_l = \begin{cases} \dfrac{\mu_0 M_0^2}{2K_1} \dfrac{h^2}{1 - h^2} & \text{for } h < 1 \\[3mm] \dfrac{\mu_0 M_0^2}{2K_1} \dfrac{1}{h - 1} & \text{for } h > 1 \end{cases}
\tag{10.83}
$$

A very large longitudinal permeability is expected therefore for $h = 1$. Note that in Fig. 10.18, the θ versus h curve has an infinite slope at the point $h = 1$ and $\theta = 90°$. The effect of the perturbing field can also be viewed as modulating the direction of the applied field by a small amount $\delta\beta$. Owing to the steep slope at the point $h = 1$, the corresponding change in θ is very much amplified. The measured value of μ_l, however, is only around 10 for a typical thin film. The discrepancy between the theoretical and experimental results is due to the dispersion in both the magnitude and direction of the effective anisotropy field.

According to Crowther,[20] the dispersion is caused by a random distribution of local strains between the film and the substrate. As discussed in Sec. 10.5, a new energy term $\mathcal{E}_r$ of Eq. (10.68) should be added to the anisotropy energy on account of the strain. The total effective anisotropy energy is thus equal to

$$
\mathcal{E} = K_1 \sin^2 \theta + K_s \sin^2 (\theta - \eta)
\tag{10.84}
$$

where η is the angle between the directions of the strain and the unstrained easy axis and the term $K_s (= 3\lambda\tau/2)$ is generally known as the *strain anisotropy*. The new apparent easy axis or the easy axis under strain has a direction shifted from the unstrained axis by an angle α such that the energy $\mathcal{E}$ is a minimum. Setting $d\mathcal{E}/d\theta = 0$ at $\theta = \alpha$ gives

$$
\tan 2\alpha = \frac{K_s \sin 2\eta}{K_1 + K_s \cos 2\eta}
\tag{10.85}
$$

The new effective anisotropy constant can be found by taking the energy difference for $\theta = \alpha$ (the new easy axis) and $\theta = \alpha + \pi/2$ (the new hard axis). Thus, we have

$$
\begin{aligned}
\frac{H_a'}{H_a} = \frac{\Delta\mathcal{E}}{K_1} &= \cos 2\alpha + \left(\frac{K_s}{K_1}\right) \cos 2(\alpha - \eta) \\
&= \left[1 + \left(\frac{K_s}{K_1}\right)^2 + 2\left(\frac{K_s}{K_1}\right) \cos 2\eta \right]^{1/2}
\end{aligned}
\tag{10.86}
$$

after using the value of 2α from Eq. (10.85).

As K_s/K_1 and η in Eqs. (10.85) and (10.86) change from place to place in a film, so do the values of α and H_a'. Consequently, a strained

[20]   T. S. Crowther, *Jour. Appl. Phys.*, vol. 34, p. 580, 1963.

film has an effective anisotropy field which is distributed (Fig. 10.20) both in magnitude and direction in a small region around H_a, the unstrained anisotropy field. The longitudinal permeability now must be averaged over this distribution, resulting in a finite value for μ_l. What is interesting is the possibility that for a given bias field, the four-leaf clover covers half the distribution. In other words, part of the film remains unswitched and part becomes switched. Since the value of K_s/K_1 is usually much smaller than unity in practical films, it only takes a field variation of $\Delta H'_a = 2H_a K_s/K_1$ to change from a situation of being on the threshold of switching to the situation of complete switching. The two situations are illustrated in Fig. 10.20. In this region of the bias field, the net magnetization of the film will

**Figure
10.20**

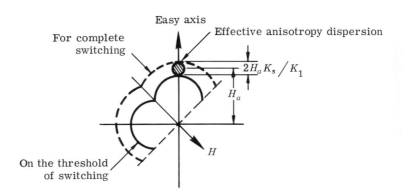

When a film is under a nonuniform strain, the anisotropy field is distributed both in magnitude and direction over a small region around H_a, the effective anisotropy field.

be very sensitive to small changes in the bias field. This fact can be utilized for the detection[21] of a very small magnetic field (of the order of few μoe).

10.7 FERROMAGNETIC, FERRIMAGNETIC, AND ANTIFERROMAGNETIC RESONANCES

The magnetization vector in ferromagnetic materials, like its counterpart in paramagnetic substances, obeys the classical equation of motion, Eq. (9.11), or

$$\frac{d\mathbf{M}}{dt} = -\gamma \mathbf{M} \times \mathbf{B} \tag{10.87}$$

Therefore, under suitable conditions, a resonance phenomenon can

[21] C. E. Frank, "A Study of the Longitudinal Permeability of Uniaxial Anisotropic Permalloy Films Using Spin Resonance," Doctoral thesis, unpublished, University of California, Berkeley, 1964.

be expected in ferromagnetic materials. As a matter of fact, ferromagnetic resonance was first observed by Griffiths[22] in Fe and Co, and only later were resonance phenomena observed in paramagnetic substances. Although Eq. (10.87) appears similar to Eq. (9.11), the induction field **B** acting on the magnetization vector is quite different from the applied field in the ferromagnetic case. Besides the external field $\mathbf{H}_{ext}$, the effective induction field **B** contains three additional terms:

$$\frac{\mathbf{B}_{eff}}{\mu_0} = \mathbf{H}_{ext} - N_{x,y,z}\mathbf{M} + \mathbf{H}_{anis} + \alpha\nabla^2\mathbf{M} \qquad (10.88)$$

The second term $N\mathbf{M}$ represents the demagnetizing field produced by the magnetization of the sample and it is similar in nature to the depolarizing field discussed in Sec. 7.5. The third and fourth terms represent the effective fields resulting from anisotropy and exchange energies respectively.

First, we shall consider the effect of the demagnetizing field alone. If we assume that the external d-c field $B_0(=\mu_0H_0)$ is applied along the z axis while the external a-c fields B_x and B_y lie in the xy plane,

[22]   J. H. E. Griffiths, *Nature*, vol. 158, p. 670, 1946.

Figure 10.21

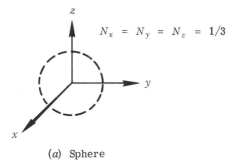

$N_x = N_y = N_z = 1/3$

(*a*) Sphere

The three sample configurations commonly used in magnetic resonance experiments: (a) sphere, (b) thin disc with the d-c field in the plane of the sample, and (c) thin disc with the d-c field perpendicular to the plane of the sample.

$N_x = N_z = 0$
$N_y = 1$

d-c field in the plane

(*b*) Disc

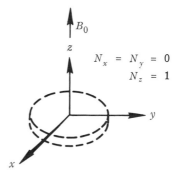

$N_x = N_y = 0$
$N_z = 1$

d-c field ⊥ to the plane

(*c*) Disc

then from Eq. (10.87) we obtain for the x and y components of $\mathbf{M}$ the following:

$$\begin{aligned}
i\omega M_x + \gamma M_y[B_0 + (N_y - N_z)\mu_0 M_z] &= \gamma M_z B_y \\
i\omega M_y - \gamma M_x[B_0 + (N_x - N_z)\mu_0 M_z] &= \gamma M_z B_x
\end{aligned} \quad (10.89)$$

Equation (10.89) is obtained by assuming a sinusoidal time variation for B_x and B_y. Resonance occurs at the angular frequency ω_0 for which the determinant of the coefficients of M_x and M_y vanishes. For small a-c fields, M_z can be taken as M_0, the spontaneous magnetization. Thus, we have

$$\omega_0{}^2 = \mu_0{}^2 \gamma^2 [H_0 + (N_x - N_z)M_0][H_0 + (N_y - N_z)M_0] \quad (10.90)$$

The three cases commonly used in experiments are illustrated in Fig. 10.21. For a spherical sample, $N_x = N_y = N_z = \frac{1}{3}$ and Eq. (10.90) yields $\omega_0 = \gamma B_0$. For a thin disc or a thin film, $\omega_0 = \mu_0 \gamma (H_0 - M_0)$ for d-c field perpendicular to the plane of the disc and $\omega_0 = \mu_0 \gamma [H_0(H_0 + M_0)]^{1/2}$ for d-c field in the plane of the disc. In all cases, H_0 is the magnetic field applied externally and $\gamma = g \times (0.88 \times 10^{11} \text{ coul/kg})$ where g is the Landé splitting factor.

Figure 10.22a shows the experimental result in supermalloy reported by Bloembergen.[23] Since most ferromagnetic materials are conductors, microwave fields can only penetrate to a depth of the order of the skin depth, corresponding to either case b or c in Fig. 10.21. The case c often requires a very high H_0 for ω_0 to be in the microwave region; therefore, the configuration shown in case b is most commonly used. The values of $B_0(= \mu_0 H_0)$ and ω_0 used in the experiment are, respectively, 0.515 weber/m² and 1.528×10^{11} rad/sec. Using the value $I_0 = \mu_0 M_0 = 0.79$ weber/m² for supermalloy, we find $g = 2.13$ as quoted in Table 10.1. Since supermalloy is magnetically soft, it has a very small anisotropy, and hence the effective anisotropy field can be neglected in the calculation. This is not generally true, however. Because a general treatment of the problem is rather involved, we shall discuss in the following only the case of uniaxial anisotropy.

Figure 10.23 shows the equilibrium position of the magnetization vector in a uniaxial thin film under the action of a d-c field which is applied in the plane of the film. It is obvious that the d-c magnetization vector lies in the xz plane. If an a-c magnetic field is applied to the film, the magnetization no longer will stay in the plane. According to Eq. (10.17) the effective anisotropy field in a uniaxial film is given by

$$H_{z\,\text{anis}} = \frac{2K_1 \cos \theta}{\mu_0 M_0} = H_a\left(\frac{M_z}{M_0}\right) = \frac{H_a(M_{z0} + m_z)}{M_0} \quad (10.91)$$

where $H_a = 2K_1/\mu_0 M_0$. The following notation is used for the various

[23] N. Bloembergen, *Phys. Rev.*, vol. 78, p. 572, 1950.

**Figure
10.22**

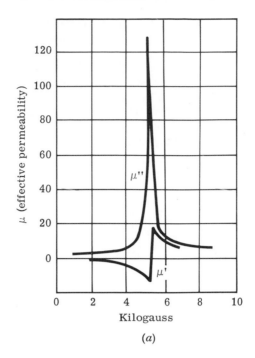

(a)

Nickel ferrite at 9,000 Mcps

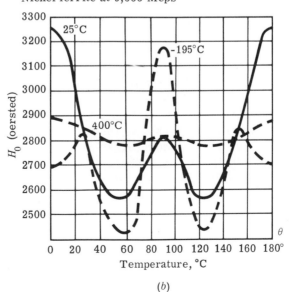

(b)

(a) *Experimental resonance curves in supermalloy. (N. Bloembergen, Phys.
Rev., vol. 78, p. 572, 1950.) (b) Experimental resonance fields plotted as a
function of the angle θ which the applied field H_0 makes with the crystal axis
in the (110) plane. (D. R. Healy, Phys. Rev., vol. 86, p. 1009, 1952.)*

magnetization components: M_z for the total z component, M_{z0} for the d-c z component, m_z for the a-c z component, and so on. The total effective field excluding the a-c applied field can thus be written as

$$\mathbf{B}_{\mathrm{eff}} = \mu_0 \begin{pmatrix} H_{x0} \\ -N_y m_y \\ H_{z0} + H_a(M_{z0} + m_z)/M_0 \end{pmatrix} \qquad (10.92)$$

Substituting Eq. (10.92) into Eq. (10.87), the natural frequency is determined by setting the determinant of the coefficients for m_x, m_y, and m_z to zero or

$$\begin{vmatrix} j\omega/\mu_0\gamma & M_{z0} + H_{z0} + H_a M_{z0}/M_0 & 0 \\ -H_{z0} - H_a M_{z0}/M_0 & j\omega/\mu_0\gamma & H_{x0} - H_a M_{x0}/M_0 \\ 0 & -M_{x0} - H_{x0} & j\omega/\mu_0\gamma \end{vmatrix}$$
$$= 0 \qquad (10.93)$$

Higher order terms involving $m_x m_y$, etc. are assumed negligible in Eq. (10.93). The natural angular frequency ω_0 is given by the roots of Eq. (10.93) or

$$\omega_0{}^2 = \mu_0{}^2\gamma^2 \left[(M_{x0} + H_{x0}) \left(H_{x0} - \frac{H_a M_{x0}}{M_0} \right) \right.$$
$$\left. + \left(H_{z0} + \frac{H_a M_{z0}}{M_0} \right) \left(M_{z0} + H_{z0} + \frac{H_a M_{z0}}{M_0} \right) \right] \qquad (10.94)$$

When the external field H_0 is along the easy axis, that is $\theta = \beta = 0$, Eq. (10.94) reduces to

$$\omega_0{}^2 = \mu_0{}^2\gamma^2 (H_0 + H_a)(H_0 + H_a + M_0) \qquad (10.95)$$

A comparison of Eqs. (10.95) and (10.90) with $N_x = N_z = 0$ and $N_y = 1$ in Eq. (10.90) shows that the effective anisotropy field can be accounted for by adding H_a to H_0. This is true, however, only for $\theta = \beta = 0$. If the applied field is in an arbitrary direction in the plane of the film, then we must know the equilibrium position of the

Figure 10.23

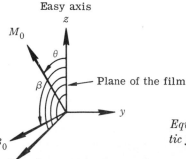

Equilibrium configuration for a thin magnetic film in a resonance experiment.

magnetization vector from Eq. (10.49) before we can calculate ω_0 from Eq. (10.94). We note that in materials with high effective anisotropy field, resonance phenomena can be observed at microwave frequencies without any d-c field, that is for $H_0 = 0$.

From the above analysis, we can see that the dependence of the resonance frequency upon the anisotropy field is a rather complicated function of the orientation of the applied d-c field with respect to the easy axis. Since in uniaxial crystals the anisotropy energy depends only on θ, the above analysis is quite general. It can be extended to include uniaxial crystals having an ellipsoidal shape with arbitrary N_x, N_y, and N_z. The situation with cubic crystals is much more complicated. Because the anisotropy energy depends on two polar angles, the effective anisotropy field needs two components in a complete description of the effect of anisotropy. These two components are usually expressed as

$$H_{x\,\text{anis}} = N_x{}^a M_x \quad \text{and} \quad H_{y\,\text{anis}} = N_y{}^a M_y \qquad (10.96)$$

in terms of two parameters $N_x{}^a$ and $N_y{}^a$ which may be called the apparent contribution to the demagnetizing factor due to anisotropy. In general, $N_x{}^a$ and $N_y{}^a$ are complicated functions of the two polar angles. One particular arrangement in which the applied d-c field H_0 is chosen to be in a (110) plane is generally favored by experimental investigators because it makes possible a relatively simple interpretation of the resonance frequency curve. Figure 10.22b shows the experimental resonance curve observed in nickel ferrite reported by Healy.[24] Values of $N_x{}^a$ and $N_y{}^a$ are given in Healy's paper. For a general discussion on how the parameters $N_x{}^a$ and $N_y{}^a$ can be found analytically, the reader should consult works[25] by Bickford and by Kittel.

Spin resonance absorption has also been observed in antiferromagnetic materials. According to the two-sublattice model discussed in Sec. 8.10, the effective fields acting on sublattice magnetization M_A and M_B are

$$\mathbf{H}_{\text{eff}\,A} = \mathbf{H} + \mathbf{H}_{aA} - \lambda\mathbf{M}_B$$
$$\mathbf{H}_{\text{eff}\,B} = \mathbf{H} + \mathbf{H}_{aB} - \lambda\mathbf{M}_A \qquad (10.97)$$

For simplicity, we include in Eq. (10.97) only the molecular field of the A-B type exchange interaction which is represented by the terms $\lambda\mathbf{M}_B$ and $\lambda\mathbf{M}_A$ since the interaction of this type is much stronger than those of the A-A and B-B types. Furthermore, the direction of the applied field is taken to be along the direction of easy magnetization, that is, the z axis. Thus, for a uniaxial crystal with an effective

[24] D. R. Healy, *Phys. Rev.*, vol. 86, p. 1009, 1952.
[25] L. R. Bickford, Jr., *Tech. Rept.* 23, Lab. for Insulation Research, MIT, Cambridge; and *Phys. Rev.*, vol. 78, p. 449, 1950; C. Kittel, *Phys. Rev.*, vol. 73, p. 155, 1948.

anisotropy field H_a, Eq. (10.87) becomes

$$\frac{d}{dt}(\mathbf{a}_x M_{Ax} + \mathbf{a}_y M_{Ay} + \mathbf{a}_z M_{Az})$$

$$= -\gamma\mu_0 \begin{vmatrix} \mathbf{a}_x & \mathbf{a}_y & \mathbf{a}_z \\ M_{Ax} & M_{Ay} & M_{Az} \\ -\lambda M_{Bx} & -\lambda M_{By} & H_0 + H_a - \lambda M_{Bz} \end{vmatrix} \qquad (10.98)$$

for M_A and the corresponding equation for M_B is similar in form except that the term $H_0 + H_a - \lambda M_{Bz}$ should be replaced by $H_0 - H_a + \lambda M_{Az}$. The sign change in the effective anisotropy and molecular fields is due to the fact that the direction of M_B is opposite to that of M_A (Fig. 10.24).

It is often convenient to use

$$M_A^+ = M_{Ax} + iM_{Ay} \qquad \text{and} \qquad M_B^+ = M_{Ax} + iM_{Ay} \qquad (10.99)$$

and rewrite the equation of motion as

$$\frac{dM_A^+}{dt} = i\mu_0\gamma[(H_0 + H_a + H_m)M_A^+ + H_m M_B^+]$$

$$\frac{dM_B^+}{dt} = i\mu_0\gamma[-H_m M_A^+ + (H_0 - H_a - H_m)M_B^+] \qquad (10.100)$$

The precessional motion of M_A and M_B is illustrated in Fig. 10.24. It is assumed in the above equation that the amplitude of the precessional motion is small so that $M_{Az} = M_A$ and $M_{Bz} = M_B$, thus the quantity H_m stands for the molecular field with $H_m = \lambda M_A = \lambda M_B$. It is evident from Fig. 10.24 that Eq. (10.100) has a solution of the form

$$M_A^+ = M \sin\theta_A \exp(i\omega t) \qquad \text{and} \qquad M_B^+ = -M \cos\theta_B \exp(i\omega t) \qquad (10.101)$$

Figure 10.24

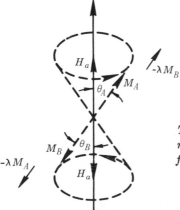

The precessional motions of the sublattice magnetization M_A and M_B in an antiferromagnet.

Since the exchange interaction between M_A and M_B is exceedingly strong, the precessional motion of M_A and M_B must oscillate in unison with the same angular frequency ω.

Substituting Eq. (10.101) into Eq. (10.100) yields

$$[\omega - \mu_0\gamma(H_0 + H_a + H_m)]\sin\theta_A - \mu_0\gamma H_m\sin\theta_B = 0$$
$$\mu_0\gamma H_m\sin\theta_A + [\omega - \mu_0\gamma(H_0 - H_a - H_m)]\sin\theta_B = 0 \qquad (10.102)$$

From Eq. (10.102), we obtain the natural angular frequency ω_0:

$$\omega_0 = \mu_0\gamma[H_0 \pm \sqrt{H_a(H_a + 2H_m)}] \qquad (10.103)$$

The above equation was derived by Kittel[26] and independently by Nagamiya.[27] The theory was further extended by Yosida[28] in his treatment of the antiferromagnetic resonance in $CuCl_2 \cdot 2H_2O$. Only for substances with a relatively weak molecular field, the resonance frequency falls in the microwave region. This happens to be the case with $CuCl_2 \cdot 2H_2O$.

10.8   THE FARADAY EFFECT

When a beam of plane-polarized light is sent through a block of transparent solid, the direction of polarization rotates upon the application of a magnetic field parallel to the direction of propagation. The effect known as the *Faraday effect* was discovered by Faraday in studying the propagation of light through glass, and since then the phenomenon has been observed in many solids, liquids, and gases. The amount of rotation is found by experiments to be proportional to the magnetic field strength and the optical path length in the medium. The optical Faraday effect is closely associated with the splitting of atomic energy levels by a magnetic field known as the Zeeman effect, but this is not the purpose of our present discussion. Instead, we shall discuss the corresponding effect observed in magnetic substances at microwave frequencies. The reason for doing this is twofold. The effect can be discussed on the basis of the classical equation of motion developed in Sec. 9.2 so that the reader can get acquainted with the physical process causing the rotation of the direction of polarization. Second, the effect is basic to the development of a large class of nonreciprocal microwave magnetic devices, such as gyrators, circulators, and isolators, which have become standard microwave components.

Consider a plane wave propagating in an infinite and lossless magnetic medium. If a d-c magnetic flux density B_0 exists in the direction of propagation which is taken to be the z axis, then the a-c magnetic flux densities in the x and y directions, B_x and B_y, are

[26]   C. Kittel, *Phys. Rev.*, vol. 82, p. 565, 1951.
[27]   T. Nagamiya, *Prog. Theor. Phys.* (Kyoto), vol. 6, p. 342, 1951.
[28]   K. Yosida, *Prog. Theor. Phys.* (Kyoto), vol. 7, pp. 25 and 425, 1952.

related to the a-c magnetic fields, H_x and H_y, as follows:

$$B_x = \mu H_x + i\kappa H_y \qquad B_y = -i\kappa H_x + \mu H_y \qquad (10.104)$$

According to Eq. (9.6), the a-c permeabilities μ and κ are given by

$$\mu = \mu_0 \left(1 + \frac{\omega_0 \omega_M}{\omega_0^2 - \omega^2}\right) \qquad (10.105a)$$

$$\kappa = \mu_0 \frac{\omega \omega_M}{\omega_0^2 - \omega^2} \qquad (10.105b)$$

where $\omega_0 = \gamma B_0$ and $\omega_M = \gamma \mu_0 M_z \simeq \gamma \mu_0 M_0$, M_0 being the spontaneous magnetization and γ the gyromagnetic ratio of the medium. Note that the a-c permeability relating the vector **B** to the vector **H** is a tensor and that the off-diagonal elements are complex conjugate of each other. Under such circumstances the problem of wave propagation can be best described in terms of two circularly polarized components.

For a right-hand circularly polarized wave, the B_x and H_x lead respectively the B_y and H_y by a phase angle of $90°$ as shown in Fig. 10.25. Hence, in complex notation, we have

$$B_x{}^R = iB_y{}^R = B^R \exp(i\omega t) \qquad H_x{}^R = iH_y{}^R = H^R \exp(i\omega t) \qquad (10.106)$$

Figure 10.25

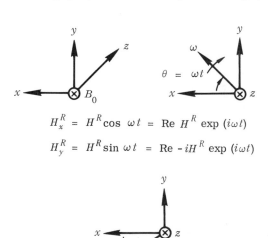

$$H_x^R = H^R \cos \omega t = \text{Re } H^R \exp(i\omega t)$$
$$H_y^R = H^R \sin \omega t = \text{Re } -iH^R \exp(i\omega t)$$

$$H_x^L = H^L \cos \omega t = \text{Re } H^L \exp(i\omega t)$$
$$H_y^L = -H^L \sin \omega t = \text{Re } iH^L \exp(i\omega t)$$

Right-hand and left-hand circularly polarized waves. The right- and left-hand senses use the direction of **B**$_0$ *as the reference axis.*

and for a left-hand circularly polarized wave,

$$B_x{}^L = -iB_y{}^L = B^L \exp(i\omega t) \qquad H_x{}^L = -iH_y{}^L = H^L \exp(i\omega t)$$

$$(10.107)$$

Substituting Eqs. (10.106) and (10.107) into Eq. (10.104) gives

$$B^R + B^L = (\mu + \kappa)H^R + (\mu - \kappa)H^L \qquad (10.108)$$

Since B^R and B^L as well as H^R and H^L are mutually independent, Eq. (10.108) gives two different permeabilities for the two circularly polarized waves:

$$\mu^R = \mu + \kappa \qquad \text{and} \qquad \mu^L = \mu - \kappa \qquad (10.109)$$

The superscripts R and L refer to the right-hand and left-hand polarized components of the plane wave. It should be emphasized that the sense of polarization is defined with respect to the direction of B_0 but not with respect to the direction of propagation.

Maxwell's field equations in rectangular coordinates read

$$\frac{\partial E_x}{\partial z} = -\frac{\partial B_y}{\partial t} \qquad \frac{\partial E_y}{\partial z} = \frac{\partial B_x}{\partial t}$$

$$\frac{\partial H_y}{\partial z} = \epsilon \frac{\partial E_x}{\partial t} \qquad \frac{\partial H_x}{\partial z} = -\epsilon \frac{\partial E_y}{\partial t} \qquad (10.110)$$

where E_x and E_y are the components of the electric field. We can again express the field equations in terms of two circularly polarized waves with

$$E_x = E_x{}^R + E_x{}^L \qquad E_y = E_y{}^R + E_y{}^L \qquad (10.111a)$$
$$H^R = H_x{}^R + iH_y{}^R \qquad H^L = H_x{}^L + iH_y{}^L \qquad (10.111b)$$
$$E^R = i(E_x{}^R + iE_y{}^R) \qquad E^L = i(E_x{}^L + iE_y{}^L) \qquad (10.111c)$$

In terms of E^R and H^R, Eq. (10.110) becomes

$$\frac{\partial E^R}{\partial z} = -i\omega\mu^R H^R \qquad \frac{\partial H^R}{\partial z} = i\omega\epsilon E^R \qquad (10.112)$$

Similar expressions are obtained for the left-hand circularly polarized waves. In Eq. (10.111c), the factor i in front accounts for the fact that **E** and **H** are mutually perpendicular in the xy plane. Assuming a plane-wave solution $E = E_0 \exp[i(\omega t - \Gamma z)]$, we find

$$\Gamma_R{}^2 = \omega^2 \mu^R \epsilon \qquad \text{and} \qquad \Gamma_L{}^2 = \omega^2 \mu^L \epsilon \qquad (10.113)$$

The two circularly polarized waves travel with two different phase velocities and the increase in the phase difference between these two waves causes a rotation of the direction of polarization. The effect is known as the *Faraday rotation*.

Consider a plane-polarized wave which has the electric field vector along the y axis at $z = 0$. The wave can be resolved into two circularly polarized waves shown graphically in Fig. 10.26a or analytically by Eqs. (10.106) and (10.107). We shall take the former approach. Since the two vectors $\mathbf{E}^L$ and $\mathbf{E}^R$ form an equilateral triangle, it is obvious from Fig. 10.26b that the resultant electric field vector makes an angle θ with the y axis,

$$\theta^+ = \frac{\theta_R^+ - \theta_L^+}{2} = \frac{(\Gamma_L - \Gamma_R)d}{2} \tag{10.114}$$

after the wave travels a distance d. The superscript $+$ refers to waves propagating in the forward or $+z$ direction. For waves propagating in the backward or $-z$ direction, $E = E_0 \exp[i(\omega t + \Gamma z)]$ and $z = -d$. Therefore, the negative sign in d compensates the change of sign in Γ and the same amount of rotation results:

$$\theta^- = \frac{\theta_R^- - \theta_L^-}{2} = \frac{(\Gamma_L - \Gamma_R)d}{2} = \theta^+ \tag{10.115}$$

Note that θ^+ and θ^- are on the same side of the y axis. In other words, if we look in the direction of the d-c field B_0, the angle of rotation is in the same sense irrespective of the direction of propagation. If we follow the direction of propagation, however, the rotations of the

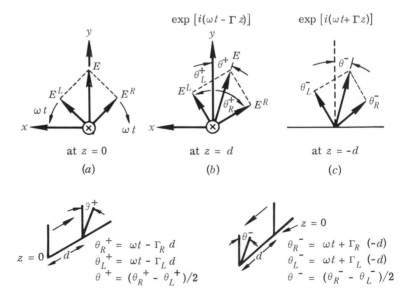

Figure
10.26

Diagrams illustrating the Faraday rotation: (a) A plane wave is resolved into two circularly polarized components. The two components propagate with different phase velocities and hence the direction of polarization changes after a distance d. Diagrams (b) and (c) are for waves propagating in the positive and negative z directions, respectively.

field vectors are in the opposite sense for waves propagating in the forward and backward directions. A wave making a round trip through the medium will find its direction of polarization shifted by an angle 2θ at the end of the trip. This basic property is used in many nonreciprocal microwave devices.

The microwave Faraday effect was first studied by Hogan[29] in ferrites after Polder[30] pointed out the tensor nature of the RF permeability. Although the Faraday effect also exists in paramagnetic materials, the effect is much larger in ferromagnetic and ferrimagnetic materials because of the larger value of magnetization. The use of the Faraday effect in ferromagnetic materials is not practical, however, because of the high electric conductivity of these materials and the consequent very small penetration depth of the microwave field. On the other hand, ferrimagnetic materials such as ferrites and garnets are almost insulators and hence they are well suited for use in many practical applications of the effect. Figure 10.27 shows the experimental arrangement in which the rotation is measured. Two rectangular waveguides are separated by a circular waveguide which contains the ferrite sample. Since the dimensions of the rectangular waveguides are such that they can only support TE_{10} mode with a particular direction of polarization as shown in Fig. 10.27, the Faraday rotation is measured by the angle of rotation of one waveguide with respect to the other for maximum energy transfer. Figure 10.28 shows the experimental result reported by Hogan in manganese zinc ferrite.

Substituting Eqs. (10.105) and (10.113) into Eq. (10.115), we find

$$\frac{\theta}{d} = \frac{\omega \sqrt{\epsilon/\epsilon_0}}{2c} \left[\left(1 + \frac{\omega_M}{\omega_0 + \omega}\right)^{\frac{1}{2}} - \left(1 + \frac{\omega_M}{\omega_0 - \omega}\right)^{\frac{1}{2}} \right] \quad (10.116)$$

where c is the velocity of light. In common practice, both ω_M and ω_0

[29]   C. L. Hogan, *Bell System Tech. J.*, vol. 31, p. 1, 1952; *Rev. Mod. Phys.*, vol. 25, p. 253, 1953.
[30]   D. Polder, *Phil. Mag.*, vol. 40, p. 99, 1949.

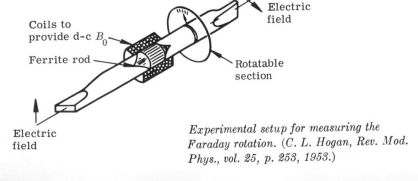

Figure 10.27

Coils to provide d-c B_0

Ferrite rod

Electric field

Rotatable section

Electric field

Experimental setup for measuring the Faraday rotation. (C. L. Hogan, Rev. Mod. Phys., vol. 25, p. 253, 1953.)

are smaller than ω. Hence, Eq. (10.116) can be approximated by

$$\frac{\theta}{d} = \sqrt{\frac{\epsilon}{\epsilon_0}}\left(\frac{\omega_M}{2c}\right) \tag{10.117}$$

For manganese zinc ferrite, $4\pi M_0 = 1500$ gauss or $\omega_M = 2.6 \times 10^{10}$ rad/sec for $g = 2$. Since the particular sample used in the experiment has a fairly large imaginary component $\epsilon_2/\epsilon_0 = 24$ as compared to a value $\epsilon_1/\epsilon_0 = 17$ for the real component, the value of ϵ is equal to $[(\epsilon_1^2 + \epsilon_2^2)^{1/2} + \epsilon_1]/2 = 23\epsilon_0$, as can be shown from Maxwell's equations. Using the values of ω_M and ϵ/ϵ_0 in Eq. (10.117), we obtain $\theta/d = 118$ degrees/cm in good agreement with the saturated value given in Fig. 10.28. When the applied field is smaller than the spontaneous magnetization, the ferrite may not attain its fully magnetized state or, in other words, ω_M in Eq. (10.117) will increase with the applied field, explaining the dependence of θ/d upon H_0 at low fields.

Finally, we will discuss briefly the optical Faraday effect. Consider the motion of an electron in a circular orbit around the nucleus. In the presence of a magnetic flux density B perpendicular to the plane of the orbit, the force equation, Eq. (1.14),

$$m_0\omega_0^2 a = \frac{ze^2}{4\pi\epsilon_0 a^2} \tag{10.118}$$

must be changed into

$$m_0\omega^2 a \pm evB = \frac{ze^2}{4\pi\epsilon_0 a^2} \tag{10.119}$$

where the negative and positive signs are respectively for right-hand

Figure 10.28

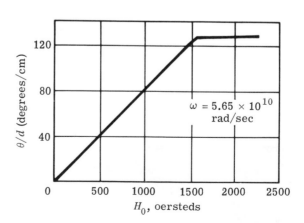

The experimental curve showing the angle of rotation θ versus the applied d-c magnetic field in manganese zinc ferrite. (C. L. Hogan, Rev. Mod. Phys., vol. 25, p. 253, 1953.)

and left-hand rotations whose senses use the direction of **B** as reference axis. Subtracting Eq. (10.118) from Eq. (10.119) gives

$$\omega^2 - \omega_0^2 = \mp \frac{e\omega B}{m_0} \quad \text{or} \quad \Delta\omega = \omega - \omega_0 = \mp \frac{eB}{2m_0} \quad (10.120)$$

in view of the fact that $\omega - \omega_0 \ll \omega$. Note that Eq. (10.120) gives the Larmor precession frequency as Eq. (8.16). However, the positive and negative signs in Eq. (10.120) are significant. The two rotations now have different angular frequencies,

$$\omega^L = \omega_0 - \frac{eB}{2m_0} \quad \text{and} \quad \omega^R = \omega_0 + \frac{eB}{2m_0} \quad (10.121)$$

respectively for left-hand and right-hand rotations. The above discussion can be extended to a general case in which the atom level is split into two levels due to the Zeeman effect. In the presence of a magnetic field, not only will a solid have two absorption peaks at ω^L and ω^R but it will also have two dispersion curves. The situation is illustrated in Fig. 10.29. The reader is referred to Eq. (7.17) or Fig. 7.1 for the relationship between the absorption and dispersion curves. It is obvious from Fig. 10.29 that there is a region in which the indices of refraction n will be different for two circularly polarized waves rotating in opposite directions. This difference in the indices of refraction is responsible for the optical Faraday effect.

Figure 10.29

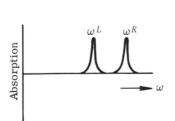

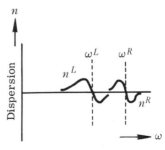

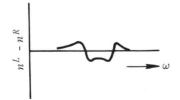

The optical Faraday effect. The absorption and dispersion curves (the imaginary and real part of the optical index of refraction) are related to each other through Eq. (7.17). Referring to Fig. 7.1, we see that if there are two resonant frequencies, ω^L and ω^R for the left-hand and right-hand waves, there also exist two dispersion curves, resulting in two indices of refraction n^L and n^R.

10.9 NONRECIPROCAL MICROWAVE FERRIMAGNETIC DEVICES

There are four principal types[31] of microwave ferrimagnetic devices using ferrites or garnets: (1) Faraday rotators, (2) resonance isolators, (3) nonreciprocal phase shifters, and (4) field-displacement devices. Since a detailed discussion of the operational principle of these devices requires a knowledge of modes of wave propagation in waveguides, we shall present first a qualitative discussion of the main features of these devices. From the discussion in the previous section, we see that the direction of polarization is rotated by the same amount and in the same sense (with respect to the direction of the d-c magnetic field H_0) for waves propagating in the forward and reverse directions. This nonreciprocal property can be put into practical use in the form of isolators and circulators. The idea is schematically illustrated in Figs. 10.30 and 10.31. The microwave transducer is a section of twisted rectangular waveguide which simply changes the direction of polarization by a specified amount. Unlike the Faraday rotation, the rotation caused by the transducer is reciprocal. A practical realization of Fig. 10.30 is actually shown in Fig. 10.27 except for the added section of microwave transducer.

According to Fig. 10.30, the total amount of rotation is 90° for the forward wave. If the two end sections of the waveguides are so chosen that their electric field vectors are spatially 90° apart, the forward wave can go through with a rotation of polarization by 90°.

[31] B. Lax, *Proc. IRE*, vol. 44, p. 1368, 1956.

Figure 10.30

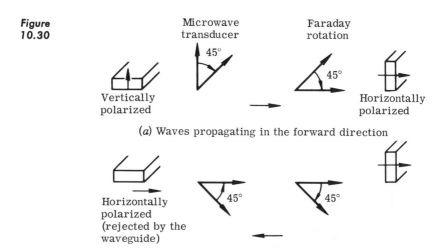

(a) Waves propagating in the forward direction

(b) Waves propagating in the reverse direction

Schematic illustrations showing the operational principle of a microwave isolator.

This, however, does not hold for the backward wave. Since the Faraday rotation is nonreciprocal, the net rotation of the backward wave is zero. As indicated in Fig. 10.30, the backward wave finds that the direction of its electric field does not fit into the propagating mode of the waveguide and hence the wave will be rejected. Therefore, the arrangement of Fig. 10.30 constitutes a one-way transmission system and is generally known as an *isolator*. We should add that in practice, it is desirable to have the nonpropagating wave absorbed instead of being reflected. This can be done by adding radial vanes which absorb energy only when the electric field is in the plane of the vane. A possible modification of the isolator arrangement leads to the structure of a *gyrator*. By changing the amount of rotation from 45° to 90°, we find that one wave will have a phase shift of 180° but not the other wave. Still another variation of Fig. 10.30 is the *circulator* of Fig. 10.31. The input and output ends of the waveguide system actually consist of two arms, one allowing the propagation of vertically polarized waves and the other that of horizontally polarized waves. These arms are labeled as a, b, c, and d in Fig. 10.31. The reader can easily verify that the wave with polarization a is turned into polarization b, b into c, c into d, and d into $-a$ which will be accepted by arm a.

The various Faraday-rotation devices just discussed are operated in a region away from the ferrimagnetic resonance to avoid absorption by the ferrimagnetic material. The *resonance isolator*, on the

Figure 10.31

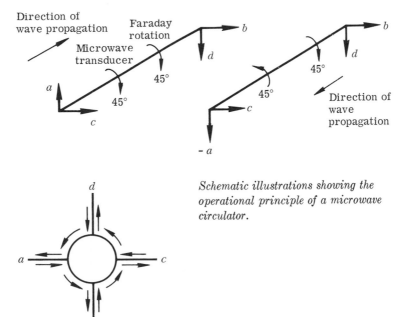

Schematic illustrations showing the operational principle of a microwave circulator.

other hand, utilizes the resonance absorption to differentiate waves propagating in opposite directions. As discussed in Sec. 9.2, only the right-hand circularly polarized wave will be strongly absorbed at resonance according to Eq. (9.10) or Fig. 9.2. The magnetic field components in an empty waveguide (Fig. 10.32a) are given by

$$H_x = (-i\beta \sin kx) \exp(-i\beta y)$$
$$H_y = (-k \cos kx) \exp(-i\beta y) \tag{10.122a}$$

for the $(TE)_{10}$ mode of propagation where $k = \pi/L$ and

$$\beta^2 = \omega^2 \mu_0 \epsilon_0 - k^2 \tag{10.122b}$$

The internal fields in a ferrimagnetic substance placed in the waveguide can be calculated from Eq. (10.122) as follows:

$$H_{ix} = H_x - N_x M_x = H_x - N_x(\chi_1 H_{ix} + i\chi H_{iy})$$
$$H_{iy} = H_y - N_y M_y = H_y - N_y(\chi_2 H_{iy} - i\chi H_{ix}) \tag{10.123}$$

where $\chi = \kappa/\mu_0$ and κ is given by Eq. (10.105). Note that Eq. (10.104) in which we have neglected demagnetizing effects is true for an infinite medium. The ferrite or garnet used in a resonance

Figure 10.32

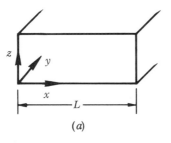

(a)

(a) A resonance isolator which is made of a ferrite (or garnet) rod placed in a waveguide and (b) the attenuation ratio plotted as a function of the position of the rod.

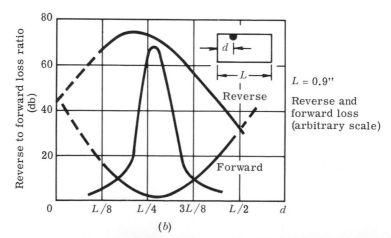

(b)

isolator is generally in the form of a slab or rod. Therefore, the χ's in Eq. (10.123) must be generalized to include demagnetizing effects with

$$\chi_1 = \frac{\omega_M \omega_1}{\omega_1 \omega_2 - (\omega - i\,\Delta\omega)^2} \qquad \chi_2 = \frac{\omega_M \omega_2}{\omega_1 \omega_2 - (\omega - i\,\Delta\omega)^2}$$

$$\chi = \frac{i\omega + \Delta\omega}{\omega_1 \omega_2 - (\omega - i\,\Delta\omega)^2} \tag{10.124a}$$

Equation (10.124a) is obtained from Eq. (9.14) by letting $1/T_2 = \Delta\omega$ and properly replacing ω_0 with either

$$\omega_1 = \mu_0 \gamma [H_0 + (N_y - N_z)M_0] \quad \text{or} \quad \omega_2 = \mu_0 \gamma [H_0 + (N_x - N_z)M_0] \tag{10.124b}$$

In principle, the internal fields H_{ix} and H_{iy} can be expressed in terms of the external fields H_x and H_y, knowing all the parameters in Eq. (10.123) from Eqs. (10.124a) and (10.124b). The position at which the internal field is circularly polarized can be found by demanding $H_{ix} = \pm iH_{iy}$. One of the two signs means a maximum absorption of the forward wave and the other a maximum absorption of the backward wave. Figure 10.32b shows the experimental results obtained by Momo and Heller[32] for a thin cylindrical ferrite rod. Actually, it is not the absolute attenuation of the reverse wave that is of interest to us but the relative attenuation of the reverse wave to that of the forward wave. Note that the positions for the optimization of these two cases do not coincide. We also should point out that the other half of the curve with $d > L/2$ is quite similar to the one with $d < L/2$ except that it interchanges the role of reverse and forward waves. In any case, Fig. 10.32b clearly shows that a large ratio of reverse to forward loss can be obtained by differential absorption of the two waves at resonance. This property constitutes another class of nonreciprocal microwave devices.

The *nonreciprocal phase shifter* is actually a variation of the Faraday rotator devices. According to the first-order perturbation calculation, the complex propagation constant Γ in a rectangular waveguide of cross section A partly filled with a ferrimagnetic material of cross section A' is given by

$$\Gamma = i\beta + i\left(\frac{A'}{A\beta}\right)[\beta^2 \chi_1 \sin^2 kx$$
$$+ k^2 \chi_2 \cos^2 kx \pm \beta k \chi \sin 2kx + \omega^2 \mu_0 \epsilon_0 \chi_e] \tag{10.125}$$

where β is the same phase constant as the one for an empty guide given by Eq. (10.122b) and χ_e is the dielectric susceptibility of the ferrimagnetic substance. The positive and negative signs correspond

[32] The result was quoted by B. Lax, *Proc. IRE*, vol. 44, p. 1379, 1956.

to the two directions of propagation. From Eq. (10.125), the differential phase shift is proportional to

$$\beta^+ - \beta^- \sim 2k \text{ (real part } \chi) \sin 2kx \qquad (10.126)$$

For optimum operation of the device, the ratio of the differential phase shift to the attenuation constant or the quantity $(\beta^+ - \beta^-)/(\alpha^+ + \alpha^-)$ is to be maximized. This procedure determines the location of the ferrite or garnet in the waveguide. It is obvious from Eq. (10.125) that the phase constant β is determined by the real parts of χ_1, χ_2, and χ while the attenuation constant α is fixed by the imaginary parts. A nonreciprocal phase shifter is operated in the region where β is large and α is small. A resonance isolator, on the other hand, makes use of the fact that the attenuation constants are different for forward and reverse waves. Therefore, Eq. (10.125) is applicable to both types of devices with χ_1, χ_2, and χ given by Eq. (10.124).

In a waveguide partly filled with ferrites or garnets, not only may the propagation constant of the wave be affected but so also may the field distribution. Another class of nonreciprocal magnetic device called the *field-displacement device* is based on differing field configurations for the two directions of propagation. Qualitatively, the energy of a wave in a nonuniform medium is not distributed evenly over the cross section but concentrated in regions where the dielectric constant and the magnetic permeability are high. The same positive and negative signs which cause two different β and α in Eq. (10.125) will result in two different effective permeabilities for the two waves. To state the physical situation differently, the presence of a ferrimagnetic slab, in a rectangular waveguide for example, creates higher modes. The situation is illustrated in Fig. 10.33. The phase angle between the dominant mode E_1 and the higher modes ΣE_n depends on the direction of propagation, resulting in two different field configurations. If a resistance sheet is placed against the face of the ferrite (Fig. 10.34a) where the electric field is zero for the forward

Figure 10.33

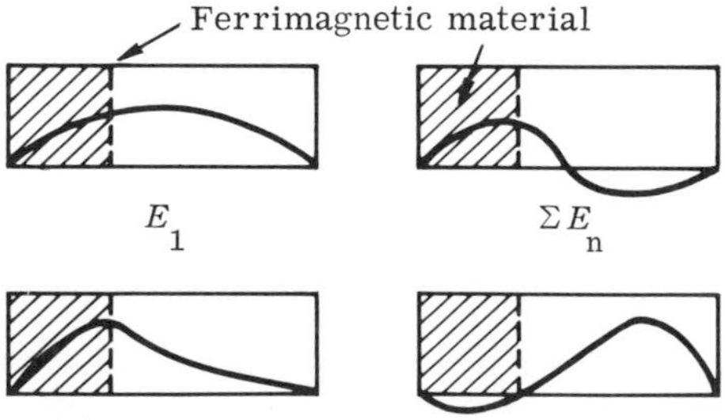

Schematic representation showing the superposition of the dominant waveguide mode (E_1) with the excited modes (ΣE_n due to the presence of the ferrimagnetic material) in a field-displacement device. Depending on the direction of wave propagation, the resultant field is given either by $E_1 + \Sigma E_n$ or by $E_1 - \Sigma E_n$.

Ferrimagnetic material

E_1

ΣE_n

$E_1 + \Sigma E_n$

$E_1 - \Sigma E_n$

wave, a large differential absorption of the two waves is obtained. The device is known as a *field-displacement isolator.*

It is also possible to make a circulator out of the field-displacement effect. Theoretical calculations show that the field displacement also happens to the longitudinal magnetic field H_y. If holes or slots are made at the top of the guide (Fig. 10.34b), energy will be coupled to the other waveguide on top through H_y. For a detailed discussion of the field-displacement devices, we must know the field distribution in a perturbed waveguide. The procedure for such a calculation is outlined in an excellent review article by Lax.[33] The reader is also referred to the article by Fox, Miller, and Weiss[34] for experimental results on the performances of these devices as well as for a general discussion of their operational principle. Specific mention should be made of the annual special issue of the *Journal of Applied Physics,* which since 1958 has been devoted to the results of research discussed at the magnetics conference.

[33] B. Lax, *op. cit.*
[34] A. G. Fox, S. E. Miller, and M. T. Weiss, *Bell System Tech. J.*, vol. 34, p. 5, 1955.

Figure 10.34

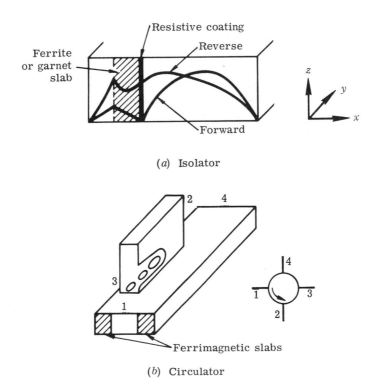

(*a*) Isolator

(*b*) Circulator

Field-displacement devices: (a) isolator and (b) circulator.

10.10 SPIN WAVES

In Sec. 8.9, we have shown that at a finite temperature, the spontaneous magnetization M_0 is smaller than the saturation magnetization M_s (Fig. 8.13). The nonperfect alignment can be understood as resulting from thermal agitation. The question of how the deviation $M_s - M_0$ will distribute itself in space is certainly interesting. From our discussion in Sec. 10.4, it is clear that any distribution which requires an abrupt change of the directions of spin on neighboring atoms is energetically unfavorable on account of the exchange energy. Furthermore, in a homogeneous medium, it is equally likely for the ith atom to have a nonparallel spin as for the jth atom. The spin wave theory which is a result of these considerations supposes that the part $M_s - M_0$ propagates like a wave through the whole magnetic medium. Analytically, $M_s - M_0$ can be thought of as consisting of many Fourier components having different angular frequencies ω and different wave vectors $\mathbf{k}$. In the traveling-wave representation

$$\frac{\mathbf{M}}{M_s} = \frac{\mathbf{M}_0}{M_s} + \sum_k \mathbf{a}_k \exp\left[i(\mathbf{k} \cdot \mathbf{r} - \omega t)\right] \tag{10.127}$$

where a_k is the normalized amplitude (with respect to M_s) of the wave with wave number k and the summation is over all possible wave numbers.

The most important information concerning a wave is the dispersion relation, that is, the relation between ω and k. To derive the dispersion relationship, we again start with the classical equation of motion. If the exchange field B_{ex} is the only field present, Eq. (10.87) becomes

$$\frac{d\mathbf{M}}{dt} = -\gamma \mathbf{M} \times \mathbf{B}_{ex} = \gamma \mathbf{M} \times \left(\frac{2A}{M_s{}^2} \nabla^2 \mathbf{M}\right) \tag{10.128}$$

where the expression for B_{ex} is obtained from Eq. (10.11) for a simple cubic lattice, $A = JS^2/l$, S is the spin quantum number, and l is the lattice constant. The factor 2 takes into account the fact that the exchange energy of Eq. (10.11) is a self-energy. For $(M_s - M_0)/M_s \ll 1$, if the spontaneous magnetization points in the z direction, M_z can be approximated by M_0 while the x and y components of M are respectively given by

$$M_x = M_s \sum_k a_{kx} \exp\left[i(\mathbf{k} \cdot \mathbf{r} - \omega t)\right] \tag{10.129}$$

and a similar expression for M_y.

Substituting Eq. (10.127) into Eq. (10.128) yields

$$i\omega a_{kx} = -\gamma \left(\frac{2A}{M_s}\right) k^2 a_{ky} \qquad \text{and} \qquad i\omega a_{ky} = \gamma \left(\frac{2A}{M_s}\right) k^2 a_{kx} \tag{10.130}$$

For nontrivial solutions,

$$\omega = \gamma \left(\frac{2A}{M_s}\right) k^2 \tag{10.131}$$

Since the saturation magnetization is related to S by $M_s = \gamma \hbar S/l^3$, the energy associated with a given spin wave mode is equal to

$$n\hbar\omega = 2JSl^2k^2n \tag{10.132}$$

according to our discussion in Sec. 1.6 where n is the occupation number of the spin wave mode. The change of spontaneous magnetization with temperature now has a new physical interpretation. As given by Eq. (1.129), n increases with temperature. At higher temperatures, more spin wave modes are excited, reducing M_0. Our next task is to establish a quantitative relationship between n and the spin wave amplitude a_k.

For this purpose, we consider the standing-wave solutions

$$\begin{aligned} a_{kx} &= a_k \sin k_x x \sin k_y y \sin k_z z \cos \omega t \\ a_{ky} &= a_k \sin k_x x \sin k_y y \sin k_z z \sin \omega t \end{aligned} \tag{10.133}$$

According to Eq. (10.11), the exchange energy involved is equal to

$$\begin{aligned} \mathcal{E}_{\text{ex}} &= -Jl^2S^2 \sum \mathbf{a}_k \cdot \nabla^2 \mathbf{a}_k \\ &= JS^2l^2 \sum_k k^2a_k^2 \sin^2 k_x x \sin^2 k_y y \sin^2 k_z z \end{aligned} \tag{10.134}$$

for a simple cube with edge l. In a volume V with $dV = dx\,dy\,dz$, the total energy is

$$\mathcal{E} = \int \frac{\mathcal{E}_{\text{ex}}dV}{l^3} = \sum_k \frac{JS^2k^2a_k^2V}{8l} \tag{10.135}$$

Equating the exchange energy associated with the spin wave mode k from Eq. (10.135) to that from Eq. (10.132) gives

$$a_k^2 = \frac{16l^3n}{VS} \tag{10.136}$$

where n is given by Eqs. (1.129) and (1.134) as follows:

$$\frac{dn}{V} = \frac{4\pi k^2\,dk/(2\pi)^3}{\exp(\hbar\omega/kT) - 1} \tag{10.137}$$

Note that for a continuous distribution, the summation over k should be replaced by a corresponding integration.

Substituting Eq. (10.137) into Eq. (10.136) yields

$$\sum \frac{a_k^2}{16} = l^3 \int \frac{dn}{VS} = \left(\frac{0.0587}{S}\right)\left(\frac{kT}{2JS}\right)^{\!3\!/\!2} \tag{10.138}$$

The constant 0.0587 in Eq. (10.138) is the result of a definite integral

$\int_0^\infty \sqrt{x}(e^x - 1)^{-1}\,dx/4\pi^2 = 0.0587$ with $x = \hbar\omega/kT$. Note that the k in Eq. (10.138) and $(\hbar\omega/kT)$ of Eq. (10.137) is the Boltzmann constant. In view of the relation $M_x{}^2 + M_y{}^2 + M_z{}^2 = M_s{}^2$, the spontaneous magnetization at a finite temperature is the spatial average of

$$M_0{}^2 = M_s{}^2 \left[1 - \sum_k (a_{kx}{}^2 + a_{ky}{}^2) \right] \tag{10.139}$$

As worked out earlier for Eq. (10.135), the spatial average of $a_{kx}{}^2 + a_{ky}{}^2$ is equal to $a_k{}^2/8$. Thus, for $(M_s - M_0)/M_s < 1$, Eq. (10.139) becomes

$$\frac{M_s - M_0}{M_s} = \sum_k \frac{a_k{}^2}{16} = \left(\frac{0.0587}{S}\right)\left(\frac{kT}{2JS}\right)^{3/2} \tag{10.140}$$

This is known as the Bloch $T^{3/2}$ law mentioned in Sec. 10.2.

Direct observation of spin wave resonance in a thin permalloy film was first reported by Seavey and Tannenwald.[35] Figure 10.35 shows the absorption due to spin wave excitation in a permalloy film 5600 Å thick with the d-c field applied perpendicular to the surface of the film. Under ordinary circumstances, only the uniform precessional mode discussed in Sec. 10.7 is excited by a uniform microwave field. However, Kittel[36] has pointed out that atoms on the surface of a specimen are in a special position. The spins can be pinned down at the boundary of the specimen due to surface anisotropy which is larger than bulk anisotropy. For further discussion of the mechanism of pinning, the reader is referred to the book by Soohoo[37] and the

[35] M. H. Seavey, Jr., and P. E. Tannenwald, *Phys. Rev. Letters*, vol. 1, p. 168, 1958; *Jour. Appl. Phys.*, vol. 30, p. 2275, 1959.
[36] C. Kittel, *Phys. Rev.*, vol. 110, p. 1295, 1958.
[37] R. F. Soohoo, "Thin Magnetic Films," chap. 11, Harper and Row, Publishers, Inc., New York, 1965.

Figure 10.35

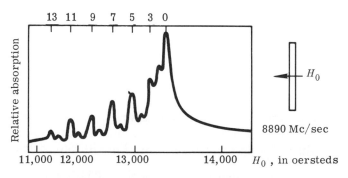

Spin wave resonance in a permalloy film at 8890 Mc/sec. The number ($p = 3$, 5,7 etc.) on top determines the spin wave number through the relation $k_p L = p\pi$, where k_p is the spin wave number and L is the thickness of the film.

article by Wigen *et al.*[38] With the field configuration of Fig. 10.35, the spin wave resonance frequency is

$$\omega = \gamma \left[\mu_0(H_0 - M_0) + \left(\frac{2A}{M_s}\right) k^2 \right]$$

(10.141)

which is a combination of the effects stated in Eqs. (10.90) and (10.131).

According to the analysis by Kittel, the spin wave varies as $\Sigma a_p \sin k_p z \cos \omega t$ where the z axis is in the direction normal to the surface of the film. For surface pinning, $k_p L = p\pi$ where L is the thickness of the film and p is an odd integer for excitation by a uniform microwave field. Eddy current effects in metallic films as in the case of permalloy films can cause a decaying field in the film and hence may permit even-mode excitation. This explains the appearance of even modes in Fig. 10.35 and their relatively small absorption. Since the direct observation of spin wave resonance, the method has been applied to a number of ferromagnetic metallic films to measure the exchange integral J and to study the mechanism of surface pinning. We should also mention that both the excitations of spin waves and magnetostatic modes by an inhomogeneous microwave field have been observed. For a discussion of these phenomena, the reader is referred to the articles by White and Solt[39] and by Walker.[40]

10.11 SPIN WAVE INSTABILITIES IN A RESONANCE EXPERIMENT

In Sec. 9.3, we discussed the saturation effect (Fig. 9.4) at high microwave power levels in a paramagnetic resonance experiment. The relevant expressions describing the saturation phenomenon are Eqs. (9.25) and (9.27b) which read at resonance

$$M_z = \frac{M_0}{1 + T_1 T_2(\gamma\mu_0 H_1)^2} \quad \text{and} \quad \chi'' = \frac{\gamma\mu_0 M_0 T_2}{1 + T_1 T_2(\gamma\mu_0 H_1)^2}$$

(10.142)

The change in the z component of magnetization M_z can be calculated by observing the voltage induced in a pickup coil at the beginning and end of the microwave pulse, while the imaginary part of susceptibility χ'' can be obtained by measuring the power absorbed in the sample. In the paper by Bloembergen and Wang,[41] a careful calibration of the system measuring M_z was made so that a detailed

[38] Wigen, Kooi, Shanabarger, and Rossig, *Phys. Rev. Letters*, vol. 5, p. 206, 1962.
[39] R. L. White and I. H. Solt, *Phys. Rev.*, vol. 104, p. 56, 1956; also J. F. Dillon, *Jour. Appl. Phys.*, vol. 31, p. 1605, 1960.
[40] L. R. Walker, *Phys. Rev.*, vol. 105, p. 390, 1957; *Jour. Appl. Phys.*, vol. 29, p. 318, 1958.
[41] N. Bloembergen and S. Wang, *Phys. Rev.*, vol. 93, p. 72, 1954.

comparison of the saturation phenomena in paramagnetic and ferrimagnetic substances was possible. Figure 10.36 shows the M_z and
χ'' curves in nickel ferrite as compared to those in $MnSO_4 \cdot 4H_2O$, a
paramagnetic salt. These curves are quite typical. While the two
curves keep in step with each other in a paramagnetic substance as
predicted by Eq. (10.142), they no longer do so in a ferrimagnetic
substance. The discrepancy between the theory and experiment is
due to the excitation of spin wave modes as first suggested by Anderson and Suhl[42] and later explained in detail by Suhl.

For a quantitative analysis of the effect, we must derive the disper-

[42]   P. W. Anderson and H. Suhl, *Phys. Rev.*, vol. 100, p. 1788, 1955; H. Suhl, *J.
Phys. Chem. Solids*, vol. 1, p. 209, 1957.

**Figure
10.36**

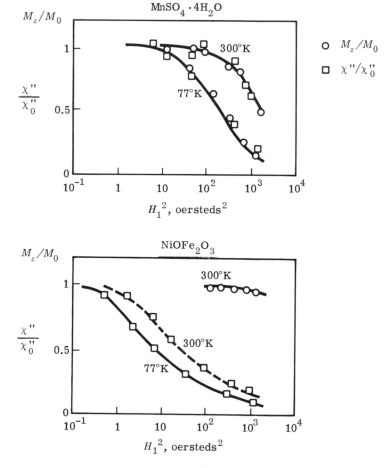

Experimental curves of M_z/M_0 and χ''/χ_0'' as functions of H_1^2 (the a-c microwave field) in paramagnetic $MnSO_4 \cdot 4H_2O$ and ferrimagnetic $NiOFe_2O_3$. (N.
Bloembergen and S. Wang, Phys. Rev., vol. 93, p. 72, 1954.)

sion relation of the normal spin wave modes appropriate to the conditions existing in a resonance experiment. The procedure follows along the same line as in the previous section. The effective exchange field $\mathbf{H}_{ke}$ due to spin waves of wave number k is given by

$$\mu_0\mathbf{H}_{ke} = \left(\frac{2A}{M_s{}^2}\right)\nabla^2\mathbf{M} = -\mu_0 H_e(kl)^2\mathbf{a}_k \qquad (10.143)$$

where $H_e = 2A/(\mu_0 M_s l^2) = 2JS/(\mu_0\gamma\hbar)$ has the dimension of a magnetic field. The effective dipolar field is much more complicated. For the uniform mode, this field is accounted for by the demagnetizing factors N_x, N_y, and N_z. For a spherical sample which is the case under consideration, the effect cancels out in Eq. (10.90) with $N_x = N_y = N_z$. By *uniform modes*, we mean that in this mode the magnetization vector will rotate as a whole with components M_x, M_y, and M_z. This is not true with the spin waves. What complicates the calculation of the effective dipolar field arising from spin waves is the propagation effects associated with spin waves. In other words, we must solve Maxwell's equations instead of the Laplace equation as was the case for the uniform mode. According to an analysis by Soohoo,[43] the dipolar field $\mathbf{H}_{kd}$ due to spin wave of wave number k is equal to

$$\frac{\mathbf{H}_{kd}}{M_s} = \frac{(\omega^2\epsilon_e/c^2)\mathbf{a}_k - \mathbf{k}(\mathbf{k}\cdot\mathbf{a}_k)}{k^2 - \omega^2\epsilon_e/c^2} \qquad (10.144)$$

where $\epsilon_e = (\epsilon/\epsilon_0)(1 + 4\pi\sigma/i\omega\epsilon)$, ϵ and σ being, respectively, the dielectric constant and conductivity of the medium, and $c = (\mu_0\epsilon_0)^{-\frac{1}{2}}$.

In a resonance experiment, we are concerned with spin waves whose wavelength is short compared with the wavelength of the electromagnetic wave in the medium. In other words, $k^2 \gg \omega^2\epsilon_e/c^2$ and hence Eq. (10.144) can be approximated by

$$\mathbf{H}_{kd} = -\frac{M_s\mathbf{k}(\mathbf{k}\cdot\mathbf{a}_k)}{k^2} \qquad (10.145)$$

Our next step is to substitute Eqs. (10.143) and (10.145) into Eq. (10.87), the equation of motion. Following the analysis by Kemanis and Wang,[44] we shall choose in this case a pair of axes u and v so that the vector $\mathbf{k}$ is in the uz plane as shown in Fig. 10.37. For spin waves propagating in different directions, the u and v axes will be different. As we shall see presently, the use of u and v axes instead of the ordinary x and y axes not only simplifies the analysis but also makes clear the physical picture of spin wave dynamics. In the new coordinate system, the total effective field $\mathbf{H}_{\text{eff}}$ can be written as

$$\mathbf{H}_{\text{eff}} = -\mathbf{u}[H_e(lk)^2 + M_s\sin^2\theta_k]a_{ku} - \mathbf{v}H_e(lk)^2 a_{kv}$$
$$- \mathbf{z}[M_s(\sin\theta_k\cos\theta_k)a_{ku} + H_0 - N_z M_0] \qquad (10.146)$$

[43] R. F. Soohoo, *Phys. Rev.*, vol. 120, p. 1978, 1960.
[44] G. Kemanis and S. Wang, *Jour. Appl. Phys.*, vol. 35, p. 1465, 1964.

where **u**, **z**, and **v** are unit vectors. As in the previous analysis, only the first-order terms are included in Eq. (10.146). In other words, terms involving $a_{kz} = -(a_{ku}^2 + a_{kv}^2)/2$ are neglected.

Substituting Eq. (10.146) into Eq. (10.87), we find

$$\frac{da_{ku}}{dt} = -(A_k - B_k)a_{kv} = -\omega_k \varepsilon_k a_{kv}$$

$$\frac{da_{kv}}{dt} = (A_k + B_k)a_{ku} = \omega_k \varepsilon_k^{-1} a_{ku}$$

(10.147)

where $\omega_k^2 = A_k^2 - B_k^2$ and $\varepsilon_k^2 = (A_k - B_k)/(A_k + B_k)$ with

$$A_k = \mu_0 \gamma \left[H_0 - N_z M_0 + H_e(lk)^2 + \left(\frac{M_s}{2}\right) \sin^2 \theta_k \right]$$

$$B_k = \left(\frac{\mu_0 \gamma M_s}{2}\right) \sin^2 \theta_k$$

(10.148)

Clearly, the solutions for the spin wave components will be sinusoids and we choose the following set of solutions for a_k:

$$a_{ku} = \varepsilon_k^{1/2} a_k \cos \omega_k t \cos \mathbf{k} \cdot \mathbf{r}$$

$$a_{kv} = \varepsilon_k^{-1/2} a_k \sin \omega_k t \cos \mathbf{k} \cdot \mathbf{r}$$

(10.149)

Figure 10.37

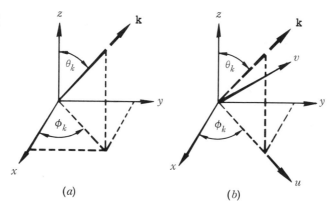

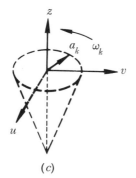

*Diagrams used in analyzing the spin wave dynamics. (a) The wave vector **k** of a given spin wave mode has polar angles θ_k and ϕ_k. (b) The new axes u and v are so chosen that the wave vector **k** lies in the uz plane. (c) In the uv plane, the locus of the spin wave vector $\mathbf{a}_k$ is described by an ellipse with **u** and **v** being the major and minor axes.*

where the angular frequency ω_k is given by[45]

$$\omega_k = \mu_0\gamma[(H_0 - N_z M_0 + H_e l^2 k^2)(H_0 - N_z M_0 + H_e l^2 k^2 + M_s \sin^2 \theta_k)]^{1/2} \quad (10.150)$$

Note that for spin waves propagating in the z direction, $\theta_k = 0$, then Eq. (10.150) reduces to Eq. (10.141) except for the term $-M_0$ which is due to the demagnetizing effect. According to Eq. (10.147), the motion of spin wave vector $\mathbf{a}_k$ is described by an ellipse in the uv plane (Fig. 10.37c).

The energy $\mathcal{E}_k$ associated with the spin wave component a_k is

$$\mathcal{E}_k = -\left(\mathbf{H}_0 + \frac{\mathbf{H}_{\text{int}}}{2}\right) \cdot (\mu_0 M_s \mathbf{a}_k) \quad (10.151)$$

The field given in Eq. (10.151) is the same as the effective field $\mathbf{H}_{\text{eff}}$ of Eq. (10.146) except for the factor $\frac{1}{2}$ which appears in the internal field $\mathbf{H}_{\text{int}}$. The factor $\frac{1}{2}$ arises from the fact that the term represents self-energy. Using the expression for $\mathbf{H}_{\text{int}}$ from Eq. (10.146) in Eq. (10.151) and realizing that $H_0 \mu_0 M_s a_{kz} = H_0 \mu_0 M_s (a_{ku}^2 + a_{kv}^2)/2$, we find

$$\mathcal{E}_k = \left(\frac{\omega_k M_s}{2\gamma}\right)[\mathcal{E}_k^{-1} a_{ku}^2 + \mathcal{E}_k a_{kv}^2] \quad (10.152)$$

Note that Eq. (10.152) reduces to Eq. (10.134) for $H_0 = 0$ and $\theta_k = 0$ as it should. Under thermal equilibrium, $\mathcal{E}_k$ is independent of time. By excitation of spin waves, we mean that under suitable conditions there will be a constant flow of power from external RF fields into the spin wave system. The power is given by the time average of $d\mathcal{E}_k/dt$

$$P_{\text{abs}} = \left\langle \frac{d\mathcal{E}_k}{dt} \right\rangle = \left(\frac{\omega_k M_s}{\gamma}\right)\left\langle \mathcal{E}_k^{-1} a_{ku}\frac{da_{ku}}{dt} + \mathcal{E}_k a_{kv}\frac{da_{kv}}{dt} \right\rangle \quad (10.153)$$

where the symbol $\langle \ \rangle$ denotes time and space average.

There are three principal types of spin wave instabilities which may take place under the simultaneous presence of d-c and a-c magnetic fields. Among the three, we shall discuss the case of parallel pumping first because it is the simplest. If we introduce an a-c field H_z parallel to the d-c field H_0

$$H_z = H_1 \cos(\omega t + \alpha) \quad (10.154)$$

into Eq. (10.146), two additional terms appear in Eq. (10.147)

$$\frac{d}{dt}(\mathbf{u}a_{ku} + \mathbf{v}a_{kv}) = \gamma\mu_0(\mathbf{v}H_z a_{ku} - \mathbf{u}H_z a_{kv}) \quad (10.155)$$

[45] The spin wave dispersion relation was first derived by T. Holstein and H. Primakoff, *Phys. Rev.*, vol. 58, p. 1098, 1940.

Substituting Eq. (10.155) into Eq. (10.153) gives

$$P_{\text{abs}} = \omega_k \mu_0 M_s \langle (\varepsilon_k - \varepsilon_k^{-1}) a_{ku} a_{kv} H_z \rangle \tag{10.156}$$

For $\langle a_{ku} a_{kv} H_z \rangle$ to be nonzero, certain conditions must be satisfied. For the spatial part, a_{ku} and a_{kv} must have the same wave number k. For the time-varying part, the condition is $2\omega_k = \omega$ and $\alpha = \pi/2$ for maximum power absorption. Unlike spin waves excited by thermal agitation, the spin waves excited by an RF field not only have a definite angular frequency $\omega_k = \omega/2$ but bear a definite phase relationship with respect to H_z. We should emphasize that this type of absorption does not need the excitation of the uniform mode and is not expected from the classical theory of resonance.

Excitation of spin waves by an RF field can also occur when the RF field is perpendicular to the d-c field. In evaluating the dipolar field in Eq. (10.145), we have neglected the contribution from the $k_z a_{kz}$ term because it is a second-order term. Under the simultaneous presence of the uniform mode a_0 and the spin wave mode a_k, the z component of magnetization becomes

$$M_z = M_s[1 - (a_{0u} + a_{ku})^2 - (a_{0v} + a_{kv})^2]^{1/2}$$
$$\simeq M_s(1 - a_{0u}a_{ku} - a_{0v}a_{kv}) \tag{10.157}$$

to the first order in a_k. Consequently, additional terms represented by $\mathbf{H}'_{kd}$ will appear in Eq. (10.146) and they are

$$\mathbf{H}'_{kd} = (\mathbf{u} \sin \theta_k + \mathbf{z} \cos \theta_k)(M_s \cos \theta_k) a_{k0}$$
$$- (\mathbf{u} \sin \theta_k + \mathbf{z} \cos \theta_k)(M_s \sin \theta_k) a_{0u} \tag{10.158}$$

where $a_{k0} = a_{0u}a_{ku} + a_{0v}a_{kv}$. Because of the presence of the uniform mode, the term $M_s(\mathbf{k} \cdot \mathbf{a}_k)$ in Eq. (10.145) should be replaced by $M_s(1 - a_{k0})[\mathbf{k} \cdot (\mathbf{z} + \mathbf{a}_k)]$, where the vector $\mathbf{z}$ actually represents $\mathbf{M}_z/M_s$ from the uniform precession. As pointed out earlier, the dipolar field due to the uniform precession has been accounted for by the demagnetizing field, thus we need not consider the $M_s\mathbf{k} \cdot \mathbf{z}$ term. However, the two first-order cross terms have not been taken into consideration. The first term in Eq. (10.158) represents the contribution from the cross-product $M_s a_{k0}(\mathbf{k} \cdot \mathbf{z})$, whereas the second term represents that from $M_s(\mathbf{k} \cdot \mathbf{a}_k)$.

Likewise, the magnetization vector can be written as

$$\mathbf{M} = M_s[\mathbf{u}(a_{0u} + a_{ku}) + \mathbf{v}(a_{0v} + a_{kv}) + \mathbf{z}(1 - a_{k0})] \tag{10.159}$$

After substituting Eqs. (10.159), (10.158), and (10.146) into the equation of motion [Eq. (10.87)], we find that the following additional terms to the first order in $a_k a_0$,

$$\frac{d}{dt} (\mathbf{u} a_{ku} + \mathbf{v} a_{kv}) = \omega_k \sin \theta_k \cos \theta_k[\mathbf{u}(a_{0u}a_{kv})$$
$$- \mathbf{v}(a_{k0} + a_{0u}a_{ku})] \tag{10.160}$$

should be added to Eq. (10.147). These terms which represent non-linear coupling terms in the equation of motion are responsible for the excitation of spin waves. Other first-order terms in $a_k a_0$, which do not contribute to a nonzero time and space average, are discarded in Eq. (10.160).

In a right-hand circularly polarized field, the uniform mode will also be circularly polarized in the xy plane with

$$\mathbf{a}_0 = a_0[\mathbf{u} \cos (\omega t + \alpha) + \mathbf{v} \sin (\omega t + \alpha)] \qquad (10.161)$$

Substituting Eqs. (10.160) and (10.161) into Eq. (10.153) and using Eq. (10.149) for a_k, we find the condition for nonzero P_{abs} to be: $\omega = 2\omega_k$ and $\alpha = \pi/2$. Experimentally, at sufficiently high microwave power levels, a new absorption appears at a lower d-c magnetic field $H_0 < \omega/(\gamma \mu_0)$ besides the normal absorption peak at $H_0 = \omega/(\gamma \mu_0)$ due to the uniform mode. This is generally referred to as the first-order instability because the excitation of spin waves is caused by the coupling term $a_0 a_k$ which is first order in a_0, the amplitude of the uniform mode. The premature saturation of χ'' (Fig. 10.36b) actually is the most complex case among the three cases of spin wave excitations because it is caused by nonlinear coupling terms of the form $a_0{}^2 a_k$. If we use the term of this form in Eq. (10.153) for da_{ku}/dt and da_{kv}/dt, we find the condition for nonzero absorption by the spin wave to be $\omega_k = 2\omega - \omega_k$ or $\omega = \omega_k$. Therefore, the excited spin wave has the same angular frequency as the uniform mode.

In summary, we have discussed three principal types of spin wave excitation by an RF field. The first case is by parallel pumping in which the a-c field is parallel to the d-c field. In this case, the spin wave is coupled directly to the RF field and hence no uniform mode is needed. When the RF field is perpendicular to the d-c field, the spin wave modes are coupled to the RF field through the uniform mode. Depending on whether the coupling terms depend on $a_0 a_k$ or $a_0{}^2 a_k$, we have either the first-order or the second-order instability. The d-c field at which these three types of instability occur is determined by the condition that a constant flow of energy goes into the spin wave system. For parallel pumping and the first-order instability,

Figure 10.38

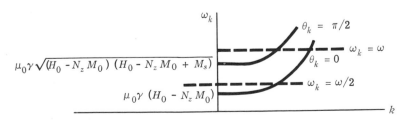

Schematic representation of the spin wave dispersion relation (ω_k versus k curve) in an ellipsoidal sample.

the d-c field must be such that $\omega_k = \omega/2$. For the second-order instability, $\omega_k = \omega$. The value of ω_k is plotted in Fig. 10.38 as a function of k for two values of $\theta_k = 0$ and $\pi/2$. For a fixed a-c frequency ω, different portions of the spin wave spectrum will intersect the ω line and the $\omega/2$ line. This also means that as we change the value of the d-c field, the wave number k of the excited spin wave modes varies with it. For a further discussion of spin wave instabilities, the reader is referred to a series of articles by Suhl[46] and by Schlömann.[47]

10.12 EXCITATION OF LATTICE WAVES BY MAGNETOSTRICTIVE COUPLING

In Sec. 10.3, we have analyzed the static magnetoelastic effect which deals with the change in the dimension of a ferromagnet as a result of magnetization changes. In the following, we shall discuss the dynamic magnetoelastic process in which the precessional motion of the magnetization is coupled to the elastic waves. If we recall our discussion in Sec. 10.3, the way to treat the effect of magnetic anisotropy is to propose an effective anisotropy field acting on the magnetization vector. This method can be extended to magnetoelastic effects. The effective field due to magnetostriction is given by

$$\mathbf{H}_{\text{eff}} = -\frac{1}{\mu_0}\left(\mathbf{x}\,\frac{\partial \mathcal{E}}{\partial M_x} + \mathbf{y}\,\frac{\partial \mathcal{E}}{\partial M_y} + \mathbf{z}\,\frac{\partial \mathcal{E}}{\partial M_z}\right) = -\frac{1}{\mu_0 M_s}\,\text{grad}_\alpha\,\mathcal{E} \quad (10.162)$$

where $\mathcal{E}$ is the magnetoelastic energy, $\mathbf{x}$, $\mathbf{y}$, and $\mathbf{z}$ are unit vectors, and α represents the direction cosine of the magnetization vector. Substituting Eq. (10.24) into Eq. (10.162), we find

$$H^z_{\text{eff}} = -\frac{1}{\mu_0 M_s}\left[2B_1\left(\frac{\partial \zeta}{\partial z}\right)\alpha_3 + \left(\frac{B_2}{2}\right)\left(\frac{\partial \eta}{\partial z} + \frac{\partial \zeta}{\partial y}\right)\alpha_2 \right.$$
$$\left. + \left(\frac{B_2}{2}\right)\left(\frac{\partial \zeta}{\partial x} + \frac{\partial \xi}{\partial z}\right)\alpha_1\right] \quad (10.163)$$

and similar expressions for H^x_{eff} and H^y_{eff}, where ξ, η, and ζ are the components of the displacement vector defined in Eqs. (7.76), (7.77), and (7.78). The reader can easily show that due to magnetostriction the following terms

$$-\gamma\mu_0(H^z_{\text{eff}}M_y - H^y_{\text{eff}}M_z) = \frac{\gamma}{M_s}\left[2B_1\left(\frac{\partial \zeta}{\partial z} - \frac{\partial \eta}{\partial y}\right)M_y \right.$$
$$\left. - \frac{B_2}{2}\left(\frac{\partial \eta}{\partial z} + \frac{\partial \zeta}{\partial y}\right)M_z - \frac{B_2}{2}\left(\frac{\partial \xi}{\partial y} + \frac{\partial \eta}{\partial x}\right)M_x\right] \quad (10.164)$$

[46] H. Suhl, *Proc. IRE*, vol. 44, p. 1270, 1956; *J. Phys. Chem. Solids*, vol. 1, p. 209, 1957; also Clogston, Suhl, Walker, and Anderson, *J. Phys. Chem. Solids*, vol. 1, p. 129, 1956.

[47] E. Schlömann, *Phys. Rev.*, vol. 116, p. 828, 1959; *Jour. Appl. Phys.*, vol. 31, p. 386S, 1960 and vol. 32, p. 165S, 1961.

should be added to the x component of $-\gamma\mu_0\mathbf{M} \times \mathbf{H}$. Similar expressions are obtained for the y and z components of $-\gamma\mu_0\mathbf{M} \times \mathbf{H}$. In obtaining Eq. (10.164), we realize that $\alpha_3 \simeq 1$, $\alpha_1 M_z \simeq \alpha_1 M_s = M_x$, and $\alpha_2 M_z \simeq \alpha_2 M_s = M_y$, and also we neglect terms involving the product $M_x M_y$.

Now let us turn our attention to lattice vibration by referring to our discussion in Sec. 7.8. Substituting Eq. (7.75) into Eq. (7.88) gives

$$\rho\frac{d^2\xi}{dt^2} = \frac{\partial T_1}{\partial x} + \frac{\partial T_6}{\partial y} + \frac{\partial T_5}{\partial z} \tag{10.165}$$

where ρ is the density of the medium, $T_1(= T_{xx})$ is the tensile stress, and $T_6(= T_{xy} = T_{yx})$ and $T_5(= T_{xz} = T_{zx})$ are the shear stresses. Similar expressions are obtained for η and ζ. In general, the stresses are related to the strains through Eq. (7.79), and the problem of wave propagation in such a medium is rather complicated. For this reason, we shall limit our discussion to isotropic cubic crystals. In crystals having cubic symmetry, there are only three independent nonvanishing elastic moduli c_{11}, c_{12}, and c_{44}, where $c_{11} = c_{ii}$, $c_{12} = c_{ij}$ with i, $j = 1, 2, 3$, and $c_{44} = c_{ii}$ with $i = 4, 5, 6$. It can be shown from symmetry considerations that the other moduli are zero. In an isotropic cubic crystal, a relation $c_{11} - c_{12} = 2c_{44}$ exists so that the velocity of wave propagation is independent of direction of propagation. Under these conditions,

$$T_1 = \frac{\partial \mathcal{E}}{\partial e_{xx}} = c_{11}e_{xx} + c_{12}(e_{yy} + e_{zz}) + B_1\alpha_1^2$$

$$T_6 = \frac{\partial \mathcal{E}}{\partial e_{xy}} = c_{44}e_{xy} + B_2\alpha_1\alpha_2 \tag{10.166}$$

$$T_5 = \frac{\partial \mathcal{E}}{\partial e_{zx}} = c_{44}e_{zx} + B_2\alpha_1\alpha_3$$

where $\mathcal{E}$ is the sum of elastic and magnetoelastic energies given, respectively, by Eqs. (10.25) and (10.24).

Substituting Eq. (10.166) into Eq. (10.165) and realizing that $c_{11} - c_{12} = 2c_{44}$, $\alpha_1^2 = M_x^2/M_s^2$, and so on, we find

$$\rho\frac{d^2\xi}{dt^2} = c_{44}\nabla^2\xi + (c_{44} + c_{12})\frac{\partial}{\partial x}\left(\frac{\partial\xi}{\partial x} + \frac{\partial\eta}{\partial y} + \frac{\partial\zeta}{\partial z}\right)$$

$$+ \frac{B_1}{M_s^2}\frac{\partial M_x^2}{\partial x} + \frac{B_2}{2M_s^2}\left(\frac{\partial M_x M_y}{\partial y} + \frac{\partial M_x M_z}{\partial z}\right) \tag{10.167}$$

and similar expressions for η and ζ. The equation of motion for the magnetization vector, in view of Eq. (10.164), can be written as

$$\frac{dM_x}{dt} = -\gamma\mu_0\left[H_z + \frac{2B_1}{\mu_0 M_s}\left(\frac{\partial\eta}{\partial y} - \frac{\partial\zeta}{\partial z}\right)\right]M_y + \gamma\mu_0 M_s H_y$$

$$- \frac{\gamma B_2}{2M_s}\left(\frac{\partial\eta}{\partial z} + \frac{\partial\zeta}{\partial y}\right)M_z - \frac{\gamma B_2}{2M_s}\left(\frac{\partial\xi}{\partial y} + \frac{\partial\eta}{\partial x}\right)M_x \tag{10.168}$$

where for simplicity H_z and H_y are the internal fields which include both the applied and demagnetizing fields. A similar expression is also obtained for dM_y/dt. It is obvious that Eqs. (10.167) and (10.168) are too involved for a general discussion. Two cases of special interest will be considered in the following discussion: the generation of microwave elastic waves in a disk by precessional motion of magnetization and the parametric excitation of magnetostatic and elastic waves.

In the experiment of Bömmel and Dransfeld,[48] a nickel film 18,000 Å thick was deposited on a single-crystal quartz rod which was an oriented AC-cut[49] crystal to allow the transmission of transverse sound waves. Both end surfaces of quartz were optically polished and parallel, and the quartz served as a link between two microwave cavities (Fig. 10.39a). When the d-c magnetic field was near the value for ferromagnetic resonance in cavity 1, a microwave power output was detected from cavity 2 (Fig. 10.39c). Acoustic waves (lattice waves) were generated by the precessional motion of magnetization in Ni and transmitted to cavity 2 through the quartz rod. The ultrasonic lattice waves were converted back into microwave power in cavity 2 by the piezoelectric effect in quartz. The same effect was also observed in a disk of yttrium-iron-garnet bonded to an AC-cut quartz rod. Since the transmission of sound waves is a separate problem, we shall only discuss the generation of acoustic waves in a ferromagnetic or ferrimagnetic medium.

For transverse acoustic waves propagating in the z direction (Fig. 10.39b), $\zeta = 0$ and ξ and η vary with z only. Thus, Eqs. (10.167) and (10.168) become

$$i\omega m^+ = i\gamma\mu_0 \left[m^+ H + \left(\frac{B_2}{2\mu_0} \right) \left(\frac{du^+}{dz} \right) - M_s h^+ \right] \quad \text{(10.169)}$$

$$-\omega^2 \rho u^+ = c_{44} \left(\frac{d^2 u^+}{dz^2} \right) + \left(\frac{B_2}{2M_s} \right) \left(\frac{dm^+}{dz} \right) \quad \text{(10.170)}$$

by assuming an exp $(i\omega t)$ time dependence. In deriving the above equations we use the notations $u^+ = \xi + i\eta$, $H = H_z$, $h^+ = H_x + iH_y$, and $m^+ = M_x + iM_y$. Furthermore, we assume that $M_z \gg M_x$ and M_y. The quantities m^+ and u^+ are circularly polarized under a circularly polarized a-c field h^+. The solution of Eq. (10.169) can be obtained by decomposing m^+ into two components: m_0^+, the spatial nonvarying part due to the uniform mode, and m_k^+, the spatial varying part due to spin waves. From Eq. (10.169), we obtain

$$m_0^+ = \frac{\gamma\mu_0 M_s}{\omega_H - \omega} h^+ \quad \text{and} \quad m_k^+ = \frac{\gamma B_2}{2(\omega - \omega_k)} \frac{du^+}{dz} \quad \text{(10.171)}$$

[48] H. E. Bömmel and K. Dransfeld, *Phys. Rev. Letters*, vol. 3, p. 83, 1959.
[49] For information on the useful orientations of quartz crystals, the reader is referred to W. P. Mason, "Piezoelectric Crystals and Their Applications to Ultrasonics," pp. 95–102, D. Van Nostrand Co., Inc., New York, 1950.

Note that the field H is the effective field and hence it is different for the uniform mode and the spin waves. According to Eq. (10.90) with $N_x = N_y = 0$ and $N_z = 1$, $\omega_H = \gamma\mu_0(H_0 - M_s)$, H_0 being the applied field. For the spin waves, ω_k is given by Eq. (10.150) with $\omega_k{}^2 = A_k{}^2 - B_k{}^2$.

The effect of the magnetization vector, $m_0^+ + m_k^+$, upon the elastic waves is twofold. From Eq. (10.171), we see that the spin waves m_k^+ constantly interact with the lattice waves du^+/dz. This interaction which is continuous throughout the medium modulates the propagation property of the lattice wave. Substituting Eq. (10.171) into Eq. (10.170), and proposing that u^+ varies as $\exp{(ikz)}$, we find

$$k^2 = \frac{\omega^2\rho}{c'_{44}} \quad \text{with } c'_{44} = c_{44} + \frac{\gamma B_2{}^2}{4M_s(\omega - \omega_k)} \qquad (10.172)$$

For the other circularly polarized components with u^- for the lattice wave and m_k^- for the spin wave, we can replace ω by $-\omega$ in the above

Figure 10.39

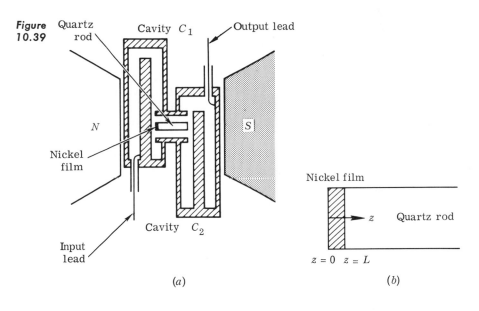

(a)

(b)

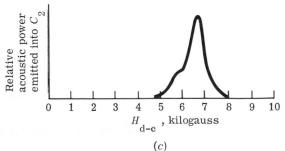

(c)

The generation of microwave elastic waves in a nickel film by the precessional motion of magnetization: (a) the experimental setup, (b) the nickel film and the quartz rod, and (c) the experimental result. (H. E. Bömmel and K. Dransfeld, Phys. Rev. Letters, vol. 3, p. 83, 1959.)

equation. As originally pointed out by Kittel,[50] the dispersion relation is now different for two circularly polarized waves with opposite senses of rotation. This can lead to nonreciprocal elastic wave propagation in the medium.

The effect which is exhibited in Fig. 10.39 does not involve the spin waves but only the uniform mode. As a matter of fact, it is desirable to operate the device away from ω_k so as to minimize power dissipation in exciting spin waves. The phenomenon of phonon-wave generation can be explained by examining Eq. (10.166) at the boundary of the specimen. At the end $z = 0$ where the specimen is free, $T_4 = T_5 = T_6 = 0$ or

$$c_{44} \frac{du^+}{dz} + \left(\frac{B_2}{2M_s}\right) m_0^+ = 0 \qquad \text{at } z = 0 \qquad (10.173)$$

At the end where the specimen is attached to the quartz, continuity of surface traction and displacement demands

$$c_{44} \frac{du^+}{dz} + \left(\frac{B_2}{2M_s}\right) m_0^+ = c_1 \frac{du_1^+}{dz}$$

$$u^+ = u_1^+ \qquad \text{at } z = L \qquad (10.174)$$

where c_1 and u_1^+ refer to the corresponding quantities in quartz. In Eqs. (10.173) and (10.174), we assume that ω is away from ω_k so the effect of spin wave is neglected. Therefore, the problem reduces to solving Eq. (10.170) subjected to the boundary conditions of Eqs. (10.173) and (10.174).

Equations (10.173) and (10.174) resemble the boundary conditions at the ends of a transmission line. If we let

$$V = -c_{44} \frac{du^+}{dz} \qquad \text{and} \qquad I = \frac{du^+}{dt} \qquad (10.175)$$

it follows that

$$-\frac{1}{c_{44}} \frac{dV}{dt} = \frac{dI}{dz} \qquad \text{and} \qquad \frac{dV}{dz} = -\rho \frac{dI}{dt} \qquad (10.176)$$

which are the transmission-line equations. Note that ρ and c_{44} are associated respectively with the kinetic and potential energies of the lattice waves so it is natural that they appear to have the place of inductance and capacitance in Eq. (10.176). If we represent $V_m = (B_2/2M_s)m_0^+$ as the generator voltage, then the transmission line analogy is complete with

$$
\begin{array}{ll}
V = V_m & \text{at } z = 0 \\
V = V_m + V_1 \quad \text{and} \quad I = I_1 & \text{at } z = L
\end{array} \qquad (10.177)
$$

A circuit representation of Eqs. (10.176) and (10.177) is shown in

[50] C. Kittel, *Phys. Rev.*, vol. 110, p. 836, 1958.

Fig. 10.40, which should look familiar to electrical engineers. For a nonzero m_0^+, an elastic wave will be set up in the magnetostrictive medium and transmitted through quartz to the load which is represented by Z. The value of Z will depend on how well the quartz rod is coupled piezoelectrically to the second waveguide. For further analysis of the problem, the reader is referred to the article by Comstock and LeCraw.[51]

Now we shall discuss a different phenomenon reported by Spencer and LeCraw.[52] In their experiment, a polished sphere of single-crystal YIG was placed loose on the bottom of a quartz tube and located at a position of maximum RF magnetic field in a microwave cavity. When the microwave power exceeded a certain critical value, the amplitude of the power absorbed by the sample was modulated and a low-frequency oscillation (8.9 Mc/sec according to Fig. 10.41) was observed. As illustrated in Fig. 10.41, the relevant frequencies obey the following relation

$$\nu_p = \nu_1 + \nu_a \tag{10.178}$$

where ν_p is the microwave (uniform mode) frequency, ν_a is the acoustic resonant frequency of the sample, and ν_1 is the resonant frequency of the magnetostatic mode. Using a g value of 2.03, the resonant field of the uniform mode is at $H = 3280$ oe for a frequency of 9320 Mc. With the excitation of acoustic waves with $\nu_a = 8.9$ Mc, the resonant field of the magnetostatic mode is at 3276.8 oe, giving a resonant frequency ν_1 of 9311 Mc so that $\nu_1 = \nu_p - \nu_a$. Note that Eq. (10.178) is the typical frequency relation in a parametric amplifier.

Referring again to Eqs. (10.167) and (10.168), we assume that two circularly polarized components of magnetization m_1^+ and m_p^+ exist at the idle and pump frequencies and that one and only one

[51] R. L. Comstock and R. C. LeCraw, *Jour. Appl. Phys.*, vol. 34, p. 3022, 1963.
[52] E. G. Spencer and R. C. LeCraw, *Phys. Rev. Letters*, vol. 1, p. 241, 1958; *Jour. Appl. Phys.*, vol. 30, p. 149S, 1959.

Figure 10.40

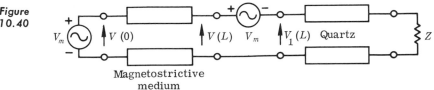

The circuit representation of Eqs. (10.176) and (10.177) used to analyze the phenomenon of generation of ultrasonic elastic waves by the precessional motion of magnetization. (R. L. Comstock and R. C. LeCraw, Jour. Appl. Phys., vol. 34, p. 3022, 1963.)

longitudinal lattice-vibrational wave propagates in the z direction ($\xi = \eta = 0$ and $\zeta(z) \neq 0$). Thus, with $\omega_M = \gamma\mu_0 M_s$ and $\omega_0 = \gamma\mu_0 H_z$, Eq. (10.168) becomes

$$\frac{dm^+}{dt} = i\omega_1 m_1^+ + i\omega_p m_p^+$$

$$= i\omega_0(m_1^+ + m_p^+) - i\omega_M h^+ - iv\frac{\partial\zeta^*}{\partial z}(m_1^+ + m_p^+) \quad (10.179)$$

Figure 10.41

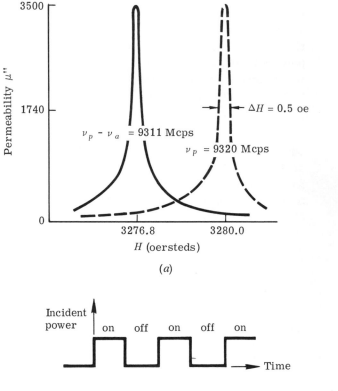

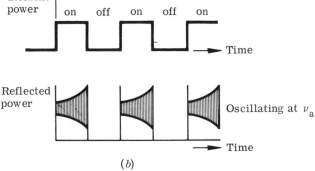

Experimental results which demonstrate the parametric excitation of elastic waves in a ferrimagnetic resonance experiment. (a) The observation of μ'' at a field below resonance. (E. G. Spencer and R. C. LeCraw, Phys. Rev. Letters, vol. 1, p. 241, 1958.) (b) The modulation of the reflected microwave power at a frequency ν_a. (R. L. Comstock, Jour. Appl. Phys., vol. 34, pp. 1461 and 1465, 1963.)

where $m^+ = M_x + iM_y$, $h^+ = h_1^+ + h_p^+ = H_x + iH_y$, and $v = 2B_1/\mu_0 M_s$. As in our discussion of parametric amplifiers in Sec. 6.6, the pump power is assumed to be sufficiently large so that m_p^+ is much larger than m_1^+ and independent of m_1^+ and ζ^*. The superscript $+$ refers to the right-hand circularly polarized wave which varies as $\exp(i\omega t)$ with time. For the lattice wave, since it is longitudinal, ζ^* simply means the complex conjugate of ζ in complex notation and it varies as $\exp(-i\omega_a t)$ with time. In view of the remarks just made about m_p^+, the quantity of interest is m_1^+. Equating the terms which vary as $\exp(i\omega_1 t)$ with time, we have

$$m_1^+ = \chi_1^+ h_1^+ + \tau m_p^+ \left(\frac{\partial \zeta^*}{\partial z} \right) \tag{10.180}$$

where h_1^+ is the microwave magnetic field induced by m_1^+ and

$$\chi_1^+ = \frac{\omega_M}{\omega_0 - \omega_1} \quad \text{and} \quad \tau = \frac{2B_1}{\mu_0 M_s(\omega_0 - \omega_1)} \tag{10.180a}$$

In the presence of a nonvanishing magnetization, the electromagnetic wave equation becomes

$$\frac{\partial^2 h_1^+}{\partial z^2} = -\omega_1^2 \mu_0 \epsilon_0 (h_1^+ + m_1^+)$$

$$= -\omega_1^2 \mu_0 \epsilon_0 \left[(1 + \chi_1^+) h_1^+ + \tau m_p^+ \left(\frac{\partial \zeta^*}{\partial z} \right) \right] \tag{10.181}$$

In the meantime, with $\xi = \eta = 0$, the lattice wave equation can be simplified into

$$-\omega_a^2 \rho \zeta = (2c_{44} + c_{12}) \left(\frac{\partial^2 \zeta}{\partial z^2} \right) + \left(\frac{B_1}{M_s^2} \right) \left(\frac{\partial M_z^2}{\partial z} \right) \tag{10.182}$$

by referring to and making appropriate changes in Eq. (10.167). Realizing that $c_{11} = c_{12} + 2c_{44}$ in an isotropic cubic medium and that $M_z^2 = M_s^2 - (M_x^2 + M_y^2) = M_s^2 - (m^+ m^-)$, Eq. (10.182) can be rewritten as

$$\frac{\partial^2 \zeta}{\partial z^2} = -\left(\frac{\omega_a^2 \rho}{c_{11}} \right) \zeta + \left(\frac{B_1}{c_{11} M_s^2} \right) \left[\frac{\partial (m_p^+ m_1^-)}{\partial z} \right] \tag{10.183}$$

Note that Eqs. (10.181) and (10.183) are quite similar to the two coupled equations discussed in Sec. 6.6 in conjunction with semiconductor parametric amplifiers. Hence, a general solution may be found, following the procedure outlined in Sec. 6.6. For concreteness, however, we shall discuss the simple case of a ferrimagnetic disk to show that excitation of lattice waves is indeed possible.

Take m_p^+ to be the uniform mode. Thus, to satisfy the phase relation throughout the medium, m_1^+ and ζ must vary spatially with the same wave number k or

$$m_1^+ = a_1 \exp(i\omega_1 t - ikz) \quad \text{and} \quad \zeta = a_2 \exp(i\omega_a t + ikz) \tag{10.184}$$

Therefore, m_1^+ must not be the uniform mode but belong to the magnetostatic modes. Furthermore, for acoustic resonance, $k = n\pi/L$ where L is the thickness of the ferrimagnetic disk and n is an integer. The applied d-c magnetic field should be so adjusted to give a value of χ_1^+ which results in $k^2 = \omega_1^2 \mu_0 \epsilon_0 (1 + \chi_1^+)$. If these conditions are satisfied, substituting Eq. (10.184) into Eqs. (10.181) and (10.183) gives

$$
\begin{aligned}
\left(\frac{\partial}{\partial z} + \alpha_1\right) a_1 &= \left(\frac{\omega_1^2 \mu_0 \epsilon_0 \tau \chi_1^+}{2}\right) a_p^+ a_2^* \\
\left(\frac{\partial}{\partial z} + \alpha_2\right) a_2 &= \left(\frac{B_1}{2c_{11} M_s^2}\right) a_p^+ a_1^*
\end{aligned}
\tag{10.185}
$$

where α_1 and α_2 are the two loss constants introduced to account for the loss mechanisms associated with the magnetostatic and lattice waves. Similar equations are obtained for the complex conjugate quantities. As the magnetic and elastic waves traverse across the medium, a continuous amplification of the waves occurs when

$$
|a_p|^2 \omega_1^2 \mu_0 \epsilon_0 \tau \chi_1^+ \frac{B_1}{4c_{11} M_s^2} > \alpha_1 \alpha_2
\tag{10.186}
$$

In obtaining Eq. (10.185), we have assumed that the magnetoelastic coupling is weak so that the amplitude of the field h_1^+ can be approximated by a_1/χ_1^+ from Eq. (10.180).

The growth of elastic waves and the observation of μ'' in a ferrimagnet at a field below resonance (Fig. 10.41) are a direct consequence of parametric excitation of elastic and magnetostatic modes by the pumping field at ν_p. The important point which we should like to stress with the present example is that parametric amplification is also possible in a system involving the interaction of two waves of different nature if suitable coupling terms are present. Therefore, the procedure developed in Sec. 6.6 for the analysis of traveling-wave parametric amplifiers can be adopted for general use in dealing with interactions of two different waves. Further examples will be discussed in Chapter 12 for the interaction of light waves with lattice waves in a nonlinear dielectric.

PROBLEMS

10.1 Consider the lattice point i. Expand the spin vector S_j on the nearest neighbors in terms of S_i in a Taylor's series expansion for simple, body-centered, and face-centered cubic structures. Show that the change in the exchange energy per unit cell is given by Eq. (10.11) with $Z = 1, 2,$ and 4, respectively, for the simple, body-centered, and face-centered structures. Note that Z is not the number of nearest neighbors.

10.2 From Fig. 10.6, find the angle θ (defined in Fig. 10.7b) in iron at $H = 200$ oe. From θ, evaluate the value of H_a. Show that the magnetization vector $\mathbf{M}$ lies in the direction of the resultant field of $\mathbf{H}$ and $\mathbf{H}_a$.

10.3 Generalizing the result in the preceding problem, show that the magnetization M_H along the [110] direction follows the equation

$$\frac{M_H}{M_0}\left[\left(\frac{M_H}{M_0}\right)^2 - \frac{1}{2}\right] = \frac{\mu_0 H M_0}{4K_1}$$

10.4 Verify the condition stated in Eq. (10.26) for the minimization of $\mathcal{E}_a + \mathcal{E}_{me} + \mathcal{E}_{el}$. Derive Eq. (10.28) from Eqs. (10.26) and (10.27).

10.5 Consider a magnetic particle having the shape of a prolate ellipsoid. Because of the demagnetizing effect (similar to the depolarizing effect discussed in Sec. 7.5), the magnetization vector lies along the direction of the polar axis to achieve minimum energy. Show that the demagnetizing energy $\mathcal{E}_d$ for other directions can be written as

$$4\mathcal{E}_d = -N_d\mu_0 M^2 \cos 2\alpha$$

where N_d is the difference between the demagnetizing factors along the polar and equatorial axes, and α is the angle between the magnetization vector and the polar axis. Show that if an external field H is applied, making an angle θ with the polar axis, the direction of the magnetization vector obeys the following equation

$$N_d M \sin 2\alpha = 2H \sin (\theta - \alpha)$$

10.6 Using the result of the preceding problem, plot the normalized M versus H curve for $\theta = 30°$ and $45°$.

10.7 Prove the condition for switching stated in Eq. (10.56). Also verify that the direction of the magnetization vector at switching is given by Eq. (10.57). In verifying these relations, it is advisable that the reader should first solve for $\sin 2\phi_0$ and $\cos 2\phi_0$ from Eqs. (10.49) and (10.55).

10.8 Following the discussion in the article by C. Kittel (*Phys. Rev.*, vol. 70, p. 965, 1946, or *Solid State Phys.*, vol. 3, p. 504, 1956), find the expression of the critical size of a spherical particle for which a single domain configuration is favored from the energy viewpoint.

10.9 Derive Eqs. (10.85) and (10.86) from Eq. (10.84).

10.10 Show that the effective torque arising from anisotropy energy $\mathcal{E}_a$ is given by

$$T_x = \sin \phi \frac{\partial \mathcal{E}_a}{\partial \theta} + \cot \theta \cos \phi \frac{\partial \mathcal{E}_a}{\partial \phi}$$

$$T_y = -\cos \phi \frac{\partial \mathcal{E}_a}{\partial \theta} + \cot \theta \sin \phi \frac{\partial \mathcal{E}_a}{\partial \phi}$$

$$T_z = -\frac{\partial \mathcal{E}_a}{\partial \phi}$$

where θ and ϕ are the polar angles of the magnetization vector.

10.11 Read the articles by L. R. Bickford (*Phys. Rev.*, vol. 78, p. 449, 1950) and by J. F. Dillon, S. Geschwind, and V. Jaccarino (*Phys. Rev.*, vol. 100, p. 750, 1955). Show that the effective anisotropy fields in a cubic crystal for H lying in a (110) plane and making an angle θ with the z axis are given by

$$H_{z \text{ anis}} = N_z{}^a M_z = \frac{K_1}{M_0} (2 - 4 \sin^2 \theta - \tfrac{3}{4} \sin^2 2\theta)$$

$$H_{y \text{ anis}} = N_y{}^a M_y = \frac{K_1}{M_0} (2 - \sin^2 \theta - 3 \sin^2 2\theta)$$

10.12 Following the discussion in the article by F. Keffer and C. Kittel (*Phys. Rev.*, vol. 85, p. 329, 1952), find the condition for antiferromagnetic resonance with the applied d-c field perpendicular to the easy direction of magnetization.

10.13 Describe the experimental condition under which the antiferromagnetic resonances are observed in Cr_2O_3 by G. S. Heller, J. J. Stickler, and J. B. Thaxter (*J. Appl. Phys.*, vol. 32, p. 307S, 1961) and in MnO by M. Tinkham (*Phys. Rev.*, vol. 124, p. 311, 1961). What are the magnitudes of the effective anisotropy and molecular fields?

10.14 Read the articles by L. R. Walker (*Phys. Rev.*, vol. 105, p. 390, 1957, and *J. Appl. Phys.*, vol. 29, p. 318, 1958) and by R. L. White (*J. Appl. Phys.*, vol. 31, p. 86S, 1960). Discuss the excitation of nonuniform precessional modes in a magnetic resonance experiment and the information obtainable from the analysis of the experimental results.

10.15 The magnetic field of a plane, linearly polarized wave varies with time as $H_x = H_1 \cos \omega t$ and $H_y = 0$ at $z = 0$. Decompose the linearly polarized wave into two circularly polarized waves, and find the expressions for $H_x{}^L$, $H_y{}^L$, $H_x{}^R$, and $H_y{}^R$. Assuming that the two waves propagate with two different phase velocities given by Eq. (10.113), show that at a distance $z = d$, the magnetic field H is polarized in a direction making an angle

$$\theta = \omega \left[\sqrt{(\mu - \kappa)} - \sqrt{(\mu + \kappa)}\right] \frac{d}{2}$$

by finding first the direction along which the magnetic field is zero.

10.16 The magnetic induction field for a left-hand circularly polarized wave is given by

$$B_x = B_1 \cos (\omega t - kz) \qquad B_y = -B_1 \sin (\omega t - kz)$$

Using the definitions given in Eq. (10.111), show that Maxwell's equation for the left-hand wave can be written as

$$\frac{\partial E^L}{\partial z} = i\omega \mu^L H^L \qquad \frac{\partial H^L}{\partial z} = -i\omega \epsilon E^L$$

10.17 Prove Eq. (10.123). The reader may consult the review article by B. Lax (*Proc. IRE*, vol. 44, p. 1379, 1956) and the relevant references quoted in Lax's article. Also outline the procedure by which Eq. (10.125) is obtained.

10.18 Write down the equation of motion for spin waves in the presence of an applied static field H. Show that if H is applied perpendicular to the plane of a ferromagnetic film, the resonance condition is given by

$$\omega = \mu_0\gamma\left(H - M_0 + \frac{2A}{\mu_0 M_0}\,k^2\right)$$

and if H is applied in the plane of the film, the resonance condition is given by

$$\omega = \mu_0\gamma\left[\left(H + \frac{2A}{\mu_0 M_0}\,k^2\right)\left(H + M_0 + \frac{2A}{\mu_0 M_0}\,k^2\right)\right]^{\frac{1}{2}}$$

Discuss the assumptions invoked in obtaining the above equations. In view of our discussion concerning Eq. (10.144), comment on the validity of the two expressions.

10.19 Derive Eq. (10.150) from the equation of motion and also verify Eq. (10.148).

10.20 Read the article by H. Suhl (*Proc. IRE*, vol. 44, p. 1270, 1956, or *J. Phys. Chem. Solids*, vol. 1, p. 209, 1957). Outline the procedure by which the condition is derived for the threshold of instability of spin waves.

REFERENCES

Chikazumi, S.: "Physics of Magnetism," John Wiley & Sons, Inc., New York, 1965.

Clarricoats, J. B.: "Microwave Ferrites," Chapman and Hall, London, 1961.

Kittel, C.: "Introduction to Solid State Physics," John Wiley & Sons, Inc., New York, 1956.

Kittel, C., and J. K. Galt: Ferromagnetic Domain Theory, *Solid State Physics*, vol. 3, p. 437, 1956.

Lax, B., and K. J. Button: "Microwave Ferrites and Ferrimagnetics," McGraw-Hill Book Company, New York, 1962.

Morrish, A. H.: "The Physical Principles of Magnetism," John Wiley & Sons, Inc., New York, 1965.

Seitz, F.: "Modern Theory of Solids," McGraw-Hill Book Company, New York, 1940.

Soohoo, R. F.: "Magnetic Thin Films," Harper and Row, Publishers, Inc., New York, 1965.

Symposiums on Magnetism and Magnetic Materials, *Jour. of Appl. Phys.*, March, 1958; April, 1959; May, 1960; March, 1961; March, 1962; March, 1963; March, 1964; March, 1965.

Ferrites Issue, *Proc. IRE*, vol. 44, pp. 1229–1468, 1956.

11 OPTICAL ABSORPTION AND EMISSION PROCESSES

11.1 INTRODUCTION

The material in this chapter deals mainly with the interaction of electromagnetic radiation with matter in the visible or near-visible region of the spectrum. Many devices in common usage employ electric effects produced by impinging photons. For example, in television pickup tubes, electrons are liberated from a photocathode, thus transforming light signals into electrical signals. Besides photoemission, photoelectric effects of major device importance include the *photovoltaic* effect and the *photoconductive* effect. In the latter two effects, electron-hole pairs are generated either near a semiconductor *p-n* junction or inside a homogeneous semiconductor, thus establishing a voltage across the junction or changing the conductivity of the semiconductor. Practical devices in which these two effects are employed include solar cells and infrared detectors.

The converse to the photon absorption process is the photon emission process. During the process, a substance absorbs energy in some form or other and converts part of the absorbed energy into photon energy. The phenomenon is commonly called *luminescence*, and basically it involves two steps: the excitation of electrons to higher energy states and the subsequent release of energy by the excited electrons in the form of electromagnetic radiation. Many excitation schemes have been used in practice; examples are: electron bombardment in cathode-ray tubes, optical excitation in ruby lasers, and the application of an electric field in gallium-arsenide lasers. The names cathode-luminescence, photoluminescence, and electro-luminescence, respectively, are often used in association with the above excitation schemes. Furthermore, if the emission process occurs during excitation, the process is called *fluorescence*. On the other hand, the emission of light may occur after excitation has ceased; in that case, the process is called *phosphorescence* or afterglow.

To understand optical processes, it is necessary to know electronic energy levels in solids, especially the energy-band structure and the impurity levels. However, our degree of understanding does not follow the historical sequence of the development of practical devices. For example, luminescent solids known as *phosphors* (such as ZnS and CdS) have been applied for many years in cathode-ray and television tubes. These phosphors need certain traces of impurities known as *activators* to produce luminescence. Although technically these activated phosphors work well in practical applications, the details of the luminescent processes in phosphors are not understood thoroughly at present. For these materials, only a qualitative explanation of the principal features of their optical properties will be presented. On the other hand, covalent semiconductors such as Ge and Si and III-V compounds such as GaAs and InSb became materials of technical importance only recently. In spite of the fact that their technical development is comparatively late, a vast amount of knowledge about their band structure is made available through many thorough and brilliant investigations. Therefore, it seems logical that we start our discussion of optical processes in materials which are suited for qualitative descriptions. This qualitative discussion will be followed by extensive and quantitative analyses of optical processes in semiconductors.

11.2 PHOTOLUMINESCENCE IN THALLIUM-ACTIVATED ALKALI HALIDES

It may be instructive to review first the energy-band diagram of KCl shown in Fig. 11.1a. As discussed in Sec. 2.3, the valence band of an ionic crystal is assumed to be due to the negative ions, which are the Cl^- ions in the case of KCl. The transition of an electron from the valence to the conduction band corresponds to the migration of the electron from the Cl^- ion to the K^+ ion which thus becomes a neutral atom. Therefore, the conduction band in KCl is ascribed to energy levels of neutral K atoms. In Fig. 11.1a, the energy χ is the electron affinity of the solid and it can be calculated from the work required in removing an electron from the bottom of the conduction band of the crystal to infinity. The energy ψ is the work required to remove an electron from the top of the valence band to infinity. The reader is referred to the book by Mott and Gurney[1] for the determination of the energies χ and ψ. For easy reference, we have kept the same notation used in their book. In KCl, the values of χ and ψ are found respectively to be 0.1 ev and 9.49 ev, yielding a value of $\psi - \chi = \mathcal{E}_g = 9.39$ ev for the energy gap.

The thallium-activated alkali-halide phosphor is an alkali halide crystal doped with thallous halide in dilute concentration. The thallous ions (Tl$^+$) are distributed randomly over lattice sites normally occupied by alkali ions. For a thallous ion to be an active luminescent center, the concentration of thallous ions must be low so that each thallous ion is essentially isolated from other thallous

[1] N. F. Mott and R. W. Gurney, "Electronic Processes in Ionic Crystals," pp. 71 and 80, Clarendon Press, Oxford, 1948. For further discussion of the energy-band structure of KCl, read L. P. Howland, *Phys. Rev.*, vol. 109, p. 1927, 1958.

ions. The electron configuration of the ground state of a Tl⁺ ion is $6s^2$ and the ground state is $^1S_0$ according to the spectroscopic notation outlined in Sec. 8.6. The lowest excited states of Tl⁺ ion have the configuration $6s\,6p$, and the excited states, in spectroscopic notation, are $^3P_0$, $^3P_1$, $^3P_2$, and $^1P_1$ with the first three states having spin parallel and the last state spin antiparallel arrangement. The

Figure 11.1

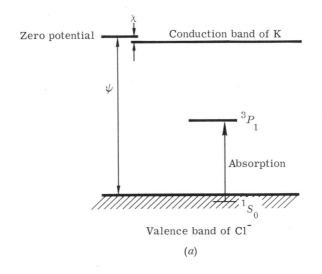

(a)

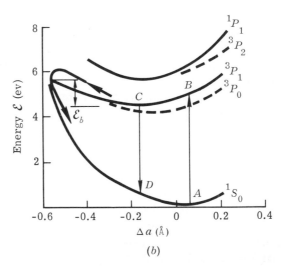

(b)

(a) *Energy-band diagram and* (b) *configuration model for the ground* ($^1S_0$), *the emitting* ($^3P_1$ *and* $^1P_1$), *and the trapping* ($^3P_2$ *and* $^3P_0$) *states in* KCl:Tl. *(P. D. Johnson and F. E. Williams, J. Chem. Phys., vol. 21, p. 125, 1953, for the* $^3P_2$ *and* $^3P_0$ *states; and Phys. Rev., vol. 113, p. 97, 1959, for the* $^3P_1$ *and* $^1P_1$ *states.)*

spectroscopic selection rules for transitions between states are: (1) for L, $\Delta L = 0$, ± 1; (2) for S, $\Delta S = 0$, and (3) for J, $\Delta J = 0$, ± 1 but $J = 0 \leftrightarrow J' = 0$ transition is forbidden. Consequently, the transitions from $^1S_0$ to $^3P_0$ and $^3P_2$ are forbidden. However, the selection rule $\Delta S = 0$ is not very effective for heavy atoms; hence both transitions from $^1S_0$ to $^3P_1$ and $^1P_1$ are allowed.

The energies of the various excited states of Tl^+ ions in KCl have been calculated by Williams and coworkers[2] and are shown in Fig. 11.1b. The abscissa is usually referred to as the *configuration coordinate* of the system, which describes the overall geometrical configuration of the nuclei surrounding the luminescent center. In the present case, only the interaction of the Tl^+ ion with the six Cl^- neighbors are of particular relevance to the state of the Tl^+ ion; therefore, the abscissa can simply be taken as the distance between the Tl^+ ion and its nearest Cl^- neighbors, using the corresponding distance in a perfect KCl crystal as reference. In the ground state, the thallous ion has a configuration coordinate close to point A, the point of lowest potential energy. Let us first consider the absorption process. The quantum-mechanical selection rule only tells us that the final state can be anywhere along the $^3P_1$ line. According to the *Franck-Condon principle*,[3] the most probable transition is the one which requires the least change in momentum and position. This transition is shown as the absorption process $A \rightarrow B$ in Fig. 11.1b.

Following excitation, the activated ion is no longer in equilibrium with the lattice because the point B does not represent the point of minimum energy on the $^3P_1$ curve. Since the lifetime of the excited state is sufficiently long for equilibrium to be established between the excited Tl^+ ion and its neighbors, the excited state first moves to point C before it returns to the ground state. The emission process is shown as transition $C \rightarrow D$ in Fig. 11.1b. The energy differences $\mathcal{E}_B - \mathcal{E}_C$ and $\mathcal{E}_D - \mathcal{E}_A$ go to the lattice. According to the calculations by Johnson and Williams and by Kristoffel,[4] the lattice vibrational energy of the ground state is $h\nu_g = 0.016$ ev and that of the excited state is $h\nu_e = 0.010$ ev. Thus, the mean number of vibrational quanta emitted is given by $S = 0.65/0.016 = 40$ during the process BC and by $S = 0.60/0.010 = 60$ during the process DA. In Fig. 11.2 are

2   For a review of these calculations, refer to F. E. Williams, Solid State Luminescence, in *Advan. Electron. Electron Phys.*, vol. 5, p. 137, 1953. For later results, read P. D. Johnson and F. E. Williams, *Phys. Rev.*, vol. 113, p. 97, 1959, and vol. 117, p. 964, 1960.

3   See, for example, D. Curie, "Luminescence in Crystals," pp. 45–51, Methuen & Co., Ltd., London, 1963; N. F. Mott and R. W. Gurney, "Electronic Processes in Ionic Crystals," p. 160, Clarendon Press, Oxford, 1948; D. L. Dexter, Theory of the Optical Properties of Imperfections in Nonmetals, in *Solid State Phys.*, vol. 6, pp. 367 and 387, 1958.

4   P. D. Johnson and F. E. Williams, *Phys. Rev.*, vol. 117, p. 964, 1960; N. N. Kristoffel, *Optics and Spectr.*, vol. 7, p. 45, 1959.

shown the experimental absorption and emission spectra of Tl-activated KCl at 298°K and the theoretical curve calculated from the energy-level diagram of Fig. 11.1. The general agreement gives conclusive evidence that the $^1S_0 \rightleftharpoons {}^3P_1$ transitions are responsible for the 2460 Å absorption and 3050 Å emission bands. The spread in the absorption and emission bands is due to perturbation of the electronic energy levels caused by thermal vibration of the lattice. Therefore, the width of the band should decrease as temperature goes down, in agreement with experiment.

Note that the states $^3P_0$ and $^3P_2$ may act as electron traps. After electrons are excited to the $^3P_1$ and $^1P_1$ states, they may be transferred to the $^3P_0$ and $^3P_2$ states through interaction with lattice vibrational modes. Since these trapping states are metastable states by virtue of the selection rule, electrons once trapped can be stored in these states at low temperatures. Upon heating, electrons may return to the emitting states $^3P_1$ and $^1P_1$, and emission of light results. This

Figure 11.2

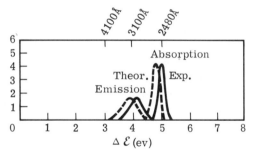

Theoretical and experimental excitation and emission spectra of KCl:Tl. (F. E. Williams, Advan. Electron. Electron Phys., vol. 5, p. 137, 1953.)

effect is known as *thermoluminescence*. We should further point out that electrons in the excited state $^3P_1$ may return to the ground state $^1S_0$ through a radiationless deexcitation process as shown in Fig. 11.1. To do this, electrons must be able to overcome a potential barrier $\mathcal{E}_b$. This radiationless process is always in direct competition with the emission process. The presence of certain impurities may make the aforementioned processes or a combination of these processes work against the emission process and hence drastically reduce the efficiency of luminescence. Such impurities are known as luminescence killers.

11.3 LUMINESCENT AND PHOTOCONDUCTIVE PROCESSES IN SULFIDE PHOSPHORS

The group II-VI compounds are an important class of materials for photoconduction and luminescence. Among the group, ZnS and CdS have been most extensively studied, the former for its luminescent properties and the latter for its photoconductive properties. As a general rule, the electrical properties of the group II-VI compounds are between those of the group I-VII and III-V compounds. For example, the energy gaps of ZnS and CdS are respectively 3.7 ev and 2.4 ev as compared to 9.4 ev for KCl and 1.4 ev for GaAs. So far as the nature of the binding is concerned, the best estimate seems to be that the ZnS lattice is about 75 per cent ionic, again intermediate between predominantly ionic KCl and predominantly covalent GaAs.

In common with most phosphors, the group II-VI compounds must be activated by suitable impurities known as activators to produce luminescence. To increase the solubility of the activator, it is necessary to introduce impurities of another kind known as *coactivator*. In Fig. 11.3, the most common activator and coactivator impurities are shown in relation to the ZnS family of phosphors in the periodic table. The role of coactivators can be explained on the basis of charge compensation as follows. To be specific, take ZnS:Cu as an example, the notation ZnS:Cu being adopted to mean Cu-activated ZnS. The monovalent Cu^+ ion is supposed to enter the ZnS crystal substitutionally and take a normal Zn lattice site. For charge compensation, a trivalent ion, say Ga^{3+} ion, must also replace a Zn^{2+} ion such that one Cu^+ and one Ga^{3+} ion are equivalent to two Zn^{2+} ions. By similar reasoning, a halide ion has to take the place of a normal S lattice site to pair with a Cu^+ ion.

Figure 11.3

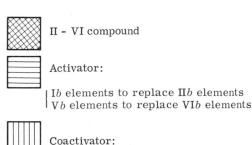

The relation of activator and coactivator impurities to the ZnS *family of phosphors with regard to their relative positions in the periodic table.* (*W. W. Piper and F. E. Williams, Solid State Physics, vol. 6, p. 96, 1958.*)

☒ II - VI compound

▤ Activator:

{ Ib elements to replace IIb elements
Vb elements to replace VIb elements

▥ Coactivator:

{ IIIb elements to replace IIb elements
VIIb elements to replace VIb elements

Although the number of publications concerning the photoconductive and luminescent properties of the group II-VI compound is
awesomely large, it is very difficult to present a definitive and unified
picture because an absolute theory simply does not exist. The following brief review of experimental observations may serve to illustrate
the complexities of the problem. Take the ZnS:Cu phosphor again.
The electroluminescent emission is usually either blue (peak at about
4550 Å) or green (peak at about 5200 Å) or a combination of both,
depending on the temperature and the amplitude and frequency of
the applied voltage. It is not at all certain whether these two emission
bands are due to two copper levels or due to the modification of the
energy levels of the S ions located nearest to the Cu ions because of
the difference in ionic charge associated with a Zn or Cu ion. The
situation is even further complicated by the presence of other emission bands under certain conditions. On the theoretical side, our
present lack of detailed knowledge about the nature of the chemical
bond and the energy band structure in the group II-VI compound
makes theoretical predictions very difficult and generally of little
use. Therefore, the following discussion is of necessity based on
phenomenological models. These models should be viewed with an
open mind.

Since II-VI compounds are situated between the ionic insulators
on the one side and the predominantly covalent semiconductors on
the other, it is natural and often tempting to adopt the existing and
established models for insulators and semiconductors. For covalent
semiconductors, the impurity atoms which appear in the periodic
table to the left of the original constituent element enter the lattice
substitutionally as acceptors; for example, Zn atoms act as acceptors,
replacing Ga atoms in GaAs crystals. By the same token, the
activator and coactivator elements become acceptors and donors
respectively in a II-VI compound. We may look in the other
direction and consider ZnS as an ionic crystal made of Zn^{2+} and
S^{2-} ions. Thus, Cu^{2+} ions replacing Zn^{2+} ions leave the crystal
electrically neutral, while Cu^+ ions replacing Zn^{2+} ions create a local
region negatively charged. Therefore, Cu^{2+} ions may be considered
as neutral or empty acceptor states while Cu^+ ions represent negatively charged or occupied acceptor states. The words empty and
occupied respectively refer to the state not occupied or occupied by
an electron.

It is also generally deduced from experimental observations that
donors have a fairly large electron-capture cross section but a very
small hole-capture cross section. In other words, empty donors are
ready to receive free electrons but occupied donors do not readily
capture free holes. On the other hand, empty acceptors are as capable
of capturing free electrons as occupied acceptors are capable of
capturing free holes. Consequently, coactivator donors behave

mainly as *electron traps* while activator acceptors act as recombination or *luminescent centers*. In the following, we shall discuss the luminescent properties of sulfide phosphors using Cu and Mn atoms as activators. The activators Cu and Mn are chosen not only because they are most extensively studied and best understood but because they represent two distinct types as well.

Luminescent centers. As mentioned earlier, in ZnS copper produces two main emission bands, a green band at about 5230 Å and a blue band at about 4450 Å. Many models of energy levels have been proposed to explain the two bands. We choose here the model proposed by Curie and Curie[5] for presentation. The energy levels which give the characteristic emission bands are shown in Fig. 11.4. As we can see, the copper impurity acts as a recombination center in both emission processes. Furthermore, the copper level has been identified as due to Cu^+ ion by magnetic susceptibility measurement. The Cu^{2+} ion is paramagnetic whereas the Cu^+ ion is diamagnetic. According to our discussion in Chapter 8, the paramagnetic susceptibility is temperature dependent while the diamagnetic susceptibility is not. Experiment by Bowers and Melamed[6] shows that the susceptibility of ZnS:Cu is independent of temperature.

In the model proposed by Prener and Williams,[7] the green emission is caused by the capture of an electron by a donor level and its subsequent recombination with a hole from the valence band through the Cu acceptor level. Since the green emission needs the presence of the donor, Williams and Prener's model is generally known as the model of associated donor-acceptor centers. The donors are believed

[5] G. Curie and D. Curie, *J. Phys. Radium*, vol. 21, p. 127, 1960.
[6] R. Bowers and N. T. Melamed, *Phys. Rev.*, vol. 99, p. 1781, 1955.
[7] J. S. Prener and F. E. Williams, *J. Phys. Radium*, vol. 17, p. 667, 1956.

Figure 11.4

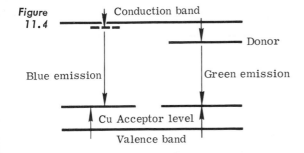

Energy levels giving the characteristic emission bands in Cu-activated ZnS. The green band is centered around 5230 Å (2.37 ev) in wurtzite structure and around 5350 Å (2.32 ev) in zinc blende structure. The blue band is centered around 4450 Å (2.79 ev) in wurtzite structure and around 4600 Å (2.70 ev) in zinc blende structure. (G. Curie and D. Curie, J. Phys. Radium, vol. 21, p. 127, 1960.)

to be due to halogen or trivalent metal (Al, Ga, and In) coactivators. In the experiment of Van Gool and Cleiren,[8] equal concentrations of copper activators and halogen coactivators were deliberately introduced into the sulfide. Under such conditions, the intensity of the green band was indeed much higher than the intensity of the blue band, supporting the model of associated donor-acceptor centers.

As shown in Fig. 11.4, the copper level also plays the role of a recombination center for the blue emission process. However, the origin of the other energy state indicated by dotted line is not at all certain. The state could be due to lattice defects or the excited states of copper. It is also possible that the blue emission is produced if an electron makes a direct transition from the conduction band to the copper acceptor state.

Figure 11.5 shows the energy levels generally present in a Mn-activated ZnS phosphor. Since Mn can be divalent, it does not need coactivators for charge compensation. In Fig. 11.5, the levels marked Mn centers represent the internal transitions in an Mn^{2+} ion similar to the thallium levels in a KCl:Tl phosphor. The levels marked L are produced by disturbances of lattice by Mn atoms while the levels V are due to Zn vacancies. The Mn^{2+} ion has a $3d^5$ configuration; therefore, the ground state is ^6S with a spin parallel arrangement. The excited states of the Mn^{2+} ion is ^4G with the spin of one electron being antiparallel to the spins of the other four electrons. The ground state of the Mn^{2+} ion has been verified by paramagnetic resonance experiments.[9] According to the theoretical calculation by Clogston,[10]

[8]   W. Van Gool and A. P. Cleiren, *Philips Res. Rept.*, vol. 15, p. 238, 1960.

[9]   See, for example, W. D. Hershberger and H. N. Leifer, *Phys. Rev.*, vol. 88, p. 714, 1952; S. P. Keller *et al.*, *Phys. Rev.*, vol. 110, p. 850, 1958; J. Lambe and G. Kikuchi, *Phys. Rev.*, vol. 119, p. 1256, 1960.

[10]   A. M. Clogston, *J. Phys. Chem. Solids*, vol. 7, p. 201, 1958.

Figure 11.5

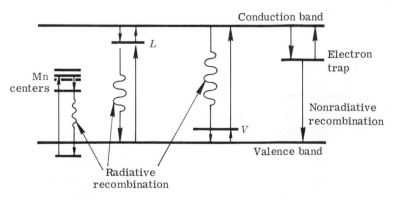

Simplified energy-level diagram for a ZnS:Mn *phosphor. (H. Gobrecht et al., Z. Physik, vol. 149, p. 504, 1957.)*

the crystal field splittings for the ^6S and ^4G states are small; therefore, as a first-order approximation, the radiative recombination in ZnS:Mn may be described as being due to transitions from the ^4G to ^6S state of a free Mn^{2+} ion.

Electron traps. Electron traps are metastable states (or energy levels) in which electrons can be captured and remain for a sufficiently long time before they are released. The metastable states $^3P_0$ and $^3P_2$ of the Tl^+ ion in KCl:Tl are examples of trapping states. Since a direct transition from the metastable state to the ground state is forbidden by quantum-mechanical selection rules, electrons captured by metastable states must be first excited to an adjacent energy state in order to participate in a luminescent process. Under the general term luminescence, however, we must distinguish between fluorescence and phosphorescence. If an electron after having been excited from the ground state into the excited state makes a spontaneous transition to the ground state, the phenomenon is known as fluorescence. If, on the contrary, emission occurs with the intervention of a metastable state, the process is known as phosphorescence.

As mentioned earlier in Sec. 5.5, the spontaneous lifetime in an atomic transition is of the order of 10^{-8} sec; therefore, the process of fluorescence is almost instantaneous. On the other hand, in the process of phosphorescence, an electron after having been excited to a higher state is captured by a metastable state (an electron trap) and the electron may remain in the trapping state for an average period of time τ before it is released. If ε is the energy separation between the metastable state and the next higher energy state from which a direct transition to the ground state is possible, then the time τ is given by

$$\tau^{-1} = s \exp\left(\frac{-\varepsilon}{kT}\right) \tag{11.1}$$

where s is a proportionality constant. Note that the dependence $\exp(-\varepsilon/kT)$ upon temperature T ($°K$) is universal for processes which require particles to overcome a potential barrier ε. Another example which has a similar temperature dependence is the diffusion constant of impurities given by Eq. (2.62).

From the preceding discussion, we see that electron traps play an important part in the process of phosphorescence. Since the physical origin of these traps is not yet ascertained, the only way to characterize them is to measure the magnitude of the energy ε. Consider a phosphor in which there exist a number of trap levels with energy $\varepsilon_1, \varepsilon_2, \ldots \varepsilon_i$. Each level has an initial density of trapped electrons n_{ti}. If it is assumed that immediately after being released from the trap an electron participates in a luminescence process, the intensity

$I(t)$ of luminescence at a time t is given by

$$I(t) = \sum C \frac{n_{ti}}{\tau_i} \exp\left(-\frac{t}{\tau_i}\right) \tag{11.2}$$

where C is a proportionality constant.

To appreciate the nature of the problem that is facing us, we shall use in Eq. (11.2) the known values for ZnS. The value of s is found to be around 10^9/sec and the energies of ε are found to be at 0.5, 0.57, 0.65, and 0.70 ev respectively. Thus, at $T = 18°C$, for $\varepsilon = 0.65$ ev, $\tau = 3$ min and for $\varepsilon = 0.57$ ev, $\tau = 5$ sec. Since τ changes very drastically with ε, a careful fit of the luminescence curve to Eq. (11.2) enables us to determine the values of n_{ti} and τ_i in Eq. (11.2). Differentiating Eq. (11.1) with respect to ε and realizing that $n_t(\tau)\,d\tau = n_t(\varepsilon)\,d\varepsilon$, we obtain

$$d\varepsilon = kT \frac{d\tau}{\tau} \qquad \text{or} \qquad n_t(\varepsilon) = n_t(\tau) \frac{\tau}{kT} \tag{11.3}$$

In Eq. (11.3), $n_t(\tau) = n_{ti}$ and $n_t(\varepsilon)$ are the distribution functions of trapped electrons in terms of τ and ε respectively. Knowing the values of $n_t(\tau)$ and τ, we can plot the product $n_t(\tau)\tau$ as a function of $\log(\tau s)$ (Fig. 11.6a). Note that Fig. 11.6a is equivalent to a plot of $n_t(\varepsilon)$ versus ε. Therefore, from the various maxima A, B, C, and D in Fig. 11.6a, we can determine the trap energies (0.5, 0.57, 0.65, and 0.70 ev).

In the preceding calculation, we have assumed a value of $s = 10^9$/sec. This value of s can be checked by an independent experiment known as thermoluminescence. Consider a phosphorescent solid which is first excited at a low temperature T_0 such that τ in Eq. (11.1) is of the order of days.[11] As we allow the temperature to rise, the traps begin to empty. For simplicity, we assume that the temperature rises at a uniform rate β degrees/sec, thus $dT = \beta\,dt$. By the definition of τ, the number of electrons released from a given trap is given by $dn_t/dt = -n_t/\tau$ and hence the electron concentration in the trap varies with time as

$$n_t = n_{t0} \exp\left[-\int_0^t \frac{dt}{\tau(T)}\right] \tag{11.4}$$

The luminescent intensity from the release of trapped electrons and their subsequent recombination with valence-band holes is given by $I(t) = Cn_t/\tau$. Using Eq. (11.4) and expressing dt in terms of dT, we find

$$I(T) = Cn_{t0}s \exp\left(-\frac{\varepsilon}{kT}\right) \exp\left[-\int_{T_0}^{T} s \exp\left(-\frac{\varepsilon}{kT}\right) \frac{dT}{\beta}\right] \tag{11.5}$$

where T_0 is the initial temperature.

[11] This is to ensure that the traps will not lose electrons at this temperature.

Equation (11.5) predicts a maximum for $I(T)$ at a temperature T_1 such that

$$\beta \frac{\mathcal{E}}{kT_1^2} = s \exp\left(-\frac{\mathcal{E}}{kT_1}\right) \tag{11.6}$$

Figure 11.6b shows a typical experimental plot of $I(T)$ as a function of temperature in ZnS:Cu. The temperatures at which the various maxima of $I(T)$ occur determine the values of the trap energies. Mainly due to the work of Randall and Wilkins[12] the thermolumi-

[12] J. T. Randall and M. H. F. Wilkins, *Proc. Roy. Soc.*, vol. A184, p. 366, 1945.

Figure
11.6

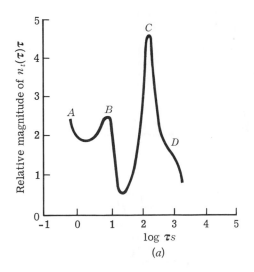

(a)

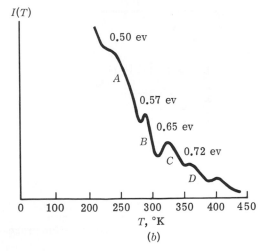

(b)

Identification of electron traps by (a) the experiment of decay of luminescence and (b) the method of thermoluminescence. (D. Curie, "Luminescence in Crystals," Methuen & Co., Ltd., London, 1963.)

nescence experiment is now extensively used in the study of electron traps. As shown in Fig. 11.6b, the values of the trap energies obtained from Eq. (11.6) agree well with the values determined from the luminescence-decay experiment if a value of 10^9/sec is chosen for s in both experiments. These trap states are associated with the introduction of copper in a sulfide phosphor. For further discussions of the luminescence-decay and thermoluminescence experiments and the proposed physical origins of electron traps, the reader is referred to Curie's book.[13]

Photoconduction and luminescence. The basic model to explain the different processes taking place in luminescence and photoconduction is shown schematically in Fig. 11.7. We should emphasize the word

[13] D. Curie, *op. cit.*

Figure 11.7

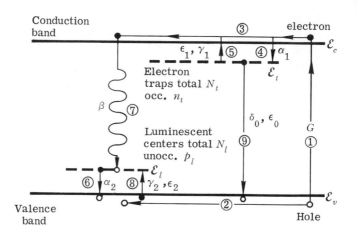

① Excitation by photons, electric field or electron bombardment
② Hole migration
③ Electron migration
④ Electron capture by traps
⑤ Release of electron by traps
⑥ Hole capture or electron release by luminescent centers
⑦ Electron capture by an empty luminescent center with subsequent emission of light
⑧ Hole release or electron capture by an empty luminescent center
⑨ Nonradiative recombination process

Basic model of a photoconductive and luminescent phosphor. (H. F. Ivey, Electroluminescense and Related Effects, Advan. Electron. Electron Phys., Suppl. 1, 1963.)

"basic" here because other models with several luminescent and trapping centers have also been proposed, but the basic processes involved are the same. It is generally deduced from experimental observations that donors have a fairly large electron-capture cross section but very small hole-capture cross section. The reader is referred to Eq. (5.67) for the definition of capture cross-sections C_n and C_p of a trap state. Since $C_n \gg C_p$ for coactivator donors, empty donors are ready to receive free electrons but occupied donors do not readily capture free holes. Consequently, coactivator donors behave mainly as electron traps. In the following discussion, we shall use the term "electron traps" to include donor levels.

A brief discussion of Fig. 11.7 is in order. The basic excitation process in electroluminescence consists of accelerating electrons under a high field, the subsequent collision of these high-energy electrons with luminescent centers or the lattice itself, and the consequent production of electron-hole pairs. Other excitation processes include photoexcitation and excitation by electron bombardment. The following discussion concerns the phenomenon of luminescence in general and does not depend on the mechanism of excitation unless stated otherwise. In Fig. 11.7, G is the rate of electron-hole pair production per second and the other symbols also represent the time rates of one kind or another associated with the following processes:

α_1 electron capture by the trap

α_2 production of an empty (active) luminescent center by capturing a hole (or releasing an electron)

γ_1 release of a trapped electron by thermal agitation

γ_2 thermal filling of an empty luminescent center by capturing a valence electron (or releasing a captured hole)

β radiative recombination through capturing an electron by an active (empty) luminescent center

δ_0 nonradiative recombination through capturing a hole from the valence band by the electron trap

The rate equations under steady-state condition read

$$0 = \frac{dn}{dt} = G - \alpha_1 n(N_t - n_t) - \beta n p_l + \gamma_1 n_t \qquad (11.7a)$$

$$0 = \frac{dn_t}{dt} = \alpha_1 n(N_t - n_t) - \delta_0 n_t p - \gamma_1 n_t \qquad (11.7b)$$

$$0 = \frac{dp_l}{dt} = \alpha_2 p(N_l - p_l) - \beta n p_l - \gamma_2 p_l \qquad (11.7c)$$

$$0 = \frac{dp}{dt} = G - \alpha_2 p(N_l - p_l) - \delta_0 n_t p + \gamma_2 p_l \qquad (11.7d)$$

where N_t and n_t are the total and occupied density of electron traps while N_l and p_l are the total and unoccupied (active) density of luminescent centers, respectively. In Eqs. (11.7a to d), all the electron-capture, hole-capture, and recombination processes are assumed to be bimolecular on account of the Pauli exclusion principle. For the processes, due to thermal agitation, of releasing a trapped electron and hole respectively by the electron trap and luminescent center, the terms $\gamma_1 n_t$ and $\gamma_2 p_l$ actually should read $\gamma_1' P_c n_t$ and $\gamma_2' N_v p_l$ where P_c and N_v are the empty conduction and occupied valence states respectively. Since the conduction and valence bands are respectively almost empty and fully occupied, the quantities P_c and N_v can be equated with the total conduction and valence band states, and hence can be incorporated into the constants γ_1 and γ_2 respectively.

Equations (11.7a–d) are nonlinear; therefore, a general solution of these equations is unavoidably very, if not hopelessly, complicated. The finding of a general solution is not only tedious algebraically but unrewarding insofar as getting physical insight into the problem is concerned. Instead, the problem will be analyzed more or less from physical intuition. A combination of Eqs. (11.7a) and (11.7b) or Eqs. (11.7c) and (11.7d) reads

$$G = \beta n p_l + \delta_0 n_t p \qquad (11.8)$$

Equation (11.8) states the physical situation very clearly. The radiative recombination process, the $\beta n p_l$ term, is in direct competition with the nonradiative process, the $\delta_0 n_t p$ term. Therefore, the luminescence efficiency can be defined as

$$\eta = \frac{\beta n p_l}{G} = \left(1 + \frac{\delta_0}{\beta} \frac{n_t}{n} \frac{p}{p_l}\right)^{-1} \qquad (11.9)$$

For low levels of excitation, it can be assumed that the condition of thermal equilibrium prevails between the electron traps and the conduction band, and between the luminescent centers and the valence band. Therefore,

$$\gamma_1 n_t \simeq \alpha_1 n (N_t - n_t) \qquad (11.10a)$$

$$\gamma_2 p_l \simeq \alpha_2 p (N_l - p_l) \qquad (11.10b)$$

Furthermore, the Fermi level is assumed to lie somewhere between $\mathcal{E}_t$ and $\mathcal{E}_l$; thus,

$$N_t - n_t \approx N_t \qquad N_l - p_l \approx N_l \qquad \frac{n_t}{p_l} \approx \left(\frac{N_t}{N_l}\right) \exp\left(\frac{\mathcal{E}_l - \mathcal{E}_t}{kT}\right)$$

$$(11.10c)$$

The luminescence efficiency[14] now becomes

$$\eta = \left(1 + \frac{\delta_0}{\beta} \frac{\alpha_1}{\alpha_2} \frac{\gamma_2}{\gamma_1} \frac{N_t}{N_l}\right)^{-1} \qquad (11.11a)$$

or, in terms of the energies,

$$\eta = \left[1 + \frac{N_t}{N_l} \frac{\delta_0}{\beta} \frac{p}{n} \exp\left(\frac{\mathcal{E}_l - \mathcal{E}_t}{kT}\right)\right]^{-1} \qquad (11.11b)$$

The temperature dependence of η comes mainly from the term $(p/n) \exp\left[(\mathcal{E}_l - \mathcal{E}_t)/kT\right]$, or in other words, the distribution of electrons between the traps and the conduction band and that of holes between the luminescent centers and the valence band. To explain the dependence of η upon the applied electric field, it is often assumed[15] that in the presence of an electric field, δ_0, γ_1, and γ_2 should be replaced by $\delta_0 + \epsilon_0$, $\gamma_1 + \epsilon_1$, and $\gamma_2 + \epsilon_2$, respectively, where ϵ_0, ϵ_1, and ϵ_2 are the field-induced components of the corresponding processes. Quenching or enhancement of luminescence by the field results, depending on whether

$$\gamma_1(\epsilon_0\gamma_2 + \delta_0\epsilon_2 + \epsilon_0\epsilon_2) < \qquad \text{or} \qquad > \delta_0\gamma_2\epsilon_1 \qquad (11.12)$$

The physical origins of the field-induced terms, however, are left unexplained in the literature.

We should point out that the luminescence process of Fig. 11.7 requires the excitation of electrons into the conduction band, and hence the luminescent phosphor is at the same time a photoconductor. It might seem illogical to discuss luminescence and photoconduction in the same material because the two processes impose conflicting requirements. Good luminescent materials require rapid recombination, whereas good photoconductors require long lifetime of free carriers or slow recombination. It is fortunate, however, that the recombination process can be speeded up or slowed down by adding the right kind of crystal imperfections. Consider the effect of introducing hole traps into the energy diagram of Fig. 11.7. The hole traps are postulated to accept holes readily from the valence band but to receive electrons from the conduction band at a much slower rate (by a factor of about 10^5 times). To account for these traps, a third recombination term must be added to Eq. (11.8). This recombination process is in direct competition with the $\beta n p_l$ process to capture holes from the valence band. The presence of this type of hole traps, often referred to as *compensated acceptors*, greatly reduces the effectiveness of the luminescent center by depleting the hole supply. The origin of these compensated acceptors has not yet

14   M. Schön, *Physica*, vol. 20, p. 430, 1954.
15   F. Matossi, *Phys. Rev.*, vol. 98, p. 434, 1955; vol. 101, p. 1835, 1956.

been positively identified. In practice, a photoconductor can be sensitized through the incorporation of halogen impurities, or excess Cd atoms in CdS and CdTe. It is suspected that either interstitial Cd atoms or anion vacancies are the cause of sensitization.

Finally, the difference between the ZnS:Cu and ZnS:Mn phosphors is worth noting. As illustrated in Fig. 11.5, excitation and luminescence in a ZnS:Mn phosphor can occur between the atomic levels of the Mn^{2+} ions, as in thallium-activated alkali halides. Therefore, photoconduction does not necessarily accompany the luminescence process. On the other hand, the other two radiative processes (Fig. 11.5) through levels L and V do need the participation of valence-band holes and conduction-band electrons. For these two processes, not only will photoconduction result but also electron and hole traps will play a part. We should add that neither quenching nor enhancement of luminescence by an applied field has been observed in the emission involving transitions between the Mn^{2+} levels. This experimental result is expected because only internal transitions are involved in the Mn^{2+} centers as illustrated in Fig. 11.5.

Thermal and optical quenching.[16]   Other interesting experimental observations which can be explained qualitatively on the simple model of Fig. 11.7 include infrared and thermal quenching of luminescence. A decrease of luminescent intensity is observed in the presence of an infrared radiation of the proper frequency. The phenomenon is usually ascribed to the filling of empty luminescent centers from the valence band similar to the γ_2 process in Fig. 11.7. The holes left in the valence band are now available for nonradiative recombination (the δ_0 process). Luminescence can also be quenched thermally. As illustrated in Fig. 11.1, an electron can make a radiationless transition from an excited state to the ground state if the electron has sufficient energy to climb over a potential barrier $\mathcal{E}_b$. This amount of energy can be supplied by the lattice at sufficiently high temperatures. It is known that thermal quenching of luminescence in ZnS:Cu occurs around 100°C for the blue band and around 300°C for the green band.

Opposite to the process of quenching is the process of activation. The release of electrons from traps in the thermoluminescence experiment is an example of thermal activation. Trapped electrons can also be released by the absorption of optical energy as well as by the utilization of thermal energy. Figure 11.8 shows the photoconductivity spectrum taken at 77°K for several p-type germanium samples doped with different impurities and their relevant energy levels. For a p-type sample, the Fermi level lies close to the valence

[16]   R. H. Bube, *Phys. Rev.*, vol. 99, p. 1105, 1955; G. F. J. Garlick, "Luminescent Materials," Clarendon Press, Oxford, 1949; H. A. Klasens, *Jour. Phys. Chem. Solids*, vol. 7, p. 175, 1958.

band at low temperatures; therefore, the impurity states shown in Fig. 11.8b are almost all empty. Under such circumstances, it is expected that the photoconductive process involves the excitation of an electron from the valence band to the impurity state as indicated by an arrow in Fig. 11.8b. The photoconductivity data shown in Fig. 11.8a confirm this view. We should point out that the photoexcitation process in sulfide phosphors is quite similar to that in germanium. The latter is chosen for presentation because an excellent compilation[17] of photoconduction data is available. For information

[17] R. Newman and W. W. Tyler, Photoconductivity in Germanium, *Solid State Phys.*, vol. 6, p. 96, 1958.

Figure 11.8

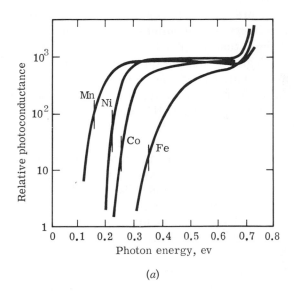

(a)

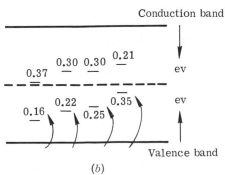

(b)

(a) *The photoconductivity spectra at 77°K in several p-type germanium samples doped with different impurities* (Mn, Ni, Co, *and* Fe) *and* (b) *the relevant energy levels.* (*R. Newman and W. W. Tyler, Solid State Physics, vol. 8, p. 49, 1959.*)

concerning the energies of luminescent centers and electron traps in sulfide phosphors, the reader is referred to the books by Curie[18] and by Bube.[19]

In conclusion, we again emphasize that although important advances have been made toward understanding the luminescent and photoconductive properties of sulfide phosphors during the past two decades, many fundamental questions still remain unanswered. The reader is well advised to search the literature before accepting any proposed model because there appear in the literature many different models. The existence of many different types of luminescent centers and electron traps makes a general analysis of such properties exceedingly complicated. In attempting a theoretical interpretation of experimental results, there appear to be two distinct approaches in the literature. The first approach takes the simplest model possible and involves the least number of parameters. The model may not reflect the actual complexity of the problem but it reduces the problem to a degree so that the approach becomes physically interpretable and mathematically manageable. Of course, it requires physical insight into the problem in deciding what can be neglected and what should be included. The other approach consists of leaving as much flexibility as possible in the proposed model. Simplifying assumptions are made later so that analytical results can fit experimental data. One great danger of the second approach is that after a few mathematical steps, the assumptions then made often become mathematical manipulations dissociated from physical processes. With so many adjustable parameters in the formulation, it becomes questionable whether curve fitting really means much. After reviewing the literature, we adopt the first approach in our presentation here.

Many aspects of photoconduction and luminescence in the group II-VI compound are left out of our discussion mainly because their picture is not clear at present. Therefore, so far as the teaching of scientific discipline is concerned, these subjects leave much to be desired. However, the rather unsatisfactory state of our understanding does not necessarily mean that these materials are not technically important. A combination of luminescent and photoconductive elements can perform a variety of functions such as light amplification, information storage, and switching. The basic circuit for these functions consists of a photoconductor and an electroluminescent crystal connected in series electrically and also coupled together optically. Emission from electroluminescence decreases the impedance of the photoconductor and hence increases the voltage applied across the electroluminescent crystal. The above processes provide the neces-

18 D. Curie, *op. cit.*
19 R. H. Bube, "Photoconductivity of Solids," John Wiley & Sons, Inc., New York, 1960.

sary feedback mechanism for amplifying, information-storage and switching circuits. For these applications, technical know-how is much more important than a clear understanding of the basic physical processes of luminescence and photoconduction.

11.4 OPTICAL CONSTANTS AND MAXWELL'S EQUATIONS

In the previous section, we have discussed optical processes in materials for which only phenomenological models are available. With other materials, however, especially covalent semiconductors, it is possible to discuss absorption and emission processes in a more definitive and coherent manner. In this and subsequent sections, we shall provide the necessary background for analysis of optical processes in these materials.

The classical theory of the optical properties of materials is based on Maxwell's field equations, which for an uncharged polarizable medium may be written:

$$\text{div}\,(\epsilon_0 \mathbf{E} + \mathbf{P}) = 0 \qquad \text{curl}\,\mathbf{E} = -\frac{\partial}{\partial t}\,\mu_0(\mathbf{H} + \mathbf{M})$$

$$\text{div}\,(\mathbf{H} + \mathbf{M}) = 0 \qquad \text{curl}\,\mathbf{H} = \sigma\mathbf{E} + \frac{\partial}{\partial t}\,(\epsilon_0\mathbf{E} + \mathbf{P}) \tag{11.13}$$

The relationship between $\mathbf{P}$ and $\mathbf{E}$ and that between $\mathbf{M}$ and $\mathbf{H}$ have been discussed in detail in Chapters 7 and 8. For all cases of practical interest in the optical region, M is small enough to be dropped, or in other words, $\mu \simeq \mu_0$. Furthermore, for most applications, we are dealing with isotropic and homogeneous media where the conductivity σ and the polarizability α are constants. Under such circumstances, Eq. (11.13) can easily be combined, yielding the well-known wave equation

$$\nabla^2\mathbf{E} - \mu_0\sigma\,\frac{\partial\mathbf{E}}{\partial t} - \mu_0\epsilon\,\frac{\partial^2\mathbf{E}}{\partial t^2} = 0 \tag{11.14}$$

where ϵ is the dielectric constant of the medium and ∇^2 is the Laplacian ($\nabla^2 = \text{grad div} - \text{curl curl}$). A similar equation is obtained for $\mathbf{H}$.

In rectangular coordinates, the Laplacian simply reduces to $\frac{\partial^2}{\partial x^2} + \frac{\partial^2}{\partial y^2} + \frac{\partial^2}{\partial z^2}$. For waves propagating in the z direction, the plane-wave solution of Eq. (11.14) is given by

$$\mathbf{E} = \mathbf{E}_0 \exp\left[i\omega\left(t - \frac{z}{v}\right)\right] \qquad \mathbf{H} = \mathbf{H}_0 \exp\left[i\omega\left(t - \frac{z}{v}\right)\right] \tag{11.15}$$

where $\mathbf{E}_0$ and $\mathbf{H}_0$ are constant vectors, mutually orthogonal to each other and also perpendicular to the direction of propagation. It is often convenient to express the velocity of propagation v in terms of velocity of light in vacuo c as

$$v = \frac{c}{N} = \frac{c}{n - i\kappa} \tag{11.16}$$

where N is called the complex index of refraction, n the ordinary *index of refraction*, and κ the *extinction coefficient*. Substituting Eq. (11.15) into Eq. (11.14) and using Eq. (11.16), we find by separating the real and imaginary parts

$$n^2 - \kappa^2 = \frac{\epsilon}{\epsilon_0} \quad \text{and} \quad 2n\kappa = \frac{\sigma}{\omega\epsilon_0} \tag{11.17}$$

It should be pointed out that in Eq. (11.17) both ϵ and σ are real. In Eq. (11.13), the last equation can be written as curl $\mathbf{H} = \mathbf{J}$ where $\mathbf{J}$ is the generalized current-density vector, and $\sigma\mathbf{E}$ and $j\omega\epsilon\mathbf{E}$ are the components of $\mathbf{J}$ in phase and in phase quadrature with $\mathbf{E}$ respectively. At high frequencies, the conceptual distinction between conduction and displacement currents becomes rather ambiguous and unnecessary. The implicit definition of σ and ϵ stated above removes this unnecessary distinction and makes the presentations to follow less wordy.

Consider a plane wave normally incident on a thin parallel sheet of a given material having a refractive index N and placed in free space as shown in Fig. 11.9. At the interface boundary, part of the incident wave is reflected and part is transmitted. According to the electromagnetic theory of propagation, the reflection and transmission coefficients r and t, defined as the ratios of the reflected and transmitted electric field $\mathbf{E}$ relative to the incident $\mathbf{E}$, are,

Figure 11.9

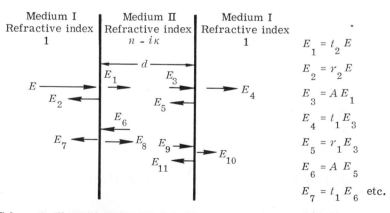

$$E_1 = t_2 E$$
$$E_2 = r_2 E$$
$$E_3 = A E_1$$
$$E_4 = t_1 E_3$$
$$E_5 = r_1 E_3$$
$$E_6 = A E_5$$
$$E_7 = t_1 E_6 \quad \text{etc.}$$

Schematic illustration showing the relationships between incident, transmitted, and reflected waves. The arrows indicate the direction of wave propagation but not the direction of electric field.

respectively,

$$r_1 = \frac{n - i\kappa - 1}{n - i\kappa + 1} \quad \text{and} \quad t_1 = \frac{2(n - i\kappa)}{n - i\kappa + 1} \tag{11.18}$$

for waves going from medium II to I, and the corresponding coefficients for waves going from medium I to II are

$$r_2 = \frac{1 - n + i\kappa}{1 + n - i\kappa} \quad \text{and} \quad t_2 = \frac{2}{n - i\kappa + 1} \tag{11.19}$$

Inside the sheet, the electromagnetic fields undergo changes in both phase and amplitude. Let A be the ratio of the initial and final values of E upon one traversal of the thin sheet of thickness d. According to Eq. (11.15), the change in the phase angle of $\mathbf{E}$ or $\mathbf{H}$ is

$$\theta = \frac{n\omega d}{c} = \frac{2\pi n d}{\lambda} \tag{11.20}$$

and the amplitude of $\mathbf{E}$ or $\mathbf{H}$ decreases by a factor of $\exp(-\omega\kappa d/c)$. The *absorption coefficient* of the medium K is defined as the reciprocal of the distance for the energy to fall by a factor of e, that means

$$K = \frac{2\omega\kappa}{c} = \frac{4\pi\kappa}{\lambda} \tag{11.21}$$

In terms of K and θ, A becomes

$$A = \exp(-i\theta) \exp\left(-\frac{Kd}{2}\right) \tag{11.22}$$

It is apparent from Fig. 11.9 that multiple reflections take place inside the sheet. The values of E_3 and E_9 differ because E_9 undergoes two more reflections at the interface boundaries and two more passages inside the sheet than E_3. The ratio E_{10}/E_4, being the same as E_9/E_3, is therefore equal to $A^2 r_1^2$. By similar reasoning applied to other components of the electric field, the total electric field of the transmitted wave E_t is

$$E_t = E_4 + E_{10} + \cdots = E_4(1 + A^2 r_1^2 + A^4 r_1^4 + \cdots)$$

$$= \frac{A t_1 t_2 E}{1 - A^2 r_1^2} \tag{11.23}$$

and the total electric field of the reflected wave E_r is

$$E_r = E_2 + E_7 + \cdots = \left(r_2 + \frac{A^2 r_1 t_1 t_2}{1 - A^2 r_1^2}\right) E \tag{11.24}$$

Similar expressions may be obtained for H_t and H_r. The power flow associated with an electromagnetic wave is given by the Poynting vector $(\mathbf{E} \times \mathbf{H}^*)$, and hence the percentage of power transmitted through the thin sheet is

$$\frac{P_t}{P_i} = \frac{|A|^2 T^4}{1 + |A|^4 R^4 - 2|A|^2 R^2 \cos 2(\theta + \psi)} \tag{11.25}$$

where R and ψ, respectively, are the magnitude and phase angle of r_1 and T^2 is the absolute magnitude of $t_1 t_2$.

From Eq. (11.25), the power transmitted has maxima at optical wavelengths such that $2(\theta + \psi) = 0$ or an integral multiple of 2π. In other words, if λ_1 and λ_2 are the wavelengths at which two successive maxima of P_t occur, then

$$n = \frac{\lambda_1 \lambda_2}{(\lambda_2 - \lambda_1)2d} \tag{11.26}$$

Equation (11.26) may be used to determine the value of n from such measurements. In order that the interference phenomenon may produce pronounced maxima and minima, it is necessary that the specimen have a sharply defined value of d. The measurement of absorption coefficient, on the other hand, does not require a uniform thickness of the same accuracy. If the maxima and minima are not resolved, meaning that the term with $\cos 2(\psi + \theta)$ in Eq. (11.25) averages to zero over the limits of d, then Eq. (11.25) becomes

$$\frac{P_t}{P_i} = \frac{T^4}{\exp(Kd) - R^4 \exp(-Kd)} \tag{11.27}$$

where $\quad T^2 = \frac{16(n^2 + \kappa^2)}{(n+1)^2 + \kappa^2} \quad$ and $\quad R^2 = \frac{(n-1)^2 + \kappa^2}{(n+1)^2 + \kappa^2}$

Knowing the values of n and P_t/P_i, the value of K or κ can be evaluated from Eqs. (11.27) and (11.21).

The absorption coefficient K of a semiconductor changes by many orders of magnitude at different wavelengths and hence it is usually more indicative to plot $\log K$ as a function of λ. Figure 11.10 shows the principal features of such a plot for a typical semiconductor.

Figure 11.10

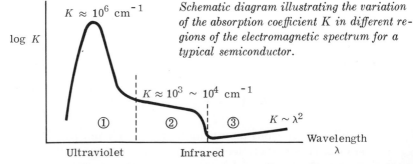

Schematic diagram illustrating the variation of the absorption coefficient K in different regions of the electromagnetic spectrum for a typical semiconductor.

Three distinctive types of absorption are shown in the figure, arising from the following physical processes: (1) In the ultraviolet region, electronic polarization as given by Eqs. (7.17) and (7.18) gives rise to an in-phase component of $\mathbf{J}$ with respect to $\mathbf{E}$, and hence contributes to a nonvanishing K. (2) The absorption in the infrared or near-visible region corresponds to an optical excitation of electrons from the valence into the conduction band. As the energy gap varies from material to material, so does the critical wavelength, at which this type of absorption sets in, usually referred to as the *absorption edge*. (3) Free carriers in a semiconductor, like free electrons in a metal, absorb electromagnetic radiation because of their conductive properties. Processes (1) and (3) can be treated classically and will be further discussed in this section. Process (2) is best treated in terms of qnautum-mechanical transitions, and furthermore, a discussion of these transitions forms the backbone of our treatment of semiconductor lasers. For these reasons, process (2) will be discussed in detail in later sections.

Let us now consider the absorption process due to electronic polarization. Using Eqs. (7.13) and (7.17), and noting that according to our new definition of σ and ϵ, α' and α'' given by Eqs. (7.17a) and (7.17b) contribute, respectively, to ϵ and σ in Eq. (11.17), we find

$$n^2 - \kappa^2 = 1 + \frac{Ne^2}{m\epsilon_0} \frac{\omega_0^2 - \omega^2}{(\omega_0^2 - \omega^2)^2 + (\omega\beta/m)^2} \qquad (11.28)$$

$$2n\kappa = \frac{Ne^2}{m\epsilon_0} \frac{\omega\beta/m}{(\omega_0^2 - \omega^2)^2 + (\omega\beta/m)^2} \qquad (11.29)$$

The behavior of $n^2 - \kappa^2$ and $2n\kappa$ is plotted in Fig. 11.11a as functions of frequency. Four distinct regions exist, depending on the solutions of n and κ from Eqs. (11.28) and (11.29). To facilitate an explanation, we shall assume that for frequencies $\omega > \omega_2$ or $\omega < \omega_1$, the value of $2n\kappa$ is much smaller than $n^2 - \kappa^2$ so that $2n\kappa$ is practically zero. That means, for $\omega > \omega_2$ or $\omega < \omega_1$, either $n = 0$ or $\kappa = 0$, depending on whether $n^2 - \kappa^2 <$ or > 0. For example, in region 3, ϵ is negative; therefore $n = 0$ and $\kappa \neq 0$. From Eq. (11.18), the absolute magnitude of the reflection coefficient is unity and hence the medium is totally reflecting. The optical behavior of the medium at other frequencies is qualitatively summarized in Fig. 11.11b.

Equation (11.28) shows that at frequencies much lower than ω_0, the value of n^2 approaches a constant value of $1 + Ne^2/m\epsilon_0\omega_0^2$. Take silicon as an example. The measured value of ω_0, at which the value of $2n\kappa$ is a maximum, is around 7.6×10^{15} rad/sec, corresponding to a wavelength of 0.25 micron and an energy of 4.9 ev. Using $N = 5 \times 10^{22}$ atoms/cm³ and four valence electrons per atom for silicon, we obtain $n^2 \approx 12$ in good agreement with the value of relative dielectric constant at low frequencies. It should be pointed out that on theoretical grounds[20] not to be discussed here, the local field correction is not needed for highly polarizable solids such as germanium and silicon. Therefore, Eq. (7.13) is used instead of the Clausius-Mossotti equation in deriving Eq. (11.28).

The problem of free-carrier absorption will be treated at the end of this chapter in Prob. 11.12. The equation of motion for free carriers is similar to that for bound electrons except for the following accounts. No restoring force is experienced by free carriers; but on the other hand, free carriers do suffer collisions due to scattering processes. That means, in Eqs. (11.28) and (11.29), $\omega_0 = 0$ and β/m should be replaced by $1/\tau$ where τ is the relaxation time of free carriers. Equation (11.21) in conjunction with Eq. (11.29) shows that the absorption coefficient K increases as λ^2 in the region where

[20] P. Nozières and D. Pines, *Phys. Rev.*, vol. 109, pp. 741, 762, and 1062, 1958.

Figure 11.11

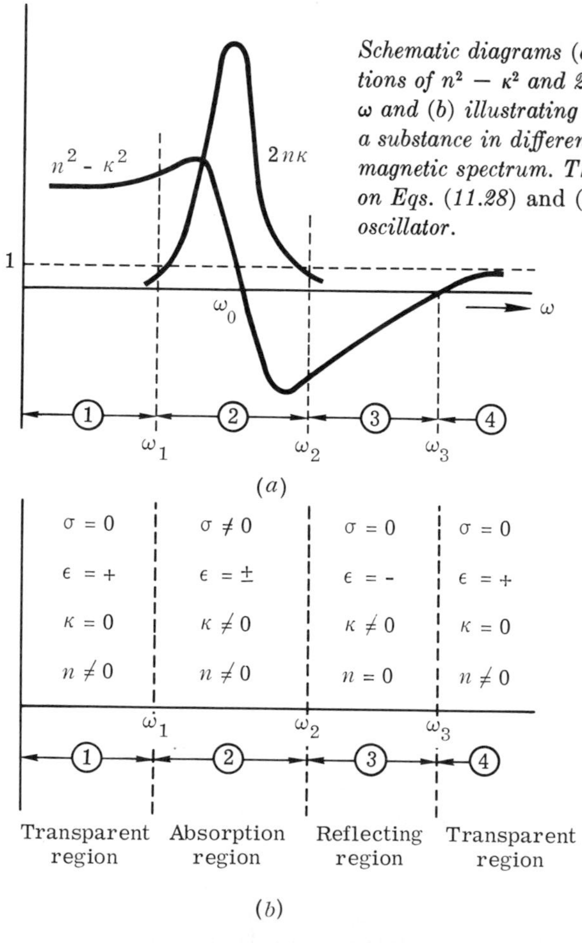

Schematic diagrams (a) showing the variations of $n^2 - \kappa^2$ and $2n\kappa$ as functions of ω and (b) illustrating the optical behavior of a substance in different regions of the electromagnetic spectrum. The analysis is based on Eqs. (11.28) and (11.29) for a single oscillator.

(a)

(1)	(2)	(3)	(4)
$\sigma = 0$	$\sigma \neq 0$	$\sigma = 0$	$\sigma = 0$
$\epsilon = +$	$\epsilon = \pm$	$\epsilon = -$	$\epsilon = +$
$\kappa = 0$	$\kappa \neq 0$	$\kappa \neq 0$	$\kappa = 0$
$n \neq 0$	$n \neq 0$	$n = 0$	$n \neq 0$
Transparent region	Absorption region	Reflecting region	Transparent region

(b)

n does not change appreciably with ω. Other modifications necessary for Eqs. (11.28) and (11.29) will be further discussed in the problem.

11.5 OPTICAL TRANSITIONS ACROSS ENERGY BANDS

The optical excitation of electrons across the energy gap from the valence band into the conduction band is of great importance in the study of the fundamental properties of semiconductors. The shape of the absorption curve in region 2 of Fig. 11.10 is determined by the band structure of a given semiconductor. Figure 11.12 shows the measured absorption coefficient of germanium as a function of photon energy at two different temperatures, and Fig. 11.13 shows the band structure of germanium. Note that in Fig. 11.12, there exist two regions of absorption coefficient of distinctly different magnitudes, corresponding to part a and part b of the curve.

In germanium, as shown in Fig. 11.13, the top of the valence band occurs at $k[000]$ in k (momentum) space while the conduction band has several valleys, one at $k[000]$, four at the edge of the [111] and its equivalent axes, and six along the [100] and its equivalent axes. In an optical transition, it is required by wave mechanics as will be

Figure 11.12

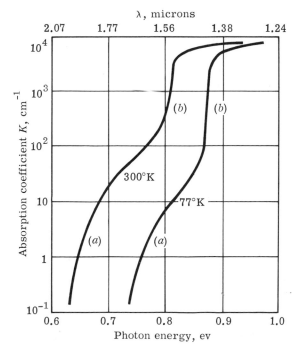

Absorption coefficient in germanium at 300°K and 77°K. (W. C. Dash and R. Newman, Phys. Rev., vol. 99, p. 1151, 1955.)

(a) Indirect optical transition   (b) Direct optical transition

shown in a later section that the total momenta should be conserved, the momentum being equal to the wave number multiplied by $\hbar$. If $\mathbf{k}_i$ and $\mathbf{k}_f$ are the initial and final wave vectors of the electron involved in an optical transition, then $\mathbf{k}_i + \mathbf{n} = \mathbf{k}_f$ where $\mathbf{n}$ is the wave vector of the electromagnetic radiation. The magnitude of k is of the order of π/a, a being the dimension of the unit lattice cell in angstrom units while that of η is equal to $2\pi/\lambda$ where λ is in the neighborhood of one micron in the infrared and near-visible regions. Therefore, the magnitude of η is very much smaller than that of k and the quantum-mechanical selection rule simply reduces to $\mathbf{k}_i \simeq \mathbf{k}_f$. This rule is observed in a direct optical transition across the bands from $k[000]$ of the valence band to the $k[000]$ valley of the conduction band, shown as process b in Figs. 11.12 and 11.13. At room temperature, the energy gap at $k[000]$, $\mathcal{E}_{g0}$, usually referred to as the *direct gap*, amounts to 0.80 ev; that means, the threshold of photon energy or the absorption edge is 0.80 ev for direct optical transition to occur. At liquid nitrogen temperature, the value of $\mathcal{E}_{g0}$ increases to 0.88 ev and so does the absorption edge as shown in Fig. 11.12.

The other absorption process, indicated as process a in Figs. 11.12 and 11.13, is often referred to as the *indirect transition*. The conduction-band minima in germanium occur at the edge of the four equivalent axes [111], [1$\bar{1}$1], [$\bar{1}$11], and [$\bar{1}$1$\bar{1}$] in k space. Note that electrons in the four conduction-band minima have different momenta or $\mathbf{k}$ values from those at the top of the valence band. To make an optical transition from the top of the valence band to one of the conduction-band minima, an electron must interact with lattice imperfections, such as phonons or impurities, to conserve momentum. In other words, an electron in the valence band first must be scattered by phonons or impurities to an intermediate state with a proper $\mathbf{k}$ value before making an optical transition to the conduction-band minimum. Since the indirect transition is a two-step process, the transition probability of such a process is much smaller than that

Figure 11.13

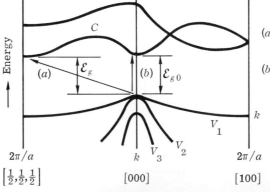

(a) Indirect optical transition

(b) Direct optical transition

Energy-band structure in germanium. (F. Herman, J. Electron., vol. 1, p. 103, 1955.)

of the direct process. This explains why the absorption coefficient in the indirect transition region is much smaller than that in the direct transition region.

Germanium is chosen as the example in our discussion for the following reasons. The fundamental properties of germanium, such as the band structure and the effective mass, have been most extensively investigated and accurately determined; therefore, a meaningful comparison between theoretical predictions and experimental results can be made. Germanium is also an indirect-gap semiconductor where the bottom of the conduction band and the top of the valence band do not occur at the same value of **k**. Hence, both the direct and indirect transitions can be discussed in the same material. Silicon is also an indirect-gap semiconductor but unlike germanium, it has six conduction-band minima along the six equivalent [100]-like axes. The direct gap of silicon is estimated around 2.5 ev; therefore, direct transitions are possible only in the visible region.

The optical properties of III-V compounds also have been studied extensively. Among these compounds, InAs, InSb, InP, GaSb, and GaAs are known to be direct-gap semiconductors whereas GaP and AlSb are known to be indirect-gap semiconductors. Figure 11.14a shows the energy-band structure of GaAs. The symbols Γ, L, and X actually denote the symmetry type of energy surfaces defined in group theory, but for our present purpose they merely refer to the various points, k[000], $k = \pi a^{-1}$ [111] and $k = 2\pi a^{-1}$ [100], respectively, in k space. The energy positions of the conduction band at L, Γ, and X are illustrated in Fig. 11.14b for several III-V compounds and the value of the various energy differences are summarized in Table 11.1.

The measured absorption coefficient in GaAs is shown in Fig. 11.15 as a function of photon energy. In contrast to Fig. 11.12, Fig. 11.15 shows only one type of absorption due to direct optical transition in accordance with the band structure of GaAs. Note that the magnitudes of the absorption coefficients for Ge and GaAs are about the same ($\sim 10^4$ cm^{-1}) in the direct transition region, indicating absorption processes of the same type.

TABLE 11.1 *Relevant energy differences in the band structures of III-V compounds*

	InSb	InAs	InP	GaSb	GaAs	GaP	AlSb
$\mathcal{E}_g$(ev)	0.236	0.360	1.20	0.813	1.41	2.2	1.6
$\mathcal{E}_{L\Gamma}$	.....	.....	....	0.09	>0.4	....	
$\mathcal{E}_{X\Gamma}$	.....	.....	0.3	0.3	0.36	....	
$\mathcal{E}_{\Gamma X}$	.....	.....	....	.....	.....	0.35	0.3

Reference: H. Ehrenreich, *J. Appl. Phys.*, vol. 32, p. 2155, 1961.

When a semiconductor is biased in the forward direction, injected minority carriers recombine with majority carriers, resulting in emission of radiation. Recombination radiation becomes coherent electromagnetic radiation in semiconductor lasers. Here the question of whether the semiconductor is a direct-gap or an indirect-gap semiconductor is an important factor. In direct-gap semiconductors, the excess injected holes and electrons, being located at the same $k[000]$ in k space, can participate in a recombination process involving a direct transition between the conduction- and valence-band states. In indirect-gap semiconductors, on the other hand, the injected holes

Figure 11.14

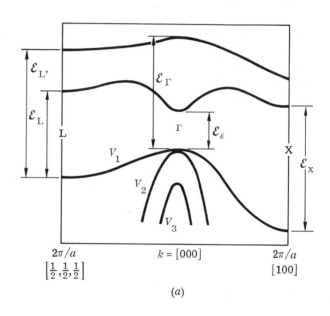

(a)

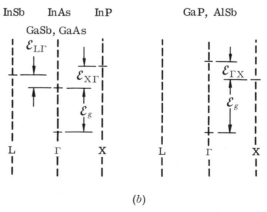

(b)

Energy-band structure of (a) GaAs and (b) other III-V compounds.

and electrons, like electrons and holes in the conduction-band minima and the valence-band maxima, respectively, have different values of k. The recombination of excess carriers, therefore, needs the assistance of phonon or impurity scattering to conserve momentum, and the transition probability of such an indirect recombination process, like the indirect absorption process, is much smaller than the corresponding probability of a direct process. This is manifested by the fact that the efficiency of recombination radiation in GaAs diodes is close to unity while that in Ge diodes is only about 10^{-4}.

Note that the emission spectrum of a semiconductor diode is determined by the energy difference of the electron and hole under transition, therefore, unlike the absorption spectrum, it is confined to energies around the gap energy. To make emission spectra available over a wide range, it is possible to use semiconductor alloys; for example, $GaSb_{1-x}As_x$ alloy should cover an energy range from 0.813 to 1.41 ev, depending on the value of x or the relative percentage of GaSb and GaAs. It is also possible with alloys to increase the gap energy of a direct-gap semiconductor by mixing it with an indirect-gap semiconductor of a larger gap energy. An interesting example is the $GaAs_{1-x}P_x$ alloy, and Fig. 11.16 shows the variation of the gap energy as a function of x. The solid curve is the meaured value of $\mathcal{E}_g$ while the dashed curves represent the energy differences between the top of the valence band and the two conduction-band valleys at $k[000]$ and $k[100]$, assuming a straight line variation with x from

Figure 11.15

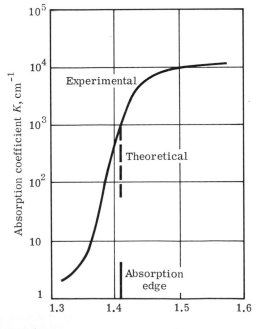

Absorption coefficient in GaAs. (*T. S. Moss, J. Appl. Phys., vol. 32, p. 2136, 1961.*)

GaAs to GaP. It is clear from Fig. 11.16 that for $x <$ about 0.5, the conduction-band valley at $k[000]$ is the lower energy valley and hence the alloy is a direct-gap semiconductor. The other curves in Fig. 11.16 are for the $Ga_{1-x}Al_xSb$ alloy. Similar curves are also obtained for the Ge-Si alloy, showing a similar shift of the conduction minimum from germanium-like to silicon-like energy-band structure.

For completeness, we should mention that inter-valence-band transitions[21] are also possible, for example, transitions between the lower valence band (V_3) and the upper two valence bands $(V_1$ and $V_2)$ in Fig. 11.13 as well as transitions between the two upper bands. Such transitions, however, occur at very long wavelengths in the range from 3 to 10 microns. Besides band-to-band transitions and the absorption processes discussed in the previous section, there are two other absorption processes which are not included here. The absorption due to lattice vibration will not be treated here. Such an absorption process occurs even at a longer wavelength than the intraband absorption; for example, in germanium, the main lattice absorption band is at 29 microns.

The other absorption process concerns the excitation of localized impurity states in impure semiconductors. Most shallow impurity states, such as Ga, In, As, P, etc., in Ge or Si have a very low activation energy, and hence at ordinary temperatures, such donor states

[21] For further discussions, the reader is referred to E. Burstein, G. Picus, and N. Sclar, Optical and Photoconductive Properties of Silicon and Germanium, in "Photoconductivity Conference," pp. 353–413, John Wiley & Sons, Inc., New York, 1956.

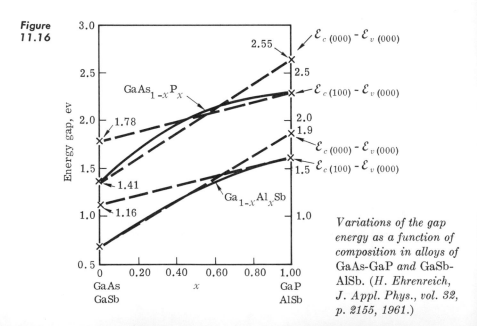

Figure 11.16

Variations of the gap energy as a function of composition in alloys of GaAs-GaP and GaSb-AlSb. (H. Ehrenreich, J. Appl. Phys., vol. 32, p. 2155, 1961.)

are completely ionized and such acceptor states are completely occupied. To give sufficient absorption, it is necessary to cool the sample to very low temperatures so that excitation of electrons from occupied donor states to the conduction band or from the valence band to empty acceptor states is possible. The same can be said about the photoconductive property of a doped semiconductor. In order that a photoconductive semiconductor may have reasonable sensitivity at moderately low temperatures, say liquid nitrogen temperature, the activation energy of a chosen impurity must be sufficiently high. In the meantime, however, the activation energy also determines the threshold of long wavelength detection of a photoconductor. Gold-doped germanium and indium-doped silicon are considered promising as infrared detectors in view of the above considerations.

In summary, this section serves to give a general discussion on various absorption processes involving transitions between different energy states in a semiconductor. In subsequent sections, we shall develop the necessary steps which lead to a quantitative analysis of the absorption process and finally to a formal treatment of semiconductor lasers. Since semiconductor lasers concern mainly transitions between conduction- and valence-band states, we shall concentrate our discussion on such transitions. Furthermore, the probability of direct band-to-band transition and the effective mass of conduction band electrons at $k[000]$ are related to the same matrix element; hence we shall express the transition probability in terms of the effective mass which can be determined from cyclotron resonance experiments. In the following section, we shall discuss the energy-band structure, the nature of the wave function and the effective mass at $k[000]$ of a semiconductor first in order to lay the foundation for the calculation of the transition probability.

11.6 BAND STRUCTURE, WAVE FUNCTIONS, AND EFFECTIVE MASSES OF COVALENT SEMICONDUCTORS

The band structure near $k[000]$ of covalent semiconductors having a diamond or zinc blende structure has been investigated theoretically by many authors. The analysis originated from the work of Shockley.[22] The electrons in the conduction and valence bands at $k[000]$ are known to possess wave functions which, in many respects, are similar to the atomic s and p functions, s wave function for the conduction band and p wave function for the valence band. Without spin-orbit interactions, the energy of the valence-band electrons is actually sixfold degenerate, corresponding to the three p functions times the two spin orientations. Spin-orbit interaction partially removes this degeneracy by lowering the two $j = \frac{1}{2}$ bands with

[22] W. Shockley, *Phys. Rev.*, vol. 78, p. 173, 1950.

respect to the four $j = 3/2$ bands, the former being represented by V_3 and the latter by V_1 and V_2 in Figs. 11.13 and 11.14.

In the absence of the spin-orbit interaction, the Schrödinger equation reads

$$\left[-\frac{\hbar^2}{2m} \nabla^2 + V(r) \right] \Phi(r) = \mathcal{E}\Phi(r) \tag{11.30}$$

where $V(r)$ represents the potential energy of the electron, which among other things, consists primarily of the periodic lattice potential. Since at present our aim is to work out some theoretical expressions and to correlate the parameters involved in these expressions with experiments, it is not necessary for us to know the explicit form of $V(r)$. In a periodic lattice, the one electron wave function can always be written as the well-known Bloch function

$$\Phi_k(r) = u_k(r) \exp (i\mathbf{k} \cdot \mathbf{r}) \tag{11.31}$$

where $u_k(r)$ is periodic from lattice cell to cell. Substituting Eq. (11.31) into Eq. (11.30), we have

$$\left[-\frac{\hbar^2}{2m} \nabla^2 - \frac{\hbar^2}{m} i\mathbf{k} \cdot \nabla + V(r) + \frac{\hbar^2 k^2}{2m} \right] u_k(r) = \mathcal{E}u_k(r) \tag{11.32}$$

Let u_s be the electronic wave function for the conduction-band electrons and u_x, u_y, and u_z be those for the valence-band electrons at $k[000]$. By definition, these wave functions should satisfy the following relations

$$\left[-\frac{\hbar^2}{2m} \nabla^2 + V(r) \right] u_s = \mathcal{E}_c u_s \tag{11.33}$$

$$\left[-\frac{\hbar^2}{2m} \nabla^2 + V(r) \right] u_{x,y,z} = \mathcal{E}_v u_{x,y,z} \tag{11.34}$$

where $\mathcal{E}_c$ and $\mathcal{E}_v$, respectively, are the energies of the conduction- and valence-band edges at $k[000]$.

For electrons near $k[000]$, the cell wave function $u_k(r)$ can be expressed in terms of u_s, u_x, u_y, u_z as

$$u_k(r) = A u_s + B u_x + C u_y + D u_z \tag{11.35}$$

The operational properties of the s and p wave functions u_s and $u_{x,y,z}$ have been discussed in Sec. 8.4. Note that the s wave function is spherically symmetrical while the p wave function is odd in x, y, or z. Substituting Eq. (11.35) into Eq. (11.32) and using Eqs. (11.33) and (11.34), we find

$$A(\mathcal{E}_c - \mathcal{E}' + h_1)u_s + B(\mathcal{E}_v - \mathcal{E}' + h_1)u_x$$
$$+ C(\mathcal{E}_v - \mathcal{E}' + h_1)u_y + D(\mathcal{E}_v - \mathcal{E}' + h_1)u_z = 0 \tag{11.36}$$

where $\mathcal{E}' = \mathcal{E} - \hbar^2 k^2/2m$ and

$$h_1 = \frac{\hbar^2}{mi}\left(k_x\frac{\partial}{\partial x} + k_y\frac{\partial}{\partial y} + k_z\frac{\partial}{\partial z}\right)$$

It is interesting to observe that the function $(\partial/\partial x)u_s$ is an odd function in x; therefore the matrix element $\int u_j^*(\partial/\partial x)u_s\, dx$ is nonzero only when $j = x$. The same is true for $\int u_s^*(\partial/\partial x)u_j\, dx$. Since the function $(\partial/\partial x)u_y$ is odd in both x and y, the integral

$$\int\!\!\int_{-\infty}^{\infty} u_j^* \frac{\partial}{\partial x} u_y\, dx\, dy$$

is always zero, $j = s, x, y,$ or z. In summary, the only nonvanishing matrix elements are the ones defined by the following expression

$$P = -\frac{\hbar^2}{m}\int_{-\infty}^{\infty} u_j\frac{\partial}{\partial j}u_s\, dq = \frac{\hbar^2}{m}\int_{-\infty}^{\infty} u_s\frac{\partial}{\partial j}u_j\, dq \qquad (11.37)^{23}$$

where $j = x, y,$ or z, and $dq = dx\, dy\, dz$.

Multiplying Eq. (11.36) by u_s and integrating over dq, we obtain a homogeneous equation in $A, B, C,$ and D. We repeat the process by multiplying Eq. (11.36) with $u_x, u_y,$ and u_z separately, and note that for nontrivial solutions, the determinant of the coefficients for $A, B, C,$ and D must be zero. Thus, we obtain:

$$\begin{vmatrix} \mathcal{E}_c - \mathcal{E}' & -ik_x P & -ik_y P & -ik_z P \\ ik_x P & \mathcal{E}_v - \mathcal{E}' & 0 & 0 \\ ik_y P & 0 & \mathcal{E}_v - \mathcal{E}' & 0 \\ ik_z P & 0 & 0 & \mathcal{E}_v - \mathcal{E}' \end{vmatrix} = 0 \qquad (11.38)$$

The four roots of Eq. (11.38) are

$$\mathcal{E}_{1,2} = \frac{\mathcal{E}_c + \mathcal{E}_v}{2} \pm \sqrt{\left(\frac{\mathcal{E}_c - \mathcal{E}_v}{2}\right)^2 + k^2 P^2 + \frac{\hbar^2 k^2}{2m}}$$
$$\mathcal{E}_{3,4} = \mathcal{E}_v + \frac{\hbar^2 k^2}{2m} \qquad (11.39)$$

In Eq. (11.39), $\mathcal{E}_1$ taking the positive sign represents the energy of conduction-band electrons while $\mathcal{E}_2$ taking the negative sign, $\mathcal{E}_3$, and $\mathcal{E}_4$ are the energies of electrons in the three valence bands. From Eq. (11.39), the effective masses can be calculated and the values thus obtained can be compared with those found from cyclotron resonance experiments. Such a comparison indicates that further refinements of the theory must be made. First, the spin-orbit interaction must be added and, second, a higher order perturbation calculation is needed for the valence bands.

The spin-orbit interaction can be incorporated into the formulation

---

[23] The wave functions $u_x, u_y,$ and u_z, being similar in nature to $\psi_A, \psi_B,$ and ψ_C of Eq. (8.26), are real and so is u_s; therefore the complex conjugate sign is not needed here.

by adding a term $\xi(r)\mathbf{l} \cdot \mathbf{s}$ into h_1 in the perturbation calculation. Realizing that the eigenvectors for the operators $\mathbf{l}$ and $\mathbf{s}$ are, respectively, u_+, u_z, and u_- and α(spin up) and β(spin down), we choose the following as the base vectors: $u_s\alpha$, $u_+\beta$, $u_z\alpha$, $u_-\beta$, $u_s\beta$, $u_-\alpha$, $u_z\beta$, and $-u_+\alpha$. The secular equation, therefore, involves an 8-by-8 matrix. It turns out that if we arrange the base vectors in the above order, the interaction matrix can be written in the form $\begin{bmatrix} \mathfrak{IC} & 0 \\ 0 & \mathfrak{IC} \end{bmatrix}$ where $\mathfrak{IC}$ is a four by four matrix. The functions u_+ and u_- are defined as $u_+ = -(u_x + iu_y)/\sqrt{2}$ and $u_- = (u_x - iu_y)/\sqrt{2}$ and the operational rules of $\mathbf{l}$ on u_s, u_+, u_-, and u_z as well as those of $\mathbf{s}$ on α and β have been discussed in Secs. 8.4 and 9.4. The calculation is much simplified if the spin-orbit interaction energy is written as

$$h_2 = -\xi(r)\mathbf{l} \cdot \mathbf{s} = -\xi(r)\left[L_z s_z + \left(\frac{L_+ s_- + L_- s_+}{2}\right)\right] \quad (11.40)$$

where

$$L_+ = L_x + iL_y, \; L_- = L_x - iL_y, \; s_+ = s_x + is_y \text{ and } s_- = s_x - is_y.$$

In general, the contribution from $\mathbf{l}$ operated on the $\exp(i\mathbf{k} \cdot \mathbf{r})$ part of $\Phi(r)$ given by Eq. (11.31) is small compared to the corresponding contribution from the u_k part, hence the former will not be considered in the following calculation.

Take the wave function $u_-\beta$ as an example. The spin operator has the following effects [Eq. (9.38)]: $s_z\beta = -\beta/2$, $s_+\beta = \alpha$, and $s_-\beta = 0$. However, since $L_-u_- = 0$, the only surviving term is $L_z s_z u_-\beta = u_-\beta/2$ and the only nonvanishing matrix element in the $u_-\beta$ column is with the $u_-\beta$ row. Let

$$4\Delta = 3\int(u_x{}^2 + u_y{}^2)\xi(r)\,dq = 6\int u_z{}^2\xi(r)\,dq \quad (11.41)$$

Then we can express the various interaction matrix elements in terms of Δ as shown in the following:

$$\begin{array}{c|cccc} & u_s\alpha & u_+\beta & u_z\alpha & u_-\beta \\ \hline u_s\alpha & \mathcal{E}_s - \mathcal{E}' & 0 & -ikP & 0 \\ u_+\beta & 0 & \mathcal{E}_p - \mathcal{E}' - \Delta/3 & \sqrt{2}\,\Delta/3 & 0 \\ u_z\alpha & ikP & \sqrt{2}\,\Delta/3 & \mathcal{E}_p - \mathcal{E}' & 0 \\ u_-\beta & 0 & 0 & 0 & \mathcal{E}_p - \mathcal{E}' + \Delta/3 \end{array} = \mathfrak{IC}$$

$$(11.42)[24]$$

To simplify the calculation, we choose the $\mathbf{k}$ vector in the z direction in Eq. (11.42). Furthermore, the spin-orbit interaction shifts the band energy so the notations $\mathcal{E}_s$ and $\mathcal{E}_p$ are used. Otherwise Eq. (11.42) with $\Delta = 0$ is identical to Eq. (11.38). The $\sqrt{2}\,\Delta/3$ terms are due to L_+s_- operated on $u_z\alpha$ and L_-s_+ operated on $u_+\beta$, and the

[24] E. O. Kane, *J. Phys. Chem. Solids*, vol. 1, p. 249, 1957.

difference in sign in $\mathcal{E}_p - \mathcal{E}' \pm \Delta/3$ is due to the fact $L_z s_z u_{+} \beta = -u_{+} \beta$ while $L_z s_z u_{-} \beta = u_{-} \beta$. Expansion of Eq. (11.42) gives

$$\mathcal{E}' = \mathcal{E}_p + \frac{\Delta}{3}$$

$$\left(\mathcal{E}' - \mathcal{E}_p + \frac{2\Delta}{3} \right) \left(\mathcal{E}' - \mathcal{E}_p - \frac{\Delta}{3} \right) (\mathcal{E}' - \mathcal{E}_s) \tag{11.43a}$$

$$- k^2 P^2 \left(\mathcal{E}' - \mathcal{E}_p + \frac{\Delta}{3} \right) = 0$$

For small values of k^2, the cubic equation can easily be solved by treating the term involving $k^2 P^2$ as a small perturbation as shown in the following example:

$$\mathcal{E}_1' = \mathcal{E}_s + \frac{k^2 P^2 (\mathcal{E}_s - \mathcal{E}_p + \Delta/3)}{(\mathcal{E}_s - \mathcal{E}_p + 2\Delta/3)(\mathcal{E}_s - \mathcal{E}_p - \Delta/3)} \tag{11.43b}$$

Letting $\mathcal{E}_v = \mathcal{E}_p + \Delta/3$, $\mathcal{E}_c = \mathcal{E}_s$, and $\mathcal{E}_c - \mathcal{E}_v = \mathcal{E}_{g0}$ the direct gap, we have

$$\mathcal{E}_1 = \mathcal{E}_c + \frac{\hbar^2 k^2}{2m} + \frac{P^2 k^2}{3} \left(\frac{2}{\mathcal{E}_{g0}} + \frac{1}{\mathcal{E}_{g0} + \Delta} \right)$$

$$\mathcal{E}_2 = \mathcal{E}_v + \frac{\hbar^2 k^2}{2m}$$

$$\mathcal{E}_3 = \mathcal{E}_v + \frac{\hbar^2 k^2}{2m} - \frac{2P^2 k^2}{3\mathcal{E}_{g0}} \tag{11.44}$$

$$\mathcal{E}_4 = \mathcal{E}_v - \Delta + \frac{\hbar^2 k^2}{2m} - \frac{P^2 k^2}{3(\mathcal{E}_{g0} + \Delta)}$$

In the above equation, $\mathcal{E}_1$ is the energy of conduction-band electrons while $\mathcal{E}_2$, $\mathcal{E}_3$, and $\mathcal{E}_4$, respectively, are the energies of electrons in the three valence bands V_1, V_2, and V_3. The composition of the wave function for conduction-band electrons and electrons in the three valence bands can be found according to the scheme discussed in Sec. 9.9 by taking the ratios of the respective minors in the determinant of Eq. (11.42). For our present discussion of the direct emission and absorption processes, however, we are mainly concerned with transitions between the conduction band and the heavy-mass valence band V_1, and furthermore, transitions across the direct gap are limited to electrons near $k[000]$. For the conduction band, the wave function near $k[000]$ can be approximated by $u_s \alpha$ and $u_s \beta$ and for the heavy-mass valence band, it can be approximated by $u_{-} \beta$ and $-u_{+} \alpha$. The functions $u_s \beta$ and $-u_{+} \alpha$ together with $u_{-} \alpha$ and $u_z \beta$ form the other four base vectors in the 8-by-8 determinant $\begin{bmatrix} \mathfrak{IC} & 0 \\ 0 & \mathfrak{IC} \end{bmatrix}$.

As will be shown in the next section, the optical transition probability between the conduction and valence bands is proportional to P^2, which can be expressed in terms of the experimentally measurable effective mass m_{c0} of the conduction-band electrons near $k[000]$ as follows:

$$P^2 = \frac{3\hbar^2}{2m_{c0}} \frac{\mathcal{E}_{g0}(\mathcal{E}_{g0} + \Delta)}{3\mathcal{E}_{g0} + 2\Delta} \frac{m - m_{c0}}{m} \qquad (11.45)$$

Equation (11.45) will be used in the next section in the calculation of the optical absorption coefficient.

The reader may wonder whether the energy expression $\mathcal{E}_2$ is correct or not because it predicts a positive mass for valence-band electrons or a negative mass for holes. It turns out that for the hole effective mass, a second-order perturbation calculation[25] of the $\mathbf{k} \cdot \nabla$ energy is needed. Then, the off-diagonal elements will have additional terms involving $k_x k_y$, $k_y k_z$, and $k_z k_x$ while the diagonal elements will have additional terms involving $k_x{}^2$, $k_y{}^2$, and $k_z{}^2$. The energy expression given in Chapter 3 for the cyclotron resonance experiment is a result of such calculations. For germanium and silicon, the energies of the three valence bands are of the form:

$$\mathcal{E}_{2,3} = \mathcal{E}_v + Ak^2 \pm [B^2k^4 + C^2(k_x{}^2k_y{}^2 + k_y{}^2k_z{}^2 + k_z{}^2k_x{}^2)]^{1/2}$$
$$\mathcal{E}_4 = \mathcal{E}_v - \Delta + Ak^2 \qquad (11.46)$$

where the constants have the approximate values, in units of $\hbar^2/2m$,

Si:	$A = -4.1$	$B = 1.6$	$C = 3.3$
Ge:	$A = -13.0$	$B = 8.9$	$C = 10.3$

The sign of A now shows positive effective masses for holes in the three valence bands and is consistent with the energy surfaces V_1, V_2, and V_3 drawn in Figs. 11.13 and 11.14. The energy $\mathcal{E}_2$ corresponding to the expression with a positive sign is the heavy-mass band V_1 while $\mathcal{E}_3$ with a negative sign is the light-mass band V_2. All these bands are doubly degenerate, giving a total of six valence bands.

11.7 CALCULATIONS OF THE DIRECT OPTICAL TRANSITION PROBABILITY AND ABSORPTION COEFFICIENT

In Sec. 9.5, we derived a general expression for the effect of a radiation field upon the relative distribution of electrons between two energy states i and j, with energy $\mathcal{E}_i > \mathcal{E}_j$. If at $t = 0$, the system is in the lower energy state j, then the probability that at a later time t the system is in the higher energy state i is given by $|a(t)|^2$ where $a(t)$ obeys the following equation

$$a(t) = \frac{1}{i\hbar} \iint \Phi_i^*(q) \exp\left(\frac{i\mathcal{E}_i t}{\hbar}\right) U\Phi_j(q) \exp\left(-\frac{i\mathcal{E}_j t}{\hbar}\right) dq \, dt \qquad (11.47)$$

where q stands for the spatial coordinates and $dq = dx \, dy \, dz$. Equa-

[25] G. Dresselhaus, A. F. Kip, and C. Kittel, *Phys. Rev.*, vol. 98, p. 368, 1955; E. O. Kane, *J. Phys. Chem. Solids*, vol. 1, p. 82, 1956, and p. 249, 1957.

tion (11.47) is identical to Eq. (9.49) except that the interaction energy U due to the radiation field now takes a different form from that given by Eq. (9.50). The classical Hamiltonian[26] in the presence of an electromagnetic field with vector potential $\mathbf{A}$ is given by

$$\mathcal{H} = \frac{1}{2m} \sum_{x,y,z} \left(p_x + \frac{eA_x}{c} \right)^2 + V \qquad (11.48)^{[27]}$$

where c is the velocity of light. It can be shown that the classical equation of motion can be derived from Eq. (11.48) following the procedure outlined in Prob. 11.18. The vector potential is generally so chosen that $\nabla \cdot \mathbf{A} = 0$; therefore expansion of Eq. (11.48) gives

$$\mathcal{H} = \mathcal{H}_0 + \frac{e\hbar}{imc} \mathbf{A} \cdot \nabla + \frac{e^2}{2mc^2} (A_x{}^2 + A_y{}^2 + A_z{}^2) \qquad (11.49)$$

The first term represents the unperturbed Hamiltonian of the system. Since the last term usually is much smaller than the middle term, the interaction energy U is given by $(e\hbar/imc)\,\mathbf{A} \cdot \nabla$.

For a plane wave propagating in the z direction and polarized in the x direction, the vector potential is of the form

$$\mathbf{A} = \mathbf{a}_x A_0 \exp\left[i(\eta z - \omega t)\right] + \text{complex conjugate} \qquad (11.50)$$

where $\mathbf{a}_x$ is a unit vector in the x direction and η is the propagation constant. Using Eq. (11.50) in the expression of U in Eq. (11.47), we have after integration over t

$$a(t) = \frac{1}{i\hbar} \frac{1 - \exp\left[i(\omega_0 - \omega)t\right]}{\omega_0 - \omega} \frac{e\hbar A_0}{mc} \int \Phi_i^* \frac{\partial}{\partial x} \Phi_j \exp\left(i\eta z\right) dq \qquad (11.51)$$

where $\omega_0 = (\mathcal{E}_i - \mathcal{E}_j)/\hbar$.

In a semiconductor, the one-electron wave function is given by the Bloch wave function of the form given by Eq. (11.31); therefore, the integration in Eq. (11.51) can be broken into a product of two parts: (1) The integration for the cell function $u_k(r)$ is over a unit cell of the crystal, and (2) the integration for $\exp(i\mathbf{k} \cdot \mathbf{r})$ can be replaced by a summation over different cells of the whole crystal.

$$\int \Phi_i^* \frac{\partial}{\partial x} \Phi_j \exp(i\eta z) \, dq = CD = \left(\exp(i\eta z_0) \int_{\text{cell}} u_i^* \frac{\partial}{\partial x} u_j \, dq \right)$$
$$\left(\sum_{\text{crystal}} \exp\left[i(\mathbf{k}_j - \mathbf{k}_i + \mathbf{n}) \cdot \mathbf{l}\right] \right) \qquad (11.52)$$

[26] See, for example, J. H. Van Vleck, "The Theory of Electric and Magnetic Susceptibilities," p. 20, Clarendon Press, Oxford, 1932; L. I. Schiff, "Quantum Mechanics," p. 135, McGraw-Hill Book Company, New York, 1949; N. F. Mott and I. N. Sneddon, "Wave Mechanics and Its Applications," p. 262, Clarendon Press, Oxford, 1948.

[27] The cgs units are used here because they are exclusively used in the literature in treating the interaction of radiation with matter.

The sum D is over the translation vector $\mathbf{l}$ of the form given in Eq. (2.20). It is unity if $\mathbf{k}_i = \mathbf{k}_j + \mathbf{n}$ and zero otherwise. This is the mathematical expression of the *momentum conservation* law stated earlier in Sec. 11.5. In Eq. (11.52), the factor exp $(i\eta z_0)$ can be taken outside the integration, making use of the fact that the wavelength λ of the light is larger by many orders of magnitude than the atomic dimensions over which the integration is carried out.

Since true monochromatic light does not exist, it is necessary to consider a radiation field having a finite bandwidth. Let $I(\nu)$ be the radiation energy density per unit frequency interval within the frequency interval between ν and $\nu + d\nu$. For a monochromatic light, the energy density equals $\epsilon \omega^2 A_0^2 / 2\pi c^2$ where ϵ is the dielectric constant of the medium. For the radiation field having a finite bandwidth, we can define an effective vector potential $\mathbf{A}_0$ such that $I(\nu)\, d\nu = \epsilon \omega^2 A_0^2 / 2\pi c^2$. Note that the population of the upper energy state is proportional to $|a(t)|^2$. Thus from Eq. (11.51), we have

$$
\begin{aligned}
|a(t)|^2 &\sim \sum \left| \frac{2 \sin \left[(\omega - \omega_0)t/2\right]}{\omega_0 - \omega} \right|^2 \frac{e^2 A_0^2}{m^2 c^2} \\
&= \int \frac{1 - \cos \left[(\omega - \omega_0)t\right]}{(\omega_0 - \omega)^2} \frac{4\pi e^2}{\epsilon m^2 c^2} I(\nu)\, d\nu
\end{aligned}
\qquad (11.53a)
$$

where the summation is over all the frequency components of A_0. Integrating $|\sin \left[(\omega_0 - \omega)t/2\right]/(\omega_0 - \omega)|^2$ over $d\nu$ and realizing that $\int_{-\infty}^{\infty} (1 - \cos x)\, dx/x^2 = \pi$, we have

$$
W = \frac{|a(t)|^2}{t} = \frac{e^2}{2\pi m^2 \nu^2 \epsilon} I(\nu) \left| \int_{cell} u_i^* \frac{\partial}{\partial x} u_j\, dq \right|^2
\qquad (11.53b)
$$

In deriving Eq. (11.53), we note that the function $(1 - \cos x)/x^2$ drops very sharply as x deviates from zero, therefore the function $I(\nu)$ can be taken outside the integral. The quotient $|a(t)|^2/t$ is known as the *transition probability* per unit time for electrons going from the lower energy state j to the higher energy state i. It should be pointed out that in transitions from state i to j, the component of the vector potential varying as exp $[i(\omega t - \eta z)]$ should be used. Since the interchange of state i and j does not change the absolute magnitude of the integral in Eq. (11.53), the probabilities for upward and downward transitions are equal.

The transition probability W given by Eq. (11.53) is identical to $B_{ij}I$ or $B_{ji}I$ given by Eq. (9.109). Therefore, the Einstein A and B coefficients in Eq. (9.109) are given by

$$
\begin{aligned}
B &= \frac{e^2}{2\pi \epsilon m^2 \nu^2} \frac{1}{3} \sum_{x,y,z} \left| \int_{cell} u_i^* \frac{\partial}{\partial x} u_j\, dq \right|^2 \\
A &= \frac{4he^2}{\epsilon \lambda^3 m^2 \nu^2} \frac{1}{3} \sum_{x,y,z} \left| \int_{cell} u_i^* \frac{\partial}{\partial x} u_j\, dq \right|^2
\end{aligned}
\qquad (11.54)
$$

The reciprocal of the Einstein coefficient of spontaneous emission (that is, $1/A$) is the *spontaneous lifetime* τ of state i in a transition to state j. In terms of P given by Eq. (11.37) or in terms of m_{c0} given by Eq. (11.45), τ is equal to

$$\tau = \frac{\epsilon\lambda^3\hbar^3\nu^2}{8\pi e^2 P^2} = \frac{n^2 c^3 m_{c0}\hbar}{12\pi e^2\nu}\frac{3\mathcal{E}_{g0} + 2\Delta}{\mathcal{E}_{g0}(\mathcal{E}_{g0} + \Delta)}\frac{m}{m - m_{c0}} \tag{11.55}$$

In using Eq. (11.55), caution must be exercised. First, it is expressed in cgs unit and second, λ and c are respectively the wavelength and the velocity of light in the dielectric medium, so they differ from the free-space values by a factor n, the refractive index. For direct transitions at $k[000]$ in germanium (process b in Fig. 11.13), we may take $h\nu = \mathcal{E}_{g0} =$ direct gap $= 0.88$ ev at 77°K, $m_{c0} = 0.037\ m$ at $k[000]$ and $\Delta = 0.28$ ev. Using these values in Eq. (11.55), we obtain $\tau = 2.4 \times 10^{-10}$ sec. In this example, we have shown that the Einstein A and B coefficients and the spontaneous lifetime in a direct transition are all related to the quantity P and in turn to the electron effective mass at $k[000]$, the value of which can be determined experimentally. A smaller effective mass means a larger transition probability and shorter lifetime.

Equations (11.53) and (11.54) can easily be extended to cover optical transitions in insulators. In a ruby laser, for example, the atomic wave functions of Cr^{3+} should be used in these equations instead of the cell functions u_i and u_j, and the integration should be carried over the full extension of these atomic wave functions. Since Φ_i and Φ_j can no longer be expressed in terms of the Bloch wave function, the momentum conservation law is automatically removed. For insulators, the dipole-moment matrix element r_{ij} is often used instead of P where $|r_{ij}|^2 = |x_{ij}|^2 + |y_{ij}|^2 + |z_{ij}|^2$ and

$$x_{ij} = \int\Phi_i^* x\Phi_j\, dq \tag{11.56}$$

Using the Schrödinger equations for Φ_i^* and Φ_j, it can be shown[28] that

$$\frac{m}{\hbar^2}(\mathcal{E}_i - \mathcal{E}_j)\int\Phi_i^* x\Phi_j\, dq = \int\Phi_i^*\frac{\partial}{\partial x}\Phi_j\, dq \tag{11.57}$$

In terms of r_{ij}, the Einstein A and B coefficients are

$$B = \frac{2\pi e^2}{3\hbar^2\epsilon}|r_{ij}|^2 \qquad A = \frac{1}{\tau} = \frac{4e^2\omega^3}{3\epsilon\hbar c^3}|r_{ij}|^2 \tag{11.58}$$

where $\hbar\omega = \mathcal{E}_i - \mathcal{E}_j$. Again c is the velocity of light in the dielectric medium.

The absorption coefficient of a semiconductor can be calculated from the optical transition probability as follows. Let $N(\nu)$ be the density of states having an energy difference $h\nu$ in a direct optical

[28] See, for example, N. F. Mott and I. N. Sneddon, *op. cit.*, pp. 170, 256.

transition. Near $k[000]$, the energies of conduction-band electrons and heavy-mass valence-band holes are given by

$$\mathcal{E}_1 = \mathcal{E}_{c0} + \frac{\hbar^2 k^2}{2m_{c0}} \qquad \mathcal{E}_2 = \mathcal{E}_v - \frac{\hbar^2 k^2}{2m_v} \qquad (11.59)$$

Since in a direct optical transition, the momentum is conserved, $k_e = k_h = k$, yielding

$$h\nu - \mathcal{E}_{g0} = \frac{\hbar^2 k^2}{2m_r} \qquad (11.60)$$

where $h\nu = \mathcal{E}_1 - \mathcal{E}_2$, and $m_r = \dfrac{m_v m_{c0}}{m_v + m_{c0}}$ is often referred to as the *reduced mass*. Using Eq. (11.60), we have

$$N(\nu)\, d\nu = \frac{8\pi}{(2\pi)^3} k^2\, dk = \frac{(2m_r)^{3/2}}{2\pi^2 \hbar^3} (h\nu - \mathcal{E}_{g0})^{1/2} h\, d\nu \qquad (11.61)$$

For a light beam of energy density $I(\nu)$, the attenuation per unit distance dx is equal to the total number of transitions per unit time multiplied by energy absorbed per transition and divided by velocity, or in other words,

$$dI(\nu) = -Wh\nu N(\nu) n \frac{dx}{c_0} \qquad (11.62)$$

where n is the index of refraction and c_0 is the velocity of light in free space. Using W given by Eq. (11.53), $N(\nu)$ by Eq. (11.61), and the definition of absorption coefficient K given by Eq. (11.22), we find

$$K = -\frac{1}{I}\frac{dI}{dx} = \frac{2e^2(2m_r)^{3/2}}{nc_0 m^2 h\nu}(h\nu - \mathcal{E}_{g0})^{1/2}\left|\int_{\text{cell}} u_i^* \frac{\partial}{\partial x} u_j\, dq\right|^2 \qquad (11.63)$$

For germanium, $\mathcal{E}_{g0} = 0.88$ ev at $77°$K, $\Delta = 0.28$ ev, $m_v = 0.20m$, $m_{c0} = 0.037m$; these values give $\left|\int_{\text{cell}} u_i^*(\partial/\partial x)u_j\, dq\right|^2 = 15m\,\mathcal{E}_{g0}/\hbar^2$ and $m_r = 0.031m$. As a check for dimension, the dimensionless constants in Eq. (11.63) are evaluated and

$$K = 0.115\left(\frac{\mathcal{E}_{g0}e^2 m^{1/2}}{h\nu c_0 \hbar^2}\right)(h\nu - \mathcal{E}_{g0})^{1/2}$$

has a dimension of $1/$length in agreement with its definition. For $h\nu - \mathcal{E}_{g0} = 0.02$ ev, $K = 4.4 \times 10^3$ cm^{-1} in good agreement with the experimental value shown in Fig. 11.12. Similar calculations can be made for III-V compound semiconductors, such as GaAs and InSb.

11.8 INDIRECT TRANSITIONS IN A SEMICONDUCTOR

Equations (11.53), (11.54), and (11.63) apply only to direct transitions in a semiconductor, process b in Fig. 11.13. Now we shall discuss the indirect transition, process a in Fig. 11.13. The indirect transition is a two-step process as shown in Fig. 11.17a. An electron in the valence band is first excited by a photon to an intermediate state a which can be a virtual state in a vertical transition (process 1) and then is scattered by a phonon into a final state f (process 2) in the conduction-band minimum away from $k = 0$. The alternate process in which the photon excites an electron into the final state f (process 3) and the phonon scatters the hole into the initial state i (process 4) will not be considered here because it is less important in germanium

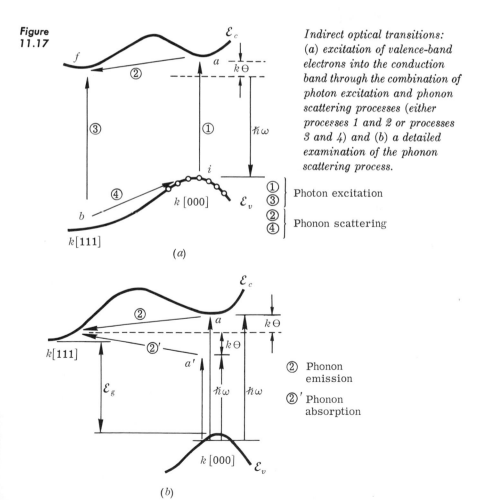

Figure 11.17

Indirect optical transitions: (a) excitation of valence-band electrons into the conduction band through the combination of photon excitation and phonon scattering processes (either processes 1 and 2 or processes 3 and 4) and (b) a detailed examination of the phonon scattering process.

and silicon. Carrying the perturbation calculation to second order[29] shows that the transition probability per unit time for state i to go to state f through state a is given by

$$W = \frac{2\pi}{\hbar} \left| \sum_{a,f} \frac{(f|h_1|a)(a|h_1|i)}{\mathcal{E}_f - \mathcal{E}_a} \right|^2 \delta(\mathcal{E}_f - \mathcal{E}_i - \mathcal{E}) \qquad (11.64)$$

where $(f|h_1|a)$ represents the usual matrix element connecting two states f and a through the perturbation potential h_1 which consists of both the interactions with photons and phonons. The factor $\delta(\mathcal{E}_f - \mathcal{E}_i - \mathcal{E})$ takes care of the energy conservation during transition and the summation is over intermediate and final states such that a net gain $\mathcal{E}$ in energy results.

For $|(a|h_1|i)|^2$, the interaction is through the photon field and $|(a|h_1|i)|^2 = \left(\dfrac{eP}{n\hbar\omega}\right)^2 2\pi I(\nu)$. For $|(f|h_1|a)|^2$, the interaction is with lattice vibrations. Actually, there are two possible ways for interaction with the lattice: phonon emission and phonon absorption processes. These are indicated as processes 2 and 2′ respectively in Fig. 11.17b. For the phonon emission process, the transition probability is proportional to $N_s + 1$ while for the phonon absorption process, it is proportional to N_s where $N_s = [\exp(\Theta/T) - 1]^{-1}$ on account of the fact that phonons obey the Bose-Einstein statistics. The factors $(N_s + 1)$ and N_s are a result of quantization of phonons. Note that similar factors appear in the photon emission and absorption processes as given in Eq. (9.112). Furthermore, for phonon emission processes, $\mathcal{E}_f - \mathcal{E}_i = \mathcal{E} = \hbar\omega - k\Theta$ while for phonon absorption processes, $\mathcal{E}_f - \mathcal{E}_i = \mathcal{E} = \hbar\omega + k\Theta$ where $k\Theta$ is the phonon energy and Θ is usually expressed in units of temperature, degrees Kelvin.

To calculate the absorption coefficient in the indirect absorption region, we need to count all possible initial and final states; that means we shall combine the summation in Eq. (11.64) and the factor $N(\nu)$ in Eq. (11.62) to form the following product:

$$N(\mathcal{E}_f)\,d\mathcal{E}_f N(\mathcal{E}_i)\,d\mathcal{E}_i = \frac{(8\pi)^2}{h^6} \sqrt{2m_c\mathcal{E}_1} \sqrt{2m_v\mathcal{E}_2}\,d(m_c\mathcal{E}_1)\,d(m_v\mathcal{E}_2) \qquad (11.65)$$

In an indirect transition, no restriction on wave number is imposed, hence the only restriction on $\mathcal{E}$ is the energy conservation relation $\mathcal{E}_f - \mathcal{E}_i = \hbar\omega \pm k\Theta$. Summing Eq. (11.65) over all possible $\mathcal{E}_i$ together with $\delta(\mathcal{E}_f - \mathcal{E}_i - \mathcal{E})$ gives

$$\int \delta(\mathcal{E}_f - \mathcal{E}_i - \mathcal{E})N(\mathcal{E}_f)\,d\mathcal{E}_f N(\mathcal{E}_i)\,d\mathcal{E}_i$$

$$= \int \frac{(4\pi)^2}{h^6}(4m_v m_c)^{3/2}\sqrt{\mathcal{E}_1(\hbar\omega \pm k\Theta - \mathcal{E}_g - \mathcal{E}_1)}\,d\mathcal{E}_1 \qquad (11.66)$$

---

[29] See, for example, *ibid.*, p. 273.

where m_c and m_v are the density-of-state effective masses of the conduction and valence bands, respectively, and $\mathcal{E}_1$ and $\mathcal{E}_2$ are the energies of the final (conduction-band) and initial (valence-band) states, in reference to the respective band edges. Therefore, the integration limit in Eq. (11.66) is from $\mathcal{E}_1 = 0$ to $\mathcal{E}_1 = \hbar\omega - \mathcal{E}_g \pm k\Theta$ or $\mathcal{E}_2 = 0$.

Substituting Eqs. (11.64) and (11.66) into Eq. (11.62), we obtain

$$K = \frac{e^2(m_c m_v)^{3/2} P^2}{2\pi n c_0 \hbar^9 \omega^2} \frac{\langle \mathcal{E}_a - \mathcal{E}_i \rangle_{av}}{\langle \mathcal{E}_a - \mathcal{E}_f \rangle_{av}^2} \sum (\hbar\omega - \mathcal{E}_g \pm k\Theta)^2 \left\{ \begin{matrix} N_s \\ N_s + 1 \end{matrix} \right\} M_s^2$$

$$(11.67)^{30}$$

In the above equation, M_s is the electron-phonon scattering matrix element, or in other words, $|(f|h_1|a)|^2 = M_s^2 N_s$ or $M_s^2(N_s + 1)$ for the phonon absorption and emission processes respectively, and the summation is meant for a sum of these two processes. In deriving Eq. (11.67), it is assumed that the quantities M_s, $\mathcal{E}_a - \mathcal{E}_f$ and $\mathcal{E}_a - \mathcal{E}_i$ vary slowly with energy so that these quantities can be taken outside the integral and replaced by their respective average values. The quantity M_s^2 has a dimension of energy² volume and using this information, the reader can check the dimension of K (cm⁻¹). The value of M_s^2 can be inferred from other experiments. For example, the indirect tunneling current in Ge depends upon phonon scattering of electrons near $k[000]$ into $k[111]$ as previously discussed in Chapter 6, and from the magnitude of the current, a value of 4×10^{-49} erg² cm³ for M_s^2 in Ge is estimated by Kane.[31] To simplify the calculation of K, we can assume the following: (1) At sufficiently low temperature, say 77°K, only the phonon-emission process is important; that means $N_s \ll 1$ in Eq. (11.67). (2) The value of $\mathcal{E}_a$ can be taken as $\mathcal{E}_c$ at $k[000]$ so $\langle \mathcal{E}_a - \mathcal{E}_f \rangle_{av} = 0.18$ ev in Ge and $\langle \mathcal{E}_a - \mathcal{E}_i \rangle_{av} \approx \hbar\omega$. For $\hbar\omega - \mathcal{E}_g - k\Theta = 0.05$ ev, the value of K is calculated from Eq. (11.67) to be 2 cm⁻¹ with $m_c = 0.55m$, $m_v = 0.37m$, and the value of P in Ge given previously. A comparison of this calculated value with experimental values shown in Fig. 11.12 shows a reasonably good agreement. Furthermore, a detailed analysis[32] of the experimental curve of K not only fits Eq. (11.67) but also yields two phonon energies having temperature equivalents of $\Theta_1 = 90°K$ and $\Theta_2 = 320°K$.

The phonon-scattering time of $k[000]$ valley electrons into conduction-band minima in Ge can be calculated as follows. The transition probability per unit time of such a process is

$$W = \frac{2\pi}{\hbar} \sum_{\text{all states}} M_s^2(2N_s + 1) = \frac{2\sqrt{2}}{\pi\hbar^4} M_s^2 m_c^{3/2} \mathcal{E}^{1/2}(2N_s + 1) \quad (11.68)$$

[30] J. Bardeen, F. J. Blatt, and L. H. Hall, Indirect Transitions from the Valence to the Conduction Bands, in "Photoconductivity Conference," p. 146, John Wiley & Sons, Inc., New York, 1956.

[31] E. O. Kane, *Jour. Appl. Phys.*, vol. 32, p. 83, 1961.

[32] G. G. Macfarlane *et al.*, *Phys. Rev.*, vol. 108, p. 1377, 1957.

where ε is the kinetic energy of the electron after it is scattered into the conduction-band minimum. The mean free path of $k[000]$ valley electrons is equal to v_0/W where v_0 is the velocity of the electron. During the scattering process, the transfer of momentum is a fixed amount; therefore $\Delta\hbar k' = \Delta\hbar k$ as shown in Fig. 11.18. Using the relation $m_{c0}^2 v_0^2 = 2m_c'\varepsilon$ [that is, $(\Delta\hbar k)^2 = (\Delta\hbar k')^2$] and assuming $N_s \ll 1$, we have the mean free path

$$l = \frac{\pi\hbar^4}{2} \frac{(m_c')^{1/2}}{M_s^2 m_c^{3/2} m_{c0}} \tag{11.69}$$

where m_c and m_c' are the density-of-states and mobility effective masses of electrons at $k[111]$ and m_{c0} is the effective mass of electrons at $k[000]$. With $m_c = 0.55\,m$, $m_c' = 0.12m$, and $m_{c0} = 0.037m$, we find $l = 1.5 \times 10^{-4}$ cm which corresponds to a mean free time of 6×10^{-12} sec at $77°$K. This value of phonon-scattering time is also in reasonable agreement with values estimated from magneto-optical absorption data[33] and acoustoelectric effect experiments.[34] The calculations given above not only show that reasonable agreements exist between various experiments, but also indicate one important consideration. Since the intervalley scattering time is shorter than the spontaneous lifetime by a factor of forty, the former may play a very important role[35] in determining whether a given material is suitable for laser operation.

[33] S. Zwerdling, B. Lax, L. M. Roth, and K. J. Button, *Phys. Rev.*, vol. 114, p. 80, 1959.

[34] G. Weinreich, T. M. Sanders, Jr., and H. G. White, *Phys. Rev.*, vol. 114, p. 33, 1959.

[35] S. Wang, *J. Appl. Phys.*, vol. 34, p. 3443, 1963.

Figure 11.18

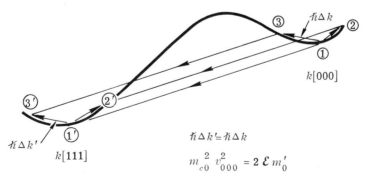

Momentum transfer during a phonon scattering process. It is assumed that the momentum transfer is of the same amount during the scattering processes of electrons from 1 to 1', from 2 to 2', and from 3 to 3'. In other words, the momentum differences between points 3 and 1 and between points 3' and 1' are the same; hence $\hbar\Delta k' = \hbar\Delta k$ where $\Delta k'$ and Δk refer to their respective minima at $k[111]$ and $k[000]$.

In concluding this section, the following comments are added. First, in calculating the optical transition probability across the energy gap, we have assumed s-like and p-like wave functions for the conduction and valence bands respectively. In certain semiconductors, the conduction band at $k[000]$ may also have p-like wave functions, hence the transition at $k[000]$ is forbidden. The problem has been treated by many authors[36] and the optical absorption coefficient for forbidden transitions shows a $(h\nu - \mathcal{E}_{g0})^{3/2}$ dependence in contrast to the $(h\nu - \mathcal{E}_{g0})^{1/2}$ dependence of Eq. (11.63). For indirect transitions, the optical absorption coefficient is proportional to $(\hbar\omega - \mathcal{E}_g \pm k\Theta)^3$ if the transition at $k[000]$ is forbidden. The second comment is about the meaning of the spontaneous lifetime. Obviously, the value calculated from Eq. (11.55) is several orders of magnitude smaller than the often-quoted lifetime of minority carriers. The reason for this discrepancy is as follows. First, band-to-band transitions may not be the only recombination process; as a matter of fact, the Hall-Shockley-Read model of recombination through intermediate states works well in explaining the behavior of minority-carrier lifetime as a function of carrier concentration and temperature. Second, in finding the spontaneous lifetime it is assumed that a vacant state in the valence band is immediately available. Even if spontaneous radiative recombination is the only recombination process, the ordinary lifetime should be longer than the spontaneous lifetime because a statistical average [Eq. (5.60)] must be taken over the availability of the occupied and vacant states.[37] The ordinary lifetime ultimately approaches the spontaneous lifetime as a limit when a semiconductor becomes degenerate. Degeneracy is needed for semiconductor laser operation to achieve a population inversion as well as to shorten the effective lifetime of carriers. Considerations of this kind will be further discussed in Chap. 12.

11.9 MEASUREMENTS OF THE DIELECTRIC CONSTANT IN THE OPTICAL AND X-RAY REGIONS

In Sec. 7.2, we discussed the behavior of the dielectric constant in the optical region on the basis of the classical treatment of Lorentz, which considers a dielectric as an assembly of oscillators which are constituted by electrons bound to the nuclei. In the presence of an applied field, the electrons are set in forced oscillation by the radiation field, giving rise to an induced polarization. In the optical and X-ray regions, the classical theory can be further developed to correlate the information obtained from the measurements of dielectric constants.

[36] See, for example, J. Bardeen, F. J. Blatt, and L. H. Hall, *op. cit.* Also H. Y. Fan, M. L. Shepherd, and W. Spitzer, *loc. cit.*, p. 184.
[37] W. P. Dumke, *Phys. Rev.*, vol. 108, p. 1377, 1957.

Since the real and imaginary parts of the dielectric constants [Eqs. (11.28) and (11.29)] for a given material are determined by the same parameters N, γ, and β, the two optical constants n and κ in Eq. (11.17) are generally related to each other. The first set of such relations is given as follows:

$$(n^2 - \kappa^2)_{\omega_0} = 1 + \frac{2}{\pi} \int_0^\infty \frac{2\omega n\kappa - \omega_0(2n\kappa)_{\omega_0}}{\omega^2 - \omega_0^2} \, d\omega \qquad (11.70a)^{38}$$

$$(2n\kappa)_{\omega_0} = -\frac{2}{\pi} \int_0^\infty \frac{\omega(n^2 - \kappa^2) - \omega_0(n^2 - \kappa^2)_{\omega_0}}{\omega^2 - \omega_0^2} \, d\omega \qquad (11.70b)$$

In Eq. (11.70), the subscript ω_0 stands for the frequency at which the quantity in parentheses is evaluated. Thus, if we have complete information about the real or the imaginary part of the dielectric constant, that is $n^2 - \kappa^2$ or $2n\kappa$, over the entire frequency spectrum, we should be able to deduce the information for the other part of the dielectric constant from Eq. (11.70) which is known as the *Kramers-Kronig relations*.[39] We should emphasize again that the symbol ω_0 used in Eqs. (11.28) and (11.29) represents the resonant frequency of the oscillator. However, the symbol ω_0 used in the following discussion does not carry this connotation unless explicitly stated.

The Kramers-Kronig relation can be stated in various forms. The form often used in the optical region relates the phase angle θ to the reflectance R as follows:

$$\theta(\omega_0) = -\frac{1}{2\pi} \int_0^\infty \frac{d \log R(\omega)}{d\omega} \log \left| \frac{\omega + \omega_0}{\omega - \omega_0} \right| \, d\omega \qquad (11.71)$$

in which the phase angle θ is given by

$$\theta = 2 \tan^{-1} \left(\frac{2\kappa}{n^2 + \kappa^2 - 1} \right) \qquad (11.72a)$$

and the reflectance R is equal to

$$R = \frac{(n - 1)^2 + \kappa^2}{(n + 1)^2 + \kappa^2} \qquad (11.72b)$$

according to the definition $r_1^2 = R \exp(-i\theta)$ given by Eq. (11.18). Figure 11.19 shows the reflectance measurement on germanium reported by Philipp and Taft.[40] As we can see, in the region from $h\nu = 0$ to 9 ev, the reflectance R (or the ratio of the reflected to the

[38] For the mathematical proof of Eq. (11.70) the reader is referred to T. S. Moss, "Optical Properties of Semiconductors," Academic Press, New York, 1959. The analogous relation known to electrical engineers can be found in texts on circuit analysis. See, for example, H. W. Bode, "Network Analysis and Feedback Design," D. Van Nostrand Co., Princeton, N.J., 1945, and D. F. Tuttle, Jr., "Network Synthesis," John Wiley & Sons, Inc., New York, 1958.

[39] H. A. Kramers and R. de L. Kronig, *Z. Physik*, vol. 30, pp. 521–522, 1929.

[40] H. R. Philipp and E. A. Taft, *Phys. Rev.*, vol. 113, p. 1002, 1959.

Figure 11.19

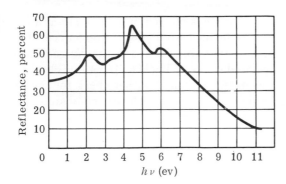

The spectral dependence of the reflectance of germanium. (H. R. Philipp and E. A. Taft, Phys. Rev., vol. 113, p. 1002, 1959.)

incident light intensity) is appreciable; consequently Eq. (11.71) provides a fairly accurate determination of θ. From the calculated values of $\theta(\omega_0)$ and the measured values of $R(\omega_0)$, the values of n and κ can be evaluated from Eq. (11.72). Figure 11.20 shows the results obtained for Ge.

From the values of n and κ given in Fig. 11.20, we can determine the value of the absorption coefficient K from Eq. (11.21). The

Figure 11.20

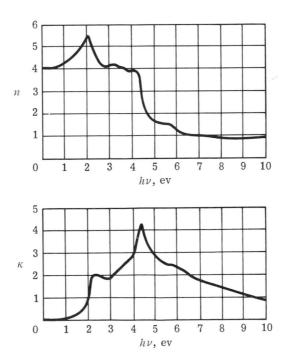

The spectral dependence of n and κ (the real and imaginary parts of the index of refraction) of Ge. (H. R. Philipp and E. A. Taft, Phys. Rev., vol. 113, p. 1002, 1959.)

calculated value of K for Ge is plotted in Fig. 11.21a as a function of the photon energy. The measured absorption curve of the type shown in Fig. 11.12 is available for Ge in the range 0.6 to 1.7 ev; therefore, the calculated curve of Fig. 11.21a can be checked against the measured curve in this region. Another check for the calculated curve can be secured from the static dielectric constant ϵ_s. According

Figure 11.21

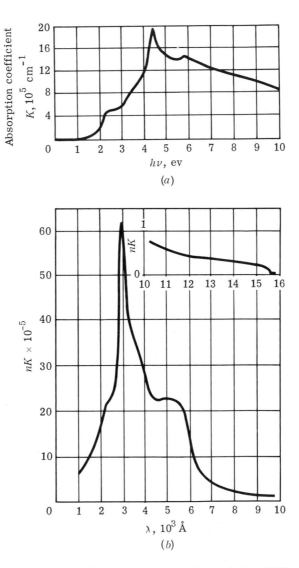

(a) The spectral dependence of the absorption coefficient K $[K = 4\pi\kappa/\lambda$, Eq. (11.21)] of Ge and (b) the function nK as a function of wavelength for Ge. (H. R. Philipp and E. A. Taft, Phys. Rev., vol. 113, p. 1002, 1959.)

to Eq. (11.70a), the static dielectric constant is given by

$$\epsilon_s = n^2(\omega_0 = 0) = 1 + \frac{1}{\pi^2} \int_0^\infty nK \, d\lambda \qquad (11.73)^{41}$$

The value of nK in Ge is plotted in Fig. 11.21b, using the value of K from Fig. 11.21a and the value of n from Fig. 11.20a. Performing the integration in Eq. (11.73), we find the value of $\epsilon_s = 15$ in good agreement with the accepted value of 16. For a similar analysis of the reflectance data for diamond using the Kramers-Kronig relations, the reader is referred to another paper by Philipp and Taft.[42]

We should point out that the curves shown in Fig. 11.11 for the real and imaginary parts of the dielectric constant are based on a single oscillator having a resonant frequency ω_0. A detailed quantum-mechanical treatment of dielectric dispersion in the optical region shows that the atomic polarizability α can be written in the general form[43]

$$\alpha = \frac{n_0 e^2}{m} \sum_{j,l} \frac{f_{jl}}{\omega_{jl}^2 - \omega^2 + i\beta_{jl}\omega/m} \qquad (11.74)$$

where the resonant frequency ω_{jl} is determined by the energy involved $(\mathcal{E}_j - \mathcal{E}_l = \hbar\omega_{jl})$ in an allowed optical transition between energy levels j and l. In Eq. (11.74) n_0 is the number of valence electrons per atom and the parameter

$$f_{jl} = \frac{2\pi}{h} \left(\frac{2m}{e^2}\right) \omega_{jl}|M_{jl}|^2 \qquad (11.75)$$

is generally referred to as the *oscillator strength* which determines the magnitude of the contribution of each transition to the optical dielectric constant. As mentioned earlier in Sec. 11.7, the matrix element M_{jl} is often expressed in terms of the electric dipole operator for insulators as

$$M_{jl} = \int \Phi_j^* er\Phi_l \, dx \, dy \, dz \qquad (11.76)$$

and expressed in terms of the momentum operator for metals and semiconductors as

$$|M_{jl}|^2 = \left(\frac{e\hbar}{\omega_{jl}m}\right)^2 \sum_{x,y,z} \left| \int \Phi_j^* \frac{\partial}{\partial x} \Phi_l \, dx \, dy \, dz \right|^2 \qquad (11.77)$$

[41]   The reader is again reminded that the equation is expressed in cgs units. Since $\epsilon_0 = 1$ for the dielectric constant of free space, $\epsilon_s/\epsilon_0 = \epsilon_s$.

[42]   H. R. Philipp and E. A. Taft, *Phys. Rev.*, vol. 136, p. A1445, 1964.

[43]   For a general discussion, see P. Nozières and D. Pines, *Phys. Rev.*, vol. 109, pp. 741, 762, and 1062, 1958; F. Seitz, "The Modern Theory of Solids," McGraw-Hill Book Company, New York, 1940; N. F. Mott and H. Jones, "Properties of Metals and Alloys," Oxford University Press, London, 1936.

Figure 11.22 shows the detailed structure of the reflectance curves of InSb, InAs, GaAs, GaP, and Ge reported in a paper by Ehrenreich, Philipp, and Phillips.[44] In the same figure, the numbers represent the gap energies (defined in Fig. 11.14) deduced from the reflectance data. The peaks in the reflectance curve can be attributed to transitions across the energy gap at different points of the k space. Therefore, the measurement of the reflectance of a given material in the optical range not only gives a complete set (real and imaginary parts) of the dielectric constant but yields information about the band structure also. In the paper by Philipp and Ehrenreich,[45] the reflectance measurement is further extended into higher photon energy from 10 to 25 ev for a number of semiconductors. A typical reflectance curve is plotted in Fig. 11.23a for InSb and the corresponding

[44] H. Ehrenreich, H. R. Philipp, and J. C. Phillips, *Phys. Rev. Letters*, vol. 8, p. 59, 1962.

[45] H. R. Philipp and H. Ehrenreich, *Phys. Rev.*, vol. 129, p. 1550, 1963.

Figure 11.22

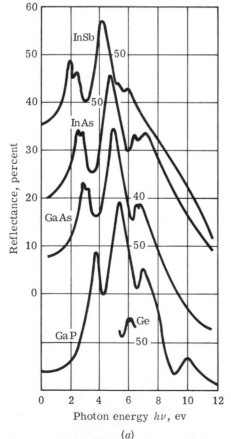

The reflectance of InSb, InAs, GaAs, GaP, *and* Ge *as a function of photon energy.* (*H. Ehrenreich, H. R. Philipp, and J. C. Phillips, Phys. Rev. Letters, vol. 8, p. 59, 1962.*)

	$\mathcal{E}_{\mathrm{L}}$	$\mathcal{E}_{\mathrm{L'}}$	$\mathcal{E}_{\Gamma}$	$\mathcal{E}_{\mathrm{X}}$
InSb	1.8 2.4	5.3 6.0	3.4	4.1
InAs	2.5 2.8	6.4 7.0	5.2	4.7
GaAs	2.9 3.1	6.6 6.9		5.0
GaP			3.7	5.3
Ge	2.1 2.3	5.9 6.1	3.1	4.5
Si	3.7	5.5	3.5	4.5
C		9	7.0	12.7
ZnSe	4.9 5.3	9.1 9.6		6.7

(*b*)

curves (deduced from Fig. 11.23a) for the real and imaginary parts of the dielectric constant are shown in Fig. 11.23b.

Another important relation useful in the analysis of the dispersion curve is the *sum rule* which can be stated mathematically as

$$\int_0^\infty \omega\epsilon''(\omega)\,d\omega = \frac{\pi}{2}\omega_p{}^2 \tag{11.78a}$$

or

$$\int_0^\infty \omega\mathrm{Im}\epsilon^{-1}(\omega)\,d\omega = \left(-\frac{\pi}{2}\right)\omega_p{}^2 \tag{11.78b}$$

where $\epsilon''(\omega)$ $(= 2n\kappa)$ is the imaginary part of the dielectric constant.

Figure 11.23

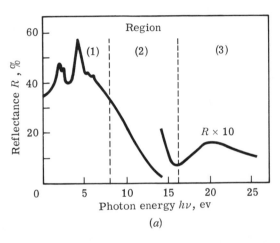

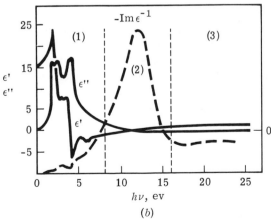

The spectral dependence of the reflectance R, the real and imaginary parts of the dielectric constant, ϵ' and ϵ'', and the energy loss function $(-\mathrm{Im}\ \epsilon^{-1})$ for InSb. (H. R. Philipp and H. Ehrenreich, Phys. Rev., vol. 129, p. 1550, 1963.)

In Eq. (11.78), the quantity ω_p defined as

$$\omega_p{}^2 = \frac{4\pi N n_0 e^2}{m} \qquad (11.79)^{46}$$

is the free-electron *plasma frequency* of a substance having a total electron density Nn_0 with n_0 being the number of electrons per atom and N being the density of atoms per unit volume. For a physical interpretation of the result, Eq. (11.78) is rewritten as

$$\left(\frac{2\pi^2 N e^2}{m}\right) n_{\text{eff}} = \int_0^{\omega_0} \omega \epsilon''(\omega) \, d\omega \qquad (11.80)^{46}$$

where n_{eff} is the effective number of electrons per atom in the energy range between $\hbar\omega = 0$ and $\hbar\omega_0$ contributing to the optical property.

Figure 11.24a shows the plot of n_{eff} as a function of $\hbar\omega_0$ reported in the aforementioned paper by Philipp and Ehrenreich. The curve for silicon appears to approach closely to a value of four valence electrons per atom as expected. We note, however, that the curves for germanium and the III-V compounds extend appreciably above four. The increase above four is associated with the onset of d-band excitation, that is, associated with the weak (oscillator strength) coupling between the conduction and d bands. Other evidences of d-band excitation come from Figs. 11.23a and 11.24b. In the region 3 of Fig. 11.23a, the reflectance R rises again, indicating the onset of a new transition. In Eq. (11.73), if the upper limit of integration is replaced by ω_0, we can define an effective static dielectric constant

$$\epsilon_{s,\text{eff}} = 1 + \frac{1}{\pi^2} \int_0^{\omega_0} n K \, d\lambda = 1 + \frac{2}{\pi} \int_0^{\omega_0} \omega^{-1} \epsilon''(\omega) \, d\omega \qquad (11.81)$$

The value of $\epsilon_{s,\text{eff}}$ is plotted in Fig. 11.24b as a function of $\hbar\omega_0$. Since there are no d electrons in silicon, the value of $\epsilon_{s,\text{eff}}$ should saturate at a value of ϵ_s corresponding to the independently measured low-frequency dielectric constant. This is indeed the case as can be seen from Fig. 11.24b. For other curves, the discrepancy between ϵ_s and $\epsilon_{s,\text{eff}}$ is due to the contribution from the d bands.

Finally, we should like to discuss briefly the measurement of optical constants in the X-ray range. In the energy region above $h\nu = 10^2$ ev, materials become relatively transparent and hence the reflectance is very small. However, absorption data are readily available in the X-ray region. In this region, the dispersion relation of interest

$$n(\omega_0) = 1 + \frac{\lambda(\omega_0)}{4\pi^2} \int_0^\infty \frac{dK(\omega)}{d\omega} \log \left| \frac{\omega + \omega_0}{\omega - \omega_0} \right| d\omega \qquad (11.82)$$

---

[46] The equations are expressed in cgs system. In mks system, Eq. (11.79) should read $\omega_p{}^2 = N n_0 e^2 / m\epsilon_0$ and Eq. (11.80) should be modified accordingly.

relates the absorption coefficient $K = 4\pi\kappa/\lambda$ to the index of refraction. Figure 11.25 shows the result of an analysis of the absorption data in Al reported by Philipp and Ehrenreich.[47] The important features are the following. According to Eq. (11.28), for $\omega \gg \omega_0$,

[47]   H. R. Philipp and H. Ehrenreich, *J. Appl. Phys.*, vol. 35, p. 1416, 1964.

Figure 11.24

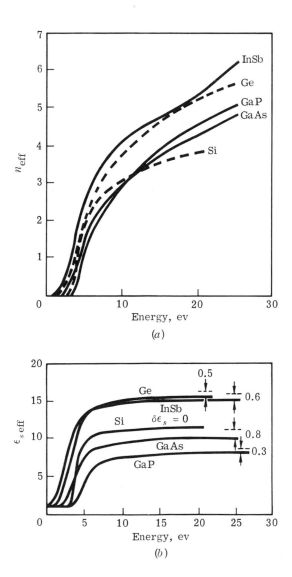

The effective number of electrons per atom and the effective dielectric constant as functions of energy in a number of semiconductors. (H. R. Philipp and H. Ehrenreich, Phys. Rev., vol. 129, p. 1550, 1963.)

$\kappa = 0$ and the index of refraction is expected to vary as

$$n^2 = 1 - \frac{\omega_p^2}{\omega^2} \qquad \text{or} \qquad n \simeq 1 - \frac{\omega_p^2}{2\omega^2} \qquad (11.83)$$

where ω_p is the plasma frequency. This relation is indeed obeyed as indicated by the dashed line in Fig. 11.25a. In the analysis of the dispersion relations [Eqs. (11.70), (11.71), and (11.73)], information concerning n in the high-energy region is not always available. Equation (11.83) provides a basis for the extrapolation of the optical data to the high-energy region. The effective number of electrons

Figure 11.25

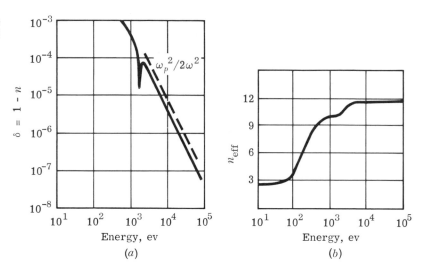

(a) The spectral dependence of $1 - n$ (the index of refraction) and (b) the effective number of electrons per atom as functions of energy in aluminum. (H. R. Philipp and H. Ehrenreich, J. Appl. Phys., vol. 35, p. 1416, 1964.)

per atom is plotted as a function of energy in Fig. 11.25b. The number 3 (with energies up to 70 ev) is associated with transitions involving $3s$ and $3p$ electrons of Al. For energies from 70 to 1.5×10^3 ev, the eight L-shell electrons also contribute to the optical properties. For energies beyond 1.5×10^3 ev, further contribution to n_{eff} comes from the two K electrons. Figure 11.25b is in qualitative agreement with this picture although the quantitative disagreement which exists between the measured value of n_{eff} and the theory is still unaccounted for.

PROBLEMS

11.1 Read and summarize the discussions in "Electronic Processes in Ionic Crystals," by N. F. Mott and R. W. Gurney, (Clarendon Press,

Oxford, 1948) concerning the determination of the energies χ and ψ in an ionic crystal (pp. 69–80).

11.2 Read Secs. 5 and 8 in the article by D. L. Dexter, Theory of the Optical Properties of Imperfections in Nonmetals (*Solid State Physics*, vol. 6, pp. 367 and 387, 1958). Explain why lattice vibrations play a part in the luminescent process of KCl:Tl. State the Franck-Condon principle in reference to the change of the vibrational modes during the luminescent process.

11.3 Read the paper by J. J. Hopfield, A Theory of Edge Emission in CdS, ZnS, and ZnO (*J. Phys. Chem. Solids*, vol. 10, p. 110, 1959). Describe the process of edge emission on the basis of the analysis by Hopfield.

11.4 Read the article by F. E. Williams, Theory of the Energy Levels of Donor-Acceptor Pairs (*J. Phys. Chem. Solids*, vol. 12, p. 265, 1960). Present the theory and experimental evidence supporting the associated donor-acceptor model.

11.5 Read the articles by R. Bowers and N. T. Melamed, Luminescent Centers in ZnS(Cu, Cl) Phosphors (*Phys. Rev.*, vol. 99, p. 1781, 1955) and by J. Lambe and G. Kikuchi, Paramagnetic Resonance of CdTe(Mn) and CdS(Mn) (*Phys. Rev.*, vol. 119, p. 1256, 1960). Discuss experimental evidences from which the ionized states of Cu and Mn can be deduced.

11.6 The recombination processes responsible for luminescence can generally be classified into two types: monomolecular and bimolecular. The rates of recombination are respectively given by

$$\frac{dn}{dt} = -\alpha n \qquad \text{and} \qquad \frac{dn}{dt} = -\alpha n^2$$

Show that the luminescence intensity in these two cases will decay, respectively, as

$$I = I_0 \exp(-\alpha t) \qquad \text{and} \qquad I = \frac{I_0}{(1 + \sqrt{I_0 \alpha}\, t)^2}$$

Compare the two decay curves if it is assumed that both curves drop to one-tenth of its initial value at $t = 10^{-3}$ sec.

11.7 Derive Eq. (11.5) and prove Eq. (11.6). If the solution of Eq. (11.6) is expressed as

$$\mathcal{E}(\text{ev}) = \frac{T_1\,(°\text{K}) - T_0(\beta/s)}{K(\beta/s)}$$

show that at $\beta/s = 10^{-4}$ and 10^{-15}, the values of $\mathcal{E}$ which satisfy Eq. (11.6) can be obtained by using in the above equation $T_0 = 35°\text{K}$ and $K = 833$ for $\beta/s = 10^{-4}$, and $T_0 = 4°\text{K}$ and $K = 312$ for $\beta/s = 10^{-15}$, respectively.

11.8 Read the discussions in the book by D. Curie, "Luminescence in Crystals," pp. 165–170, Methuen & Company, Ltd., London, 1963. Explain the physical model used to calculate the energies of the various traps (0.50, 0.57, 0.65, and 0.72 ev of Fig. 11.6*b*) in ZnS:Cu phosphors.

11.9 Read Secs. 4, 5, and 6 in the article by W. W. Piper and F. E. Williams, Electroluminescence (*Solid State Physics*, vol. 6, p. 95, 1958). Discuss the processes of excitation, energy transfer, and emission in an electroluminescent material.

11.10 Show that the reflection coefficient for oblique incidence at an air-dielectric boundary is equal to

$$r = \frac{\cos \theta - A}{\cos \theta + A}$$

for a plane wave with the electric field normal to the plane of incidence. In the above equation, $A = \sqrt{(n - i\kappa)^2 - \sin^2 \theta}$ and θ is the angle of incidence. Simplify the expression of r for the case $n^2 \gg \kappa^2$ and $n^2 > 1$.

11.11 The problem of wave propagation of the kind shown in Fig. 11.9 is formally solved by proposing that (1) two waves propagating in opposite directions exist in the dielectric medium and (2) an incident wave and a reflected wave exist in the air medium on the left and a transmitted wave exists in the air medium on the right. By applying the boundary conditions to **E** and **H** at the air-dielectric boundaries, prove Eq. (11.24) and derive an expression for the ratio of the reflected to the incident light for normal incidence.

11.12 Starting from Eq. (3.14), the equation of motion for free electrons, show that the contribution from free carriers to the dielectric constant is given by

$$n^2 - \kappa^2 = -\frac{Ne^2}{m^*\epsilon_0}\frac{\tau^2}{1 + \omega^2\tau^2} \qquad 2n\kappa = \frac{Ne^2}{m^*\epsilon_0}\frac{\tau}{\omega(1 + \omega^2\tau^2)}$$

Find the absorption coefficient due to free carrier absorption and discuss its dependence on carrier concentration and wavelength. Read the article by H. Y. Fan, W. Spitzer, and R. J. Collins (*Phys. Rev.*, vol. 101, p. 566, 1956), and compare the theoretical and experimental results.

11.13 Read the articles by F. Matossi and F. Stern (*Phys. Rev.*, vol. 111, p. 472, 1958) and by E. O. Kane (*J. Phys. Chem. Solids*, vol. 1, p. 94, 1956). Explain the mechanisms for the absorption observed in p-type Ge and InSb in the micron wavelength region.

11.14 Read the article by E. Burstein, G. Picus, and N. Sclar ("Photoconductivity Conference," p. 353, John Wiley & Sons, Inc., 1956). Discuss the various absorption processes in Ge and Si for photon energies smaller than 1.5 ev.

11.15 Read the article by G. G. Macfarlane et al. (*Phys. Rev.*, vol. 111, p. 1245, 1958) and discuss how the phonon energy is determined in an indirect transition in germanium. Obtain the lattice-vibration spectrum of germanium from the article by B. N. Brockhouse and P. K. Iyengar (*Phys. Rev.*, vol. 108, p. 894, 1957; vol. 111, p. 747, 1958). Compare the phonon energies involved in the optical transition and in the tunneling process (Ref: J. J. Tiemann and H. Fritzsche, *Phys. Rev.*, vol. 132, p. 2506, 1963) with the energy obtained from the vibrational spectrum.

11.16 Read the article by R. Braunstein, A. R. Moore, and F. Herman (*Phys. Rev.*, vol. 109, p. 695, 1958) and discuss how the optical energy gap observed in germanium-silicon alloy can be explained on the basis of the energy-band structure of the alloy.

11.17 Show that in Eq. (11.42),

$$(u_s\alpha|h_1 + h_2|u_z\alpha) = -ikP \qquad \text{and} \qquad (u_+\beta|h_1 + h_2|u_z\alpha) = \sqrt{2}\,\Delta/3$$

where h_1 and h_2 are the perturbation potentials defined in the text.

11.18 Show from Eq. (11.48) that

$$\frac{dx}{dt} = \frac{\partial\mathcal{H}}{\partial p_x} = \frac{1}{m}\left(p_x + \frac{eA_x}{c}\right)$$

$$\frac{dp_x}{dt} = -\frac{\partial\mathcal{H}}{\partial x} = -\frac{e}{mc}\left[\sum_{j=x,y,z}\left(p_j + \frac{eA_j}{c}\right)\frac{\partial A_j}{\partial x}\right] - \frac{\partial V}{\partial x}$$

From the above two equations, derive the following equation of motion:

$$m\frac{d^2x}{dt^2} = -\frac{\partial V}{\partial x} - eE_x - \frac{e}{c}(\mathbf{v} \times \mathbf{H})_z$$

where $\mathbf{H} = \text{curl }\mathbf{A}$ and $\mathbf{E} = -(1/c)(\partial\mathbf{A}/\partial t)$.

11.19 Derive the $(\hbar\omega - \mathcal{E}_g \pm k\Theta)^2$ dependence in Eq. (11.67) by noting the following

$$\tfrac{1}{2}\int\sqrt{x(a - x)}\,dx = -\frac{\sqrt{x(a - x)^3}}{4} + \frac{a}{8}\left(\sqrt{x(a - x)} + a\sin^{-1}\sqrt{\frac{x}{a}}\right)$$

Check the dimension of K in Eq. (11.67).

11.20 Read the article by W. P. Dumke (*Phys. Rev.*, vol. 105, p. 139, 1957). Outline the procedure in calculating the average lifetime of excess carriers due to direct transitions across the energy gap in nondegenerate germanium. Make clear how the averaging procedure is carried out.

11.21 Repeat Prob. 11.20 for indirect transitions.

11.22 Based on the analysis given in Sec. 11.6, especially Eq. (11.42), find the composition of wave functions near $k = 0$ for conduction-band and valence-band states.

11.23 Read the article by F. Stern, (*Solid State Physics*, vol. 15, p. 328, sec. 17, 1963). Outline the procedure in deriving the dielectric dispersion relations, Eq. (11.70).

11.24 Discuss how Figs. 11.20 and 11.22 can be explained on the basis of Eq. (11.74).

11.25 Read the article by F. Stern (*Phys. Rev.*, vol. 133, p. A1653, 1964). Show that the change in the index of refraction near an absorption edge can be approximated by the following expression

$$\Delta n = \begin{cases} \dfrac{Bc}{2\nu}\,[\sqrt{\mathcal{E}_g + h\nu} - \sqrt{\mathcal{E}_g - h\nu}] & \text{for } h\nu < \mathcal{E}_g \\[2em] \dfrac{Bc}{2\nu}\,\sqrt{\mathcal{E}_g + h\nu} & \text{for } h\nu > \mathcal{E}_g \end{cases}$$

Explain how the parameter B is related to the absorption coefficient K.

REFERENCES

Bube, R. H.: "Photoconductivity of Solids," John Wiley & Sons, Inc., New York, 1960.

Curie, D.: "Luminescence in Crystals," Methuen & Company, Ltd., London, 1963.

Ivey, H. F.: Electroluminescence and Related Effects, *Advan. Electron. Electron Phys.*, Suppl. 1, 1963.

Moss, T. S.: "Optical Properties of Semiconductors," Academic Press Inc., New York, 1959.

Seitz, F.: "Modern Theory of Solids," McGraw-Hill Book Company, New York, 1940.

Stern, F.: Elementary Theory of Optical Properties of Solids, *Solid State Physics*, vol. 15, p. 300, 1963.

Williams, F. E.: Solid State Luminescence, *Advan. Electron. Electron Phys.*, vol. 5, p. 137, 1953.

"Photoconductivity Conference," John Wiley & Sons, Inc., New York, 1956.

12 LASERS AND RELATED OPTICAL EFFECTS

12.1 LASERS AND CONVENTIONAL OPTICAL RADIATORS

The main characteristics of a laser beam in contrast to those of an ordinary incandescent light are coherence, directionality, monochromaticity, and concentrated intensity. In the optical (visible and infrared) region, lasers are the only source of coherent electromagnetic radiation so far available. Masers, on the other hand, do not enjoy such a unique position. First, coherent sources such as klystrons and magnetrons are available in the microwave region. Even for low-noise applications, masers face serious competition from parametric amplifiers. This explains why maser applications so far have been limited to radio and radar astronomy and microwave spectroscopy while many varied laser applications have been either demonstrated or suggested. These include microsurgery such as welding the loose retina tissue of an eye back to the supporting tissue, precision micro-welding in metalwork, and plasma diagnostics, to name a few. Lasers can also be valuable tools in scientific research such as Raman spectroscopy, photochemistry, and studies of chemical bonds and reactions. Needless to say, interstellar and space communication and optical radar systems are among other possible applications of lasers being contemplated. Because the laser development is still in its infancy, uses which have not yet been foreseen will undoubtedly arise and systems contemplated today will probably become outdated tomorrow. For these reasons, we shall concentrate only on those aspects of engineering applications which are intimately connected with the basic properties of materials. Among the many possible laser applications, the use of lasers in a communication system is closest to the heart of electrical engineering, and among the functions of various component parts in a communication system, modulation and harmonic generation of signals are ranked most essential. Consequently, in concluding this chapter, we shall review the various schemes suggested for modulating the amplitude and frequency of a laser beam and discuss the nonlinear properties of solids for detection and harmonic generation of optical signals.

Before we discuss the various types of lasers, it is instructive to compare the optical properties of a coherent laser beam with those of a conventional incoherent beam. For a plane wave, we can write the electric field as

$$\mathbf{E} = \mathbf{E}_0 \exp\left[i(\omega t - kz + \beta)\right] \tag{12.1}$$

where z is the direction of wave propagation, $\mathbf{E}_0$ is the electric field vector lying in a plane perpendicular to the z axis, and β is the phase angle. Generally speaking, a coherent source is characterized by a clearly defined ω and by a fixed phase angle β. In contrast, for the radiation field from an incoherent source, not only does ω spread over a relatively wide frequency spectrum but also β is random. Examples of incoherent radiation include black-body radiation (caused by thermal excitations) and mercury-arc lamp (involving atomic transitions).

A familiar source of coherent radiation in the radio frequency region is a dipole antenna. The far-zone electric field from a short antenna of length b, which lies in the z direction, is given by

$$\mathbf{E} = \boldsymbol{\theta}\, \frac{i\omega\mu_0 I_0 b \sin\theta}{4\pi r} \exp\left[i(\omega t - kr + \beta)\right] \tag{12.2}[1]$$

where I_0 is the amplitude of the antenna current and (r,θ,ϕ) is the polar coordinate of the point at which $\mathbf{E}$ is measured. The wave described by Eq. (12.2) is generally referred to as a spherical wave. From Eqs. (12.1) and (12.2), we see that if β is independent of both time and space coordinates, the wave front [planar in the case of Eq. (12.1) and spherical in the case of Eq. (12.2)] is also a surface of constant phase angle. In general, however, β may be fixed in time but varies spatially, Therefore, in dealing with the coherence of a laser beam, we are concerned with both the degree of temporal coherence which describes phase correlation in the direction of propagation and the degree of spatial coherence which describes phase correlation across the wave front.

[1] S. Ramo and J. R. Whinnery, "Fields and Waves in Modern Radio," p. 498, John Wiley & Sons, Inc., New York, 1958. The equation is in mks units.

Figure 12.1

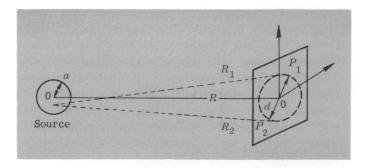

The coherence problem considered by van Cittert and Zernike. (M. Born and E. Wolf, "Principles of Optics," p. 508, Pergamon Press, London, 1964.)

To discuss the coherence characteristics of a light beam obtained from a conventional source, we shall present the problem originally considered by Van Cittert and Zernike.[2] Figure 12.1 shows a source S emitting quasimonochromatic light, that is, light consisting of spectral components which cover a frequency range $\Delta \nu$ small compared to the mean frequency. It is further assumed that the electromagnetic radiation arising from different elements of the source is mutually incoherent. Under such circumstances, the correlation function which defines the degree of coherence of the two light beams received at points P_1 and P_2 is given by

$$\mu(P_1, P_2) = \frac{1}{\sqrt{I(P_1)}\sqrt{I(P_2)}} \int_{\text{source}} I(S) \frac{\exp\left[ik(R_1 - R_2)\right]}{R_1 R_2} dS \qquad (12.3)$$

where

$$I(P_1) = \int \frac{I(S)}{R_1^2} dS \qquad I(P_2) = \int \frac{I(S)}{R_2^2} dS \qquad (12.4)$$

are the light intensities at P_1 and P_2. In Eqs. (12.3) and (12.4), $I(S)$ is the intensity per unit area of the source, R_1 and R_2 are respectively the distances from P_1 and P_2 to the source, and k is the average wave number of the emitted radiation.

For complete coherence, the value of μ is unity. As the point P_1 moves away from P_2, the degree of coherence decreases. The analysis by Van Cittert and Zernike shows that the value of μ drops to 0.88 when

$$d = \frac{0.16\lambda R}{a} \qquad (12.5)$$

where a is the radius of the source (assumed to be circular) and d is the distance between P_1 and P_2. If we set the tolerance limit of coherence at $\mu \geq 0.88$, then the diameter of the circular region illuminated almost coherently is equal to $0.16\lambda R/a$. To give a coherence region of 1 mm in diameter at $\lambda = 8000$ Å, the ratio a/R must be smaller than 1.3×10^{-4}. According to Lambert's law, the power densities I_s at the source and I_p at the receiving point are related to each other through

$$I_p = \frac{\pi a^2 I_s}{\pi R^2} \qquad (12.6)$$

Therefore, in order to obtain a coherence region of 1 mm in diameter at $\lambda = 8000$ Å, we must place the source so far away that only 1 part in 10^8 of the flux density emitted by the source falls into this coherence region.

Next, we shall discuss the focusing property of a laser beam. Figure 12.2a shows schematically the image formation by a lens.

[2] M. Born and E. Wolf, "Principles of Optics," p. 502, Pergamon Press, London, 1959.

According to the Abbe theory,[3] if $F(x,y,z)$ is the amplitude of the electric field at the point (x,y,z) in the plane of the object (plane II of Fig. 12.2a) the amplitude of the electric field at a point (ξ,η,ζ) in the image plane (plane II$'$ of Fig. 12.2a) is given by

$$G(\xi,\eta) = \frac{k}{2\pi f} \int F(x,y) \exp\left[-i\frac{k}{f}(x\xi + y\eta) \right] dx\, dy \qquad (12.7)$$

where f is the distance of the focal plane. As an illustration, we shall use the experimental results of Boyd and coworkers.[4] The laser beam from a confocal resonator in which phase-correcting optics are employed to eliminate diffraction loss has been shown to have a Gaussian intensity distribution and to operate in the uniphase TEM_{00} mode. In other words, the source function $F(x,y)$ in Eq. (12.7) can be taken as

$$F(x,y) = \frac{1}{\alpha} \sqrt{\frac{P}{2\pi}} \exp\left(-\frac{\rho^2}{4\alpha^2} \right) \qquad (12.8)$$

where P is the laser power, α is the standard deviation of the laser beam intensity in x and y directions, and $\rho^2 = x^2 + y^2$.

[3] M. Born and E. Wolf, *op. cit.*, pp. 420 and 481.
[4] G. D. Boyd and J. P. Gordon, *Bell System Tech. J.*, vol. 40, p. 489, 1961; G. D. Boyd and H. Kogelnik, *Bell System Tech. J.*, vol. 41, p. 1347, 1962.

Figure 12.2

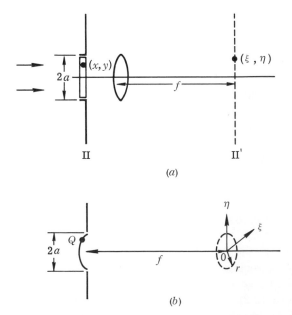

Illustrations showing (a) the image formation by a lens and (b) the diffraction at a circular aperture of a spherical wave converging toward the focal point 0.

In cylindrical coordinates, we can write

$$x\xi + y\eta = \rho\rho_1 \cos (\phi_1 - \phi) \qquad \text{and} \qquad dx\, dy = \rho\, d\rho\, d\phi \qquad (12.9)$$

where $\rho_1{}^2 = \xi^2 + \eta^2$, $\phi_1 = \tan^{-1} (\eta/\xi)$ and $\phi = \tan^{-1} (y/x)$. Substituting Eq. (12.8) into Eq. (12.7) and using Eq. (12.9), we find

$$G(\xi,\eta) = \frac{k}{f\alpha} \sqrt{\frac{P}{2\pi}} \int_0^a J_0 \left(\frac{\rho\rho_1 k}{f}\right) \exp\left(-\frac{\rho^2}{4\alpha^2}\right) \rho\, d\rho \qquad (12.10)$$

At the focal point $\xi = \eta = 0$, integrating Eq. (12.6) gives

$$G(0,0) = \frac{ka}{f} \sqrt{\frac{P}{2\pi}} \frac{1 - \exp(-\beta)}{\sqrt{\beta}} \qquad (12.11)$$

where $\beta = (a/2\alpha)^2$ and a is the radius of the aperture. Equation (12.11) has a maximum at $\beta = 1.14$. Using this value of β in Eq. (12.11), we find the maximum intensity at the focal point to be

$$I_{\max} = |G(0,0)|^2 = 0.066 \left(\frac{ka}{f}\right)^2 P \qquad (12.12)$$

Consider a similar problem (Fig. 12.2b) analyzed by Wolf,[5] in which a spherical monochromatic wave emerges from a circular aperture of radius a and converges toward the axial focal point O. If P denotes the total power incident on the aperture, the amount of power P' received in the focal plane within a circular region of radius r around the focal point can be expressed in terms of two Bessel functions J_0 and J_1 as follows:

$$P' = P[1 - J_0{}^2(v) - J_1{}^2(v)] \qquad (12.13)$$

where $v = (ka/f)r$. For $v = 3.83$, $J_1 = 0$ and $J_0 = 0.402$; thus we have $P' = 0.838P$. In other words, the average intensity near the focal point in the focal plane is given by

$$I_{av} = \frac{P'}{\pi r^2} = \frac{0.838P}{\pi(3.83)^2} \left(\frac{ka}{f}\right)^2 = 0.018 \left(\frac{ka}{f}\right)^2 P \qquad (12.14)$$

Equation (12.13) is well known in the theory of diffraction as the Rayleigh formula, and the radius $r = 3.83f/ka$ is the radius of the first dark ring. Therefore, Eqs. (12.12) and (12.14) state physically the same limitation imposed by the diffraction theory. Since the size of the first ring is given by $3.83f/(ka)$ and can not be arbitrarily reduced, the light intensity at or near the focal point is also limited.

Finally, we should mention that so far as the maximum power density obtainable from a light beam is concerned, a coherent beam with power P (watts) is approximately equivalent (Prob. 12.1) to an incoherent beam with brightness b (watts/cm^2) such that $b = P/\lambda^2$. Thus, with $P = 2$ milliwatts easily obtainable from a helium-neon

[5] M. Born and E. Wolf, op. cit., p. 442.

gas laser at $\lambda = 1.118 \times 10^{-6}$ m, we have $b = 1.6 \times 10^5$ watts/cm². The commercially available mercury-arc lamp has a brightness about 250 watts/cm² over the entire visible spectrum. In other words, the brightness of a helium-neon laser beam exceeds that of a mercury-arc lamp by a factor of about 6000. We should emphasize that the helium-neon laser beam has a frequency spread $\Delta\nu \sim 20$ Kc corresponding to a spread in wavelength $\Delta\lambda \sim 10^{-6}$ Å. Therefore, the difference in the brightness per unit frequency interval of a laser beam and that of a conventional light beam is even much more impressive than the difference in the total brightness.

12.2 SEMICONDUCTOR LASERS

Condition for population inversion. When a junction diode made of degenerate semiconductors is biased sufficiently in the forward direction, a state of population inversion is established near the p-n junction. For clarity, the energy-band diagram of a degenerate junction is shown in Fig. 12.3. The quasi-Fermi levels on both sides of the junction are given as $\mathcal{E}_{fn}$ and $\mathcal{E}_{fp}$ in the diagram. At absolute zero temperature, the conduction-band states are occupied up to $\mathcal{E}_{fn}$ while the valence-band states are empty down to $\mathcal{E}_{fp}$. Clearly a state of complete population inversion exists in the region bounded by 1 and 2. The total rate of radiative recombination depends on both the availability of electrons in the conduction band and the availability of holes in the valence band. From Fig. 12.3, it is clear that radiative recombination will be most intense in a narrow region near the physical junction. In the following, we shall first discuss the condition of population inversion, and then the laser oscillation condition.

The condition for population inversion in a maser system has been discussed fully in Chapter 9. As we recall, the population distribution under equilibrium condition is described by Boltzmann statistics and the magnetic ions are treated as a system of isolated ions. If a higher energy state is more populated than a lower energy state, a state of population inversion is established. In a semiconductor, the electron distribution obeys Fermi-Dirac statistics, hence the state of population inversion needs a new definition. Consider two energy states, one in the conduction band and the other in the valence band, as

Figure 12.3

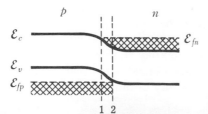

A degenerate semiconductor junction biased in the forward direction.

shown in Fig. 12.4a. Let $N_c(\mathcal{E})$ and $N_v(\mathcal{E})$ be the total density of states and $n_c(\mathcal{E})$ and $n_v(\mathcal{E})$ be the density of occupied states having energies between $\mathcal{E}$ and $\mathcal{E} + d\mathcal{E}$ in the conduction and valence band, respectively, then

$$n_c(\mathcal{E}) = N_c(\mathcal{E})f_c(\mathcal{E}) \qquad \text{and} \qquad n_v(\mathcal{E}) = N_v(\mathcal{E})f_v(\mathcal{E}) \qquad (12.15)$$

In Eq. (12.15), $f_c(\mathcal{E})$ and $f_v(\mathcal{E})$ are the usual Fermi-Dirac probability functions with

$$f_c = \left[1 + \exp\left(\frac{\mathcal{E}_1 + \mathcal{E}_c - \mathcal{E}_{fn}}{kT}\right) \right]^{-1}$$

and

$$f_v = \left[1 + \exp\left(\frac{\mathcal{E}_v - \mathcal{E}_2 - \mathcal{E}_{fp}}{kT}\right) \right]^{-1}$$

$$(12.16)$$

where $\mathcal{E}_{fn}$ and $\mathcal{E}_{fp}$ are the quasi-Fermi levels for conduction- and valence-band electrons and $\mathcal{E}_1$ and $\mathcal{E}_2$ are the respective energies of these electrons measured from the band edges. The net power of recombination radiation resulting from optical transitions between these energy states is proportional to

$$\text{Power} \sim (\mathcal{E}_1 + \mathcal{E}_2 + \mathcal{E}_g) W[n_c(N_v - n_v) \\ - n_v(N_c - n_c)] \, \delta(k) \, \delta(\mathcal{E}) \qquad (12.17)$$

Figure 12.4

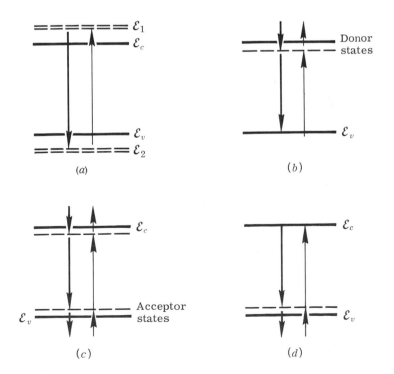

(a)

(b)

(c)

(d)

Transitions between various energy states.

In the above equation, the two delta functions $\delta(k)$ and $\delta(\mathcal{E})$ appear on account of momentum and energy conservation, and W is the transition probability. To obtain a net emission of radiation, we must have more downward than upward transitions, that means

$$n_c(N_v - n_v) > n_v(N_c - n_c) \qquad \text{or} \qquad \frac{n_c}{N_c} > \frac{n_v}{N_v} \qquad (12.18)$$

Under equilibrium, $\mathcal{E}_{fn} = \mathcal{E}_{fp}$; thus we find

$$\frac{n_c(N_v - n_v)}{n_v(N_c - n_c)} = \exp\left(-\frac{\mathcal{E}_1 + \mathcal{E}_2 + \mathcal{E}_g}{kT}\right) \ll 1 \qquad (12.19)$$

meaning that a nonequilibrium situation must be created to achieve a population inversion. Equation (12.18) further stipulates that the percentage of occupied conduction-band states must be greater than the percentage of occupied valence-band states for population inversion. In Eqs. (12.17) and (12.19) as well as in subsequent discussions, it is understood that $\mathcal{E}_g$ represents the direct-gap energy.

Oscillation condition. The consideration of energy and momentum conservation limits the optical transition only to states which obey Eqs. (11.59), (11.60), and (11.61). Therefore, the increase in $I(\nu)$ per unit length or the amplification constant K' is equal to

$$K'I(\nu) = \frac{dI(\nu)}{dx} = Wh\nu N(\nu)(f_c - f_v)\left(\frac{n}{c_0}\right) - \alpha I(\nu) \qquad (12.20)$$

where $\mathcal{E}_1$ in f_c and $\mathcal{E}_2$ in f_v are related by $\mathcal{E}_1 m_{c0} = \mathcal{E}_2 m_v$. In the above equation, the first term is from stimulated emission while the second term $\alpha I(\nu)$ represents the effective loss in the medium mostly due to free-carrier absorption discussed in Sec. 11.4. To achieve a condition of sustained oscillation, that is, a laser operating condition, two requirements must be met. First, a resonant structure must be provided so that the electromagnetic radiation released through recombination may make several passages to be further amplified in the active medium before leaving the medium. The simplest resonant structure consists of two parallel reflecting surfaces perpendicular to the plane of a semiconductor junction. Such a configuration is known as a *Fabry-Perot interferometer*. The semiconductor-air interface boundary may serve adequately as a reflecting surface if the refractive index of the semiconductor is large enough as is the case for GaAs.

The other requirement is that the overall amplification constant per round trip through the active medium must be positive. Let R be the reflection coefficient at the interface boundary and d be the separation between the two reflecting boundaries. For each traversal (forth and back) between the boundaries, the radiation energy is reduced by a factor R twice at the interface boundaries and augmented by a factor exp $(2K'd)$ inside the active medium. Hence, the laser oscillation condition becomes $R^2 \exp (2K'd) > 1$ or

$$\frac{32\pi^4 e^2 (2m_r)^{3/2}}{n c_0 h^5 \nu} (h\nu - \mathcal{E}_g)^{1/2} |P|^2 (f_c - f_v) \geq \alpha - d^{-1} \ln R \quad (12.21)[6]$$

Note that the left-hand side of Eq. (12.21) can be expressed in terms of the absorption coefficient K defined by Eq. (11.63) as $K(f_c - f_v)$.

A brief and qualitative discussion of Eq. (12.21) follows. A semiconductor diode is in an absorption or an emission mode, depending on the sign of $f_c - f_v$. For a nondegenerate semiconductor used in an absorption experiment, the conduction band is almost empty and the valence band fully occupied; therefore, in deriving the expression for the absorption coefficient, we can assume $f_c = 0$ and $f_v = 1$. In a laser operation, however, we hope that the opposite may be true, or in other words, $f_c = 1$ and $f_v = 0$ for a complete population inversion. This is the same condition used in defining the spontaneous lifetime. It should be pointed out that the rate of spontaneous recombination is proportional to $f_c(1 - f_v)$ instead of $f_c - f_v$. Therefore, three distinct regions exist for a semiconductor diode as a function of the diode current. As soon as a nonequilibrium situation is created by injection, spontaneous emission begins. As the injection current is increased, a condition $f_c > f_v$ is reached, which marks the beginning of stimulated emission. On further increase of the injected current, the difference $f_c - f_v$ increases. When the threshold condition given by Eq. (12.21) is met, the semiconductor diode goes into an oscillating mode of operation, emitting coherent electromagnetic radiation. Since the spread in the energy of electrons decreases as the operating temperature is lowered, the difference $f_c - f_v$ increases for the same amount of injection current. In other words, the threshold current decreases with decreasing temperature. This is borne out by experiments.

Figures 12.5 and 12.6 are the experimental curves[7] of emission intensity from a GaAs diode at 4.2°K and 77°K, showing a greatly improved performance at 4.2°K from both viewpoints of low threshold current and narrow bandwidth. The abrupt increase in radiation intensity may indicate the onset of stimulated emission. The linewidth of the emission spectrum of a different GaAs diode is plotted as a function of injection current[8] in Fig. 12.7, showing the narrowing of the spectral line as stimulated emission sets in. Direct evidence of laser action, however, must come from either the spatial coherence or the polarization of the emitted radiation. Below threshold, the light is unpolarized; above threshold, it is found experi-

[6] The expression is similar to the one given in the paper by A. L. Schawlow and C. H. Townes, *Phys. Rev.*, vol. 112, p. 1940, 1958.
[7] T. M. Quist et al., *Appl. Phys. Letters*, vol. 1, p. 92, 1962.
[8] M. I. Nathan et al., *Appl. Phys. Letters*, vol. 1, p. 62, 1962.

mentally thatt he electric field is predominantly in the plane of the junction. Figure 12.8 shows the radiation pattern[9] from a GaAs diode photographed as an image on the screen of an infrared converter tube which converts the infrared radiation into a visible image. Below threshold, the picture in Fig. 12.8a does not show any characteristic pattern, indicating scattered light from an incoherent light source. Above threshold, vertical bands appear as shown in Fig. 12.8b. These bands are due to interference between the waves emitted at various points along the edge of the diode junction; therefore, the interference pattern is a clear evidence that a definite phase relation or spatial coherence exists. The diodes used in early experiments have two sides polished optically flat and nearly parallel, and experimental results regarding threshold and bandwidth have varied, depending on the preparation of the two sides. It turns out that the zinc blende structure of GaAs has a cleavage plane (110) and the two sides of a laser diode can be made optically flat and parallel by cleaving. It is also possible to have a structure with four faces made flat and perpendicular to the junction plane through cleavage so that internally reflecting modes may exist with incident angles at four interface boundaries larger than the critical angle for total reflection. Since the transmission loss of an internally reflecting structure is extremely small, laser diodes using such a structure have a very high quality factor and hence a low threshold.

[9]   R. N. Hall et al., *Phys. Rev. Letters*, vol. 9, p. 366, 1962.

Figure 12.5

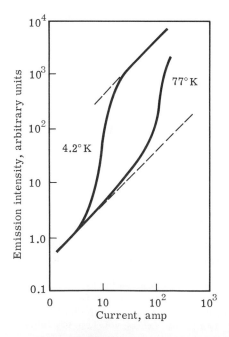

Relative emission intensity from a GaAs *diode as a function of diode current. (T. M. Quist* et al., *Appl. Phys. Letters, vol. 1, p. 91, 1962.)*

Figure 12.6

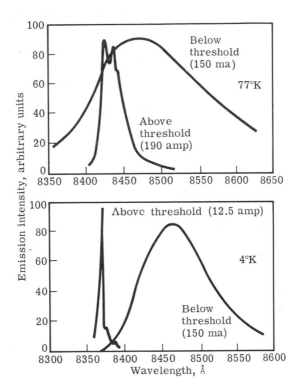

Spectra of emitted radiation from a GaAs diode before and after threshold of oscillation is reached. (*T. M. Quist et al., Appl. Phys. Letters, vol. 1, p. 91, 1962.*)

Figure 12.7

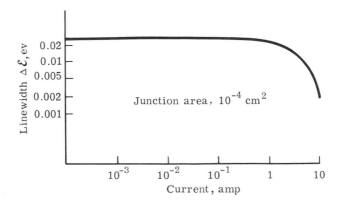

Linewidth of the emitted radiation from a GaAs laser diode as a function of diode current. (*M. I. Nathan et al., Appl. Phys. Letters, vol. 1, p. 62, 1962.*)

The reader may have noticed that in the above discussion of experimental results, only a qualitative but not a quantitative reference to Eq. (12.21) has been made. The reason is simple. In deriving Eq. (12.21), two basic assumptions are made: First, a plane wave mode of propagation is used. Second, the assumed transitions are between conduction- and valence-band states. The gap energy of intrinsic GaAs is 1.51 ev while the strong emission line of a GaAs diode, made by diffusing Zn into Te (3×10^{17} atoms/cm³) doped GaAs, occurs around 1.47 ev. Had transitions occurred between valence-band and conduction-band states, as shown in Fig. 12.4a, the emission spectrum should be on the higher energy side of the gap energy as indicated by Eq. (12.21). Figure 12.4b, c, and d shows other possible laser transitions involving donor and valence-band states, donor and acceptor states, and conduction-band and acceptor states respectively. All these transitions involve an energy smaller than the gap energy. The problem has not been entirely settled to this writing, but it looks quite certain that impurities of either or both kinds are involved in a laser transition.

Role of impurity states.   In discussing the role which impurities play in an optical transition, two main considerations come to the fore regarding the magnitude of the transition probability: (1) the nature of the electron wave function pertaining to impurity centers and (2) the spread of the wave function. At low impurity concentrations, information on both questions is available from the effective mass approximation for impurity states[10] mentioned in Sec. 3.10. To make use of the effective mass approximation, the following assumptions are needed: (1) The impurity states are discrete, shallow, and non-overlapping and (2) the valence- or conduction-band states are not

---

[10]   For a detailed discussion, the reader is referred to W. Kohn, Shallow Impurity States in Silicon and Germanium, in *Solid State Physics*, vol. 5, p. 258, 1957.

Figure 12.8

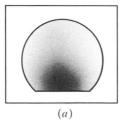
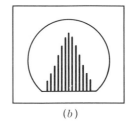

(a)   (b)

Below threshold   Above threshold

Radiation pattern from a GaAs diode. The existence of an interference pattern above the threshold indicates spatial coherence between rays emitted from different points along the edge of the junction. (R. N. Hall et al., Phys. Rev. Letters, vol. 9, p. 366, 1962.)

affected by the presence of these impurities. For definiteness, we shall consider transitions between donor and valence-band states. In the effective mass approximation, the donor-state wave function can be expressed in terms of the conduction-band cell periodic function $u_{c0}(r)$ at $k[000]$ as

$$\Phi_d = \frac{u_{c0}(r) \exp{(-r/a)}}{\sqrt{\pi}\, a^{3/2}} \tag{12.22}$$

where a is the effective donor-orbit radius defined as

$$a = \frac{\epsilon\hbar^2}{ze^2 m_{c0}} = \frac{ze^2}{2\epsilon\mathcal{E}_d} \tag{12.23}$$

$\mathcal{E}_d$ being the donor binding energy, ze the effective charge of the ionized donor, and m_{c0} the effective mass of conduction-band electrons at $k[000]$. Substituting Eq. (12.22) into Eq. (11.51) and following the same procedure used in deriving Eqs. (11.52) and (11.53), we find

$$\int \Phi_i^* \frac{\partial}{\partial x} \Phi_j \exp{(i\mathbf{n}\cdot\mathbf{r})}\, dq$$

$$= \int_{\text{cell}} u_v^* \frac{\partial}{\partial x} u_{c0}\, dq \int_{\text{crystal}} \frac{\exp{[i(\mathbf{k}-\mathbf{n})\cdot\mathbf{r}]} \exp{(-r/a)}\, dq}{V^{1/2}\pi^{1/2}a^{3/2}}$$

$$\simeq \frac{8\pi^{1/2}a^{3/2}mP}{(1+k^2a^2)^2\hbar^2 V^{1/2}} \tag{12.24}$$

and hence the transition probability W to be

$$W = \frac{32e^2a^3|P|^2 I(\nu)}{(1+k^2a^2)^4\hbar^4\nu^2 n^2 V} \tag{12.25}$$

In Eqs. (12.24) and (12.25), P is as defined in Eq. (11.37) and $dq = dx\, dy\, dz$. Since $|\eta| \ll |k|$, the contribution of η to Eqs. (12.24) and (12.25) has been neglected. In Eq. (12.25), V is the volume of the crystal and it enters into the equation because the valence-band electron wave function spreads over the whole crystal according to the one-electron Bloch wave function.

The spontaneous lifetime of an electron in an occupied donor state through recombination with a hole in the valence band is equal to[11]

$$\tau_s = \frac{(1+k^2a^2)^4\epsilon c^3 V\hbar^4}{256\pi\hbar\omega e^2 P^2 a^3} \tag{12.26}$$

Since in volume V, the total number of valence-band states having energy between $\mathcal{E}$ and $\mathcal{E}+d\mathcal{E}$ is $V4\pi(2m_v)^{3/2}\mathcal{E}^{1/2}\, d\mathcal{E}/h^3$, the lifetime of an occupied donor state is given by

$$\frac{1}{\tau} = \int \frac{256\pi\hbar\omega e^2 P^2 a^3}{(1+k^2a^2)^4\epsilon c^3 V\hbar^4} \frac{V4\pi(2m_v)^{3/2}}{h^3}(1-f_v)\mathcal{E}^{1/2}\, d\mathcal{E} \tag{12.27}$$

[11] W. P. Dumke, *Phys. Rev.*, vol. 132, p. 1998, 1963; vol. 133, p. 553A, 1964.

Note that from Eq. (12.23), $a = \hbar/(2m_{c0}\mathcal{E}_d)^{1/2}$; therefore, the quantity k^2a^2 can be expressed in terms of the energies of states under transition, that is, $\mathcal{E}_d$ for the donor state and $\mathcal{E}_{vk}$, being equal to $\mathcal{E}$ in Eq. (12.27), for the valence-band states with wave number k. Since $\mathcal{E}_{vk} = \hbar^2 k^2/2m_v$,

$$k^2a^2 = \frac{m_v \mathcal{E}_{vk}}{m_{c0}\mathcal{E}_d} \tag{12.28}$$

Thus, Eq. (12.27) becomes

$$\frac{1}{\tau} = \int \frac{32\sqrt{2}\, e^2 P^2 (2m_v)^{3/2}(m_{c0}\mathcal{E}_d)^{5/2}(\mathcal{E}_g - \mathcal{E}_d + \mathcal{E}_{vk})}{(m_{c0}\mathcal{E}_d + m_v\mathcal{E}_{vk})^4 \epsilon c^3 \pi \hbar^4}(1 - f_v)\mathcal{E}_{vk}^{1/2}\, d\mathcal{E}_{vk} \tag{12.29}$$

In general, the lifetime given by Eq. (12.29) involves a rather complicated integration over $\mathcal{E}_{vk}$. For our present purpose, to get an order-of-magnitude estimate of τ we shall make the following simplifying assumptions: (1) The valence band is degenerate so $1 - f_v = 1$ from $\mathcal{E}_{vk} = 0$ to the quasi-Fermi level $\mathcal{E}_{pf}$ and (2) the empty valence-band states have k values not far from $k = 0$ so that the term $\mathcal{E}_{vk}$ can be neglected in comparison with other terms in Eq. (12.29). Under these circumstances, Eq. (12.29) becomes

$$\tau = \frac{(m_{c0}\mathcal{E}_d)^{3/2}\epsilon c^3 \hbar}{64\sqrt{2}\,\pi e^2 P^2(\mathcal{E}_c - \mathcal{E}_d)p} \tag{12.30}$$

where p is the concentration of holes in the valence band. A comparison of Eq. (12.30) with Eq. (11.55) shows that the ratio of the lifetime for the donor to valence-band transition over that for the conduction-band to valence-band transition is given by

$$\frac{\tau_{dv}}{\tau_{cv}} = \frac{(m_{c0}\mathcal{E}_d)^{3/2}h\nu}{16\sqrt{2}\,\pi \hbar^3(\mathcal{E}_g - \mathcal{E}_d)p} \tag{12.31}$$

In GaAs, $m_{c0} = 0.072m$, $\mathcal{E}_d = 0.01$ ev. Thus, if we take $p = 10^{17}$ cm^{-3} and assume $h\nu \approx \mathcal{E}_g - \mathcal{E}_d$, then we find $\tau_{dv} = 0.1\,\tau_{cv}$. The value of $p = 10^{17}$ cm^{-3} is chosen because much below this value, the first assumption is no longer true while above this value, the second assumption becomes questionable. Even at $p = 10^{17}$ cm^{-3}, the valence band will have holes with enough energy to make $m_v\mathcal{E}_{vk} \simeq m_{c0}\mathcal{E}_d$ for the heavy-mass valence band with $m_v = 0.5m$. Therefore, the actual value of τ should be longer than the value predicted by Eqs. (12.30) and (12.31). A similar equation for the lifetime of a hole in an unoccupied acceptor state through recombination with conduction-band electrons can be obtained from Eqs. (12.30) or (12.31) by substituting $\mathcal{E}_a$ and m_v, respectively, for $\mathcal{E}_d$ and m_{c0}.

In applying the above calculations to lasers, the situation is complicated by the fact that at sufficiently high impurity densities, the impurity levels broaden and merge with the adjacent band edges. Furthermore, the impurity energy levels are shifted as a whole toward the center of the gap, accompanied by a tail extending into the forbidden gap as a result of random impurity distributions. Under such circumstances, it is even difficult to calculate the density of states as a function of energy in the impurity energy band, not to mention the matrix elements connecting the impurity states with valence- and conduction-band states. Note that the shallow donor and acceptor states respectively become fully occupied and empty before the conduction- and valence-band states from energy consideration. Even though these states may suffer in transition probability, they certainly gain in the Fermi-Dirac distribution function. For the above reasons, Eq. (12.30) or the more general expression Eq. (12.29) serves a very useful purpose in obtaining an order-of-magnitude estimate of the transition probability through impurity states. In general, a higher binding energy means a smaller spread in the impurity-state wave function, or a smaller Fourier component of the wave function near $k = 0$ and hence a smaller transition probability.

Modes of wave propagation. Now let us discuss the mode of light propagation in a semiconductor diode. Figure 12.9 shows (a) the experimental setup in measuring the light output from a laser junction and (b) the experimental curve showing the light energy distribution across the junction. The energy envelope of Fig. 12.9b indicates that the wave attenuates in the base regions of the diode as expected from a lossy dielectric. For simplicity, consider the ideal model of Fig. 12.10 in which region 2 is characterized by a lossy dielectric constant $\epsilon_2 = \epsilon_e + \sigma_2/i\omega$. Region 1 of thickness $2d$ has an effective dielectric constant $\epsilon_1 = \epsilon - \sigma_1/i\omega$. Region 1 represents the region of population inversion which is in the depletion layer of the junction; therefore, the negative conductivity in ϵ_1 indicates an active region and ϵ in ϵ_1 is the regular dielectric constant of the semiconductor (12.5 for GaAs). One possible origin for the difference between ϵ_e and ϵ lies in the fact that there is a free-carrier contribution to the dielectric constant in region 2, that is, the p and n base regions of the laser diode. For the TE mode of Fig. 12.10, the field components are

$$E_y = A_1 \cos{(kx)} \exp{(-i\beta z)} \qquad |x| < d \qquad \text{(12.32a)}$$

$$E_y = A_2 \exp{(-\alpha x - i\beta z)} \qquad |x| > d \qquad \text{(12.32b)}$$

$$H_x = -\frac{\beta E_y}{\omega \mu_0} \quad \text{and} \quad H_z = \frac{i}{\omega \mu_0} \frac{\partial E_y}{\partial x} \qquad \text{(12.32c)}$$

The continuity of E_y and H_z at $|x| = d$ requires

$$\alpha d = kd \tan{kd} \qquad \text{(12.33)}$$

Substituting Eq. (12.32) into the wave equation further yields

$$\beta^2 = \omega^2\mu_0\epsilon_1 - k^2 = \omega^2\mu_0\epsilon_2 + \alpha^2 \qquad (12.34)$$

From Fig. 12.9b, it is clear that $kd \ll 1$ so $\alpha \simeq k^2 d$ from Eq. (12.33) and $\alpha \ll k$. Thus from Eq. (12.34), we have

$$\frac{\alpha}{d} \simeq k^2 \simeq \omega^2\mu_0\left[(\epsilon - \epsilon_e) - \frac{(\sigma_1 + \sigma_2)}{i\omega}\right] \qquad (12.35)$$

Figure 12.9

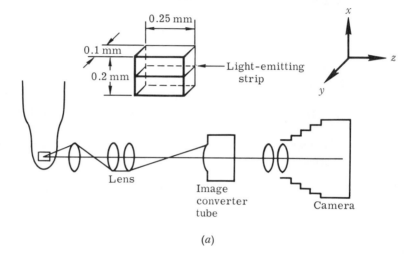

(a)

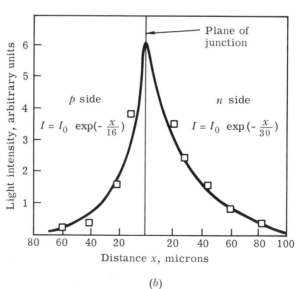

(b)

(a) *The experimental setup measuring the light output from a semiconductor laser diode and (b) the experimental curve showing the light energy distribution across the junction.* (W. L. Bond et al., Appl. Phys. Letters, vol. 2, p. 57, 1963.)

The real part of α is the attenuation constant which determines the envelope of the curves shown in Fig. 12.9b. From Eq. (12.35) we find

$$d_0 = \frac{1}{\text{Real part } \alpha} = \frac{\epsilon}{d\omega^2 \mu_0 \epsilon(\epsilon - \epsilon_e)} = \frac{c^2}{d\omega_p^2} \qquad (12.36)^{12}$$

In Eq. (12.36), the quantity $(\epsilon - \epsilon_e)/\epsilon$ is obtained by assuming that the difference $\epsilon - \epsilon_e$ is due solely to the free carrier contribution of the dispersion. Following the discussion at the end of Sec. 11.4, we have

$$\frac{\epsilon - \epsilon_e}{\epsilon} = \frac{Ne^2}{m^* \epsilon_0} \frac{1}{\omega^2} = \frac{\omega_p^2}{\omega^2} \qquad (12.37)$$

where N is the concentration of free carriers, m^* the effective mass of these carriers, and ω_p is generally referred to as the plasma frequency. For a typical GaAs degenerate junction with $2d = 0.1$ micron and $N = 10^{18}$ cm^{-3}, Eqs. (12.36) and (12.37) yield a value of $d_0 = 45$ microns in reasonable agreement with experimental values shown in Fig. 12.9b.

[12] It is not yet certain that the difference $\epsilon - \epsilon_e$ is necessarily due entirely to the free carrier contribution of the dispersion as suggested by Eq. (12.37). However, the experimental decay constant α is determined by the differential $\Delta\epsilon = \epsilon - \epsilon_e$ whatever its origin.

Figure 12.10

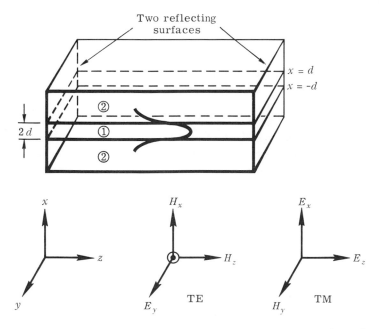

Idealized model used in analyzing wave propagation in a semiconductor laser diode.

Next we shall consider the loading of the lossy dielectric on the laser mode in the depletion region. For this purpose, solutions of Eqs. (12.33) and (12.35) must be carried to the second order, yielding

$$\alpha = k^2 d \left(1 + \frac{k_0^2 d^2}{3} \right)$$

$$k^2 = \frac{k_0^2}{1 + k_0^2 d^2} \simeq k_0^2 (1 - k_0^2 d^2) \qquad (12.38)$$

where $k_0^2 = \omega^2 \mu_0 (\epsilon_1 - \epsilon_2)$. Substituting Eq. (12.38) into Eq. (12.34), we find

$$\beta^2 = \omega^2 \mu_0 \left[\epsilon_2 + k_0^2 d^2 (\epsilon_1 - \epsilon_2) - 4(k_0 d)^4 \frac{\epsilon_1 - \epsilon_2}{3} \right] \qquad (12.39)$$

To obtain the laser oscillation condition, the contribution of free carriers to the dielectric constant can be neglected; thus $k_0^2 = i\omega \mu_0 (\sigma_1 + \sigma_2)$. Assuming $\beta = \omega(\mu_0 \epsilon)^{1/2}(1 + i\eta/2)$, we find

$$\eta = - \frac{\sigma_2}{\omega \epsilon} + \frac{8}{3} \left(\frac{\sigma_1 + \sigma_2}{\omega \epsilon} \right)^3 (\omega^2 \mu_0 \epsilon d^2)^2 \qquad (12.40)[13]$$

The condition for oscillation is that $\eta > 0$. The oscillation condition given by Eq. (12.21) assumes a plane-wave mode of propagation as if the electromagnetic radiation would not go into region 2, or in other words, $\sigma_2 = 0$ in Eq. (12.40). Obviously, consideration of the lossy dielectric which acts as a dielectric waveguide for the electromagnetic radiation raises the threshold for laser oscillation.

In summary, we can say that the major part of the theory concerning semiconductor lasers has been worked out quite satisfactorily, but some unanswered problems still remain. Most important, the energy levels involved in a laser transition have not been ascertained, and for that matter, laser transitions may occur at several wavelengths if the injected current is raised above the threshold current density. The question as to whether the TE or TM mode will have a lower threshold for oscillation also has not been thoroughly investigated. Notice that two different languages are used in Eqs. (12.21) and (12.40), the former in terms of the amplification constant K' and the latter in terms of a negative conductivity $- \sigma_1$. There must be a way to tie these two descriptions together. This problem is also a good exercise for the reader. The treatment given above is by no means exhaustive, but provides a basic and working knowledge for those who may want to carry out the analyses further and work out the details. Finally, we should like to mention that different schemes[14] have been proposed to achieve direct radiative recombination in indirect-gap semiconductors. However, the intervalley scattering time is about two orders of

[13] A. L. McWhorter, H. J. Zeiger, and B. Lax, *J. Appl. Phys.*, vol. 34, p. 235, 1963.
[14] S. Wang, *J. Appl. Phys.*, vol. 34, p. 3443, 1963.

magnitude faster than the spontaneous lifetime; therefore, electrons in the $k[000]$ valley are scattered back to the conduction-band minimum before they have a chance to recombine with valence-band holes.

12.3 GASEOUS LASERS

Besides semiconductor lasers, there are two other principal types: the gaseous and the solid-state lasers. In the gaseous type, transitions between the various excited states and the ground state of free atoms are used. The energy-level schemes for gaseous lasers such as the He-Ne system are comparatively easy to understand because both theoretical analyses and experimental data of the atomic spectra of noble and alkali vapors are generally available. In the solid-state laser, however, the energy-level schemes, for example those of Cr^{3+} ion in a ruby laser, are much more complicated because of the presence of the crystalline field. As discussed in Chapter 8, the crystalline field affects the energy levels of ions in varying degrees, depending on the electronic configuration of a given ion. For example, the energy levels of ions having $4f$ electrons are much less affected by the crystalline field than those of $3d$ electrons because the $4f$ electrons are electrically shielded by the outer $5s^2$ and $5p^6$ electrons. In order to understand the technical aspect of these lasers, it is necessary to make a spectroscopic analysis of optical absorption curves in these materials. We shall start our discussion with gaseous lasers, follow this treatment by considering solid-state lasers using the energy levels of rare-earth elements, and then turn to the ruby laser. The reason for this sequence is that the spectroscopic aspects of the problem become progressively more complicated in the same order. This order of presentation, however, does not exactly follow the history of laser development.

Consider the energy-level scheme used in the He-Ne laser system as shown in Fig. 12.11. The ground state of He atom has an electronic configuration $1s^2$ and using the spectroscopic notation adopted in Chapter 8, it is an S state. The first group of excited states, having an electronic configuration $1s2s$, consists of two levels which correspond to two possible spin arrangements, with the parallel arrangement shown as ^3S in Fig. 12.11. The electronic configurations of the ground state and various excited states of the Ne atom are also included in the same figure. The ^3S state of He is a metastable state because any direct optical transition to or from the ground state is forbidden by the quantum-mechanical selection rule. Excitations of He atoms are achieved through inelastic collisions with electrons in a gaseous discharge. The He metastable states are used as carriers of energy to excite the Ne atoms when excited He atoms collide with Ne atoms in the ground state. Due to the resonant nature of this energy transfer process, transfer of excitation occurs only to those levels of Ne which fall within a few kT of the He ^3S level. In other words, only the group $2p^54s$ levels are excited. Radiative transitions may then take place from the four $2p^54s$ levels to the ten $2p^53p$ levels. A detailed calculation by Koster and Statz[15] shows that only sixteen transitions out of a possible forty show nonvanishing transition probabilities. Of these, five have been observed in laser action, at 1.118, 1.153, 1.160, 1.199, and 1.207 microns.

A brief discussion of the energy-level diagram shown in Fig. 12.11 is in order. First, consider the five $2p$ electrons of Ne. Because of the Pauli exclusion principle, the two configurations $2p^5$ and $2p$ are almost equivalent so far as the values of the resultant S and L are concerned. Following the discussion given in Chapter 8, the $2p^5$ configuration has $S = \frac{1}{2}$ and $L = 1$, yielding two possible values for J, $\frac{1}{2}$ and $\frac{3}{2}$. Since the shell is more than half-filled, the $\frac{3}{2}$ state should be the lower energy state. The remaining question is how to couple the angular momentum values of the excited electron to the core $2p^5$ configuration. One obvious way is to use the scheme of Russell-Saunders coupling; that is, the value of S and that of L should be combined first, or, symbolically,

$$(l_1 s_1)(l_2 s_2) \cdots \Rightarrow (LS) \Rightarrow J \qquad (12.41)$$

with $\mathbf{L} = \mathbf{l}_1 + \mathbf{l}_2$ and $\mathbf{S} = \mathbf{s}_1 + \mathbf{s}_2$. Take the $2p^54s$ configuration. The possible values for S are 1 and 0 out of any $(s_1 = \frac{1}{2}, s_2 = \frac{1}{2})$ combination, and the only possible value for L is 1 out of the $(l_1 = 1, l_2 = 0)$ combination. Therefore, the Russell-Saunders coupling scheme gives the following states: ^3P, ^1P. Another possible

[15] G. F. Koster and H. Statz, *J. Appl. Phys.*, vol. 32, p. 2054, 1961.

Figure 12.11

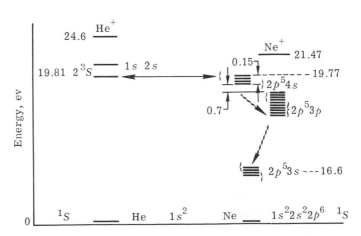

Energy-level diagram for a He-Ne laser. The arrow indicates the resonant transfer of excitation between He and Ne atoms.

scheme is the *j-j* coupling scheme in which the resultant values of J are coupled together, or, symbolically,

$$(l_1s_1)(l_2s_2) \cdot \cdot \cdot \Rightarrow (j_1, j_2, \ldots) \Rightarrow J \qquad (12.42)$$

For the $2p^5$ electrons, we have two values of j_1, $\frac{3}{2}$ and $\frac{1}{2}$; and for the $4s$ electron, we have $j_2 = \frac{1}{2}$. Therefore, the excited state $2p^54s$ has four levels according to four possible combinations of J, arranged in two groups which may be schematically shown as follows:

$$(j_1j_2) \begin{Bmatrix} \frac{1}{2} & \frac{1}{2} \to \\ \frac{3}{2} & \frac{1}{2} \to \end{Bmatrix} J \begin{Bmatrix} 1 & 0 \\ 2 & 1 \end{Bmatrix} \quad \begin{matrix} \text{(upper group of two levels)} \\ \text{(lower group of two levels)} \end{matrix}$$

with column headings $2p^5 \quad 4s$ and $2p^54s$.

The questions now arise as to what causes the differences in the two schemes and which scheme corresponds to the actual situation.

For a system of N electrons moving about a nucleus of charge ze, the Hamiltonian is approximately

$$\mathcal{H} = \sum_{i=1}^{N} \left[\left(\frac{1}{2m_0} p_i^2 - \frac{ze^2}{r_i} \right) + \xi(r_i) \mathbf{l}_i \cdot \mathbf{s}_i \right] + \sum_{i>j=1}^{N} \frac{e^2}{r_{ij}} \qquad (12.43)^{16}$$

where r_i represents the distance of the *i*th electron from the nucleus and r_{ij} is the mutual distance between the *i*th and *j*th electrons. Based on the model of an electron as a spinning top, Thomas and Frenkel have obtained an expression for the spin-orbit interaction energy of an electron in a central field with potential $V(r)$ to be

$$\mathcal{H}_1 = \frac{\hbar^2}{2m_0^2c^2} \left(\frac{1}{r} \frac{\partial V(r)}{\partial r} \right) \mathbf{l} \cdot \mathbf{s} = \xi(r) \mathbf{l} \cdot \mathbf{s} \qquad (12.44)$$

where c is the velocity of light. Therefore, the third term in Eq. (12.43) represents the spin-orbit interaction energy mentioned in Chapter 8. Here, we intend only to give a qualitative explanation of the various interactions, hence we need not be concerned with the exact form of $V(r)$. In Eq. (12.43) the first and second terms represent symbolically the kinetic and potential energies of the *i*th electron, without taking into account the necessary refinements for a many-body system. The fourth term is the electrostatic interaction energy of the *i*th and *j*th electrons.

Knowing the individual atomic wave functions of the N electrons, the spin-orbit interaction energy and the electrostatic repulsion energy between any two states can be calculated, using the terms $\xi(r)\mathbf{l} \cdot \mathbf{s}$ and e^2/r_{ij} as perturbation potentials. Furthermore, the energy difference between two states having the same n, l, and s but a different combination of l and s can be determined. For a detailed discussion of the calculation of these energies, the reader should refer to an advanced book on atomic spectroscopy, and our present

---

[16] The equation is in cgs units.

discussion is an outline of the treatment given in the book by Condon and Shortley.[17] For example, the energy of an electron in the presence of spin-orbit interaction is given by

$$W = W_{nl} + \frac{l\zeta_{nl}}{2} \qquad \text{for } j = l + \tfrac{1}{2}$$

$$W = W_{nl} - \frac{(l+1)\zeta_{nl}}{2} \qquad \text{for } j = l - \tfrac{1}{2}$$

(12.45)

where W_{nl} is the energy of the electron in the central field of $V(r)$ and the parameter ζ_{nl}, describing the strength of the spin-orbit interaction, is given by

$$\zeta_{nl} = \int R^2(n,l)\xi(r)\,dr \tag{12.46}$$

where $R^2(n,l)$ is the radial part of the electron wave function.

Similarly, the electrostatic interaction energy can be calculated from the wave functions of any two given states. By expanding $1/r_{ij}$ in a series of Legendre polynomials, the interaction can be expressed as a sum of the following functions:

$$F_k(n,l;n',l') = e^2 \int\!\!\int_0^\infty \frac{r_<^k}{r_>^{k+1}} R_1^2(n,l)R_2^2(n',l')\,dr_1\,dr_2$$

(12.47)

$$G_k(n,l;n',l') = e^2 \int\!\!\int_0^\infty \frac{r_<^k}{r_>^{k+1}} R_1(n,l)R_1(n',l')R_2(n,l)R_2(n',l')\,dr_1\,dr_2$$

in which $r_<$ is the lesser and $r_>$ the greater of r_1 and r_2 and these are a direct result in the expansion of $1/r_{ij}$. Here, we are not concerned with the actual evaluation of these integrals. The expressions[18] for ζ, F, and G are given not only because they will be referred to later for concreteness but also because they are often used in the literature.

When the spin-orbit interaction is much larger than the electrostatic interaction, the individual **l** is coupled to **s** to form a resultant **j** and we have the *j-j* coupling. When the opposite is true, the individual orbit- and spin-angular momenta are coupled separately to form a resultant **L** and **S**, and then **L** and **S** are coupled together to form **J**. The reason for selecting this set of **L** and **S** as a description of the atomic states is because without the spin-orbit interaction terms, the Hamiltonian commutes with **L** and **S**. That means, the off-diagonal matrix elements are zero and this offers a great simplifica-

[17] E. U. Condon and G. H. Shortley, "The Theory of Atomic Spectra," Cambridge University Press, London, 1959.

[18] For the definition of these integrals, the reader is referred to E. U. Condon and G. H. Shortley, *op. cit.*, pp. 122 and 177.

tion in the calculation of energies for the atomic states as the calculation given in Sec. 9.9 can testify. This latter coupling scheme is
known as the Russell-Saunders coupling. Experience shows that in
many atoms, the spin-orbit interaction is much weaker than the
electrostatic interaction and hence the Russell-Saunders coupling is
a good approximation of the actual situation.

Figure 12.12 shows the transition from the L-S (Russell-Saunders)
to the j-j coupling as a function of $\chi = 3\zeta/4G_1$, that is, the relative
strength of the electrostatic and spin-orbit interactions. The ordinate
shows the energy difference ε between the various atomic states.
The scales of the abscissa and ordinate are so chosen that their
values fall in a finite range as χ goes from zero to infinity. At the left
end where $\chi = 0$, the two states, ^1P and ^3P, are the direct consequence of Russell-Saunders coupling scheme. As χ differs from zero,
or in other words, the spin-orbit coupling is introduced, the state
^3P is split into three separate states because the three possible combinations of L and S involve different spin-orbit interaction energies.
At the right end where $\chi = \infty$ or the spin-orbit interaction prevails,
we have the j-j coupling. Note that in the $2p^5ns$ configuration of Ne,
as n increases, the distance between the ns electron in the excited
state and the core $2p^5$ electrons becomes larger, resulting in a smaller
electrostatic interaction energy. This is indicated in Fig. 12.12 by the
progressive shift of Ne I ns levels toward the right. Furthermore,
we see that the number of terms and the values of J are the same for

Figure 12.12

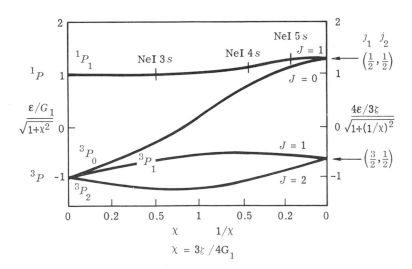

$$\chi = 3\zeta/4G_1$$

*Schematic representation showing the transition from the Russell-Saunders
coupling to the j-j coupling in the p⁵s configuration. (E. U. Condon and G. H.
Shortley, "The Theory of Atomic Spectra," p. 304, Cambridge University
Press, London, 1959.)*

both types of coupling, and hence an unambiguous correlation between the levels in the two coupling schemes is possible as indicated by the curves shown in Fig. 12.12.

Now let us turn to the $2p^53p$ configuration. In the j-j coupling scheme with $j_1 = \frac{1}{2}$ or $\frac{3}{2}$ for the core $(2p^5)$ electrons and $j_2 = \frac{1}{2}$ or $\frac{3}{2}$ for the excited $(3p)$ electron, possible combinations are: $J = 1$, 0 for $(j_1, j_2) = (\frac{1}{2}, \frac{1}{2})$; $J = 2$, 1 for $(j_1, j_2) = (\frac{1}{2}, \frac{3}{2})$; $J = 2$, 1 for $(j_1, j_2) = (\frac{3}{2}, \frac{1}{2})$; and $J = 3, 2, 1, 0$ for $(j_1, j_2) = (\frac{3}{2}, \frac{3}{2})$. Altogether, there are ten levels as shown in Fig. 12.11. In the Russell-Saunders coupling scheme, $L_1 = 1$ and $S_1 = \frac{1}{2}$ for the core $(2p^5)$ electrons and $L_2 = 1$ and $S_2 = \frac{1}{2}$ for the excited $(3p)$ electron. Possible combinations of L and S give the following: ^3D, ^1D, ^3P, ^1P, ^3S, and ^1S. Due to spin-orbit coupling, the ^3D and ^3P states are further split into $^3D_1$, $^3D_2$, $^3D_3$, and $^3P_0$, $^3P_1$, $^3P_2$ respectively. Again, we have ten states. The energies of the ten levels can be calculated from either coupling scheme in terms of the spin-orbit interaction energy ζ and the electrostatic interaction energy F and G, and these energy levels are plotted in Fig. 12.13 in terms of λ^{-1} together with the four energy levels of the $2p^54s$ configuration. The values of J associated with the atomic states as well as the observed laser transitions between some of these states are also indicated in the diagram.

Not all transitions between the four states of the $2p^54s$ configuration and the ten states of the $2p^53p$ configuration are allowed by the quantum-mechanical selection rules. For electric dipole transitions, the general selection rule is $\Delta J = 0$, ± 1 with $J = 0$ to $J' = 0$ transition forbidden. According to this general selection rule, ten out of the forty transitions are forbidden, with two examples shown by dotted lines in Fig. 12.13. For pure Russell-Saunders coupling, additional selection rules are $\Delta S = 0$ and $\Delta L = 0$ or ± 1. For the intermediate coupling as it is the case here, the selection rule is generally relaxed as compared to that for pure Russell-Saunders coupling. The identification of atomic spectral lines is a difficult task and it is outside the scope of the present book. The discussion given above merely serves to give a better understanding of the He-Ne laser system. For a thorough, theoretical discussion on atomic spectroscopy, the reader should refer to the book by Condon and Shortley and a series of articles by Racah.[19] The book by Herzberg[20] provides an excellent background for these advanced readings.

Figure 12.14 shows schematically the experimental arrangement for a gaseous laser. The RF generator is to cause gas discharge in the discharge tube containing He-Ne mixture, with He at 1 mm and Ne at 0.1 mm Hg pressure in the original work of Javan, Bennett, and

[19] G. Racah, *Phys. Rev.*, vol. 61, p. 537, 1942; vol. 62, p. 438, 1942; vol. 63, p. 367, 1943.
[20] H. G. Herzberg, "Atomic Spectra and Atomic Structure," Dover Publications, Inc., New York, 1944.

Figure 12.13

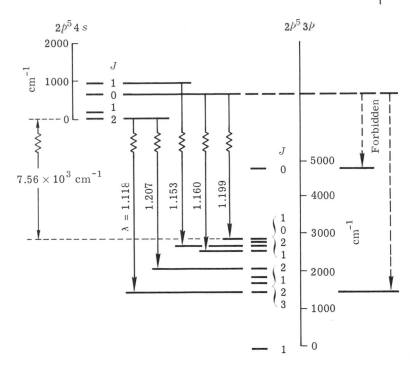

Energy separation between the various states of the $2p^5 4s$ configuration and the various states of the $2p^5 3p$ configuration of Ne. Solid arrows indicate the reported laser transitions (with λ expressed in microns) and broken arrows illustrate the transitions forbidden by quantum-mechanical selection rules.

Figure 12.14

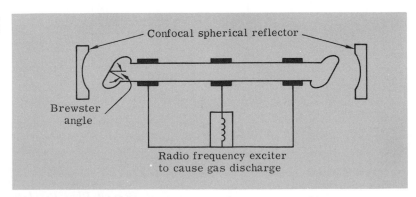

Experimental arrangement for a gaseous laser.

Herriott.[21] The quartz windows at the end of the tube are optically flat and are oriented at the Brewster angle to minimize undesired reflection at these windows. Reflectors are mounted outside the discharge tube instead of being a part of it to provide flexibility in

[21] A. Javan, W. R. Bennett, Jr., and D. R. Herriott, *Phys. Rev. Letters*, vol. 6, p. 106, 1961.

Figure 12.15

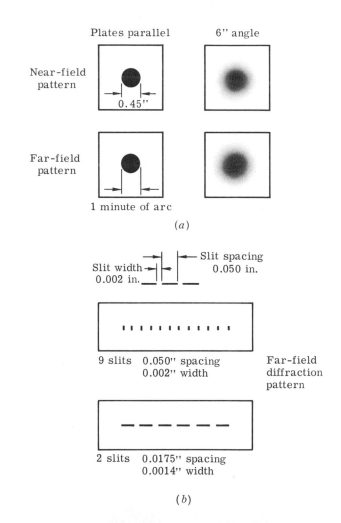

(a) *Near-field and far-field patterns for two Fabry-Perot plate angles, one parallel and the other with a 6″ angle inclination; and (b) the far-field diffraction pattern of a helium-neon laser. The diffraction mask covers the full 0.45-in. aperture of the laser beam. The diagrams are drawn for illustration purposes only, and they correspond to the negatives of the photographs shown in the paper by A. Javan, W. R. Bennett, Jr., and D. R. Herriott, Phys. Rev. Letters, vol. 6, p. 106, 1961.*

adjustment; furthermore, spherical reflectors are preferable to plane reflectors for noncritical adjustment. The overall tolerance for the spherical mirrors, that is, the change of angle between the axes of the two mirrors at two half-power points, is about one half degree as compared with only a few seconds of arc (Fig. 12.15a) for the plane mirrors. A discussion of the performance of optical cavities using spherical and plane mirrors can be found in articles by Boyd and Gordon[22] and by Fox and Li.[23] Figure 12.15 shows (a) the near-field and far-field pattern of a He-Ne laser beam and (b) the diffraction pattern observed by placing a multiple slit in the beam, the reflectors in the experimental setup being parallel plane mirrors. If the mirrors are adjusted in perfect parallelism, the far-field pattern of the beam has a width of 1 minute of arc. The diffraction pattern is due to interference of light from different slits, and the fact that the intensity of the minima is very low indicates a phase coherence over the aperture.

A brief discussion concerning laser modes now follows. First, let us consider the width of the optical cavity in relation to the natural width of the emission spectrum. Since the velocities of the light-emitting atoms are random in direction and widely distributed in magnitude, the frequency of emitted light will be shifted up and down by varying amounts. This is known as the *Doppler effect* and the spectrum linewidth due to this effect is given by $\Delta \nu / \nu = v/c$. For a thermal velocity v of 10^5 cm/sec, a width about 10^9 cps results at

[22]   G. D. Boyd and J. P. Gordon, *Bell System Tech. J.*, vol. 40, p. 489, 1961.
[23]   A. G. Fox and Tingyi Li, *Bell System Tech. J.*, vol. 40, p. 453, 1961.

Figure
12.16

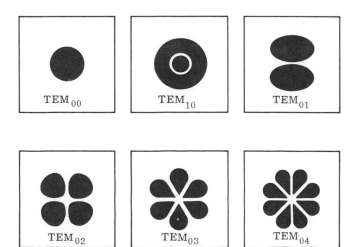

Axisymmetrical modes of a concave mirror interferometer. (W. W. Rigrod, Appl. Phys. Letters, vol. 2, p. 51, 1963.)

Figure 12.17

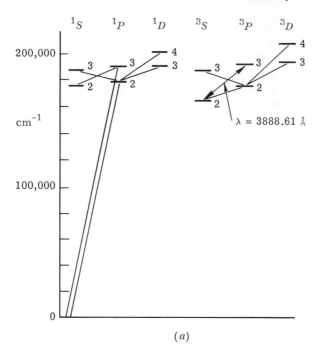

(a)

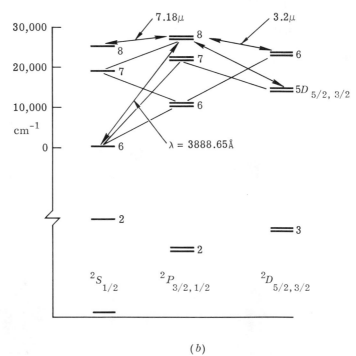

(b)

Relevant atomic-energy levels for a He-Cs *laser with* (a) *for* He *and* (b) *for* Cs.

$\lambda = 1.2$ microns. For simple TEM_{00n} modes propagating between two parallel plates of separation d, the resonance condition is that $2d = \lambda_n n = cn/\nu_n$ where n represents the number of standing wave patterns between the two reflector plates. Therefore, the separation in frequency between two adjacent modes is equal to $\Delta\nu = \nu_{n+1} - \nu_n = c/2d$.

For appreciable gain, the discharge tube is fairly long, say of the order of 1 m. This gives a mode separation of $\Delta\nu = 150$ Mc, which is much smaller than the width of the emission spectrum. Therefore, many modes can be excited simultaneously inside the optical cavity system and beat notes between these modes may be observed through a photo-mixing device. Techniques for isolating the various modes are discussed in a paper by Rigrod[24] and as a result of isolation, each individual mode can be separately excited and observed. Figure 12.16 shows some of the modes observed by Rigrod where the second subscript n in the notation TEM_{mn} indicates the azimuthal wave number and the first subscript m, the radial variation in terms of the associated Laguerre polynomials $\mathcal{L}_m{}^n$.

There are other gas systems which have been either reported successful or investigated for laser operation and these include potassium vapor in the original proposal of Schawlow and Townes,[25] Hg-Na mixture and He-Cs mixture with the last proven to be successful. All elements under study have closed-shell or near closed-shell configurations because for these elements, the atomic spectra are relatively simple. For example, in the case of potassium and cesium, the spectra are due to the excited states of the $4s$ and $6s$ valence electron respectively. Figure 12.17 shows the sketch of relevant atomic energy levels in the He-Cs system with possible transitions between states joined by lines. The triplet states of He are metastable states. The close coincidence[26] between the $6^2S_{1/2} \leftrightarrow 8^2P_{1/2}$ Cs transition and the $2^3S_1 \leftrightarrow 3^3P_{2,1}$ He line permits a large amount of excitation power transfer to the Cs vapor. Laser operation[27] has been reported between the $8^2P_{1/2}$ and $8^2S_{1/2}$ states at 7.18 microns and population inversion has also been demonstrated between the $8^2P_{1/2}$ and $6^2D_{3/2}$ states at 3.2 microns.

12.4 SOLID-STATE LASERS

Although both semiconductor and ruby lasers use solids as the medium for laser action, they are fundamentally different according to the way electrons behave in these materials. The electronic states of solids can be treated theoretically from two drastically different

24 W. W. Rigrod, *Appl. Phys. Letters*, vol. 2, p. 51, 1963; also H. Kogelnik and W. W. Rigrod, *Proc. IRE*, vol. 50, p. 220, 1962.
25 A. L. Schawlow and C. H. Townes, *Phys. Rev.*, vol. 112, p. 1940, 1958.
26 H. Z. Cummins et al., "Quantum Electronics," p. 12, Columbia University Press, New York, 1961.
27 S. Jacobs, G. Gould, and P. Rabinowitz, *Phys. Rev. Letters*, vol. 7, p. 415, 1961.

approaches which differ in the choice of electronic wave function in the starting zeroth order approximation: one assuming an extremely localized and the other an extremely widespread electronic wave function. Metals and semiconductors are best described, starting with wave functions of free electrons. As the discussion in Sec. 11.6 shows, the electronic states inside a semiconductor are shared by valence electrons throughout the crystal in the Bloch formulation in terms of running electron waves with a wave vector $\mathbf{k}$ and an effective mass m_c or m_v. Molecular and ionic crystals, on the other hand, are best described in terms of localized states of the free molecule or ion. The lasers to be discussed in this section use ionic crystals as the host lattice and the energy levels of added impurities as the laser transition levels. Examples include CaF_2 doped with U^{3+}, Dy^{2+}, Sm^{2+}, or Tm^{2+}, $CaWO_4$ doped with Er^{3+} or Nd^{3+}, and Al_2O_3 doped with Cr^{3+}, to name a few. The optical spectra of these ions are best explained on the model of tight binding (localized wave function), and the electronic wave functions of these ions can be expressed in terms of atomic wave functions similar to those discussed in the previous section. Therefore, a discussion of solid-state lasers requires a basic knowledge of atomic states, and consequently the solid-state lasers to be presented in this section are more closely related to the gaseous lasers than to the semiconductor lasers. This explains why the present sequence of introducing the subject materials is adopted instead of following the history of laser development.

The difference in the behavior of transition metal ions such as Cr^{3+} and the behavior of ions in the lanthanide and actinide series such as Tm^{2+} and U^{3+} lies in the relative effect of the crystalline field upon the energy levels of these ions. For concreteness, consider the $3d^2$ electrons of V^{3+} situated in the center of a regular octahedron. The electric potential in the vicinity of the central ion due to the six point charges located at the corners of the octahedron is given by

$$V = C + D(x^4 + y^4 + z^4 - \tfrac{3}{5}r^4) + \text{higher terms} \qquad (12.48)$$

The crystalline potential is in direct competition with the electrostatic interaction and the spin-orbit interaction in determining the composition of the electron wave functions. In an optical transition, only the low-lying excited states of an ion are involved and for these energy states, the electrostatic interaction energy is much stronger than the spin-orbit interaction as pointed out in the previous section in reference to Fig. 12.12. Therefore, in the following discussion, we shall consider only the effects of crystalline potential and electrostatic interaction.

As discussed in Sec. 9.8, the energies of the five $3d$-orbitals are split into two groups, the two $d\gamma$ orbitals and the three $d\epsilon$ orbitals, in the presence of the octahedral potential V. The energy difference between the $d\epsilon$ and $d\gamma$ orbitals can be calculated from Eq. (12.48) and is

TABLE 12.1 The possible values of resultant M_L and M_S from various combinations of individual m_l and m_s for V^{3+} ions

M_S \ M_L	4	3	2	1	0	-1	-2	-3	-4
1	X	$(2^+,1^+)$	$(2^+,0^+)$	$(2^+,-1^+)$ $(1^+,0^+)$	$(2^+,-2^+)$ $(1^+,-1^+)$	$(1^+,-2^+)$ $(0^+,-1^+)$	$(0^+,-2^+)$	$(-1^+,-2^+)$	X
0	$(2^+,2^-)$	$(2^+,1^-)$ $(2^-,1^+)$	$(2^+,0^-)$ $(2^-,0^+)$ $(1^+,1^-)$	$(2^+,-1^-)$ $(2^-,-1^+)$ $(1^+,0^-)$ $(1^-,0^+)$	$(2^+,-2^-)$ $(2^-,-2^+)$ $(1^+,-1^-)$ $(1^-,-1^+)$ $(0^+,0^-)$	$(1^+,-2^-)$ $(1^-,-2^+)$ $(0^+,-1^-)$ $(0^-,-1^+)$	$(0^+,-2^-)$ $(0^-,-2^+)$ $(-1^+,-1^-)$	$(-1^+,-2^-)$ $(-1^-,-2^+)$	$(-2^+,-2^-)$
-1	X	$(2^-,1^-)$	$(2^-,0^-)$	$(2^-,-1^-)$ $(1^-,0^-)$	$(2^-,-2^-)$ $(1^-,-1^-)$	$(1^-,-2^-)$ $(0^-,-1^-)$	$(0^-,-2^-)$	$(-1^-,-2^-)$	X

designated as $\Delta = 10D(2e\bar{r}^4/105)$ where $\bar{r}^4$ is the average of r^4 over the normalized $3d$ wave functions. If the crystalline-field splitting is much larger than the electrostatic interaction energy, then the $d\gamma$ and $d\epsilon$ orbitals are the proper wave functions. If the opposite is true, the electronic states of a given ion can be expressed in terms of wave functions of free ions obtained from the Russell-Saunders coupling scheme as discussed in the previous section. For the $3d^2$ configuration of V^{3+} ion, possible combinations of m_l and m_s (orbital and spin magnetic quantum number) in accordance with the Pauli exclusion principle are given in Table 12.1.

In Table 12.1, the positive and negative signs in the superscript indicate the spin states. It is obvious that the combinations $(2^+,1^+)$ and $(2^+,2^-)$ respectively must belong to the ^3F and ^1G states. The two states with $M_L = 3$ and $M_S = 0$ must come from ^3F ($M_L = 3$, $M_S = 0$) and $^1G(3,0)$. The addition of a new term with $M_L = 2$ and $M_S = 0$, however, must be due to the beginning of a new series, that is ^1D. Following similar reasoning, Table 12.1 can be transformed into another table consisting of ^3F, ^1G, etc. as in Table 12.2. A comparison of Table 12.2 with Table 12.1 gives the identity $^3F(1,1) + {}^3P(1,1) = (2^+,-1^+) + (1^+,0^+)$ and similar expressions. The functional relationship between the various ^3F states, for example between $^3F(1,1)$ and $^3F(1,0)$, and between $^3F(1,1)$ and $^3F(2,1)$, can be obtained from the quantum-mechanical operational relations governing momentum operators L and S. The method[28] which is originally due to Gray and Willis enables us to express the various states $^3F(1,1)$ and so on in terms of a linear combination of the products of single electron wave functions such as $(2^+,-1^+) = \psi_{\text{I}}(2,+)\psi_{\text{II}}(-1,+)$.

Knowing the compositions of the various states, ^3F and so on, the electrostatic interaction energy of these states can be calculated in terms of F_k and G_k given by Eq. (12.47). Figure 12.18 shows the change of the electronic states of the $3d^2$ electrons of V^{3+} ion from

[28]   See E. U. Condon and G. H. Shortley, *op. cit.*, p. 226.

TABLE 12.2   *The various states in Table 12.1 expressed in terms of the atomic states* $(a)^3F$, $(b)^3P$, $(c)^1G$, $(d)^1D$, and $(e)^1S$

M_S \ M_L	4	3	2	1	0	-1	-2	-3	-4
1	$\times$	a	a	$a\ b$	$a\ b$	$a\ b$	a	a	$\times$
0	c	a c	a $c\ d$	$a\ b$ $c\ d$	$a\ b$ $c\ d$ e	$a\ b$ $c\ d$	a $c\ d$	a c	c
-1	$\times$	a	a	$a\ b$	$a\ b$	$a\ b$	a	a	$\times$

the weak to the strong crystalline field case, the parameter C/D being used to express the relative strength of the crystalline-field splitting and the electrostatic interaction energy with $B = F_2 - 5F_4$, $C = 35\,F_4$, and D given in Eq. (12.48). Figure 12.18 serves the same purpose as Fig. 12.12 and it is arranged in a similar manner. The notations A, E, and T in Fig. 12.18 are used in group theory to indicate the character and symmetry property of the wave functions belonging to a given representation. Nondegenerate representations are designated by the letters A and B, the A_1's being totally symmetrical and the A_2's and B's being respectively symmetrical and antisymmetrical with respect to a rotation about the principal axis in systems which have such an axis, for example, the axis of a distorted octahedron. Doubly degenerate representations are denoted by E and triply degenerate representations by T or sometimes F. These designations are determined by both the crystal symmetry and the electronic configuration of a given ion. For a detailed discussion of the subject, the reader should consult the book on quantum chemistry by Eyring, Walter, and Kimball[29] and two review articles by Koster[30] and by McClure.[31]

[29] H. Eyring, J. Walter, and G. E. Kimball, "Quantum Chemistry," John Wiley & Sons, Inc., New York, 1963.
[30] G. F. Koster, Space Groups and Their Representations, in *Solid State Phys.*, vol. 5, p. 174, 1957.
[31] D. S. McClure, Electronic Spectra of Molecules and Ions in Crystals, in *Solid State Phys.* vol. 8, p. 1, and vol. 9, p. 400, 1959.

Figure
12.18

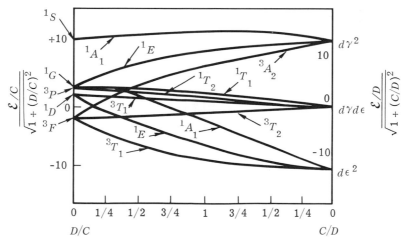

*The correlation between the states of weak and strong crystal fields for the 3d²
electrons of V³⁺ ion in an octahedral field. The curves are calculated for
B/C = 4.5. (D. S. McClure, Solid State Phys., vol. 9, p. 399, 1959.)*

Figure 12.19

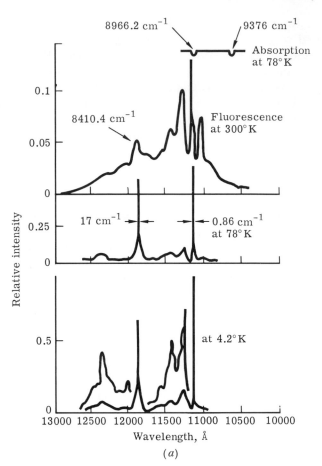

(a)

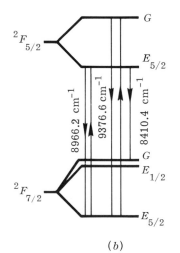

(b)

(a) Fluorescence and absorption lines of the Tm²⁺ ion in CaF₂ and (b) the relevant energy levels, with the upward arrow indicating the absorption process and the downward arrow indicating the fluorescence process. (Z. J. Kiss, Phys. Rev., vol. 127, p. 718, 1962.)

Weak crystal-field case. After a general discussion of the role played by the crystalline field and electrostatic interaction between orbitals, we can now discuss in some detail the optical transitions in solid-state lasers. Among the solid-state lasers using rare-earth elements, the $CaF_2:Tm^{2+}$ laser will be chosen as the example in our discussion because the energy levels of divalent Tm in CaF_2 have been extensively studied by Kiss.[32] Figure 12.19a shows the absorption and fluorescent lines having a very narrow linewidth of 0.86 cm^{-1} and 17 cm^{-1} at 78°K. Figure 12.19b shows the energy-level scheme of the $4f^{13}$ configuration of Tm^{2+}. For the ions of rare-earth elements, the magnitudes of the various interaction energies are generally arranged in the following order: (1) the electrostatic interaction energy, (2) the spin-orbit interaction and (3) the crystalline field splitting. Thus, the energy levels of Tm^{2+} ion in CaF_2 can be best described starting from the energy levels of the free Tm^{2+} ion. According to the Russell-Saunders coupling scheme, the ground state of the $4f^{13}$ configuration of Tm^{2+} is ^2F with $L = 3$ and $S = \frac{1}{2}$, and it is split into $^2F_{5/2}$ and $^2F_{7/2}$ by the spin-orbit interaction. By the Hund rule, the lower energy state of the two levels is $^2F_{7/2}$ as shown in Fig. 12.19b. The cubic crystalline field of CaF_2 splits the two states further, the upper $^2F_{5/2}$ being split into a doubly degenerate state $E_{5/2}$ and a fourfold degenerate state G, and the lower $^2F_{7/2}$ into $E_{1/2}$, $E_{5/2}$ and G states. By studying the temperature variation of the absorption and fluorescence spectra in connection with the lattice-vibration spectrum, Kiss has concluded the following: (1) The two strong and narrow fluorescent lines at 8410.4 cm^{-1} and 8966.2 cm^{-1} correspond to magnetic dipole transitions from the upper $E_{5/2}$ level to the lower G and $E_{5/2}$ levels respectively, with the transition to the lower $E_{1/2}$ forbidden by the selection rule. (2) The satellite structures around these two lines are due to the various vibrational modes of the CaF_2 crystal.

Figure 12.20 shows the relevant energy levels in the laser scheme used by Kiss. The strong, broad-band absorption spectrum in the visible region corresponds to the $4f$ to $5d$ or other allowed interconfigurational transitions. Because of its strongly absorbing and wide-band properties, these interconfigurational transitions are used as the pump in a multilevel laser system to depopulate the ground state. The 1.116 micron fluorescent line (ν_3 transition) breaks into oscillation as a three-level laser at 4.2°K. The 1.189 micron radiation (ν_4 transition) is reabsorbed by transitions from the excited $^2F_{5/2}$ state into the $5d$ band and hence it is prevented from breaking into oscillation. Had it not been for this reason, the ν_4 transition should have a lower oscillation threshold than the ν_3 transition. In general, the terminal state in a four-level laser system, if it is not the ground state, is almost depopulated even at room temperature and hence it

[32] Z. J. Kiss, *Phys. Rev.*, vol. 127, p. 718, 1962; *Proc. IRE*, vol. 50, p. 1532, 1962; *Appl. Phys. Letters*, vol. 2, p. 61, 1963.

is easier to achieve a population inversion in a four-level than in a three-level system.

Many other laser systems using the energy levels of rare-earth elements have been reported in the literature, with some being capable of continuous operation.[33] Among these, we shall use the $CaF_2:U^{3+}$ system reported by Boyd et al.[34] as a further illustration. Figure 12.21 shows the relevant energy levels of such a system and Fig. 12.22 the experimental setup for continuous laser operation at

[33] Continuous laser operations have been reported in $CaWO_4:Nd^{3+}$ and in $CaF_2:Dy^{2+}$ by L. F. Johnson et al., *Phys. Rev.*, vol. 126, p. 1406, 1962, and *Proc. IRE*, vol. 50, p. 1691, 1962.

[34] G. D. Boyd, *Phys. Rev. Letters*, vol. 8, p. 269, 1962.

Figure 12.20

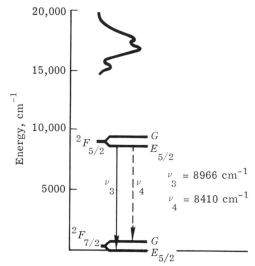

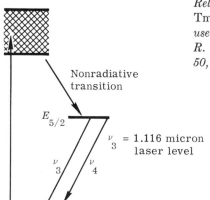

Relevant energy levels of the Tm^{2+} ion in the laser scheme used by Kiss. (Z. J. Kiss and R. C. Duncan, Proc. IRE, vol. 50, p. 1532, 1962.)

77°K. It is worth mentioning that the threshold energy input into the flash tube changes from 2.0 joules at 20°K to 4.35 joules at 90°K and around 1200 joules at 300°K. The increase in the threshold energy is due to a depopulation of the metastable state at higher temperatures as a result of thermal equalization of population among the various states belonging to the $^4I_{1\frac{1}{2}}$ group because the energy separation between these states due to crystalline field splitting is relatively small. The energy levels of U^{3+} and Nd^{3+} in solids have been analyzed in great detail by Carlson and Dieke[35] and by Satten.[36] Valuable information concerning the composition of the ground state can be obtained from paramagnetic resonance experiment as discussed in the papers by Bleaney et al.,[37] and by Judd and Wong.[38]

[35] E. H. Carlson and G. H. Dieke, *J. Chem. Phys.*, vol. 34, p. 1602, 1961; vol. 29, p. 229, 1958; vol. 28, p. 1097, 1958.
[36] R. A. Satten, *J. Chem. Phys.*, vol. 21, p. 637, 1953.
[37] B. Bleaney, P. M. Llewellyn, and D. A. Jones, *Proc. Phys. Soc.* (London), vol. B69, p. 858, 1956.
[38] B. R. Judd and E. Wong, *J. Chem. Phys.*, vol. 28, p. 1097, 1958.

Figure 12.21

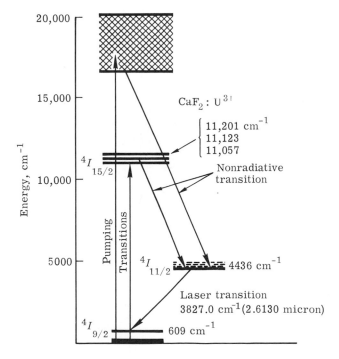

The relevant energy levels in $CaF_2:U^{3+}$ involved in the absorption-fluorescence transitions. (G. D. Boyd et al., *Phys. Rev. Letters*, vol. 8, p. 269, 1962.)

Strong crystal-field case. Now we shall discuss the ruby laser and the energy levels of Cr^{3+} ions in Al_2O_3. The sharp absorption and emission bands of ruby have attracted the attention of many investigators. Figure 12.23 shows the splitting of states of the d^3 configuration by an octahedral field calculated by Tanabe and Sugano[39] as a function of the strength of the crystalline-field splitting relative to the electrostatic interaction energy. Similar to Fig. 12.18, the extreme left end of Fig. 12.23 corresponds to zero crystalline field and the various energy states ^4F, ^4P, ^2G, ^2H, etc., are the energy states of a free Cr^{3+} ion with ^4F being the ground state. The extreme right side of Fig. 12.23 corresponds to the strong field situation where the energy states are best described by $d\epsilon$ and $d\gamma$ orbitals. Under a strong crystalline field (cubic), the ground state is an orbital singlet with three electrons having three different $d\epsilon$ orbitals and a resultant

---

[39]   Y. Tanabe and S. Sugano, *J. Phys. Soc.* (Japan), vol. 9, pp. 753 and 766, 1954.

Figure 12.22

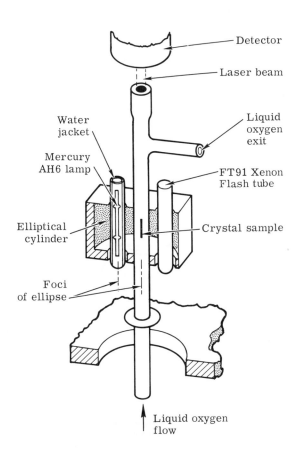

Detector

Laser beam

Water jacket

Liquid oxygen exit

Mercury AH6 lamp

FT91 Xenon Flash tube

Elliptical cylinder

Crystal sample

Foci of ellipse

Liquid oxygen flow

Experimental setup for continuous laser operation in $CaF_2:U^{3+}$. (G. D. Boyd et al., Phys. Rev. Letters, vol. 8, p. 269, 1962.)

Figure 12.23

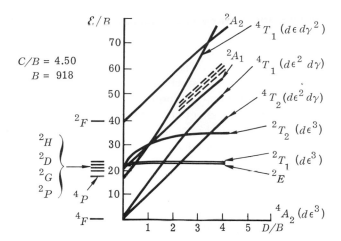

Splitting of the states of the d^3 configuration by an octahedral field. The parameters B and D are the same as the ones in Fig. 12.18. (D. S. McClure, Solid State Phys., vol. 9, p. 400, 1959.)

$S = \frac{3}{2}$, and it is represented by $^4A_2$, a notation used in group theory. The excited states of Cr^{3+} ion require either pairing of $d\epsilon$ orbitals or lifting one electron from the $d\epsilon$ to the $d\gamma$ orbital, the former at the expense of electrostatic interaction energy and the latter at the expense of crystalline-field splitting energy.

Figure 12.24 is the absorption spectrum of Cr^{3+} in Al_2O_3 at 77°K. The six nearest neighbors (oxygen ions) of Cr form a distorted

Figure 12.24

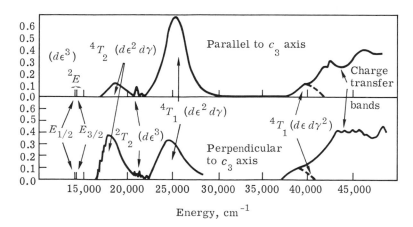

Polarized absorption spectrum (from the ground states to the various excited states) of the Cr^{3+} ion in Al_2O_3 at 77°K. (D. S. McClure, Solid State Phys., vol. 9, p. 400, 1959.)

octahedron as shown in Fig. 12.25 and the C_3 axis in the figure refers to the threefold symmetry axis of the distorted octahedron. Figure 12.26a is the energy-level diagram of ruby which summarizes the information given in Fig. 12.24 for easy comparison with Fig. 12.23. The effect of the distortion in the octahedral arrangement is to introduce a trigonal component of the crystalline field. A detailed calculation of the effect of the trigonal field and the spin-orbit interaction can be found in the article by Sugano and Tanabe.[40] The red lines (R_1 and R_2) which are observable both in absorption and emission spectra of ruby can be assigned to the $^2E(d\epsilon^3)$ and $^2T_1(d\epsilon^3)$ states and the group of absorption lines B in the blue region is due to transitions from the ground state to the excited $^2T_2(d\epsilon^3)$ states. The splitting of these lines is due to the combined effect of the trigonal field and the spin-orbit interaction. Figure 12.26b shows the excitation scheme for laser action. It is preferable that radiation from a high-intensity excitation source such as a flashlamp be centered around the green band (5500Å) for reasons of achieving a high power efficiency and avoiding the depopulation of the ^2E states. The blue radiation from the lamp if the radiation is centered around the violet band (4100Å) may depopulate the ^2E states by lifting electrons from these states into the $^4T_1(d\epsilon\,d\gamma^2)$ and charge transfer bands. The red line R_1 is favored over the R_2 line for stimulated emission because of a somewhat larger population inversion in the lower than in the upper states of ^2E on account of the Boltzmann factor.

The examples given above serve to illustrate the different energy-level schemes in solid-state lasers using rare-earth and transition metal ions. It is not our intent to give an exhaustive survey of various solid-state laser materials because it is almost an impossible task to do so in a field so rapidly developing. For a survey of the spectra of divalent and trivalent rare-earth ions, the reader is referred to two

[40] S. Sugano and Y. Tanabe, *J. Phys. Soc.* (Japan), vol. 13, p. 880, 1958.

Figure 12.25

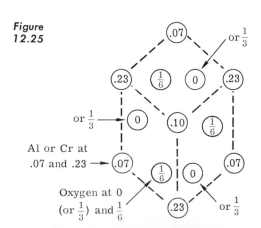

Al or Cr at .07 and .23

Oxygen at 0 (or $\frac{1}{3}$) and $\frac{1}{6}$

Projection of the atoms in Al$_2$O$_3$ upon a plane normal to the threefold c_3 axis and passing through the apex of a unit rhombohedron. The numbers in circles indicate the height of each atom.

articles by Dieke[41] and by McClure and Kiss[42]; and for a general discussion of the absorption spectra of transition metal ions, a review article by McClure. The reader may also find abundant literature in spectroscopic analyses of these ions in the Journal of Chemical Physics and in paramagnetic resonance studies in the Report on Progress in Physics. The paramagnetic resonance studies yield information not only about the composition of the ground state but also about the possibility of doing optical and microwave experiments simultaneously as in the modulation of an optical signal by a microwave field. For a theoretical discussion of the energy levels, the composition of wave functions, and the quantum-mechanical selection rules governing electric and magnetic dipole and electric quad-

[41] G. H. Dieke and B. Pandey, Symposia series XIII, Optical Masers, p. 327. Brooklyn Polytechnic Institute, 1963.

[42] D. S. McClure and Z. J. Kiss, Symposia series XIII, Optical Masers, p. 357, Brooklyn Polytechnic Institute, 1963.

Figure 12.26

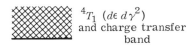

$^4T_1\ (d\epsilon\, d\gamma^2)$ and charge transfer band

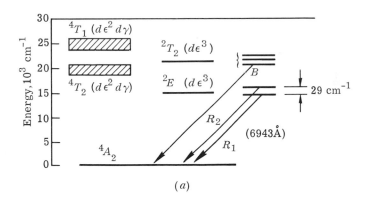

(a)

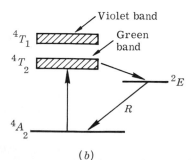

(b)

Relevant energy levels in a ruby laser.

rupole transitions, the reader is again referred to the review article by McClure[43] for further references.

Needless to say, spectroscopic analysis of the absorption and fluorescence data is a difficult and painstaking task if we want to understand fully the quantum-mechanical transitions involved. This is a job mainly for physicists and chemists. So far as laser work is concerned, valuable information can still be derived from the absorption and fluorescence studies without a thorough analysis if we know how to interpret the experimental data correctly. The transition probability between different energy levels is often expressed in spectroscopy in terms of a dimensionless quantity f, called the oscillator strength of a transition, given by Eq. (11.75) as follows:

$$\hbar^2 \left| \int \Phi_i^* \frac{\partial}{\partial x} \Phi_j \, dq \right|^2 = m^2 \omega^2 \left| \int \Phi_i^* x \Phi_j \, dq \right|^2 = \frac{m\hbar\omega f}{2} \qquad (12.49)$$

Substituting Eq. (12.49) into Eq. (11.58) gives

$$f = \frac{\epsilon mc^3}{2\omega^2 e^2 \tau} \qquad (12.50)$$

where τ is the spontaneous lifetime of the transition, and ϵ and c respectively are the dielectric constant and velocity of light in the medium. The oscillator strength is also related to the absorption coefficient in the following manner. From Eqs. (11.58), (11.62), and (11.63), we have

$$K(\nu) = \frac{n2\pi e^2}{\hbar^2 \epsilon c} |x_{ij}|^2 \hbar \nu N(\nu) = \frac{n\pi e^2}{m\epsilon c} N(\nu) f \qquad (12.51)$$

Integrating Eq. (12.51) with respect to ν and noting that $\int N(\nu) \, d\nu = N$, the number of active ions per cm³, we have

$$f = \frac{nmc^2}{\pi N e^2} \int K(\nu) \frac{d\nu}{c} \qquad (12.52)$$

where n is the refractive index of the medium and $d\nu/c$ is expressed in cm⁻¹. Therefore, with N known, the oscillator strength of a given transition can be calculated from the area of the absorption curve. Once f is determined, the spontaneous lifetime of a given transition can be calculated from Eq. (12.50).

It should be pointed out that Eq. (12.52) is just another way of stating the sum rule expressed in Eq. (11.80). Here, we are interested in the oscillator strength associated with a particular transition which occurs in an impurity atom. Therefore the limits of integration in Eq. (12.52) should run across the width of a particular absorption line under investigation. Furthermore, because the absorption spectrum associated with impurity atoms in an insulator

---

[43] D. S. McClure, "Electronic Spectra of Molecules and Ions in Crystals," *Solid State Physics*, vol. 8, p. 1 and vol. 9, p. 400, 1959.

is discrete whereas the absorption spectrum in a semiconductor is continuous for $h\nu > \mathcal{E}_g$, Eq. (12.52) is useful in analyzing the absorption data in insulators, whereas Eq. (11.80) is useful in analyzing the absorption spectra in semiconductors and metals.

The reason why τ must be obtained from Eqs. (12.52) and (12.50) instead of the fluorescence lifetime is because there may be other processes than the spontaneous radiative transition. Let τ' be the lifetime of the upper state due to other processes. The fluorescence lifetime τ_f, therefore, is given by $\tau\tau'/(\tau + \tau')$, and the fluorescent quantum efficiency η is given by

$$\eta = \frac{\tau^{-1}}{\tau'^{-1} + \tau^{-1}} = \frac{\tau'}{\tau + \tau'} = \frac{\tau_f}{\tau} \tag{12.53}$$

which is the percentage of electrons in the fluorescent level undergoing radiative transition. In solid-state lasers, the quantum efficiency is much less than unity in contrast to a value close to unity in semiconductor lasers.

The condition for laser oscillation can be derived in a similar manner as Eq. (12.21), and it is given by[44]

$$\frac{\Delta N}{V} \geq 8\pi \frac{\Delta\nu}{\lambda^2} \frac{(\alpha d - \ln R)}{d} \tau \tag{12.54}$$

where ΔN is the excess population in the upper state in a volume V and $\Delta\nu$ the fluorescence linewidth. Both the values of η and τ_f can be measured experimentally, as described in the articles by Kiss[45] and by Maiman et al.[46] The value of τ obtained from Eq. (12.53) then can be compared with that from Eq. (12.50) for use in Eq. (12.54). If it is assumed that the nonradiative transitions from the pump band to the ground state are negligible and that nonradiative decay from the pump band to the emitting laser level is very rapid, the power required to maintain the laser operation is approximately

$$P = \frac{h\nu_{\text{pump}}}{\eta\tau}\left(\frac{N + \Delta N}{3}\right) \text{ watts} \tag{12.55}$$

where N is the total number of active impurities involved in the laser transition.

It may be worth discussing aspects concerning modes of light propagation in solid-state lasers. The R_1 fluorescent line of ruby at 300°K has a natural linewidth of 4Å which corresponds to an energy separation of 8 cm^{-1} or 240 Gc. For simplicity, let us assume that a ruby rod of length d has reflectors at the two ends. The condition for plane-wave resonant mode is that $m\lambda = 2d$ where m is an in-

[44] A. L. Schawlow and C. H. Townes, *Phys. Rev.*, vol. 112, p. 1940, 1958.
[45] Z. J. Kiss, *Phys. Rev.*, vol. 127, p. 718, 1962.
[46] T. H. Maiman et al., *Phys. Rev.*, vol. 123, p. 1151, 1961.

teger. Therefore, the separation between two adjacent modes is $\lambda_m - \lambda_{m+1} = \lambda^2/2d$. For a rod 6 cm long, $\Delta\lambda$ is of the order of 0.04 Å. If the ruby rod is driven far above the threshold condition, all the modes within the natural linewidth can go into oscillation. A quick calculation shows that there are approximately 100 (4/0.04) modes and hence mode selection is an urgent problem. Consider a Fabry-Perot etalon shown in Fig. 12.27a. The ratio of the transmitted wave to the incident wave is given by

$$\frac{I_t}{I_0} = \left[1 + \left(\frac{2F}{\pi}\right)^2 \sin^2 \left(\frac{\delta}{2}\right) \right]^{-1} \qquad (12.56)$$

where $\delta/2 = 2\pi(d/\lambda)$ cos θ and $F = \pi \sqrt{R}/(1 - R)$. Maximum transmission occurs when $\delta/2$ is an integral multiple of π. This provides not only a selection in λ but also in θ, the direction of beam propagation. Figure 12.27b shows the arrangement of Fabry-Perot etalons used by Collins and White[47] as mode selectors. The improvement in the laser beam is demonstrated by a sharpening of the diffraction rings and a narrowing of the beam angle. For lasers having

[47]   S. A. Collins and G. R. White, *Appl. Optics*, vol. 2, p. 448, 1963; W. S. C. Chang (ed.), "Lasers and Applications," p. 96, Ohio State University, Columbus, Ohio, 1963; *Quantum Electronics*, vol. 3, p. 1291, Columbia University Press, New York, 1964.

Figure 12.27

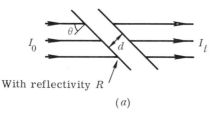

With reflectivity R

(a)

Scheme used to separate different laser modes. (S. A. Collins and G. R. White, Appl. Optics, vol. 2, p. 448, 1963.)

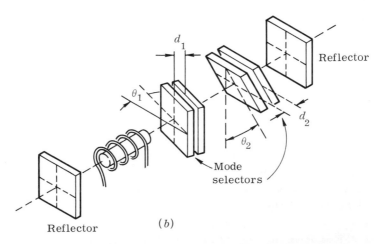

Reflector

Mode selectors

Reflector

(b)

a narrower natural linewidth, the problem of mode selection becomes less serious as in the case of $CaF_2:Tm^{2+}$ and $Al_2O_3:Cr^{3+}$ at low temperature.

12.5 MODULATION OF A LASER BEAM

We shall start the discussion with the scheme used by McClung and Hellwarth[48] to generate giant optical pulsations from a ruby rod as shown in Fig. 12.28. When a polar liquid such as nitrobenzene is placed in an electric field E_0, it behaves optically like a uniaxial crystal having two optic dielectric constants, one perpendicular and the other parallel to the field direction. Such a device is known as a *Kerr cell*. In Fig. 12.28, the output laser beam has the electric field vector **E** perpendicular to the optic axis C_3 of the ruby rod and the Kerr cell is oriented 45° with respect to the optic axis. Because of the difference in refractive indices, the laser beam is broken into two components parallel and perpendicular to the biasing field E_0. In the original scheme of McClung and Hellwarth, the magnitude of the biasing field E_0 is so chosen that the relative phase change of these two components is 180° in passing through the Kerr cell and being reflected back through the cell. In other words, the direction of polarization of the laser beam changes by 90° reentering the ruby rod as compared to that leaving the ruby rod. The direction of polarization will be restored after another roundtrip passage through the Kerr cell. That means, the effective amplification constant of the laser medium is reduced by a factor of 2 with the field on as compared to that with the field off. Suppose that with the biasing field turned on, the ruby rod is pumped to a level above the threshold

48   F. J. McClung and R. W. Hellwarth, *J. Appl. Phys.*, vol. 33, p. 828, 1962; *Proc. IEEE*, vol. 51, p. 46, 1963.

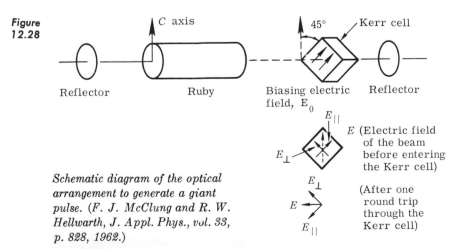

Figure 12.28

Reflector   Ruby   Biasing electric field, E_0 Reflector

$E_{||}$

$E_\perp$ E (Electric field of the beam before entering the Kerr cell)

$E_\perp$ (After one round trip through the Kerr cell)

E $E_{||}$

Schematic diagram of the optical arrangement to generate a giant pulse. (F. J. McClung and R. W. Hellwarth, J. Appl. Phys., vol. 33, p. 828, 1962.)

condition for laser oscillation with the field off but below the threshold with the field on. If the biasing field is suddenly turned off, all the stored energy in the rod becomes available for laser action, creating a giant optical pulse. The peak power of the laser beam reported by McClung and Hellwarth is around 600 kw as compared to 6 kw obtainable in a comparable unit without the Q switch to control the laser-firing. Later results show that peak power up to 75 megawatts can be obtained for a duration of about 10 nsec. It should be pointed out that many ideas worked out for microwave circuitry can be adopted for uses in optical circuits. The Kerr cell in the optical path of the laser cavity is to control the feedback loop or the effective Q of the optical cavity and hence the laser using such an arrangement is known as the Q-switched laser. The interferometric etalon discussed in the previous section in connection with Fig. 12.27 is the optical equivalent of a bandpass filter. Wave propagation in dielectric fibers has been studied for their use as optical wave guides.

The amplitude of a laser beam can also be modulated outside the optical cavity as shown in Fig. 12.29, using the scheme first developed by Kaminow[49] and later applied to laser beam by Harris et al.[50] Ways to produce plane-polarized light are many; one method uses Brewster's law. If an unpolarized beam is incident on a dielectric medium at the polarizing or Brewster angle, the component of the wave having its electric field in the plane of incidence passes into the dielectric medium without reflection and hence the reflected wave is plane-polarized with electric field normal to the plane of incidence.

[49] I. P. Kaminow, *Phys. Rev. Letters*, vol. 6, p. 528, 1961.
[50] S. E. Harris, B. J. McMurry, and A. E. Siegman, *Appl. Phys. Letters*, vol. 1, p. 37, 1962; also S. E. Harris, *Appl. Phys. Letters*, vol. 2, p. 47, 1963.

Figure 12.29

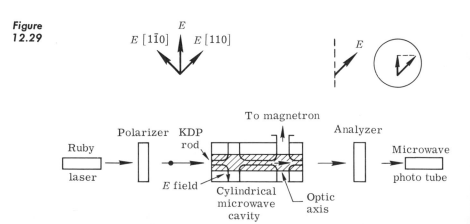

Scheme for amplitude modulation of a laser beam using the linear electro-optic effect. (*I. P. Kaminow, Phys. Rev. Letters, vol. 6, p. 528, 1961.*)

Other common types of polarizer and analyzer use doubly refracting crystals or dichroic organic compounds to resolve an unpolarized wave into two separate plane-polarized components (as a polarizer) or to allow the passage of a particular plane-polarized component (as an analyzer). The function of the KDP (KH_2PO_4) is similar to that of the Kerr cell in Fig. 12.28. According to a model theory of KDP, the electronic polarization and hence the optical property of the crystal depend upon the motion of protons in a double-trough potential minimum. Without an applied field, KDP is optically uniaxial with the optic axis along its tetragonal z axis. Light propagating along the z axis travels with the same velocity irrespective of the plane of polarization. Upon the application of an electric field, the crystal becomes orthorhombic and for light propagating along the z axis, a difference in refractive indices results:

$$n_1 - n_2 = n^3 r E_z \qquad (12.57)$$

where n_1 and n_2 respectively are the indices of refraction for light polarizing along [110] and [1$\bar{1}$0] directions, r the electro-optic modulus in cm/volt, n the normal index of refraction, and E_z the applied field. The effect is called the *linear electro-optic effect* which is subject to the same symmetry requirement as the piezoelectric effect, that is, the lack of a crystallographic center of inversion. A laser beam, in passing through the KDP crystal as shown in Fig. 12.29, suffers a change in the direction of polarization on account of the phase difference between the two polarization components $E[110]$ and $E[1\bar{1}0]$. The electro-optic modulus in KDP is around 10^{-9} cm/volt. For a microwave field of $E_z = 2000$ volts/cm and a length of KDP of $L = 5$ cm, the phase difference is calculated to be about 1.4 rad with $\lambda = 7000$ Å and $n = 1.47$. Since the electro-optic effect is produced by electronic polarization, it is expected that the effect will stay constant in the microwave region. In the experiment of Harris et al., approximately 300 watts of microwave power produce almost 100 per cent amplitude modulation of the transmitted light from a ruby laser, the microwave frequency being centered at 2700 Mc. The modulated light can then be converted into a microwave signal by a microwave phototube.

Amplitude modulation of a light beam is also possible by cross magnetic fields. Figure 12.30 shows two modulation schemes proposed by Bloembergen et al.[51] The electronic polarizability based on classical models as given by Eq. (7.17) is no longer adequate for the present discussion. Quantum-mechanical calculations similar to but more complicated than those given in Sec. 9.6 show that the polarizability of a medium is not only a tensor but also composed of two parts, a Hermitian and an anti-Hermitian part. The Hermitian part

[51] N. Bloembergen, P. S. Pershan, and L. R. Wilcox, *Phys. Rev.*, vol. 120, p. 2014, 1960.

has off-diagonal elements which are complex conjugates of each other. Analogous to the magnetic case discussed in Sec. 10.8, the Hermitian part causes a rotation of the plane of polarization known as the Faraday rotation. The anti-Hermitian part determines the absorptive property of the medium. Both the absorption and Faraday rotation of a light beam can be modulated by a microwave signal if the frequency of the microwave signal corresponds to the transition energy between any two magnetic levels of the ground state. Consider the energy-level scheme of the ground state $^4A_2$ of ruby, as shown in Fig. 9.9 and in the article of Bloembergen et al. Let β_j and β_l be the levels involved in a microwave transition and α_i and β_j those involved in an optical transition. The optical property of the medium, by the quantum-mechanical perturbation calculation, is shown to be proportional to

$$\sum_{\alpha_i \beta_j \beta_l} (\beta_j|er|\alpha_i)(\alpha_i|er|\beta_l)\rho(\beta_j\beta_l)$$

where $\rho(\beta_j\beta_l)$ is the density matrix similar to the AA^*, AB^*, etc. defined in Sec. 9.6 and $(\beta_j|er|\alpha_i)$ is the electric dipole-moment matrix element. Note that $\rho(\beta_j\beta_l)$ if $j \neq l$ behaves like u_1 and u_2 in Eq. (9.60) and hence it varies with the microwave field. If a light beam propa-

Figure 12.30

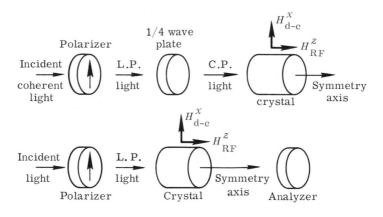

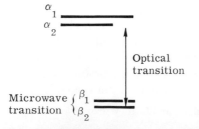

Scheme for modulation of a laser beam using the magneto-optic effect. L. P. = linearly polarized and C. P. = circularly polarized. (N. Bloembergen, P. S. Pershan, and L. R. Wilcox, Phys. Rev., vol. 120, p. 2014, 1960.)

gates in the z direction, the d-c magnetic field should be applied in the xy plane so that the magnetization has an a-c component in the direction of light propagation as shown in the scheme of Fig. 12.30. A detailed analysis of microwave modulation of the optical absorption and Faraday rotation by Bloembergen et al. shows that these effects are relatively small in paramagnetic and dielectric solids at ordinary temperatures. The Kerr effect in KDP and similar piezo-electric crystals discussed in the previous paragraph seems to be much superior to the magneto-optic effects so far as modulation of the amplitude of a light beam is concerned.

The Zeeman splitting of the energies of laser levels by a d-c magnetic field, however, can be used for another purpose, namely, the tuning of the frequency of a laser beam. Take the energy level scheme of $CaF_2:Tm^{2+}$ discussed in a paper by Kiss.[52] In the presence of a d-c magnetic field, the two levels $^2F_{5/2}$ and $^2F_{7/2}$ involved in the laser transition are further split as shown in Fig. 12.31. Since the g values of the two Kramers doublets are different, four fluorescent lines are observed, with the inner components ($\Delta m = \pm 1$) being circularly polarized and the outer components ($\Delta m = 0$) linearly polarized. At liquid neon temperature (27°K), the two inner components have been reported to break into laser oscillations with an energy separation equal to $u_B g B_0$ where u_B is the Bohr magneton, B_0 the d-c magnetic field and g has an effective value of 2. For a magnetic field of 9000 oersteds, an energy separation of the order of 1 cm^{-1} is expected. We should mention that the emission spectrum of a semiconductor laser[53] is also affected by a magnetic field. As shown in Fig. 12.32, as the magnetic field increases, not only does the emission line at 31,125 Å of an InAs diode increase at the expense of the

[52] Z. J. Kiss, *Appl. Phys. Letters*, vol. 2, p. 61, 1963.
[53] I. Melngailis and R. H. Rediker, *Appl. Phys. Letters*, vol. 2, p. 202, 1963.

Figure 12.31

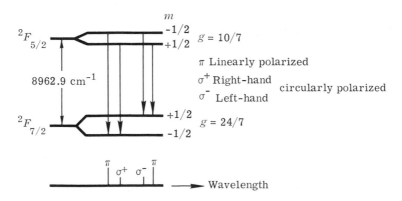

Splitting of the energy levels in $CaF_2:Tm^{2+}$ *by a magnetic field. (Z. J. Kiss, Appl. Phys. Letters, vol. 2, p. 61, 1963.)*

line at 31,170 Å but also the whole emission spectrum shifts toward shorter wavelengths. Two tentative explanations have been offered. The dielectric constant may decrease in a magnetic field due to the magneto-optic effect discussed in the preceding paragraph, resulting in a shift toward shorter wavelength modes. The observed change could also be due to the Zeeman splitting of an exciton level or that of the Landau levels of conduction-band electrons if either the exciton or the conduction-band states were the initial states in a laser transition.

Figure 12.32

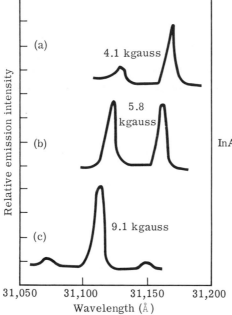

InAs diode at 4.2°K

Emission spectrum from an InAs diode as the d-c magnetic field (in kilogauss) increases. (I. Melngailis and R. H. Rediker, Appl. Phys. Letters, vol. 2, p. 202, 1963.)

12.6 NONLINEAR OPTICAL EFFECTS AND SECOND-HARMONIC GENERATION

Harmonic generation of laser beams has been observed in quartz, triglycine, potassium dihydrogen phosphate, gallium arsenide, and many other piezoelectric crystals except in cases where there is absorption of the radiation involved. There are several physical mechanisms which can give rise to nonlinear dielectric properties of solids and among these, the nonlinearity resulting from anharmonicity can be treated classically. Take the electronic polarizability as an example. If an anharmonic component δx^2 of the restoring force is introduced into Eq. (7.10), then Eq. (7.10) becomes

$$m \frac{d^2x}{dt^2} = -\beta \frac{dx}{dt} - \gamma x - \delta x^2 - eE_1 \exp{(i\omega t)} \tag{12.58}$$

Assuming that $x = x_1 \exp(i\omega t) + x_2 \exp(i2\omega t)$ and that $x_2 \ll x_1$, we find

$$ex_2 = -\frac{(e/m)^3 \delta E_1^2}{(\omega_0^2 - \omega^2 + i\omega\beta/m)^2(\omega_0^2 - 4\omega^2 + 2i\omega\beta/m)} \tag{12.59}$$

where $\omega_0 = \sqrt{\gamma/m}$. If the field term in Eq. (12.58) contains two frequency components ω_1 and ω_2, then mixing of these two frequencies, that is $\omega_1 \pm \omega_2$, is also possible through the anharmonic term δx^2.

The generalization of Eq. (12.58) to include other physical mechanisms requires the introduction of the internal energy U of the system. In the presence of an electric field,

$$U = (b_1 x^2 + 2b_2 xE + b_3 E^2)/2 \tag{12.60}$$

Here, we again take the one-dimensional case to preserve consistency. By definition, the function U satisfies the following equations:

$$P = \frac{\partial U}{\partial E} \quad \text{and} \quad m\frac{\partial^2 x}{\partial t^2} = -\frac{\partial U}{\partial x} \tag{12.61}$$

where P is the electric polarization. A comparison of Eq. (12.61) with Eq. (12.58) and the definition of P shows that in Eq. (12.60), b_1 represents the force constant γ, b_2 the effective charge e, and b_3 the polarizability resulting from ionic or other contributions. The expression, Eq. (12.60), now can be generalized to include higher-order terms which consist of

$$U' = \frac{C_0 x^3 + 3C_1 x^2 E + 3C_2 xE^2 + C_3 E^3}{3} \tag{12.62}$$

The first term arises from the anharmonic restoring force, the second term, the second-order electric movements, and the third term, the Raman effect. A general discussion of the physical origins of these terms can be found in the original article by Kleinman.[54] If still higher order terms are added to U, the electric polarization P can be expressed as a function of the electric field E at optical frequencies as

$$P = a_0 E(1 + a_1 E + a_2 E^2 + \cdots) \tag{12.63}$$

The coefficients a_1 and a_2 of the nonlinear terms can be estimated from quantum-mechanical calculations, and they are so small that only laser light sources have sufficient intensity to produce detectable optical harmonics.

For second-harmonic generation, symmetry considerations can rule out a large number of crystal classes. For crystals possessing a center of inversion, the polarization must reverse sign for a reversal of the field; that means $a_1 = 0$. Therefore, the symmetry restrictions

[54] D. A. Kleinman, *Phys. Rev.*, vol. 128, p. 1761, 1962; also vol. 126, p. 1977, 1962.

are the same as those for piezoelectric effects. For a three-dimensional lattice, the relationship between P and E must be replaced by a tensor. For simplicity, we shall limit our discussion to terms which will contribute to the generation of second harmonics, namely,

$$P_i(2\omega) = \chi_{i1}E_x^2 + \chi_{i2}E_y^2 + \chi_{i3}E_z^2 + \chi_{i4}E_yE_z$$
$$+ \chi_{i5}E_zE_x + \chi_{i6}E_xE_y \quad (12.64)$$

where $i = x$, y, or z. Note that Eq. (12.64) has a similar functional relationship between P_i and the various E_jE_k with j, $k = x$, y, or z as that between the electric displacement D_i and the various stresses T_j of Eq. (7.82). Since the χ's in Eq. (12.64) observe the same symmetry restrictions as the d's in Eq. (7.82), the nonvanishing elements of χ can be read directly from tabulations of the piezoelectric coefficients published in the Proceedings of IRE.[55] Examples are given below. The KDP which has point group symmetry $(\overline{4}2m)$ has nonvanishing d_{36}, and $d_{14} = d_{25}$. Thus, for KDP, we have

$$P_x = aE_yE_z \qquad P_y = aE_zE_x \qquad P_z = bE_xE_y \quad (12.65)$$

For alpha-quartz which has point group symmetry (32) of the D_3 class, we can express the second-harmonic component of P as

$$P_x = a(E_x^2 - E_y^2) + bE_yE_z$$
$$P_y = -bE_xE_z - aE_xE_y \quad (12.66)$$
$$P_z = 0$$

It should be emphasized, however, that the magnitudes of the second-harmonic and piezoelectric tensors are not related. Strictly speaking, Eqs. (12.65) and (12.66) are true only for second-harmonic generation. For frequency mixing, a general expression for P_x of Eq. (12.65), for example, should read

$$P_x = a_1E_y(\omega_1)E_z(\omega_2) + a_2E_z(\omega_1)E_y(\omega_2)$$

In other words, the magnitude of P_x at $(\omega_1 \pm \omega_2)$ depends on whether $E(\omega_1)$ is applied to the y axis and $E(\omega_2)$ is applied to the z axis or vice versa. Therefore the number of independent constants in Eqs. (12.65) and (12.66) will increase when these equations are used for frequency mixing.

Next, we shall discuss aspects of electromagnetic wave propagation associated with second harmonic waves. Consider the case of a plane wave of frequency ω_1 propagating through a nonlinear medium of thickness l as shown in Fig. 12.33. If the electric field at the fundamental frequency is given by $E_1 \exp[i(\omega_1 t - k_1 x)]$, then according to Eq. (12.65) or Eq. (12.66), the second harmonics generated between x and $x + dx$ must vary as $\exp[i2(\omega_1 t - k_1 x)]$ or

$$dE(2\omega_1) \sim \exp[i2(\omega_1 t - k_1 x)]\, dx \quad (12.67)$$

[55] Piezoelectric Crystals, *Proc. IRE*, vol. 46, p. 766, 1958.

If the refractive indices of the medium at ω_1 and ω_2 are different, then the second harmonic wave will propagate with a different phase velocity, $v_2 = \omega_2/k_2 = c/n_2$ where $\omega_2 = 2\omega_1$. The time delay required for the second harmonic wave to reach the exit surface of the nonlinear crystal is equal to $t' = (l - x)/v_2 = (l - x)k_2/\omega_2$. Therefore, at time t, the total second-harmonic electric field of the wave coming out from the crystal is given by

$$E \sim \int_0^l \exp\left[i2(\omega_1 t - k_1 x) - i\omega_2 t'\right] dx$$

$$\sim \frac{\lambda \sin\left[2\pi l(n_2 - n_1)/\lambda\right]}{2\pi(n_2 - n_1)} \exp\left\{i\left[\omega_2 t - \frac{2\pi(n_1 + n_2)l}{\lambda}\right]\right\} \quad (12.68)$$

Note that the intensity of the second-harmonic radiation is a maximum when $n_1 = n_2$. The condition $n_1 = n_2$ is referred to as *index-matching* at the fundamental and second-harmonic frequencies.

Figure 12.34 shows (a) the experimental setup for the production and (b) the measured output of second harmonics published in a paper by Maker et al.[56] As the thin quartz plate is rotated, l in Eq. (12.68) becomes $l/\cos\theta$, or, in other words, the effective optical length of the plate increases, thus generating successive maxima and minima shown in Fig. 12.34b. The characteristic length l' of a medium defined as

$$l' = \frac{\lambda}{4(n_2 - n_1)} \quad (12.69)$$

is often referred to as the *coherence length* and a maximum occurs if $l/\cos\theta$ is an odd integral multiple of l'. The value of l' amounts to the order of twenty wavelengths for quartz at the red radiation and its second harmonic of a ruby laser beam. Knowing l, the spacing between successive maxima gives the value of l'. Figure 12.34b shows a good agreement between the theoretical value of the spacing cal-

[56]   P. D. Maker et al., *Phys. Rev. Letters*, vol. 8, p. 21, 1962.

Figure 12.33

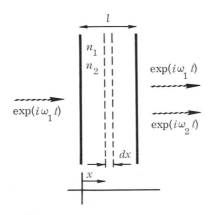

Second-harmonic generation in a dieletric slab.

culated from Eq. (12.69) and the observed value expressed in terms of thickness changes.

It should be pointed out that the refractive index matching is equivalent to a matching of the wave vectors, which is well-known in the theory of traveling-wave parametric amplifiers. The simple index-matching condition can be generalized to include two or more incident waves at ω_i, ω_j, etc., with progagation vectors $\mathbf{k}_i$, $\mathbf{k}_j$, etc. In order to obtain a maximum output at frequency $\omega = \omega_i \pm \omega_j \pm \cdots$, it is necessary to satisfy the momentum conservation relation $\mathbf{k} = \mathbf{k}_i \pm \mathbf{k}_j \pm \cdots$. A clever method has been developed independently by Giordmaine[57] and by Maker et al.[58] In certain uniaxial crystals such as quartz, KDP, and calcite, there are two refractive indices, the ordinary n^o which is independent of θ and the extraordinary n^e

[57] J. A. Giordmaine, *Phys. Rev. Letters*, vol. 8, p. 19, 1962.
[58] P. D. Maker et al., *Phys. Rev. Letters*, vol. 8, p. 21, 1962.

Figure 12.34

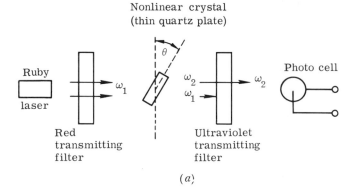

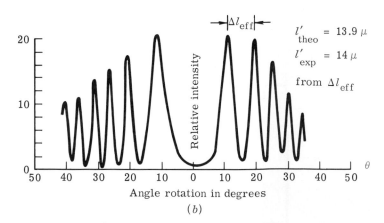

(a) *Experimental arrangement in the detection of second-harmonic generation of light and* (b) *the second-harmonic intensity as a function of the angle of rotation θ.* (P. D. Maker et al., Phys. Rev. Letters, vol. 8, p. 21, 1962.)

Figure 12.35

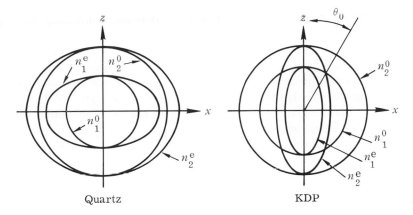

Quartz   KDP

Refractive index surfaces for quartz and KDP. *The superscripts* o *and* e *refer to the ordinary and extraordinary rays, respectively, and the subscripts 1 and 2 refer to the fundamental and second-harmonic frequencies, respectively. The figures are not drawn to scale. (P. A. Franken and J. F. Ward, Rev. Mod. Phys., vol. 35, p. 23, 1963.)*

which is a function of θ, the angle θ being the angle between the wave vector **k** and the uniaxial axis. The variations with θ of the refractive indices of quartz and KDP are shown schematically in Fig. 12.35. Note that in KDP, the index-matching condition is satisfied when $n_1^o = n_2^e(\theta_0)$. Figure 12.36 shows the experimental curve of the second-harmonic intensity as a function of crystal orientation reported by Maker et al.

There are other phase or index matching schemes to enhance the second-harmonic intensity. It is also worth mentioning that if two

Figure 12.36

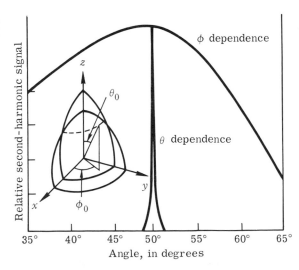

The second-harmonic signal (blue light) intensity as a function of the orientation for KDP. *Maximum intensity occurs at* $\theta_0 = 52°$, $\phi_0 = 45°$. *(P. D. Maker et al., Phys. Rev. Letters, vol. 8, p. 21, 1962.)*

plane waves of ω_1 and ω_2 are incident on a nonlinear medium, the reflected wave must also contain components at sum and difference frequencies $\omega_1 \pm \omega_2$ to satisfy the boundary condition. For a general discussion of harmonic generation and wave propagation in nonlinear crystals as well as for a quantum-mechanical estimate of the non-linear effect, the reader is referred to articles by Franken,[59] by Bloembergen,[60] and by Kleinman.[61] Figure 12.37 shows the conversion efficiency of second- and third-harmonic generation using a giant ruby laser and index-matched ADP and calcite reported in a paper by Terhune et al.[62] A peak second-harmonic power of 200 kw was achieved, using a 30 nsec pulse. From such a measurement, the coefficient a in Eq. (12.65) is estimated to be 3×10^{-10} esu for ADP. Second-harmonic generation from GaAs injection laser itself has also been observed by Armstrong et al.[63] GaAs has point group symmetry $(\bar{4}3m)$; therefore, the second-order polarization is related to the electric field by an equation similar to Eq. (12.65) but

[59] P. A. Franken and J. F. Ward, *Rev. Mod. Phys.*, vol. 35, p. 23, 1963.
[60] N. Bloembergen and P. S. Pershan, *Phys. Rev.*, vol. 128, p. 606, 1962; J. A. Armstrong et al., *Phys. Rev.*, vol. 127, p. 1918, 1962.
[61] D. A. Kleinman, *Phys. Rev.*, vol. 126, p. 1977, 1962; vol. 128, p. 1761, 1962.
[62] W. R. Terhune, P. D. Maker, and C. M. Savage, *Appl. Phys. Letters*, vol. 2, p. 54, 1963.
[63] J. A. Armstrong, M. I. Nathan, and A. W. Smith, *Appl. Phys. Letters*, vol. 3, p. 68, 1963.

Figure 12.37

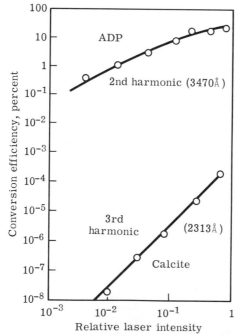

Conversion efficiency of ADP for second-harmonic generation and that of calcite for third-harmonic generation using a giant-pulse ruby laser beam. (R. W. Terhune, P. D. Maker, and C. M. Savage, Appl. Phys. Letters, vol. 2, p. 54, 1963.)

with $a = b$. Experimental results seem to indicate even a larger value of a for GaAs than that for KDP. A compilation of data on second-harmonic generation in various piezoelectric materials may be found in the book by Bloembergen.[64]

12.7 BRILLOUIN AND RAMAN SCATTERING

If a material is elastically deformed, the index of refraction of the medium changes as a result of changing α_e or N or both in Eq. (7.13). Brillouin predicted that a substance under periodic elastic deformation would diffract light in a way similar to the diffraction of light by a grating. The phenomenon is known as *Brillouin scattering*.[65] Consider a monochromatic plane wave (of frequency ν and wave vector $\mathbf{k}$) incident on a dielectric slab (Fig. 12.38). In the meantime, we also let a plane compression wave (of frequency ν_a and wave vector $\mathbf{q}$) progress through the slab along the direction of $\mathbf{q}$. Since the velocity of sound waves is much smaller than the velocity of light, we may consider the sound wave to be stationary insofar as the propagation of the light wave is concerned. Therefore, to a first-order approximation, the dielectric slab can be pictured as being made of regions having different indices of refraction.

In order to obtain appreciable intensity of the diffracted light, the optical path difference (Fig. 12.38) between the rays from two successive planes of distance $2\pi/q$ apart must satisfy the following relation

$$BD - AC = \left(\frac{2\pi}{q}\right)(\sin \phi' - \sin \theta') = l\left(\frac{2\pi}{k'}\right) \qquad (12.70)$$

where k' is the wave number of the light wave in the dielectric

[64] N. Bloembergen, "Nonlinear Optics," W. A. Benjamin, Inc., New York, 1965.
[65] See, for example, M. Born and E. Wolf, *op. cit.*, p. 590; M. Born and K. Huang, "Dynamic Theory of Crystal Lattices," p. 373, Clarendon Press, Oxford, 1954.

Figure 12.38

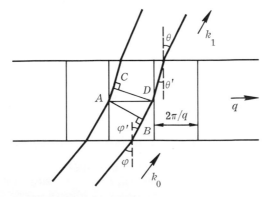

Schematic illustration of the diffraction of light by an ultrasonic wave (Brillouin scattering).

medium and l is an integer. Using Snell's law of refraction

$$\frac{\sin \phi'}{\sin \phi} = \frac{\sin \theta'}{\sin \theta} = \frac{c'}{c} = \frac{k}{k'} \qquad (12.71)$$

and letting $l = 1$ in Eq. (12.70), we have

$$k(\sin \phi - \sin \theta) = q \qquad (12.72)$$

In terms of wave mechanics, the Brillouin scattering can be considered as a collision process between a photon and a phonon. During such a process, the conservation of energy and momentum requires

$$\nu_0 = \nu_1 \pm \nu_a \qquad \text{and} \qquad \mathbf{k}_0 = \mathbf{k}_1 + \mathbf{q} \qquad (12.73)$$

where the subscripts 0 and 1 refer to the incident and diffracted waves respectively, and the positive and negative signs refer respectively to the process in which a phonon is emitted and to the process in which a phonon is absorbed.

Note that the optical frequencies ν_0 and ν_1 are many orders of magnitude larger than the acoustic frequency ν_a. Thus, the magnitudes of the wave vectors, $\mathbf{k}_0$ and $\mathbf{k}_1$, before and after scattering are nearly the same. Referring to Fig. 12.38, we see that Eq. (12.72) is actually a statement of momentum conservation along the direction of the acoustic wave vector $\mathbf{q}$. Next, let us discuss how the direction of the wave vector $\mathbf{k}$ is affected by the acoustic wave. The velocity of acoustic waves is around 5×10^5 cm/sec in most solids. Thus, at $\nu_a = 5$ Gc/sec, the wavelength of the acoustic wave is of the order of 10^{-4} cm which is close to the optical wavelength in the visible and near-infrared regions. In other words, the magnitudes of $\mathbf{q}$ and $\mathbf{k}$ in Eq. (12.73) are comparable for ultrasonic waves. Therefore, in the normal Brillouin scattering experiment, we expect to have a very small frequency shift but a significant scattering angle of the light beam.

Figure 12.39 shows the experimental observation of Brillouin scattering with a He-Ne laser beam reported by Siegman et al.[66] Because the quartz bar (excited by an ultrasonic wave) is placed inside the optical cavity where the interaction between acoustic and optic waves is enhanced, an appreciable amount of the primary laser beam is converted into the secondary scattered beam. When the intensity of the radiation from a laser is sufficiently high, a dielectric medium may be driven into coherent acoustic oscillations without external acoustic excitation. The phenomenon is known as the stimulated Brillouin scattering. In the experiment reported by Chiao, Townes, and Stoicheff,[67] the intense radiation from a giant-pulse ruby laser ($\lambda = 6940$ Å) was used with a power output of about 50 megawatts

[66] A. E. Siegman et al., *Appl. Phys. Letters*, vol. 5, p. 1, 1964.
[67] R. Y. Chiao, C. H. Townes, and B. P. Stoicheff, *Phys. Rev. Letters*, vol. 12, p. 592, 1964.

and a power density of the order of 10^6 megawatts/cm² at the focus. The scattered light was shifted from the incident beam by about 1 cm⁻¹, corresponding to an acoustic wave of frequency near 3×10^{10} cps from Eq. (12.73). Because the acoustic loss in a solid medium is generally high at ordinary temperatures, high-intensity laser beam is needed for stimulated Brillouin scattering.

Now let us discuss the Raman effect. When a light beam is passing through a gaseous, liquid or solid medium, the scattered light contains, in addition to the frequency of the incident light, a series of other frequencies. The classical explanation is as follows. If a light beam with an electric field $\mathbf{E} = \mathbf{E}_0 \cos \omega t$ falls on a molecule, the induced electric dipole moment is given by

$$\mathbf{p} = \alpha \mathbf{E}_0 \cos \omega t \qquad (12.74)$$

The quantity α, according to the definition given in Sec. 7.2, is called the polarizability and for a three-dimensional molecule, it is a tensor. At present, we are interested in the effects produced by the rotational and vibrational state of the molecule on the induced polarizability. In the derivation of α in Sec. 7.2 we have assumed that the nuclei are stationary. Furthermore, because we take the one-dimensional

Figure 12.39

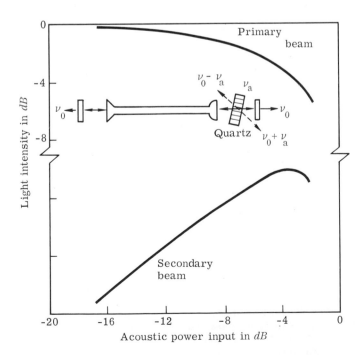

Experimental observation of Brillouin scattering with a He-Ne laser beam. (A. E. Siegman et al., Appl. Phys. Letters, vol. 5, p. 1, 1964.)

model, no rotation of molecules is admitted. When a molecule rotates or vibrates, the direction and magnitude of the induced polarization change with the rotational and vibrational motions of the molecule. Phenomenologically, the polarizability tensor, to include the vibrational and rotational effects, can be written as

$$\alpha = \alpha_0 + \alpha_1 \cos (\omega_v t + \delta) \qquad (12.75)$$

where ω_v is the angular frequency of one particular Fourier component of the vibrational or rotational spectra of the molecule. Substituting Eq. (12.75) into Eq. (12.74) shows that the induced dipole moment and hence the scattered light contain three frequency components, $\nu \pm \nu_v$ in addition to frequency ν of the incident light where $\nu = \omega/2\pi$.

The two additional lines at $\nu_S = \nu - \nu_v$ and $\nu_A = \nu + \nu_v$ are generally referred to as the *Stokes* and *anti-Stokes lines*, a terminology used in fluorescence. Note that $\hbar\omega_v$ represents the quantized energy associated with a particular mode of the vibrational and rotational motion of the molecule. A light quantum with energy $\hbar\omega$ when hitting a molecule may lift the molecule from the ground state to an excited vibrational or rotational state by giving the molecule an energy $\hbar\omega_v$. The light quantum after scattering has an energy $\hbar(\omega - \omega_v)$, corresponding to the Stokes line. It is also possible but less probable that a light quantum may receive an energy of the amount $\hbar\omega_v$ from a molecule in an excited vibrational or rotational state, resulting in the emission of anti-Stokes line and the returning of the molecule to the ground state. The observation of stimulated emission by Woodbury and Ng[68] near 7670 Å from a nitrobenzene Q-switch Kerr cell using the experimental arrangement of Fig. 12.28 has prompted attempts to obtain *stimulated Raman scattering* from other substances. By stimulated scattering, we mean that the phase angle δ in Eq. (12.75) is no longer random at sufficiently high electric fields. So far, stimulated emission of Stokes and anti-Stokes lines has been observed from various organic liquids (such as benzene, nitrobenzene, toluene),[69] gases (H_2, D_2, and CH_4),[70] and solids (diamond, calcite, and α-sulfur).[71] A giant-pulse laser beam from a ruby rod is needed in these experiments to provide a high enough field so that the condition for stimulated Raman scattering may be achieved.

Equation (12.75) is no longer adequate to explain the phenomenon of stimulated Raman scattering. At high electric-field intensity, we must consider effects produced by nonlinear polarization. Thus, in

[68] E. J. Woodbury and W. K. Ng, *Proc. IRE*, vol. 50, p. 2367, 1962.
[69] G. Eckhardt et al., *Phys. Rev. Letters*, vol. 9, p. 455, 1962.
[70] R. W. Minck, R. W. Terhune, and W. G. Rado, *Appl. Phys. Letters*, vol. 3, p. 181, 1963.
[71] G. Eckhardt, D. P. Bortfeld, and M. Geller, *Appl. Phys. Letters*, vol. 3, p. 137, 1963.

the auxiliary Maxwell equation

$$\mathbf{D} = \epsilon\mathbf{E} + 4\pi\mathbf{P}'$$ (12.76)

The component P'_a of the polarization vector $\mathbf{P}'$ should be expressed in a power series of E as

$$P'_a = \sum_{\alpha,\beta} \chi_{a\alpha\beta} E_\alpha E_\beta + \sum_{\alpha,\beta,\gamma} \chi_{a\alpha\beta\gamma} E_\alpha E_\beta E_\gamma + \cdots$$ (12.77)

where a, α, β, and $\gamma = x$, y, or z. For convenience, the linear component of the polarization is incorporated in $\epsilon\mathbf{E}$ in Eq. (12.76); therefore the component $\mathbf{P}'$ represents only the nonlinear polarization. Equation (12.77), which is in essence a restatement of Eq. (12.63), shows explicitly the tensorial nature of the nonlinear susceptibilities $\chi_{a\alpha\beta}$ and $\chi_{a\alpha\beta\gamma}$. As will be shown later, the phenomenon of stimulated Raman scattering arises from the cubic polarization; that is, the second term in Eq. (12.77). For a discussion of the symmetry property of the susceptibility tensor, the reader is referred to the book by Bloembergen[72] and the research report by Butcher.[73] In view of Eq. (12.76), Maxwell's field equations now become

$$\nabla \times \mathbf{E} = -\frac{i\omega}{c}\mathbf{H}$$ (12.78a)[74]

$$\nabla \times \mathbf{H} = \frac{i\omega\epsilon}{c}\mathbf{E} + \frac{4\pi i\omega}{c}\mathbf{P}'$$ (12.78b)[74]

For simplicity, we shall assume a plane wave having an $\exp[i(\omega t - \mathbf{k}\cdot\mathbf{r})]$ dependence; thus we have for the wave equation

$$\nabla \times \nabla \times \mathbf{E} - \left(\frac{\omega}{c}\right)^2 \epsilon\mathbf{E} = 4\pi\left(\frac{\omega}{c}\right)^2 \mathbf{P}'$$ (12.79)

where $k = \omega/c$.

Before analyzing the phenomenon of stimulated Raman scattering, we should point out that distinctions are often made between external and internal modes of vibrations. Internal modes are due to vibrations of the molecule itself or a group of molecules in the unit cell of a solid, whereas external modes arise from the relative movements of molecules or groups of molecules occupying different lattice sites. In gases and liquids, only internal modes exist; hence the angular frequency ω_v in CH_4, for example, is determined by the molecular structure and the symmetry property of the CH_4 molecule. In solids, however, both internal and external vibrational modes may exist, depending upon the nature of the solid. In $CaCO_3$ and α-sulphur

[72] N. Bloembergen, "Nonlinear Optics," W. J. Benjamin, Inc., New York, 1965.
[73] P. N. Butcher, "Nonlinear Optical Phenomena," *Ohio State Univ., Bulletin* 200, *Eng. Expt. Sta.* Columbus, Ohio, 1965.
[74] The reader is again reminded that the equations in this chapter are expressed in cgs units.

(which is a molecular crystal), the observed Raman effect is due to the internal modes of vibration associated with the $(CO_3)^{--}$ and sulphur molecules. In diamond, however, the observed Raman shift corresponds to the triply degenerate vibration of the two inter-penetrating cubic sublattices of the diamond structure against one another. In other words, the shift is caused by the interaction of the optical branch of lattice vibration with the electromagnetic wave through the nonlinear polarization. We shall treat the Raman effect due to external modes of vibrations first simply, because the treat-ment follows a similar procedure as the one developed for parametric amplifiers which we have already discussed in Sec. 6.6.

Let us now discuss the nonlinear polarization $\mathbf{P}'$ in connection with the Raman effect. In Sec. 7.2, we have assumed a stationary nucleus in the derivation of the electronic polarization. According to the model developed by Placzek,[75] when a nucleus vibrates about its equilibrium position, the electronic polarizability depends para-metrically on the nuclear vibrational coordinate R, as

$$\alpha = \alpha_0 + \frac{\partial \alpha}{\partial R} R + \cdots \qquad (12.80)[76]$$

For ease of discussion, we shall not be concerned with the tensor nature of the polarizability and simply treat $\partial \alpha / \partial R$ as a constant. Since R is associated with the optical vibration waves, we can write

$$R = 2 \operatorname{Re} A_v \exp \left[i(\omega_v t - \mathbf{k}_v \cdot \mathbf{r}) \right] \qquad (12.81)$$

where A_v is the amplitude of the vibrational mode (with angular frequency ω_v and wave vector $\mathbf{k}_v$), and the symbol Re stands for the real part. In the presence of two light waves of the form

$$\mathbf{E} = \mathbf{E}_0 \exp \left[i(\omega_0 t - \mathbf{k}_0 \cdot \mathbf{r}) \right] + \mathbf{E}_1 \exp \left[i(\omega_1 t - \mathbf{k}_1 \cdot \mathbf{r}) \right] \qquad (12.82)$$

the component of the nonlinear polarization $\mathbf{P}'$ at frequency ω_1 is given by

$$P'(\omega_1) = N \frac{\partial \alpha}{\partial R} A_v^* E_0 \exp \left[i(\omega_1 t - \mathbf{k}_1 \cdot \mathbf{r}) \right] \qquad (12.83)$$

with $\omega_0 = \omega_1 + \omega_v$ and $\mathbf{k}_0 = \mathbf{k}_1 + \mathbf{k}_v$. In Eq. (12.83), N is the number of molecules per unit volume. A similar expression is obtained for the component of nonlinear polarization at ω_0.

To obtain the differential equation for the amplitude E_1 and A_v, we follow the same procedure as outlined in Sec. 6.6 for parametric amplification. Substituting Eqs. (12.82) and (12.83) into Eq. (12.79) and assuming that the variation of E_1 over one wavelength is small,

[75] See N. Bloembergen, *op. cit.*, pp. 39 and 104.
[76] Since in the optical region, the only contribution to polarizability is electronic, the subscript e in α_e is dropped.

that is, $|\partial^2 E_1/\partial z^2| \ll k_1|\partial E_1/\partial z|$, we have

$$2ik_1 \frac{\partial E_1}{\partial z} = 4\pi N \frac{\partial \alpha}{\partial R} \left(\frac{\omega_1}{c}\right)^2 E_0 A_v^* \qquad (12.84)$$

In Eq. (12.84), the direction of propagation is taken to be along the z axis. The other equation, that is, the equation for A_v, can be obtained by considering the energy exchange between the photon and phonon fields. According to Eq. (12.80), the free energy U of the medium for N molecules in the presence of electric fields is

$$U = \frac{N}{2} \alpha E^2 + \frac{N}{2} \frac{\partial \alpha}{\partial R} RE^2 \qquad (12.85)$$

Note that the quantity $(N\alpha/2)E^2$ is the ordinary polarization energy. The term $(N/2)(\partial\alpha/\partial R)RE^2$, however, is new; it represents the additional free energy due to the *photoelastic coupling*. We further note that the quantity R is the displacement of the molecule with respect to its equilibrium position; hence R is proportional to the strain S in Eq. (7.77). According to Eq. (7.83), the additional stress due to the photoelastic coupling is given by

$$T = \frac{\partial U}{\partial S} = \frac{1}{B} \frac{N}{2} \frac{\partial \alpha}{\partial R} E^2 \qquad (12.86)$$

where the factor B is a normalization constant as a result of changing the variable from R to S.

According to Eq. (12.86), the stress component at frequency ω_v is given by

$$T(\omega_v) = \frac{1}{B} \frac{N}{2} \frac{\partial \alpha}{\partial R} E_0 E_1^* \exp\{i[\omega_v t - (\mathbf{k}_0 - \mathbf{k}_1) \cdot \mathbf{r}]\} \qquad (12.87)$$

Adding the component $T(\omega_v)$ of Eq. (12.87) to Eq. (7.89) and realizing that $S = \partial\zeta/\partial z$, we have

$$\rho \frac{\partial^2 S}{\partial t^2} - \frac{1}{s} \frac{\partial^2 S}{\partial z^2} = \frac{\partial^2 T(\omega_v)}{\partial z^2} \qquad (12.88)$$

where s is the elastic constant defined in Eq. (7.80) and ρ is the mass density of the medium. In Eq. (12.88), the vibrational mode is assumed to be a longitudinal wave propagating in the z direction. Substituting Eq. (12.87) into Eq. (12.88) and using $S = BR$, we obtain

$$\rho \frac{\partial^2 R}{\partial t^2} - \frac{1}{s} \frac{\partial^2 R}{\partial z^2} = k_{vg}{}^2 \frac{E_0 E_1^*}{B^2} \frac{N}{2} \frac{\partial \alpha}{\partial R} \exp[i(\omega_v t - \mathbf{k}_v \cdot \mathbf{r})] \qquad (12.89)$$

Using the expression for R given by Eq. (12.81) in Eq. (12.89) and following the same procedure as used in deriving Eq. (12.84) from

Eq. (12.79), we find

$$2ik_{vg}\left(\frac{dA_v}{dz} + \frac{A_v}{v_g\tau}\right) = N\frac{\partial\alpha}{\partial R}\frac{E_0 E_1^*}{2\rho\omega_v^2}k_{vg}^2 \tag{12.90}$$

with $B = k_v$. A similar equation is obtained for A_v^*.

Several comments about Eq. (12.90) are in order. First, in our earlier discussion of the Brillouin scattering, we are concerned with the acoustic phonons whereas in the Raman scattering, we are dealing with the optical branch of the vibrational mode. Even for long-wavelength optical phonons, the dispersion relation (ω vs. k) is not linear. Therefore, in Eq. (12.90), $v_g(= \partial\omega_v/\partial k_v)$ is the group velocity of the vibrational wave and $k_{vg} = \omega_v/v_g = \omega_v(\partial k_v/\partial\omega_v)$. The term $A_z/(v_g\tau)$ represents the loss associated with the vibrational mode and τ is the characteristic damping time. The quantity R given by Eq. (12.81) actually is the Fourier component of the vibrational mode having a wave vector $\mathbf{k}_v$. Equation (12.90) can be formally derived by using the normalized vibrational coordinate instead of R. The step by letting $B = k_v$ in Eq. (12.89) is taken so that Eq. (12.90) agrees with the result obtained from the formal analysis.

Note the similarities between Eqs. (12.84) and (12.90) and Eqs. (6.86) and (6.87). Coherent generation of vibrational waves results when the power received from the laser beam overcomes the loss associated with the vibrational mode or

$$\frac{2\pi}{\rho\omega_v^2}\left(N\frac{\partial\alpha}{\partial R}\right)^2\left(\frac{\omega_1}{c}\right)^2\left|\frac{E_0}{2}\right|^2\frac{k_{vq}}{k_1} \geq \frac{1}{(v_g\tau)^2} \tag{12.91}$$

In the derivation of Eqs. (12.84) and (12.90), we have assumed $\nu_0 = \nu_1 + \nu_v$. Thus, the secondary beam has a frequency $\nu_1 = \nu_0 - \nu_v$, corresponding to the frequency of the Stokes line $\nu_S = \nu_0 - \nu_v$. Once the first Stokes line (with $\nu_1 = \nu_0 - \nu_v$) goes into coherent oscillation, its intensity can build up to serve as the pump for the second Stokes line (with $\nu_2 = \nu_1 - \nu_v = \nu_0 - 2\nu_v$). As the power of the laser pump increases, more and more Stokes lines will go into oscillation. The excitation of higher-order Stokes lines could be prevented by selective antireflective coating in the Raman cell.

Now let us discuss the stimulated Raman effect associated with the internal modes of molecular oscillations. Referring to Eq. (7.42), the energy U arising from the electrostatic interaction between two dipole moments $\mathbf{p}_1$ and $\mathbf{p}_2$ induced on two adjacent molecules can be written as

$$U = \frac{A\mathbf{p}_1\cdot\mathbf{p}_2}{d^3} = \frac{A\alpha_1\alpha_2}{d^3}\mathbf{E}\cdot\mathbf{E} \tag{12.92}[77]$$

[77] The equation is expressed in cgs units. Here we follow the treatment given by E. Garmire, F. Pandarese, and C. H. Townes, *Phys. Rev. Letters*, vol. 11, p. 160, 1963.

where d is the intermolecular distance, α_1 and α_2 are the polarizabilities of the two molecules, and A is a constant which depends on the relative orientations of $\mathbf{p}_1$ and $\mathbf{p}_2$ with respect to $\mathbf{d}$. If both polarizabilities depend on the vibration coordinate R of the molecule as stated in Eq. (12.80), then the molecule will experience a force of magnitude F

$$F = \frac{\partial U}{\partial R} = \frac{A\alpha}{d^3}\frac{\partial \alpha}{\partial R}\,\mathbf{E}\cdot\mathbf{E} = f\mathbf{E}\cdot\mathbf{E} \tag{12.93}$$

Since the two molecules are assumed identical and, hence, have identical polarizabilities, the subscripts on α_1 and α_2 can be dropped after differentiation.

For molecular vibrations, the force equation reads

$$\frac{d^2x}{dt^2} + \Gamma\frac{dx}{dt} + \omega_v{}^2 x = \frac{F}{M} \tag{12.94}$$

where ω_v is the characteristic frequency of a given internal mode of molecular vibration. For easy comparison with Eq. (12.58), Eq. (12.94) is also written in the one-dimensional form. In order to use Eq. (12.93) in Eq. (12.94), we assume that two light waves of the form given by Eq. (12.82) exist in the interaction medium. For the Raman effect, only the frequency-difference term is important and, hence, will be retained in Eq. (12.93). After substituting Eq. (12.93) into Eq. (12.94), we obtain

$$x(\omega_1 - \omega_0) = \frac{f\mathbf{E}_1\cdot\mathbf{E}_0^*\,\exp\{i[(\omega_1 - \omega_0)t - (\mathbf{k}_1 - \mathbf{k}_0)\cdot\mathbf{r}]\}}{M[\omega_v{}^2 - (\omega_1 - \omega_0)^2 + i(\omega_1 - \omega_0)\Gamma]} \tag{12.95}$$

Using the value of $x(\omega_1 - \omega_0)$ in Eq. (12.80), we find that the driven molecular vibration produces an oscillating electric dipole moment

$$\begin{aligned}
\mathbf{p} &= \frac{\partial \alpha}{\partial x}\,x\mathbf{E}\\[2mm]
&= \frac{f\mathbf{E}_1\cdot\mathbf{E}_0^*\,\exp\{i[(\omega_1 - \omega_0)t - (\mathbf{k}_1 - \mathbf{k}_0)\cdot\mathbf{r}]\}}{M[\omega_v{}^2 - (\omega_1 - \omega_0)^2 + i(\omega_1 - \omega_0)\Gamma]}\frac{\partial \alpha}{\partial x}\,\mathbf{E}
\end{aligned} \tag{12.96}$$

If ω_0 is taken as the angular frequency of the laser beam, obviously it is the polarization at ω_1 (the first Stokes line) that is of interest to us. A quick reference to Eqs. (12.95) and (12.96) shows that intense molecular oscillation occurs and, hence, a large nonlinear polarization results only when $\omega_1 - \omega_0 = \pm\omega_v$. This is the frequency condition for the first Stokes and anti-Stokes lines. At resonance, the nonlinear polarization $\mathbf{P}'$ having angular frequency ω_1 is given by

$$\mathbf{P}'(\omega_1) = \frac{Nf\mathbf{E}_1\cdot\mathbf{E}_0^*\,\exp[i(\omega_1 t - \mathbf{k}_1\cdot\mathbf{r})]}{iM(\omega_1 - \omega_0)\Gamma}\frac{\partial \alpha}{\partial x}\,\mathbf{E}_0 \tag{12.97}$$

where N is the density of molecules under consideration. Thus the average power delivered to the Raman line at ω_1 is equal to

$$\text{Power} = -\operatorname{Re}\left(\frac{d\mathbf{P}'}{dt}\right)^* \cdot \mathbf{E}(\omega_1) = \frac{Nf|(\mathbf{E}_1 \cdot \mathbf{E}_0^*)|^2}{M\Gamma}\frac{\partial\alpha}{\partial x}\frac{\omega_1}{\omega_0 - \omega_1} \qquad (12.98)$$

Note that for stable molecules, $\partial\alpha/\partial R$ is always positive from energy considerations. According to Eq. (12.98), the Stokes radiation at $\omega_1 = \omega_0 - \omega_v$ will be amplified while the anti-Stokes radiation at $\omega_1 = \omega_0 + \omega_v$ will be attenuated. Before discussing the mechanism by which the anti-Stokes lines can also be amplified, we shall generalize the result expressed in Eq. (12.96) so that it can be used more effectively.

First, we notice that the nonlinear polarization responsible for the Raman effect is proportional to E^3. The most general expression relating the polarization to the field is through a third-order susceptibility tensor of the form

$$P'_a = \sum_{\alpha,\beta,\gamma} \chi_{a\alpha\beta\gamma}(\omega_1,\omega_2,\omega_3)E_\alpha(\omega_1)E_\beta(\omega_2)E_\gamma(\omega_3) \qquad (12.99)$$

Obviously the cubic polarization will give rise to third-harmonic generation with $\omega_1 = \omega_2 = \omega_3$. When a medium is subject to a d-c electric field with $\omega_2 = \omega_3 = 0$, we see from Eq. (12.99) that the optical index of refraction of the medium will have a varying part which is proportional to the square of the d-c field. This is known as the Kerr effect, mentioned earlier in connection with the Q-switch giant-pulse laser.

For a discussion of the Raman effect, it is more convenient to express Eq. (12.99) in complex notations as

$$P'_a(\omega_1) = \sum_{\alpha,\beta,\gamma} \chi_{a\alpha\beta\gamma}(\omega_0, -\omega_0, \omega_1)E_\alpha(\omega_0)E_\beta^*(-\omega_0)E_\gamma(\omega_1) \qquad (12.100)$$

Note that according to the notation adopted in Eq. (12.82), taking a complex conjugate means replacing ω_0 by $-\omega_0$. This is exactly what is meant by $-\omega_0$ in Eq. (12.100). As shown in Eq. (12.95), the frequencies must obey the relation $\omega_1 - \omega_0 = \pm\omega_v$ in order to have an intense molecular oscillation. Therefore, to study the effect produced by the nonlinear polarization at ω_1, we must choose E_α at ω_0 such that $\omega_0 - \omega_0 + \omega_1 = \omega_1$. This relation is implicitly stated in Eq. (12.100). We also should point out that in Eq. (12.83), A_v^* is proportional to $E_1(\omega_1)E_0^*(-\omega_0)$. This implicit relation can be deduced from Eq. (12.95). The polarization of Eq. (12.83) and the $3C_2xE^2$ term in Eq. (12.62) are all due to the cubic dependence of the polarization given by Eq. (12.100), although this fact is not explicitly stated in these two equations.

Now let us return to Eqs. (12.96) and (12.100) to make further use of the results of our analysis. Phenomenologically, the important factor contributing to the stimulated Raman effect is the fact that, due to nonlinear cubic polarization, the radiation field experiences a positive imaginary susceptibility at ω_1. In other words, the dielectric constant in the presence of a pump field E_0 can be written as

$$\epsilon(E_0) = \epsilon(1 + ia|E_0|^2) \qquad (12.101)$$

where ϵ is the dielectric constant of the medium with $E_0 = 0$ and a is a proportionality constant which, among other things, contains a factor $(\omega_v \Gamma)^{-1}(\partial \alpha / \partial R)^2$. In view of Eq. (12.101), the wave equation becomes

$$\frac{\partial^2 E_1}{\partial z^2} = -\left(\frac{\omega_1}{c}\right)^2 \epsilon(1 + ia|E_0|^2) \qquad (12.102)$$

For ease of discussion, we assume that the wave propagates in the z direction and that the electric field E_1 is in a direction along which Eq. (12.101) applies. It is instructive to compare Eq. (12.102) with the wave equation in a conductive medium

$$\frac{\partial^2 E_1}{\partial z^2} = -\left(\frac{\omega_1}{c}\right)^2 \epsilon\left(1 - \frac{i\sigma}{\omega\epsilon}\right) \qquad (12.103)$$

where σ is the conductivity of the medium. In both Eqs. (12.102) and (12.103), a periodic time dependence of the form $\exp(i\omega_1 t)$ is assumed for E_1. Therefore a positive imaginary susceptibility is equivalent to a negative conductivity, which means an active medium at ω_1.

Since the susceptibility at the anti-Stokes frequency $\omega_2 = \omega_0 + \omega_v$ is a negative imaginary quantity, stimulated Raman radiation at the anti-Stokes frequency occurs only in the presence of the stimulated radiation at the Stokes frequency. If we write

$$E = E_0 \exp[i(\omega_0 t - k_0 \cdot r)] + E_1 \exp[i(\omega_1 t - k_1 \cdot r)] \\ + E_2 \exp[i(\omega_2 t - k_2 \cdot r)] \qquad (12.104)$$

with $\omega_1 = \omega_0 - \omega_v$ and $\omega_2 = \omega_0 + \omega_v$, we see from Eq. (12.100) that a nonlinear polarization at ω_2 results from $E_\alpha(\omega_0)E_\beta(\omega_0)E_\gamma^*(-\omega_1)$. Rearranging the fields in Eq. (12.96) in this fashion, we obtain for the nonlinear dipole moment at ω_2

$$p(\omega_2) = \frac{fE_1^* \cdot E_0 \exp\{i[\omega_2 t - (2k_0 - k_1) \cdot r]\}}{iM(\omega_0 - \omega_1)\Gamma} \frac{\partial \alpha}{\partial x} E_0 \qquad (12.105)$$

This component of $p(\omega_2)$ from Eq. (12.105) should be added to the contribution from the $E_\alpha(\omega_0)E_\beta^*(-\omega_0)E_\gamma(\omega_2)$ term in Eq. (12.100). Adding the two components together and following a similar procedure to the one used in deriving Eq. (12.98), we find the average power delivered to the anti-Stokes line to be

$$\text{Power} = \frac{N f \omega_2}{(\omega_0 - \omega_1)\Gamma} \frac{\partial \alpha}{\partial x} \{ -|\mathbf{E}_1 \cdot \mathbf{E}_0| \, |\mathbf{E}_0 \cdot \mathbf{E}_2| \cos [(2\mathbf{k}_0 \\ - \mathbf{k}_1 - \mathbf{k}_2) \cdot \mathbf{r} + \varphi_1 + \varphi_2] - |\mathbf{E}_2 \cdot \mathbf{E}_0|^2 \} \quad (12.106)$$

where φ_1 and φ_2 are, respectively, the phase angles of $\mathbf{E}_1$ and $\mathbf{E}_2$ with respect to $\mathbf{E}_0$.

To obtain amplification for E_2, the first term in Eq. (12.106) must have a positive sign and must be greater than the second term. Therefore the amplitude of E_1 should remain greater than that of E_2. For Eq. (12.106) to have nonzero spatial average, index matching $2\mathbf{k}_0 = \mathbf{k}_1 + \mathbf{k}_2$ is required. Furthermore, the phase relation between $\mathbf{E}_0$, $\mathbf{E}_1$, and $\mathbf{E}_2$ should be such that $\varphi_1 + \varphi_2 = \pi$ for maximum gain. Anti-Stokes radiation was first observed by Terhune[78] in benzene. In liquids and solids, the radiation is emitted in cones in the forward direction around the initial laser beam. In gases, however, all anti-Stokes cones overlap. For experimental results on anti-Stokes radiation, the reader is referred to the articles by Minck et al.,[79] and by Chiao and Stoicheff.[80] It is apparent from Eq. (12.106) that a theoretical analysis of the wave propagation of the anti-Stokes radiation would involve a set of coupled-wave equations at ω_1, ω_2, and ω_v even though the amplitude of E_0 (the laser beam) is assumed to be constant. The reader is referred to the book by Bloembergen[81] for such an analysis.

Finally, we should mention that other schemes[82] for obtaining stimulated Raman radiation have been suggested. Moreover, in the phenomenological theory described previously, no mention has been made of the physical origin of the $\partial \alpha / \partial R$ term. The physical mechanisms which give rise to the $\partial \alpha / \partial R$ term may be quite different in the two cases just discussed. Even in the case involving external modes of vibration, there are two possible mechanisms through which photons and phonons are coupled indirectly via the electrons of the crystal, using the electron-photon and electron-phonon interactions. In covalent crystals such as diamond, the electron-phonon interaction through deformation potential may be important, whereas in crystals which are either piezoelectric or ionic, the long-range electromagnetic interaction (which is proportional to the electric field associated with the phonons) may dominate.

[78] R. W. Terhune, *Solid State Design*, vol. 4, p. 38, 1963.
[79] R. W. Minck, R. W. Terhune, and W. R. Rado, *Appl. Phys. Letters*, vol. 3, p. 181, 1963.
[80] R. Chiao and B. P. Stoicheff, *Phys. Rev. Letters*, vol. 12, p. 290, 1964.
[81] N. Bloembergen, "Nonlinear Optics," W. A. Benjamin Inc., New York, 1965.
[82] See, for example, H. J. Zeiger and P. E. Tannenwald, *Proc. 3rd International Symp. on Quantum Electronics, Paris*, p. 1589; R. J. Elliott and R. Loudon, *Phys. Letters*, vol. 3, p. 189, 1963; R. Loudon, *Proc. Phys. Soc.* (London), vol. 82, p. 393, 1963.

PROBLEMS

12.1 Supposing that a black body is placed in a resonator of length L, show that the energy stored in all the axial modes is equal to

$$\mathcal{E} = \frac{2\pi^2 (kT)^2}{3hc} \quad \text{joule/m}$$

The above equation is obtained by using the black-body radiation law of Eq. (1.130) and by using the definite integral

$$\int_0^\infty \frac{x \, dx}{\exp (x) - 1} = \frac{\pi^2}{6}$$

As will be shown in the next problem, the black-body radiation has a pronounced maximum around $hc/\lambda_m = 3kT$. Show that the power P associated with the waves set up in the resonator by the radiation from the black body can be expressed in terms of λ_m as

$$P = \frac{c\mathcal{E}}{4} = \sigma T^4 \frac{45\lambda_m{}^2}{4\pi^3} \quad \text{watts}$$

where $\sigma = 2\pi^5 k^4 / 15 h^3 c^2$.

12.2 From the black-body radiation law of Eq. (1.130), show that the total power per unit area radiated by a black body is given by

$$F(T) = \int_0^\infty \frac{c}{4} \frac{8h\nu^3 \, d\nu}{c^3 \left[\exp \left(\dfrac{h\nu}{kT} \right) - 1 \right]} = \sigma T^4 \quad \text{joule/m}^2$$

which is known as Stefan's law. Find the expression for the Stefan-Boltzmann constant σ. Referring to Prob. 1.25, explain the factor 1/4 which appears in both $F(T)$ and the expression of P of the preceding problem.

12.3 Refer to the article by W. P. Dumke (*Phys. Rev.*, vol. 132, p. 1998, 1963). Derive the expression for the absorption coefficient due to transitions between the valence-band and donor states.

12.4 Read the article by G. Lasher and F. Stern (*Phys. Rev.*, vol. 133, p. A553, 1964). Discuss how one obtains the expression for the spontaneous emission spectrum of a semiconductor diode.

12.5 Repeat Prob. 12.4 for a discussion of the spectrum from the stimulated emission of a semiconductor diode. Explain physically the difference between the spontaneous and the stimulated emission spectra.

12.6 Read the articles by F. L. Galeener et al. (*Phys. Rev. Letters*, vol. 10, p. 472, 1963), and by M. I. Nathan and G. Burns (*Appl. Phys. Letters*, vol. 1, p. 89, 1962). Discuss the essential features of the experiments which give clue to the nature of the transition involved in GaAs electroluminescence.

12.7 Describe the band-filling model proposed by D. F. Nelson et al. (*Appl. Phys. Letters*, vol. 2, p. 182, 1963) for GaAs injection luminescence and the experimental evidence supporting the model. Also discuss the luminescence experiment reported by D. K. Wilson (*Appl. Phys. Letters*, vol. 3, p. 127, 1963) on GaAs *p-n* junction with a lightly doped *n*-type epitaxial layer.

12.8   Relate the negative conductivity σ_1 which appears in Eq. (12.35) to the transition probability W and the difference in the Fermi functions $(f_c - f_v)$ of Eq. (12.20). For simplicity, assume a plane wave solution in deriving this relationship.

12.9   Read the article by B. Lax (*Proc. of the Symposium on Optical Masers*, p. 119, Brooklyn Polytechnic Institute, 1963). Discuss other schemes being investigated in semiconductors for possible laser action besides the direct transition laser discussed in the text.

12.10   Read the articles by A. L. Schawlow and C. H. Townes (*Phys. Rev.*, vol. 112, p. 1940, 1958) and C. K. N. Patel ("Lasers and Applications," p. 49, Ohio State University, 1963). Show that for gaseous lasers where the width of the spontaneous emission spectrum is determined by the Doppler effect, the oscillation condition can be expressed as

$$\Delta N = N_1 - N_2 = \frac{\nu}{f}\left(\frac{\pi e^2}{mc^2}\right)^{-1}\sqrt{\frac{2\pi kT}{M}} \times (\text{loss})$$

In the above equation, ΔN is the difference in the population density between the upper and lower laser levels, f is the oscillator strength defined in Eq. (12.49), and the term (loss) represents the percentage loss for a single pass including reflection losses at the mirrors and diffraction losses.

12.11   Read the article by A. G. Fox and T. Li (*Bell System Tech. J.*, vol. 40, p. 453, 1961). Outline the procedure by which the steady-state field distribution in a Fabry-Perot interferometer may be obtained. Discuss the diffraction loss mechanism and its dependence on $a^2/b\lambda$ where a is the radius of the end circular plates, b is the separation between the plates, and λ is the wavelength.

12.12   Read the articles by G. D. Boyd and J. P. Gordon (*Bell System Tech. J.*, vol. 40, p. 489, 1961) and by G. D. Boyd and H. Kogelnik (*Bell System Tech. J.*, vol. 41, p. 1347, 1961). Discuss how the spot size for the lowest mode of a confocal system is related to the confocal resonator spacing and the wavelength. Compare the diffraction loss for a confocal system with that for the plane-parallel mirrors.

12.13   Describe the power-dependent frequency-pulling effect observed by W. R. Bennett, Jr. (*Phys. Rev.*, vol. 126, p. 580, 1962) in a He-Ne laser. Also derive the expression for the separation between adjacent modes in a semiconductor laser. (References: G. Burns et al., *Proc. IEEE*, vol. 51, p. 947, 1963; and G. Birnbaum, "Optical Masers," Supplement 2, *Advances in Electronics and Electron Physics*, p. 165, 1964).

12.14   Read the article by W. E. Lamb (*Phys. Rev.*, vol. 134, p. A1429, 1964). Explain the effects of mode coupling in a gaseous laser.

12.15   Read the articles by W. Kaiser, C. G. B. Garrett, and D. L. Wood (*Phys. Rev.*, vol. 123, p. 766, 1961), and by Z. J. Kiss and R. C. Duncan

(*Proc. IRE*, vol. 50, p. 1531, 1962). Describe the energy-level schemes used in $CaF_2:Dy^{2+}$ and $CaF_2:Sm^{2+}$ lasers.

12.16 Most ruby lasers oscillate in random spikes which occur in somewhat random fashion. Read the article by K. Shimoda (*Proc. of the Symposium on Optical Masers*, pp. 100–105, Brooklyn Polytechnic Institute, 1963) and outline the theory proposed by Shimoda in explaining the observed relaxation oscillations, both damped and undamped.

12.17 Read the articles by D. F. Nelson and F. K. Reinhart (*Appl. Phys. Letters*, vol. 5, p. 148, 1964), by A. Frova and P. Handler (*Appl. Phys. Letters*, vol. 5, p. 11, 1964), by B. O. Seraphin and D. A. Orton (*J. Appl. Phys.*, vol. 34, p. 1743, 1963), and by G. E. Fenner (*Appl. Phys. Letters*, vol. 5, p. 198, 1964). Discuss the possible use of these reported effects in the modulation of a semiconductor laser beam.

12.18 Read the articles by D. F. Holshouser et al. (*J. Optic. Soc. Am.*, vol. 51, p. 1360, 1961) and by R. H. Blumenthal (*Proc. IRE*, vol. 50, p. 452, 1962). Discuss the use of Kerr and Pockel effects in the modulation of a laser beam.

12.19 Read the article by E. I. Gordon and J. D. Rigden (*Bell System Tech. J.*, vol. 42, p. 155, 1963). Discuss the use of electro-optic effects in the modulation of a laser beam.

12.20 Refer to the table for piezoelectric constants (*Proc. IRE*, vol. 46, p. 767, 1958). Write down the second-order dielectric tensor similar to Eqs. (12.65) and (12.66) for calcite, $BaTiO_3$, CdS, and GaAs.

12.21 The $(\overline{4}2m)$ crystal group to which KDP belongs has the following symmetry properties: (1) an inversion fourfold axis (chosen as the z axis) along the tetragonal axis, (2) two orthogonal twofold axes (chosen as the x and y axes) perpendicular to the tetragonal axis, and (3) two mirror planes containing the inversion fourfold axis and inclined 45° to the two-fold axes. These symmetry properties are designated by the symbol $(\overline{4}2m)$ in the order described above, and all the symmetry operations can be represented mathematically by a transformation matrix $[T]$ such that

$$\begin{bmatrix} x' \\ y' \\ z' \end{bmatrix} = [T] \begin{bmatrix} x \\ y \\ z \end{bmatrix}$$

The unprimed and primed coordinates represent, respectively, the coordinates of a lattice point before and after the symmetry operation.

Identify the following matrices with the respective symmetry operation of the $(\overline{4}2m)$ crystal group:

$$\begin{bmatrix} 1 & 0 & 0 \\ 0 & -1 & 0 \\ 0 & 0 & -1 \end{bmatrix} \begin{bmatrix} -1 & 0 & 0 \\ 0 & 1 & 0 \\ 0 & 0 & -1 \end{bmatrix} \begin{bmatrix} 0 & -1 & 0 \\ 1 & 0 & 0 \\ 0 & 0 & -1 \end{bmatrix}$$

$$\begin{bmatrix} 0 & 1 & 0 \\ 1 & 0 & 0 \\ 0 & 0 & 1 \end{bmatrix} \begin{bmatrix} 0 & -1 & 0 \\ -1 & 0 & 0 \\ 0 & 0 & 1 \end{bmatrix}$$

12.22 According to Eq. (12.77), the quadratic polarization can be expressed explicitly in terms of the electric fields through a susceptibility tensor $\chi_{a\alpha\beta}$ as

$$[P] = [\chi_{a\alpha\beta}][EE]$$

where

$$[\chi_{a\alpha\beta}] = \begin{bmatrix} xxx & xyy & xzz & xyz & xzy & xzx & xxz & xxy & xyx \\ yxx & yyy & yzz & yyz & yzy & yzx & yxz & yxy & yyx \\ zxx & zyy & zzz & zyz & zzy & zzx & zxz & zxy & zyx \end{bmatrix}$$

and $[P]$ is a column matrix arranged in the order P_x, P_y, P_z, whereas $[EE]$ is also a column matrix arranged in the order E_xE_x, E_yE_y, E_zE_z, E_yE_z, E_zE_y, E_zE_x, E_xE_z, E_xE_y, E_yE_x. In the susceptibility matrix, the element zyx, for example, actually represents χ_{zyx} and the symbol χ is dropped for compactness.

If a crystal has a certain symmetry property, the susceptibility tensor should remain unchanged after a symmetry operation associated with a given symmetry property is applied to $[P]$ and $[EE]$. Therefore if an element in the susceptibility matrix changes sign after a symmetry operation, the element must be zero.

Show that in crystals having three orthogonal twofold axes, the elements of the second-order susceptibility tensor are zero, except those with nonidentical indices ($a \neq \alpha \neq \beta \neq a$).

12.23 Consider an ionic crystal in which an anharmonic force δx^2 exists. Starting from Eq. (7.23), derive the following equation of motion

$$\frac{d^2x}{dt^2} + \Gamma \frac{dx}{dt} + \omega_v{}^2 x + \frac{\delta}{M} x^2 = \frac{F}{M}$$

where $x = u - v$ is the relative displacement of neighboring positive and negative ions, $M = M_1 M_2/(M_1 + M_2)$ is the reduced mass, Γ is a damping constant, and F is the force due to applied fields.

Treating the δx^2 term as a small perturbation, find the amplitude of oscillation $x(\omega_1 - \omega_0)$ at the difference frequency $\omega_1 - \omega_0$ under the applied field condition of Eq. (12.82). If it is assumed that the dipole-dipole interaction expressed in Eq. (12.92) is responsible for the anharmonic force, find the expression for $\partial\alpha/\partial x$ by comparing the value of $x(\omega_1 - \omega_0)$ just calculated with that given in Eq. (12.95).

12.24 In the interaction of light waves with lattice waves of the optical branch, the amplification factor of the stimulated Raman radiation is usually limited by the damping associated with the phonon waves. Find the expression of the amplification factor for E_1 from Eqs. (12.84) and (12.90). Compare it with the corresponding expression obtained from Eq. (12.102), regarding their functional dependence on such factors as $|E_0|^2$, $\partial\alpha/\partial R$, the various frequencies, and the damping associated with vibrational losses.

REFERENCES

Bloembergen, N.: "Nonlinear Optics," W. A. Benjamin, Inc., New York, 1965.

Born, M., and E. Wolf: "Principle of Optics," Pergamon Press, New York, 1964.

Condon, E. U., and G. H. Shortley: "The Theory of Atomic Spectra," Cambridge University Press, London, 1959.

Heavens, O. S.: "Optical Masers," Methuen & Co., Ltd., London, 1964.

Herzberg, G.: "Atomic Spectra and Atomic Structure," Dover Publications, Inc., New York, 1944.

Lengyel, B. A.: "Lasers," John Wiley & Sons, Inc., New York, 1962.

McClure, D. S.: Electronic Spectra of Molecules and Ions in Crystals, *Solid State Physics*, vol. 8, p. 1, and vol. 9, p. 399, 1959.

Townes, C. H.: 1964 Nobel Lecture presented in Stockholm, Sweden, *IEEE Spectrum*, August, 1965.

"Optical Masers," Symposia series XIII, Brooklyn Polytechnic Institute, 1963.

"Quantum Electronics," *Proc. IEEE*, vol. 51, January, 1963.

"Quantum Electronics," Proceedings of the first, second, and third International Congress on Quantum Electronics, Columbia University, New York, 1960, 1961, and 1964.

"Lasers and Applications," W. S. C. Chang (ed.), The Ohio State University, Columbus, 1963.

AUTHOR INDEX

SUBJECT INDEX

Absorption and emission spectra, of
Al_2O_3:Cr^{3+}, 725, 727
of CaF_2:Tm^{2+}, 720
of GaAs, 657, 697
of KCl:Tl$^+$, 632
Absorption coefficient, 649
for atomic transitions, 728
and band structure, 653–655
determination of, 650–651
for direct optical transition, 668
in GaAs, 657
in Ge, 653
for indirect optical transition, 671
and oscillation strength, 728
and static dielectric constant, 677,
680
Acceptor impurities, 133
ionization energy of, 133, 192
in III-V compounds, 193
in II-VI compounds, 196, 633, 634
Acoustic branch, 120–121
Acoustoelectric current, 256
effect on V-I relation, 256
Activators, 629, 633, 634
(*See also* Luminescent centers)
Allowed bands, 161
Alloying process, 115, 340, 342
phase diagram, 115, 116
Angular momentum, 29, 464–467
in j-j coupling scheme, 707
and magnetic moment, 457, 467
orbital, 29
effect of crystalline field, 476
operator, 29, 509
quantization, 29–30, 465–466
in Russell-Saunders coupling
scheme, 471–472, 706
spin, 33
operators, 511
quantization, 466–467
vector model for addition of, 466,
470–471
Anisotropy, magnetic, 559
caused by strain, 585
constants, 560, 562

Anisotropy, magnetic, easy axis, 559
energy, 560
field, 560–561, 591
hard axis, 559
origin of, 562
Annihilation operator, 23, 223
Antiferroelectric material, 434
physical model, 433
Antiferromagnetic resonance, 591
Antiferromagnetism, 484
and exchange interaction, 481–482,
484
superexchange coupling, 497, 557
Atomic number, 37
Atomic scattering factor, 91
Avalanche breakdown, 359
breakdown voltage and impurity
concentration, 361
and impact ionization, 359
Avalanche transistor, 362
condition for negative-resistance
region, 365
V-I characteristic, 363
Azimuthal (orbital) quantum number, 29–30

Barium titanate, 431–432
Black-body radiation, 47
Einstein's interpretation of, 538–
539
Planck's law of, 47–48
Bloch equation, 504
Bloch wave function, 163
Bohr magneton, 459
magnetic moment expressed in, 483
Bohr radius, 11
Bohr theory, 9
of spectral lines of hydrogen atom, 9
Boltzmann constant, 44
Boltzmann distribution function, 44
Boltzmann relation, 105, 143
Boltzmann transport equation, 204,
241–242
solution, for electric conduction,
205–206